Inverse properties For each real number a there is a single real number $-a$ such that

$$a + (-a) = 0 \quad \text{and} \quad (-a) + a = 0.$$

For each nonzero real number a, there is a single real number $\frac{1}{a}$ such that

$$a \cdot \frac{1}{a} = 1 \quad \text{and} \quad \frac{1}{a} \cdot a = 1.$$

Distributive property $a(b + c) = ab + ac$

2.2 LINEAR EQUATION

A linear equation can be written in the form $ax + b = c$, for real numbers a, b, and c, with $a \neq 0$.

2.2 ADDITION PROPERTY OF EQUALITY

For all real numbers a, b, and c,

$$\text{if } a = b, \text{ then } a + c = b + c.$$

2.2 MULTIPLICATION PROPERTY OF EQUALITY

For all real numbers a, b, and $c \neq 0$,

$$\text{if } a = b, \text{ then } ac = bc.$$

2.3 SOLVING LINEAR EQUATIONS

Step 1 Combine like terms to simplify each side.

Step 2 Get the variable term on one side, and a number on the other.

Step 3 Get the equation into the form $x =$ a number.

Step 4 Check by substituting into the original equation.

2.9 SOLVING LINEAR INEQUALITIES

To solve a linear inequality, follow the same steps as those for solving a linear equation. However, when each side is multiplied or divided by a negative number, the direction of the inequality symbol must be reversed.

3.2, 3.5 RULES FOR EXPONENTS

For any integers m and n:

Product rule $a^m \cdot a^n = a^{m+n}$

Quotient rule $\frac{a^m}{a^n} = a^{m-n} \quad (a \neq 0)$

Zero exponent $a^0 = 1 \quad (a \neq 0)$

Negative exponent $a^{-n} = \frac{1}{a^n} \quad (a \neq 0)$

Power rules

(a) $(a^m)^n = a^{mn}$

(b) $(ab)^m = a^m b^m$

(c) $\left(\frac{a}{b}\right)^m = \frac{a^m}{b^m} \quad (b \neq 0)$

(d) $\frac{a^{-n}}{b^{-n}} = \frac{b^n}{a^n} \quad (a, b \neq 0)$

(e) $\left(\frac{a}{b}\right)^{-n} = \left(\frac{b}{a}\right)^n \quad (a, b \neq 0)$

SIXTH EDITION

BEGINNING ALGEBRA

BEGINNING ALGEBRA

SIXTH EDITION

MARGARET L. LIAL
American River College

CHARLES D. MILLER

E. JOHN HORNSBY, JR.
University of New Orleans

HarperCollins*Publishers*

TO THE STUDENT

If you need further help with algebra, you may want to obtain a copy of the *Student's Solutions Manual* that goes with this book. It contains solutions to all the odd-numbered exercises and all the chapter test exercises. You also may want the *Student's Study Guide*. It has extra examples and exercises to complete, corresponding to each learning objective of the book. In addition, there is a practice test for each chapter. Your college bookstore either has these books or can order them for you.

On the cover: Flushing salts from contoured fields in California's Coachella Valley, farmers flood fallow land with from one to five acre-feet of water, sometimes more than they use to irrigate crops. Soluble salts leach through the soil into buried drains. Caused mainly by use of saline Colorado River water for irrigation, salt buildup can significantly reduce soil productivity.

Sponsoring Editor: Anne Kelly
Developmental Editors: Liz Lee and Adam Bryer
Project Editor: Janet Tilden
Assistant Art Director: Julie Anderson
Text and Cover Design: Lucy Lesiak Design, Inc.: Lucy Lesiak
Cover Photo: Comstock/George Gerster
Photo Researchers: Kelly Mountain and Sandy Schneider
Production: Jeanie Berke, Linda Murray, and Helen Driller
Compositor: York Graphic Services
Printer and Binder: R. R. Donnelley & Sons Company
Cover Printer: Lehigh Press Lithographers

For permission to use copyrighted material, grateful acknowledgment is made to the copyright holders on page 536, which are hereby made part of this copyright page.

Beginning Algebra, Sixth Edition

Library of Congress Cataloging-in-Publication Data

Lial, Margaret L.
Beginning algebra/Margaret L. Lial, Charles D. Miller, E. John Hornsby, Jr.—6th ed.
p. cm.
Includes index.
ISBN 0-673-46459-8
1. Algebra. I. Miller, Charles David. II. Hornsby, E. John. III. Title.
QA152.2.L5 1992 91-19376
512.9—dc20 CIP

92 93 94 9 8 7 6 5 4 3 2

PREFACE

In the sixth edition of *Beginning Algebra,* we have maintained the strengths of past editions, while enhancing the pedagogy, readability, and attractiveness of the text. Many features have been added to make the text easier and more enjoyable for students and teachers to use, including new exercises, Quick Reviews, and the use of full color. We continue to provide an extensive supplemental package. For students, we offer a solutions manual, a study guide, interactive tutorial software, videotapes, and audiotapes. For instructors, we present the Annotated Instructor's Edition with conveniently placed answers and teaching tips. Further, for the instructor, we provide alternative forms of tests and additional exercises, computer-generated tests, complete solutions to all exercises, and transparencies.

All the successful features of the previous edition are carried over in the new edition: learning objectives for each section, careful exposition, fully developed examples with comments printed at the side (almost 700 examples), and carefully graded section, chapter review, and chapter test exercises (more than 5000 exercises, with more than 50 percent of the applications exercises new to this edition). Screened boxes set off important definitions, formulas, rules, and procedures to further aid in learning and reviewing the course material.

NEW FEATURES

Conceptual and Writing Exercises

To complement the drill exercises, several exercises requiring an understanding of the concepts introduced in a section are included in almost every exercise set. There are almost 300 of these conceptual exercises. Also, more than 130 exercises require the student to respond by writing a few sentences. (Note that some of these writing exercises are also labeled as conceptual exercises.) Directions to conceptual and writing exercises include references to specific learning objectives to help students achieve a broader perspective should they need to turn back to the explanations and examples.

Focus on Problem Solving

An application from the exercise set will be featured at the beginning of selected sections. Each application is chosen from among the most interesting in the exercise set and is intended to help motivate the study of the section.

> The winner of the 1988 Indianapolis 500 (mile) race was Rick Mears, who drove his Penske-Chevy V8 at an average speed of 144.8 miles per hour. What was Mears' driving time?
>
> The relationship between distance, rate, and time is one that is used quite frequently in everyday life. In order to solve this problem, we need to know how to find time if we are given distance and rate. In this section we look at some applications of the distance, rate, and time relationship.
>
> This problem is Exercise 1 in the exercises for this section. After working through this section, you should be able to solve this problem.
>
> FOCUS ON PROBLEM SOLVING

Problem-Solving Strategies

In special paragraphs clearly distinguished by the heading ''Problem Solving,'' we have expanded our discussion of strategies to include connections to techniques learned earlier.

PROBLEM SOLVING

A common sight in supermarkets is shoppers carrying hand-held calculators to assist them in their job of budgeting. While the most common use is to make sure that the shopper does not go over budget, another use is to see which size of an item offered in different sizes produces the best price per unit. In order to do this, simply divide the price of the item by the unit of measure in which the item is labelled. The next example illustrates this idea. ■

Quick Review

A Quick Review at the end of each chapter provides a capsule summary of key ideas and is set in tabular form to enable students to find the important concepts easily and review them more effectively. In addition, worked-out examples accompany each section-referenced key idea.

CHAPTER 3 QUICK REVIEW

SECTION	CONCEPTS	EXAMPLES
3.1 POLYNOMIALS	**Addition:** Add like terms.	Add: $\begin{array}{r} 2x^2 + 5x - 3 \\ 5x^2 - 2x + 7 \\ \hline 7x^2 + 3x + 4 \end{array}$
	Subtraction: Change the signs of the terms in the second polynomial and add to the first polynomial.	Subtract: $(2x^2 + 5x - 3) - (5x^2 - 2x + 7)$ $= (2x^2 + 5x - 3) + (-5x^2 + 2x - 7)$ $= -3x^2 + 7x - 10$
3.2 EXPONENTS	For any integers m and n with no denominators zero:	
	Product rule $a^m \cdot a^n = a^{m+n}$	$2^4 \cdot 2^5 = 2^9$

Design Use of full color and changes in format help to create a fresh look. We have enhanced the book's appeal and increased its usefulness.

Cautionary Remarks Common student errors and difficulties are now highlighted graphically and identified with the heading "Caution." **Notes** have a similar graphic treatment.

CAUTION Remember that a sum or difference of radicals can be simplified only if the radicals are *like radicals*. For example, $\sqrt{5} + 3\sqrt{5} = 4\sqrt{5}$, but $\sqrt{5} + 5\sqrt{3}$ cannot be simplified further. Also, $2\sqrt{3} + 5\sqrt[3]{3}$ cannot be simplified further.

NOTE In the next section we will learn to add and subtract rational expressions, and this will often require the skill illustrated in Example 5. While it will often be beneficial to leave the denominator in factored form, we have multiplied the factors in the denominator in Example 5 to give the answer in the form the original problem was presented.

Geometry Exercises Review of geometry is a thread that runs through the text. We have increased the number of exercises that relate geometric concepts to the new algebraic concepts. A brief review of geometry is included in Appendix A, expanded from the fifth edition to include a greater number of basic definitions and relationships.

Applications These exercises have been extensively rewritten and updated to more closely reflect the student's world. More than 50 percent are new to the sixth edition.

Glossary A glossary of key terms, followed by a description of new symbols, is provided at the end of each chapter. A comprehensive glossary is placed at the end of the book. Each term in the glossary is defined and then cross-referenced to the appropriate section, where students may find a more detailed explanation of the term.

CHAPTER 9 GLOSSARY

KEY TERMS

9.3 **standard form (of a quadratic equation)** A quadratic equation written as $ax^2 + bx + c = 0$ $(a \neq 0)$ is in standard form.

9.4 **complex number** A complex number is a number of the form $a + bi$, where a and b are real numbers.

imaginary number The complex number $a + bi$ is imaginary if $b \neq 0$.

real part The real part of $a + bi$ is a.

imaginary part The imaginary part of $a + bi$ is b.

standard form (of a complex number) A complex number written in the form $a + bi$ (or $a + ib$) is in standard form.

conjugate The conjugate of $a + bi$ is $a - bi$.

9.5 **parabola** The graph of a quadratic equation of the form $y = ax^2 + bx + c$ $(a \neq 0)$ is called a parabola.

vertex The vertex of a parabola is the highest or lowest point on the graph.

axis The axis of a parabola is a vertical line through the vertex.

quadratic function A function of the form $f(x) = ax^2 + bx + c$ $(a \neq 0)$ is called a quadratic function.

Example Titles Each example now has a title to help students see the purpose of the example. The titles also facilitate working the exercises and studying for examinations.

EXAMPLE 1 USING THE ZERO-FACTOR PROPERTY

Solve the equation $(x + 3)(2x - 1) = 0$.

The product $(x + 3)(2x - 1)$ is equal to zero. By the zero-factor property, the only way that the product of these two factors can be zero is if at least one of the factors is zero. Therefore, either $x + 3 = 0$ or $2x - 1 = 0$. Solve each of these two linear equations as in Chapter 2.

$$x + 3 = 0 \quad \text{or} \quad 2x - 1 = 0$$

$$x = -3 \quad \text{or} \quad 2x = 1 \qquad \text{Add to both sides.}$$

$$x = \frac{1}{2} \qquad \text{Divide by 2.}$$

Calculator Usage The use of a scientific calculator is explained and referred to at appropriate points in the text.

Success in Algebra This foreword to the student provides additional support by offering suggestions for studying the course material.

Preview Exercises Formerly called Review Exercises, these are intended to sharpen the basic skills needed to do the work in the next section. Students need to review material presented earlier, even though it seems very basic. These exercises also help to show how earlier material connects with and is needed for later topics.

NEW CONTENT HIGHLIGHTS

We have endeavored to stress throughout the text the difference between an expression and an equation. This emphasis begins in Chapter 1. We have expanded the introduction of the operations with real numbers, working in interpretation of words and phrases indicating these operations. Each of the properties of real numbers is preceded by an intuitive explanation.

Chapter 2 has been rewritten to present applied problems in a more effective way, grouping them by type and keeping these first ones simpler. More examples are given and the exercises are carefully cross-referenced to the examples. A number of applications from geometry are included here.

The topics in Chapter 3 have been reorganized to help improve students' work with exponents. We begin with addition, subtraction, and multiplication of polynomials. Then the laws of exponents are discussed in two sections, followed by polynomial division, which reinforces the product and quotient rules for exponents. The chapter ends with scientific notation.

In Chapter 4, the discussion of factoring trinomials has been expanded. Factoring general trinomials by grouping now precedes the trial-and-error method.

Working with rational expressions is difficult for students, so we have added many cautions and notes throughout Chapter 5 to help them avoid common errors. New examples and exercises have been added, including some that point out the distinction between expressions and equations.

Chapter 9 now includes more applications. The derivation of the quadratic formula is given in a parallel development with an equation having numerical coefficients. The presentation of graphing parabolas has been rewritten to make it more appropriate for students at this level.

Throughout the book, summaries of procedures have been moved earlier in the section, so students can refer to the summary while working through the examples.

SUPPLEMENTS

Our extensive supplemental package includes an annotated instructor's edition, testing materials, solutions, software, videotapes, and audiotapes.

Annotated Instructor's Edition

With this, instructors have immediate access to the answers to every drill and conceptual exercise in the text: each answer is printed in color next to the corresponding text exercise or in the margin. In addition, challenging exercises, which will require most students to stretch beyond the concepts discussed in the text, are marked with the symbol ▲. The conceptual (◉) and writing (✎) exercises are also marked in this edition so instructors may use discretion in assigning these problems. (Calculator exercises will be marked by ▦ in both the student's and instructor's editions.) This edition also includes Teaching Tips that point out rough spots or give additional hints and warnings for students. Further, the instructor's edition includes Byways, which provide additional information, interesting sidelights, historical notes, and other material that an instructor can use to introduce concepts or provoke classroom conversation. Resources, a set of cross-references to all the supplements, appears at the head of each section. The student's edition does not contain the on-page answers; conceptual, writing, and challenging exercise symbols; Teaching Tips; Byways; or Resources references.

Instructor's Test Manual

Included are two versions of a pretest—one open-response and one multiple-choice; six versions of a chapter test for each chapter—four open-response and two multiple-choice; three versions of a final examination—two open-response and one multiple-choice; and an extensive set of additional exercises, providing ten to twenty exercises for each textbook objective, which can be used as an additional source of questions for tests, quizzes, or student review of difficult topics. Answers to all tests and additional exercises are provided.

Instructor's Solutions Manual

This manual includes solutions to all exercises and a list of all conceptual, writing, and challenging exercises.

HarperCollins Test Generator for Mathematics

The *HarperCollins Test Generator* is one of the top testing programs on the market for IBM and Macintosh computers. It enables instructors to select questions for any section in the text or to use a ready-made test for each chapter. Instructors may generate tests in multiple-choice or open-response formats, scramble the order of questions while printing, and produce twenty-five versions of each test. The system features printed graphics and accurate mathematical symbols. The program also allows instructors to choose problems randomly from a section or problem type or to choose questions manually while viewing them on the screen, with the option to regenerate variables if desired. The editing feature allows instructors to customize the chapter data disks by adding their own problems.

Transparencies Nearly 100 color overhead transparencies of figures, examples, definitions, procedures, properties, and problem-solving methods will help instructors present important points during lecture.

Student's Solutions Manual Solutions are given for odd-numbered exercises and chapter test questions.

Student's Study Guide Written in a semiprogrammed format, the *Study Guide* provides additional practice problems and reinforcement for each learning objective in the textbook. A self-test is given at the end of each chapter.

Computer-Assisted Tutorials The tutorials offer self-paced, interactive review in IBM, Apple, and Macintosh formats. Solutions are given for all examples and exercises, as needed.

Videotapes A new videotape series has been developed to accompany *Beginning Algebra,* Sixth Edition. The tapes cover all objectives, topics, and problem-solving techniques within the text.

Audiotapes A set of audiotapes, one tape per chapter, guides students through each topic, allowing individualized study and additional practice for troublesome areas. These tapes are especially helpful for visually impaired students.

ACKNOWLEDGMENTS

We wish to thank the many users of the fifth edition for their insightful suggestions on improvements for this book. We also wish to thank our reviewers for their contributions:

Carol Achs, *Mesa Community College*
Peter Arvanites, *Rockland Community College*
Alan Bishop, *Western Illinois University*
Rodney E. Chase, *Oakland Community College*
Kenneth R. Crowe, *Rio Salado Community College*
David Dudley, *Phoenix College*
Carol A. Edwards, *St. Louis Community College at Florissant Valley*
Kathy Fink, *Moorpark College*
Al Giambrone, *Sinclair Community College*
John Haldi, *Spokane Community College*
Brian Hayes, *Triton College*
Harold Hiken, *University of Wisconsin–Milwaukee*
Donald K. Hostetler, *Mesa Community College*
Tracey Hoy, *College of Lake County*
Keith Jorgensen, *Orange Coast College*
Phyllis Kester, *Liberty University*
Mary Lois King, *Tallahassee Community College*
Molly S. Krejewski, *Daytona Beach Community College*
Arthur H. Litka, *Umpqua Community College*
Carl Mancuso, *William Paterson College*
Philip M. Meyer, *Skyline College*
Richard Quint, *Ventura College*
William P. Robertson, *Charles County Community College*
Kenneth S. Ross, *Broward Community College–Central Campus*

Carol Roush, *Catonsville Community College*
Samuel Sargis, *Modesto Junior College*
Mary Kay Schippers, *Fort Hays State University*
Vic Schneider, *University of Southwestern Louisiana*
Melissa J. Simpkins, *Columbus State Community College*
Judy Staver, *Florida Community College–South Campus*
Raymond H. Tanner, *Jackson County Community College*
Beverly Weatherwax, *Southwest Missouri State University*
Gail G. Wiltse, *Saint Johns River Community College*

Paul Eldersveld, College of DuPage, deserves our gratitude for an excellent job coordinating all of the print ancillaries for us, an enormous and time-consuming task. Careful work by Barbara Wheeler, of California State University–Sacramento, has helped ensure the accuracy of the answers. Jim Walker and Paul Van Erden, of American River College, have done a superb job creating the indexes, and Paul Van Erden and Mike Karelius, also of American River College, have provided the excellent set of Teaching Tips found in the Annotated Instructor's Edition of this text and of our *Intermediate Algebra,* Sixth Edition. We also want to thank Tommy Thompson of Seminole Community College for his suggestions for the feature "To the Student: Success in Algebra" that follows this preface.

Special thanks go to Harold Hiken for his many valuable suggestions, and to those among the staff at HarperCollins whose assistance and contributions have been very important: Jack Pritchard, Anne Kelly, Linda Youngman, Adam Bryer, Liz Lee, Ellen Keith, Janet Tilden, and Cathy Wacaser.

Margaret L. Lial
E. John Hornsby, Jr.

CONTENTS

CHAPTER 3 POLYNOMIALS AND EXPONENTS

CHAPTER 4 FACTORING AND APPLICATIONS

CHAPTER 5 RATIONAL EXPRESSIONS

CHAPTER 6
GRAPHING LINEAR EQUATIONS

CHAPTER 7
LINEAR SYSTEMS

CHAPTER 8
ROOTS AND RADICALS

CHAPTER 9
QUADRATIC EQUATIONS

TO THE STUDENT: SUCCESS IN ALGEBRA

The main reason students have difficulty with mathematics is that they don't know how to study it. Studying mathematics *is* different from studying subjects like English or history. The key to success is regular practice.

This should not be surprising. After all, can you learn to play the piano or to ski well without a lot of regular practice? The same thing is true for learning mathematics. Working problems nearly every day is the key to becoming successful. Here is a list of things you can do to help you succeed in studying algebra.

1. *Attend class regularly*. Pay attention in class to what your teacher says and does, and make careful notes. In particular, note the problems the teacher works on the board and copy the complete solutions. Keep these notes separate from your homework to avoid confusion when you read them over later.
2. Don't hesitate to ask questions in class. It is not a sign of weakness, but of strength. There are always other students with the same question who are too shy to ask.
3. *Read your text carefully*. Many students read only enough to get by, usually only the examples. Reading the complete section will help you to be successful with the homework problems. Most exercises are keyed to specific examples or objectives that will explain the procedures for working them.
4. Before you start on your homework assignment, rework the problems the teacher worked in class. This will reinforce what you have learned. Many students say, "I understand it perfectly when you do it, but I get stuck when I try to work the problem myself."
5. Do your homework assignment only *after* reading the text and reviewing your notes from class. Check your work with the answers in the back of the book. If you get a problem wrong and are unable to see why, mark that problem and ask your instructor about it. Then practice working additional problems of the same type to reinforce what you have learned.
6. Work as neatly as you can. Write your symbols clearly, and make sure the problems are clearly separated from each other. Working neatly will help you to think clearly and also make it easier to review the homework before a test.

7. After you have completed a homework assignment, look over the text again. Try to decide what the main ideas are in the lesson. Often they are clearly highlighted or boxed in the text.
8. Use the chapter test at the end of each chapter as a practice test. Work through the problems under test conditions, without referring to the text or the answers until you are finished. You may want to time yourself to see how long it takes you. When you have finished, check your answers against those in the back of the book and study those problems that you missed. Answers are referenced to the appropriate sections of the text.
9. Keep any quizzes and tests that are returned to you and use them when you study for future tests and the final exam. These quizzes and tests indicate what your instructor considers most important. Be sure to correct any problems on these tests that you missed, so you will have the corrected work to study.
10. Don't worry if you do not understand a new topic right away. As you read more about it and work through the problems, you will gain understanding. Each time you look back at a topic you will understand it a little better. No one understands each topic completely right from the start.

CHAPTER ONE

THE REAL NUMBER SYSTEM

The Italian astronomer Galileo Galilei (1564–1643) once wrote: *Mathematics is the language with which God has written the Universe*. Can you imagine trying to function in our world without numbers? The most familiar set of numbers is the set of *natural numbers,* the numbers used for counting. But the natural numbers form only a part of a much more extensive set of numbers called the *real numbers*.

In this chapter we set out to examine properties of the real numbers that will be used extensively in this book and in all other mathematics courses you will take. These properties will enable us, in later chapters, to solve equations, work with algebraic fractions and radicals, and graph sets of points.

Numbers were developed in the early history of the human race for the purpose of counting. Early humans learned that mathematical concepts allowed them to solve a large variety of problems, and our study of algebra will allow us to do the same.

1.1 FRACTIONS

OBJECTIVES

1 LEARN THE DEFINITION OF *FACTOR*.

2 WRITE FRACTIONS IN LOWEST TERMS.

3 MULTIPLY AND DIVIDE FRACTIONS.

4 ADD AND SUBTRACT FRACTIONS.

5 SOLVE PROBLEMS THAT INVOLVE OPERATIONS WITH FRACTIONS.

FOCUS ON PROBLEM SOLVING

A hardware store sells a 40-piece socket wrench set. The measure of the largest socket is 3/4 inch, while the measure of the smallest socket is 3/16 inch. What is the difference between these measures?

A common use of fractions is the measurement of dimensions in tools, building materials, and so on. In order to work the problem above, we must know how to subtract fractions. Operations with fractions are covered in this section. This problem is Exercise 75 in the exercises for this section. After working through this section, you should be able to solve this problem.

As preparation for the study of algebra, this section begins with a brief review of arithmetic. In everyday life the numbers seen most often are the **natural numbers,**

$$1, 2, 3, 4, \ldots,$$

the **whole numbers,**

$$0, 1, 2, 3, 4, \ldots,$$

and the **fractions,** such as

$$\frac{1}{2}, \quad \frac{2}{3}, \quad \frac{11}{12},$$

and so on. In a fraction, the top number is called the **numerator** and the bottom number is called the **denominator.**

1 In the statement $2 \times 9 = 18$, the numbers 2 and 9 are called **factors** of 18. Other factors of 18 include 1, 3, 6, and 18. The result of the multiplication, 18, is called the **product.**

The number 18 is **factored** by writing it as the product of two or more numbers. For example, 18 can be factored in several ways, as $6 \cdot 3$, or $18 \cdot 1$, or $9 \cdot 2$, or $3 \cdot 3 \cdot 2$. In algebra, raised dots are used instead of the $\times$ symbol to indicate multiplication.

A natural number (except 1) is **prime** if it has only itself and 1 as factors. "Factors" are understood here to mean natural number factors. (By agreement, the number 1 is not a prime number.) The first dozen primes are listed here.

$$2, 3, 5, 7, 11, 13, 17, 19, 23, 29, 31, 37$$

It is often useful to find all the **prime factors** of a number—those factors that are prime numbers. For example, the only prime factors of 18 are 2 and 3.

EXAMPLE 1
FACTORING NUMBERS

Write each number as the product of prime factors.

(a) 35

Write 35 as the product of the prime factors 5 and 7, or as

$$35 = 5 \cdot 7.$$

(b) 24

One way to begin is to divide by the smallest prime, 2, to get

$$24 = 2 \cdot 12.$$

Now divide 12 by 2 to find factors of 12.

$$24 = 2 \cdot 2 \cdot 6$$

Since 6 can be written as $2 \cdot 3$,

$$24 = 2 \cdot 2 \cdot 2 \cdot 3,$$

where all factors are prime. ■

NOTE It is not necessary to start with the smallest prime factor, as shown in Example 1(b). In fact, no matter which prime factor we start with, we will *always* obtain the same prime factorization.

2 Prime factors are used to write fractions in **lowest terms.** A fraction is in lowest terms when the numerator and the denominator have no factors in common (other than 1). Use the following steps to write a fraction in lowest terms.

WRITING A FRACTION IN LOWEST TERMS

Step 1 Write the numerator and the denominator as the product of prime factors.

Step 2 Divide the numerator and the denominator by the **greatest common factor,** the product of all factors common to both.

EXAMPLE 2
WRITING FRACTIONS IN LOWEST TERMS

Write each fraction in lowest terms.

(a) $\frac{10}{15} = \frac{2 \cdot 5}{3 \cdot 5} = \frac{2}{3}$

Since 5 is the greatest common factor of 10 and 15, dividing both numerator and denominator by 5 gives the fraction in lowest terms.

(b) $\frac{15}{45} = \frac{3 \cdot 5}{3 \cdot 3 \cdot 5} = \frac{1 \cdot 3 \cdot 5}{3 \cdot 3 \cdot 5} = \frac{1}{3}$

The factored form shows that 3 and 5 are the common factors of both 15 and 45. Dividing both 15 and 45 by $3 \cdot 5 = 15$ gives 15/45 in lowest terms as 1/3. ■

NOTE When you are factoring to write a fraction in lowest terms, you can simplify the process if you can find the greatest common factor in the numerator and denominator by inspection. For example, in Example 2(b), we can use 15 rather than $3 \cdot 5$.

3 The basic operations on whole numbers are addition, subtraction, multiplication, and division. These same operations apply to fractions. We multiply two fractions by first multiplying their numerators and then multiplying their denominators. This rule is written in symbols as follows.

MULTIPLYING FRACTIONS

If $\frac{a}{b}$ and $\frac{c}{d}$ are fractions, then $\frac{a}{b} \cdot \frac{c}{d} = \frac{a \cdot c}{b \cdot d}$.

EXAMPLE 3
MULTIPLYING FRACTIONS

Find the product of 3/8 and 4/9, and write it in lowest terms.

First, multiply 3/8 and 4/9.

$$\frac{3}{8} \cdot \frac{4}{9} = \frac{3 \cdot 4}{8 \cdot 9}$$ Multiply numerators; multiply denominators.

It is easiest to write a fraction in lowest terms while the product is in factored form. Factor 8 and 9 and then divide out common factors in the numerator and denominator.

$$\frac{3 \cdot 4}{8 \cdot 9} = \frac{3 \cdot 4}{2 \cdot 4 \cdot 3 \cdot 3} = \frac{1 \cdot 3 \cdot 4}{2 \cdot 4 \cdot 3 \cdot 3}$$ Factor. Introduce a factor of 1.

$$= \frac{1}{2 \cdot 3}$$ 3 and 4 are common factors.

$$= \frac{1}{6}$$ Lowest terms ■

Two fractions are **reciprocals** of each other if their product is 1. For example, 3/4 and 4/3 are reciprocals since

$$\frac{3}{4} \cdot \frac{4}{3} = \frac{12}{12} = 1.$$

Also, 7/11 and 11/7 are reciprocals of each other. The reciprocal is used to divide fractions. To *divide* two fractions, multiply the first fraction and the reciprocal of the second one.

DIVIDING FRACTIONS

For the fractions $\frac{a}{b}$ and $\frac{c}{d}$, $\frac{a}{b} \div \frac{c}{d} = \frac{a}{b} \cdot \frac{d}{c}$.

The reason this method works will be explained in Chapter 5. The answer to a division problem is called a **quotient.** For example, the quotient of 20 and 10 is 2, since $20 \div 10 = 2$.

EXAMPLE 4 DIVIDING FRACTIONS

Find the following quotients, and write them in lowest terms.

(a) $\frac{3}{4} \div \frac{8}{5} = \frac{3}{4} \cdot \frac{5}{8} = \frac{3 \cdot 5}{4 \cdot 8} = \frac{15}{32}$ Multiply by the reciprocal of $\frac{8}{5}$.

(b) $\frac{3}{4} \div \frac{5}{8} = \frac{3}{4} \cdot \frac{8}{5} = \frac{3 \cdot 8}{4 \cdot 5} = \frac{3 \cdot 4 \cdot 2}{4 \cdot 5} = \frac{6}{5}$ ■

4 The **sum** of two fractions having the same denominator is found by adding the numerators, and keeping the same denominator.

ADDING FRACTIONS

If $\frac{a}{b}$ and $\frac{c}{b}$ are fractions, then $\frac{a}{b} + \frac{c}{b} = \frac{a+c}{b}$.

EXAMPLE 5 ADDING FRACTIONS WITH THE SAME DENOMINATOR

Add.

(a) $\frac{3}{7} + \frac{2}{7} = \frac{3+2}{7} = \frac{5}{7}$ Add numerators and keep the same denominator.

(b) $\frac{2}{10} + \frac{3}{10} = \frac{2+3}{10} = \frac{5}{10} = \frac{1}{2}$ ■

If two fractions to be added do not have the same denominators, the rule above can still be used, but only after the fractions are rewritten with a common denominator. For example, to rewrite 3/4 as a fraction with a denominator of 32,

$$\frac{3}{4} = \frac{?}{32},$$

find the number that can be multiplied by 4 to give 32. Since $4 \cdot 8 = 32$, use the number 8. Multiplying a number by 1 does not change its value. The value of 3/4 will be the same if it is multiplied by 8/8, which equals 1.

$$\frac{3}{4} = \frac{3}{4} \cdot \frac{8}{8} = \frac{3 \cdot 8}{4 \cdot 8} = \frac{24}{32}$$

EXAMPLE 6
CHANGING THE DENOMINATOR

Write 5/8 as a fraction with a denominator of 72.

Since 8 must be multiplied by 9 to get 72, multiply both numerator and denominator by 9.

$$\frac{5}{8} = \frac{5}{8} \cdot \frac{9}{9} = \frac{5 \cdot 9}{8 \cdot 9} = \frac{45}{72} \quad \blacksquare$$

EXAMPLE 7
ADDING FRACTIONS WITH DIFFERENT DENOMINATORS

Add the following fractions.

(a) $\frac{1}{2} + \frac{1}{3}$

These fractions cannot be added until both have the same denominator. Use 6 as a common denominator, since both 2 and 3 divide into 6. Write both 1/2 and 1/3 as fractions with a denominator of 6.

$$\frac{1}{2} = \frac{1}{2} \cdot \frac{3}{3} = \frac{1 \cdot 3}{2 \cdot 3} = \frac{3}{6} \quad \text{and} \quad \frac{1}{3} = \frac{1}{3} \cdot \frac{2}{2} = \frac{1 \cdot 2}{3 \cdot 2} = \frac{2}{6}$$

Now add.

$$\frac{1}{2} + \frac{1}{3} = \frac{3}{6} + \frac{2}{6} = \frac{3 + 2}{6} = \frac{5}{6}$$

(b) $3\frac{1}{2} + 2\frac{3}{4}$

These numbers are called mixed numbers. A **mixed number** is understood to be the sum of a whole number and a fraction. We can add mixed numbers using either of two methods.

Method 1

Rewrite both numbers as follows.

$$3\frac{1}{2} = 3 + \frac{1}{2} = \frac{3}{1} + \frac{1}{2} = \frac{6}{2} + \frac{1}{2} = \frac{6 + 1}{2} = \frac{7}{2}$$

$$2\frac{3}{4} = 2 + \frac{3}{4} = \frac{8}{4} + \frac{3}{4} = \frac{8 + 3}{4} = \frac{11}{4}$$

Now add. The common denominator is 4.

$$3\frac{1}{2} + 2\frac{3}{4} = \frac{7}{2} + \frac{11}{4} = \frac{14}{4} + \frac{11}{4} = \frac{25}{4} \text{ or } 6\frac{1}{4}$$

Method 2

Write 3 1/2 as 3 2/4. Then add vertically.

$$\begin{array}{r} 3\frac{1}{2} \\ +2\frac{3}{4} \\ \hline \end{array} \longrightarrow \begin{array}{r} 3\frac{2}{4} \\ +2\frac{3}{4} \\ \hline 5\frac{5}{4} \end{array}$$

Since 5/4 = 1 1/4,

$$5\frac{5}{4} = 5 + 1\frac{1}{4} = 6\frac{1}{4}, \text{ or } \frac{25}{4}. \quad ■$$

The **difference** of two numbers is found by subtraction. For example, 9 − 5 = 4 so the difference of 9 and 5 is 4. Subtraction of fractions is similar to addition. Just subtract the numerators instead of adding them, according to the following definition. Again, keep the same denominator.

SUBTRACTING FRACTIONS

$$\frac{a}{b} - \frac{c}{b} = \frac{a-c}{b}$$

EXAMPLE 8 SUBTRACTING FRACTIONS

Subtract.

(a) $\frac{5}{8} - \frac{3}{8} = \frac{5-3}{8} = \frac{2}{8} = \frac{1}{4}$ Subtract numerators and keep the same denominator.

(b) $\frac{3}{4} - \frac{1}{3}$

These numbers must be written with a common denominator to use the rule for subtraction. Here a common denominator is 12.

$$\frac{3}{4} - \frac{1}{3} = \frac{9}{12} - \frac{4}{12} = \frac{9-4}{12} = \frac{5}{12} \quad ■$$

5 Sometimes applied problems require work with fractions.

PROBLEM SOLVING

When a carpenter reads diagrams and plans, he or she often must work with fractions whose denominators are 2, 4, 8, 16, or 32. Therefore, operations with fractions are sometimes necessary to solve such problems. The next example shows a typical diagram. ■

EXAMPLE 9
ADDING FRACTIONS TO SOLVE A WOODWORKING PROBLEM

The diagram shown appears in the book *Woodworker's 39 Sure-Fire Projects*. It is the front view of a corner bookcase/desk. Add the fractions shown in the diagram to find the approximate height of the bookcase/desk.

We must add the following measures (in inches):

$$\frac{3}{4},\ 4\frac{1}{2},\ 9\frac{1}{2},\ \frac{3}{4},\ 9\frac{1}{2},\ \frac{3}{4},\ 4\frac{1}{2}.$$

Begin by changing 4 1/2 to 4 2/4 and 9 1/2 to 9 2/4, since the common denominator is 4. Then, use Method 2 from Example 7.

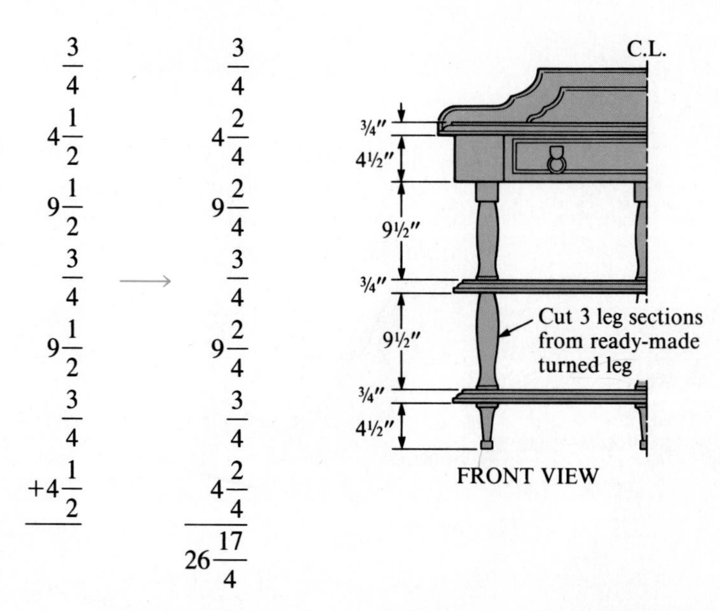

$$\begin{array}{r} \frac{3}{4} \\ 4\frac{1}{2} \\ 9\frac{1}{2} \\ \frac{3}{4} \\ 9\frac{1}{2} \\ \frac{3}{4} \\ +4\frac{1}{2} \\ \hline \end{array} \quad \longrightarrow \quad \begin{array}{r} \frac{3}{4} \\ 4\frac{2}{4} \\ 9\frac{2}{4} \\ \frac{3}{4} \\ 9\frac{2}{4} \\ \frac{3}{4} \\ 4\frac{2}{4} \\ \hline 26\frac{17}{4} \end{array}$$

Since 17/4 = 4 1/4, 26 17/4 = 26 + 4 1/4 = 30 1/4. The approximate height is 30 1/4 inches. It is best to give answers as mixed numbers in applications like this. ■

1.1 EXERCISES

Write each number as the product of prime factors. See Example 1.

1. 30 **2.** 40 **3.** 50 **4.** 72
5. 65 **6.** 85 **7.** 100 **8.** 110
9. 17 **10.** 13 **11.** 124 **12.** 120

Write each fraction in lowest terms. See Example 2.

13. $\frac{7}{14}$ **14.** $\frac{3}{9}$ **15.** $\frac{10}{12}$ **16.** $\frac{8}{10}$ **17.** $\frac{16}{18}$

18. $\frac{14}{20}$ **19.** $\frac{50}{75}$ **20.** $\frac{32}{48}$ **21.** $\frac{72}{108}$ **22.** $\frac{96}{120}$

23. Which one of the following is the correct way to write 16/24 in lowest terms? (See Objective 2.*)

(a) $\frac{16}{24} = \frac{8+8}{8+16} = \frac{8}{16} = \frac{1}{2}$

(b) $\frac{16}{24} = \frac{4 \cdot 4}{4 \cdot 6} = \frac{4}{6}$

(c) $\frac{16}{24} = \frac{8 \cdot 2}{8 \cdot 3} = \frac{2}{3}$

(d) $\frac{16}{24} = \frac{14+2}{21+3} = \frac{2}{3}$

24. If p/q and r/s are fractions, which one of the following can serve as a common denominator? (See Objective 3.)

(a) $q \cdot s$

(b) $q + s$

(c) $p \cdot r$

(d) $p + r$

Find the products or quotients, and write answers in lowest terms. In Exercises 39–42, begin by writing the mixed numbers as fractions. See Examples 3 and 4.

25. $\frac{3}{4} \cdot \frac{3}{5}$ **26.** $\frac{3}{8} \cdot \frac{5}{7}$ **27.** $\frac{1}{10} \cdot \frac{6}{5}$ **28.** $\frac{6}{7} \cdot \frac{1}{3}$

29. $\frac{9}{4} \cdot \frac{8}{15}$ **30.** $\frac{3}{5} \cdot \frac{20}{15}$ **31.** $\frac{3}{8} \div \frac{5}{4}$ **32.** $\frac{9}{16} \div \frac{3}{8}$

33. $\frac{5}{12} \div \frac{15}{4}$ **34.** $\frac{15}{16} \div \frac{30}{8}$ **35.** $\frac{15}{32} \div \frac{25}{8}$ **36.** $\frac{24}{25} \div \frac{3}{50}$

37. $\frac{5}{9} \cdot \frac{7}{10}$ **38.** $\frac{21}{30} \cdot \frac{5}{7}$ **39.** $1\frac{3}{10} \cdot 1\frac{2}{3}$ **40.** $1\frac{5}{16} \cdot 1\frac{1}{7}$

41. $9\frac{1}{3} \div 1\frac{1}{6}$ **42.** $13\frac{4}{9} \div \frac{11}{18}$ **43.** $\frac{28}{15} \div \frac{14}{5}$ **44.** $\frac{120}{7} \div \frac{45}{3}$

Add or subtract the following fractions. Write answers in lowest terms. Work from left to right in Exercises 63–66. See Examples 5–8.

45. $\frac{1}{12} + \frac{5}{12}$ **46.** $\frac{2}{5} + \frac{1}{5}$ **47.** $\frac{1}{10} + \frac{7}{10}$ **48.** $\frac{3}{8} + \frac{1}{8}$

49. $\frac{4}{9} + \frac{2}{3}$ **50.** $\frac{3}{5} + \frac{2}{15}$ **51.** $\frac{8}{11} + \frac{3}{22}$ **52.** $\frac{9}{10} + \frac{3}{5}$

53. $\frac{2}{3} - \frac{3}{5}$ **54.** $\frac{7}{12} - \frac{5}{9}$ **55.** $\frac{5}{6} - \frac{3}{10}$ **56.** $\frac{11}{4} - \frac{5}{8}$

57. $3\frac{1}{4} + \frac{1}{8}$ **58.** $5\frac{2}{3} + \frac{1}{4}$ **59.** $4\frac{1}{2} + 3\frac{2}{3}$ **60.** $7\frac{5}{8} + 3\frac{3}{4}$

*For help with this exercise, read the discussion of Objective 2 in this section.

61. $6\frac{2}{3} - 5\frac{1}{4}$ **62.** $8\frac{8}{9} - 7\frac{4}{5}$ **63.** $\frac{2}{5} + \frac{1}{3} + \frac{9}{10}$ **64.** $\frac{3}{8} + \frac{5}{6} + \frac{2}{3}$

65. $\frac{5}{14} + \frac{1}{6} - \frac{4}{9}$ **66.** $\frac{2}{15} + \frac{1}{6} - \frac{1}{10}$

67. A cent is equal to 1/100 of one dollar. Two dimes added to three dimes gives an amount equal to that of one half-dollar. Give an arithmetic problem using fractions that describes this equality, using 100 as a denominator throughout.

68. Three nickels added to twelve nickels gives an amount equal to that of three quarters. Give an arithmetic problem using fractions that describes this equality, using 100 as a denominator throughout.

Work each of the following problems. See Example 9.

69. Wing's favorite recipe for barbecue sauce calls for 2 1/3 cups of tomato sauce. The recipe makes enough barbecue sauce to serve 7 people. How much tomato sauce is needed for 1 serving?

70. If an upholsterer needs 2 1/4 yards of fabric to re-cover a reclining chair, how many chairs can be re-covered with 27 yards of fabric?

71. Last month, the Salvage Recycling Center received 3 1/4 tons of newspaper, 2 3/8 tons of aluminum cans, 7 1/2 tons of glass, and 1 5/16 tons of used writing paper. Find the total number of tons of material received by the center during the month.

72. A motel owner has decided to expand his business by buying a piece of property next to the motel. The property has an irregular shape, with five sides as shown in the figure. Find the total distance around the piece of property.

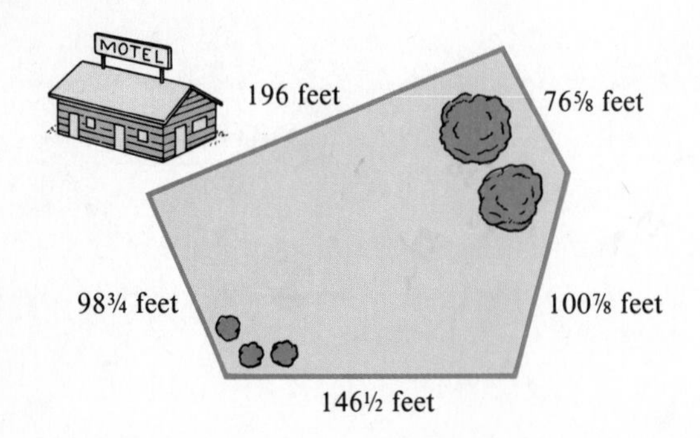

73. Martha Wailes works 40 hours per week in a stationery store. She worked 8 1/4 hours on Monday, 6 3/8 hours on Tuesday, 7 2/3 hours on Wednesday, and 8 3/4 hours on Thursday. How many hours did she work on Friday?

74. A concrete truck is loaded with 9 7/8 cubic yards of concrete. The driver delivers 1 1/2 cubic yards at the first stop and 2 3/4 cubic yards at the second stop. At the third stop, the customer receives 3 5/12 cubic yards. How much concrete is left in the truck?

75. A hardware store sells a 40-piece socket wrench set. The measure of the largest socket is 3/4 inch, while the measure of the smallest socket is 3/16 inch. What is the difference between these measures?

76. Two sockets in a socket wrench set have measures of 9/16 inch and 3/8 inch. What is the difference between these two measures?

77. In your own words, write an explanation of how to add two fractions that have different denominators. (See Objective 4.)

78. In your own words, write an explanation of how to divide two fractions. (See Objective 3.)

1.2 EXPONENTS, ORDER OF OPERATIONS, AND INEQUALITY

OBJECTIVES

1 USE EXPONENTS.
2 USE THE ORDER OF OPERATIONS.
3 USE MORE THAN ONE GROUPING SYMBOL.
4 KNOW THE MEANINGS OF $\neq, <, >, \leq$, AND $\geq$.
5 TRANSLATE WORD STATEMENTS TO SYMBOLS.
6 WRITE STATEMENTS THAT CHANGE THE DIRECTION OF INEQUALITY SYMBOLS.

1 It is common for a multiplication problem to have the same factor appearing several times. For example, in the product

$$3 \cdot 3 \cdot 3 \cdot 3 = 81$$

the factor 3 appears four times. In algebra, repeated factors are written with an *exponent*. For example, in $3 \cdot 3 \cdot 3 \cdot 3$, the number 3 appears as a factor four times, so the product is written as 3^4, and is read "3 to the fourth power."

$$3 \cdot 3 \cdot 3 \cdot 3 = 3^4$$

The number 4 is the **exponent** and 3 is the **base.** A natural number exponent, then, tells how many times the base is used as a factor. A number raised to the first power is simply that number. For example, $5^1 = 5$ and $(1/2)^1 = 1/2$.

EXAMPLE 1 EVALUATING A NUMBER RAISED TO A POWER

Find the values of the following.

(a) 5^2

$$\underbrace{5 \cdot 5}_{} = 25$$

5 is used as a factor 2 times.

Read 5^2 as "5 squared."

(b) 6^3

$$\underbrace{6 \cdot 6 \cdot 6}_{} = 216$$

6 is used as a factor 3 times.

Read 6^3 as "6 cubed."

(c) $2^5 = 2 \cdot 2 \cdot 2 \cdot 2 \cdot 2 = 32$ 2 is used as a factor 5 times.
Read 2^5 as "2 to the fifth power."

(d) $\left(\frac{2}{3}\right)^3 = \frac{2}{3} \cdot \frac{2}{3} \cdot \frac{2}{3} = \frac{8}{27}$ $\frac{2}{3}$ is used as a factor 3 times. ■

2 Many problems involve more than one operation of arithmetic. To indicate the order in which the operations should be performed, we often use **grouping symbols.** If no grouping symbols are used, we apply the order of operations rules, which are discussed below.

Suppose we consider the expression $5 + 3 \cdot 2$. If we wish to show that the multiplication should be performed before the addition, parentheses can be used to write

$$5 + (2 \cdot 3) = 5 + 6 = 11.$$

If addition is to be performed first, the parentheses should group $5 + 2$ as follows.

$$(5 + 2) \cdot 3 = 7 \cdot 3 = 21$$

Other grouping symbols used in more complicated expressions are

brackets, [ ], braces, { },

and fraction bars.

The most useful way to work problems with more than one operation is to use the following **order of operations.** This order is used by most calculators and computers.

ORDER OF OPERATIONS

If grouping symbols are present, simplify within them, innermost first (and above and below fraction bars separately), in the following order.

Step 1 Apply all exponents.

Step 2 Do any multiplications or divisions in the order in which they occur, working from left to right.

Step 3 Do any additions or subtractions in the order in which they occur, working from left to right.

If no grouping symbols are present, start with Step 1.

A dot has been used to show multiplication; another way to show multiplication is with parentheses. For example, 3(7), (3)7, and (3)(7) each mean $3 \cdot 7$ or 21. The next example shows the use of parentheses for multiplication.

EXAMPLE 2 **USING THE ORDER OF OPERATIONS**

Find the values of the following.

(a) $9(6 + 11)$

Using the order of operations given above, work first inside the parentheses.

$$\begin{aligned} 9(6 + 11) &= 9(17) && \text{Work inside parentheses.} \\ &= 153 && \text{Multiply.} \end{aligned}$$

(b) $6 \cdot 8 + 5 \cdot 2$

Do any multiplications, working from left to right, and then add.

$$\begin{aligned} 6 \cdot 8 + 5 \cdot 2 &= 48 + 10 && \text{Multiply.} \\ &= 58 && \text{Add.} \end{aligned}$$

(c) $2(5 + 6) + 7 \cdot 3 = 2(11) + 7 \cdot 3$ Work inside parentheses.
$= 22 + 21$ Multiply.
$= 43$ Add.

(d) $9 + 2^3 - 5$
Find 2^3 first.

$$9 + 2^3 - 5 = 9 + 2 \cdot 2 \cdot 2 - 5 \quad \text{Use the exponent.}$$
$$= 9 + 8 - 5 \quad \text{Multiply.}$$
$$= 17 - 5 \quad \text{Add.}$$
$$= 12 \quad \text{Subtract.}$$

(e) $16 - 3^2 + 4^2 = 16 - 3 \cdot 3 + 4 \cdot 4$ Use the exponents.
$= 16 - 9 + 16$ Multiply.
$= 7 + 16$ Subtract.
$= 23$ Add.

Notice that 3^2 must be evaluated before subtracting. ■

NOTE Parentheses and fraction bars are used as grouping symbols to indicate an expression that represents a single number. That is why we must first simplify within parentheses and above and below fraction bars.

3 An expression with double parentheses, such as $2(8 + 3(6 + 5))$, can be confusing. To eliminate this, square brackets, [ ], often are used instead of one of the pairs of parentheses, as shown in the next example.

EXAMPLE 3 USING BRACKETS

Simplify $2[8 + 3(6 + 5)]$.

Work first within the parentheses, and then simplify until a single number is found inside the brackets.

$$2[8 + 3(6 + 5)] = 2[8 + 3(11)]$$
$$= 2[8 + 33]$$
$$= 2[41]$$
$$= 82 \quad ■$$

Sometimes fraction bars are grouping symbols, as the next example shows.

EXAMPLE 4 USING A FRACTION BAR AS A GROUPING SYMBOL

Simplify $\dfrac{4(5 + 3) + 3}{2(3) - 1}$.

The expression can be written as the quotient

$$[4(5 + 3) + 3] \div [2(3) - 1],$$

which shows that the fraction bar serves to group the numerator and denominator separately. Simplify both numerator and denominator, then divide, if possible.

$$\frac{4(5+3)+3}{2(3)-1} = \frac{4(8)+3}{2(3)-1} \quad \text{Work inside parentheses.}$$
$$= \frac{32+3}{6-1} \quad \text{Multiply.}$$
$$= \frac{35}{5} \quad \text{Add and subtract.}$$
$$= 7 \quad \text{Divide.}$$ ■

4 So far, we have used the symbols for the operations of arithmetic and the symbol for equality (=). The equality symbol with a slash through it, ≠, means "is not equal to." For example,

$$7 \neq 8$$

indicates that 7 is not equal to 8.

If two numbers are not equal, then one of the numbers must be smaller than the other. The symbol < represents "is less than," so that "7 is less than 8" is written

$$7 < 8.$$

Also, write "6 is less than 9" as $6 < 9$.

The symbol > means "is greater than." Write "8 is greater than 2" as

$$8 > 2.$$

The statement "17 is greater than 11" becomes $17 > 11$.

Keep the meanings of the symbols < and > clear by remembering that the symbol always points to the smaller number. For example, write "8 is less than 15" by pointing the symbol toward the 8:

$$8 < 15.$$

Two other symbols, ≤ and ≥, also represent the idea of inequality. The symbol ≤ means "is less than or equal to," so that

$$5 \leq 9$$

means "5 is less than or equal to 9." This statement is true, since $5 < 9$ is true. If either the < part or the = part is true, then the inequality ≤ is true.

The symbol ≥ means "is greater than or equal to." Again,

$$9 \geq 5$$

is true because $9 > 5$ is true. Also, $8 \leq 8$ is true since $8 = 8$ is true. But it is not true that $13 \leq 9$ because neither $13 < 9$ nor $13 = 9$ is true.

EXAMPLE 5
USING INEQUALITY SYMBOLS

Determine whether each statement is true or false.

(a) $6 \neq 6$
The statement is false because 6 *is equal to* 6.

(b) $5 < 19$
Since 5 represents a number that is indeed less than 19, this statement is true.

(c) $15 \leq 20$
The statement $15 \leq 20$ is true, since $15 < 20$.

(d) $25 \geq 30$
Both $25 > 30$ and $25 = 30$ are false. Because of this, $25 \geq 30$ is false.

(e) $12 \geq 12$
Since $12 = 12$, this statement is true. ■

5 An important part of algebra deals with translating words into algebraic notation.

PROBLEM SOLVING

As we will see throughout this book, the ability to solve problems using mathematics is based on translating the words of the problem into symbols. The next example is the first of many that will be included to illustrate translations from words to symbols. ■

EXAMPLE 6
TRANSLATING FROM WORDS TO SYMBOLS

Write each word statement in symbols.

(a) Twelve **equals** ten **plus** two.

$$12 = 10 + 2$$

(b) Nine **is less than** ten.

$$9 < 10$$

(c) Fifteen **is not equal to** eighteen.

$$15 \neq 18$$

(d) Seven **is greater than** four.

$$7 > 4$$

(e) Thirteen **is less than or equal to** forty.

$$13 \leq 40$$

(f) Eleven **is greater than or equal to** eleven.

$$11 \geq 11$$ ■

6 Any statement with $<$ can be converted to one with $>$, and any statement with $>$ can be converted to one with $<$. Do this by reversing the order of the numbers and the direction of the symbol. For example, the statement $6 < 10$ can be written with $>$ as $10 > 6$. Similarly, the statement $4 \leq 10$ can be changed to $10 \geq 4$.

EXAMPLE 7
CONVERTING BETWEEN INEQUALITY SYMBOLS

The following examples show the same statement written in two equally correct ways.

(a) $9 < 16$ $\quad 16 > 9$

(b) $5 > 2$ $\quad 2 < 5$

(c) $3 \leq 8$ $\quad 8 \geq 3$

(d) $12 \geq 5$ $\quad 5 \leq 12$ ■

Here is a summary of the symbols discussed in this section.

SYMBOLS OF EQUALITY AND INEQUALITY		
	$=$ is equal to	$\neq$ is not equal to
	$<$ is less than	$>$ is greater than
	$\leq$ is less than or equal to	$\geq$ is greater than or equal to

1.2 EXERCISES

Find the values of the following. See Example 1.

1. 6^2 **2.** 9^2 **3.** 8^2 **4.** 10^2

5. 17^2 **6.** 22^2 **7.** 5^3 **8.** 7^3

9. 6^4 **10.** 3^4 **11.** 2^5 **12.** 4^5

13. 3^6 **14.** 2^6 **15.** $\left(\frac{1}{2}\right)^2$ **16.** $\left(\frac{3}{4}\right)^2$

17. $\left(\frac{2}{5}\right)^3$ **18.** $\left(\frac{3}{7}\right)^3$ **19.** $\left(\frac{4}{5}\right)^3$ **20.** $\left(\frac{2}{3}\right)^5$

---

21. Is it possible to square a fraction and get a smaller number than the one you started with? If so, give an example. (See Objective 1.)

22. Explain in your own words what it means to raise a number to a power. (See Objective 1.)

23. If you are asked to evaluate $9 + 4 \cdot 3$, which operation is performed first? Why? (See Objective 2.)

24. If you are asked to evaluate $16 \div 8 \cdot 4$, which operation is performed first? Why? (See Objective 2.)

Find the values of the following expressions. See Examples 2–4.

25. $4 + 6 \cdot 2$

26. $9 + 3 \cdot 4$

27. $12 - 5 \cdot 2$

28. $16 - 3 \cdot 5$

29. $3 \cdot 8 - 4 \cdot 6$

30. $2 \cdot 20 - 8 \cdot 5$

31. $6 \cdot 5 + 3 \cdot 10$

32. $5 \cdot 8 + 10 \cdot 4$

33. $5[8 + (2 + 3)]$

34. $9[(14 + 5) - 10]$

35. $(6 - 3)[8 - (2 + 1)]$

36. $(7 - 1)[9 + (6 - 3)]$

37. $\dfrac{2(5 + 3) + 2 \cdot 2}{2(4 - 1)}$

38. $\dfrac{9(7 - 1) - 8 \cdot 2}{4(6 - 1)}$

39. $\dfrac{8^2 + 2}{5 - 2^2}$

40. $\dfrac{4^2 - 8}{15 - 3^2}$

Which of the symbols $\neq$, $<$, $>$, $\leq$, and $\geq$ make the following statements true? Give all possible correct answers.

41. 6 ____ 9 **42.** 18 ____ 12 **43.** 51 ____ 50 **44.** 0 ____ 12

45. 5 ____ 5

46. 10 ____ 10

47. 48 ____ 0

48. 100 ____ 1000

49. 16 ____ 10

50. 5 ____ 3

51. $\frac{1}{4}$ ____ $\frac{2}{5}$

52. $\frac{2}{3}$ ____ $\frac{5}{8}$

53. .609 ____ .61

54. .5 ____ .499

55. $3\frac{1}{2}$ ____ 4

56. $5\frac{7}{8}$ ____ 6

---

57. What English language phrase is used to express the fact that one person's age *is less than* another person's age? (See Objective 5.)

58. What English language phrase is used to express the fact that one person's height *is greater than* another person's height? (See Objective 5.)

Tell whether each statement is true or false. In Exercises 71–76, evaluate the expression first. See Example 5.

59. $8 + 2 = 10$

60. $8 \neq 9 - 1$

61. $12 \geq 10$

62. $45 < 45$

63. $0 < 15$

64. $16 \geq 10$

65. $1\frac{2}{3} + 2\frac{3}{4} = \frac{53}{12}$

66. $3\frac{2}{5} < 6\frac{1}{4}$

67. $\frac{25}{3} \geq \frac{19}{2}$

68. $\frac{18}{5} < \frac{5}{4}$

69. $9 < 0$

70. $15 \leq 32$

71. $[3 \cdot 4 + 5(2)] \cdot 3 > 72$

72. $2 \cdot [7 \cdot 5 - 3(2)] \leq 58$

73. $\frac{3 + 5(4 - 1)}{2 \cdot 4 + 1} \geq 3$

74. $\frac{7(3 + 1) - 2}{3 + 5 \cdot 2} \leq 2$

75. $3 \geq \frac{2(5 + 1) - 3(1 + 1)}{5(8 - 6) - 4 \cdot 2}$

76. $7 \leq \frac{3(8 - 3) + 2(4 - 1)}{9(6 - 2) - 11(5 - 2)}$

Write the following word statements in symbols. See Example 6.

77. Seven equals five plus two.

78. Nine is greater than the product of four and two.

79. Three is less than the quotient of fifty and five.

80. Five equals ten minus five.

81. Twelve is not equal to five.

82. Fifteen does not equal sixteen.

83. Zero is greater than or equal to zero.

84. Six is less than or equal to six.

Rewrite the following true statements so the inequality symbol points in the opposite direction. See Example 7.

85. $6 < 14$

86. $8 \leq 9$

87. $15 \geq 3$

88. $9 > 8$

89. $12 < 17$

90. $0 \leq 6$

---

91. $12 \geq 12$ is a true statement. Suppose that someone tells you the following: "$12 \geq 12$ is false, because even though 12 is equal to 12, 12 is not greater than 12." How would you respond to this? (See Objective 4.)

92. The symbol $\neq$ means "is not equal to." How do you think we read the symbol $\ngtr$? How do you think we read $\nless$? (See Objective 4.)

1.3 VARIABLES, EXPRESSIONS, AND EQUATIONS

OBJECTIVES

1. DEFINE *VARIABLE*, AND FIND THE VALUE OF AN ALGEBRAIC EXPRESSION, GIVEN THE VALUES OF THE VARIABLES.
2. CONVERT PHRASES FROM WORDS TO ALGEBRAIC EXPRESSIONS.
3. IDENTIFY SOLUTIONS OF EQUATIONS.
4. IDENTIFY SOLUTIONS OF EQUATIONS FROM A SET OF NUMBERS.
5. DISTINGUISH BETWEEN AN *EXPRESSION* AND AN *EQUATION*.

1 A **variable** is a symbol, usually a letter, such as x, y, or z, used to represent any unknown number. An **algebraic expression** is a collection of numbers, variables, symbols for operations, and symbols for grouping (such as parentheses). For example,

$$6(x + 5), \qquad 2m - 9, \qquad \text{and} \qquad 8p^2 + 6p + 2$$

are all algebraic expressions. In the algebraic expression $2m - 9$, the expression $2m$ indicates the product of 2 and m, just as $8p^2$ shows the product of 8 and p^2. Also, $6(x + 5)$ means the product of 6 and $x + 5$. An algebraic expression takes on different numerical values as the variables take on different values.

EXAMPLE 1 EVALUATING EXPRESSIONS

Find the numerical values of the following algebraic expressions when $m = 5$.

(a) $8m$

Replace m with 5, to get

$$8m = 8 \cdot 5 = 40.$$

(b) $3m^2$

For $m = 5$,

$$3m^2 = 3 \cdot 5^2 = 3 \cdot 25 = 75. \quad \blacksquare$$

CAUTION In Example 1(b), it is important to notice that $3m^2$ means $3 \cdot m^2$; it *does not* mean $3m \cdot 3m$. The product $3m \cdot 3m$ is indicated by $(3m)^2$.

EXAMPLE 2 EVALUATING EXPRESSIONS

Find the value of each expression when $x = 5$ and $y = 3$.

(a) $2x + 7y$

Replace x with 5 and y with 3. Do the multiplication first, and then add.

$2x + 7y = 2 \cdot 5 + 7 \cdot 3$	Let $x = 5$ and $y = 3$.
$= 10 + 21$	Multiply.
$= 31$	Add.

(b) $\dfrac{9x - 8y}{2x - y}$

Replace x with 5 and y with 3.

$$\frac{9x - 8y}{2x - y} = \frac{9 \cdot 5 - 8 \cdot 3}{2 \cdot 5 - 3} \qquad \text{Let } x = 5 \text{ and } y = 3.$$

$$= \frac{45 - 24}{10 - 3} \qquad \text{Multiply.}$$

$$= \frac{21}{7} \qquad \text{Subtract.}$$

$$= 3 \qquad \text{Divide.}$$

(c) $x^2 - 2y^2 = 5^2 - 2 \cdot 3^2$ Let $x = 5$ and $y = 3$.

$= 25 - 2 \cdot 9$ Use the exponents.

$= 25 - 18$ Multiply.

$= 7$ Subtract. ■

2 In Section 1.2 we saw how to translate from words to symbols.

PROBLEM SOLVING

Sometimes variables must be used in changing word phrases into algebraic expressions in order to solve problems. The next example illustrates this. Such translations will be used extensively in problem solving. ■

EXAMPLE 3

CHANGING WORD PHRASES TO ALGEBRAIC EXPRESSIONS

Change the following word phrases to algebraic expressions. Use x as the variable.

(a) The **sum** of a number and 9

"**Sum**" is the answer to an addition problem. This phrase translates as

$$x + 9 \quad \text{or} \quad 9 + x.$$

(b) 7 **minus** a number

"**Minus**" indicates subtraction, so the answer is $7 - x$.

CAUTION Here $x - 7$ would *not* be correct; this statement translates as "a number minus 7," not "7 minus a number." The expressions $7 - x$ and $x - 7$ are rarely equal. For example, if $x = 10$, $10 - 7 \neq 7 - 10$. ($7 - 10$ is a *negative number*, covered in Section 1.4.)

(c) 7 taken from a number

Since 7 is taken *from* a number, write $x - 7$. In this case $7 - x$ would not be correct, because "taken from" means "subtracted from."

(d) The product of 11 and a number

$$11 \cdot x \quad \text{or} \quad 11x$$

As mentioned above, $11x$ means 11 times x. No symbol is needed to indicate the product of a number and a variable.

(e) 5 divided by a number
This translates as $\frac{5}{x}$.

The expression $\frac{x}{5}$ would *not* be correct here.

(f) The product of 2, and the sum of a number and 8

$$2(x + 8)$$ ■

3 An **equation** is a statement that two algebraic expressions are equal. Therefore, an equation always includes the equality symbol, =. Examples of equations are

$$x + 4 = 11, \quad 2y = 16, \quad \text{and} \quad 4p + 1 = 25 - p.$$

SOLVING AN EQUATION

To **solve** an equation means to find the values of the variable that make the equation true. The values of the variable that make the equation true are called the **solutions** of the equation.

EXAMPLE 4 **DECIDING WHETHER A NUMBER IS A SOLUTION**

Decide whether the given number is a solution of the equation.

(a) $5p + 1 = 36$; 7
Replace p with 7.

$$5p + 1 = 36$$
$$5 \cdot 7 + 1 = 36 \quad ? \qquad \text{Let } p = 7.$$
$$35 + 1 = 36 \quad ?$$
$$36 = 36 \qquad \text{True}$$

The number 7 is a solution of the equation.

(b) $9m - 6 = 32$; 4

$$9m - 6 = 32$$
$$9 \cdot 4 - 6 = 32 \quad ? \qquad \text{Let } m = 4.$$
$$36 - 6 = 32 \quad ?$$
$$30 = 32 \qquad \text{False}$$

The number 4 is not a solution of the equation. ■

4 A **set** is a collection of objects. In mathematics, these objects are most often numbers. The objects that belong to the set, called **elements** of the set, are written between **set braces.** For example, the set containing the numbers 1, 2, 3, 4, and 5 is written as

$$\{1, 2, 3, 4, 5\}.$$

For more information about sets, see Appendix D at the back of this book.

PROBLEM SOLVING

In some cases, the set of numbers from which the solutions of an equation must be chosen is specifically stated. In an application, this set is often determined by the natural restrictions of the problem. For example, if the answer to a problem is a number of people, only whole numbers would make sense, so the set would be the set of whole numbers. In other situations the set of restrictions may be an arbitrary choice. ■

EXAMPLE 5
FINDING A SOLUTION FROM A GIVEN SET

Change each word statement to an equation. Use x as the variable. Then find all solutions for the equation from the set

$$\{0, 2, 4, 6, 8, 10\}.$$

(a) The sum of a number and four is six.

The word "is" suggests "equals." If x represents the unknown number, then translate as follows.

The sum of a number and four	is	six.
↓	↓	↓
$x + 4$	$=$	6

Try each number from the given set $\{0, 2, 4, 6, 8, 10\}$, in turn, to see that 2 is the only solution of $x + 4 = 6$.

(b) 9 more than five times a number is 49.

Use x to represent the unknown number.

9	more than	five times a number	is	49.
↓	↓	↓	↓	↓
9	$+$	$5x$	$=$	49

Try each number from $\{0, 2, 4, 6, 8, 10\}$. The solution is 8, since $9 + 5 \cdot 8 = 49$. ■

5 Students often have trouble distinguishing between equations and expressions. Remember that an equation is a sentence; an expression is a phrase.

EXAMPLE 6
DISTINGUISHING BETWEEN EQUATIONS AND EXPRESSIONS

Decide whether each of the following is an equation or an expression.

(a) $2x - 5y$

There is no equals sign, so this is an expression.

(b) $2x = 5y$

Because of the equals sign, this is an equation. ■

1.3 EXERCISES

Find the numerical values of the following when (a) $x = 3$ *and* (b) $x = 15$. *See Example 1.*

1. $x + 9$

2. $x - 1$

3. $5x$

4. $7x$

5. $2x + 8$

6. $9x - 5$

7. $\frac{x + 1}{3}$

8. $\frac{x - 2}{5}$

9. $\frac{3x - 5}{2x}$

10. $\frac{x + 2}{x - 1}$

11. $3x^2 + x$

12. $2x + x^2$

---

13. Explain in your own words why, when evaluating the expression $4x^2$ for $x = 3$, 3 must be squared *before* multiplying by 4. (See Objective 1.)

14. What value of x would cause the expression $2x + 3$ to equal 9? (See Objective 1.)

15. There are many pairs of values of x and y for which $2x + y$ will equal 6. Name two such pairs. (See Objective 1.)

16. Suppose that for the equation $3x - y = 9$, the value of x is given to be 4. What would be the corresponding value of y? (See Objective 4.)

Find the numerical values of the following when (a) $x = 4$ *and* $y = 2$ *and* (b) $x = 1$ *and* $y = 5$. *See Example 2.*

17. $3(x + 2y)$

18. $2(2x + y)$

19. $x + \frac{4}{y}$

20. $y + \frac{8}{x}$

21. $\frac{x}{3} + \frac{5}{y}$

22. $\frac{x}{5} + \frac{y}{4}$

23. $5(4x + 7y)$

24. $8(5x + 9y)$

25. $\frac{2x + 3y}{x + y + 1}$

26. $\frac{5x + 3y + 1}{2x}$

27. $\frac{2x + 4y - 6}{5y + 2}$

28. $\frac{4x + 3y - 1}{x}$

29. $\frac{x^2 + y^2}{x + y}$

30. $\frac{9x^2 + 4y^2}{3x^2 + 2y}$

31. $x + y(3 + x)$

32. $x + y(5 - y)$

Change the word phrases to algebraic expressions. Use x to represent the variable. See Example 3.

33. Eight times a number

34. Fifteen divided by a number

35. The quotient of five and a number

36. Six added to a number

37. A number subtracted from eight

38. Nine subtracted from a number

39. Eight added to the product of a number and three

40. The difference between five times a number and six

41. Eight times a number, added to fifty-two

42. Six added to two-thirds of a number

Decide whether the given number is a solution of the equation. See Example 4.

43. $p - 5 = 12$; 17
44. $x + 6 = 15$; 9
45. $5m + 2 = 7$; 2
46. $3r + 5 = 8$; 2
47. $2y + 3(y - 2) = 14$; 4
48. $6a + 2(a + 3) = 14$; 1
49. $6p + 4p - 9 = 11$; 2
50. $2x + 3x + 8 = 38$; 6
51. $3r^2 - 2 = 46$; 4
52. $2x^2 + 1 = 19$; 3
53. $\dfrac{z + 4}{z - 2} = 2$; 8
54. $\dfrac{x + 6}{x - 2} = 9$; 3

---

55. In the phrase "Four more than the product of a number and 6," does the word *and* signify the operation of addition? Explain. (See Objective 2.)

56. Suppose that someone says the following: "The solution of $2x = 4$ is 2, and the solution of $2 + x = 4$ is 2. So it really doesn't matter whether we add or multiply on the left; we will always get the same solution in a simple equation like this." Is this a correct conclusion? If not, give an example of a pair of simple equations of this type that have different solutions. (See Objectives 3 and 4.)

Change the word statements to equations. Use x as the variable. Find the solutions from the set {0, 1, 2, 3, 4, 5, 6}. See Example 5.

57. The sum of a number and 8 is 12.
58. A number minus three equals two.
59. Twice a number plus two is ten.
60. The sum of twice a number and 6 is 18.
61. Five more than twice a number is 13.
62. The product of a number and 8 is 24.
63. Three times a number is equal to two more than twice the number.
64. Twelve divided by a number equals three times that number.
65. The quotient of twenty and five times a number is 2.
66. A number divided by 2 is 0.

Decide whether each of the following is an equation or an expression. See Example 6.

67. $2x + 5y - 7$
68. $2x + 5 = 7$
69. $\dfrac{x}{y - 3} = 4x$
70. $\dfrac{3x - 1}{5}$
71. $x^2 + 6x + 9$
72. $x^2 + 6x + 9 = 0$

---

73. Explain in your own words the difference between an *expression* and an *equation*. (See Objective 5.)

74. Suppose that the directions on a test read "Solve the following expressions." How would you politely correct the person who wrote these directions? (See Objective 5.)

Find the value of the following when $x = 4$.

75. $\dfrac{(2x - 3)(5x + 2)}{x - 1}$
76. $\dfrac{7x - 3}{(x + 2)(x - 1)}$
77. $\dfrac{2[4(x + 3) - x]}{2(x + 1)}$
78. $\dfrac{3[x(x - 1) + 2]}{5(2x - 5)}$

1.4 REAL NUMBERS AND THE NUMBER LINE

OBJECTIVES

1. SET UP NUMBER LINES.
2. IDENTIFY NATURAL NUMBERS, WHOLE NUMBERS, INTEGERS, RATIONAL NUMBERS, IRRATIONAL NUMBERS, AND REAL NUMBERS.
3. TELL WHICH OF TWO DIFFERENT REAL NUMBERS IS SMALLER.
4. FIND ADDITIVE INVERSES OF REAL NUMBERS.
5. FIND ABSOLUTE VALUES OF REAL NUMBERS.

1 In Section 1.1 we introduced two important sets of numbers, the *natural numbers* and the *whole numbers*.

NATURAL NUMBERS {1, 2, 3, 4, . . .} is the set of **natural numbers.**

WHOLE NUMBERS {0, 1, 2, 3, . . .} is the set of **whole numbers.**

NOTE The three dots show that the list of numbers continues in the same way indefinitely.

These numbers, along with many others, can be represented on **number lines** like the one pictured in Figure 1.1. We draw a number line by locating any point on the line and calling it 0. Choose any point to the right of 0 and call it 1. The distance between 0 and 1 gives a unit of measure used to locate other points, as shown in Figure 1.1. The points labeled in Figure 1.1 and those continuing in the same way to the right correspond to the set of whole numbers.

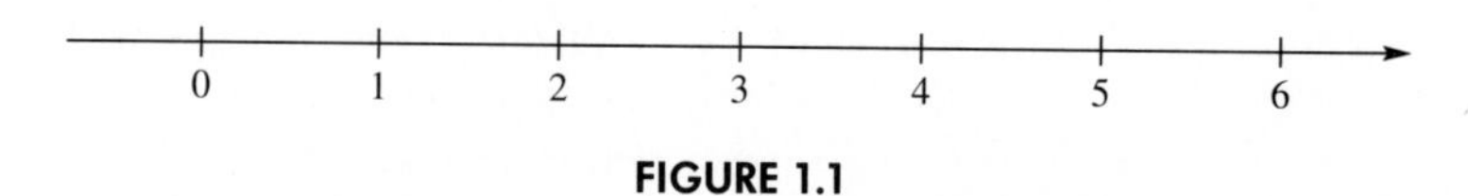

FIGURE 1.1

All the whole numbers starting with 1 are located to the right of 0 on the number line. But numbers may also be placed to the left of 0. These numbers, written -1, -2, -3, and so on, are shown in Figure 1.2. (The minus sign is used to show that the numbers are located to the *left* of 0.)

The numbers to the *left* of 0 are **negative numbers.** The numbers to the *right* of 0 are **positive numbers.** The number 0 itself is neither positive nor negative. Positive numbers and negative numbers are called **signed numbers.**

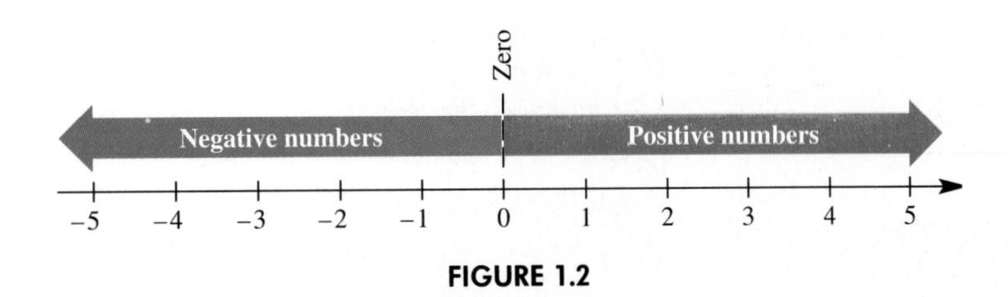

FIGURE 1.2

There are many practical applications of negative numbers. For example, temperatures sometimes fall below zero. The lowest temperature ever recorded in meteorological records was $-128.6°$ F at Vostok, Antarctica, on July 22, 1983. A business that spends more than it takes in has a negative "profit." Altitudes below sea level can be represented by negative numbers. The shore surrounding the Dead Sea is 1312 feet below sea level; this can be represented as -1312 feet.

2 The set of numbers marked on the number line in Figure 1.2, including positive and negative numbers and zero, is part of the set of *integers*.

INTEGERS $\{\ldots, -3, -2, -1, 0, 1, 2, 3, \ldots\}$ is the set of **integers.**

Not all numbers are integers. For example, 1/2 is not; it is a number halfway between the integers 0 and 1. Also, 3 1/4 is not an integer. Several numbers that are not integers are *graphed* in Figure 1.3. The **graph** of a number is a point on the number line. Think of the graph of a set of numbers as a picture of the set.

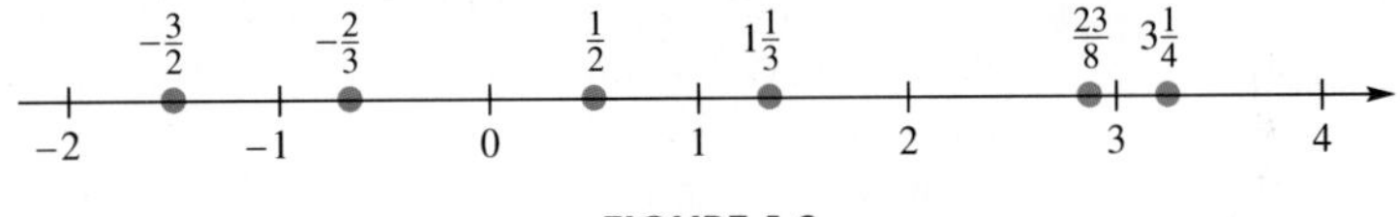

FIGURE 1.3

All the numbers in Figure 1.3 can be written as quotients of integers. These numbers are examples of *rational numbers*.

RATIONAL NUMBERS $\{x \mid x$ is a quotient of two integers, with denominator not $0\}$ is the set of **rational numbers.**
(Read the part in the braces as "the set of all numbers x such that x is a quotient of two integers, with denominator not 0.")

NOTE The set symbolism used in the definition of rational numbers,

$$\{x \mid x \text{ has a certain property}\},$$

is called **set-builder notation.** This notation is convenient to use when it is not possible to list all the elements of the set.

Since any integer can be written as the quotient of itself and 1, all integers also are rational numbers.

All numbers that can be represented by points on the number line are called *real numbers*.

REAL NUMBERS $\{x \mid x$ is a number that can be represented by a point on the number line$\}$ is the set of **real numbers.**

Although a great many numbers are rational, not all are. For example, a floor tile one foot on a side has a diagonal whose length is the square root of 2 (written $\sqrt{2}$). It can be shown that $\sqrt{2}$ cannot be written as a quotient of integers. Because of this, $\sqrt{2}$ is not rational; it is *irrational.*

IRRATIONAL NUMBERS $\{x \mid x$ is a real number that is not rational$\}$ is the set of **irrational numbers.**

Examples of irrational numbers include $\sqrt{3}$, $\sqrt{7}$, $-\sqrt{10}$, and π, which is the ratio of the distance around a circle to the distance across it.

Real numbers can be written as decimal numbers. Any rational number will have a decimal that will come to an end (terminate), or repeat in a fixed "block" of digits. For example, $2/5 = .4$ and $27/100 = .27$ are rational numbers with terminating decimals; $1/3 = .3333\ldots$ and $3/11 = .27272727\ldots$ are repeating decimals. The decimal representation of an irrational number will neither terminate nor repeat. (A review of decimal numbers can be found in Appendix C.)

An example of a number that is not a real number is the square root of a negative number. These numbers are discussed in the last chapter of this book.

Two ways to represent the relationships among the various types of numbers are shown in Figure 1.4. Part (a) also gives some examples.

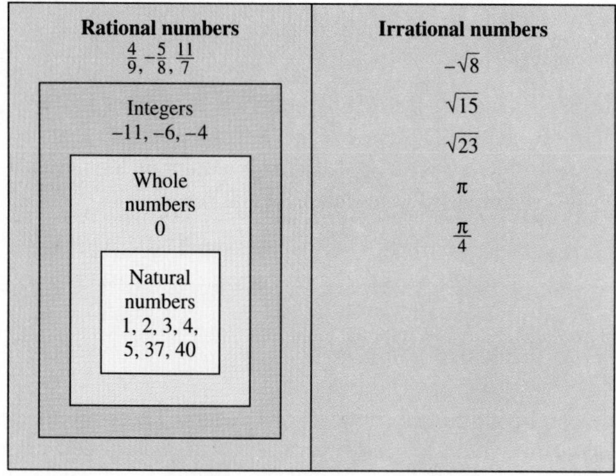

All numbers shown are real numbers.

(a)

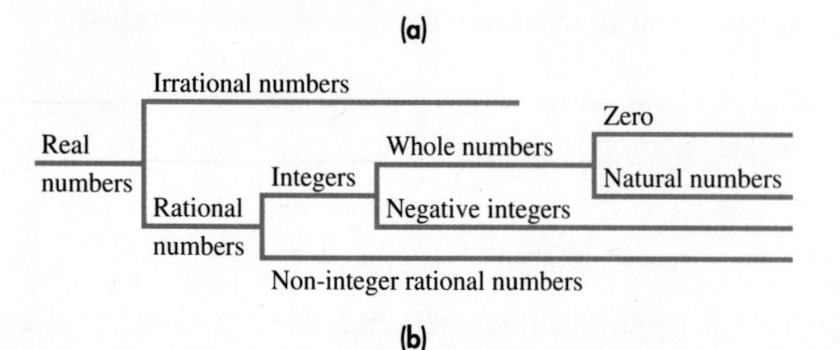

(b)

FIGURE 1.4

EXAMPLE 1
DETERMINING WHETHER A NUMBER BELONGS TO A SET

List the numbers in the set

$$\left\{-5, \quad -\frac{2}{3}, \quad 0, \quad \sqrt{2}, \quad 3\frac{1}{4}, \quad 5, \quad 5.8\right\}$$

that belong to each of the following sets of numbers.

(a) Natural numbers
The only natural number in the set is 5.

(b) Whole numbers
The whole numbers consist of the natural numbers and 0. So the elements of the set that are whole numbers are 0 and 5.

(c) Integers
The integers in the set are -5, 0, and 5.

(d) Rational numbers
The rational numbers are -5, $-2/3$, 0, 3 1/4, 5, 5.8, since each of these numbers *can* be written as the quotient of two integers. For example, $5.8 = 58/10$.

(e) Irrational numbers
The only irrational number in the set is $\sqrt{2}$.

(f) Real numbers
All the numbers in the set are real numbers. ■

3 Given any two whole numbers, you probably can tell which number is smaller. But what happens with negative numbers, as in the set of integers? Positive numbers decrease as the corresponding points on the number line go to the left. For example, $8 < 12$, and 8 is to the left of 12 on the number line. This ordering is extended to all real numbers by definition.

THE ORDERING OF REAL NUMBERS

For any two real numbers a and b, ***a* is less than *b*** if a is to the left of b on the number line.

This means that any negative number is smaller than 0, and any negative number is smaller than any positive number. Also, 0 is smaller than any positive number.

EXAMPLE 2
DETERMINING THE ORDER OF REAL NUMBERS

Is it true that $-3 < -1$?

To decide whether the statement $-3 < -1$ is true, locate both numbers, -3 and -1, on a number line, as shown in Figure 1.5. Since -3 is to the left of -1 on the number line, -3 is smaller than -1. The statement $-3 < -1$ is true. ■

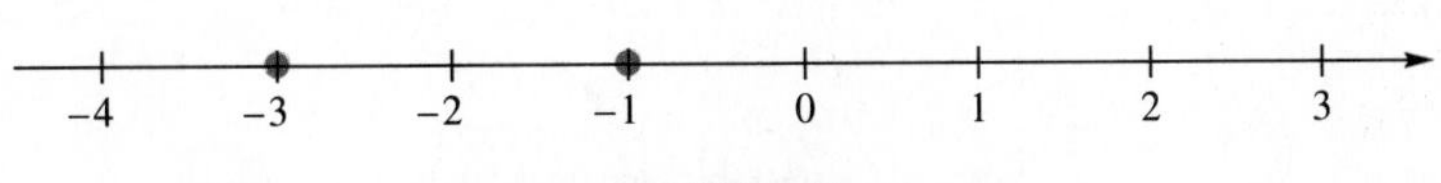

FIGURE 1.5

NOTE In Section 1.2 we saw how it is possible to rewrite a statement involving $<$ as an equivalent statement involving $>$. The question in Example 2 can also be worded as follows: Is it true that $-1 > -3$? This is, of course, also a true statement.

We can say that for any two real numbers a and b, ***a* is greater than *b*** if a is to the right of b on the number line.

4 By a property of the real numbers, for any real number x (except 0), there is exactly one number on the number line the same distance from 0 as x but on the opposite side of 0. For example, Figure 1.6 shows that the numbers 3 and -3 are each the same distance from 0 but are on opposite sides of 0. The numbers 3 and -3 are called **additive inverses,** or **opposites,** of each other.

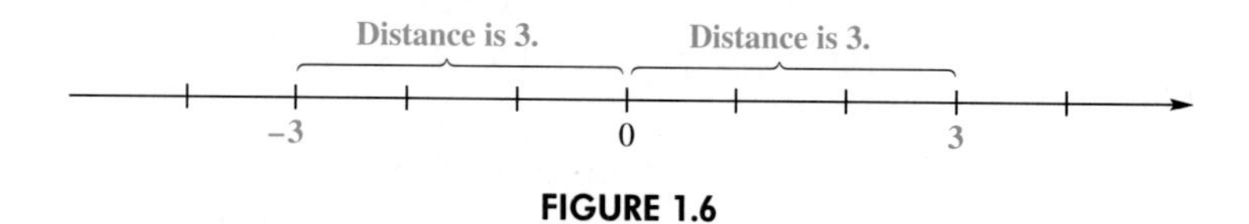

FIGURE 1.6

ADDITIVE INVERSE The **additive inverse** of a number x is the number that is the same distance from 0 on the number line as x, but on the opposite side of 0.

The additive inverse of the number 0 is 0 itself. This makes 0 the only real number that is its own additive inverse. Other additive inverses occur in pairs. For example, 4 and -4, and 5 and -5, are additive inverses of each other. Several pairs of additive inverses are shown in Figure 1.7.

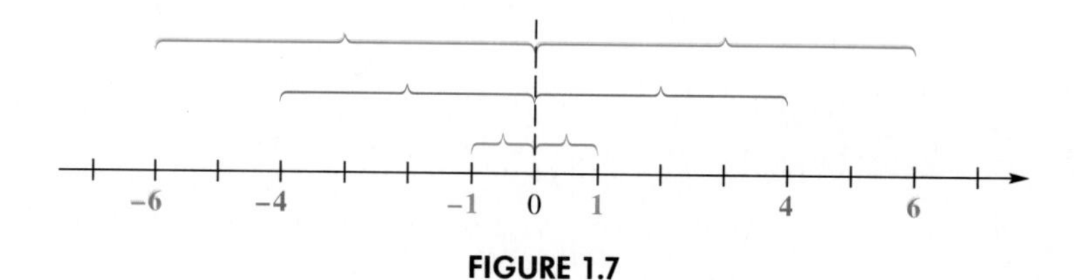

FIGURE 1.7

The additive inverse of a number can be indicated by writing the symbol $-$ in front of the number. With this symbol, the additive inverse of 7 is written -7. The additive inverse of -4 is written $-(-4)$, and can be read "the opposite of -4" or "the negative of -4." Figure 1.7 suggests that 4 is an additive inverse of -4. Since a number can have only one additive inverse, the symbols 4 and $-(-4)$ must represent the same number, which means that

$$-(-4) = 4.$$

This idea can be generalized as follows.

DOUBLE NEGATIVE RULE

For any real number x, $-(-x) = x$.

EXAMPLE 3
FINDING THE ADDITIVE INVERSE OF A NUMBER

The following chart shows several numbers and their additive inverses.

NUMBER	ADDITIVE INVERSE
-4	$-(-4)$, or 4
0	0
-2	2
19	-19
3	-3 ■

NOTE Example 3 suggests that the additive inverse of a number is found by changing the sign of the number.

An important property of additive inverses will be studied in more detail in a later section of this chapter: $a + (-a) = (-a) + a = 0$ for all real numbers a.

5 As mentioned above, additive inverses are numbers that are the same distance from 0 on the number line. This idea can also be expressed by saying that a number and its additive inverse have the same absolute value. The **absolute value** of a real number can be defined as the distance between 0 and the number on the number line. The symbol for the absolute value of the number x is $|x|$, read "the absolute value of x." For example, the distance between 2 and 0 on the number line is 2 units, so that

$$|2| = 2.$$

Because the distance between -2 and 0 on the number line is also 2 units,

$$|-2| = 2.$$

Since distance is a physical measurement, which is never negative, **the absolute value of a number is never negative**. For example, $|12| = 12$ and $|-12| = 12$, since both 12 and -12 lie at a distance of 12 units from 0 on the number line. Also, since 0 is a distance 0 units from 0, $|0| = 0$.

In symbols, the absolute value of x is defined as follows.

FORMAL DEFINITION OF ABSOLUTE VALUE

$$|x| = \begin{cases} x & \text{if } x \geq 0 \\ -x & \text{if } x < 0 \end{cases}$$

By this definition, if x is a positive number or 0, then its absolute value is x itself. For example, since 8 is a positive number, $|8| = 8$. However, if x is a negative number, then its absolute value is the additive inverse of x. This means that if $x = -9$, then $|-9| = -(-9) = 9$, since the additive inverse of -9 is 9.

CAUTION The formal definition of absolute value can be confusing if it is not read carefully. The "$-x$" in the second part of the definition *does not* represent a negative number; since x is negative in the second part, $-x$ represents a positive number. Remember that the absolute value of a number is never negative.

EXAMPLE 4 **FINDING ABSOLUTE VALUE**

Simplify by removing absolute value symbols.

(a) $|5| = 5$

(b) $|-5| = -(-5) = 5$

(c) $-|5| = -(5) = -5$

(d) $-|-14| = -(14) = -14$

(e) $|8 - 2| = |6| = 6$

(f) $|3 - 15| = |-12| = -(-12) = 12$ ■

Parts (e) and (f) of Example 4 show that absolute value bars are also grouping symbols. You must perform any operations that appear inside absolute value symbols before finding the absolute value.

1.4 EXERCISES

For each of the following, (a) *find the additive inverse, and* (b) *find the absolute value of the number. See Examples 3 and 4.*

1. 8 **2.** 12 **3.** -9 **4.** -11

5. -2 **6.** -3 **7.** 0 **8.** -1

9. A commonly heard statement from students is "Absolute value is always positive." Is this true? If not, explain. (See Objective 5.)

10. If a is a negative number, then is $-|-a|$ positive or negative? (See Objectives 4 and 5.)

11. Match each expression in Column I with its value in Column II. Some choices in Column II may not be used. (See Objectives 4 and 5.)

I	II
(a) $\lvert -7\rvert$	A. 7
(b) $-(-7)$	B. -7
(c) $-\lvert -7\rvert$	C. neither A nor B
(d) $-\lvert -(-7)\rvert$	D. both A and B

12. Fill in the blanks with the correct values. The opposite of -2 is _____, while the absolute value of -2 is _____. The additive inverse of -2 is _____, while the additive inverse of the absolute value of -2 is _____. (See Objectives 4 and 5.)

Select the smaller of the two given numbers. See Examples 2 and 4.

13. $-12, -4$ **14.** $-9, -14$ **15.** $-8, -1$ **16.** $-15, -16$

17. $3, |-4|$ **18.** $5, |-2|$ **19.** $|-3|, |-4|$ **20.** $|-8|, |-9|$

21. $-|-6|, -|-4|$ **22.** $-|-2|, -|-3|$ **23.** $|5 - 3|, |6 - 2|$ **24.** $|7 - 2|, |8 - 1|$

Write true *or* false *for each statement in Exercises 25–46. See Examples 2 and 4.*

25. $-2 < -1$ **26.** $-8 < -4$ **27.** $-3 \geq -7$

28. $-9 \geq -12$

29. $-15 \leq -20$

30. $-21 \leq -27$

31. $-8 \leq -(-4)$

32. $-9 \leq -(-6)$

33. $0 \leq -(-4)$

34. $0 \geq -(-6)$

35. $6 > -(-2)$

36. $-8 > -(-2)$

37. $-4 < -(-5)$

38. $-6 \leq -0$

39. $|-6| < |-9|$

40. $|-12| < |-20|$

41. $-|8| > |-9|$

42. $-|12| > |-15|$

43. $-|-5| \geq -|-9|$

44. $-|-12| \leq -|-15|$

45. $|6 - 5| \geq |6 - 2|$

46. $|13 - 8| \leq |7 - 4|$

For Exercises 47 and 48, see Example 1.

47. List all numbers from the set

$$\left\{-9, -\sqrt{7}, -1\frac{1}{4}, -\frac{3}{5}, 0, \sqrt{5}, 3, 5.9, 7\right\}$$

that are

(a) natural numbers; **(b)** whole numbers;
(c) integers; **(d)** rational numbers;
(e) irrational numbers; **(f)** real numbers.

48. List all numbers from the set

$$\left\{-5.3, -5, -\sqrt{3}, -1, -\frac{1}{9}, 0, 1.2, 1.8, 3, \sqrt{11}\right\}$$

that are

(a) natural numbers; **(b)** whole numbers;
(c) integers; **(d)** rational numbers;
(e) irrational numbers; **(f)** real numbers.

Graph each group of numbers on a number line.

49. $0, 3, -5, -6$

50. $2, 6, -2, -1$

51. $-2, -6, |-4|, 3, -|4|$

52. $-5, -3, -|-2|, -0, |-4|$

53. $\frac{1}{4}, 2\frac{1}{2}, -3\frac{4}{5}, -4, -1\frac{5}{8}$

54. $5\frac{1}{4}, 4\frac{5}{9}, -2\frac{1}{3}, 0, -3\frac{2}{5}$

55. $|3|, -|3|, -|-4|, -|-2|$

56. $|6|, -|6|, -|-8|, -|-3|$

In Exercises 57–62, give three examples of numbers that satisfy the given condition. (See Objective 2.)

57. Positive real numbers but not integers

58. Real numbers but not positive numbers

59. Real numbers but not whole numbers

60. Rational numbers but not integers

61. Real numbers but not rational numbers

62. Rational numbers but not negative numbers

In Exercises 63–72, write true *or* false *for each statement. (See Objective 2.)*

63. Every rational number is a real number.

64. Every integer is a rational number.

65. Some integers are not real numbers.

66. Every integer is positive.

67. Every whole number is positive.

68. Some irrational numbers are negative.

69. Some real numbers are not rational.

70. Not every rational number is positive.

71. Some whole numbers are not integers.

72. The number 0 is irrational.

For each of the following, give values of a and b that make the following statements true. Then give values of a and b that make the statement false. (See Objective 5.)

73. $|a + b| = |a - b|$

74. $|a - b| = |b - a|$

75. $|a + b| = -|a + b|$

76. $|-(a + b)| = -(a + b)$

1.5 ADDITION OF REAL NUMBERS

OBJECTIVES

1 ADD TWO NUMBERS WITH THE SAME SIGN.
2 ADD POSITIVE AND NEGATIVE NUMBERS.
3 USE THE ORDER OF OPERATIONS WITH REAL NUMBERS.
4 INTERPRET WORDS AND PHRASES THAT INDICATE ADDITION.
5 INTERPRET GAINS AND LOSSES AS POSITIVE AND NEGATIVE NUMBERS.

FOCUS ON PROBLEM SOLVING

On January 23, 1943, the temperature rose 49° F in two minutes in Spearfish, South Dakota. If the starting temperature was $-4°$ F, what was the temperature two minutes later?

In order to solve a problem like this one, we must be able to add signed numbers. In this section we develop rules for addition of signed numbers. This problem is Exercise 71 in the exercises for this section. After working through this section, you should be able to solve this problem.

1 The number line can be used to show the addition of real numbers, as in the following examples.

EXAMPLE 1 ADDING WITH THE NUMBER LINE

Use the number line to find the sum $2 + 3$.

Add the positive numbers 2 and 3 on the number line by starting at 0 and drawing an arrow two units to the *right,* as shown in Figure 1.8. This arrow represents the number 2 in the sum $2 + 3$. Then, from the right end of this arrow draw another arrow three units to the right. The number below the end of this second arrow is 5, so $2 + 3 = 5$. ■

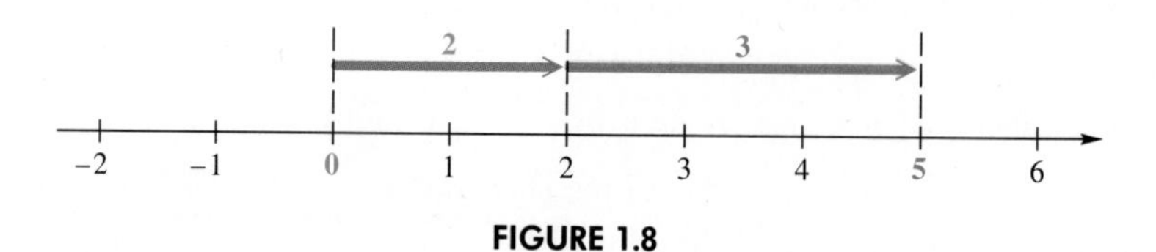

FIGURE 1.8

EXAMPLE 2 ADDING WITH THE NUMBER LINE

Use the number line to find the sum $-2 + (-4)$. (Parentheses are placed around the -4 to avoid the confusing use of $+$ and $-$ next to each other.)

Add the negative numbers -2 and -4 on the number line by starting at 0 and drawing an arrow two units to the *left,* as shown in Figure 1.9. The arrow is drawn to the left to represent the addition of a *negative* number. From the left end of this first arrow, draw a second arrow four units to the left. The number below the end of this second arrow is -6, so $-2 + (-4) = -6$. ■

FIGURE 1.9

In Example 2, the sum of the two negative numbers -2 and -4 is a negative number whose distance from 0 is the sum of the distance of -2 from 0 and the distance of -4 from 0. That is, *the sum of two negative numbers is the negative of the sum of their absolute values.*

$$-2 + (-4) = -(|-2| + |-4|) = -(2 + 4) = -6$$

ADDING NUMBERS WITH THE SAME SIGNS

Add two numbers having the *same* signs by adding the absolute values of the numbers. The result has the same sign as the numbers being added.

EXAMPLE 3
ADDING TWO NEGATIVE NUMBERS

Find the sums.

(a) $-2 + (-9) = -(|-2| + |-9|) = -(2 + 9) = -11$

(b) $-8 + (-12) = -20$

(c) $-15 + (-3) = -18$ ■

2 We can use the number line again to give meaning to the sum of a positive number and a negative number.

EXAMPLE 4
ADDING NUMBERS WITH DIFFERENT SIGNS

Use the number line to find the sum $-2 + 5$.

Find the sum $-2 + 5$ on the number line by starting at 0 and drawing an arrow two units to the left. From the left end of this arrow, draw a second arrow five units to the right, as shown in Figure 1.10. The number below the end of the second arrow is 3, so $-2 + 5 = 3$. ■

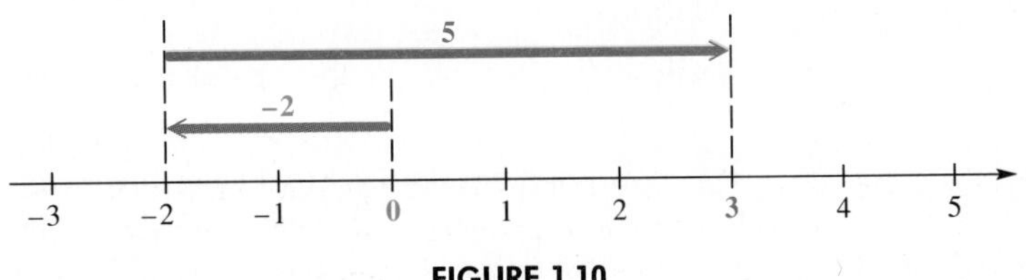

FIGURE 1.10

Addition of numbers with different signs can also be defined using absolute value.

ADDING NUMBERS WITH DIFFERENT SIGNS

Add two numbers with *different* signs by subtracting the smaller absolute value from the larger absolute value. The answer is given the sign of the number with the larger absolute value.

For example, to add -12 and 5, find their absolute values: $|-12| = 12$ and $|5| = 5$. Then find the difference between these absolute values: $12 - 5 = 7$. Since $|-12| > |5|$, the sum will be negative, so that the final answer is $-12 + 5 = -7$.

While a number line is useful in showing the rules for addition, it is important to be able to do the problems mentally.

EXAMPLE 5 ADDING MENTALLY

Check each answer, trying to work the addition mentally. If you get stuck, use a number line.

(a) $7 + (-4) = 3$

(b) $-8 + 12 = 4$

(c) $-\frac{1}{2} + \frac{1}{8} = -\frac{4}{8} + \frac{1}{8} = -\frac{3}{8}$ Remember to get a common denominator first.

(d) $\frac{5}{6} + \left(-\frac{4}{3}\right) = -\frac{1}{2}$

(e) $-4.6 + 8.1 = 3.5$

(f) $-16 + 16 = 0$

(g) $42 + (-42) = 0$ ■

Parts (f) and (g) in Example 5 suggest that the sum of a number and its additive inverse is 0. This is always true, and this property will be discussed further in Section 1.9.

The rules for adding signed numbers are summarized below.

ADDING SIGNED NUMBERS

Like signs Add the absolute values of the numbers. The sum has the same sign as the given numbers.

Unlike signs Find the difference between the larger absolute value and the smaller. Give the answer the sign of the number having the larger absolute value.

3 Sometimes an addition problem involves adding more than two numbers. As mentioned earlier, do the calculations inside the brackets or parentheses until a single number is obtained. Remember to use the order of operations given in Section 1.2 when adding more than two numbers.

EXAMPLE 6
ADDING WITH BRACKETS

Find the sums.

(a) $-3 + [4 + (-8)]$

First work inside the brackets. Follow the rules for the order of operations given in Section 1.2.

$$-3 + [4 + (-8)] = -3 + (-4) = -7$$

(b) $8 + [(-2 + 6) + (-3)] = 8 + [4 + (-3)] = 8 + 1 = 9$ ■

4 We now look at the interpretation of words and phrases that involve addition.

PROBLEM SOLVING

Problem solving often requires translating words and phrases into symbols. The word *sum* is one of the words that indicates addition. The chart below lists some of the words and phrases that signify that addition is to be performed. ■

WORD OR PHRASE	EXAMPLE	NUMERICAL EXPRESSION AND SIMPLIFICATION
Sum of	The *sum of* -3 and 4	$-3 + 4 = 1$
Added to	5 *added to* -8	$-8 + 5 = -3$
More than	12 *more than* -5	$-5 + 12 = 7$
Increased by	-6 *increased by* 13	$-6 + 13 = 7$
Plus	3 *plus* 14	$3 + 14 = 17$

EXAMPLE 7
INTERPRETING WORDS AND PHRASES INVOLVING ADDITION

Write a numerical expression for each phrase, and simplify the expression.

(a) The sum of -8 and 4 and 6

$$-8 + 4 + 6 = [-8 + 4] + 6 = -4 + 6 = 2$$

Notice that brackets were placed around $-8 + 4$ and this addition was done first, using the order of operations given earlier. The same result would be obtained if the brackets were placed around $4 + 6$. (This idea will be discussed further in Section 1.9.)

$$-8 + 4 + 6 = -8 + [4 + 6] = -8 + 10 = 2$$

(b) 3 more than -5, increased by 12

$$-5 + 3 + 12 = [-5 + 3] + 12 = -2 + 12 = 10$$ ■

5 Gains (or increases) and losses (or decreases) sometimes appear in applied problems.

PROBLEM SOLVING

When problems deal with gains and losses, the gains may be interpreted as positive numbers and the losses as negative numbers. The next example illustrates this idea. ■

EXAMPLE 8
INTERPRETING GAINS AND LOSSES

A football team gained 3 yards on the first play from scrimmage, lost 12 yards on the second play, and then gained 13 yards on the third play. How many yards did the team gain or lose altogether?

The gains are represented by positive numbers and the loss by a negative number.

$$3 + (-12) + 13$$

Add from left to right.

$$3 + (-12) + 13 = [3 + (-12)] + 13 = (-9) + 13 = 4$$

The team gained 4 yards altogether. ■

1.5 EXERCISES

Find the sums. See Examples 1–6.

1. $5 + (-3)$
2. $11 + (-8)$
3. $6 + (-8)$
4. $3 + (-7)$
5. $-6 + (-2)$
6. $-8 + (-3)$
7. $-3 + (-9)$
8. $-11 + (-5)$
9. $12 + (-8)$
10. $10 + (-2)$
11. $4 + [13 + (-5)]$
12. $6 + [2 + (-13)]$
13. $8 + [-2 + (-1)]$
14. $12 + [-3 + (-4)]$
15. $-2 + [5 + (-1)]$
16. $-8 + [9 + (-2)]$
17. $-6 + [6 + (-9)]$
18. $-3 + [4 + (-8)]$
19. $[(-9) + (-14)] + 12$
20. $[(-8) + (-6)] + 10$
21. $-\frac{1}{6} + \frac{2}{3}$
22. $\frac{9}{10} + \left(-\frac{3}{5}\right)$
23. $\frac{5}{8} + \left(-\frac{17}{12}\right)$
24. $-\frac{6}{25} + \frac{19}{20}$
25. $2\frac{1}{2} + \left(-3\frac{1}{4}\right)$
26. $-4\frac{3}{8} + 6\frac{1}{2}$
27. $-6.1 + [3.2 + (-4.8)]$
28. $-9.4 + [-5.8 + (-1.4)]$
29. $[-3 + (-4)] + [5 + (-6)]$
30. $[-8 + (-3)] + [-7 + (-6)]$
31. $[-4 + (-3)] + [8 + (-1)]$
32. $[-5 + (-9)] + [16 + (-21)]$
33. $[-4 + (-6)] + [(-3) + (-8)] + [12 + (-11)]$
34. $[-2 + (-11)] + [12 + (-2)] + [18 + (-6)]$

whats diff.

35. Is it possible to add a negative number to another negative number and get a positive number? If so, give an example. (See Objective 1.)

36. Under what conditions will the sum of a positive number and a negative number be a number which is neither negative nor positive? (See Objective 2.)

Write a numerical expression for each phrase and simplify. See Example 7.

37. The sum of -9 and 2 and 6
38. The sum of 4 and -7 and -3
39. 12 added to the sum of -17 and -6
40. -3 added to the sum of 15 and -1
41. The sum of -11 and -4 increased by -5
42. The sum of -8 and -15 increased by -3

Write true *or* false *for each statement.*

43. $-9 + 5 + 6 = -2$

44. $-6 + (8 - 5) = -3$

45. $-3 + 5 = 5 + (-3)$

46. $11 + (-6) = -6 + 11$

47. $|-8 + 3| = 8 + 3$

48. $|-4 + 2| = 4 + 2$

49. $|12 - 3| = 12 - 3$

50. $|-6 + 10| = 6 + 10$

51. $[4 + (-6)] + 6 = 4 + (-6 + 6)$

52. $[(-2) + (-3)] + (-6) = 12 + (-1)$

53. $-7 + [-5 + (-3)] = [(-7) + (-5)] + 3$

54. $6 + [-2 + (-5)] = [(-4) + (-2)] + 5$

55. $-5 + (-|-5|) = -10$

56. $|-3| + (-5) = -2$

57. $-2 + |-5| + 3 = 8 + (-2)$

Find all solutions for the following equations from the set $\{-3, -2, -1, 0, 1, 2, 3\}$. *Use guessing or trial and error.*

58. $x + 3 = 0$

59. $x + 1 = -2$

60. $x + 2 = -1$

61. $14 + x = 12$

62. $x + 8 = 7$

63. $x + (-4) = -6$

64. $x + (-2) = -5$

65. $-8 + x = -6$

66. $-2 + x = -1$

Solve each of the following problems by writing a sum of real numbers and adding. No variables are needed. See Example 8.

67. A college student received a \$50 check in the mail from her parents. She then spent \$36 for a sociology textbook. How much money does she have left?

68. Eryn's checking account balance is \$24.00. She then takes a gamble by writing a check for \$39.00. What is her new balance? (Write the balance as a signed number.)

69. A pilot announces to his passengers that the current altitude of their plane is 34,000 feet. Because of some unexpected turbulence, he is forced to descend 3500 feet. What is the new altitude of the plane? (Write the altitude as a signed number.)

70. The surface, or rim, of a canyon is at altitude 0. On a hike down into the canyon, a party of hikers stops for a rest at 120 meters below the surface. They then descend another 48 meters. What is their new altitude? (Write the altitude as a signed number.)

71. On January 23, 1943, the temperature rose 49° F in two minutes in Spearfish, South Dakota. If the starting temperature was −4°, what was the temperature two minutes later?

72. The lowest temperature ever recorded in Little Rock, Arkansas, was −5° F. The highest temperature ever recorded there was 117° F more than the lowest. What was this highest temperature?

73. On a series of three consecutive running plays, Herschel Walker of the Minnesota Vikings gained 4 yards, lost 3 yards, and lost 2 yards. What positive or negative number represents his total net yardage for the series of plays?

74. On three consecutive passes, Joe Montana of the San Francisco 49ers passed for a gain of 6 yards, was sacked for a loss of 12 yards, and passed for a gain of 43 yards. What positive or negative number represents the total net yardage for the plays?

Photo for Exercise 71

75. A welder working with stainless steel must use precise measurements. Suppose the welder attaches two pieces of steel that are each 3.589 inches in length, then attaches an additional three pieces that are each 9.089 inches long, and finally cuts off a piece that is 7.612 inches long. Find the length of the welded piece of steel.

76. Sarah Long owes $983.72 on her Visa credit card. She returns for credit items costing $74.18 and $12.53. She makes two purchases of $11.79 each and further purchases of $106.58, $29.81, and $73.24. She makes a payment of $186.50. Find the amount that she then owes.

(For Exercises 77–80, see Objectives 1 and 2.)

77. What number must be added to -9 to get 8?

78. What number must be added to -15 to get 3?

79. The sum of what number and 5 is -11?

80. The sum of what number and -6 is -9?

1.6 SUBTRACTION OF REAL NUMBERS

OBJECTIVES

1 FIND A DIFFERENCE ON A NUMBER LINE.
2 USE THE DEFINITION OF SUBTRACTION.
3 WORK SUBTRACTION PROBLEMS THAT INVOLVE GROUPING.
4 INTERPRET WORDS AND PHRASES THAT INVOLVE SUBTRACTION.
5 SOLVE PROBLEMS THAT INVOLVE SUBTRACTION.

FOCUS ON PROBLEM SOLVING

The top of Mount Whitney, visible from Death Valley, has an altitude of 14,494 feet above sea level. The bottom of Death Valley is 282 feet below sea level. Using zero as sea level, find the difference between these two elevations.

In order to solve this problem, we must know how to subtract signed numbers. In the previous section we learned how to add signed numbers. In this section, we learn how to subtract by changing subtraction to addition. This problem is Exercise 65 in the exercises for this section. After working through this section, you should be able to solve this problem.

1 Recall that the answer to a subtraction problem is a *difference*. Differences between signed numbers can be found by using a number line. Since *addition* of a positive number on the number line is shown by drawing an arrow to the *right, subtraction* of a positive number is shown by drawing an arrow to the *left*.

EXAMPLE 1
SUBTRACTING WITH THE NUMBER LINE

Use the number line to find the difference $7 - 4$.

To find the difference $7 - 4$ on the number line, begin at 0 and draw an arrow seven units to the right. From the right end of this arrow, draw an arrow four units to the left, as shown in Figure 1.11. The number at the end of the second arrow shows that $7 - 4 = 3$. ■

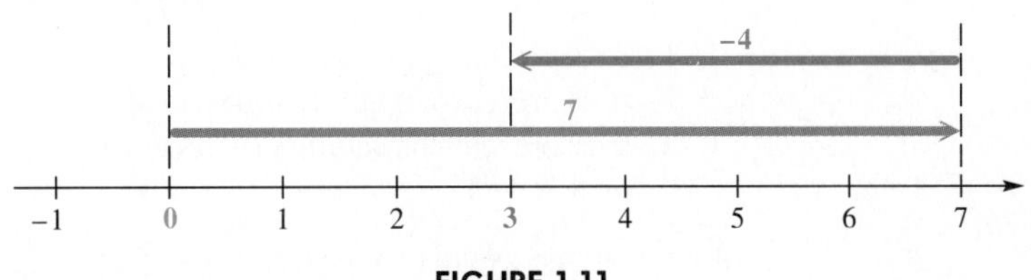

FIGURE 1.11

The procedure used in Example 1 to find $7 - 4$ is exactly the same procedure that would be used to find $7 + (-4)$, so that

$$7 - 4 = 7 + (-4).$$

2 Example 1 suggests that *subtraction* of a positive number from a larger positive number is the same as *adding* the additive inverse of the smaller number to the larger. This result is extended as the definition of subtraction for all real numbers.

DEFINITION OF SUBTRACTION

For any real numbers x and y, $\quad x - y = x + (-y).$

That is, to **subtract** y from x, *add the additive inverse* (or opposite) of y to x. This definition gives the following procedure for subtracting signed numbers.

SUBTRACTING SIGNED NUMBERS

Step 1 Change the subtraction symbol to addition.
Step 2 Change the sign of the number being subtracted.
Step 3 Add, as in the previous section.

EXAMPLE 2
USING THE DEFINITION OF SUBTRACTION

Subtract.

(No change; Change − to +.; Additive inverse of 3)

(a) $12 - 3 = 12 + (-3) = 9$

(b) $5 - 7 = 5 + (-7) = -2$

(No change; Change − to +.; Additive inverse of −5)

(c) $-3 - (-5) = -3 + (5) = 2$

(d) $-6 - (-9) = -6 + (9) = 3$ ■

Subtraction can be used to reverse the result of an addition problem. For example, if 4 is added to a number and then subtracted from the sum, the original number is the result:

$$12 + 4 = 16 \quad \text{and} \quad 16 - 4 = 12.$$

The symbol $-$ has now been used for three purposes:

1. to represent subtraction, as in $9 - 5 = 4$;
2. to represent negative numbers, such as -10, -2, and -3;
3. to represent the additive inverse of a number, as in "the additive inverse of 8 is -8."

More than one use may appear in the same problem, such as $-6 - (-9)$, where -9 is subtracted from -6. The meaning of the symbol depends on its position in the algebraic expression.

3 As before, with problems that have grouping symbols, first do any operations inside the parentheses and brackets. Work from the inside out.

EXAMPLE 3 SUBTRACTING WITH GROUPING SYMBOLS

Work each problem.

(a)
$$\begin{aligned} -6 - [2 - (8 + 3)] &= -6 - [2 - 11] && \text{Add.} \\ &= -6 - [2 + (-11)] && \text{Use the definition of subtraction.} \\ &= -6 - (-9) && \text{Add.} \\ &= -6 + (9) && \text{Use the definition of subtraction.} \\ &= 3 && \text{Add.} \end{aligned}$$

(b)
$$\begin{aligned} 5 - [(-3 - 2) - (4 - 1)] &= 5 - [(-3 + (-2)) - 3] \\ &= 5 - [(-5) - 3] \\ &= 5 - [(-5) + (-3)] \\ &= 5 - (-8) \\ &= 5 + 8 \\ &= 13 \end{aligned}$$

(c)
$$\begin{aligned} \frac{2}{3} - \left[\frac{1}{12} - \left(-\frac{1}{4}\right)\right] &= \frac{8}{12} - \left[\frac{1}{12} - \left(-\frac{3}{12}\right)\right] && \text{Get a common denominator.} \\ &= \frac{8}{12} - \left[\frac{1}{12} + \frac{3}{12}\right] && \text{Use the definition of subtraction.} \\ &= \frac{8}{12} - \frac{4}{12} && \text{Add.} \\ &= \frac{4}{12} && \text{Subtract.} \\ &= \frac{1}{3} && \text{Lowest terms} \end{aligned}$$

■

4 Let us now look at how we interpret words and phrases that involve subtraction.

PROBLEM SOLVING

In order to solve problems that involve subtraction, we must be able to interpret key words and phrases that indicate subtraction. *Difference* is one of them. Some of these are given in the chart below. ■

WORD OR PHRASE	EXAMPLE	NUMERICAL EXPRESSION AND SIMPLIFICATION
Difference between	The *difference between* -3 and -8	$-3-(-8)=-3+8=5$
Subtracted from	12 *subtracted from* 18	$18-12=6$
Less than	6 *less than* 5	$5-6=5+(-6)=-1$
Decreased by	9 *decreased by* -4	$9-(-4)=9+4=13$
Minus	8 *minus* 5	$8-5=3$

CAUTION When you are subtracting two numbers, it is important that you write them in the correct order, because, in general, $a-b \neq b-a$. For example, $5-3 \neq 3-5$. For this reason, it is important to *think carefully before interpreting an expression involving subtraction.* (This problem did not arise for addition.)

EXAMPLE 4 **INTERPRETING WORDS AND PHRASES INVOLVING SUBTRACTION**

Write a numerical expression for each phrase, and simplify the expression.

(a) The **difference between** -8 and 5

It is conventional to write the numbers in the order they are given when "difference between" is used.

$$-8-5=-8+(-5)=-13$$

(b) 4 **subtracted from** the sum of 8 and -3

Here the operation of addition is also used, as indicated by the word *sum*. First, add 8 and -3. Next, subtract 4 from this sum.

$$[8+(-3)]-4=5-4=1$$

(c) 4 **less than** -6

Be careful with order here. 4 must be taken *from* -6.

$$-6-4=-6+(-4)=-10$$

Notice that "4 less than -6" differs from "4 *is less than* -6." The second of these is symbolized as $4 < -6$ (which is a false statement).

(d) 8, **decreased by** 5 **less than** 12

First, write "5 less than 12" as $12-5$. Next, subtract $12-5$ from 8.

$$8-(12-5)=8-7=1 \quad ■$$

5 Recall from Section 1.5 that gains and losses may be interpreted as signed numbers.

PROBLEM SOLVING

Other applications of signed numbers include profit and loss in business, temperatures above and below 0°, and altitudes above and below sea level. The next example illustrates subtraction of signed numbers. ■

EXAMPLE 5 SOLVING A PROBLEM INVOLVING SUBTRACTION

The record high temperature of 134° F in the United States was recorded at Death Valley, California, in 1913. The record low was −80° F at Prospect Creek, Alaska, in 1971. What is the difference between the highest and the lowest temperatures?

We must find the value of the highest temperature minus the lowest temperature.

$$134 - (-80) = 134 + 80 \quad \text{Use the definition of subtraction.}$$
$$= 214 \quad \text{Add.}$$

The difference between the highest and the lowest temperatures is 214° F. ■

1.6 EXERCISES

Subtract. See Examples 1–3.

1. $3 - 6$
2. $7 - 12$
3. $5 - 9$
4. $8 - 13$
5. $-6 - 2$
6. $-11 - 4$
7. $-9 - 5$
8. $-12 - 15$
9. $6 - (-3)$
10. $12 - (-2)$
11. $-6 - (-2)$
12. $-7 - (-5)$
13. $2 - (3 - 5)$
14. $-3 - (4 - 11)$
15. $\frac{1}{2} - \left(-\frac{1}{4}\right)$
16. $\frac{1}{3} - \left(-\frac{4}{3}\right)$
17. $-\frac{3}{4} - \frac{5}{8}$
18. $-\frac{5}{6} - \frac{1}{2}$
19. $\frac{5}{8} - \left(-\frac{1}{2} - \frac{3}{4}\right)$
20. $\frac{9}{10} - \left(\frac{1}{8} - \frac{3}{10}\right)$
21. $3.4 - (-8.2)$
22. $5.7 - (-11.6)$
23. $-6.4 - 3.5$
24. $-4.4 - 8.6$
25. $-4.2 - (7.4 - 9.8)$
26. $(-1.8 - 3.7) - (-9.8)$

27. Explain in your own words how to subtract signed numbers. (See Objective 2.)
28. We know that in general, $a - b \neq b - a$. Can you give values for a and b such that $a - b$ does equal $b - a$? If so, give two such pairs. (See Objective 2.)

Work each problem. See Example 3.

29. $(4 - 6) + 12$
30. $(3 - 7) + 4$
31. $(8 - 1) - 12$
32. $(9 - 3) - 15$
33. $6 - (-8 + 3)$
34. $8 - (-9 + 5)$
35. $2 + (-4 - 8)$
36. $6 + (-9 - 2)$

37. $(-5 - 6) - (9 + 2)$

38. $(-4 + 8) - (6 - 1)$

39. $(-8 - 2) - (-9 - 3)$

40. $(-4 - 2) - (-8 - 1)$

41. $-9 + [(3 - 2) - (-4 + 2)]$

42. $-8 - [(-4 - 1) + (9 - 2)]$

43. $-3 + [(-5 - 8) - (-6 + 2)]$

44. $-4 + [(-12 + 1) - (-1 - 9)]$

In Exercises 45–48, suppose that a represents a positive number and b represents a negative number. Determine whether the given expression must represent a positive number or a negative number. (See Objective 2.)

45. $a - b$

46. $b - a$

47. $a + |b|$

48. $b - |a|$

Write a numerical expression for each phrase, and simplify. See Example 4.

49. The difference between 3 and -7

50. The difference between 8 and -2

51. The sum of -11 and the difference between -4 and 2

52. The sum of 8 and the difference between 1 and -3

53. -2 subtracted from the sum of 8 and -13

54. 4 subtracted from the sum of -12 and -3

55. 8 less than the difference between -3 and -7

56. 12, decreased by 9 less than 4

Find all solutions for the following equations from the set $\{-3, -2, -1, 0, 1, 2, 3\}$. Use guessing or trial and error.

57. $x - 1 = -3$

58. $x - 2 = -1$

59. $3 - x = 6$

60. $2 - x = 3$

61. $3 - (-x) = 0$

62. $1 - (-x) = 0$

Solve each of the following problems by writing a difference between real numbers and subtracting. No variables are needed. See Example 5.

63. On a cold winter day, the temperature reached $-5°$ F in Glenview, Illinois. The following day it was even colder, and the temperature dropped 10° F below $-5°$ F. What was the temperature on the following day?

64. A chemist is running an experiment under precise conditions. At first, she runs it at $-174.6°$ F. She then lowers it by 2.3° F. What is the new temperature for the experiment?

65. The top of Mount Whitney, visible from Death Valley, has an altitude of 14,494 feet above sea level. The bottom of Death Valley is 282 feet below sea level. Using zero as sea level, find the difference between these two elevations.

66. The highest point in Louisiana is Driskill Mountain, at an altitude of 535 feet. The lowest point is at Spanish Fort, 8 feet below sea level. Using zero as sea level, find the difference between these two elevations.

67. Josh owed his brother \$10. He later borrowed \$7. What positive or negative number represents his present financial status?

68. Francesca has \$15 in her purse, and Emilio has a debt of \$12. Find the difference between these amounts.

Photo for Exercise 65

69. For the year of 1990, one exercise facility showed a profit of \$76,000, while another showed a loss of \$29,000. Find the difference between these.

70. At 1:00 A.M., a plant worker found that a dial reading was 7.904. At 2:00 A.M., she found the reading to be -3.291. By how much had the reading declined?

Let $x = -5$, $y = -4$, and $z = 8$. Find the value of each expression.

71. $x + y$

72. $y - x$

73. $x + y - z$

74. $z - y - x$

1.7 MULTIPLICATION OF REAL NUMBERS

OBJECTIVES

1 FIND THE PRODUCT OF A POSITIVE NUMBER AND A NEGATIVE NUMBER.
2 FIND THE PRODUCT OF TWO NEGATIVE NUMBERS.
3 IDENTIFY FACTORS OF INTEGERS.
4 USE THE ORDER OF OPERATIONS IN MULTIPLICATION WITH SIGNED NUMBERS.
5 EVALUATE EXPRESSIONS INVOLVING VARIABLES.
6 INTERPRET WORDS AND PHRASES THAT INDICATE MULTIPLICATION.

You already know the rule for multiplying two positive numbers. The product of two positive numbers is also a positive number. But what about multiplying other real numbers? Any rules for multiplication of real numbers ought to be consistent with the rules for multiplication from arithmetic. For example, the product of 0 and any real number (positive or negative) should be 0.

MULTIPLICATION BY ZERO

For any real number x, $\quad x \cdot 0 = 0.$

1 In order to define the product of a positive and a negative number so that the result is consistent with the multiplication of two positive numbers, look at the following pattern.

$$\begin{aligned} 3 \cdot 5 &= 15 \\ 3 \cdot 4 &= 12 \\ 3 \cdot 3 &= 9 \\ 3 \cdot 2 &= 6 \\ 3 \cdot 1 &= 3 \\ 3 \cdot 0 &= 0 \\ 3 \cdot (-1) &= ? \end{aligned}$$

Numbers decrease by 3.

What should $3(-1)$ equal? The product $3(-1)$ represents the sum

$$-1 + (-1) + (-1) = -3,$$

so the product should be -3. Also,

$$3(-2) = -2 + (-2) + (-2) = -6$$

and

$$3(-3) = -3 + (-3) + (-3) = -9.$$

These results maintain the pattern in the list, which suggests the following rule.

MULTIPLYING NUMBERS WITH DIFFERENT SIGNS

For any positive real numbers x and y,

$$x(-y) = -(xy) \quad \text{and} \quad (-x)y = -(xy).$$

That is, the product of two numbers with opposite signs is negative.

EXAMPLE 1 **MULTIPLYING A POSITIVE NUMBER AND A NEGATIVE NUMBER**

Find the products using the multiplication rule given above.

(a) $8(-5) = -(8 \cdot 5) = -40$

(b) $5(-4) = -(5 \cdot 4) = -20$

(c) $(-7)(2) = -(7 \cdot 2) = -14$

(d) $(-9)(3) = -(9 \cdot 3) = -27$ ■

2 The product of two positive numbers is positive, and the product of a positive and a negative number is negative. What about the product of two negative numbers? Look at another pattern.

$$(-5)(4) = -20$$
$$(-5)(3) = -15$$
$$(-5)(2) = -10$$
$$(-5)(1) = -5$$
$$(-5)(0) = 0$$
$$(-5)(-1) = ?$$

Numbers increase by 5.

The numbers on the left of the equals sign (in color) decrease by 1 for each step down the list. The products on the right increase by 5 for each step down the list. To maintain this pattern, $(-5)(-1)$ should be 5 more than $(-5)(0)$, or 5 more than 0, so

$$(-5)(-1) = 5.$$

The pattern continues with

$$(-5)(-2) = 10$$
$$(-5)(-3) = 15$$
$$(-5)(-4) = 20$$
$$(-5)(-5) = 25,$$

and so on. This pattern suggests the rule at the top of the next page.

MULTIPLYING TWO NEGATIVE NUMBERS

For any positive real numbers x and y,

$$(-x)(-y) = xy.$$

In words, the product of two negative numbers is positive.

EXAMPLE 2
MULTIPLYING TWO NEGATIVE NUMBERS

Find the products using the multiplication rule given above.

(a) $(-9)(-2) = 9 \cdot 2 = 18$ **(b)** $(-6)(-12) = 6 \cdot 12 = 72$

(c) $(-8)(-1) = 8 \cdot 1 = 8$ **(d)** $(-15)(-2) = 15 \cdot 2 = 30$ ■

A summary of the results for multiplying signed numbers is given here.

MULTIPLYING SIGNED NUMBERS

The product of two numbers having the *same* sign is *positive,* and the product of two numbers having *different* signs is *negative*.

3 In Section 1.1 the definition of a *factor* was given for whole numbers. (For example, since $9 \cdot 5 = 45$, both 9 and 5 are factors of 45.) The definition can now be extended to integers.

If the product of two integers is a third integer, then each of the two integers is a **factor** of the third. For example, $(-3)(-4) = 12$, so -3 and -4 are both factors of 12. The factors of 12 are the numbers -12, -6, -4, -3, -2, -1, 1, 2, 3, 4, 6, and 12.

EXAMPLE 3
IDENTIFYING FACTORS OF AN INTEGER

The following chart shows several integers and the factors of those integers.

INTEGER	FACTORS
18	$-18, -9, -6, -3, -2, -1, 1, 2, 3, 6, 9, 18$
20	$-20, -10, -5, -4, -2, -1, 1, 2, 4, 5, 10, 20$
15	$-15, -5, -3, -1, 1, 3, 5, 15$
7	$-7, -1, 1, 7$
1	$-1, 1$ ■

4 The next example shows the order of operations discussed earlier, used with the multiplication of positive and negative numbers.

EXAMPLE 4
USING THE ORDER OF OPERATIONS

Perform the indicated operations.

(a) $(-9)(2) - (-3)(2)$

First find all products, working from left to right.

$$(-9)(2) - (-3)(2) = -18 - (-6)$$

Now perform the subtraction.

$$-18 - (-6) = -18 + 6 = -12$$

(b) $(-6)(-2) - (3)(-4) = 12 - (-12) = 12 + 12 = 24$

(c) $-5(-2 - 3) = -5(-5) = 25$ ■

5 The next examples show numbers substituted for variables where the rules for multiplying signed numbers must be used.

EXAMPLE 5 **EVALUATING AN EXPRESSION FOR NUMERICAL VALUES**

Evaluate the expression

$$(3x + 4y)(-2m)$$

given the following values.

(a) $x = -1, \quad y = -2, \quad m = -3$

First substitute the given values for the variables. Then find the value of the expression.

$$\begin{aligned}(3x + 4y)(-2m) &= [3(-1) + 4(-2)][-2(-3)] && \text{Put parentheses around the number for each variable.}\\ &= [-3 + (-8)][6] && \text{Find the products.}\\ &= (-11)(6) && \text{Use order of operations.}\\ &= -66\end{aligned}$$

(b) $x = 7, \quad y = -9, \quad m = 5$

Substitute.

$$\begin{aligned}(3x + 4y)(-2m) &= [3 \cdot 7 + 4(-9)][-2(5)] && \text{Use parentheses around } -9.\\ &= [21 + (-36)](-10) && \text{Find the products.}\\ &= (-15)(-10)\\ &= 150\end{aligned}$$ ■

EXAMPLE 6 **EVALUATING AN EXPRESSION FOR NUMERICAL VALUES**

Evaluate $2x^2 - 3y^2$ if $x = -3$ and $y = -4$.

Use parentheses as shown.

$$\begin{aligned}2(-3)^2 - 3(-4)^2 &= 2(9) - 3(16) && \text{Square } -3 \text{ and } -4.\\ &= 18 - 48 && \text{Multiply.}\\ &= -30 && \text{Subtract.}\end{aligned}$$ ■

6 Just as there are words and phrases that indicate addition and subtraction, certain ones also indicate multiplication.

PROBLEM SOLVING

The word *product* refers to multiplication. The chart on the next page gives some of the key words and phrases that indicate multiplication. ■

WORD OR PHRASE	EXAMPLE	NUMERICAL EXPRESSION AND SIMPLIFICATION
Product of	The *product of* -5 and -2	$(-5)(-2) = 10$
Times	13 *times* -4	$13(-4) = -52$
Twice (meaning "2 times")	*Twice* 6	$2(6) = 12$
Of (used with fractions)	$\frac{1}{2}$ *of* 10	$\frac{1}{2}(10) = 5$
Percent of	12% *of* -16	$.12(-16) = -1.92$

EXAMPLE 7 INTERPRETING WORDS AND PHRASES INVOLVING MULTIPLICATION

Write a numerical expression for each phrase and simplify. Use the order of operations.

(a) The product of 12 and the sum of 3 and -6
Here 12 is multiplied by "the sum of 3 and -6."

$$12[3 + (-6)] = 12(-3) = -36$$

(b) Twice the difference between 8 and -4

$$2[8 - (-4)] = 2[8 + 4] = 2(12) = 24$$

(c) Two-thirds of the sum of -5 and -3

$$\frac{2}{3}[-5 + (-3)] = \frac{2}{3}[-8] = -\frac{16}{3}$$

(d) 15% of the difference between 14 and -2
Remember that 15% = .15.

$$.15[14 - (-2)] = .15(14 + 2) = .15(16) = 2.4 \quad ■$$

1.7 EXERCISES

Find the products. See Examples 1 and 2.

1. $(-3)(-4)$ **2.** $(-3)(4)$ **3.** $3(-4)$ **4.** $-2(-8)$

5. $(-10)(-12)$ **6.** $9(-5)$ **7.** $0(-11)$ **8.** $3(-15)$

9. $(15)(-11)$ **10.** $(-9)(-4)$ **11.** $-\frac{3}{8}\cdot\left(-\frac{10}{9}\right)$ **12.** $-\frac{5}{4}\cdot\left(-\frac{5}{8}\right)$

13. $\left(-1\frac{1}{4}\right)\left(\frac{2}{15}\right)$ **14.** $\left(\frac{3}{7}\right)\left(-1\frac{5}{9}\right)$ **15.** $(-8)\left(-\frac{3}{4}\right)$ **16.** $(-6)\left(-\frac{5}{3}\right)$

17. $(-5.1)(.02)$ **18.** $(-3.7)(-2.1)$ **19.** $(1.8)(-7.2)$ **20.** $(-4.7)(-6.8)$

21. $(3.4)(-3.5)$ **22.** $(-5.2)(-7.4)$

Find all integer factors of each number. See Example 3.

23. 36 **24.** 32 **25.** 25 **26.** 40 **27.** 17 **28.** 29

29. What, if anything, is wrong with this statement: A prime number is a natural number greater than 1 that has only 1 and itself as integer factors. (See Objective 3.)

30. Multiplying three negative numbers will give a ___?___ (positive/negative) product. (See Objectives 1 and 2.)

31. True or False: Multiplying a negative number by a positive number will always give a negative number. If false, explain. (See Objective 1.)

32. True or False: Multiplying a negative number by a nonnegative number will always give a negative number. If false, explain. (See the introduction to this section and Objective 1.)

Perform the indicated operations. See Example 4.

33. $6 - 4 \cdot 5$
34. $3 - 2 \cdot 9$
35. $-9 - (-2) \cdot 3$
36. $-11 - (-7) \cdot 4$
37. $9(6 - 10)$
38. $5(12 - 15)$
39. $-6(2 - 4)$
40. $-9(5 - 8)$
41. $(4 - 9)(2 - 3)$
42. $(6 - 11)(3 - 6)$
43. $(2 - 5)(3 - 7)$
44. $(5 - 12)(2 - 6)$
45. $(-4 - 3)(-2) + 4$
46. $(-5 - 2)(-3) + 6$
47. $5(-2) - 4$
48. $9(-6) - 8$
49. $3(-4) - (-2)$
50. $5(-2) - (-9)$

Evaluate the following expressions, given $x = -2$, $y = 3$, and $a = -4$. See Examples 5 and 6.

51. $5x - 2y + 3a$
52. $6x - 5y + 4a$
53. $(2x + y)(3a)$
54. $(5x - 2y)(-2a)$
55. $(3x - 4y)(-5a)$
56. $(6x + 2y)(-3a)$
57. $(-5 + x)(-3 + y)(2 - a)$
58. $(6 - x)(5 + y)(3 + a)^2$
59. $-2y^2 + 3(a + 2)$
60. $5(x - 4) - 4a^2$
61. $3a^2 - (x - 3)^2$
62. $(y - 4)^2 - 2x^3$

---

63. If you square a negative number, will the product be positive or negative? (See Objective 2.)

64. If you cube a negative number, will the product be positive or negative? (See Objectives 1 and 2.)

Write a numerical expression for each phrase and simplify. Use the order of operations. See Example 7.

65. The product of -9 and 2, added to 6

66. The product of 4 and -7, added to -9

67. The difference between -9 and the product of -1 and 6

68. The difference between -4 and twice the product of -8 and 2

69. Nine subtracted from the product of 7 and -6

70. Three subtracted from the product of -2 and 3

71. Three-fifths of the sum of -8 and 3

72. 35% of the difference between 12 and -4

Find the solution for the following equations from the set $\{-3, -2, -1, 0, 1, 2, 3\}$. Use guessing or trial and error.

73. $2x = 4$
74. $-4m = 0$
75. $-9y = 0$
76. $-8p = 16$
77. $-9r = 27$
78. $2x + 1 = 3$
79. $3w + 3 = -3$
80. $-4a + 2 = -10$
81. $-5t + 6 = 11$

Replace a, b, and c with various integers to decide whether the following statements are true or false. For each false statement, give an example showing it is false. (See Objectives 1 and 2.)

82. $ab = ba$

83. $a(bc) = (ab)c$

84. $a^2 = 2a$

85. $a(b + c) = ab + ac$

Perform the indicated operations.

86. $(-8 - 2)(-4)^2 - (-5)$

87. $(-9 - 1)(-2)^2 - (-6)$

88. $|-4(-2)| + |-4|$

89. $|8(-5)| + |-2|$

90. $|2|(-4)^2 + |6| \cdot |-4|$

91. $|-3|(-2) + |-8| \cdot |5|$

1.8 DIVISION OF REAL NUMBERS

OBJECTIVES

1 FIND THE RECIPROCAL, OR MULTIPLICATIVE INVERSE, OF A NUMBER.

2 DIVIDE USING SIGNED NUMBERS.

3 SIMPLIFY NUMERICAL EXPRESSIONS INVOLVING QUOTIENTS.

4 INTERPRET WORDS AND PHRASES THAT INDICATE DIVISION.

5 TRANSLATE SIMPLE SENTENCES INTO EQUATIONS.

1 In Section 1.6 we saw that the difference of two numbers is found by adding the additive inverse of the second number to the first. Similarly, the *quotient* of two numbers is found by *multiplying* by the *multiplicative inverse*. By definition, since

$$8 \cdot \frac{1}{8} = \frac{8}{8} = 1 \qquad \text{and} \qquad \frac{5}{4} \cdot \frac{4}{5} = \frac{20}{20} = 1,$$

the multiplicative inverse of 8 is 1/8, and of 5/4 is 4/5.

MULTIPLICATIVE INVERSE

Pairs of numbers whose product is 1 are **multiplicative inverses,** or **reciprocals,** of each other.

EXAMPLE 1 FINDING THE MULTIPLICATIVE INVERSE

The following chart shows several numbers and their multiplicative inverses.

NUMBER	MULTIPLICATIVE INVERSE (RECIPROCAL)
4	$\frac{1}{4}$
-5	$\frac{1}{-5}$ or $-\frac{1}{5}$
$-\frac{5}{8}$	$-\frac{8}{5}$
0	None ■

Why is there no multiplicative inverse for the number 0? Suppose that k is to be the multiplicative inverse of 0. Then $k \cdot 0$ should equal 1. But $k \cdot 0 = 0$ for any number k. Since there is no value of k that is a solution of the equation $k \cdot 0 = 1$, the following statement can be made.

0 has no multiplicative inverse.

2 By definition, the quotient of x and y is the product of x and the multiplicative inverse of y.

DEFINITION OF DIVISION

For any real numbers x and y, with $y \neq 0$, $\frac{x}{y} = x \cdot \frac{1}{y}$.

The definition of division indicates that y, the number to divide by, cannot be 0. The reason is that 0 has no multiplicative inverse, so that $1/0$ is not a number. Because 0 has no multiplicative inverse, *division by 0 is undefined.* If a division problem turns out to involve division by 0, write "undefined."

NOTE While division by zero is undefined, we may divide 0 by any nonzero number. In fact, if $a \neq 0$,

$$\frac{0}{a} = 0.$$

Since division is defined in terms of multiplication, all the rules of multiplication of signed numbers also apply to division.

EXAMPLE 2 USING THE DEFINITION OF DIVISION

Find the quotients using the definition of division.

(a) $\frac{12}{3} = 12 \cdot \frac{1}{3} = 4$

(b) $\frac{-10}{2} = -10 \cdot \frac{1}{2} = -5$

(c) $\frac{-14}{-7} = -14\left(\frac{1}{-7}\right) = 2$

(d) $\frac{42}{-7} = 42\left(\frac{1}{-7}\right) = -6$

(e) $\frac{0}{13} = 0\left(\frac{1}{13}\right) = 0$ $\quad \frac{0}{a} = 0 \quad (a \neq 0)$

(f) $\frac{-10}{0}$ undefined ■

The following rule for division with signed numbers follows from the definition of division and the rules for multiplication with signed numbers.

DIVIDING SIGNED NUMBERS The quotient of two numbers having the *same* sign is *positive;* the quotient of two numbers having *different* signs is *negative.*

EXAMPLE 3 **DIVIDING SIGNED NUMBERS**

Find the quotients.

(a) $\frac{8}{-2} = -4$

(b) $\frac{-45}{-9} = 5$

(c) $-\frac{1}{8} \div \left(-\frac{3}{4}\right) = -\frac{1}{8} \cdot \left(-\frac{4}{3}\right) = \frac{1}{6}$ ■

From the definitions of multiplication and division of real numbers,

$$\frac{-40}{8} = -40 \cdot \frac{1}{8} = -5,$$

and

$$\frac{40}{-8} = 40\left(\frac{1}{-8}\right) = -5,$$

so that

$$\frac{-40}{8} = \frac{40}{-8}.$$

Based on this example, the quotient of a positive and a negative number can be expressed in any of the following three forms.

For any positive real numbers x and y, $\frac{-x}{y} = \frac{x}{-y} = -\frac{x}{y}$.

The form $x/-y$ is seldom used.

The quotient of two negative numbers can be expressed as a quotient of two positive numbers.

For any positive real numbers x and y, $\frac{-x}{-y} = \frac{x}{y}$.

3 The next example shows how to simplify numerical expressions involving quotients.

EXAMPLE 4 SIMPLIFYING EXPRESSIONS INVOLVING DIVISION

Simplify $\dfrac{5(-2) - (3)(4)}{2(1 - 6)}$.

Simplify the numerator and denominator separately. Then divide.

$$\frac{5(-2) - (3)(4)}{2(1 - 6)} = \frac{-10 - 12}{2(-5)} = \frac{-22}{-10} = \frac{11}{5} \quad \blacksquare$$

The rules for operations with signed numbers are summarized here.

OPERATIONS WITH SIGNED NUMBERS

Addition

Like signs Add the absolute values of the numbers. The result has the same sign as the numbers.

Unlike signs Subtract the number with the smaller absolute value from the one with the larger. Give the result the sign of the number having the larger absolute value.

Subtraction

Add the additive inverse, or negative, of the second number.

Multiplication and Division

Like signs The product or quotient of two numbers with like signs is positive.

Unlike signs The product or quotient of two numbers with unlike signs is negative.

Division by 0 is undefined.

0 divided by a nonzero number equals 0.

4 Certain words and phrases indicate the operation of division.

PROBLEM SOLVING

The word *quotient* refers to the result obtained in a division problem. In algebra, quotients are usually represented with a fraction bar. The symbol ÷ is seldom used. When solving stated problems involving division, interpret using the fraction bar. The following chart gives some key phrases associated with division.

PHRASE	EXAMPLE	NUMERICAL EXPRESSION AND SIMPLIFICATION
Quotient of	The *quotient of* −24 and 3	$\frac{-24}{3} = -8$
Divided by	−16 *divided by* −4	$\frac{-16}{-4} = 4$ ■

It is customary to write the first number named as the numerator and the second as the denominator when interpreting a phrase involving division. This is shown in the next example.

EXAMPLE 5
INTERPRETING WORDS AND PHRASES INVOLVING DIVISION

Write a numerical expression for each phrase, and simplify the expression.

(a) The quotient of 14 and the sum of -9 and 2

"Quotient" indicates division. The number 14 is the numerator and "the sum of -9 and 2" is the denominator.

$$\frac{14}{-9+2} = \frac{14}{-7} = -2$$

(b) The product of 5 and -6, divided by the difference between -7 and 8

The numerator of the fraction representing the division is obtained by multiplying 5 and -6. The denominator is found by subtracting -7 and 8.

$$\frac{5(-6)}{-7-8} = \frac{-30}{-15} = 2 \quad \blacksquare$$

5 In this section and the preceding three sections, important words and phrases involving the four operations of arithmetic have been introduced.

PROBLEM SOLVING

We can use words and phrases involving arithmetic operations to interpret sentences that translate into equations. The ability to do this will help us to solve the types of problems found in later chapters. ■

EXAMPLE 6
TRANSLATING WORDS INTO AN EQUATION

Write the following in symbols, using x as the variable, and use guessing or trial and error to find the solution. All solutions come from the list of integers between -12 and 12, inclusive.

(a) Three times a number is -18.

The word *times* indicates multiplication, and the word *is* translates as the equals sign (=).

$$3x = -18$$

Since the integer between -12 and 12, inclusive, that makes this statement true is -6, the solution of the equation is -6.

(b) The sum of a number and 9 is 12.

$$x + 9 = 12$$

Since $3 + 9 = 12$, the solution of this equation is 3.

(c) The difference between a number and 5 is 0.

$$x - 5 = 0$$

Since $5 - 5 = 0$, the solution of this equation is 5.

(d) The quotient of 24 and a number is -2.

$$\frac{24}{x} = -2$$

Here, x must be a negative number, since the numerator is positive and the quotient is negative. Since $\frac{24}{-12} = -2$, the solution is -12. ■

CAUTION It is important to recognize the distinction between the types of problems found in Example 5 and Example 6. In Example 5, the phrases translate as *expressions,* while in Example 6, the sentences translate as *equations*. Remember that an equation is a sentence with an = sign, while an expression is a phrase.

1.8 EXERCISES

Find the multiplicative inverse. If it is undefined, say so. See Example 1.

1. 9
2. 8
3. -4
4. -10

5. 0
6. $\frac{3}{4}$
7. $-\frac{9}{10}$
8. $-\frac{4}{5}$

---

9. Explain in your own words what is meant by *multiplicative inverse*. (See Objective 1.)

10. Explain in your own words why 0 has no multiplicative inverse. (See Objective 1.)

Find each quotient. If the expression is undefined, say so. See Examples 2 and 3.

11. $\frac{-10}{5}$
12. $\frac{-12}{3}$
13. $\frac{18}{-3}$
14. $\frac{-280}{-20}$

15. $\frac{-180}{0}$
16. $\frac{-350}{0}$
17. $\frac{0}{-2}$
18. $\frac{0}{12}$

19. $-\frac{1}{2} \div \left(-\frac{3}{4}\right)$
20. $-\frac{5}{8} \div \left(-\frac{3}{16}\right)$
21. $(-4.2) \div (-2)$
22. $(-9.8) \div (-7)$

23. $\frac{4}{-.8}$
24. $\frac{-6}{.3}$
25. $\frac{12}{2-5}$
26. $\frac{15}{3-8}$

27. $\frac{50}{2-7}$
28. $\frac{30}{5-5}$
29. $\frac{-40}{8-(-8)}$
30. $\frac{-72}{6-(-2)}$

31. $\frac{-120}{-3-(-5)}$
32. $\frac{-200}{-6-(-4)}$
33. $\frac{-30-(-8)}{-11}$
34. $\frac{-17-(-12)}{5}$

Simplify the numerators and denominators separately. Then find the quotients. See Example 4.

35. $\frac{-8(-2)+4}{3-(-1)}$
36. $\frac{-12(-3)-6}{-15-(-3)}$
37. $\frac{-15(2)-10}{-7-3}$

38. $\dfrac{-20(6) + 6}{-5 - 1}$

39. $\dfrac{-2(6)^2 + 3}{2 - (-1)}$

40. $\dfrac{3(-8)^2 + 3}{-6 + 1}$

41. $\dfrac{6^2 + 4^2}{5(2 + 13)}$

42. $\dfrac{4^2 - 5^2}{3(6 - 9 + 2)}$

43. $\dfrac{3^2 + 5^2 + 1}{4^2 - 1^2}$

For Exercises 44–49, assume that a is negative, b is positive, and c is positive. Tell whether the value of the given expression is positive or negative. (See Objective 2.)

44. $\dfrac{a \cdot b}{c}$

45. $\dfrac{a}{b \cdot c}$

46. $a^2 \cdot b \cdot c$

47. $a \cdot b^2 \cdot c$

48. $b \cdot (c - a)$

49. $\dfrac{a - c}{b}$

Find the solution of each equation from the set $\{-8, -6, -4, -2, 0, 2, 4, 6, 8\}$. *Use trial and error or guessing.*

50. $\dfrac{x}{4} = -2$

51. $\dfrac{x}{2} = -1$

52. $\dfrac{n}{-2} = 3$

53. $\dfrac{t}{-2} = -2$

54. $\dfrac{q}{-3} = 0$

55. $\dfrac{p}{5} = 0$

56. $\dfrac{m}{-2} = -4$

57. $\dfrac{y}{-1} = 2$

Write each statement in symbols and find the solution. Use x as the variable. The solutions come from the set of integers from −12 to 12, inclusive. Use guessing or trial and error. See Example 6.

58. Six times a number is −42.

59. Four times a number is −32.

60. When a number is divided by 3, the result is −3.

61. When a number is divided by −3, the answer is −4.

62. The quotient of a number and 2 is −6. $\left(\text{Write the quotient as } \dfrac{x}{2}.\right)$

63. The quotient of a number and 5 is −2.

64. The quotient of 6 and one more than a number is 3.

65. When the square of a number is divided by 3, the result is 12.

Write a numerical expression and simplify. See Example 5.

66. Add −9 to the quotient of 15 and −3.

67. Add 4 to the quotient of −28 and 4.

68. Subtract 3 from the quotient of 8 and −2.

69. Subtract 3 from the quotient of −16 and 4.

70. Find the product of −10 and 2. Then find the quotient of this product and 5.

Find each quotient.

71. $\dfrac{-5(2) + [3(-2) - 4]}{-3 - (-1)}$

72. $\dfrac{[4(-1) + 3](-2)}{-2 - 3}$

73. $\dfrac{-9(-2) - [(-4)(-2) + 3]}{-2(3) - 2(2)}$

74. $\dfrac{10^2 - [5^2 - 15]}{2(8^2 + 3^2) + 2}$

1.9 PROPERTIES OF ADDITION AND MULTIPLICATION

OBJECTIVES

1. IDENTIFY THE USE OF THE COMMUTATIVE PROPERTIES.
2. IDENTIFY THE USE OF THE ASSOCIATIVE PROPERTIES.
3. IDENTIFY THE USE OF THE IDENTITY PROPERTIES.
4. IDENTIFY THE USE OF THE INVERSE PROPERTIES.
5. IDENTIFY THE USE OF THE DISTRIBUTIVE PROPERTY.

If you were asked to find the sum $3 + 89 + 97$, it is likely that you would mentally add $3 + 97$ to get 100, and then add $100 + 89$ to get 189. While the rule for order of operations says to add from left to right, it is a fact that we may change the order of the terms and group them in any way we choose without affecting the sum. These are examples of shortcuts that we use in everyday mathematics. These shortcuts are justified by the basic properties of addition and multiplication, which are discussed in this section. In the following statements, a, b, and c represent real numbers.

1 The word *commute* means to go back and forth. Many people commute to work or to school. If you travel from home to work and follow the same route from work to home, you travel the same distance each time. The commutative properties say that if two numbers are added or multiplied in any order, they give the same result.

COMMUTATIVE PROPERTIES

$$a + b = b + a$$
$$ab = ba$$

EXAMPLE 1 USING THE COMMUTATIVE PROPERTIES

Use a commutative property to complete each statement.

(a) $-8 + 5 = 5 +$ _____

By the commutative property for addition, the missing number is -8, since $-8 + 5 = 5 + (-8)$.

(b) $(-2)(7) =$ _____(-2)

By the commutative property for multiplication, the missing number is 7, since $(-2)(7) = (7)(-2)$. ■

2 When we *associate* one object with another, we tend to think of those objects as being grouped together. The associative properties say that when we add or multiply three numbers, we can group the first two together or the last two together and get the same answer.

ASSOCIATIVE PROPERTIES

$$(a + b) + c = a + (b + c)$$
$$(ab)c = a(bc)$$

EXAMPLE 2
USING THE ASSOCIATIVE PROPERTIES

Use an associative property to complete each statement.

(a) $8 + (-1 + 4) = (8 + \_\_\_\_) + 4$
The missing number is -1.

(b) $[2 \cdot (-7)] \cdot 6 = 2 \cdot \_\_\_\_$
The completed expression on the right should be $2 \cdot [(-7) \cdot 6]$. ■

By the associative property of addition, the sum of three numbers will be the same no matter which way the numbers are "associated" in groups. For this reason, parentheses can be left out in many addition problems. For example, both

$$(-1 + 2) + 3 \quad \text{and} \quad -1 + (2 + 3)$$

can be written as

$$-1 + 2 + 3.$$

In the same way, parentheses also can be left out of many multiplication problems.

EXAMPLE 3
DISTINGUISHING BETWEEN THE ASSOCIATIVE AND COMMUTATIVE PROPERTIES

(a) Is $(2 + 4) + 5 = 2 + (4 + 5)$ an example of the associative property or the commutative property?

The order of the three numbers is the same on both sides of the equals sign. The only change is in the grouping, or association, of the numbers. Therefore, this is an example of the associative property.

(b) Is $6(3 \cdot 10) = 6(10 \cdot 3)$ an example of the associative property or the commutative property?

The same numbers, 3 and 10, are grouped on each side. On the left, the 3 appears first, but on the right, the 10 appears first. Since the only change involves the order of the numbers, this statement is an example of the commutative property.

(c) Is $(8 + 1) + 7 = 8 + (7 + 1)$ an example of the associative property or the commutative property?

In the statement, both the order and the grouping are changed. On the left the order of the three numbers is 8, 1, and 7. On the right it is 8, 7, and 1. On the left the 8 and 1 are grouped, and on the right the 7 and 1 are grouped. Therefore, both the associative and the commutative properties are used. ■

EXAMPLE 4
USING THE COMMUTATIVE AND ASSOCIATIVE PROPERTIES

Find the sum: $23 + 41 + 2 + 9 + 25$.

The commutative and associative properties make it possible to choose pairs of numbers whose sums are easy to add.

$$\begin{aligned} 23 + 41 + 2 + 9 + 25 &= (41 + 9) + (23 + 2) + 25 \\ &= 50 + 25 + 25 = 100 \end{aligned}$$ ■

3 If a child wears a sheet to masquerade as a ghost on Halloween, the child's appearance is changed, but his or her *identity* is unchanged. The identity of a real number is left unchanged when identity properties are applied. The identity properties say that the sum of 0 and any number equals that number, and the product of 1 and any number equals that number.

IDENTITY PROPERTIES

$$a + 0 = a \quad \text{and} \quad 0 + a = a$$
$$a \cdot 1 = a \quad \text{and} \quad 1 \cdot a = a$$

The number 0 leaves the identity, or value, of any real number unchanged by addition. For this reason, 0 is called the **identity element for addition.** Since multiplication by 1 leaves any real number unchanged, 1 is the **identity element for multiplication.**

EXAMPLE 5
USING THE IDENTITY PROPERTIES

These statements are examples of the identity properties.

(a) $-3 + 0 = -3$

(b) $0 + \frac{1}{2} = \frac{1}{2}$

(c) $-\frac{3}{4} \cdot 1 = -\frac{3}{4} \cdot \frac{8}{8} = -\frac{24}{32} \quad \left(\frac{8}{8} = 1\right)$

We use the identity property of multiplication to change a fraction to an equivalent one with a larger denominator. This is often done in adding or subtracting fractions.

(d) $1 \cdot 25 = 25$ ■

4 Each day before you go to work or school, you probably put on your shoes before you leave. Before you go to sleep at night, you probably take them off, and this leads to the same situation that existed before you put them on. These operations from everyday life are examples of inverse operations. The inverse properties of addition and multiplication lead to the additive and multiplicative identities, respectively. Recall that $-a$ is the **additive inverse** of a and $1/a$ is the **multiplicative inverse** of the nonzero number a. The sum of the numbers a and $-a$ is 0, and the product of the nonzero numbers a and $1/a$ is 1.

INVERSE PROPERTIES

$$a + (-a) = 0 \quad \text{and} \quad -a + a = 0$$
$$a \cdot \frac{1}{a} = 1 \quad \text{and} \quad \frac{1}{a} \cdot a = 1 \quad (a \neq 0)$$

EXAMPLE 6
USING THE INVERSE PROPERTIES

The following statements are examples of the inverse properties.

(a) $\frac{2}{3} \cdot \frac{3}{2} = 1$ **(b)** $(-5)\left(-\frac{1}{5}\right) = 1$ **(c)** $-\frac{1}{2} + \frac{1}{2} = 0$ **(d)** $4 + (-4) = 0$ ■

5 The everyday meaning of the word *distribute* is "to give out from one to several." An important property of real number operations involves this idea.

Look at the following statements.

$$2(5 + 8) = 2(13) = 26$$
$$2(5) + 2(8) = 10 + 16 = 26$$

Since both expressions equal 26,

$$2(5 + 8) = 2(5) + 2(8).$$

This result is an example of the *distributive property,* the only property involving *both* addition and multiplication. With this property, a product can be changed to a sum or difference.

The distributive property says that multiplying a number a by a sum of numbers $b + c$ gives the same result as multiplying a by b and a by c and then adding the two products.

DISTRIBUTIVE PROPERTY

$$a(b + c) = ab + ac \quad \text{and} \quad (b + c)a = ba + ca$$

As the arrows show, the a outside the parentheses is "distributed" over the b and c inside. Another form of the distributive property is valid for subtraction.

$$a(b - c) = ab - ac \quad \text{and} \quad (b - c)a = ba - ca$$

The distributive property also can be extended to more than two numbers.

$$a(b + c + d) = ab + ac + ad$$

EXAMPLE 7

USING THE DISTRIBUTIVE PROPERTY

Use the distributive property to rewrite each expression.

(a) $5(9 + 6) = 5 \cdot 9 + 5 \cdot 6$ Distributive property

$= 45 + 30$ Multiply.

$= 75$ Add.

(b) $4(x + 5 + y) = 4x + 4 \cdot 5 + 4y$ Distributive property

$= 4x + 20 + 4y$ Multiply.

(c) $-2(x + 3) = -2x + (-2)(3)$ Distributive property

$= -2x - 6$ Multiply.

(d) $3(k - 9) = 3k - 3 \cdot 9$ Distributive property

$= 3k - 27$ Multiply.

(e) $6 \cdot 8 + 6 \cdot 2 = 6(8 + 2)$ Distributive property

$= 6(10) = 60$ Add, then multiply.

(f) $4x - 4m = 4(x - m)$ Distributive property

(g) $8(3r + 11t + 5z) = 8(3r) + 8(11t) + 8(5z)$ Distributive property

$= (8 \cdot 3)r + (8 \cdot 11)t + (8 \cdot 5)z$ Associative property

$= 24r + 88t + 40z$ ■

The symbol $-a$ may be interpreted as $-1 \cdot a$. Similarly, when a minus sign precedes an expression within parentheses, it may also be interpreted as a factor of -1. The distributive property is used to remove parentheses from expressions such as $-(2y + 3)$. We do this by first writing $-(2y + 3)$ as $-1 \cdot (2y + 3)$.

$$\begin{aligned} -(2y + 3) &= -1 \cdot (2y + 3) \\ &= -1 \cdot (2y) + (-1) \cdot (3) && \text{Distributive property} \\ &= -2y - 3 && \text{Multiply.} \end{aligned}$$

EXAMPLE 8 USING THE DISTRIBUTIVE PROPERTY TO REMOVE PARENTHESES

Write without parentheses.

(a) $-(7r - 8) = -1(7r) + (-1)(-8)$ Distributive property

$= -7r + 8$ Multiply.

(b) $-(-9w + 2) = 9w - 2$ ■

The properties of addition and multiplication of real numbers are summarized below.

PROPERTIES OF ADDITION AND MULTIPLICATION

For any real numbers a, b, and c, the following properties hold.

Commutative properties $a + b = b + a$ $ab = ba$

Associative properties $(a + b) + c = a + (b + c)$

$(ab)c = a(bc)$

Identity properties There is a real number 0 such that

$a + 0 = a$ and $0 + a = a$.

There is a real number 1 such that

$a \cdot 1 = a$ and $1 \cdot a = a$.

Inverse properties For each real number a, there is a single real number $-a$ such that

$a + (-a) = 0$ and $(-a) + a = 0$.

For each nonzero real number a, there is a single real number $\frac{1}{a}$ such that

$a \cdot \frac{1}{a} = 1$ and $\frac{1}{a} \cdot a = 1$.

Distributive property $a(b + c) = ab + ac$

$(b + c)a = ba + ca$

1.9 EXERCISES

Label each statement as an example of the commutative, associative, identity, inverse, *or* distributive *property. See Examples 1–7.*

1. $5(15 \cdot 8) = (5 \cdot 15)8$

2. $(23)(9) = (9)(23)$

3. $12(-8) = (-8)(12)$

4. $(-9)[6(-2)] = [-9(6)](-2)$

5. $2 + (p + r) = (p + r) + 2$

6. $(4m)n = 4(mn)$

7. $(2 + p) + r = 2 + (p + r)$

8. $(-9)(-11) = (-11)(-9)$

9. $6 + (-6) = 0$

10. $9 + (11 + 4) = (9 + 11) + 4$

11. $-4 + 0 = -4$

12. $0 + (-9) = -9$

13. $3\left(\frac{1}{3}\right) = 1$

14. $-7\left(-\frac{1}{7}\right) = 1$

15. $\frac{2}{3} \cdot 1 = \frac{2}{3}$

16. $-\frac{9}{4} \cdot 1 = -\frac{9}{4}$

17. $6(5 - 2x) = 6 \cdot 5 - 6(2x)$

18. $5(2m) + 5(7n) = 5(2m + 7n)$

---

19. Give an example of an everyday situation that illustrates the concept of an inverse property. (See Objective 4.)

20. The following conversation actually took place between a mathematics teacher and his four-year-old son, who was learning some basic number facts.

Dad: "Jack, what is 3 + 0?"
Jack: "3."
Dad: "Jack, what is 4 + 0?"
Jack: "4. And, Daddy, *string* plus zero equals *string*!"

What property of addition did Jack recognize? (See Objective 3.)

Use the indicated property to write a new expression that is equal to the given expression. Simplify the new expression if possible. See Examples 1, 2, and 5–7.

21. $9k$; commutative

22. $z + 5$; commutative

23. $m + 0$; identity

24. $(-9) + 0$; identity

25. $3(r + m)$; distributive

26. $11(k + z)$; distributive

27. $8 \cdot \frac{1}{8}$; inverse

28. $\frac{1}{6} \cdot 6$; inverse

29. $12 + (-12)$; inverse

30. $-8 + 8$; inverse

31. $5 + (-5)$; commutative

32. $-9\left(-\frac{1}{9}\right)$; commutative

33. $-3(r + 2)$; distributive

34. $4(k - 5)$; distributive

35. $9 \cdot 1$; identity

36. $1(-4)$; identity

37. $5k(-6)$; associative

38. $(m + 4) + (-2)$; associative

39. Write a paragraph explaining in your own words the following properties of addition: commutative, associative, identity, inverse. (See Objectives 1–4.)

40. Write a paragraph explaining in your own words the following properties of multiplication: commutative, associative, identity, inverse. (See Objectives 1–4.)

Use the distributive property to rewrite each expression, and simplify if possible. See Example 7.

41. $5(m + 2)$
42. $6(k + 5)$
43. $-4(r + 2)$
44. $-3(m + 5)$
45. $-8(k - 2)$
46. $-4(z - 5)$
47. $-9(a + 3)$
48. $-3(p + 5)$
49. $(r + 8)4$
50. $(m + 12)6$
51. $(8 - k)(-2)$
52. $(9 - r)(-3)$
53. $2(5r + 6m)$
54. $5(2a + 4b)$
55. $-4(3x - 4y)$
56. $-9(5k - 12m)$
57. $5 \cdot 8 + 5 \cdot 9$
58. $4 \cdot 3 + 4 \cdot 9$
59. $7 \cdot 2 + 7 \cdot 8$
60. $6x + 6m$
61. $9p + 9q$
62. $8(2x) + 8(3p)$
63. $5(7z) + 5(8w)$
64. $11(2r) + 11(3s)$

Use the distributive property to rewrite each expression. See Example 8.

65. $-(3k + 5)$
66. $-(2z + 12)$
67. $-(4y - 8)$
68. $-(3r - 15)$
69. $-(-4 + p)$
70. $-(-12 + 3a)$
71. $-(-1 - 15r)$
72. $-(-14 - 6y)$

73. The distributive property holds for multiplication with respect to addition. Does the distributive property hold for addition with respect to multiplication? That is, is $a + (b \cdot c)$ equal to $(a + b) \cdot (a + c)$ for all values of a, b, and c? (*Hint:* Let $a = 2$, $b = 3$, and $c = 4$.) (See Objective 5.)

74. Suppose that a student shows you the following work.

$$-3(4 - 6) = -3(4) - 3(6) = -12 - 18 = -30$$

The student has made a very common error. Write a short paragraph explaining the student's mistake, and work the problem correctly. (See Objective 5.)

Decide whether or not the events in Exercises 75–78 are commutative. (See Objective 1.)

75. Getting out of bed and taking a shower
76. Putting on your right shoe or your left shoe
77. Taking English or taking history
78. Putting on your shoe or putting on your sock

79. Evaluate $25 - (6 - 2)$ and evaluate $(25 - 6) - 2$. Do you think subtraction is associative?

80. Evaluate $180 \div (15 \div 3)$ and evaluate $(180 \div 15) \div 3$. Do you think division is associative?

Replace a, b, and c with various integers to decide whether the following statements are true or false. For each false statement, give an example showing it is false. (See Objectives 1–5.)

81. $(-a) + (-b) = -(a + b)$
82. $(-a) + b = b - a$
83. $a + [b + (-a)] = b$
84. $(a + b) + (-b) = a$

CHAPTER 1 GLOSSARY

KEY TERMS

1.1 **natural numbers** The set of natural numbers is $\{1, 2, 3, 4, \ldots\}$.

whole numbers The set of whole numbers is $\{0, 1, 2, 3, 4, \ldots\}$.

numerator The numerator of a fraction is the number above the fraction bar.

denominator The denominator of a fraction is the number below the fraction bar.

factor If $a \cdot b = c$, then a and b are called factors of c.

product The answer to a multiplication problem is called the product.

factored A number is factored when it is written as the product of two or more numbers.

prime A natural number (except 1) is prime if it has only itself and 1 as natural number factors.

lowest terms A fraction is in lowest terms when the numerator and denominator have no factors in common (other than 1).

quotient The answer to a division problem is called the quotient.

sum The answer to an addition problem is called the sum.

mixed number A mixed number is understood to be the sum of a whole number and a fraction. For example, $3\frac{1}{4} = 3 + \frac{1}{4}$.

difference The answer to a subtraction problem is called the difference.

1.2 **exponent** A natural number exponent is a number that indicates how many times a base is used as a factor.

base The base is the number that is a repeated factor when written with an exponent.

grouping symbols Grouping symbols, such as parentheses and brackets, are symbols that are used to indicate the order in which operations should be performed.

1.3 **variable** A variable is a symbol, usually a letter, such as x, y, or z, used to represent any unknown number.

algebraic expression An algebraic expression is a collection of numbers, variables, symbols for operations, and symbols for grouping (such as parentheses).

equation An equation is a statement that two algebraic expressions are equal.

solution A solution of an equation is any replacement for the variable that makes the equation true.

set A set is a collection of objects, such as numbers.

element An object that belongs to a set is an element of the set.

1.4 **number line** A line used to picture a set of numbers is called a number line.

negative numbers The numbers to the left of 0 on a number line are negative numbers.

positive numbers The numbers to the right of 0 on a number line are positive numbers.

signed numbers Positive numbers and negative numbers are called signed numbers.

integers The set of integers is $\{\ldots, -3, -2, -1, 0, 1, 2, 3, \ldots\}$.

graph The graph of a number is a point on the number line that indicates the number.

rational numbers The set of rational numbers is $\{x \mid x \text{ is a quotient of integers, with denominator not } 0\}$.

real numbers The set of real numbers is $\{x \mid x \text{ is a number that can be represented by a point on the number line}\}$.

irrational numbers The set of irrational numbers is $\{x \mid x \text{ is a real number that is not rational}\}$.

additive inverse (opposite) The additive inverse of a number x is the number that is the same distance from 0 on the number line as x, but on the opposite side of 0.

absolute value The absolute value of a real number can be defined as the distance between 0 and the number on the number line.

1.8 **multiplicative inverse (reciprocal)** Pairs of numbers whose product is 1 are multiplicative inverses, or reciprocals, of each other.

identity element for addition 0 is called the identity element for addition.

identity element for multiplication 1 is called the identity element for multiplication.

NEW SYMBOLS

Symbol	Meaning
a^n	n factors of a
[ ]	square brackets (used as grouping symbols)
$=$	is equal to
$\neq$	is not equal to
$<$	is less than (read from left to right)
$>$	is greater than (read from left to right)
$\leq$	is less than or equal to (read from left to right)
$\geq$	is greater than or equal to (read from left to right)
{ }	set braces
$\{x \mid x$ **has a certain property**$\}$	set-builder notation
$-x$	the additive inverse, or opposite, of x
$\lvert x \rvert$	absolute value of x
$\frac{1}{x}$ or $1/x$	the multiplicative inverse, or reciprocal, of the nonzero number x
$a(b)$, $(a)(b)$, $a \cdot b$, or ab	a times b
$\frac{a}{b}$ or a/b	a divided by b

CHAPTER 1 QUICK REVIEW

SECTION	CONCEPTS	EXAMPLES
1.1 FRACTIONS	**Operations with Fractions**	Perform the operations.
	Addition: Add numerators; the denominator is the same.	$\frac{2}{5} + \frac{7}{5} = \frac{9}{5}$
	Subtraction: Subtract numerators; the denominator is the same.	$\frac{5}{4} - \frac{1}{4} = \frac{4}{4} = 1$
	Multiplication: Multiply numerators and multiply denominators.	$\frac{4}{3} \cdot \frac{5}{6} = \frac{20}{18} = \frac{10}{9}$
	Division: Multiply the first fraction by the reciprocal of the second fraction.	$\frac{6}{5} \div \frac{1}{4} = \frac{6}{5} \cdot \frac{4}{1} = \frac{24}{5}$

SECTION	CONCEPTS	EXAMPLES
1.2 EXPONENTS, ORDER OF OPERATIONS, AND INEQUALITY	a^n indicates that a is used as a factor n times.	$6^2 = 6 \cdot 6 = 36$ $5^3 = 5 \cdot 5 \cdot 5 = 125$
	Order of Operations *If grouping symbols are present,* simplify within them, innermost first (and above and below fraction bars separately), in the following order.	
	Step 1 Apply all exponents. *Step 2* Do any multiplications or divisions in the order in which they occur, working from left to right. *Step 3* Do any additions or subtractions in the order in which they occur, working from left to right. *If no grouping symbols are present,* start with Step 1.	$36 - 4(2^2 + 3) = 36 - 4(4 + 3)$ $= 36 - 4(7)$ $= 36 - 28$ $= 8$
1.3 VARIABLES, EXPRESSIONS, AND EQUATIONS	Evaluate an expression with a variable by substituting a given number for the variable.	Evaluate $2x + y^2$ if $x = 3$ and $y = -4$. $2x + y^2 = 2(3) + (-4)^2$ $= 6 + 16$ $= 22$
	Values of a variable that make an equation true are solutions of the equation.	Is 2 a solution of $5x + 3 = 18$? $5(2) + 3 = 18$? $13 = 18$ False 2 is not a solution.
1.4 REAL NUMBERS AND THE NUMBER LINE	**The Ordering of Real Numbers** For any two real numbers a and b, a is less than b if a is to the left of b on the number line.	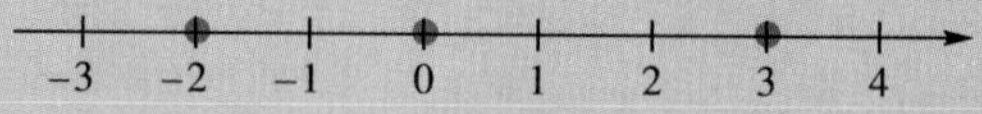 $-2 < 3$ $3 > 0$ $0 < 3$
	The additive inverse of x is $-x$.	$-(5) = -5$ $-(-7) = 7$ $-0 = 0$
	The absolute value $\lvert x \rvert$ of x is the distance between x and 0 on the number line.	$\lvert 13 \rvert = 13$ $\lvert 0 \rvert = 0$ $\lvert -5 \rvert = 5$

SECTION	CONCEPTS	EXAMPLES
1.5 ADDITION OF REAL NUMBERS	We add two numbers having the same sign by adding the absolute values of the numbers. The result has the same sign as the numbers being added. We add two numbers having different signs by first finding the difference between the absolute values of the numbers. Give the answer the same sign as the number with the larger absolute value.	Add. $9 + 4 = 13$ $-8 + (-5) = -13$ $7 + (-12) = -5$ $-5 + 13 = 8$
1.6 SUBTRACTION OF REAL NUMBERS	**Definition of Subtraction** $x - y = x + (-y)$ **Procedure for Subtracting Signed Numbers** **1.** Change the subtraction symbol to addition. **2.** Change the sign of the number being subtracted. **3.** Add, as in the previous section.	Subtract. $5 - (-2) = 5 + 2 = 7$ $-3 - 4 = -3 + (-4) = -7$ $-2 - (-6) = -2 + 6$ $= 4$ $13 - (-8) = 13 + 8$ $= 21$
1.7, 1.8 MULTIPLICATION AND DIVISION OF REAL NUMBERS	**Definition of Division** $\frac{x}{y} = x \cdot \frac{1}{y}, \quad y \neq 0$ **Multiplying and Dividing Signed Numbers** The product (or quotient) of two numbers having the *same sign* is *positive;* the product (or quotient) of two numbers having *different signs* is *negative.* Division by 0 is undefined. 0 divided by a nonzero number equals 0.	Multiply or divide. $\frac{10}{2} = 10 \cdot \frac{1}{2} = 5$ $6 \cdot 5 = 30$ $\quad (-7)(-8) = 56$ $\frac{20}{4} = 5$ $\quad \frac{-24}{-6} = 4$ $(-6)(5) = -30$ $\quad (6)(-5) = -30$ $\frac{-18}{9} = -2$ $\quad \frac{49}{-7} = -7$ $\frac{5}{0}$ is undefined. $\frac{0}{5} = 0$

SECTION	CONCEPTS	EXAMPLES
1.9 PROPERTIES OF ADDITION AND MULTIPLICATION	**Commutative**	
	$a + b = b + a$	$7 + (-1) = -1 + 7$
	$ab = ba$	$5(-3) = (-3)5$
	Associative	
	$(a + b) + c = a + (b + c)$	$3 + (4 + 8) = (3 + 4) + 8$
	$(ab)c = a(bc)$	$[(-2)(6)](4) = (-2)[(6)(4)]$
	Identity	
	$a + 0 = a \quad 0 + a = a$	$-7 + 0 = -7 \quad 0 + (-7) = -7$
	$a \cdot 1 = a \quad 1 \cdot a = a$	$9 \cdot 1 = 9 \quad 1 \cdot 9 = 9$
	Inverse	
	$a + (-a) = 0 \quad -a + a = 0$	$7 + (-7) = 0 \quad -7 + 7 = 0$
	$a \cdot \frac{1}{a} = 1 \quad \frac{1}{a} \cdot a = 1 \quad (a \neq 0)$	$-2\left(-\frac{1}{2}\right) = 1 \quad -\frac{1}{2}(-2) = 1$
	Distributive	
	$a(b + c) = ab + ac$	$5(4 + 2) = 5(4) + 5(2)$
	$(b + c)a = ba + ca$	$(4 + 2)5 = 4(5) + 2(5)$
	$a(b - c) = ab - ac$	$9(5 - 4) = 9(5) - 9(4)$

CHAPTER 1 REVIEW EXERCISES

If you have trouble with any of these exercises, look in the section given in brackets.

[1.1] *Write each fraction in lowest terms.*

1. $\frac{18}{54}$ **2.** $\frac{60}{72}$ **3.** $\frac{22}{110}$ **4.** $\frac{120}{150}$

Perform each operation. Write the answers in lowest terms.

5. $\frac{7}{10} \cdot \frac{1}{5}$ **6.** $\frac{32}{15} \div \frac{8}{5}$ **7.** $\frac{3}{8} + \frac{1}{2}$ **8.** $6\frac{1}{4} - 5\frac{3}{4}$

Work the following problems.

9. John and Gwen are painting a bedroom for their new baby. They painted $\frac{1}{4}$ of the room on Saturday and $\frac{1}{3}$ of the room on Sunday. How much of the room is still unpainted?

10. A recipe for lemon sherbet calls for $1\frac{1}{3}$ cups of lemon juice. How many cups of lemon juice would be needed for 9 recipes?

[1.2] *Find the following.*

11. 4^3

12. $\left(\frac{7}{3}\right)^3$

Which of the symbols $\neq$, $<$, $>$, $\leq$, *and* $\geq$ *make the following statements true? Give all possible correct answers.*

13. .87 ____ .865

14. .94 ____ .904

15. $\frac{2}{3}$ ____ .7

16. $\frac{3}{4}$ ____ $\frac{4}{5}$

Simplify each statement and then decide whether the statement is true or false.

17. $70 < 4 - 9(6 + 2)$

18. $2 \cdot 3 + 5(4 + 8) \leq 68$

19. $\frac{2(1 + 3)}{3(2 + 1)} < 1$

20. $6^2 - 4^2 \geq 5$

21. $5^2 + 4^2 \leq 40$

[1.3] *Find the numerical value of the given expression when* $x = 2$ *and* $y = 5$.

22. $\frac{3x - y}{2x}$

23. $\frac{2x}{y + 1}$

24. $x^2 + y^2$

25. $\frac{3x^2 + 5y^2}{7x^2 - y^2}$

Write each word phrase using algebraic expressions. Use x *as the variable.*

26. Six subtracted from three times a number

27. The difference between 4 and a number

[1.4] *Select the smaller of the two given numbers.*

28. $5, |-8|$

29. $9, |-7|$

30. $-|-2|, -|-9|$

31. $-|-7|, -|-4|$

Decide whether each statement is true or false.

32. $3 \leq -(-5)$

33. $9 \geq -(-10)$

34. $-3 > -2$

35. $-5 < -4$

36. $-|7| \leq -|-2|$

37. $-|4| > -|-3|$

38. $|-(-2)| > -|-2|$

39. $|-5| < |5|$

Graph each group of numbers on a number line.

40. $-1, -3, |-4|, |-1|$

41. $3\frac{1}{4}, -2\frac{4}{5}, -1\frac{1}{8}, \frac{2}{3}$

42. $|-2|, -|-5|, -|3|, -|0|$

[1.5] *Find the sums.*

43. $(-9 + 6) + (-10)$

44. $[-3 + (-5)] + (-8)$

45. $\frac{7}{8} + \left(-\frac{3}{10}\right)$

46. $\frac{7}{12} + \left(-\frac{2}{9}\right)$

47. $14.2 + (-8.6)$

48. $-11.3 + (-2.9)$

---

49. Write in symbols: 5 more than the sum of 12 and -7. Evaluate the expression.

50. The altitude of Mount Hunter in Alaska is 14,573 feet. If a hiker descends 2573 feet from the top, what is the altitude of the hiker?

[1.6] *Find any sums or differences.*

51. $15 - (-3)$

52. $-8 - 9$

53. $\frac{3}{4} - \left(-\frac{2}{3}\right)$

54. $-\frac{1}{5} - \left(-\frac{7}{10}\right)$

55. $(-9 + 6) - (-3)$

56. $(-15 - 7) - (-9)$

---

57. Write in symbols: 12 less than the difference between 9 and -3. Evaluate the expression.

58. Chris owed his brother \$12. He repaid \$9 and later borrowed \$13. What positive or negative number represents his present financial status?

[1.7] *Perform the indicated operations.*

59. $-\frac{4}{5}\left(-\frac{10}{7}\right)$

60. $-\frac{3}{8}\left(-\frac{16}{15}\right)$

61. $(-11.3)(2.5)$

62. $(-9.4)(-2.8)$

63. $8(-9) - (6)(-2)$

64. $-6(-2) - (-4)(3)$

Evaluate the following expressions, given $x = -5$, $y = 4$, *and* $z = -3$.

65. $y^2 - 2z^2$

66. $(2x - 8y)(z^2)$

67. $3z^2 - 4x^2$

---

68. Write in symbols: The sum of 9 and the product of -3 and 5. Evaluate the expression.

69. If you raise a negative number to the fourth power, will the product be positive or negative?

[1.8] *Find the quotients.*

70. $\frac{-8 - (-9)}{10}$

71. $\frac{-7 - (-21)}{8}$

72. $\frac{8 - 4(-2)}{-5(3) - 1}$

---

73. If a is positive and b and c are negative, is $\frac{a \cdot b^2}{c}$ positive or negative?

74. Write in symbols: The quotient of -12 and the product of -2 and -3. Evaluate the expression.

[1.9] *Decide whether each statement is an example of the commutative, associative, identity, inverse, or distributive property.*

75. $8.974 \cdot 1 = 8.974$

76. $-\frac{2}{3} + \frac{2}{3} = 0$

77. $7 + 4m = 4m + 7$

78. $8(4 \cdot 3) = (8 \cdot 4)3$

79. $\frac{5}{8} \cdot \frac{8}{5} = 1$

80. $9p + 0 = 9p$

Simplify using the distributive property.

81. $6(3k + 5r)$

82. $-3(2m - 7k)$

MIXED REVIEW EXERCISES*

Perform the indicated operations.

83. $[(-4) + 6 + (-9)] + [-3 + (-5)]$

84. $\left(\frac{5}{8}\right)^2$

85. $-|(-9)(4)| - |-2|$

86. 6^3

87. $\dfrac{5(-3) - 8(3)}{(-5)(-4) + (-7)}$

88. $\dfrac{7}{10} - \dfrac{1}{4}$

89. $\dfrac{11^2 - 5^2 + 2}{5^2 + 5^2 - 1^2}$

90. $\dfrac{7^2 - 3^2}{2^2 + 4^2}$

91. $-11(-4) - (3)(-7)$

92. $2\frac{3}{4} + 5\frac{5}{12}$

93. $-3 + [(-12 + 15) - (-4 - 5)]$

94. $\dfrac{8}{5} \div \dfrac{32}{15}$

95. $(-8 - 2) - [(-1 - 4) - (-1)]$

96. $[(-6) + (-7) + 8] + [8 + (-15)]$

97. $\dfrac{3}{7} \cdot \dfrac{5}{6}$

98. $|6(-8)| - |-3|$

---

99. Write a short paragraph explaining the special considerations involving zero when dividing.

100. ''Two negatives give a positive'' is often heard from students. Is this correct? Can you explain it in more precise language?

101. Use x as the variable, and write an expression for ''the product of 5 and the sum of a number and 6.'' Then use the distributive property to rewrite the expression.

102. The highest temperature ever recorded in Albany, New York, was 99° F. The lowest temperature was 112 degrees less than the highest. What was the lowest temperature ever recorded in Albany?

CHAPTER 1 TEST

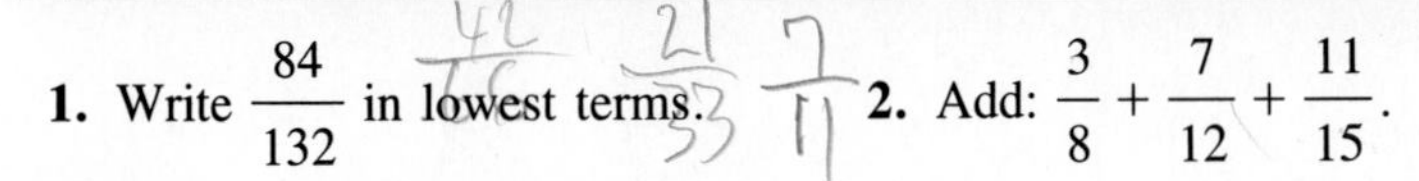

1. Write $\dfrac{84}{132}$ in lowest terms.

2. Add: $\dfrac{3}{8} + \dfrac{7}{12} + \dfrac{11}{15}$.

3. Divide: $\dfrac{6}{5} \div \dfrac{19}{15}$.

Answer true *or* false *for the following.*

4. $6 - |-4| \geq 10$

5. $4[-20 + 7(-2)] \leq 135$

6. $-2 \geq \dfrac{-36 - 3(-6)}{(-2)(-4) - (-1)}$

7. $-14 < \dfrac{32 + 4(1 + 5)}{6(3 + 5) - 2}$

8. $(-3)^2 + 2^2 = 5^2$

---

*The order of exercises in this final group does not correspond to the order in which topics occur in the chapter. This random ordering should help you prepare for the chapter test in yet another way.

Find the numerical value of the given expression when $m = 6$ and $p = 2$.

9. $5m + 2p^3$

10. $\dfrac{7m^2 - p^2}{m + 4}$

Select the smaller number from each pair.

11. $6, \quad -|-8|$

12. $.742, \quad .705$

---

13. Write in symbols: The quotient of -6 and the product of 8 and -3. Evaluate the expression.

14. If a and b are both negative, is $\dfrac{a + b}{a \cdot b}$ positive or negative?

Decide whether the given number is a solution for the equation.

15. $6m + m + 2 = 37; \quad 5$

16. $8(y + 3) + 2y = 18; \quad 4$

Perform the indicated operations wherever possible.

17. $-9 - (4 - 11) + (-5)$

18. $-2\frac{1}{5} + 5\frac{1}{4}$

19. $-6 - [-5 + (8 - 9)]$

20. $3^2 + (-7) - (2^3 - 5)$

21. $\dfrac{-7 - (-5 + 1)}{-4 - (-3)}$

22. $\dfrac{-6[5 - (-1 + 4)]}{-9[2 - (-1)] - 6(-4)}$

23. $\dfrac{15(-4 - 2)}{16(-2) + (-7 - 1)(-3 - 1)}$

---

24. The cruising altitude of a jet bound from San Francisco to New York is 36,000 feet. If it descends 5500 feet due to some unexpected turbulence, what is its new altitude?

25. The highest temperature ever recorded in Idaho was 118° F, while the lowest was $-60°$ F. What is the difference between the highest and the lowest temperature?

26. Find the solution for the equation $x - 4 = -1$ from the set $\{-3, -2, -1, 0, 1, 2, 3\}$. Use trial and error or guessing.

Match the property in Column I with the example of it from Column II.

Column I	*Column II*
27. Commutative	**A.** $2x + (-2x) = 0$
28. Associative	**B.** $(4 + 3) + 2 = 2 + (4 + 3)$
29. Identity	**C.** $6x + 6y = 6(x + y)$
30. Inverse	**D.** $(4 + 3) + 2 = 4 + (3 + 2)$
31. Distributive	**E.** $-12 + 0 = -12$

32. Use the distributive property to simplify $-(3 - 4m)$.

33. Explain in your own words how the commutative property differs from the associative property.

CHAPTER TWO

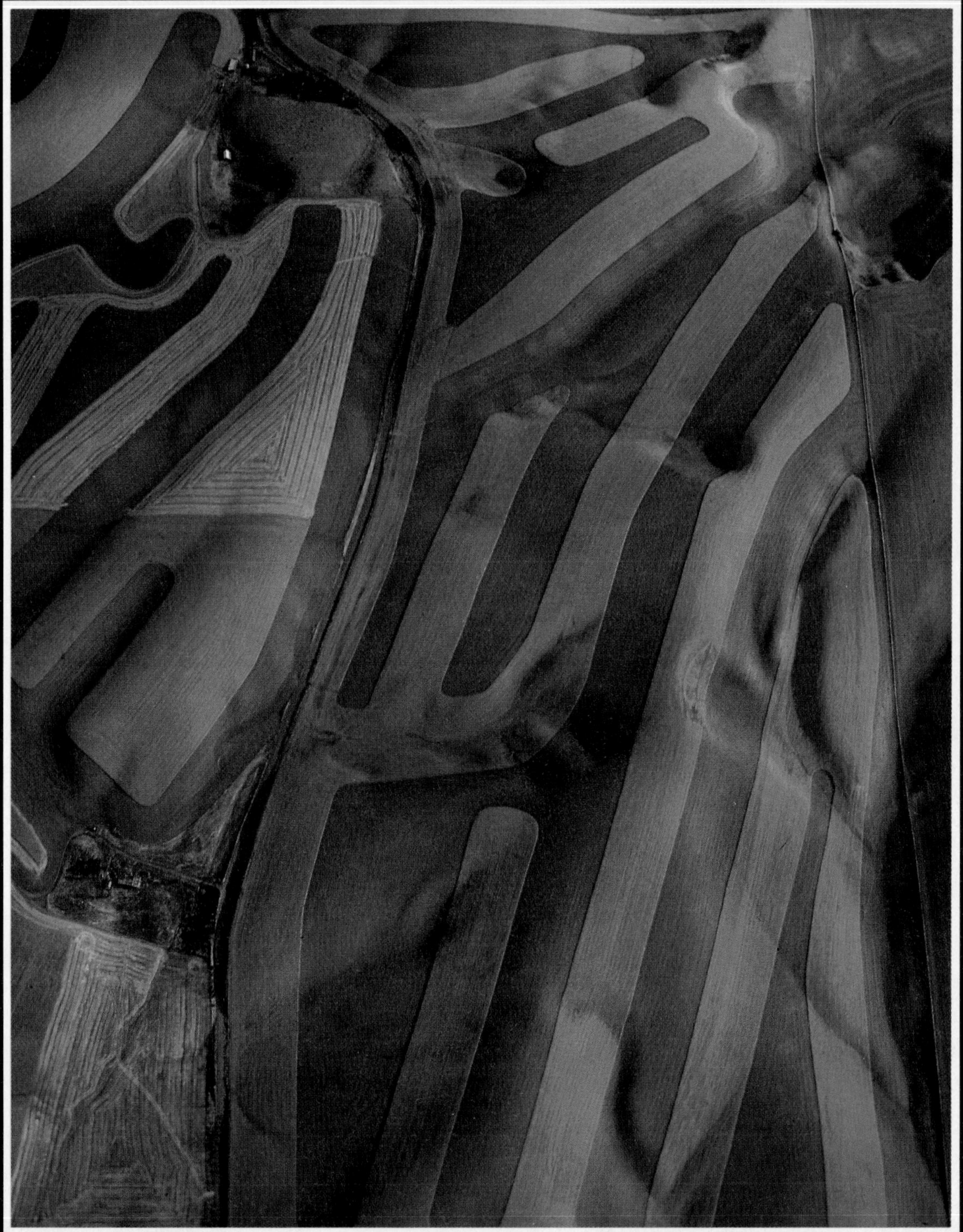

EQUATIONS AND INEQUALITIES

In this chapter we begin our study of the solution of equations and inequalities, and how they are applied to problem solving. The very word *algebra* is derived from the treatise *Hisâb al-jabr w'al muquâbalah,* written in the ninth century by the Arab Muhammed ibn Mûsâ al-Khowârizmî. The title means "the science of reunion and reduction," or more generally, "the science of transposition and cancellation." These ideas are used in equation solving. Several centuries later the book was translated into Latin under the title *Liber Algebrae et Almucabola*. Later, in Spain, the word *algebrista* came to mean "a bonesetter" and on occasion one would see a sign reading *Algebrista y Sangradoe* ("bonesetting and bloodletting") at barbershops, for barbers did this sort of work at that time.

2.1 SIMPLIFYING EXPRESSIONS

OBJECTIVES

1 SIMPLIFY EXPRESSIONS.

2 IDENTIFY TERMS AND NUMERICAL COEFFICIENTS.

3 IDENTIFY LIKE TERMS.

4 COMBINE LIKE TERMS.

5 SIMPLIFY EXPRESSIONS FROM WORD PHRASES.

1 It is often necessary to simplify the expressions on either side of an equation as the first step in solving it. This section shows how to simplify expressions using the properties of addition and multiplication introduced in Chapter 1.

EXAMPLE 1 SIMPLIFYING EXPRESSIONS

Simplify the following expressions.

(a) $4x + 8 + 9$
Since $8 + 9 = 17$,

$$4x + 8 + 9 = 4x + 17.$$

(b) $4(3m - 2n)$
Use the distributive property first.

$$\begin{aligned} 4(3m - 2n) &= 4(3m) - 4(2n) && \text{Arrows denote distributive property.} \\ &= (4 \cdot 3)m - (4 \cdot 2)n && \text{Associative property} \\ &= 12m - 8n \end{aligned}$$

(c)
$$\begin{aligned} 6 + 3(4k + 5) &= 6 + 3(4k) + 3(5) && \text{Distributive property} \\ &= 6 + (3 \cdot 4)k + 3(5) && \text{Associative property} \\ &= 6 + 12k + 15 \\ &= 6 + 15 + 12k && \text{Commutative property} \\ &= 21 + 12k \end{aligned}$$

(d) $5 - (2y - 8) = 5 - 1 \cdot (2y - 8)$ Replace − with −1.
$= 5 - 2y + 8$ Distributive property
$= 5 + 8 - 2y$ Commutative property
$= 13 - 2y$ ■

NOTE In Example 1, parts (c) and (d), a different use of the commutative property would have resulted in answers of $12k + 21$ and $-2y + 13$. These answers also would be acceptable.

The steps using the commutative and associative properties will not be shown in the rest of the examples, but you should be aware that they are usually involved.

2 A **term** is a number, a variable, or a product or quotient of numbers and variables raised to powers. Examples of terms include

$$-9x^2, \quad 15y, \quad -3, \quad 8m^2n, \quad \frac{2}{p}, \quad \text{and} \quad k.$$

The **numerical coefficient** of the term $9m$ is 9, the numerical coefficient of $-15x^3y^2$ is -15, the numerical coefficient of x is 1, and the numerical coefficient of 8 is 8.

CAUTION It is important to be able to distinguish between *terms* and *factors*. For example, in the expression $8x^3 + 12x^2$, there are two terms. They are $8x^3$ and $12x^2$. On the other hand, in the expression $(8x^3)(12x^2)$, $8x^3$ and $12x^2$ are *factors*.

EXAMPLE 2
IDENTIFYING THE NUMERICAL COEFFICIENT OF A TERM

Give the numerical coefficient of each of the following terms.

TERM	NUMERICAL COEFFICIENT
$-7y$	-7
$8p$	8
$34r^3$	34
$-26x^5yz^4$	-26
$-k$	-1 ■

3 Terms with exactly the same variables that have the same exponents are **like terms.** For example, $9m$ and $4m$ have the same variable and are like terms. Also, $6x^3$ and $-5x^3$ are like terms. The terms $-4y^3$ and $4y^2$ have different exponents and are **unlike terms.**

Here are some examples of like terms.

$$5x \text{ and } -12x \qquad 3x^2y \text{ and } 5x^2y$$

Here are some examples of unlike terms.

$$4xy^2 \text{ and } 5xy \qquad -7w^3z^3 \text{ and } 2xz^3$$

4 Recall the distributive property:

$$x(y + z) = xy + xz.$$

This statement can also be written "backward" as

$$xy + xz = x(y + z).$$

This form of the distributive property may be used to find the sum or difference of terms. For example,

$$3x + 5x = (3 + 5)x = 8x.$$

This process is called **combining like terms.**

NOTE It is important to remember that only like terms may be combined.

EXAMPLE 3 COMBINING LIKE TERMS

Combine like terms in the following expressions.

(a) $9m + 5m$

Use the distributive property as given above to combine the like terms.

$$9m + 5m = (9 + 5)m = 14m$$

(b) $6r + 3r + 2r = (6 + 3 + 2)r = 11r$ Distributive property

(c) $4x + x = 4x + 1x = (4 + 1)x = 5x$ (Note: $x = 1x$.)

(d) $16y^2 - 9y^2 = (16 - 9)y^2 = 7y^2$

(e) $32y + 10y^2$ cannot be simplified because $32y$ and $10y^2$ are unlike terms. The distributive property cannot be used here to combine coefficients. ■

When an expression involves parentheses, the distributive property is used both "forward" and "backward" to simplify the expression by combining like terms, as shown in the following example.

EXAMPLE 4 SIMPLIFYING EXPRESSIONS INVOLVING LIKE TERMS

Combine like terms in the following expressions.

(a)

$$\begin{aligned} 14y + 2(6 + 3y) &= 14y + 2(6) + 2(3y) && \text{Distributive property} \\ &= 14y + 12 + 6y && \text{Multiply.} \\ &= 20y + 12 && \text{Combine like terms.} \end{aligned}$$

(b)

$$\begin{aligned} 9k - 6 - 3(2 - 5k) &= 9k - 6 - 3(2) - 3(-5k) && \text{Distributive property} \\ &= 9k - 6 - 6 + 15k && \text{Multiply.} \\ &= 24k - 12 && \text{Combine like terms.} \end{aligned}$$

(c)

$$\begin{aligned} -(2 - r) + 10r &= -1(2 - r) + 10r && \text{Replace } - \text{ with } -1. \\ &= -1(2) - 1(-r) + 10r && \text{Distributive property} \\ &= -2 + r + 10r && \text{Multiply.} \\ &= -2 + 11r && \text{Combine like terms.} \end{aligned}$$

(d) $5(2a - 6) - 3(4a - 9)$

$= 10a - 30 - 12a + 27$	Distributive property
$= 10a + (-30) + (-12a) + 27$	Definition of subtraction
$= 10a + (-12a) + (-30) + 27$	Commutative property
$= -2a - 3$	Combine like terms. ■

Example 4(d) shows that the commutative property can be used with subtraction by treating the subtracted terms as the addition of their additive inverses.

NOTE Examples 3 and 4 suggest that like terms may be combined by combining the coefficients of the terms and keeping the same variable factors.

5 The next example reviews simplifying the result of converting a word phrase into a mathematical expression.

PROBLEM SOLVING

In Chapter 1 we saw how to translate words, phrases, and statements into expressions and equations. This was done mainly to prepare for solving applied problems. This idea will be used extensively in the later sections of this chapter. ■

EXAMPLE 5
CONVERTING WORDS TO A MATHEMATICAL EXPRESSION

Convert to a mathematical expression, and simplify: The sum of 9, five times a number, four times the number, and six times the number.

The word "sum" indicates that the terms should be added. Use x to represent the number. Then the phrase translates as follows.

$9 + 5x + 4x + 6x$	Write as a mathematical expression.
$= 9 + 15x$	Combine like terms. ■

CAUTION In Example 5, we are dealing with an expression to be simplified, and *not* an equation to be solved.

2.1 EXERCISES

Give the numerical coefficient of the following terms. See Example 2.

1. $15y$ **2.** $7z$ **3.** $-22m^4$ **4.** $-2k^7$

5. $35a^4b^2$ **6.** $12m^5n^4$ **7.** -9 **8.** 21

9. y^2 **10.** x^4 **11.** $-r$ **12.** $-z$

Write like *or* unlike *for the following groups of terms.*

13. $6m, -14m$ **14.** $-2a, 5a$ **15.** $7z^3, 7z^2$ **16.** $10m^5, 10m^6$

17. $25y, -14y, 8y$ **18.** $-11x, 5x, 7x$ **19.** $2, 5, -2$ **20.** $-8, 3, 9$

Simplify the following expressions by combining terms. See Examples 1 and 3.

21. $9y + 8y$

22. $15m + 12m$

23. $-4a - 2a$

24. $-3z - 9z$

25. $12b + b$

26. $30x + x$

27. $2k + 9 + 5k + 6$

28. $2 + 17z + 1 + 2z$

29. $-5y + 3 - 1 + 5 + y - 7$

30. $2k - 7 - 5k + 7k - 3 - k$

31. $-2x + 3 + 4x - 17 + 20$

32. $r - 6 - 12r - 4 + 6r$

33. $16 - 5m - 4m - 2 + 2m$

34. $6 - 3z - 2z - 5 + z - 3z$

35. $-10 + x + 4x - 7 - 4x$

36. $-p + 10p - 3p - 4 - 5p$

37. $1 + 7x + 11x - 1 + 5x$

38. $-r + 2 - 5r + 3 + 4r$

39. $6y^2 + 11y^2 - 8y^2$

40. $-9m^3 + 3m^3 - 7m^3$

41. $2p^2 + 3p^2 - 8p^3 - 6p^3$

42. $5y^3 + 6y^3 - 3y^2 - 4y^2$

---

43. Explain in your own words what is meant by *like terms*. Give examples of like terms and unlike terms. (See Objective 3.)

44. You have probably heard the saying "You can't add apples and oranges." How does this saying apply to this section? (See Objective 3.)

Use the distributive property to simplify the following expressions. See Example 4.

45. $6(5t + 11)$

46. $2(3x + 4)$

47. $-3(n + 5)$

48. $-4(y - 8)$

49. $-3(2r - 3) + 2(5r + 3)$

50. $-4(5y - 7) + 3(2y - 5)$

51. $8(2k - 1) - (4k - 3)$

52. $6(3p - 2) - (5p + 1)$

53. $-2(-3k + 2) - (5k - 6) - 3k - 5$

54. $-2(3r - 4) - (6 - r) + 2r - 5$

Convert the following word phrases into mathematical expressions. Use x as the variable. Combine terms whenever possible. See Example 5.

55. The difference between the sum of a number and 3, and three times the number

56. The sum of four times a number and the sum of the number and -12

57. Five times a number subtracted from 8 times the number, with the result subtracted from 9 times the number

58. The difference between a number and 4 times the number, with the result subtracted from the sum of 8 and the product of 12 and the number

59. The sum of nine multiplied by the sum of six times a number and 3, and the difference between 3 and the number

60. The sum of seven times a number and 12, and four times the sum of three times the number and 5

Combine terms.

61. $5 - 3x + 2y + x - 7 + 4y$

62. $3x^2 + 17 - 4x + 3x^2 + 9x + 5x$

63. $-2p + p^2 - p + 4p^2 + 6p^2 + 8p + 1$

64. $-3a + 2 + 5b - a - b - 3a$

65. $6(r - 5) + 3r - (4 - r)$

66. $-(m + 6) - 4(m + 1) + 3$

67. $z + 2(5 - z) - (2z + 3) - z^2$

68. $2k^2 + 4(k - 1) - 3(k + 2) + k$

PREVIEW EXERCISES

Most of the exercise sets in the rest of the book end with brief sets of "preview exercises." These exercises are designed to help you review ideas needed for the next few sections in the chapter. If you need help with these preview exercises, look in the sections indicated each time.

Find the additive inverse of each number. For help see Section 1.5.

69. 6 **70.** 15 **71.** -4 **72.** -8

Add a number to each expression so that the result is just x. See Section 1.9.

73. $x - 7$ **74.** $x + 10$ **75.** $x + 6$ **76.** $x - 3$

Multiply each expression by a number so that the result is just x. See Section 1.9.

77. $\frac{1}{3}x$ **78.** $\frac{2}{5}x$ **79.** $-3x$ **80.** $-7x$

2.2 THE ADDITION AND MULTIPLICATION PROPERTIES OF EQUALITY

OBJECTIVES

1. IDENTIFY LINEAR EQUATIONS.
2. USE THE ADDITION PROPERTY OF EQUALITY.
3. SIMPLIFY EQUATIONS, AND THEN USE THE ADDITION PROPERTY OF EQUALITY.
4. USE THE MULTIPLICATION PROPERTY OF EQUALITY.
5. SIMPLIFY EQUATIONS, AND THEN USE THE MULTIPLICATION PROPERTY OF EQUALITY.

1 In algebra, an equation is said to be *linear* if the variable is raised to the first power. Linear equations are the simplest types of equations that we will solve.

LINEAR EQUATION

A **linear equation** can be written in the form

$$ax + b = 0$$

for real numbers a and b, with $a \neq 0$.

Methods of solving linear equations will be introduced in this section and in Section 2.3. As discussed in Section 1.3, a solution of an equation is a number that when substituted for the variable makes the equation a true statement. If two equations have exactly the same solutions, they are **equivalent equations.** Linear equations are solved by using a series of steps to produce equivalent equations until an equation of the form

$$x = \text{a number}$$

is obtained.

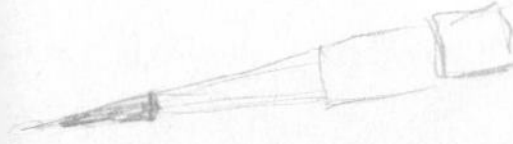

2 According to the equation $x - 5 = 2$, both $x - 5$ and 2 represent the same number, since this is the meaning of the equals sign. Solve the equation by changing the left side from $x - 5$ to just x. This is done by adding 5 to $x - 5$. Keep the two sides equal by also adding 5 on the right side.

$$x - 5 = 2 \qquad \text{Given equation}$$

$$x - 5 + 5 = 2 + 5 \qquad \text{Add 5 on both sides.}$$

Here 5 was added on both sides of the equation. Now simplify each side separately to get

$$x = 7.$$

The solution of the given equation is 7. Check by replacing x with 7 in the given equation.

$$x - 5 = 2 \qquad \text{Given equation}$$

$$7 - 5 = 2 \quad ? \qquad \text{Let } x = 7.$$

$$2 = 2 \qquad \text{True}$$

Since this final result is true, 7 checks as the solution.

The equation above was solved by adding the same number to both sides, as justified by the **addition property of equality.**

ADDITION PROPERTY OF EQUALITY

If A, B, and C are real numbers, then the equations

$$A = B \quad \text{and} \quad A + C = B + C$$

have exactly the same solution. In words, the same number may be added to both sides of an equation without changing the solution.

In the addition property, C represents a real number. That means that numbers or terms with variables, or even sums of terms that represent real numbers, can be added to both sides of an equation.

EXAMPLE 1

USING THE ADDITION PROPERTY OF EQUALITY

Solve the equation $x - 16 = 7$.

If the left side of this equation were just x, the solution would be found. Get x alone by using the addition property of equality and adding 16 on both sides.

$$x - 16 = 7$$

$$(x - 16) + 16 = 7 + 16 \qquad \text{Add 16 on both sides.}$$

$$x = 23$$

Check by substituting 23 for x in the original equation.

$$x - 16 = 7 \qquad \text{Given equation}$$

$$23 - 16 = 7 \quad ? \qquad \text{Let } x = 23.$$

$$7 = 7 \qquad \text{True}$$

Since the check results in a true statement, 23 is the solution. ■

In this example, why was 16 added to both sides of the equation $x - 16 = 7$? The equation would be solved if it could be rewritten so that one side contained only the variable and the other side contained only a number. Since $x - 16 + 16 = x + 0 = x$, adding 16 on the left side simplifies that side to just x, the variable, as desired.

The addition property of equality says that the same number may be *added* to both sides of an equation. As was shown in Chapter 1, subtraction is defined in terms of addition. Because of the way subtraction is defined, the addition property also permits *subtracting* the same number on both sides of an equation.

EXAMPLE 2
SUBTRACTING A VARIABLE EXPRESSION TO SOLVE AN EQUATION

Solve the equation $3k + 12 + k - 8 - 4 = 15 + 3k + 2$.

Begin by combining like terms on each side of the equation to get

$$4k = 17 + 3k.$$

Next, get all terms that contain variables on the same side of the equation. One way to do this is to subtract $3k$ from both sides.

$$4k = 17 + 3k$$

$$4k - 3k = 17 + 3k - 3k \qquad \text{Subtract } 3k \text{ from both sides.}$$

$$k = 17$$

Check by substituting 17 for k in the original equation.

$$3k + 12 + k - 8 - 4 = 15 + 3k + 2 \qquad \text{Given equation}$$

$$3(17) + 12 + 17 - 8 - 4 = 15 + 3(17) + 2 \quad ? \qquad \text{Let } k = 17.$$

$$51 + 12 + 17 - 8 - 4 = 15 + 51 + 2 \quad ? \qquad \text{Multiply.}$$

$$68 = 68 \qquad \text{True}$$

The check results in a true statement, so the solution is 17. ■

NOTE Subtracting $3k$ in Example 2 is the same as adding $-3k$, since by the definition of subtraction, $a - b = a + (-b)$.

3 Sometimes an equation must be simplified as a first step in its solution. This often requires the use of the distributive property, as shown in the next example.

EXAMPLE 3
USING THE DISTRIBUTIVE PROPERTY IN SOLVING AN EQUATION

Solve the equation $3(2 + 5x) - (1 + 14x) = 6$.

Use the distributive property first to simplify the equation.

$$3(2 + 5x) - (1 + 14x) = 6$$

$$6 + 15x - 1 - 14x = 6 \qquad \text{Distributive property}$$

$$x + 5 = 6 \qquad \text{Combine like terms.}$$

Subtract 5 on both sides of the equation to get the variable term alone on one side.

$$x = 1$$

Check this answer by substituting 1 for x in the original equation. ■

4 The addition property of equality by itself is not enough to solve an equation like $3x + 2 = 17$.

$$3x + 2 = 17$$
$$3x + 2 - 2 = 17 - 2 \qquad \text{Subtract 2 from both sides.}$$
$$3x = 15$$

The variable x is not alone on one side of the equation: the equation has $3x$ instead. Another property is needed to change $3x = 15$ to $x =$ a number.

If $3x = 15$, then $3x$ and 15 both represent the same number. Multiplying both $3x$ and 15 by the same number will also result in an equality. The **multiplication property of equality** states that both sides of an equation can be multiplied by the same number.

MULTIPLICATION PROPERTY OF EQUALITY

If A, B, and C are real numbers, then the equations

$$A = B \qquad \text{and} \qquad AC = BC$$

have exactly the same solution. (Assume that $C \neq 0$.) In words, both sides of an equation may be multiplied by the same nonzero number without changing the solution.

This property can be used to solve $3x = 15$. The $3x$ on the left must be changed to $1x$, or x, instead of $3x$. To get x, multiply both sides of the equation by $1/3$. Use $1/3$ since it is the reciprocal of the coefficient of x. This works because $3 \cdot 1/3 = 3/3 = 1$.

$$3x = 15$$
$$\frac{1}{3}(3x) = \frac{1}{3} \cdot 15 \qquad \text{Multiply both sides by } \frac{1}{3}.$$
$$\left(\frac{1}{3} \cdot 3\right)x = \frac{1}{3} \cdot 15 \qquad \text{Associative property}$$
$$1x = 5 \qquad \text{Multiply.}$$
$$x = 5 \qquad \text{Identity property}$$

The solution of the equation is 5. Check this by substituting 5 for x in the given equation.

Just as the addition property of equality permits subtracting the same number from both sides of an equation, the multiplication property of equality permits dividing both sides of an equation by the same nonzero number. For example, the equation $3x = 15$, solved above by multiplication, could also be solved by dividing both sides by 3, as follows.

$$3x = 15$$
$$\frac{3x}{3} = \frac{15}{3} \qquad \text{Divide by 3.}$$
$$x = 5 \qquad \text{Simplify.}$$

NOTE In practice, it is usually easier to multiply on each side if the coefficient of the variable is a fraction, and divide on each side if the coefficient is an integer. For example, to solve

$$-\frac{3}{4}x = 12$$

it is easier to multiply by $-4/3$ than to divide by $-3/4$. On the other hand, to solve

$$-5x = -20$$

it is easier to divide by -5 than to multiply by $-1/5$.

EXAMPLE 4
DIVIDING EACH SIDE OF AN EQUATION BY A NONZERO NUMBER

Solve the equation $25p = 30$.

Get p (instead of $25p$) by using the multiplication property of equality and dividing both sides of the equation by 25, the coefficient of p.

$$25p = 30$$

$$\frac{25p}{25} = \frac{30}{25} \qquad \text{Divide by 25.}$$

$$p = \frac{30}{25} = \frac{6}{5}$$

The solution is 6/5. Check by substituting 6/5 for p in the given equation.

$$25p = 30 \qquad \text{Given equation}$$

$$25\left(\frac{6}{5}\right) = 30 \quad ? \qquad \text{Let } p = \frac{6}{5}.$$

$$30 = 30 \qquad \text{True}$$

The solution is 6/5. ■

In the next two examples, multiplication produces the solution more quickly than division.

EXAMPLE 5
USING THE MULTIPLICATION PROPERTY OF EQUALITY

Solve the equation $\frac{a}{4} = 3$.

Replace $\frac{a}{4}$ by $\frac{1}{4}a$ since division by 4 is the same as multiplication by $\frac{1}{4}$. Get a alone by multiplying both sides by 4, the reciprocal of the coefficient of a.

$$\frac{a}{4} = 3$$

$$\frac{1}{4}a = 3 \qquad \text{Change } \frac{a}{4} \text{ to } \frac{1}{4}a.$$

$$4 \cdot \frac{1}{4}a = 4 \cdot 3 \qquad \text{Multiply by 4.}$$

$$1a = 12 \qquad \text{Inverse property}$$

$$a = 12 \qquad \text{Identity property}$$

Check the answer.

$$\frac{a}{4} = 3 \qquad \text{Given equation}$$

$$\frac{12}{4} = 3 \quad ? \qquad \text{Let } a = 12.$$

$$3 = 3 \qquad \text{True}$$

The solution 12 is correct. ■

EXAMPLE 6
USING THE MULTIPLICATION PROPERTY OF EQUALITY

Solve the equation $\frac{3}{4}h = 6$.

We will show two ways to solve this equation.

Method 1

To get h alone, multiply both sides of the equation by $\frac{4}{3}$, the reciprocal of $\frac{3}{4}$. Use $\frac{4}{3}$ because $\frac{4}{3} \cdot \frac{3}{4}h = 1 \cdot h = h$.

$$\frac{3}{4}h = 6$$

$$\frac{4}{3}\left(\frac{3}{4}h\right) = \frac{4}{3} \cdot 6 \qquad \text{Multiply by } \frac{4}{3}.$$

$$1 \cdot h = \frac{4}{3} \cdot \frac{6}{1} \qquad \text{Inverse property}$$

$$h = 8 \qquad \text{Identity property}$$

Method 2

Begin by multiplying both sides of the equation by 4 to eliminate the denominator.

$$\frac{3}{4}h = 6$$

$$4\left(\frac{3}{4}h\right) = 4 \cdot 6 \qquad \text{Multiply by 4.}$$

$$3h = 24$$

Now divide both sides by 3.

$$\frac{3h}{3} = \frac{24}{3} \qquad \text{Divide by 3.}$$

$$h = 8$$

Using either method, the solution is 8. Check the answer by substitution in the given equation. ■

5 The final example of this section requires simplification of one side of the equation before the multiplication property of equality is used.

EXAMPLE 7 **SIMPLIFYING AN EQUATION AND USING THE MULTIPLICATION PROPERTY**

Solve the equation $2(3 + 7x) - (1 + 15x) = 2$.

Use the distributive property to first simplify the equation.

$$2(3 + 7x) - (1 + 15x) = 2$$
$$6 + 14x - 1 - 15x = 2 \quad \text{Distributive property}$$
$$5 - x = 2 \quad \text{Combine like terms.}$$

To get the variable alone, subtract 5 from both sides.

$$5 - x - 5 = 2 - 5$$
$$-x = -3$$

The variable is alone on the left, but its coefficient is -1, since $-x = -1 \cdot x$. When this occurs, simply multiply both sides by -1 (which is the reciprocal of -1).

$$-x = -3$$
$$-1 \cdot x = -3 \quad -x = -1 \cdot x$$
$$-1(-1 \cdot x) = -1(-3) \quad \text{Multiply by } -1.$$
$$x = 3$$

The solution is 3. Check by substituting into the original equation. (Incidentally, *dividing* by -1 would also allow us to solve an equation of the form $-x = a$.) ■

NOTE From the final steps in Example 7, we can see that the following is true.

If $-x = a$, then $x = -a$.

2.2 EXERCISES

Solve each equation by using the addition property of equality. Check each solution. See Examples 1 and 2.

1. $x - 3 = 7$
2. $x + 5 = 13$
3. $7 + k = 5$
4. $9 + m = 4$
5. $3r - 10 = 2r$
6. $2p = p + 3$
7. $7z = -8 + 6z$
8. $4y = 3y - 5$
9. $m + 5 = 0$
10. $k - 7 = 0$
11. $2 + 3x = 2x$
12. $10 + r = 2r$
13. $2p + 6 = 10 + p$
14. $5r + 2 = -1 + 4r$
15. $2k + 2 = -3 + k$
16. $6 + 7x = 6x + 3$
17. $x - 5 = 2x + 6$
18. $-3r + 7 = -4r - 19$
19. $6z + 3 = 5z - 3$
20. $6t + 5 = 5t + 7$

21. Refer to the definition of *linear equation* given in this section. Why is the restriction $a \neq 0$ necessary? (See Objective 1.)

22. Which of the following equations are not linear equations? (See Objective 1.)
(a) $x^2 + 2x - 3 = 0$ **(b)** $x^3 = x$
(c) $3x - 4 = 0$ **(d)** $2x - 5x = 3 + 9x$

Solve the following equations. First simplify each side of the equation as much as possible and check each solution. You may need to refer to the "Note" following Example 7. See Example 2.

23. $3x + 2x - 6 + x - 5x = 9 + 4$

24. $9r + 4r + 6 - 8 = 10r + 6 + 2r$

25. $-3t + 5t - 6t + 4 - 3 = -3t + 2$

26. $11z + 2 + 4z - 3z = 5z - 8 + 6z$

27. $2k + 8k + 6k - 4k - 8 + 2 = 3k + 2 + 10k$

28. $4m + 8m - 9m + 2 - 5 = 4m + 6$

29. $15y - 4y + 8 - 2 + 7 - 4 = 4y + 2 + 8y$

30. $-9p + 4p - 3p + 2p - 6 = -5p - 6$

31. $5x - 2x + 3x - 4x + 8 - 2 + 4 = x + 10$

Solve the following equations. Check each solution. See Examples 3 and 7.

32. $5(m - 1) - 6m = -8$

33. $2(k + 5) - 3k = 8$

34. $-4(3y + 2) = -13y - 10$

35. $-5(2q - 3) = -11q + 19$

36. $(11y + 10) - 3(3 + 4y) = 2$

37. $(15p - 3) - 2(7p + 1) = -3$

38. $2(r + 5) - (9 + r) = -1$

39. $4(y - 6) - (3y + 2) = 8$

40. $-6(2a + 1) + (13a - 7) = 4$

41. $-5(3k - 3) + (1 + 16k) = 2$

42. $4(7x - 1) + 3(2 - 5x) = 4(3x + 5)$

43. $9(2m - 3) - 4(5 + 3m) = 5(4 + m)$

44. $-2(8p + 7) - 3(4 - 7p) = 2(3 + 2p) - 6$

45. $-5(8 - 2z) + 4(7 - z) = 7(8 + z) - 3$

Solve each equation. See Examples 4–6.

46. $5x = 25$

47. $7x = 28$

48. $3a = -24$

49. $5k = -60$

50. $8s = -56$

51. $10t = -36$

52. $-6x = 16$

53. $-6x = 24$

54. $-18z = 108$

55. $-11p = 77$

56. $5r = 0$

57. $2x = 0$

58. $2x + 3x = 20$

59. $3k + 4k = 14$

60. $7r - 13r = -24$

61. $12a - 18a = -36$

62. $5m + 6m - 2m = 72$

63. $11r - 5r + 6r = 84$

64. $k + k + 2k = 80$

65. $4z + z + 2z = 28$

66. $\frac{p}{5} = 3$

67. $\frac{x}{7} = 7$

68. $\frac{2}{3}t = 6$

69. $\frac{4}{3}m = 18$

70. $-\frac{15}{2}z = 20$

71. $-\frac{12}{5}r = 18$

72. $\frac{3}{4}p = -60$

73. $\frac{5}{8}z = -40$

74. $\frac{2}{3}k = 5$

75. $\frac{5}{3}m = 6$

76. $-\frac{2}{7}p = -7$

77. $-\frac{3}{11}y = -2$

---

78. In the statement of the multiplication property of equality in this section, there is a restriction that $C \neq 0$. What would happen if you should multiply both sides of an equation by 0? (See Objective 4.)

79. Which one of the equations that follow does not require the use of the multiplication property of equality? (See Objective 4.)

(a) $3x - 5x = 6$ **(b)** $-\frac{1}{4}x = 12$

(c) $5x - 4x = 7$ **(d)** $\frac{x}{3} = -2$

Solve each equation.

80. $3m + 5 - 2m - 1 = 4m + 2 + 7$

81. $6a + 2a - 4 = 1 + 8a - 6 + 2a$

82. $-p + 3p - 6 = 7p + 3 - 5 + p$

83. $2z - z + 4z + 5 = -7 + 3 + 8z$

84. $4k - 1 + 3 + 2k = 6 - 5 - 13k + 9k$

85. $12x - 2x - 4 - x + 1 = 4x + 7 + 2x + 2$

Write an equation using the information given in the problem. Use x as the variable. Then solve the equation.

86. Three times a number is 17 more than twice the number. Find the number.

87. If six times a number is subtracted from seven times the number, the result is -9. Find the number.

88. If five times a number is added to three times the number, the result is the sum of seven times the number and 9. Find the number.

89. When a number is multiplied by 4, the result is 6. Find the number.

90. When a number is divided by -5, the result is 2. Find the number.

91. If twice a number is divided by 5, the result is 4. Find the number.

PREVIEW EXERCISES

Use the distributive property and then combine like terms. See Sections 1.9 and 2.1.

92. $9(2q + 7)$ **93.** $4(3m - 5)$ **94.** $-4(5p - 1) + 6$ **95.** $-3(2y + 7) - 9$

96. $-(2 - 5r) + 6r$ **97.** $-(12 - 3y) - 2$ **98.** $6 - 3(4a + 3)$ **99.** $9 - 5(7 - 8p)$

2.3 MORE ON SOLVING LINEAR EQUATIONS

OBJECTIVES

1 LEARN THE FOUR STEPS FOR SOLVING A LINEAR EQUATION AND HOW TO USE THEM.

2 SOLVE EQUATIONS WITH FRACTIONS AS COEFFICIENTS.

3 SOLVE EQUATIONS WITH DECIMALS AS COEFFICIENTS.

4 RECOGNIZE EQUATIONS WITH NO SOLUTIONS OR INFINITELY MANY SOLUTIONS.

5 WRITE A SENTENCE OR SENTENCES USING AN ALGEBRAIC EXPRESSION.

1 In the last section we began our study of linear equations and developed techniques for solving them. In this section we will learn more about solving linear equations, and prepare for solving the types of problems that we will encounter in the next section.

In order to solve linear equations in general, a four-step method can be applied.

SOLVING A LINEAR EQUATION

Step 1 Simplify each side of the equation by combining like terms. Use the commutative, associative, and distributive properties as needed.

Step 2 If necessary, use the addition property of equality to simplify further, so that the variable is on one side of the equals sign and the number is on the other.

Step 3 If necessary, use the multiplication property of equality to simplify further. This gives an equation of the form $x =$ a number.

Step 4 Check the solution by substituting into the original equation. (Do *not* substitute into an intermediate step.)

EXAMPLE 1
USING THE FOUR STEPS TO SOLVE AN EQUATION

Solve the equation $3r + 4 - 2r - 7 = 4r + 3$.

Step 1 $3r + 4 - 2r - 7 = 4r + 3$

$r - 3 = 4r + 3$ Combine like terms.

Step 2 $r - 3 - r = 4r + 3 - r$ Use the addition property of equality. Subtract r.

$-3 = 3r + 3$

$-3 - 3 = -3 + 3r + 3$ Add -3.

$-6 = 3r$

Step 3 $\dfrac{-6}{3} = \dfrac{3r}{3}$ Use the multiplication property of equality. Divide by 3.

$-2 = r$ or $r = -2$

Step 4 Substitute -2 for r in the original equation.

$$3r + 4 - 2r - 7 = 4r + 3$$

$$3(-2) + 4 - 2(-2) - 7 = 4(-2) + 3 \quad ?$$

$$-6 + 4 + 4 - 7 = -8 + 3 \quad ?$$

$$-5 = -5 \quad \text{True}$$

The solution of the equation is -2. ■

In Step 2 of Example 1, the terms were added and subtracted in such a way that the variable term ended up on the right. Choosing differently would lead to the variable term being on the left side of the equation. Usually there is no advantage either way.

EXAMPLE 2
USING THE FOUR STEPS TO SOLVE AN EQUATION

Solve the equation $4(k - 3) - k = k - 6$.

Step 1 Before combining like terms, use the distributive property to simplify $4(k - 3)$.

$4(k - 3) - k = k - 6$

$4 \cdot k - 4 \cdot 3 - k = k - 6$ Distributive property

$4k - 12 - k = k - 6$

$3k - 12 = k - 6$ Combine like terms.

Step 2 $3k - 12 + 12 = k - 6 + 12$ Add 12.

$3k = k + 6$

$3k - k = k + 6 - k$ Subtract k.

$2k = 6$

Step 3 $\dfrac{2k}{2} = \dfrac{6}{2}$ Divide by 2.

$k = 3$

Step 4 Check your answer by substituting 3 for k in the given equation. Remember to do the work inside the parentheses first.

$$\begin{aligned} 4(k-3)-k &= k-6 & & \\ 4(3-3)-3 &= 3-6 & ? & \quad \text{Let } k=3. \\ 4(0)-3 &= 3-6 & ? & \\ 0-3 &= 3-6 & ? & \\ -3 &= -3 & & \quad \text{True} \end{aligned}$$

The solution of the equation is 3. ■

EXAMPLE 3
USING THE FOUR STEPS TO SOLVE AN EQUATION

Solve the equation $8a - (3 + 2a) = 3a + 1$.

Step 1 Simplify.

$$\begin{aligned} 8a-(3+2a) &= 3a+1 & & \\ 8a-3-2a &= 3a+1 & & \text{Distributive property} \\ 6a-3 &= 3a+1 & & \end{aligned}$$

Step 2

$$\begin{aligned} 6a-3+3 &= 3a+1+3 & & \text{Add 3.} \\ 6a &= 3a+4 & & \\ 6a-3a &= 3a+4-3a & & \text{Subtract } 3a. \\ 3a &= 4 & & \end{aligned}$$

Step 3

$$\begin{aligned} \frac{3a}{3} &= \frac{4}{3} & & \text{Divide by 3.} \\ a &= \frac{4}{3} & & \end{aligned}$$

Step 4 Check that the solution is 4/3.

$$\begin{aligned} 8a-(3+2a) &= 3a+1 & & \\ 8\left(\frac{4}{3}\right)-\left[3+2\left(\frac{4}{3}\right)\right] &= 3\left(\frac{4}{3}\right)+1 & ? & \quad \text{Let } a=\frac{4}{3}. \\ \frac{32}{3}-\left[3+\frac{8}{3}\right] &= 4+1 & ? & \\ \frac{32}{3}-\left[\frac{9}{3}+\frac{8}{3}\right] &= 5 & ? & \\ \frac{32}{3}-\frac{17}{3} &= 5 & ? & \\ 5 &= 5 & & \quad \text{True} \end{aligned}$$

The check shows that 4/3 is the solution. ■

CAUTION Be very careful with signs when solving equations like the one in Example 3. When a subtraction sign appears immediately in front of a quantity in parentheses, such as in the expression

$$8 - (3 + 2a),$$

remember that the $-$ sign acts like a factor of -1, and has the effect of changing the sign of *every* term within the parentheses. Thus,

$$8 - (3 + 2a) = 8 - 3 - 2a.$$

↑ ↑ Change to $-$ in both terms.

EXAMPLE 4 USING THE FOUR STEPS TO SOLVE AN EQUATION

Solve the equation $4(8 - 3t) = 32 - 8(t + 2)$.

Step 1 Use the distributive property.

$$4(8 - 3t) = 32 - 8(t + 2) \quad \text{Given equation}$$
$$32 - 12t = 32 - 8t - 16 \quad \text{Distributive property}$$
$$32 - 12t = 16 - 8t$$

Step 2

$$32 - 12t + 12t = 16 - 8t + 12t \quad \text{Add } 12t.$$
$$32 = 16 + 4t$$
$$32 - 16 = 16 + 4t - 16 \quad \text{Subtract 16.}$$
$$16 = 4t$$

Step 3

$$\frac{16}{4} = \frac{4t}{4} \quad \text{Divide by 4.}$$
$$4 = t \quad \text{or} \quad t = 4$$

Step 4 Check this solution in the given equation.

$$4(8 - 3t) = 32 - 8(t + 2)$$
$$4(8 - 3 \cdot 4) = 32 - 8(4 + 2) \quad ? \quad \text{Let } t = 4.$$
$$4(8 - 12) = 32 - 8(6) \quad ?$$
$$4(-4) = 32 - 48 \quad ?$$
$$-16 = -16 \quad \text{True}$$

The solution, 4, checks. ■

2 Sometimes equations will involve fractions as coefficients. There is a way that we can avoid dealing with fractions in equations. Since the multiplication property of equality states that we can multiply on both sides of an equation without changing the solution, if we multiply both sides by the least common denominator of all the fractions in the equation, an equivalent equation with only integers as coefficients will be obtained.

EXAMPLE 5
SOLVING AN EQUATION WITH FRACTIONS AS COEFFICIENTS

Solve the equation $\frac{2}{3}x - \frac{1}{2}x = -\frac{1}{6}x - 2$.

The least common denominator of all the fractions in the equation is 6. Start by multiplying both sides of the equation by 6.

$$\frac{2}{3}x - \frac{1}{2}x = -\frac{1}{6}x - 2$$

$$6\left(\frac{2}{3}x - \frac{1}{2}x\right) = 6\left(-\frac{1}{6}x - 2\right) \qquad \text{Multiply by 6.}$$

$$6\left(\frac{2}{3}x\right) + 6\left(-\frac{1}{2}x\right) = 6\left(-\frac{1}{6}x\right) + 6(-2) \qquad \text{Distributive property}$$

$$4x - 3x = -x - 12$$

Now use the four steps to solve this equivalent equation.

Step 1 $\quad x = -x - 12 \qquad$ Combine like terms.

Step 2 $\quad x + x = x - x - 12 \qquad$ Add x.

$$2x = -12$$

Step 3 $\quad \dfrac{2x}{2} = \dfrac{-12}{2} \qquad$ Divide by 2.

$$x = -6$$

Step 4 Check by substituting -6 for x in the original equation.

$$\frac{2}{3}(-6) - \frac{1}{2}(-6) = -\frac{1}{6}(-6) - 2 \quad ? \qquad \text{Let } x = -6.$$

$$-4 + 3 = 1 - 2 \quad ?$$

$$-1 = -1 \qquad \text{True}$$

The solution of the equation is -6. ■

NOTE The equation in Example 5 could have been solved without multiplying through to clear fractions; however, most students make fewer errors working with integers as coefficients rather than fractions as coefficients.

3 Just as we can clear fractions from equations, we can also clear decimals.

PROBLEM SOLVING

In Section 2.7 we will learn how to solve problems that involve mixture, interest, and denominations of money. In all of these types of problems, rates that appear as decimals will be found; the equations that we will use to solve these problems will have decimal numbers as coefficients. Just as we can clear fractions in an equation (as shown in Example 5), we can also clear decimals by multiplying both sides of the equation by a power of 10 that will give us only integer coefficients. A typical equation for solving these types of problems is given in the next example. ■

EXAMPLE 6
SOLVING AN EQUATION WITH DECIMAL COEFFICIENTS

Solve the equation $.20t + .10(20 - t) = .18(20)$.

A number can be multiplied by 100 by moving the decimal point two places to the right. Start the solution by multiplying both sides of the equation by 100.

$$.20t + .10(20 - t) = .18(20)$$

$$.20t + .10(20 - t) = .18(20) \qquad \text{Multiply by 100.}$$

$$20t + 10(20 - t) = 18(20)$$

Now use the four steps.

Step 1 $\quad 20t + 10(20) + 10(-t) = 360 \qquad$ Distributive property

$$20t + 200 - 10t = 360$$

$$10t + 200 = 360 \qquad \text{Combine like terms.}$$

Step 2 $\quad 10t + 200 - 200 = 360 - 200 \qquad$ Subtract 200.

$$10t = 160$$

Step 3 $\quad \dfrac{10t}{10} = \dfrac{160}{10} \qquad$ Divide by 10.

$$t = 16$$

Step 4 Check to see that 16 is the solution of the equation by substituting into the original equation. ■

4 Each of the equations that we have solved so far has had exactly one solution. As the next examples show, linear equations may have no solutions or infinitely many solutions as well. (The four steps are not identified in these examples. See if you can identify them.)

EXAMPLE 7
SOLVING AN EQUATION THAT HAS INFINITELY MANY SOLUTIONS

Solve $5x - 15 = 5(x - 3)$.

$$5x - 15 = 5(x - 3)$$

$$5x - 15 = 5x - 15 \qquad \text{Distributive property}$$

$$5x - 15 + 15 = 5x - 15 + 15 \qquad \text{Add 15 to each side.}$$

$$5x = 5x \qquad \text{Combine terms.}$$

$$5x - 5x = 5x - 5x \qquad \text{Subtract 5x from each side.}$$

$$0 = 0$$

The variable has ''disappeared.'' When this happens, look at the resulting statement $(0 = 0)$. Since the statement is a *true* one, *any* real number is a solution. Indicate the solution as ''all real numbers.'' ■

CAUTION When you are solving an equation like the one in Example 7, do not write ''0'' as the solution. While 0 is a solution, there are infinitely many other solutions.

EXAMPLE 8
SOLVING AN EQUATION THAT HAS NO SOLUTION

Solve $2x + 3(x + 1) = 5x + 4$.

$$2x + 3(x + 1) = 5x + 4$$
$$2x + 3x + 3 = 5x + 4 \quad \text{Distributive property}$$
$$5x + 3 = 5x + 4 \quad \text{Combine terms.}$$
$$5x + 3 - 5x = 5x + 4 - 5x \quad \text{Subtract 5x from each side.}$$
$$3 = 4 \quad \text{Combine terms.}$$

Again, the variable has disappeared, but this time a *false* statement ($3 = 4$) results. When this happens, the equation has no solution. Indicate this by writing "no solution." ■

5 We continue our work with translating from words to symbols.

PROBLEM SOLVING

The next example illustrates a type of translation that is important in many types of problem-solving situations. Very often we are given that the sum of two quantities is a particular number and we are asked to find the values of the two quantities. Example 9 shows how to express the unknown quantities in terms of a single variable. ■

EXAMPLE 9
TRANSLATING PHRASES INTO AN ALGEBRAIC EXPRESSION

Two numbers have a sum of 23. If one of the numbers is represented by k, find an expression for the other number.

First, suppose that the sum of two numbers is 23, and one of the numbers is 10. How would you find the other number? You would subtract 10 from 23 to get 13: $23 - 10 = 13$. So instead of using 10 as one of the numbers, use k as stated in the problem. The other number would be obtained in the same way. You must subtract k from 23. Therefore, an expression for the other number is $23 - k$. ■

2.3 EXERCISES

Solve each equation. Check your solutions. See Examples 1–4.

1. $4h + 8 = 16$
2. $3x - 15 = 9$
3. $6k + 12 = -12 + 7k$
4. $2m - 6 = 6 + 3m$
5. $12p + 18 = 14p$
6. $10m - 15 = 7m$
7. $3x + 9 = -3(2x + 3)$
8. $4z + 2 = -2(z + 2)$
9. $2(2r - 1) = -3(r + 3)$
10. $3(3k + 5) = 2(5k + 5)$
11. $2(3x + 4) = 8(2 + x)$
12. $4(3p + 3) = 3(3p - 1)$

Combine terms as necessary. Then solve the equations. See Examples 2–4, 7 and 8.

13. $-4 - 3(2x + 1) = 11$
14. $8 - 2(3x - 4) = 2x$
15. $-5k - 8 = 2(k + 6) + 1$
16. $4a - 7 = 3(2a + 5) - 2$
17. $5(2m - 1) = 4(2m + 1) + 7$
18. $3(3k - 5) = 4(3k - 1) - 17$
19. $5(4t + 3) = 6(3t + 2) - 1$
20. $7(2y + 6) = 9(y + 3) + 5$
21. $5(x - 3) + 2 = 5(2x - 8) - 3$
22. $6(2v - 1) - 5 = 7(3v - 2) - 24$

23. $-2(3s + 9) - 6 = -3(3s + 11) - 6$

24. $-3(5z + 24) + 2 = 2(3 - 2z) - 10$

25. $6(2p - 8) + 24 = 3(5p - 6) - 6$

26. $2(5x + 3) - 3 = 6(2x - 3) + 15$

27. $3(m - 4) - (4m - 11) = -5$

28. $4(2a + 6) - (a + 1) = 2$

29. $-(4m + 2) - (-3m - 5) = 3$

30. $-(6k - 5) - (-5k + 8) = -4$

31. $2x + 2(3x + 2) - 9 = 3x - 9 + 3$

32. $3(z - 2) + 4z = 8 + z + 1 - z$

33. $2(r - 3) + 5(r + 4) = 9$

34. $-4(m - 8) + 3(2m + 1) = 6$

35. $5(4x - 1) = 2(10x - 3)$

36. $3(8x + 1) = 6(4x + 2)$

37. $9(2x - 4) = 18(x - 2)$

38. $5(6 - 3x) = 15(2 - x)$

39. $8(p - 3) + 4p = 6(2p + 1) - 3$

40. $9(x + 1) - 3x = 2(3x + 1) - 4$

---

41. After correctly working through several steps of an equation, a student obtains the equation $5x = 3x$. Then the student divides both sides by x to get $5 = 3$ and gives "no solution" as the answer. Is this correct? If not, explain why. (See Objectives 1 and 4.)

42. Which one of the following equations does not have all real numbers as solutions? (See Objective 4.)

(a) $5x = 4x + x$ **(b)** $2(x + 6) = 2x + 12$

(c) $\frac{1}{2}x = .5x$ **(d)** $\frac{3}{x - 2} = \frac{3}{x - 2}$

Solve each of the following equations. See Examples 5 and 6.

43. $\frac{3}{5}t - \frac{1}{10}t = t - \frac{5}{2}$

44. $-\frac{2}{7}r + 2r = \frac{1}{2}r + \frac{17}{2}$

45. $-\frac{1}{4}(x - 12) + \frac{1}{2}(x + 2) = x + 4$

46. $\frac{1}{9}(y + 18) + \frac{1}{3}(2y + 3) = y + 3$

47. $\frac{2}{3}k - \left(k + \frac{1}{4}\right) = \frac{1}{12}(k + 4)$

48. $-\frac{5}{6}q - \left(q - \frac{1}{2}\right) = \frac{1}{4}(q + 2)$

49. $.20(60) + .05x = .10(60 + x)$

50. $.30(30) + .15x = .20(30 + x)$

51. $1.00x + .05(12 - x) = .10(12)$

52. $.92x + .98(12 - x) = .96(12)$

53. $.06(10{,}000) + .08x = .072(10{,}000 + x)$

54. $.02(5000) + .03x = .025(5000 + x)$

Write the answer to each of the following problems as an algebraic expression. See Example 9.

55. Two numbers have a sum of 11. One of the numbers is q. Find the other number.

56. The product of two numbers is 9. One of the numbers is k. What is the other number?

57. Yesterday Walt bought x apples. Today he bought 7 apples. How many apples did he buy altogether?

58. Joann has 15 books. She donated p books to the library. How many books does she have left?

59. Mary is a years old. How old will she be in 12 years?

60. Tom has r quarters. Find the value of the quarters in cents.

61. A bank teller has t dollars, all in five-dollar bills. How many five-dollar bills does the teller have?

62. A plane ticket costs b dollars for an adult and d dollars for a child. Find the total cost for 3 adults and 2 children.

PREVIEW EXERCISES

Write each phrase as a mathematical expression. Use x as the variable. See Sections 1.5, 1.6, 1.7, and 1.8.

63. Three-fifths of a number

64. Triple a number

65. A number added to twice the number

66. A nonzero number subtracted from twice its reciprocal

67. The difference between a nonzero number and 7, divided by the number

68. The sum of a number and 6, divided by -4

69. The quotient of -9 and the sum of a number and 3

70. The product of 8 and the sum of a number and 3

2.4 AN INTRODUCTION TO APPLICATIONS OF LINEAR EQUATIONS

OBJECTIVES

1. LEARN THE SIX STEPS TO BE USED TO SOLVE AN APPLIED PROBLEM.
2. SOLVE PROBLEMS INVOLVING UNKNOWN NUMBERS.
3. SOLVE PROBLEMS INVOLVING SUMS OF QUANTITIES.
4. SOLVE PROBLEMS INVOLVING SUPPLEMENTARY AND COMPLEMENTARY ANGLES.

FOCUS ON PROBLEM SOLVING

In the 1960 United States presidential election, John F. Kennedy received 84 more electoral votes than Richard M. Nixon. Together the two men received 522 electoral votes. How many votes did each of the candidates receive?

In order to solve this problem, we must devise a strategy for letting a variable represent one of the numbers of votes, and then express the other number of votes in terms of the same variable. After an equation is written, the problem may then be solved. In this section we will develop some strategies about problem solving in general. These strategies can then be applied to many kinds of applications.

This problem is Exercise 5 in the exercises for this section. After working through this section, you should be able to solve this problem.

1 We now begin to look at how algebra is used to solve applied problems.

PROBLEM SOLVING

In earlier sections we learned how to translate words, phrases, and sentences into mathematical expressions and equations. In this section and in many other sections in this book, we will use these translations in solving applied problems using algebra. It must be emphasized that many *meaningful* applications of mathematics require concepts that are beyond the level of this book. Some of the problems you will encounter will seem "contrived," and to some extent they are. But the skills you will develop in solving simple problems will help you in solving more realistic problems in chemistry, physics, biology, business, and other fields.

While there is no specific method that will enable you to solve all kinds of applied problems, the following general method is suggested. It consists of six steps, and the steps will be specified by number in the examples in this section.

SOLVING AN APPLIED PROBLEM

Step 1 Read the problem carefully, and choose a variable to represent the numerical value that you are asked to find—the unknown number. *Write down* what the variable represents.
Step 2 *Write down* a mathematical expression using the variable for any other unknown quantities. Draw figures or diagrams if they apply.
Step 3 Translate the problem into an equation.
Step 4 Solve the equation.
Step 5 Answer the question asked in the problem.
Step 6 Check your solution by using the original words of the problem. Be sure that your answer makes sense.

2 Some of the simplest applied problems involve unknown numbers, and we will look at this type in the next example.

PROBLEM SOLVING

The third step in solving a word problem is often the hardest. Begin to translate the problem into an equation by writing the given phrases as mathematical expressions. Since equal mathematical expressions are names for the same number, translate any words that mean *equal* or *same* as =. The = sign leads to an equation to be solved. ■

EXAMPLE 1 **FINDING THE VALUE OF AN UNKNOWN NUMBER**

The product of 4, and a number decreased by 7, is 100. Find the number.

Step 1 After reading the problem carefully, decide on what you are being told to find, and then choose a variable to represent the unknown quantity. In this problem, we are told to find a number, so we write

Let x = the number.

Step 2 There are no other unknown quantities to find.
Step 3 Translate as follows.

Because of the comma in the given sentence, writing the equation as $4x - 7 = 100$ is incorrect. The equation $4x - 7 = 100$ corresponds to the statement "The product of 4 and a number, decreased by 7, is 100."

Step 4 Solve the equation.

$$4(x - 7) = 100$$
$$4x - 28 = 100 \quad \text{Distributive property}$$
$$4x = 128 \quad \text{Add 28 to both sides.}$$
$$x = 32 \quad \text{Divide by 4.}$$

Step 5 The number is 32.
Step 6 Check the solution by using the original words of the problem. When 32 is decreased by 7, we get $32 - 7 = 25$. If 4 is multiplied by 25, we get 100, as the problem required. The answer, 32, is correct. ■

NOTE The commas in the statement of the problem in Example 1 are used in translating correctly.

3 A common type of problem that occurs in elementary algebra is the type that involves finding two quantities when the sum of the quantities is known.

PROBLEM SOLVING

In general, to solve such problems, choose a variable to represent one of the unknowns and then represent the other quantity in terms of the same variable, using information obtained in the problem. Then write an equation based upon the words of the problem. The next example illustrates these ideas. ■

EXAMPLE 2 **FINDING THE NUMBERS OF MEN AND WOMEN AT A CONCERT**

At a concert, there were 25 more women than men. The total number of people at the concert was 139. Find the number of men and the number of women at the concert.

Step 1 Let $x =$ the number of men.

Step 2 Let $x + 25 =$ the number of women.

Step 3 Now write an equation.

The total	is	the number of men	plus	the number of women.
↓	↓	↓	↓	↓
139	$=$	x	$+$	$(x + 25)$

Step 4 Solve the equation.

$$139 = 2x + 25 \quad \text{Combine terms.}$$
$$139 - 25 = 2x + 25 - 25 \quad \text{Subtract 25.}$$
$$114 = 2x \quad \text{Simplify.}$$
$$57 = x \quad \text{Divide by 2.}$$

Step 5 Because x represents the number of men, there were 57 men at the concert. Because $x + 25$ represents the number of women, there were $57 + 25 = 82$ women at the concert.

Step 6 Since there were 57 men and 82 women present, there were $57 + 82 = 139$ people there. Because $82 - 57 = 25$, there were 25 more women than men. This information agrees with what is given in the problem, so the answers check. ■

NOTE The problem in Example 2 could also have been solved by letting x represent the number of women. Then $x - 25$ would represent the number of men. The equation would then be

$$139 = x + (x - 25).$$

The solution of this equation is 82, which is the number of women. The number of men would then be $82 - 25 = 57$. You can see that the answers are the same, no matter which approach is used.

EXAMPLE 3
FINDING THE NUMBER OF ORDERS FOR TEA

The owner of Café du Monde found that on one day the number of orders for tea was 1/3 the number of orders for coffee. If the total number of orders for the two drinks was 76, how many orders were placed for tea?

Step 1 Let $x =$ the number of orders for coffee.

Step 2 Let $\frac{1}{3}x =$ the number of orders for tea.

Step 3 Use the fact that the total number of orders was 76 to write an equation.

The total	is	orders for coffee	plus	orders for tea.
↓	↓	↓	↓	↓
76	$=$	x	$+$	$\frac{1}{3}x$

Step 4 Now solve the equation.

$$76 = \frac{4}{3}x \qquad \text{Combine like terms.}$$

$$\frac{3}{4}(76) = \frac{3}{4}\left(\frac{4}{3}x\right) \qquad \text{Multiply by } \frac{3}{4}.$$

$$57 = x$$

Step 5 In this problem, *x does not represent the quantity that we are asked to find.* The number of orders placed for tea was $\frac{1}{3}x$. So $\frac{1}{3}(57) = 19$ is the number of orders for tea.

Step 6 The number of coffee orders (x) was 57 and the number of tea orders was 19. 19 is one-third of 57, and $19 + 57 = 76$. Since this agrees with the information given in the problem, the answer is correct. ■

PROBLEM SOLVING

In Example 3, it was easier to let the variable represent the quantity that was *not* asked for. This required an extra step in Step 5 to find the number of orders for tea. In some cases, this approach is easier than letting the variable represent the quantity that we are asked to find. Experience in solving problems will indicate when this approach is useful, and experience comes only from solving many problems!

Sometimes it is necessary to find three unknown quantities in an applied problem. Frequently the three unknowns are compared in *pairs*. When this happens, it is usually easiest to let the variable represent the unknown found in both pairs. ■

The next example illustrates how we can find more than two unknown quantities in an application.

EXAMPLE 4
DIVIDING A BOARD INTO PIECES

Maria Gonzales has a piece of board 70 inches long. She cuts it into three pieces. The longest piece is twice the length of the middle-sized piece, and the shortest piece is 10 inches shorter than the middle-sized piece. How long are the three pieces?

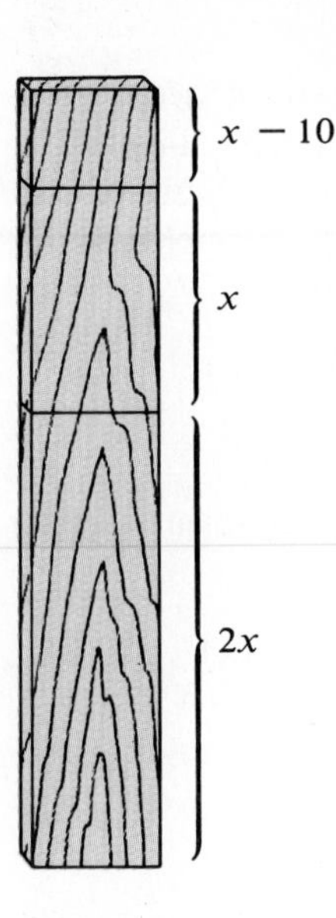

Steps 1 and 2 Since the middle-sized piece appears in both pairs of comparisons, let x represent the length of the middle-sized piece. We have

$$x = \text{the length of the middle-sized piece}$$
$$2x = \text{the length of the longest piece}$$
$$x - 10 = \text{the length of the shortest piece.}$$

A sketch is helpful here. (See the illustration at the side.)

Longest, Middle-sized, Shortest, Total length

Step 3 $$2x + x + (x - 10) = 70$$

Step 4

$$4x - 10 = 70 \quad \text{Combine terms.}$$
$$4x - 10 + 10 = 70 + 10 \quad \text{Add 10 to each side.}$$
$$4x = 80 \quad \text{Combine terms.}$$
$$\frac{4x}{4} = \frac{80}{4} \quad \text{Divide by 4 on each side.}$$
$$x = 20$$

Step 5 The middle-sized piece is 20 inches long, the longest piece is $2(20) = 40$ inches long, and the shortest piece is $20 - 10 = 10$ inches long.

Step 6 Check to see that the sum of the lengths is 70 inches, and that all conditions of the problem are satisfied. ■

EXAMPLE 5

ANALYZING A GASOLINE/OIL MIXTURE

A lawn trimmer uses a mixture of gasoline and oil. For each ounce of oil the mixture contains 16 ounces of gasoline. If the tank holds 68 ounces of the mixture, how many ounces of oil and how many ounces of gasoline does it require when it is full?

Step 1 Let x = the number of ounces of oil required when full.

Step 2 Let $16x$ = the number of ounces of gasoline required when full.

Amount of gasoline, Amount of oil, Total amount in tank

Step 3 $$16x + x = 68$$

Step 4

$$17x = 68 \quad \text{Combine terms.}$$
$$\frac{17x}{17} = \frac{68}{17} \quad \text{Divide by 17.}$$
$$x = 4$$

Step 5 The trimmer requires 4 ounces of oil and $16(4) = 64$ ounces of gasoline when full.

Step 6 Since $4 + 64 = 68$, and 64 is 16 times 4, the answers check. ■

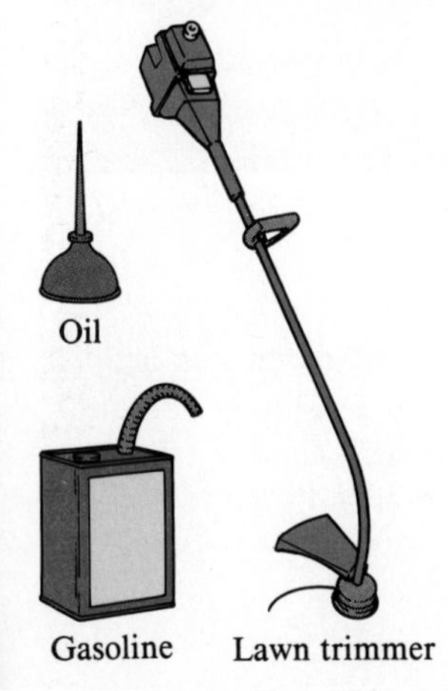

4 The final example of problem solving in this section deals with concepts from geometry. An angle can be measured by a unit called the **degree** (°). See Figure 2.1. Two angles whose sum is 90° are said to be **complementary,** or complements of each other. Two angles whose sum is 180° are said to be **supplementary,** or supplements of each other. If x represents the degree measure of an angle, then

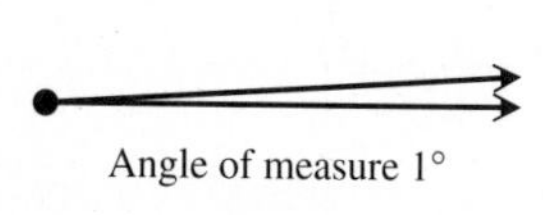

FIGURE 2.1

$90 - x$ represents the degree measure of its complement, and

$180 - x$ represents the degree measure of its supplement.

EXAMPLE 6
FINDING THE MEASURE OF AN ANGLE

Find the measure of an angle whose supplement is 10 degrees more than twice its complement.

Step 1 Let $x =$ the degree measure of the angle.

Step 2 Let $90 - x =$ the degree measure of its complement;
$180 - x =$ the degree measure of its supplement.

Step 3

Supplement	is	10	more than	twice	its complement.
↓	↓	↓	↓	↓	↓
$180 - x$	$=$	10	$+$	2	$\cdot\ (90 - x)$

Step 4 Solve the equation.

$$180 - x = 10 + 180 - 2x \qquad \text{Distributive property}$$
$$180 - x = 190 - 2x \qquad \text{Combine terms.}$$
$$180 - x + 2x = 190 - 2x + 2x \qquad \text{Add } 2x.$$
$$180 + x = 190 \qquad \text{Combine terms.}$$
$$180 + x - 180 = 190 - 180 \qquad \text{Subtract 180.}$$
$$x = 10$$

Step 5 The measure of the angle is 10 degrees.

Step 6 The complement of 10° is 80° and the supplement of 10° is 170°. 170° is equal to 10° more than twice 80° ($170 = 10 + 2(80)$ is true); therefore, the answer is correct. ■

2.4 EXERCISES

The steps for solving applied problems are repeated here. Follow these steps in working all problems in this exercise set.

Step 1 Read the problem carefully, and choose a variable to represent the numerical value that you are asked to find—the unknown number. *Write down* what the variable represents.

Step 2 *Write down* a mathematical expression using the variable for any other unknown quantities. Draw figures or diagrams if they apply.

Step 3 Translate the problem into an equation.

Step 4 Solve the equation.

Step 5 Answer the question asked in the problem.

Step 6 Check your solution by using the original words of the problem. Be sure that your answer makes sense.

Solve the following problems. See Example 1.

1. If 4 is added to a number and this sum is doubled, the result is 1 less than 3 times the number. Find the number.

2. If 8 is subtracted from a number and this difference is tripled, the result is -3 times the number. Find the number.

3. If the sum of a number and 5 is multiplied by 3, the result is the same as if 16 were added to twice the number. Find the number.

4. If 5 is multiplied by the difference between a number and 4, the result is 20 less than the number. Find the number.

Solve the following problems. See Examples 2–5.

5. In the 1960 United States presidential election, John F. Kennedy received 84 more electoral votes than Richard M. Nixon. Together the two men received 522 electoral votes. How many votes did each of the candidates receive?

6. In 1989 the state of Florida had a total of 120 members in its House of Representatives, consisting of only Democrats and Republicans. There were 30 more Democrats than Republicans. How many representatives of each party were there?

7. In their daily workout, Katherine did 28 more sit-ups than Brian. The total number of sit-ups for both of them was 80. Find the number of sit-ups that each did.

8. On a geometry test, the highest grade was 66 points higher than the lowest grade. The sum of the two grades was 140. Find the highest and the lowest grades.

9. Captain Tom Tupper gives deep-sea fishing trips. One day he noticed that the boat contained (not counting himself) 2 fewer men than women. If he had 20 customers on the boat, how many men and how many women were there?

10. A farmer has 28 more hens than roosters, with 170 chickens in all. Find the number of hens and the number of roosters on the farm.

11. In a given amount of time, Larry drove twice as far as Jean. Altogether they drove 90 miles. Find the number of miles driven by each.

12. Mark is two years older than Linda. The sum of their ages is 70 years. Find the age of each.

13. A piece of string is 40 centimeters long. It is cut into three pieces. The longest piece is 3 times as long as the middle-sized piece, and the shortest piece is 23 centimeters shorter than the longest piece. Find the lengths of the three pieces.

14. A strip of paper is 56 inches long. It is cut into three pieces. The longest piece is 12 inches longer than the middle-sized piece, and the shortest piece is 16 inches shorter than the middle-sized piece. Find the lengths of the three pieces.

15. During the 1987 baseball season, Wade Boggs had 10 more at-bats than his teammate Dwight Evans. Evans had 17 fewer at-bats than another teammate, Ellis Burks. The three players had a total of 1650 at-bats. How many times did each player come to bat?

16. During the 1987 baseball season, Bret Saberhagen pitched 9 fewer innings than Jack Morris. Mark Langston pitched 6 more innings than Morris. How many innings did each player pitch, if their total number of innings pitched was 795?

17. A pharmacist found that at the end of the day she had 4/3 as many prescriptions for antibiotics as she did for tranquilizers. She had 42 prescriptions altogether for these two types of drugs. How many did she have for tranquilizers?

18. In a mixture of concrete, there are 3 pounds of cement mix for every 1 pound of gravel. If the mixture contains a total of 140 pounds of these two ingredients, how many pounds of gravel are there?

19. A mixture of nuts contains only peanuts and cashews. For every ounce of cashews there are 5 ounces of peanuts. If the mixture contains a total of 27 ounces, how many ounces of each type of nut does the mixture contain?

20. An insecticide contains 95 centigrams of inert ingredient for every 1 centigram of active ingredient. If a quantity of the insecticide weighs 336 centigrams, how much of each type of ingredient does it contain?

21. If the sum of two numbers is k, and one of the numbers is m, how can you express the other number? (See Objective 3.)

22. If the product of two numbers is r, and one of the numbers is s $(s \neq 0)$, how can you express the other number?

23. Is there an angle whose supplement is equal to its complement? If so, what is the measure of the angle? (See Objective 4.)

24. Is there an angle that is equal to its supplement? Is there an angle that is equal to its complement? If the answer is yes to either question, give the measure of the angle. (See Objective 4.)

Solve the following problems. See Example 6.

25. Find the measure of an angle whose supplement measures 10 times the measure of its complement.

26. Find the measure of an angle whose supplement measures 4 times the measure of its complement.

27. Find the measure of an angle, if its supplement measures 38° less than three times its complement.

28. Find the measure of an angle, if its supplement measures 39° more than twice its complement.

29. Find the measure of an angle such that the sum of the measures of its complement and its supplement is 160°.

30. Find the measure of an angle such that the difference between the measures of its supplement and three times its complement is 10°.

The following problems are a bit different from those described in the examples of this section. Solve these using the general method described. (In Exercises 31–34, consecutive integers and consecutive even integers are mentioned. Some examples of consecutive integers are 5, 6, 7, and 8; some examples of consecutive even integers are 4, 6, 8, and 10.)

31. If x represents an integer, then $x + 1$ represents the next larger consecutive integer. If the sum of two consecutive integers is 137, find the integers.

32. (See Exercise 31.) If the sum of two consecutive integers is -57, find the integers.

33. If x represents an even integer, then $x + 2$ represents the next larger even integer. Find two consecutive even integers such that the smaller added to three times the larger equals 46.

34. (See Exercise 33.) Find two consecutive even integers such that six times the smaller added to the larger gives a sum of 86.

35. If x represents a quantity, then $.06x$ represents 6% of the quantity. Louise has 6% of her salary deducted for her daughter's educational fund. After the deduction, one week her check was for \$423. How much did she earn before the deduction?

36. (See Exercise 35.) At the end of a day, the owner of a gift shop had \$2394 more in the cash register than she had when the shop opened that day. This included sales tax of 5% on all sales. Find the amount of her sales.

37. A store has 39 quarts of milk, some in pint cartons and some in quart cartons. There are six times as many quart cartons as pint cartons. How many quart cartons are there? (*Hint:* 1 quart = 2 pints.)

38. Marin is three times as old as Kasey. Three years ago the sum of their ages was 22 years. How old is each now? (*Hint:* First write an expression for the age of each now, then for the age of each three years ago.)

PREVIEW EXERCISES

Use the given values to evaluate each expression. See Section 1.3.

39. LW; $L = 6$, $W = 7$

40. rt; $r = 25$, $t = 2.5$

41. prt; $p = 4000$, $r = .08$, $t = 2$

42. $\frac{1}{2}Bh$; $B = 27$, $h = 4$

43. $2L + 2W$; $L = 18$, $W = 12$

44. $\frac{1}{2}(B + b)h$; $B = 10$, $b = 7$, $h = 5$

2.5 FORMULAS AND APPLICATIONS FROM GEOMETRY

OBJECTIVES

1. SOLVE A FORMULA FOR ONE VARIABLE GIVEN THE VALUES OF THE OTHER VARIABLES.
2. SOLVE A FORMULA FOR A SPECIFIED VARIABLE.
3. USE A FORMULA TO SOLVE A GEOMETRIC APPLICATION.
4. SOLVE PROBLEMS ABOUT ANGLE MEASURES.

FOCUS ON

The survey plat shown in Figure 2.2 shows two lots that form a figure called a trapezoid. The measures of the parallel sides are 115.80 feet and 171.00 feet. The height of the trapezoid is 165.97 feet. Find the combined area of the two lots. Round your answer to the nearest hundredth of a square foot.

In order to solve this problem, we must know the formula for finding the area of a trapezoid. This section introduces a number of formulas from mathematics, with many of them from geometry. Formulas provide information for writing equations necessary to solve applied problems (see Step 3 in the method described in Section 2.4).

This problem is Exercise 47 in the exercises for this section. After working through this section, you should be able to solve this problem.

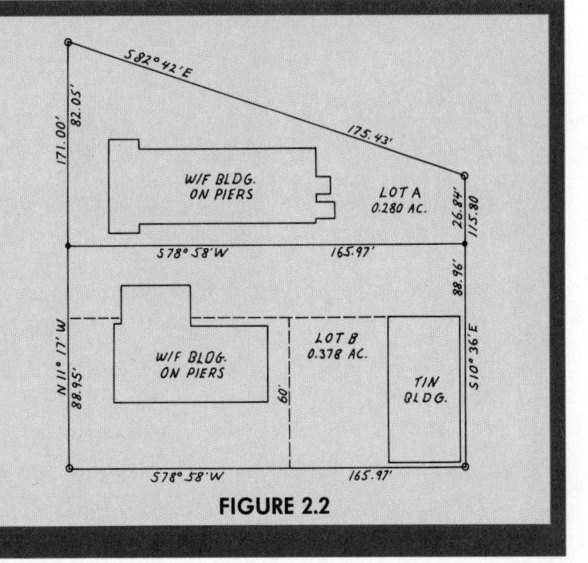

FIGURE 2.2

PROBLEM SOLVING

PROBLEM SOLVING

Many applied problems can be solved with formulas. Formulas exist for geometric figures such as squares and circles, for distance, for money earned on bank savings, and for converting English measurements to metric measurements, for example. The formulas used in this book are given in Appendices A and B. ■

1 Given the values of all but one of the variables in a formula, the value of the remaining variable can be found by using the methods introduced in this chapter for solving equations.

EXAMPLE 1
USING A FORMULA TO EVALUATE A VARIABLE

Find the value of the remaining variable in each of the following.

(a) $A = LW$; $A = 64$, $L = 10$

As shown in Figure 2.3(a), this formula gives the area of a rectangle with length L and width W. Substitute the given values into the formula and then solve for W.

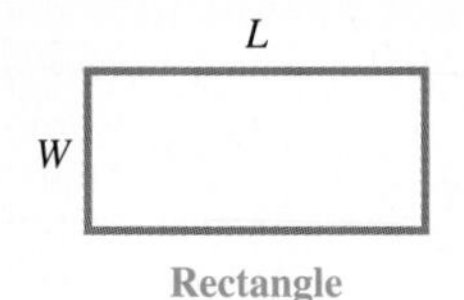

Rectangle
$A = LW$

(a)

$$A = LW$$
$$64 = 10W \qquad \text{Let } A = 64, L = 10.$$
$$6.4 = W \qquad \text{Divide by 10.}$$

Check that the width of the rectangle is 6.4.

(b) $A = \frac{1}{2}(b + B)h$; $A = 210$, $B = 27$, $h = 10$

This formula gives the area of a trapezoid with parallel sides of lengths b and B and distance h between the parallel sides. See Figure 2.3(b). Again, begin by substituting the given values into the formula.

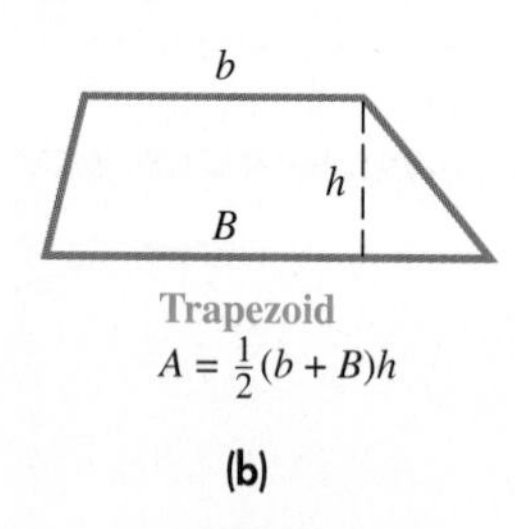

Trapezoid
$A = \frac{1}{2}(b + B)h$

(b)

FIGURE 2.3

$$A = \frac{1}{2}(b + B)h$$
$$210 = \frac{1}{2}(b + 27)(10) \qquad A = 210, B = 27, h = 10$$

Now solve for b.

$$210 = \frac{1}{2}(10)(b + 27) \qquad \text{Commutative property}$$
$$210 = 5(b + 27)$$
$$210 = 5b + 135 \qquad \text{Distributive property}$$
$$75 = 5b \qquad \text{Subtract 135.}$$
$$15 = b \qquad \text{Divide by 5.}$$

Check that the length of the shorter parallel side, b, is 15. ■

2 In the next two examples we will see how problem solving can be simplified in some cases.

PROBLEM SOLVING

Sometimes it is necessary to solve a large number of problems that use the same formula. For example, a surveying class might need to solve several problems that involve the formula for the area of a rectangle, $A = LW$. Suppose that in each problem the area (A) and the length (L) of a rectangle are given and the width (W) must be found. Rather than solving for W each time the formula is used, it would be simpler to rewrite the *formula* so that it is solved for W. This process is called **solving for a specified variable.** As the following examples will show, solving a formula for a specified variable requires the same steps used earlier to solve equations with just one variable. ■

EXAMPLE 2 **SOLVING FOR A SPECIFIED VARIABLE**

Solve $A = LW$ for W.

Think of undoing what has been done to W. Since W is multiplied by L, undo the multiplication by dividing both sides of $A = LW$ by L.

$$A = LW$$

$$\frac{A}{L} = \frac{LW}{L} \qquad \text{Divide by } L.$$

$$\frac{A}{L} = W \qquad \text{or} \qquad W = \frac{A}{L}$$

The formula is now solved for W. ■

NOTE When solving a formula for a specified variable, treat that variable as if it were the only variable in the equation, and treat all others as if they were constants. Use the method of solving equations described in Sections 2.2 and 2.3 to solve for the specified variable.

The formula for converting temperatures given in degrees Celsius to degrees Fahrenheit is

$$F = \frac{9}{5}C + 32.$$

The next example shows how to solve this formula for C.

EXAMPLE 3 **SOLVING FOR A SPECIFIED VARIABLE**

Solve $F = \frac{9}{5}C + 32$ for C.

First undo the addition of 32 to $(9/5)C$ by subtracting 32 from both sides.

$$F = \frac{9}{5}C + 32$$

$$F - 32 = \frac{9}{5}C + 32 - 32 \qquad \text{Subtract 32.}$$

$$F - 32 = \frac{9}{5}C$$

Now multiply both sides by 5/9. Use parentheses on the left.

$$\frac{5}{9}(F - 32) = \frac{5}{9} \cdot \frac{9}{5}C \qquad \text{Multiply by } \frac{5}{9}.$$

$$\frac{5}{9}(F - 32) = C$$

This last result is the formula for converting temperatures from Fahrenheit to Celsius. ■

EXAMPLE 4
SOLVING FOR A SPECIFIED VARIABLE

Solve $A = \frac{1}{2}(b + B)h$ for B.

This is the formula for the area of a trapezoid. Begin by multiplying both sides by 2.

$$A = \frac{1}{2}(b + B)h \qquad \text{Given formula}$$

$$2A = 2 \cdot \frac{1}{2}(b + B)h \qquad \text{Multiply by 2.}$$

$$2A = (b + B)h$$

$$2A = bh + Bh \qquad \text{Distributive property}$$

Undo what was done to B by first subtracting bh on both sides. Then divide both sides by h.

$$2A - bh = Bh \qquad \text{Subtract } bh.$$

$$\frac{2A - bh}{h} = B \quad \text{or} \quad B = \frac{2A - bh}{h} \qquad \text{Divide by } h.$$

The result can be written in a different form as follows.

$$B = \frac{2A - bh}{h} = \frac{2A}{h} - \frac{bh}{h} = \frac{2A}{h} - b$$

Either form is correct. ■

3 As the next examples show, formulas can be used to solve applications involving geometric figures. In Step 2, we use a drawing or a diagram to visualize.

EXAMPLE 5
FINDING THE LENGTH OF A SIDE OF A SQUARE SIGN

A homemade sign advertising a yard sale is posted on a telephone pole. It is in the shape of a square with perimeter of 96 inches. Find the length of a side of the square sign.

The problem states that the sign is square, so we will need to use the appropriate formula for a square. The list of formulas in Appendix A gives the formula $P = 4s$ for the **perimeter,** or the distance around a square. We now use the six steps from Section 2.4 for problem solving.

Step 1 Let s = the length of a side of the square, in inches.

Step 2 Draw a square. Label each side s.

Step 3 Use the formula $P = 4s$, and let $P = 96$.

$$96 = 4s$$

Step 4 Solve the equation.

$$\frac{96}{4} = \frac{4s}{4} \qquad \text{Divide by 4.}$$

$$24 = s$$

Step 5 The length of a side of the sign is 24 inches.

Step 6 Since the perimeter, 96 inches, equals 4 times 24 inches, the answer is correct. ■

EXAMPLE 6
FINDING THE WIDTH OF A RECTANGULAR LOT

A rectangular lot has perimeter 80 meters and length 25 meters. Find the width of the lot. (See Figure 2.4.)

Step 1 We are told to find the width of the lot, so

let W = the width of the lot in meters.

Step 2 See Figure 2.4.

Step 3 The distance around a rectangle is called the perimeter of the rectangle. As found in Appendix A, the formula for the perimeter of a rectangle is

$$P = 2L + 2W.$$

Find the width by substituting 80 for P and 25 for L in the formula.

$$80 = 2(25) + 2W \qquad P = 80, \quad L = 25$$

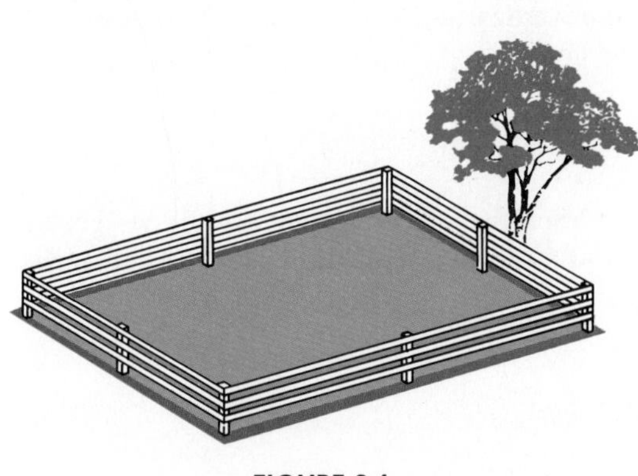

FIGURE 2.4

Step 4 Solve the equation.

$$80 = 50 + 2W \qquad \text{Multiply.}$$
$$80 - 50 = 50 + 2W - 50 \qquad \text{Subtract 50.}$$
$$30 = 2W$$
$$15 = W \qquad \text{Divide by 2.}$$

Step 5 The width is 15 meters.

Step 6 Check this result. If the width is 15 meters and the length is 25 meters, the distance around the rectangular lot (perimeter) is $2(25) + 2(15) = 50 + 30 = 80$ feet, as required. ■

EXAMPLE 7
FINDING THE HEIGHT OF A TRIANGULAR SAIL

The area of a triangular sail of a sailboat is 126 square meters. The base of the sail is 21 meters. Find the height of the sail.

Step 1 Since we must find the height of the triangular sail,

let h = the height of the sail in meters.

Step 2 See Figure 2.5.

Step 3 The formula for the area of a triangle is $A = (1/2)bh$, where A is the area, b is the base, and h is the height. Using the information given in the problem, substitute 126 for A and 21 for b in the formula.

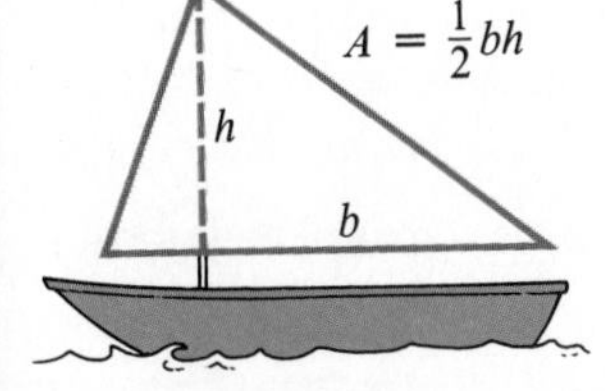

FIGURE 2.5

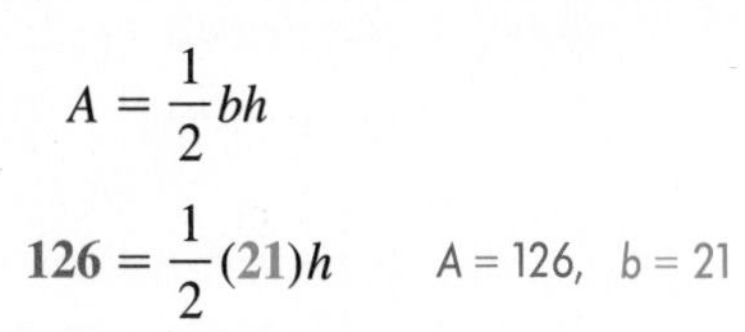

$$A = \frac{1}{2}bh$$

$$126 = \frac{1}{2}(21)h \qquad A = 126, \quad b = 21$$

Step 4 Solve the equation.

$$126 = \frac{21}{2}h$$

To find h, multiply both sides by 2/21.

$$\frac{2}{21}(126) = \frac{2}{21} \cdot \frac{21}{2}h$$

$$12 = h \qquad \text{or} \qquad h = 12$$

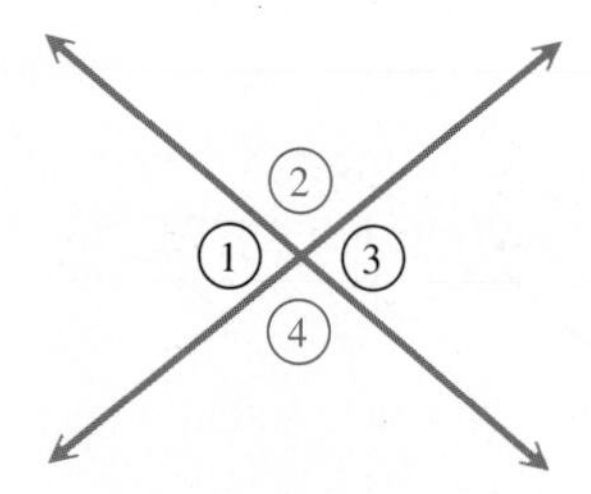

FIGURE 2.6

Step 5 The height of the sail is 12 meters.

Step 6 Check to see that the values $A = 126$, $b = 21$, and $h = 12$ satisfy the formula for the area of a triangle. ■

4 Refer to Figure 2.6, which shows two intersecting lines forming angles that are numbered ①, ②, ③, and ④. Angles ① and ③ lie "opposite" each other. They are called **vertical angles.** Another pair of vertical angles are ② and ④. In geometry, it is shown that the following property holds.

VERTICAL ANGLES Vertical angles have equal measures.

Now look at angles ① and ②. When their measures are added, we get the measure of a **straight angle,** which is 180°. There are three other such pairs of angles: ② and ③, ③ and ④, ① and ④.

The next example uses these ideas.

EXAMPLE 8 FINDING ANGLE MEASURES

Refer to the appropriate figures in each part.

(a) Find the measure of each marked angle in Figure 2.7.

Since the marked angles are vertical angles, they have the same measures. Set $4x + 19$ equal to $6x - 5$ and solve.

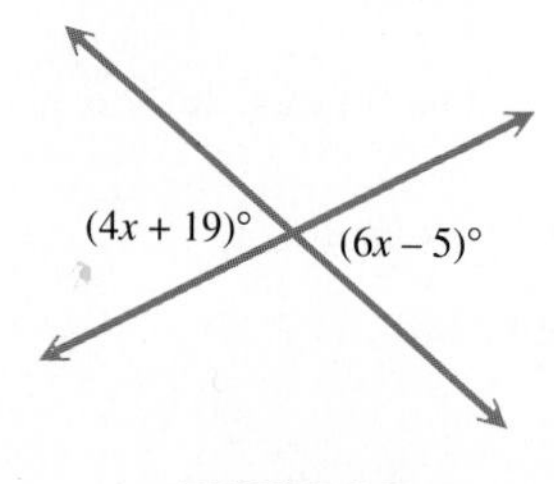

FIGURE 2.7

$$4x + 19 = 6x - 5$$

$$-4x + 4x + 19 = -4x + 6x - 5 \qquad \text{Add } -4x.$$

$$19 = 2x - 5$$

$$19 + 5 = 2x - 5 + 5 \qquad \text{Add 5.}$$

$$24 = 2x$$

$$12 = x \qquad \text{Divide by 2.}$$

Since $x = 12$, one angle has measure $4(12) + 19 = 67$ degrees. The other has the same measure, since $6(12) - 5 = 67$ as well. Each angle measures 67°.

(b) Find the measure of each marked angle in Figure 2.8.

The measures of the marked angles must add to 180° since together they form a straight angle. The equation to solve is

$$(3x - 30) + 4x = 180.$$

$$7x - 30 = 180 \qquad \text{Combine like terms.}$$

$$7x - 30 + 30 = 180 + 30 \qquad \text{Add 30.}$$

$$7x = 210$$

$$x = 30 \qquad \text{Divide by 7.}$$

(3x – 30)° (4x)°

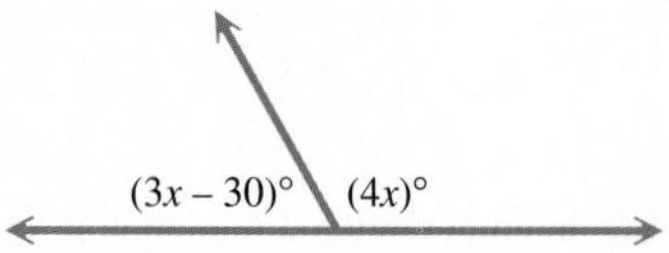

FIGURE 2.8

To find the measures of the angles, replace x with 30 in the two expressions.

$$3x - 30 = 3(30) - 30 = 90 - 30 = 60$$
$$4x = 4(30) = 120$$

The two angle measures are 60° and 120°. ■

2.5 EXERCISES

In the following exercises a formula is given, along with the values of all but one of the variables in the formula. Find the value of the variable that is not given. See Example 1.

1. $V = \frac{1}{3}Bh$; $B = 20$, $h = 9$

2. $V = \frac{1}{3}Bh$; $B = 82$, $h = 12$

3. $d = rt$; $d = 8$, $r = 2$

4. $d = rt$; $d = 100$, $t = 5$

5. $A = \frac{1}{2}bh$; $A = 20$, $b = 5$

6. $A = \frac{1}{2}bh$; $A = 30$, $b = 6$

7. $P = 2L + 2W$; $P = 40$, $W = 6$

8. $P = 2L + 2W$; $P = 180$, $L = 50$

9. $V = \frac{1}{3}Bh$; $V = 80$, $B = 24$

10. $V = \frac{1}{3}Bh$; $V = 52$, $h = 13$

11. $A = \frac{1}{2}(b + B)h$; $b = 6$, $B = 8$, $h = 3$

12. $A = \frac{1}{2}(b + B)h$; $b = 10$, $B = 12$, $h = 3$

13. $C = 2\pi r$; $C = 9.42$, $\pi = 3.14$*

14. $C = 2\pi r$; $C = 25.12$, $\pi = 3.14$

15. $A = \pi r^2$; $r = 9$, $\pi = 3.14$

16. $A = \pi r^2$; $r = 15$, $\pi = 3.14$

17. $V = \frac{4}{3}\pi r^3$; $r = 3$, $\pi = 3.14$

18. $V = \frac{4}{3}\pi r^3$; $r = 6$, $\pi = 3.14$

19. $I = prt$; $I = 100$, $p = 500$, $r = .10$

20. $I = prt$; $I = 60$, $p = 150$, $r = .08$

21. $V = LWH$; $V = 150$, $L = 10$, $W = 5$

22. $V = LWH$; $V = 800$, $L = 40$, $W = 10$

23. $A = \frac{1}{2}(b + B)h$; $A = 42$, $b = 5$, $B = 7$

24. $A = \frac{1}{2}(b + B)h$; $A = 70$, $b = 15$, $B = 20$

25. $V = \frac{1}{3}\pi r^2 h$; $V = 9.42$, $\pi = 3.14$, $r = 3$

26. $V = \frac{1}{3}\pi r^2 h$; $V = 37.68$, $\pi = 3.14$, $r = 6$

---

27. If a formula contains exactly five variables, how many values would you need to be given in order to find values for the remaining one(s)? (See Objective 1.)

28. The formula for changing Celsius to Fahrenheit is given in Example 3 as $F = \frac{9}{5}C + 32$. Sometimes it is seen as $F = \frac{9C}{5} + 32$. These are both correct. Why is it true that $\frac{9}{5}C$ is equal to $\frac{9C}{5}$?

*Actually, π is *approximately* equal to 3.14, not *exactly* equal to 3.14.

Solve the given formulas for the indicated variables. See Examples 2–4.

29. $A = LW$; for L

30. $d = rt$; for t

31. $V = LWH$; for H

32. $I = prt$; for t

33. $C = 2\pi r$; for r

34. $A = \frac{1}{2}bh$; for b

35. $P = 2L + 2W$; for W

36. $a + b + c = P$; for b

37. $A = \frac{1}{2}(b + B)h$; for b

38. $C = \frac{5}{9}(F - 32)$; for F

39. $S = 2\pi rh + 2\pi r^2$; for h

40. $A = p + prt$; for r

---

41. In your own words, write a definition of perimeter of a geometric figure. (See Objective 3.)

42. In order to purchase fencing to go around a rectangular yard, would you need to use perimeter or area to decide how much to buy? (See Objective 3.)

43. In order to purchase fertilizer for the lawn of a yard, would you need to use perimeter or area to decide how much to buy? (See Objective 3.)

44. What is the length of a side of a square whose perimeter is numerically equal to its area? Answer this question *without* solving an equation. (See Objective 3.)

For each of the following geometric applications, use the six-step method of solving problems explained in Section 2.4. Draw a sketch if it will help. The necessary formulas are given in Appendix A. Use 3.14 as an approximation for π, if necessary. See Examples 5–7.

45. A children's playroom has an area of 61.2 square meters, and the width is 6 meters. Find the length of the room.

46. The perimeter of a square backyard patio is 80 feet. Find the length of a side of the patio.

47. The survey plat shown in Figure 2.2 shows two lots that form a figure called a trapezoid. The measures of the parallel sides are 115.80 feet and 171.00 feet. The height of the trapezoid is 165.97 feet. Find the combined area of the two lots. Round your answer to the nearest hundredth of a square foot.

48. Lot A in Figure 2.2 is in the shape of a trapezoid. The parallel sides measure 26.84 feet and 82.05 feet. The height of the trapezoid is 165.97 feet. Find the area of Lot A. Round your answer to the nearest hundredth of a square foot.

49. The radius of a circular above-ground swimming pool is 6 feet. Find the circumference of the pool. (Circumference is the distance around the pool.)

50. A triangular lot has perimeter 72 meters. One side measures 16 meters and another side measures 32 meters. Find the length of the third side.

Find the measure of each marked angle. See Example 8.

51.

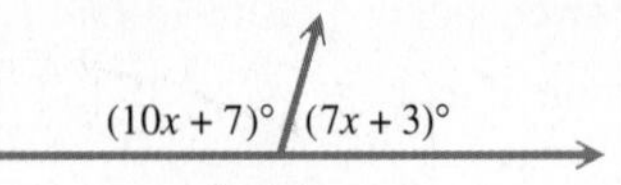

52.

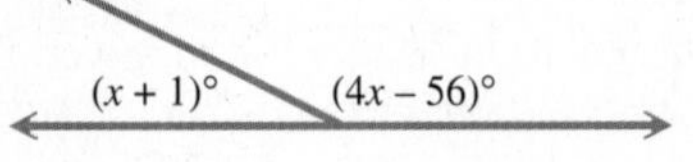

53.

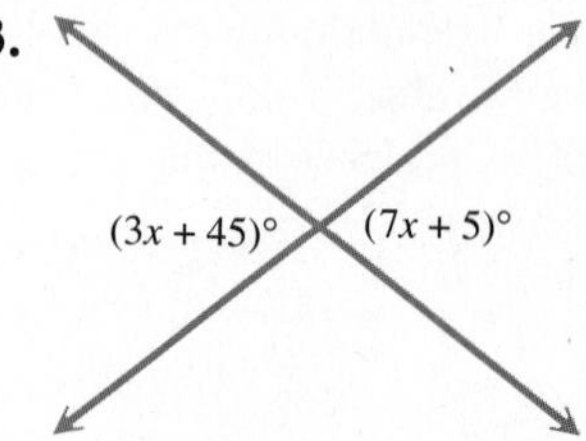

54.
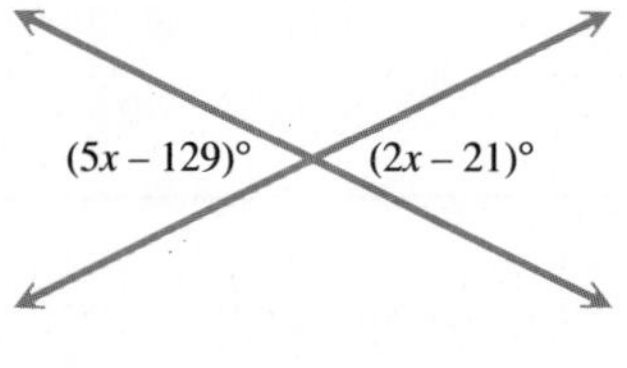

55.
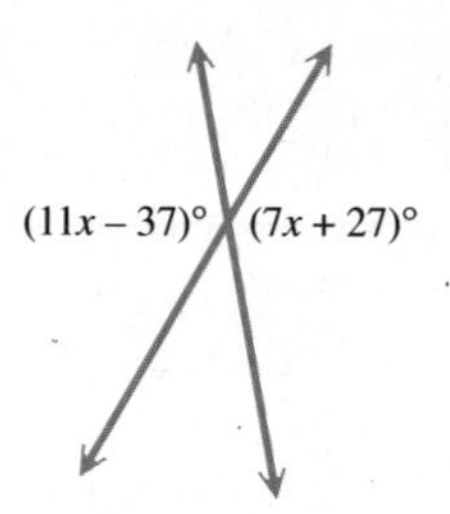

56.
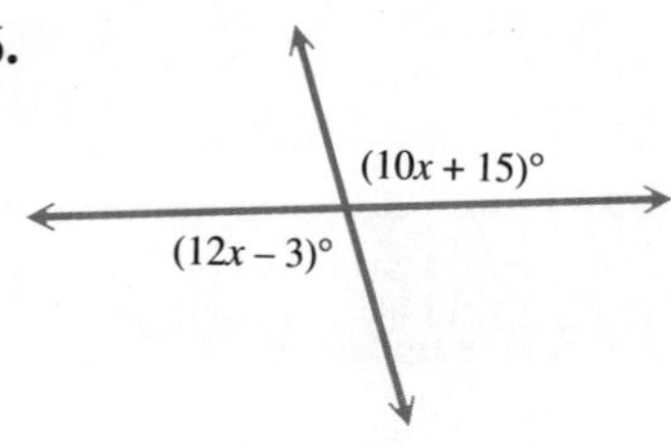

Solve each equation for x.

57. $y = 6 - 5x$ **58.** $y = 8 + 3x$ **59.** $2x + 3y = 9$

60. $5x - 4y = 12$ **61.** $3x - 5y = 15$ **62.** $-6x + 2y = 9$

PREVIEW EXERCISES

Solve each equation. See Section 2.2.

63. $3p = 8$ **64.** $2q = 11$ **65.** $-7z = 15$ **66.** $-5m = 3$ **67.** $-\frac{3}{5}x = 12$ **68.** $-\frac{2}{3}m = 4$

2.6 RATIOS AND PROPORTIONS

OBJECTIVES

1 WRITE RATIOS.
2 DECIDE WHETHER PROPORTIONS ARE TRUE.
3 SOLVE PROPORTIONS.
4 SOLVE APPLIED PROBLEMS USING PROPORTIONS.
5 SOLVE PROBLEMS INVOLVING UNIT PRICING.

FOCUS ON PROBLEM SOLVING

The local supermarket charges the following prices for different sizes of a popular name-brand cereal.

10-ounce size: \$1.49
13-ounce size: \$1.85
19-ounce size: \$2.81

Based on price per ounce, which size is the best buy for the consumer?

Unit pricing is an example of a rate, or a ratio of price to unit of measure. Solving simple proportions is one of the most common everyday occurrences of applying mathematics. In this section we learn how to solve proportions. This problem is Exercise 59 in the exercises for this section. After working through this section, you should be able to solve this problem.

PROBLEM SOLVING

An example of a type of problem that often occurs is given below.

> A carpet cleaning service charges \$45.00 to clean 2 similarly sized rooms of carpet. How much would it cost to clean 5 rooms of carpet?

Assuming that the cleaning service does not discount its prices for cleaning additional rooms after the first two, the reasoning for solving this problem might be as follows: if it costs \$45.00 to clean 2 rooms, then it would cost \$45.00/2 = \$22.50 per room. So, the total cost for cleaning 5 rooms would be 5 × \$22.50 = \$112.50. ■

1 The quotient \$45.00/2 is an example of a ratio of price to number of rooms. Ratios provide a way of comparing two numbers or quantities. A **ratio** is a quotient of two quantities. The ratio of the number a to the number b is written as follows.

RATIO

$$a \text{ to } b, \quad \frac{a}{b}, \quad \text{or} \quad a:b$$

When ratios are used in comparing units of measure, the units should be the same. This is shown in Example 1.

EXAMPLE 1
WRITING A RATIO

Write a ratio for each word phrase.

(a) The ratio of 5 hours to 3 hours

This ratio can be written as $\frac{5}{3}$.

(b) The ratio of 5 hours to 3 days

First convert 3 days to hours: 3 days $= 3 \cdot 24 = 72$ hours. The ratio of 5 hours to 3 days is thus $\frac{5}{72}$. ■

2 A ratio is used to compare two numbers or amounts. A **proportion** is a statement that two ratios are equal. For example,

$$\frac{3}{4} = \frac{15}{20}$$

is a proportion that says that the ratios 3/4 and 15/20 are equal. In the proportion

$$\frac{a}{b} = \frac{c}{d},$$

a, b, c, and d are the **terms** of the proportion. Beginning with the proportion

$$\frac{a}{b} = \frac{c}{d}$$

and multiplying both sides by the common denominator, bd, gives

$$bd \cdot \frac{a}{b} = bd \cdot \frac{c}{d}$$
$$ad = bc.$$

The products ad and bc can be found by multiplying diagonally.

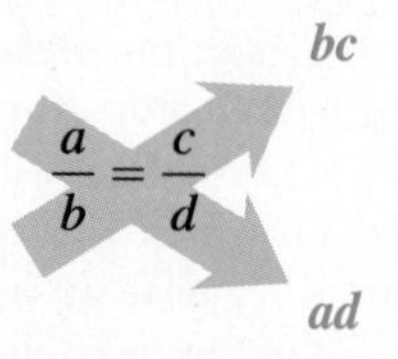

This is called **cross multiplication** and ad and bc are called **cross products.**

CROSS PRODUCTS

If $\frac{a}{b} = \frac{c}{d}$, then the cross products ad and bc are equal.

Also, if $ad = bc$, then $\frac{a}{b} = \frac{c}{d}$.

From the rule given above,

$$\text{if} \quad \frac{a}{b} = \frac{c}{d} \quad \text{then} \quad ad = bc.$$

However, if $\frac{a}{c} = \frac{b}{d}$, then $ad = cb$, or $ad = bc$. This means that the two proportions are equivalent and

$$\text{the proportion } \frac{a}{b} = \frac{c}{d} \text{ can also be written as } \frac{a}{c} = \frac{b}{d}.$$

Sometimes one form is more convenient to work with than the other.

EXAMPLE 2 **DECIDING WHETHER A PROPORTION IS TRUE**

Decide whether the following proportions are true or false.

(a) $\frac{3}{4} = \frac{15}{20}$

Check to see whether the cross products are equal.

$$4 \cdot 15 = 60$$

$$\frac{3}{4} = \frac{15}{20}$$

$$3 \cdot 20 = 60$$

The cross products are equal, so the proportion is true.

(b) $\frac{6}{7} = \frac{30}{32}$

The cross products are $6 \cdot 32 = 192$ and $7 \cdot 30 = 210$. The cross products are different, so the proportion is false. ■

CAUTION The cross product method cannot be used directly if there is more than one term on either side. For example, you cannot use the method directly to solve the equation

$$\frac{4}{x} + 3 = \frac{1}{9},$$

because there are two terms on the left side.

3 Four numbers are used in a proportion. If any three of these numbers are known, the fourth can be found.

EXAMPLE 3 FINDING AN UNKNOWN IN A PROPORTION

(a) Find x in the proportion

$$\frac{63}{x} = \frac{9}{5}.$$

The cross products must be equal, so

$$63 \cdot 5 = 9x$$
$$315 = 9x.$$

Divide both sides by 9 to get

$$35 = x.$$

(b) Solve for r in the proportion $\frac{8}{5} = \frac{12}{r}$.

Set the cross products equal to each other.

$$8r = 5 \cdot 12$$
$$8r = 60$$
$$r = \frac{60}{8} = \frac{15}{2} \quad ■$$

EXAMPLE 4 SOLVING AN EQUATION USING CROSS PRODUCTS

Solve the equation

$$\frac{m-2}{5} = \frac{m+1}{3}.$$

Find the cross products, and set them equal to each other.

$$3(m-2) = 5(m+1) \qquad \text{Be sure to use parentheses.}$$
$$3m - 6 = 5m + 5 \qquad \text{Distributive property}$$
$$3m = 5m + 11 \qquad \text{Add 6.}$$
$$-2m = 11 \qquad \text{Subtract } 5m.$$
$$m = -\frac{11}{2} \qquad \text{Divide by } -2. \quad ■$$

4 Proportions occur in many practical applications.

PROBLEM SOLVING

If we know the price of a number of similarly priced items, proportions can be used to determine the price of some other number of these items (assuming no discounts are given). ■

EXAMPLE 5
APPLYING PROPORTIONS

A local drugstore is offering 3 packs of toothpicks for $.87. How much would it charge for 10 packs?

Let x = the cost of 10 packs of toothpicks.

Set up a proportion. One ratio in the proportion can involve the number of packs, and the other can involve the costs. Make sure that the corresponding numbers appear in the numerator and the denominator.

$$\frac{\text{Cost of 3}}{\text{Cost of 10}} = \frac{3}{10}$$

$$\frac{.87}{x} = \frac{3}{10}$$

$$3x = .87(10) \qquad \text{Cross products}$$

$$3x = 8.7$$

$$x = 2.90 \qquad \text{Divide by 3.}$$

The 10 packs should cost $2.90. As shown earlier, the proportion could also be written as $\frac{3}{.87} = \frac{10}{x}$, and the same result would be obtained. ■

NOTE Many people would solve the problem in Example 5 mentally as follows: Three packs cost $.87, so one pack costs $.87/3 = $.29. Then ten packs will cost 10($.29) = $2.90. If you would do this problem this way, you would be using proportions and probably not even realizing it!

5 We now look at the idea of *unit pricing*.

PROBLEM SOLVING

A common sight in supermarkets is shoppers carrying hand-held calculators to assist them in their job of budgeting. While the most common use is to make sure that the shopper does not go over budget, another use is to see which size of an item offered in different sizes produces the best price per unit. In order to do this, simply divide the price of the item by the unit of measure in which the item is labelled. The next example illustrates this idea. ■

EXAMPLE 6
DETERMINING UNIT PRICE TO OBTAIN THE BEST BUY

The local supermarket charges the following prices for a popular brand of pancake syrup:

SIZE	PRICE
36-ounce	$3.89
24-ounce	$2.79
12-ounce	$1.89.

Which size is the best buy? That is, which size has the lowest unit price?

To find the best buy, divide the price by the number of units to get the price per ounce. Each result in the following table was found by using a calculator and rounding the answer to three decimal places.

SIZE	UNIT COST (dollars per ounce)
36-ounce	$\frac{\$3.89}{36} = \$.108$ ⟵ The best buy
24-ounce	$\frac{\$2.79}{24} = \$.116$
12-ounce	$\frac{\$1.89}{12} = \$.158$

Since the 36-ounce size produces the lowest price per unit, it would be the best buy. (Be careful: Sometimes the largest container *does not* produce the lowest price per unit.) ■

2.6 EXERCISES

Determine the following ratios. Write each ratio in lowest terms. See Example 1.

1. 30 miles to 20 miles
2. 50 feet to 90 feet
3. 72 dollars to 110 dollars
4. 120 people to 80 people
5. 6 feet to 5 yards
6. 10 yards to 8 feet
7. 30 inches to 4 feet
8. 100 inches to 5 yards
9. 12 minutes to 2 hours
10. 8 quarts to 5 pints
11. 4 dollars to 10 quarters
12. 35 dimes to 6 dollars
13. 20 hours to 5 days
14. 6 days to 9 hours
15. 80¢ to $3

Decide whether the following proportions are true. See Example 2.

16. $\frac{5}{8} = \frac{35}{56}$
17. $\frac{4}{7} = \frac{12}{21}$
18. $\frac{9}{10} = \frac{18}{20}$
19. $\frac{6}{8} = \frac{15}{20}$
20. $\frac{12}{18} = \frac{8}{12}$
21. $\frac{7}{10} = \frac{82}{120}$
22. $\frac{19}{30} = \frac{57}{90}$
23. $\frac{110}{18} = \frac{160}{27}$

24. Explain the distinction between *ratio* and *proportion*. (See Objectives 1 and 2.)

25. Suppose that someone told you to use cross products in order to multiply fractions. How would you explain to the person what is wrong with his or her thinking? (See Objective 3.)

26. How would you explain the concept of unit price in terms of ratios? (See Objective 5.)

27. Write an application of ratios that uses the ratio 5/2.

Solve the following equations. See Examples 3 and 4.

28. $\frac{7}{12} = \frac{a}{24}$
29. $\frac{35}{4} = \frac{k}{20}$
30. $\frac{z}{56} = \frac{7}{8}$
31. $\frac{m}{32} = \frac{3}{24}$
32. $\frac{x}{6} = \frac{18}{4}$
33. $\frac{z}{80} = \frac{20}{100}$
34. $\frac{m}{8} = \frac{100}{25}$
35. $\frac{2}{3} = \frac{y}{7}$
36. $\frac{5}{8} = \frac{m}{5}$
37. $\frac{5}{9} = \frac{z}{15}$
38. $\frac{3}{4} = \frac{r}{10}$
39. $\frac{k}{7} = \frac{5}{8}$

40. $\frac{p}{5} = \frac{4}{3}$ **41.** $\frac{m}{5} = \frac{m-3}{3}$ **42.** $\frac{r+1}{3} = \frac{r}{2}$ **43.** $\frac{k+3}{4} = \frac{3k-2}{10}$

Solve the following problems involving proportions. See Example 5.

44. Two slices of bacon contain 85 calories. How many calories are there in twelve slices of bacon?

45. Three ounces of liver contain 22 grams of protein. How many ounces of liver provide 121 grams of protein?

46. If 6 gallons of premium unleaded gasoline cost \$9.72, how much would 9 gallons cost?

47. If 3 music cassettes cost \$5.94, how much would 5 of them cost?

48. The sales tax on a \$24 headset radio is \$1.68. How much would the sales tax be on a pair of binoculars that costs \$36?

49. A piece of property assessed at \$42,000 requires an annual property tax of \$273. How much property tax would be charged for a similar piece of property assessed at \$52,000?

50. The distance between Kansas City, Missouri, and Denver is 600 miles. On a certain wall map, this is represented by a length of 2.4 feet. On the map, how many feet would there be between Memphis and Philadelphia, two cities that are actually 1000 miles apart?

51. A small tree 15 feet high casts a shadow 12 feet long. If Loc Ho is 5 feet, 6 inches tall and is standing next to the tree, how long is his shadow?

52. If 9 pairs of slacks cost \$283.50, how much would 11 pairs of the slacks cost?

53. The distance between Singapore and Tokyo is 3300 miles. On a certain wall map, this distance is represented by 11 inches. The actual distance between Mexico City and Cairo is 7700 miles. How far apart are they on the same map?

54. A recipe for green salad for 70 people calls for 18 heads of lettuce. How many heads of lettuce would be needed if 175 people were to be served?

55. A recipe for oatmeal macaroons calls for 1 2/3 cups of flour to make four dozen cookies. How many cups of flour would be needed for six dozen cookies?

56. If 4 pounds of fertilizer will cover 50 square feet of garden, how many pounds would be needed for 225 square feet?

A supermarket was surveyed to find the prices charged for items in various sizes. Find the best buy (based on price per unit) for each of the following. See Example 6.

57. Black pepper
4-ounce size: \$1.57
8-ounce size: \$2.27

58. Trash bags
20-count: \$2.49
30-count: \$4.29

59. Breakfast cereal
10-ounce size: \$1.49
13-ounce size: \$1.85
19-ounce size: \$2.81

60. Spaghetti sauce
15 1/2-ounce size: \$1.19
32-ounce size: \$1.69
48-ounce size: \$2.69

61. Extra crunchy peanut butter
12-ounce size: \$1.49
28-ounce size: \$1.99
40-ounce size: \$3.99

62. Tomato ketchup
14-ounce size: \$.93
32-ounce size: \$1.19
44-ounce size: \$2.19

Photo for Exercise 59

Solve.

63. $\frac{3y-2}{5} = \frac{6y-5}{11}$ **64.** $\frac{2p+7}{3} = \frac{p-1}{4}$ **65.** $\frac{2r+8}{4} = \frac{3r-9}{3}$ **66.** $\frac{5k+1}{6} = \frac{3k-2}{3}$

67. The amount of material needed for 5 dresses is 7.2 yards less than the amount needed for 8 dresses. How much is needed for 8 dresses?

68. To prepare a medication, a nurse knows he must mix 4 ounces more than twice as much medicine with 50 ounces of water than he uses with 20 ounces. How much medicine should be used in each case?

PREVIEW EXERCISES

Solve each of the following using the proper formula. See Section 2.5.

69. If an investment of \$8000 earns \$2560 in simple interest in 4 years, what rate of interest is being earned?

70. If \$200 earned \$75 in simple interest at 5%, for how many years was the money earning interest?

71. Seamus earned \$5700 in interest on a deposit of \$19,000 at 12%. For how long was the money deposited?

72. What is the monetary value of 34 quarters? (*Hint:* Monetary value = number of coins × denomination.)

Solve each of the following equations. See Section 2.3.

73. $.15x + .30(3) = .20(3 + x)$

74. $.20(60) + .05x = .10(60 + x)$

75. $.92x + .98(12 - x) = .96(12)$

76. $.10(7) + 1.00x = .30(7 + x)$

2.7 APPLICATIONS OF PERCENT: MIXTURE, INTEREST, AND MONEY

OBJECTIVES

1 LEARN HOW TO USE PERCENT IN PROBLEMS INVOLVING RATES.

2 LEARN HOW TO SOLVE PROBLEMS INVOLVING MIXTURES.

3 LEARN HOW TO SOLVE PROBLEMS INVOLVING SIMPLE INTEREST.

4 LEARN HOW TO SOLVE PROBLEMS INVOLVING DENOMINATIONS OF MONEY.

FOCUS ON PROBLEM SOLVING

Minoxidil is a drug that has recently proven to be effective in treating male pattern baldness. A pharmacist wishes to mix a solution that is 2% minoxidil. She has on hand 50 milliliters of a 1% solution, and she wishes to add some 4% solution to it to obtain the desired 2% solution. How much 4% solution should she add?

A strategy that can be used to solve this mixture problem is explained in this section. It is a valuable technique in problem solving, since it can be applied to other problems which *seem* to be different, but are actually quite similar: simple interest investment problems, and problems involving different denominations of money.

This problem is Exercise 15 in the exercises for this section. After working through this section, you should be able to solve this problem.

1 Recall that percent means "per hundred."

PROBLEM SOLVING

Percents are often used in problems that involve concentrations or rates. In general, we multiply the rate by the total amount to get the percentage. (The percentage may be an amount of pure substance, or an amount of money, as seen in the examples in this section.) In order to prepare to solve mixture, investment, and money problems, the first example illustrates this basic idea. ■

EXAMPLE 1
USING PERCENT TO FIND A PERCENTAGE

(a) If a chemist has 40 liters of a 35% acid solution, then the amount of pure acid in the solution is

40	×	.35	=	14 liters.
↑		↑		↑
Amount of solution		Rate of concentration		Amount of pure acid

(b) If \$1300 is invested for one year at 7% simple interest, the amount of interest earned in the year is

\$1300	×	.07	=	\$91.
↑		↑		↑
Principal		Interest rate		Interest earned

(c) If a jar contains 37 quarters, the monetary amount of the coins is

37	×	\$.25	=	\$9.25.
↑		↑		↑
Number of coins		Denomination		Monetary value

■

PROBLEM SOLVING

In the examples that follow, we will use *box diagrams* to organize the information in the problems. (Some students may prefer to use charts.) Either method enables us to more easily set up the equation for the problem, which is usually the most difficult part of the problem-solving process. The six steps as described in Section 2.4 will be used, but will not specifically be numbered. ■

2 In the next example, we will use percent to solve a mixture problem.

EXAMPLE 2
SOLVING A MIXTURE PROBLEM

A chemist needs to mix 20 liters of 40% acid solution with some 70% solution to get a mixture that is 50% acid. How many liters of the 70% solution should be used?

Let x = the number of liters of 70% solution that are needed.

Recall from part (a) of Example 1 that the amount of pure acid in this solution will be given by the product of the percent of strength and the number of liters of solution, or

$$\text{liters of pure acid in } x \text{ liters of 70\% solution} = .70x.$$

The amount of pure acid in the 20 liters of 40% solution is

$$\text{liters of pure acid in the 40\% solution} = .40(20) = 8.$$

The new solution will contain $20 + x$ liters of 50% solution. The amount of pure acid in this solution is

$$\text{liters of pure acid in the 50\% solution} = .50(20 + x).$$

The given information can be summarized in the box diagram below.

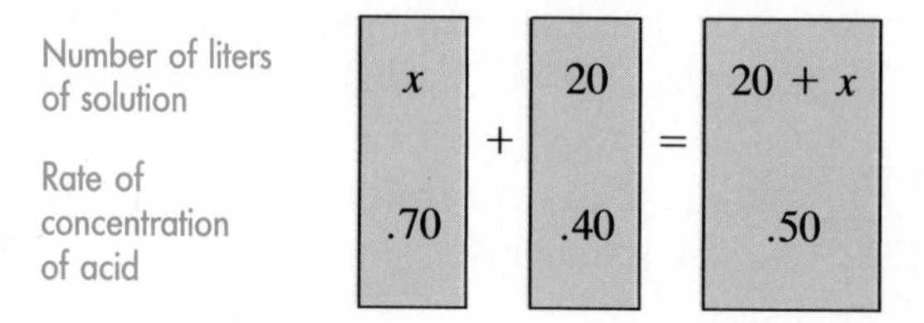

The number of liters of pure acid in the 70% solution added to the number of liters of pure acid in the 40% solution will equal the number of liters of pure acid in the final mixture, so the equation is

Pure acid in 70%	plus	pure acid in 40%	is	pure acid in 50%.
↓	↓	↓	↓	↓
$.70x$	$+$	$.40(20)$	$=$	$.50(20 + x)$.

Multiply by 100 to clear decimals.

$$70x + 40(20) = 50(20 + x)$$

Solve for x.

$70x + 800 = 1000 + 50x$	Distributive property
$20x + 800 = 1000$	Subtract $50x$.
$20x = 200$	Subtract 800.
$x = 10$	Divide by 20.

Check this solution to see that the chemist needs to use 10 liters of 70% solution. ■

3 The next example uses the formula for simple interest, $I = prt$. Remember that when $t = 1$, the formula becomes $I = pr$, and once again the idea of multiplying the total amount (principal) by the rate (rate of interest) gives the percentage (amount of interest).

NOTE In most real-life applications of interest, compound interest is used; that is, interest is paid upon interest. Compound interest involves concepts that are not covered in this book, so our examples must be limited to investments that pay simple interest.

EXAMPLE 3
SOLVING A SIMPLE INTEREST PROBLEM

Elizabeth Thornton receives an inheritance. She invests part of it at 9% and $2000 more than this amount at 10%. Altogether, she makes $1150 per year in interest. How much does she have invested at each rate?

Let $x =$ the amount invested at 9% (in dollars);

$x + 2000 =$ the amount invested at 10% (in dollars).

Use box diagrams to arrange the information given in the problem.

Amount invested (in dollars)	x	+	$x + 2000$	=	1150
Rate of interest	.09		.10		

In each box on the left side, multiply amount by rate to get the interest earned. Since the sum of the interest amounts is $1150, the equation is

Interest at 9%	plus	interest at 10%	is	total interest.
↓	↓	↓	↓	↓
$.09x$	$+$	$.10(x + 2000)$	$=$	$1150.$

Multiply by 100 to clear decimals.

$$9x + 10(x + 2000) = 115{,}000$$

Now solve for x.

$$9x + 10x + 20{,}000 = 115{,}000 \quad \text{Distributive property}$$
$$19x + 20{,}000 = 115{,}000 \quad \text{Combine terms.}$$
$$19x = 95{,}000 \quad \text{Subtract 20,000.}$$
$$x = 5000 \quad \text{Divide by 19.}$$

She has $5000 invested at 9% and $5000 + $2000 = $7000 invested at 10%. ■

NOTE Although decimals were cleared in Examples 2 and 3, the equations also can be solved without clearing decimals.

4 The final example is a problem that can be solved using the same ideas as those in Examples 2 and 3. It deals with different denominations of money.

EXAMPLE 4
SOLVING A PROBLEM ABOUT MONEY

A bank teller has 25 more five-dollar bills than ten-dollar bills. The total value of the money is $200. How many of each denomination of bill does he have?

We must find the number of each denomination of bill that the teller has.

Let $x =$ the number of ten-dollar bills;

$x + 25 =$ the number of five-dollar bills.

The information given in the problem can once again be organized in a box diagram.

Number of bills	$x + 25$	+	x	=	200
Denomination (in dollars)	5		10		

Multiplying the number of bills by the denomination gives the monetary value. The value of the tens added to the value of the fives must be \$200:

Value of fives	plus	value of tens	is	\$200.
↓	↓	↓	↓	↓
$5(x + 25)$	$+$	$10x$	$=$	$200.$

Solve this equation.

$$5x + 125 + 10x = 200 \quad \text{Distributive property}$$
$$15x + 125 = 200 \quad \text{Combine terms.}$$
$$15x = 75 \quad \text{Subtract 125.}$$
$$x = 5 \quad \text{Divide by 15.}$$

Since x represents the number of tens, the teller has 5 tens and $5 + 25 = 30$ fives. Check that the value of this money is $5(\$10) + 30(\$5) = \$200$. ■

CAUTION Most difficulties in solving problems of these types occur because students do not take enough time to read the problem carefully and organize the information given in the problem. By organizing the information in a box diagram (or chart), the chances of solving the problem correctly are greatly increased.

2.7 EXERCISES

Work each of the following. See Example 1.

1. How much pure alcohol is in 50 milliliters of a 20% alcohol solution?
2. How much pure acid is in 30 liters of a 55% acid solution?
3. If \$5000 is invested for one year at 6% simple interest, how much interest is earned?
4. If \$20,000 is invested for one year at 5% simple interest, how much interest is earned?
5. What is the monetary value of 54 nickels?
6. What is the monetary value of 29 half-dollars?
7. Express the amount of alcohol in r liters of pure water. (See Objective 1.)
8. Express the amount of alcohol in k liters of pure alcohol. (See Objective 1.)

Work each of the following mixture problems. See Example 2.

9. How many liters of 25% acid solution must be added to 80 liters of 40% solution to get a solution that is 30% acid?

Liters of solution	x	+	80	=	$x + 80$
Rate of concentration	.25		.40		.30

10. How many gallons of 50% antifreeze must be mixed with 80 gallons of 20% antifreeze to get a mixture that is 40% antifreeze?

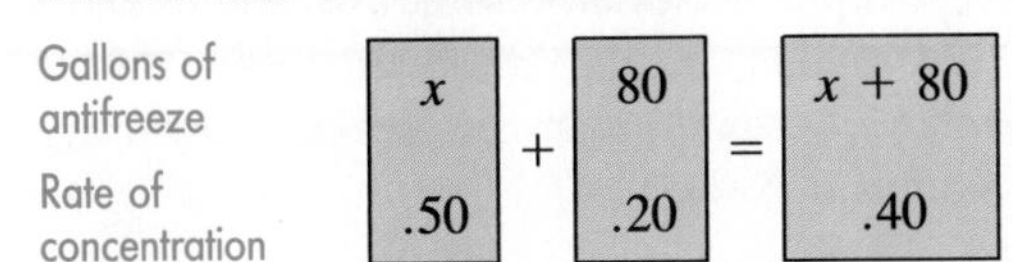

11. Ink worth \$100 per barrel will be mixed with 30 barrels of ink worth \$60 per barrel to get a mixture worth \$75 per barrel. How many barrels of \$100 ink should be used?

12. A certain metal is 40% tin. How many kilograms of this metal must be mixed with 80 kilograms of a metal that is 70% tin to get a metal that is 50% tin?

13. How many gallons of a 12% indicator solution must be mixed with a 20% indicator solution to get 10 gallons of a 14% solution? (*Hint:* Let x represent the number of gallons of 12% solution. Then $10 - x$ represents the number of gallons of 20% solution.)

14. How many liters of a 60% acid solution must be mixed with a 75% acid solution to get 20 liters of a 72% solution?

15. Minoxidil is a drug that has recently proven to be effective in treating male pattern baldness. A pharmacist wishes to mix a solution that is 2% minoxidil. She has on hand 50 milliliters of a 1% solution, and she wishes to add some 4% solution to it to obtain the desired 2% solution. How much 4% solution should she add?

16. Water must be added to 20 milliliters of a 4% minoxidil solution to dilute it to a 2% solution. How many milliliters of water should be used?

Work each of the following investment problems. Assume that simple interest is being paid. See Example 3.

17. Choi Mee invested some money at 18%, and \$3000 less than that amount at 20%. The two investments produce a total of \$3200 per year interest. How much is deposited at 18%?

18. Tri Pham inherited a sum of money from a relative. He deposits some of the money at 16%, and \$4000 more than this amount at 12%. He earns \$3840 in interest per year. Find the amount he has invested at 16%.

19. Annie Boyle invested some money at 10%, and invested \$5000 more than this at 14%. Her total annual income from these investments is \$3100. How much does she have invested at each rate?

20. Two investments produce an annual interest income of \$4200. The amount invested at 14% is \$6000 less than the amount invested at 10%. Find the amount invested at each rate.

21. An actor invests his earnings in two ways: some goes into a 5% tax-free bond, and \$5000 more than twice as much goes into an apartment house paying 10%. His total annual income from the investments is \$4600. Find the amount he has invested at 5%.

22. With income earned by selling a patent, an engineer invests some money at 8%, and \$3000 more than twice as much at 10%. The total annual income from the investments is \$2540. Find the amount invested at 8%.

Work each of the following problems involving denominations of money. See Example 4.

23. A woman has \$1.70 in dimes and nickels; she has 2 more dimes than nickels. How many nickels does she have?

24. A bank teller has some five-dollar bills and some twenty-dollar bills. The teller has 5 more twenties than fives. The total value of the money is \$725. Find the number of five-dollar bills that the teller has.

25. A convention manager finds that she has \$1290, made up of twenties and fifties. She has a total of 42 bills. How many of each kind does she have?

26. A cashier has a total of 126 bills, made up of fives and tens. The total value of the money is \$840. How many of each kind does he have?

27. Suppose that a chemist is mixing two acid solutions, one of 20% concentration and the other of 30% concentration. Which one of the following concentrations could *not* be obtained? (See Objective 2.)
(a) 22% **(b)** 24% **(c)** 28% **(d)** 32%

28. Read Example 2. Can a problem of this type have a fraction as an answer? Now read Example 4. Can a problem of this type have a fraction as an answer? Explain.

29. A teacher once commented that the method of solving problems of the type found in this section could be interpreted as "stuff plus stuff equals stuff." Refer to Examples 2, 3, and 4, and determine exactly what the "stuff" is in each problem. (See Objective 1.)

30. Imagine that you need to make up an application involving denominations of money. In order to avoid fractional answers, begin with the answers (in whole numbers), and then write the problem. Solve your own problem. (See Objective 4.)

Work the following miscellaneous problems using the ideas of this section.

31. A stamp collector buys some 16¢ stamps and some 29¢ stamps, paying \$8.68 for them. He buys 2 more 29¢ stamps than 16¢ stamps. How many 16¢ stamps does he buy?

32. For a retirement party, a person buys some 32¢ favors and some 50¢ favors, paying \$46 in total. If she buys 10 more of the 50¢ favors, how many of the 32¢ favors were bought?

33. At Vern's Grill, hamburgers cost 90 cents each, and a bag of french fries costs 40 cents. How many hamburgers and how many bags of french fries can a customer buy with \$8.80 if he wants twice as many hamburgers as bags of french fries?

34. A merchant wishes to mix candy worth \$5 per pound with 40 pounds of candy worth \$2 per pound to get a mixture that can be sold for \$3 per pound. How many pounds of \$5 candy should be used?

35. A pharmacist has 20 liters of a 10% drug solution. How many liters of 5% solution must be added to get a mixture that is 8%?

36. How many gallons of milk that is 2% butterfat must be mixed with milk that is 3.5% butterfat to get 10 gallons of milk that is 3% butterfat?

PREVIEW EXERCISES

Solve each of the following formulas for the indicated variable. See Section 2.5.

37. $d = rt$; for r **38.** $d = rt$; for t **39.** $P = 2L + 2W$; for L

40. $P = a + b + c$; for c **41.** $A = \frac{1}{2}bh$; for h **42.** $180 = A + B + C$; for A

43. Tom travels 520 miles in 13 hours. What is his rate?

44. Joann goes 49 miles per hour, and covers 367.5 miles. How many hours did she travel?

2.8 MORE ABOUT PROBLEM SOLVING

OBJECTIVES

1 USE THE FORMULA $d = rt$ TO SOLVE PROBLEMS.

2 SOLVE PROBLEMS INVOLVING DISTANCE, RATE, AND TIME.

3 SOLVE PROBLEMS ABOUT GEOMETRIC FIGURES.

FOCUS ON PROBLEM SOLVING

The winner of the 1988 Indianapolis 500 (mile) race was Rick Mears, who drove his Penske-Chevy V8 at an average speed of 144.8 miles per hour. What was Mears' driving time?

The relationship between distance, rate, and time is one that is used quite frequently in everyday life. In order to solve this problem, we need to know how to find time if we are given distance and rate. In this section we look at some applications of the distance, rate, and time relationship.

This problem is Exercise 1 in the exercises for this section. After working through this section, you should be able to solve this problem.

1 If an automobile travels at an average rate of 50 miles per hour for two hours, then it travels $50 \times 2 = 100$ miles. This is an example of the basic relationship between distance, rate, and time:

$$\text{distance} = \text{rate} \times \text{time}.$$

This relationship is given by the formula $d = rt$. By solving, in turn, for r and t in the formula, we obtain two other equivalent forms of the formula. The three forms are given below.

DISTANCE, RATE, TIME RELATIONSHIP

$$d = rt \qquad r = \frac{d}{t} \qquad t = \frac{d}{r}$$

The first example illustrates the uses of these formulas.

EXAMPLE 1
FINDING DISTANCE, RATE, OR TIME

(a) The speed of sound is 1,088 feet per second at sea level at 32° F. In 5 seconds under these conditions, sound travels

$$\underset{\text{Rate}}{1088} \times \underset{\text{Time}}{5} = \underset{\text{Distance}}{5440} \text{ feet.}$$

Here, we found distance given rate and time, using $d = rt$.

(b) Over a short distance, an elephant can travel at a rate of 25 miles per hour. In order to travel 1/4 mile, it would take an elephant

$$\frac{\frac{1}{4}}{25} = \frac{1}{4} \times \frac{1}{25} = \frac{1}{100} \text{ hour.}$$

(Distance → $\frac{1}{4}$; Rate → 25; ← Time)

Here, we find time given rate and distance, using $t = d/r$. To convert 1/100 hour to minutes, multiply 1/100 by 60 to get 60/100 or 3/5 minute. To convert 3/5 minute to seconds, multiply 3/5 by 60 to get 36 seconds.

(c) In the 1988 Olympic Games, the USSR won the 400-meter relay with a time of 38.19 seconds. The rate of the team was

$$\frac{400}{38.19} = 10.47 \text{ (rounded) meters per second.}$$

(Distance → 400; Time → 38.19; ← Rate)

This answer was obtained using a calculator. Here, we found rate given distance and time, using $r = d/t$. ■

2 Many applied problems use the formulas just discussed.

PROBLEM SOLVING

The next example shows how to solve a typical application of the formula $d = rt$. A strategy for solving such problems involves two major steps:

SOLVING MOTION PROBLEMS

Step 1 Set up a sketch showing what is happening in the problem.
Step 2 Make a chart using the information given in the problem, along with the unknown quantities.

The chart will help you organize the information, and the sketch will help you set up the equation. ■

EXAMPLE 2 SOLVING A MOTION PROBLEM

Two cars leave Baton Rouge, Louisiana, at the same time and travel east on Interstate 12. One travels at a constant speed of 55 miles per hour and the other travels at a constant speed of 63 miles per hour. In how many hours will the distance between them be 24 miles?

Since we are looking for time,

let t = the number of hours until the distance between them is 24 miles.

FIGURE 2.9

The sketch in Figure 2.9 shows what is happening in the problem.
Now, construct a chart like the one below.

	RATE	TIME	DISTANCE
Faster car			
Slower car			

Fill in the information given in the problem, and use t for the time traveled by each car. Multiply rate by time to get the expressions for distances traveled.

	RATE ×	TIME	= DISTANCE
Faster car	63	t	$63t$
Slower car	55	t	$55t$

Difference is 24 miles.

The quantities $63t$ and $55t$ represent the different distances. Refer to Figure 2.9 and notice that the *difference* between the larger distance and the smaller distance is 24 miles. Now write the equation and solve it.

$$63t - 55t = 24$$

$$8t = 24 \qquad \text{Combine terms.}$$

$$t = 3 \qquad \text{Divide by 8.}$$

After 3 hours the faster car will have traveled $63 \times 3 = 189$ miles, and the slower car will have traveled $55 \times 3 = 165$ miles. Since $189 - 165 = 24$, the conditions of the problem are satisfied. It will take 3 hours for the distance between them to be 24 miles. ■

NOTE In motion problems like the one in Example 2, once you have filled in two pieces of information in each row of the chart, you should automatically fill in the third piece of information, using the appropriate form of the formula relating distance, rate, and time. Set up the equation based upon your sketch and the information in the chart.

3 In Section 2.5 we saw some applications of geometric formulas. Example 3 shows another such application.

PROBLEM SOLVING

Remember that a sketch is very helpful when solving geometric applications. ■

EXAMPLE 3
FINDING THE LENGTH AND THE WIDTH OF A ROOM

A couple wishes to add a laundry room onto their house. Due to construction limitations, the length of the room must be 2 feet more than the width. Find the length and the width of the room, if the perimeter is 40 feet.

Start by drawing a rectangle to represent the floor of the room. See Figure 2.10.

Let x = width of the room in feet;

$x + 2$ = length of the room in feet.

The formula for the perimeter of a rectangle is $P = 2L + 2W$. Substitute x for W, $x + 2$ for L, and 40 for P.

$L = x + 2$

$W = x$

$P = 40$

FIGURE 2.10

$$P = 2L + 2W$$

$$40 = 2(x + 2) + 2x \qquad P = 40,\ W = x,\ L = x + 2$$

$$40 = 2x + 4 + 2x \qquad \text{Distributive property}$$

$$40 = 4x + 4 \qquad \text{Combine terms.}$$

$$36 = 4x \qquad \text{Subtract 4 from both sides.}$$

$$9 = x \qquad \text{Divide by 4.}$$

The width of the room is 9 feet, and the length is $9 + 2 = 11$ feet. Check these answers in the formula for the perimeter. Since $2(11) + 2(9) = 40$, the answers are correct. ■

2.8 EXERCISES

Solve each of the following problems, using $d = rt$, $r = d/t$, or $t = d/r$, as necessary. In Exercises 1–4, round answers to the nearest thousandth. See Example 1.

1. The winner of the 1988 Indianapolis 500 (mile) race was Rick Mears, who drove his Penske-Chevy V8 at an average speed of 144.8 miles per hour. What was Mears' driving time?

2. In 1989, Emerson Fitipaldi won the Indianapolis 500 (mile) race in 2.984 hours. What was his average speed?

3. The record-holder for men's freestyle swimming for 50 meters is 22.120 seconds, held by Tom Jager. What was Jager's average speed?

4. In 1976, the Indianapolis 500 race covered a distance of only 255 miles. The winner, Johnny Rutherford, averaged 148.725 miles per hour. What was his driving time?

5. A driver averaged 53 miles per hour and took 10 hours to travel from Memphis to Chicago. What is the distance between Memphis and Chicago?

6. A small plane traveled from Warsaw to Rome, averaging 164 miles per hour. The trip took 2 hours. What is the distance from Warsaw to Rome?

7. Suppose that an automobile averages 45 miles per hour, and travels for 30 minutes. Is the distance traveled $45 \times 30 = 1350$ miles? If not, explain why not, and give the correct distance. (See Objectives 1 and 2.)

8. Which of the following choices is the best *estimate* for the average speed of a trip of 405 miles that lasted 8.2 hours? (See Objective 1.)

(a) 50 miles per hour **(b)** 30 miles per hour
(c) 60 miles per hour **(d)** 40 miles per hour

Solve each of the following problems. See Example 2.

9. Atlanta and Cincinnati are 440 miles apart. John leaves Cincinnati, driving toward Atlanta at an average speed of 60 miles per hour. Pat leaves Atlanta at the same time, driving toward Cincinnati in her antique auto, averaging 28 miles per hour. How long will it take them to meet?

	r	t	d
John	60	t	$60t$
Pat	28	t	$28t$

Pat
Atlanta
John
Cincinnati
440 miles

10. St. Louis and Portland are 2060 miles apart. A small plane leaves Portland, traveling toward St. Louis at an average speed of 90 miles per hour. Another plane leaves St. Louis at the same time, traveling toward Portland, averaging 116 miles per hour. How long will it take them to meet?

	r	t	d
Plane leaving Portland	90	t	$90t$
Plane leaving St. Louis	116	t	$116t$

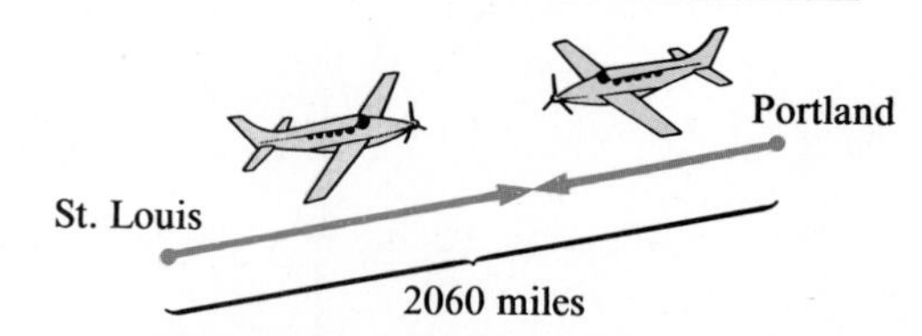

11. Two planes leave an airport at the same time. One flies east at 300 miles per hour, and the other flies west at 450 miles per hour. In how many hours will they be 2250 miles apart?

	r	t	d
Eastbound	300	t	
Westbound	450	t	

12. Two trains leave a city at the same time. One travels north at 60 miles per hour, and the other travels south at 80 miles per hour. In how many hours will they be 280 miles apart?

	r	t	d
Northbound	60	t	
Southbound	80	t	

13. From a point on a straight road, Lupe and Maria ride bicycles in opposite directions. Lupe rides 10 miles per hour and Maria rides 12 miles per hour. In how many hours will they be 55 miles apart?

14. At a given hour, two steamboats leave a city in the same direction on a straight canal. One travels at 18 miles per hour, and the other travels at 25 miles per hour. In how many hours will the boats be 35 miles apart?

Work the following problems involving geometric figures. See Example 3.

15. A stained-glass window in a church is in the shape of a square. The perimeter of the square is 7 times the length of a side in meters, decreased by 12. Find the length of a side of the window.

16. A video rental establishment displayed a rectangular cardboard standup advertisement for the movie *Field of Dreams*. The length was 20 inches more than the width, and the perimeter was 176 inches. What were the dimensions of the rectangle?

17. A lot is in the shape of a triangle. One side is 100 feet longer than the shortest side, while the third side is 200 feet longer than the shortest side. The perimeter of the lot is 1200 feet. Find the lengths of the sides of the lot.

18. A wall pennant is in the shape of an isosceles triangle. (Two sides have the same length.) Each of the two equal sides measures 18 inches more than the third side, and the perimeter of the triangle is 54 inches. What are the lengths of the sides of the pennant?

19. The Peachtree Plaza Hotel in Atlanta is in the shape of a cylinder, with a circular foundation. The circumference of the foundation is 6 times the radius, increased by 12.88 feet. Find the radius of the circular foundation. (Use 3.14 as an approximation for π.)

20. If the radius of a certain circle is tripled, with 8.2 centimeters then added, the result is the circumference of the circle. Find the radius of the circle. (Use 3.14 as an approximation for π.)

The remaining applications in this exercise set are not grouped by type. Use the problem-solving techniques described in this chapter to solve each of the following.

21. The sum of the measures of the angles of any triangle is 180 degrees. In triangle ABC, angles A and B have the same measure, while the measure of angle C is 24 degrees larger than each of A and B. What are the measures of the three angles?

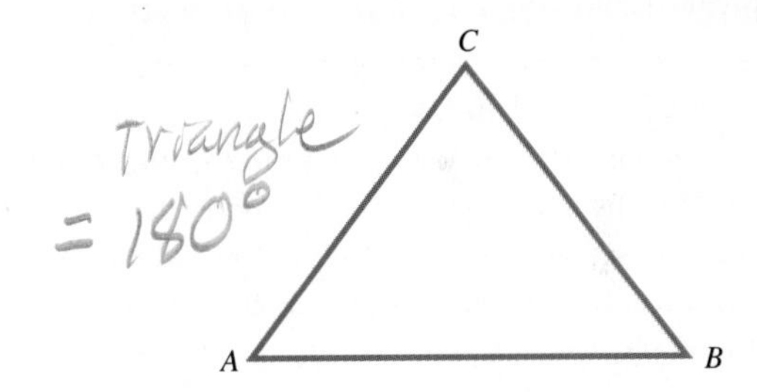

22. (See Exercise 21.) In triangle ABC, the measure of angle A is 30 degrees more than the measure of angle B. The measure of angle B is the same as the measure of angle C. Find the measure of each angle.

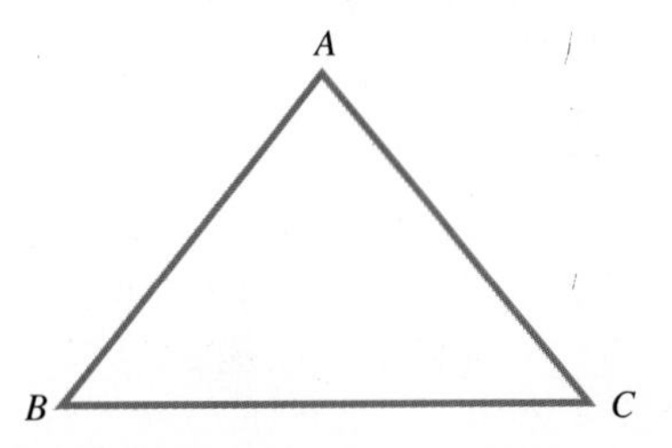

23. In an automobile race, a driver was 120 miles from the finish line after 5 hours. Another driver, who was in a later race, traveled at the same speed as the first driver. After 3 hours, the second driver was 250 miles from the finish. Find the speed of each driver.

24. Two cars are 400 miles apart. Both start at the same time and travel toward one another. They meet 4 hours later. If the speed of one car is 20 miles per hour faster than the other, what is the speed of each car?

25. A merchant has 120 pounds of candy worth \$1.50 per pound. She wishes to upgrade the candy to sell for \$3.50 per pound by mixing it with candy worth \$5.50 per pound. How many pounds of the \$5.50 candy should she use?

26. How many liters of water must be added to 4 liters of pure acid to obtain a 10% acid solution?

27. The length of a rectangle is 8 inches more than the width. The perimeter is 6 inches more than 5 times the width. Find the width of the rectangle.

28. The perimeter of a rectangle is 16 times its width. The length is 12 centimeters more than the width. Find the width of the rectangle.

29. In 1988, the United States exported to the Bahamas 315 million dollars more in goods than it imported. Together, the two amounts totalled 1167 million dollars. How much were the exports and how much were the imports?

30. In the 1988 U.S. presidential election, Sherman County in Oregon registered 120 more votes for George Bush than it did for Michael Dukakis. If the two men together received 990 votes, how many did each receive?

31. If three gallons of gasoline cost \$4.65, how much will 11 gallons cost?

32. Tulsa and Toledo are 850 miles apart. On a certain map, this distance is represented by 14 inches. Houston and Kansas City are 710 miles apart. How far apart are they on the same map? Round your answer to the nearest tenth of an inch.

33. Find the measure of each marked angle.

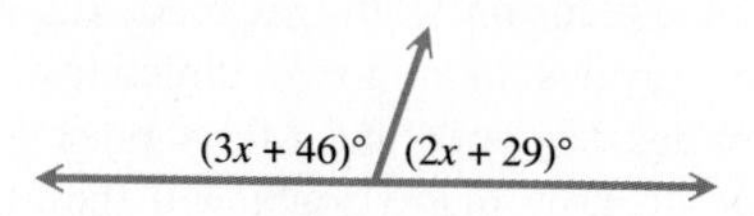

34. Find the measure of each marked angle.

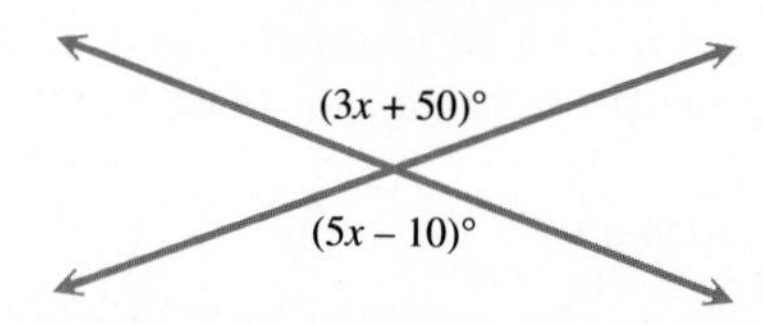

35. Which is the best buy for a popular breakfast cereal? 15-ounce size: \$2.69
20-ounce size: \$3.29
25.5-ounce size: \$3.49

36. Which is the best buy for apple juice?
32-ounce size: \$1.19
48-ounce size: \$1.79
64-ounce size: \$1.99

37. Write a short paragraph explaining the general method for problem solving as explained in this chapter.

38. Of all the different types of problems presented in this chapter, which type is your favorite? Which type is your least favorite? Explain.

PREVIEW EXERCISES

Place $<$ or $>$ in each blank to make a true statement. See Section 1.2.

39. -7 ____ 3 **40.** -9 ____ 6 **41.** -11 ____ -4 **42.** -10 ____ -12

43. -8 ____ -10 **44.** -5 ____ -3 **45.** 5 ____ -9 **46.** 8 ____ -1

2.9 THE ADDITION AND MULTIPLICATION PROPERTIES OF INEQUALITY

OBJECTIVES

1 GRAPH INTERVALS ON A NUMBER LINE.
2 USE THE ADDITION PROPERTY OF INEQUALITY.
3 USE THE MULTIPLICATION PROPERTY OF INEQUALITY.
4 SOLVE LINEAR INEQUALITIES.
5 SOLVE APPLIED PROBLEMS BY USING INEQUALITIES.
6 SOLVE THREE-PART INEQUALITIES.

FOCUS ON PROBLEM SOLVING

The formula for converting from Fahrenheit to Celsius temperature is $C = \frac{5}{9}(F - 32)$. What temperature range in degrees Fahrenheit corresponds to 0° to 35° Celsius?

In order to solve this problem we must know how to solve inequalities, since the Celsius temperature must be *greater than* 0° and *less than* 35°. Solving linear inequalities is accomplished in much the same way as solving linear equations (with one important extra consideration).

This problem is Exercise 57 in the exercises for this section. After working through this section, you should be able to solve this problem.

Inequalities are statements with algebraic expressions related by

$<$ "is less than"
$\leq$ "is less than or equal to"
$>$ "is greater than"
$\geq$ "is greater than or equal to."

An inequality is solved by finding all real number solutions for it. For example, the solution of $x \leq 2$ includes all real numbers that are less than or equal to 2, and not just the integers less than or equal to 2. For example, -2.5, -1.7, -1, $7/4$, $1/2$, $\sqrt{2}$, and 2 are all real numbers less than or equal to 2, and are therefore solutions of $x \leq 2$.

1 A good way to show the solution of an inequality is by graphing. Graph all real numbers satisfying $x \leq 2$ by placing a dot at 2 on a number line and drawing an arrow extending from the dot to the left (to represent the fact that all numbers less than 2 are also part of the graph). The graph is shown in Figure 2.11.

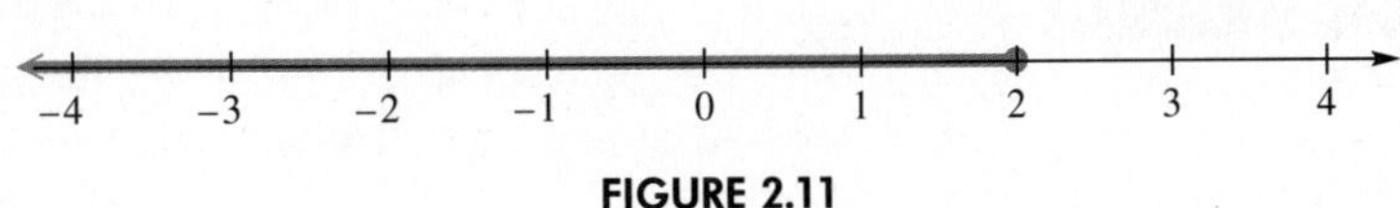

FIGURE 2.11

EXAMPLE 1 GRAPHING AN INTERVAL ON A NUMBER LINE

Graph $x > -5$.

The statement $x > -5$ says that x can represent any number greater than -5, but x cannot equal -5 itself. Show this on a graph by placing an open circle at -5 and drawing an arrow to the right, as in Figure 2.12. The open circle at -5 shows that -5 is not part of the graph. ■

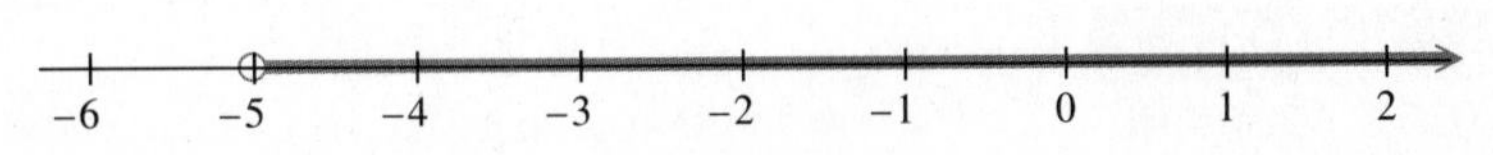

FIGURE 2.12

NOTE Some texts use parentheses rather than open circles and square brackets rather than closed circles (dots). We have chosen to use open and closed circles throughout this text.

EXAMPLE 2 **GRAPHING AN INTERVAL ON A NUMBER LINE**

Graph $-3 \le x < 2$.

The statement $-3 \le x < 2$ is read "-3 is less than or equal to x *and* x is less than 2." Graph this inequality by placing a solid dot at -3 (because -3 is part of the graph) and an open circle at 2 (because 2 is not part of the graph). Then draw a line segment between the two circles, as in Figure 2.13. ■

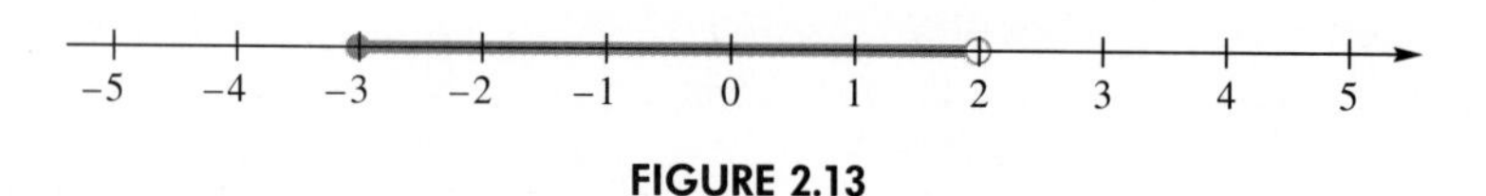

FIGURE 2.13

2 Inequalities such as $x + 4 \le 9$ can be solved in much the same way as equations. Consider the inequality $2 < 5$. If 4 is added to both sides of this inequality, the result is

$$2 + 4 < 5 + 4$$
$$6 < 9,$$

a true sentence. Now subtract 8 from both sides:

$$2 - 8 < 5 - 8$$
$$-6 < -3.$$

The result is again a true sentence. These examples suggest the following **addition property of inequality,** which states that the same real number can be added to both sides of an inequality without changing the solutions.

ADDITION PROPERTY OF INEQUALITY

For any real numbers A, B, and C, the inequalities

$$A < B \qquad \text{and} \qquad A + C < B + C$$

have exactly the same solutions. In words, the same real number may be added to both sides of an inequality without changing the solutions.

The addition property of inequality also works with $>$, $\le$, or $\ge$. Just as with the addition property of equality, the same number may also be *subtracted* from both sides of an inequality.

The following examples show how the addition property is used to solve inequalities.

EXAMPLE 3 **USING THE ADDITION PROPERTY OF INEQUALITY**

Solve the inequality $7 + 3k > 2k - 5$.

Use the addition property of inequality twice, once to get the terms containing k on one side of the inequality and a second time to get the integers together on the other side. (These steps can be done in either order.)

$$7 + 3k > 2k - 5$$

$$7 + 3k - 2k > 2k - 5 - 2k \quad \text{Subtract } 2k.$$

$$7 + k > -5 \quad \text{Combine terms.}$$

$$7 + k - 7 > -5 - 7 \quad \text{Subtract 7.}$$

$$k > -12$$

The graph of the solutions $k > -12$ is shown in Figure 2.14. ■

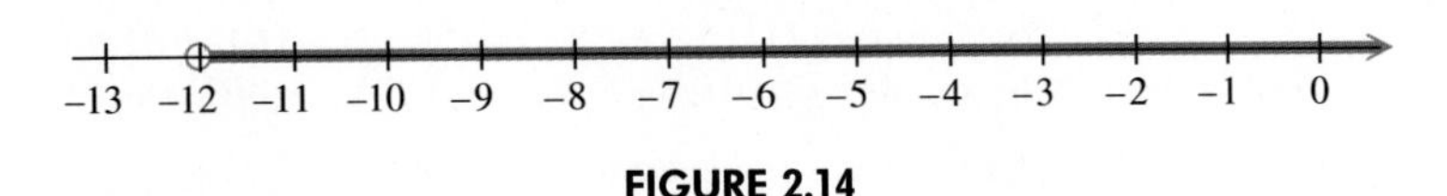

FIGURE 2.14

3 The addition property of inequality alone cannot be used to solve inequalities such as $4y \geq 28$. These inequalities require the multiplication property of inequality. To see how this property works, it will be helpful to look at some examples.

First, start with the inequality $3 < 7$ and multiply both sides by the positive number 2.

$$3 < 7$$

$$2(3) < 2(7) \quad \text{Multiply both sides by 2.}$$

$$6 < 14 \quad \text{True}$$

Now multiply both sides of $3 < 7$ by the negative number -5.

$$3 < 7$$

$$-5(3) < -5(7) \quad \text{Multiply both sides by } -5.$$

$$-15 < -35 \quad \text{False}$$

To get a true statement when multiplying both sides by -5 requires reversing the direction of the inequality symbol.

$$3 < 7$$

$$-5(3) > -5(7) \quad \text{Multiply by } -5\text{; reverse the symbol.}$$

$$-15 > -35 \quad \text{True}$$

Take the inequality $-6 < 2$ as another example. Multiply both sides by the positive number 4.

$$-6 < 2$$

$$4(-6) < 4(2) \quad \text{Multiply by 4.}$$

$$-24 < 8 \quad \text{True}$$

Multiplying both sides of $-6 < 2$ by -5 and at the same time reversing the direction of the inequality symbol gives

$$-6 < 2$$

$$(-5)(-6) > (-5)(2) \quad \text{Multiply by } -5\text{; change } < \text{ to } >.$$

$$30 > -10. \quad \text{True}$$

The two parts of the **multiplication property of inequality** are stated below.

MULTIPLICATION PROPERTY OF INEQUALITY

For any real numbers A, B, and C $(C \neq 0)$,
(1) if C is *positive,* then the inequalities

$$A < B \quad \text{and} \quad AC < BC$$

have exactly the same solutions;
(2) if C is *negative,* then the inequalities

$$A < B \quad \text{and} \quad AC > BC$$

have exactly the same solutions.

In other words, both sides of an inequality may be multiplied by the same positive number without changing the solutions. If the number is negative, we must reverse the direction of the inequality symbol.

The multiplication property of inequality works with $>$, $\leq$, or $\geq$, as well. The multiplication property of inequality also permits *division* of both sides of an inequality by the same nonzero number.

It is important to remember the differences in the multiplication property for positive and negative numbers.

1. When both sides of an inequality are multiplied or divided by a positive number, the direction of the inequality symbol *does not change.* Adding or subtracting terms on both sides also does not change the symbol.
2. When both sides of an inequality are multiplied or divided by a negative number, the direction of the symbol *does change. Reverse the symbol of inequality only when you multiply or divide both sides by a negative number.*

The next examples show how to solve inequalities with the multiplication property.

EXAMPLE 4 USING THE MULTIPLICATION PROPERTY OF INEQUALITY

Solve the inequality $3r < -18$.

Simplify this inequality by using the multiplication property of inequality and dividing both sides by 3. Since 3 is a positive number, the direction of the inequality symbol does not change.

$$3r < -18$$

$$\frac{3r}{3} < \frac{-18}{3} \qquad \text{Divide by 3.}$$

$$r < -6$$

The graph of the solutions is shown in Figure 2.15. ■

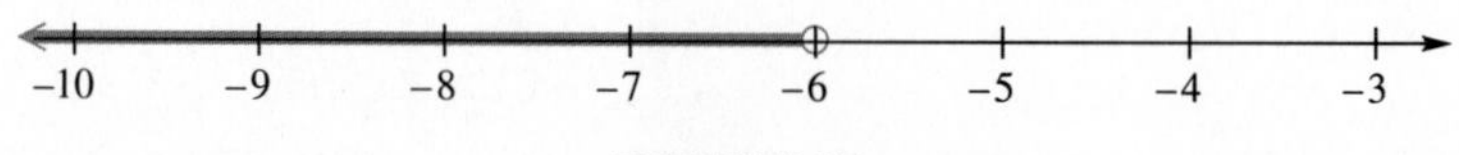

FIGURE 2.15

CAUTION Even though the number on the right side of the inequality in Example 4 is negative (-18), *do not reverse the direction of the inequality symbol.* Reverse the direction only when multiplying or dividing *by* a negative number, as shown in Example 5.

EXAMPLE 5 **USING THE MULTIPLICATION PROPERTY OF INEQUALITY**

Solve the inequality $-4t \geq 8$.

Here both sides of the inequality must be divided by -4, a negative number, which *does* change the direction of the inequality symbol.

$$-4t \geq 8$$

$$\frac{-4t}{-4} \leq \frac{8}{-4} \qquad \text{Divide by } -4\text{; symbol is reversed.}$$

$$t \leq -2$$

The solutions are graphed in Figure 2.16. ■

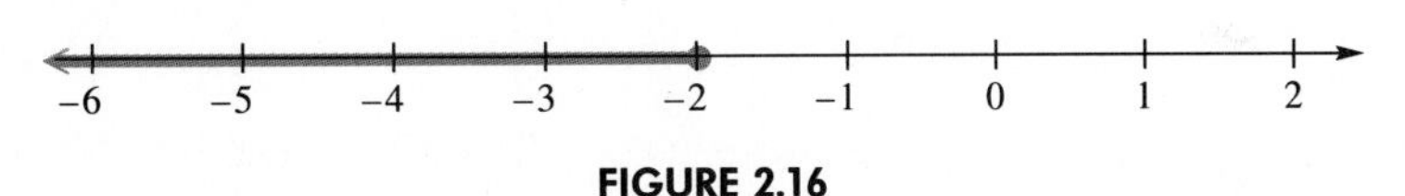

FIGURE 2.16

4 A **linear inequality** is an inequality that can be written in the form $ax + b < 0$, for real numbers a and b, with $a \neq 0$. ($<$ may be replaced with $>$, $\leq$, or $\geq$ in this definition.) In order to solve a linear inequality, go through the following steps.

SOLVING A LINEAR INEQUALITY

Step 1 Use the associative, commutative, and distributive properties to combine like terms on each side of the inequality.

Step 2 Use the addition property of inequality to simplify the inequality to one of the form $ax < b$ or $ax > c$, where a, b, and c are real numbers.

Step 3 Use the multiplication property of inequality to simplify further to an inequality of the form $x < d$ or $x > d$, where d is a real number.

Notice how these steps are used in the next example.

EXAMPLE 6 **SOLVING A LINEAR INEQUALITY**

Solve the inequality $3z + 2 - 5 > -z + 7 + 2z$.

Step 1 Combine like terms and simplify.

$$3z + 2 - 5 > -z + 7 + 2z$$

$$3z - 3 > z + 7$$

Step 2 Use the addition property of inequality.

$$3z - 3 + 3 > z + 7 + 3 \qquad \text{Add 3.}$$

$$3z > z + 10$$

$$3z - z > z + 10 - z \qquad \text{Subtract } z.$$

$$2z > 10$$

Step 3 Use the multiplication property of inequality.

$$\frac{2z}{2} > \frac{10}{2} \quad \text{Divide by 2.}$$

$$z > 5$$

Since 2 is positive, the direction of the inequality symbol was not changed in the third step. A graph of the solutions is shown in Figure 2.17. ■

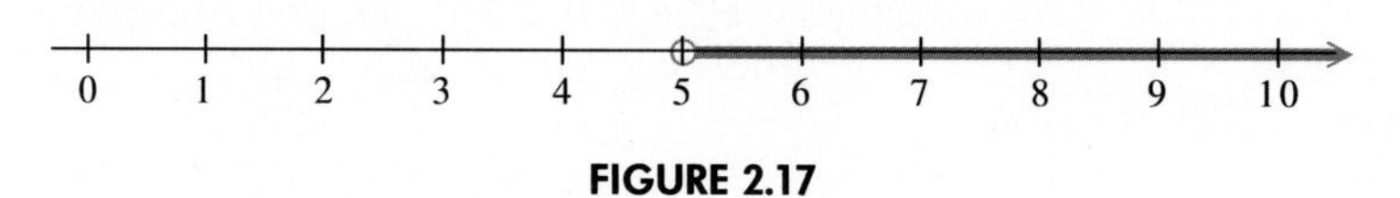

FIGURE 2.17

EXAMPLE 7
SOLVING A LINEAR INEQUALITY

Solve $5(k - 3) - 7k \geq 4(k - 3) + 9$.

Step 1 Simplify and combine like terms.

$$5(k - 3) - 7k \geq 4(k - 3) + 9$$

$$5k - 15 - 7k \geq 4k - 12 + 9 \quad \text{Distributive property}$$

$$-2k - 15 \geq 4k - 3 \quad \text{Combine like terms.}$$

Step 2 Use the addition property.

$$-2k - 15 - 4k \geq 4k - 3 - 4k \quad \text{Subtract } 4k.$$

$$-6k - 15 \geq -3$$

$$-6k - 15 + 15 \geq -3 + 15 \quad \text{Add 15.}$$

$$-6k \geq 12$$

Step 3 Divide both sides by -6, a negative number. Change the direction of the inequality symbol.

$$\frac{-6k}{-6} \leq \frac{12}{-6} \quad \text{Divide by } -6\text{; symbol is reversed.}$$

$$k \leq -2$$

A graph of the solutions is shown in Figure 2.18. ■

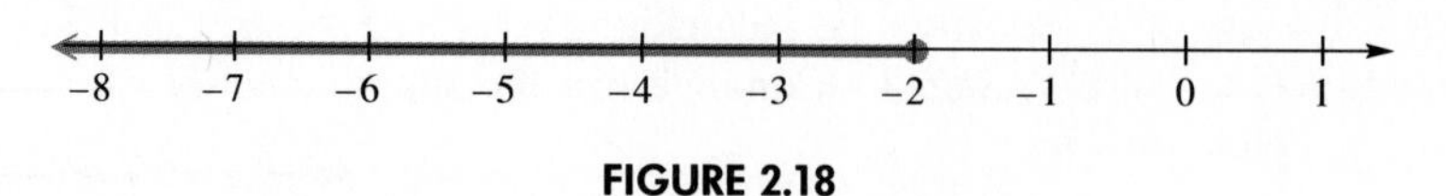

FIGURE 2.18

5 Until now, the applied problems that we have studied have all led to equations.

PROBLEM SOLVING

Inequalities can be used to solve applied problems involving phrases that suggest inequality. The following chart gives some of the more common such phrases, along with examples and translations.

PHRASE	EXAMPLE	INEQUALITY
Is more than	A number *is more than* 4	$x > 4$
Is less than	A number *is less than* -12	$x < -12$
Is at least	A number *is at least* 6	$x \geq 6$
Is at most	A number *is at most* 8	$x \leq 8$ ■

CAUTION Do not confuse statements like "5 is more than a number" with the phrase "5 more than a number." The first of these is expressed as "$5 > x$" while the second is expressed with addition, as "$x + 5$."

PROBLEM SOLVING

The next example shows an application of algebra that is important to anyone who has ever asked himself or herself "What score can I make on my next test and have a (particular grade) in this course?" It uses the idea of finding the average of a number of grades. In general, to find the average of n numbers, add the numbers, and divide by n. ■

EXAMPLE 8 FINDING AN AVERAGE TEST SCORE

Brent has test grades of 86, 88, and 78 on his first three tests in geometry. If he wants an average of at least 80 after his fourth test, what are the possible scores he can make on his fourth test?

Let x = Brent's score on his fourth test. To find his average after 4 tests, add the test scores and divide by 4.

Average ↓ is at least ↓ 80.

$$\frac{86 + 88 + 78 + x}{4} \geq 80$$

$$\frac{252 + x}{4} \geq 80 \qquad \text{Add the known scores.}$$

$$4\left(\frac{252 + x}{4}\right) \geq 4(80) \qquad \text{Multiply by 4.}$$

$$252 + x \geq 320$$

$$252 - 252 + x \geq 320 - 252 \qquad \text{Subtract 252.}$$

$$x \geq 68 \qquad \text{Combine terms.}$$

He must score 68 or more on the fourth test to have an average of *at least* 80. ■

CAUTION Errors often occur when the phrases "at least" and "at most" appear in applied problems. Remember that

at least translates as **is greater than or equal to**

and

at most translates as **is less than or equal to.**

6 Inequalities that say that one number is *between* two other numbers are *three-part inequalities*. For example,

$$-3 < 5 < 7$$

says that 5 is between -3 and 7. The inequality translates in words as "-3 is less than 5 *and* 5 is less than 7." It would be *wrong* to write $7 < 5 < -3$, since this would imply that $7 < 5$ and $5 < -3$, which are both false statements.

Three-part inequalities can also be solved by using the addition and multiplication properties of inequality. The idea is to get the inequality in the form

a number $< x <$ another number.

The solutions can then easily be graphed.

EXAMPLE 9 SOLVING THREE-PART INEQUALITIES

(a) Solve $4 \leq 3x - 5 < 6$ and graph the solutions.

If we were to solve either of the two inequalities alone, the first step would be to add 5 to each part. Do this for each of the three parts.

$$4 \leq 3x - 5 < 6$$

$$4 + 5 \leq 3x - 5 + 5 < 6 + 5 \qquad \text{Add 5.}$$

$$9 \leq 3x < 11$$

Now divide each part by the positive number 3.

$$\frac{9}{3} \leq \frac{3x}{3} < \frac{11}{3} \qquad \text{Divide by 3.}$$

$$3 \leq x < \frac{11}{3}$$

A graph of the solutions is shown in Figure 2.19.

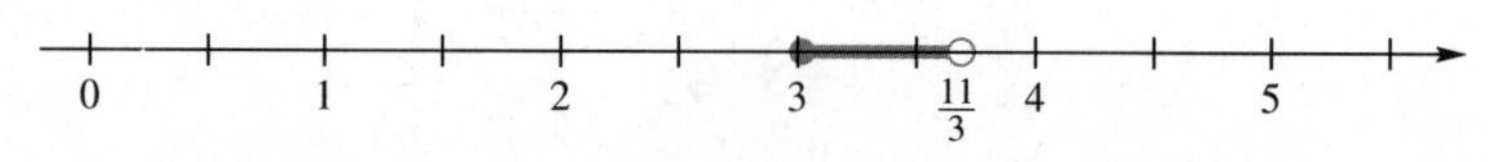

FIGURE 2.19

(b) Solve $-4 \leq \frac{2}{3}m - 1 < 8$ and graph the solutions.

Recall from Section 2.3 that fractions as coefficients in equations can be eliminated by multiplying both sides by the least common denominator of the fractions. The same is true for inequalities. One way to begin is to multiply all three parts by 3.

$$-4 \leq \frac{2}{3}m - 1 < 8$$

$$3(-4) \leq 3\left(\frac{2}{3}m - 1\right) < 3(8) \qquad \text{Multiply by 3.}$$

$$-12 \leq 2m - 3 < 24 \qquad \text{Distributive property}$$

Now add 3 to each part.

$$-12 + 3 \leq 2m - 3 + 3 < 24 + 3 \qquad \text{Add 3.}$$
$$-9 \leq 2m < 27$$

Finally, divide by 2 to get

$$-\frac{9}{2} \leq m < \frac{27}{2}.$$

A graph of the solutions is shown in Figure 2.20.

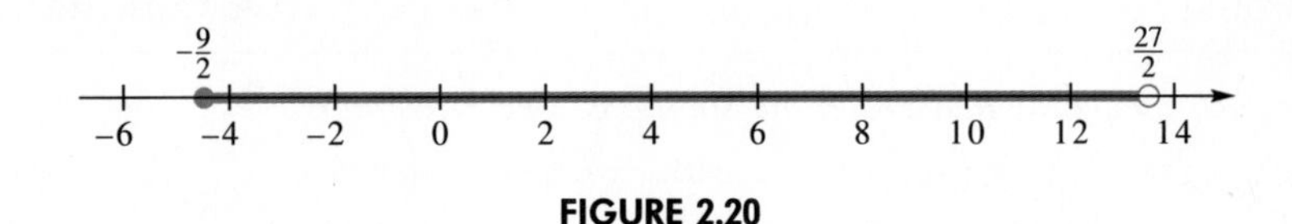

FIGURE 2.20

This inequality could also have been solved by first adding 1 to each part, and then multiplying each part by 3/2. ■

2.9 EXERCISES

Graph each inequality on a number line. See Examples 1 and 2.

1. $x \leq 4$ **2.** $k \geq -5$ **3.** $a < 3$ **4.** $p > 4$

5. $-2 \leq x \leq 5$ **6.** $8 \leq m \leq 10$ **7.** $3 \leq y < 5$ **8.** $0 < y \leq 10$

---

9. Explain why it is wrong to write $3 < x < -2$ to indicate that x is between -2 and 3. (See Objective 6.)

10. If $p < q$ and $r < 0$, which one of the following statements is false? (See Objectives 2 and 3.)
(a) $pr < qr$ **(b)** $pr > qr$
(c) $p + r < q + r$ **(d)** $p - r < q - r$

Solve each inequality. See Examples 3–5.

11. $z - 3 \geq -2$ **12.** $p + 2 \geq -6$ **13.** $-3 + k \geq 2$ **14.** $x + 5 > 5$

15. $3x < 27$ **16.** $5h \geq 20$ **17.** $-2k \leq 12$ **18.** $-3v > 6$

19. $-8y < -72$ **20.** $-9a \geq -63$

Solve each inequality and graph the solutions. In Exercises 27–30, you may wish to start by multiplying both sides by the least common denominator of the fractions. See Examples 6 and 7.

21. $3n + 5 \leq 2n - 6$ **22.** $5x - 2 < 4x - 5$ **23.** $2z - 8 > z - 3$

24. $4x + 6 \leq 3x - 5$ **25.** $4k + 1 \geq 2k - 9$ **26.** $5y + 3 < 2y + 12$

27. $3 + \frac{2}{3}r > \frac{5}{3}r - 27$ **28.** $8 + \frac{6}{5}t \leq \frac{8}{5}t + 12$

29. $4q + 1 - \frac{5}{3} < 8q + \frac{4}{3}$

30. $5x - \frac{2}{3} \leq \frac{2}{3}x + 6 - x$

31. $10p + 20 - p > p + 3 - 23$

32. $-3v + 6 + 3 - 2 > -5v - 19$

33. $-k + 4 + 5k \leq -1 + 3k + 5$

34. $6y - 2y - 4 + 7y > 3y - 4 + 7y$

35. $5 - (2 - r) \leq 3r + 5$

36. $-9 + (8 + y) > 7y - 4$

37. Write a short explanation of the additional rule that must be remembered when solving an inequality (as opposed to solving an equation). (See Objective 3.)

38. Write an explanation of how to decide when a closed circle and when an open circle should be used in graphing the solutions of an inequality. (See Objective 1.)

Solve each inequality and graph the solutions. See Example 9.

39. $-5 \leq 2x - 3 \leq 9$

40. $-7 \leq 3x - 4 \leq 8$

41. $5 < 1 - 6m < 12$

42. $-1 \leq 1 - 5q \leq 16$

43. $10 < 7p + 3 < 24$

44. $-8 \leq 3r - 1 \leq -1$

45. $-12 \leq \frac{1}{2}z + 1 \leq 4$

46. $-6 \leq 3 + \frac{1}{3}a \leq 5$

47. $1 \leq 3 + \frac{2}{3}p \leq 7$

48. $2 < 6 + \frac{3}{4}y < 12$

49. $-7 \leq \frac{5}{4}r - 1 \leq -1$

50. $-12 \leq \frac{3}{7}a + 2 \leq -4$

For each of the following problems, write an inequality and then solve the problem. See Example 8.

51. When four times a number is subtracted from 8, the result is less than 15. Find all numbers that satisfy this condition.

52. If half a number is added to 5, the result is greater than or equal to -3. Find all such numbers.

53. Maggie has scores of 98, 86, and 88 on her first three tests in algebra. If she wants an average of at least 90 after her fourth test, what possible scores can she make on her fourth test?

54. Inkie has grades of 75 and 82 on her first two computer science tests. What possible scores on a third test would give her an average of at least 80?

55. Greg Jackson earned \$200 at odd jobs during July, \$300 during August, and \$225 during September. If his average salary for the four months from July through October is to be at least \$250, what possible amounts could he earn during October?

56. In order to qualify for a company pension plan, an employee must average at least \$1000 per month in earnings. During the first four months of the year, an employee made \$900, \$1200, \$1040, and \$760. What possible amounts earned during the fifth month will qualify the employee?

57. The formula for converting from Fahrenheit to Celsius temperature is $C = \frac{5}{9}(F - 32)$. What temperature range in degrees Fahrenheit corresponds to 0° to 35° Celsius? (*Hint:* Write a three-part inequality.)

58. The formula for converting from Celsius to Fahrenheit temperature is $F = \frac{9}{5}C + 32$. What temperature range in degrees Celsius corresponds to 41° to 113° Fahrenheit?

59. A product will break even or produce a profit only if the revenue R from selling the product is at least the cost C of producing it. Suppose that the cost to produce x units of carpet is $C = 50x + 5000$, while the revenue is $R = 60x$. For what values of x is R at least equal to C?

60. Refer to Exercise 59. Suppose that the cost to produce x units of books is $C = 100x + 6000$, while the revenue is $R = 500x$. For what values of x is R at least equal to C?

61. The perimeter of a rectangular playground must be no greater than 120 meters. The width of the playground must be 22 meters. Find the possible lengths of the playground.

62. One side of a rectangular solar collector is 8 meters long. The area of the collector must be at least 240 square meters. Find the possible lengths of the collector.

Solve each inequality.

63. $5(2k + 3) - 2(k - 8) > 3(2k + 4) + k - 2$

64. $2(3z - 5) + 4(z + 6) \geq 2(3z + 2) + 3(z - 5)$

65. $3(p + 1) - 2(p - 4) \geq 5(2p - 3) + 2$

66. $-5(m - 3) + 4(m + 6) < 2(m - 3) + 4$

PREVIEW EXERCISES

Evaluate each expression for $x = 2$. See Sections 1.3 and 1.7.

67. $2x^2 - 3x + 9$

68. $3x^2 - 3x + 2$

69. $4x^3 - 5x^2 + 2x - 6$

70. $-4x^3 + 2x^2 - 9x - 3$

Simplify, combining like terms. See Section 2.1.

71. $3(2x + 4) + 4(2x - 6)$

72. $8(-3x + 7) - 4(2x + 3)$

73. $5x^3 - 2x^2 + 3x - 10 - 2x^3 + 9x^2 - 3x + 12$

74. $-8x^3 - 4x^2 + 12x - 3 + 9x^3 - 8x^2 + 6x - 14$

CHAPTER 2 GLOSSARY

KEY TERMS

2.1 **term** A term is a number, a variable, or a product or quotient of numbers and variables raised to powers.

numerical coefficient The numerical factor in a term is its numerical coefficient.

like terms Terms with exactly the same variables that have the same exponents are like terms.

2.2 **linear equation** A linear equation can be written in the form $ax + b = 0$, for real numbers a and b, with $a \neq 0$.

equivalent equations If two equations have exactly the same solutions, they are equivalent equations.

2.4 **degree** A degree is a unit of measure of an angle.

complementary angles Two angles whose sum is 90° are called complementary angles.

supplementary angles Two angles whose sum is 180° are called supplementary angles.

2.5 **vertical angles** Vertical angles are formed by intersecting lines, and they have the same measures (see Figure 2.6).

straight angle An angle that measures 180° is called a straight angle.

perimeter The distance around a geometric figure is called its perimeter.

2.6 **ratio** A ratio is a quotient of two quantities.

proportion A proportion is a statement that two ratios are equal.

2.9 **linear inequality** A linear inequality is an inequality that can be written in the form $ax + b < 0$, for real numbers a and b, with $a \neq 0$. ($<$ may be replaced with $>$, $\leq$, or $\geq$.)

NEW SYMBOLS

1° one degree

***a* to *b*, *a*:*b*, or $\frac{a}{b}$** the ratio of a to b

CHAPTER 2 QUICK REVIEW

SECTION	CONCEPTS	EXAMPLES
2.1 SIMPLIFYING EXPRESSIONS	Only like terms may be combined.	$-3y^2 + 6y^2 + 14y^2 = 17y^2$ $4(3 + 2x) - 6(5 - x)$ $= 12 + 8x - 30 + 6x$ Distributive property $= 14x - 18$
2.2 THE ADDITION AND MULTIPLICATION PROPERTIES OF EQUALITY	The same number may be added to (or subtracted from) each side of an equation without changing the solution. Each side of an equation may be multiplied (or divided) by the same nonzero number without changing the solution.	Solve $x - 6 = 12$. $x - 6 + 6 = 12 + 6$ Add 6. $x = 18$ Combine terms. Solve $\frac{3}{4}x = -9$. $\frac{4}{3} \cdot \frac{3}{4}x = \frac{4}{3}(-9)$ Multiply by $\frac{4}{3}$. $x = -12$
2.3 MORE ON SOLVING LINEAR EQUATIONS	**Solving a Linear Equation** **1.** Combine like terms to simplify each side. **2.** Get the variable term on one side, a number on the other. **3.** Get the equation into the form $x =$ a number. **4.** Check by substituting the result into the original equation.	Solve the equation $2x + 3x + 3 = 38$. **1.** $2x + 3x + 3 = 38$ $5x + 3 = 38$ Combine like terms. **2.** $5x + 3 - 3 = 38 - 3$ Subtract 3. $5x = 35$ Combine terms. **3.** $\frac{5x}{5} = \frac{35}{5}$ Divide by 5. $x = 7$ Reduce. **4.** $2x + 3x + 3 = 38$ Check. $2(7) + 3(7) + 3 = 38$? Let $x = 7$. $14 + 21 + 3 = 38$? Multiply. $38 = 38$ True
2.4 AN INTRODUCTION TO APPLICATIONS OF LINEAR EQUATIONS	**Solving an Applied Problem** **1.** Choose a variable to represent the unknown. **2.** Determine expressions for any other unknown quantities, using the variable. Draw figures or diagrams if they apply. **3.** Translate the problem into an equation.	One number is 5 more than another. Their sum is 21. Find both numbers. **1.** Let x be the smaller number. **2.** Let $x + 5$ be the larger number. **3.** $x + (x + 5) = 21$

SECTION	CONCEPTS	EXAMPLES
	4. Solve the equation.	**4.** $2x + 5 = 21$ Combine terms. $2x + 5 - 5 = 21 - 5$ Subtract 5. $2x = 16$ Combine terms. $\frac{2x}{2} = \frac{16}{2}$ Divide by 2. $x = 8$
	5. Answer the question asked in the problem. **6.** Check your solution by using the original words of the problem. Be sure that the answer is appropriate and makes sense.	**5.** The numbers are 8 and 13. **6.** 13 is 5 more than 8, and $8 + 13 = 21$. It checks.
2.5 FORMULAS AND APPLICATIONS FROM GEOMETRY	Given the values of all but one of the variables in a formula, the value of the remaining variable can be found. Solve for a specified variable in a formula by treating that variable as if it were the only one, and all others as if they were constants.	Find L if $A = LW$, given that $A = 24$ and $W = 3$. $24 = L \cdot 3$ $A = 24, W = 3$ $\frac{24}{3} = \frac{L \cdot 3}{3}$ Divide by 3. $8 = L$ Solve $A = \frac{1}{2}bh$ for b. $2A = 2\left(\frac{1}{2}bh\right)$ Multiply by 2. $2A = bh$ $\frac{2A}{h} = b$ Divide by h.
2.6 RATIOS AND PROPORTIONS	To write a ratio, express quantities in the same units. To solve a proportion, use the method of cross products.	Express as a ratio: 4 feet to 8 inches. 4 feet to 8 inches = 48 inches to 8 inches $= \frac{48}{8} = \frac{6}{1}$ or 6 to 1 or 6:1 Solve $\frac{x}{12} = \frac{35}{60}$. $60x = 12 \cdot 35$ Cross products $60x = 420$ Multiply. $\frac{60x}{60} = \frac{420}{60}$ Divide by 60. $x = 7$

<table>
<tr><th>SECTION</th><th>CONCEPTS</th><th>EXAMPLES</th></tr>
<tr>
<td>2.7 APPLICATIONS OF PERCENT: MIXTURE, INTEREST, AND MONEY</td>
<td>Problems involving applications of percent can be solved using box diagrams or charts.</td>
<td>A sum of money is invested at simple interest in two ways. Part is invested at 12%, and $20,000 less than that amount is invested at 10%. If the total interest for one year is $9000, find the amount invested at each rate.

Let x = amount invested at 12%;
$x - 20{,}000$ = amount invested at 10%.

<table>
<tr><td>Amount invested</td><td>x</td><td rowspan="2">+</td><td>$x - 20{,}000$</td><td rowspan="2">=</td><td rowspan="2">9000</td></tr>
<tr><td>Rate</td><td>.12</td><td>.10</td></tr>
</table>

$.12x + .10(x - 20{,}000) = 9000$
$12x + 10(x - 20{,}000) = 900{,}000$ Multiply by 100.
$12x + 10x - 200{,}000 = 900{,}000$ Distributive property
$22x - 200{,}000 = 900{,}000$ Combine terms.
$22x = 1{,}100{,}000$ Add 200,000.
$x = 50{,}000$ Divide by 22.

$50,000 is invested at 12% and $30,000 is invested at 10%.</td>
</tr>
<tr>
<td>2.8 MORE ABOUT PROBLEM SOLVING</td>
<td>The three forms of the formula relating distance, rate, and time are $d = rt$, $r = \frac{d}{t}$, and $t = \frac{d}{r}$.

To solve a problem about distance, set up a sketch showing what is happening in the problem.

Make a chart using the information given in the problem, along with the unknown quantities.</td>
<td>Two cars leave from the same point, traveling in opposite directions. One travels at 45 miles per hour and the other at 60 miles per hour. How long will it take them to be 210 miles apart?

Let t = time it takes for them to be 210 miles apart.
The sketch shows what is happening in the problem.
210 miles
The chart gives the information from the problem, with expressions for distance obtained by using $d = rt$.

<table>
<tr><th></th><th>r</th><th>t</th><th>d</th></tr>
<tr><td>One car</td><td>45</td><td>t</td><td>$45t$</td></tr>
<tr><td>Other car</td><td>60</td><td>t</td><td>$60t$</td></tr>
</table>

The sum of the distances, $45t$ and $60t$, must be 210 miles.</td>
</tr>
</table>

SECTION	CONCEPTS	EXAMPLES
		$45t + 60t = 210$ $105t = 210$ Combine like terms. $t = 2$ Divide by 2. It will take them 2 hours to be 210 miles apart.
2.9 THE ADDITION AND MULTIPLICATION PROPERTIES OF INEQUALITY	The same number may be added or subtracted on each side of an inequality without changing the solutions. Each side of an inequality may be multiplied or divided by the same nonzero number.	Solve each inequality and graph. $x - 8 < -4$ $x - 8 + 8 < -4 + 8$ Add 8. $x < 4$ Combine terms. −4 0 4
	(a) If the number is positive, the direction of the inequality symbol *does not* change.	$5x \geq 10$ $\frac{5x}{5} \geq \frac{10}{5}$ Divide by 5. $x \geq 2$ −4 −2 0 2 4
	(b) If the number is negative, the direction of the inequality symbol *must be reversed*.	$-\frac{1}{4}y < 3$ $(-4)\left(-\frac{1}{4}y\right) > -4(3)$ Multiply by −4; reverse symbol. $y > -12$ −12 −6 0 6
	To solve an inequality: **1.** Combine like terms to simplify each side. **2.** Get the variable term on one side, a number on the other. **3.** Get the inequality into the form $x < a$ or $x > a$ where a is a number. (The symbols $\leq$ or $\geq$ may also appear.)	Solve: $4x - 3 + 2x - 5 > 10 - 6$ $6x - 8 > 4$ Combine terms. $6x - 8 + 8 > 4 + 8$ Add 8. $6x > 12$ Combine terms. $\frac{6x}{6} > \frac{12}{6}$ Divide by 6. $x > 2.$

SECTION	CONCEPTS	EXAMPLES
	To solve an inequality such as $4 < 2x + 6 < 8$ work with all three expressions at the same time.	Solve $4 < 2x + 6 < 8$. $4 - 6 < 2x + 6 - 6 < 8 - 6$ Subtract 6. $-2 < 2x < 2$ Combine terms. $\frac{-2}{2} < \frac{2x}{2} < \frac{2}{2}$ Divide by 2. $-1 < x < 1$

CHAPTER 2 REVIEW EXERCISES

[2.1] *Combine terms whenever possible.*

1. $2m + 9m$ **2.** $15p^2 - 7p^2 + 8p^2$ **3.** $5p^2 - 4p + 6p + 11p^2$

4. $-2(3k - 5) + 2(k + 1)$ **5.** $7(2m + 3) - 2(8m - 4)$ **6.** $-(2k + 8) - (3k - 7)$

[2.2–2.3] *Solve each equation.*

7. $m - 5 = 1$ **8.** $y + 8 = -4$ **9.** $3k + 1 = 2k + 8$

10. $5k = 4k + \frac{2}{3}$ **11.** $(4r - 2) - (3r + 1) = 8$ **12.** $3(2y - 5) = 2 + 5y$

13. $7k = 35$ **14.** $12r = -48$ **15.** $2p - 7p + 8p = 15$

16. $\frac{m}{12} = -1$ **17.** $\frac{5}{8}k = 8$ **18.** $12m + 11 = 59$

19. $3(2x + 6) - 5(x + 8) = x - 22$ **20.** $5x + 9 - (2x - 3) = 2x - 7$

21. $\frac{1}{2}r - \frac{r}{3} = \frac{r}{6}$ **22.** $.10(x + 80) + .20x = 14$

23. $3x - (-2x + 6) = 4(x - 4) + x$ **24.** $2(y - 3) - 4(y + 12) = -2(y + 27)$

[2.4] *Solve the following problems.*

25. If 7 is added to six times a number, the result is 22. Find the number.

26. If 4 is subtracted from twice a number, the result is 16. Find the number.

27. The land area of Hawaii is 5213 square miles greater than the area of Rhode Island. Together, the areas total 7637 square miles. What is the area of each of the two states?

28. The height of Seven Falls in Colorado is 5/2 the height (in feet) of Twin Falls in Idaho. The sum of the heights is 420 feet. Find the height of each.

29. The supplement of an angle measures 10 times the measure of its complement. What is the measure of the angle?

[2.5] *A formula is given in the following exercises, along with the values of some of the variables. Find the value of the variable that is not given.*

30. $A = \frac{1}{2}bh$; $A = 22, b = 4$ **31.** $A = \frac{1}{2}(b + B)h$; $b = 9, B = 12, h = 8$

32. $C = 2\pi r$; $C = 12.56$, $\pi = 3.14$

33. $V = \frac{4}{3}\pi r^3$; $\pi = 3.14$, $r = 1$

Solve each formula for the specified variable.

34. $A = LW$; for W

35. $A = \frac{1}{2}(b + B)h$; for h

---

36. The perimeter of a rectangular picture frame is 84 inches. The width is 17 inches. Find the length.

37. Find the measure of each marked angle.

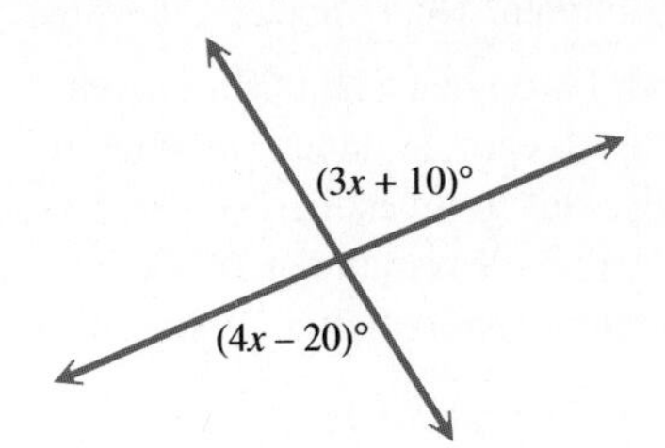

38. Find the measure of each marked angle.

(8x – 1)° [3(x + 3) – 15]°

[2.6] *Write the following ratios and reduce them to lowest terms.*

39. 50 centimeters to 30 centimeters

40. 6 days to 1 week

41. 45 inches to 5 feet

42. 2 months to 3 years

Decide whether or not the following proportions are true.

43. $\frac{15}{18} = \frac{45}{54}$

44. $\frac{11}{19} = \frac{55}{105}$

45. $\frac{38}{51} = \frac{722}{1020}$

46. $\frac{29}{72} = \frac{899}{2232}$

Solve each proportion.

47. $\frac{p}{5} = \frac{21}{30}$

48. $\frac{2 + m}{3} = \frac{2 - m}{4}$

Solve each problem.

49. The distance between Milwaukee and Boston is 1050 miles. On a certain map this distance is represented by 21 inches. On the same map, Seattle and Cincinnati are 46 inches apart. What is the actual distance between Seattle and Cincinnati?

50. A recipe for biscuit tortoni calls for 2/3 cup of macaroon cookie crumbs. The recipe is for 8 servings. How many cups of macaroon cookie crumbs would be needed for 30 servings?

51. Which is the better buy for processed cheese slices: 8 slices for \$2.19, or 16 slices for \$4.48?

[2.7]

52. A person has \$250 in fives and tens. He has twice as many tens as fives. How many fives does he have?

53. For a party, Joann bought some 15¢ candy and some 30¢ candy, paying \$15 in all. If there are 25 more pieces of 15¢ candy, how many pieces of 30¢ candy did she buy?

54. How many liters of 40% dye solution should be added to 72 liters of 80% dye solution to get a 70% mixture?

55. A nurse must mix 15 liters of a 10% solution of a drug with some 60% solution to get a 20% mixture. How many liters of the 60% solution would be needed?

56. Gwen Boyle has money invested in two accounts, both paying simple interest. She has $4000 more invested at 8% a year than she has at 12% a year. She receives a total of $1520 a year from the two investments. How much is invested at each rate?

57. Habib Paroo has $10,000 to invest from which he wishes to earn an annual income of $700 per year. He decides to invest part of it at 5% annual interest and invest the remainder in bonds paying 9% interest. How much should he invest at each rate?

[2.8] *Solve each problem.*

58. In 1846, the vessel Yorkshire traveled from Liverpool to New York, a distance of 3150 miles, in 384 hours. What was the Yorkshire's average speed? Round your answer to the nearest tenth.

59. Sue Fredine drove from Louisville to Dallas, a distance of 819 miles, averaging 63 miles per hour. What was her driving time?

60. Two planes leave St. Louis at the same time. One flies north at 350 miles per hour and the other flies south at 420 miles per hour. In how many hours will they be 1925 miles apart?

61. Jim leaves his house on his bicycle and averages 5 miles per hour. His wife, Annie, leaves 1/2 hour later, following the same path and averaging 8 miles per hour. How long will it take for Annie to catch up with Jim?

62. The perimeter of a rectangle is ten times the width. The length is 9 meters more than the width. Find the width of the rectangle.

63. The longest side of a triangle is 11 meters longer than the shortest side. The medium side is 15 meters long. The perimeter of the triangle is 46 meters. Find the length of the shortest side of the triangle.

[2.9] *Graph each inequality on a number line.*

64. $m \geq -2$ **65.** $a < -3$ **66.** $-5 \leq p < 6$ **67.** $1 \leq m < 4$

Solve each inequality.

68. $y + 5 \geq 2$

69. $5y > 4y + 8$

70. $9(k - 5) - (3 + 8k) \geq 5$

71. $3(2z + 5) + 4(8 + 3z) \leq 5(3z + 2) + 2z$

72. $-6 \leq x + 2 \leq 0$

73. $3 < y - 4 < 5$

74. $6k \geq -18$

75. $-11y < 22$

76. $2 - 4p + 7p + 8 < 6p - 5p$

77. $-(y + 2) + 3(2y - 7) \leq 4 - 5y$

78. $-3 \leq 2m + 1 \leq 4$

79. $9 < 3m + 5 \leq 20$

---

80. Carlotta Valdez has grades of 94 and 88 on her first two calculus tests. What possible scores on a third test will give her an average of at least 90?

81. If nine times a number is added to 6, the result is at most 4. Find all such numbers.

MIXED REVIEW EXERCISES*

Solve each of the following.

82. $\frac{y}{7} = \frac{y-5}{2}$

83. $I = prt$; for r

84. $-z > -4$

85. $\frac{3}{5} = \frac{k}{12}$

86. $2k - 5 = 4k + 7$

87. $7m + 3 \leq 5m - 9$

88. $a + b + c = P$; for a

89. $2 - 3(y - 5) = 4 + y$

---

90. If 1 quart of oil must be mixed with 24 quarts of gasoline, how much oil would be needed for 192 quarts of gasoline?

91. Two trains are 390 miles apart. They start at the same time and travel toward one another, meeting 3 hours later. If the speed of one train is 30 miles per hour faster than the speed of the other train, find the speed of each train.

92. One side of a triangle is 3 centimeters longer than the shortest side. The third side is twice as long as the shortest side. If the perimeter of the triangle cannot exceed 39 centimeters, find all possible lengths for the shortest side.

93. An inheritance is invested two ways—some at 5% and $6000 more at 10%. The total annual income from interest is $2100. Find the amount invested at each rate.

94. The shorter base of a trapezoid is 42 centimeters long, and the longer base is 48 centimeters long. The area of the trapezoid is 360 square centimeters. Find the height of the trapezoid.

95. The area of a triangle is 25 square meters. The base is 10 meters in length. Find the height.

96. On a test in geometry, the highest grade was 35 points more than the lowest. The sum of the highest and lowest grades was 157. Find the lowest score.

97. The perimeter of a square cannot be greater than 200 meters. Find the possible values for the length of a side.

98. The distance between two cities on a road map is 16 centimeters. The two cities are actually 150 kilometers apart. The distance on the map between two other cities is 40 centimeters. How far apart are these cities?

CHAPTER 2 TEST

Simplify by combining like terms.

1. $9r + 3r - 4r - r - 8r$

2. $3z - 7z + 8 - 9 - (-5) + 4z$

3. $4(2m - 1) - (m + 5) - 3(m - 5)$

Solve each equation.

4. $3(a + 12) = 1 - 2(a - 5)$

5. $4k - 6k + 8(k - 3) = -2(k + 12)$

6. $\frac{5}{4}m = -3$

7. $\frac{1}{2}p + \frac{1}{3} = \frac{5}{2}p - \frac{4}{3}$

8. $-(y + 3) + 2y - 5 = 4 - 3y$

9. $5(2x + 3) - 3x = 7x + 9$

---

*The order of the exercises in this final group does not correspond to the order in which topics occur in the chapter. This random ordering should help you in your preparation for the chapter test.

10. If nine times a number is subtracted from 8, the result is -10. Find the number.

11. Vern paid \$93 more to tune up his Bronco than his Oldsmobile. He paid \$313 in all. How much did it cost to tune up each car?

12. Solve the formula $P = 2L + 2W$ for L.

13. A triangular lot has an area of 1825 square yards. The height of the triangle measures 50 yards. What is the length of its base?

14. Suppose that in the equation $\frac{a}{b} = \frac{c}{d}$, a and d are positive numbers and b is a negative number. Must c be positive or negative? Explain your answer.

Solve each equation.

15. $\dfrac{z}{16} = \dfrac{3}{48}$

16. $\dfrac{y+5}{3} = \dfrac{y-2}{4}$

Solve each problem.

17. If 5 pineapples cost \$9.75, how much will 9 pineapples cost?

18. A woman invests some money at 6%, with \$6000 more than this amount invested at 7.5%. Her total annual income from interest is \$1935. How much is invested at each rate?

19. How many liters of a 20% chemical solution must be mixed with 30 liters of a 60% solution to get a 50% mixture?

20. Two cars leave from the same point, traveling in opposite directions. One travels at a constant rate of 50 miles per hour, while the other travels at a constant rate of 65 miles per hour. How long will it take for them to be 460 miles apart?

Solve each inequality. Graph each solution.

21. $-2m < -14$

22. $5(k-2) + 3 \le 2(k-3) + 2k$

23. $-4r + 2(r-3) \ge 5r - (3 + 6r) - 7$

24. $-8 < 3k - 2 \le 12$

25. Penny Heath has grades of 86 and 98 on her first two history of mathematics tests. What possible scores could she make on her third test so that her average would be at least 90?

CHAPTER THREE

POLYNOMIALS AND EXPONENTS

Just as the integers are the basic real numbers, polynomials are the basic algebraic expressions. Have you ever wondered how your scientific calculator evaluates a number like $\sqrt{.5}$ when you touch the $\sqrt{}$ key? The value of $\sqrt{.5}$ is approximated using the basic operations of addition, subtraction, multiplication, and division by writing $\sqrt{.5}$ as a polynomial, a sum of terms.

In Chapter 2 we worked with linear expressions, which are the simplest polynomials. In this chapter we discuss polynomial expressions that include variables with whole number exponents greater than one. The properties and techniques in this chapter will be important in the subjects to be discussed later in Chapters 4 and 5 and again in Chapter 9.

3.1 POLYNOMIALS

OBJECTIVES

1 IDENTIFY TERMS AND COEFFICIENTS.
2 ADD LIKE TERMS.
3 KNOW THE VOCABULARY FOR POLYNOMIALS.
4 EVALUATE POLYNOMIALS.
5 ADD POLYNOMIALS.
6 SUBTRACT POLYNOMIALS.

1 Recall that in an expression such as

$$4x^3 + 6x^2 + 5x + 8,$$

the quantities $4x^3$, $6x^2$, $5x$, and 8 are called *terms*. In the term $4x^3$, the number 4 is called the **numerical coefficient,** or simply the **coefficient,** of x^3. In the same way, 6 is the coefficient of x^2 in the term $6x^2$, 5 is the coefficient of x in the term $5x$, and 8 is the coefficient in the term 8. A constant term, like 8 in the polynomial above, can be thought of as $8x^0$, where x^0 is defined to equal 1. We will explain the reason for this definition later in this chapter.

EXAMPLE 1 **IDENTIFYING COEFFICIENTS**

Name the (numerical) coefficient of each term in these expressions.

(a) $4x^3$

The coefficient is 4.

(b) $x - 6x^4$

The coefficient of x is 1 because $x = 1 \cdot x$. The coefficient of x^4 is -6 since $x - 6x^4$ can be written as the sum $x + (-6x^4)$.

(c) $5 - v^3$

The coefficient of the term 5 is 5 since $5 = 5v^0$. By writing $5 - v^3$ as a sum, $5 + (-v^3)$, or $5 + (-1v^3)$, the coefficient of v^3 can be identified as -1. ■

2 Recall that **like terms** have exactly the same combination of variables with the same exponents on the variables. Only the coefficients may be different. Examples of like terms are

$$19m^5 \quad \text{and} \quad 14m^5,$$
$$6y^9, \quad -37y^9, \quad \text{and} \quad y^9,$$
$$3pq \quad \text{and} \quad -2pq,$$
$$2xy^2 \quad \text{and} \quad -xy^2.$$

Like terms are added by using the distributive property.

EXAMPLE 2
ADDING LIKE TERMS

Simplify each expression by adding like terms.

(a) $-4x^3 + 6x^3 = (-4 + 6)x^3 = 2x^3$ Distributive property

(b) $9x^6 - 14x^6 + x^6 = (9 - 14 + 1)x^6 = -4x^6$

(c) $12m^2 + 5m + 4m^2 = (12 + 4)m^2 + 5m = 16m^2 + 5m$

(d) $3x^2y + 4x^2y - x^2y = (3 + 4 - 1)x^2y = 6x^2y$ ■

Example 2(c) shows that it is not possible to combine $16m^2$ and $5m$. These two terms are unlike because the exponents on the variables are different. **Unlike terms** have different variables or different exponents on the same variables.

3 *Polynomials* are basic to algebra. A **polynomial in *x*** is a term or the sum of a finite number of terms of the form ax^n, for any real number a and any whole number n. For example,

$$16x^8 - 7x^6 + 5x^4 - 3x^2 + 4$$

is a polynomial in x (the 4 can be written as $4x^0$). This polynomial is written in **descending powers** of the variable, since the exponents on x decrease from left to right. On the other hand,

$$2x^3 - x^2 + \frac{4}{x}$$

is not a polynomial in x, since $4/x$ is not a *product*, ax^n, for a *whole number n*. (Of course, we could define *polynomial* using any variable or variables, and not just x.)

The **degree** of a term is the sum of the exponents on the variables. For example, $3x^4$ has degree 4, while $6x^{17}$ has degree 17. The term $5x$ has degree 1, -7 has degree 0 (since -7 can be written as $-7x^0$), and $2x^2y$ has degree $2 + 1 = 3$ (y has an exponent of 1). The **degree of a polynomial** is the highest degree of any nonzero term of the polynomial. For example, $3x^4 - 5x^2 + 6$ is of degree 4, the polynomial $5x + 7$ is of degree 1, 3 (or $3x^0$) is of degree 0, and $x^2y + xy - 5xy^2$ is of degree 3.

Three types of polynomials are very common and are given special names. A polynomial with exactly three terms is called a **trinomial.** (*Tri*- means "three," as in *tri*angle.) Examples are

$$9m^3 - 4m^2 + 6, \qquad 19y^2 + 8y + 5, \qquad \text{and} \qquad -3m^5 - 9m^2 + 2.$$

A polynomial with exactly two terms is called a **binomial.** (*Bi*- means "two," as in *bi*cycle.) Examples are

$$-9x^4 + 9x^3, \qquad 8m^2 + 6m, \qquad \text{and} \qquad 3m^5 - 9m^2.$$

A polynomial with only one term is called a **monomial.** (*Mon(o)*- means "one," as in *mono*rail.) Examples are

$$9m, \quad -6y^5, \quad a^2, \quad \text{and} \quad 6.$$

EXAMPLE 3
CLASSIFYING POLYNOMIALS

For each polynomial, first simplify if possible by combining like terms. Then give the degree and tell whether it is a monomial, a binomial, a trinomial, or none of these.

(a) $2x^3 + 5$

The polynomial cannot be simplified. The degree is 3. The polynomial is a binomial.

(b) $4x - 5x + 2x$

Add like terms to simplify: $4x - 5x + 2x = x$, which is a monomial of degree 1. ■

4 A polynomial usually represents different numbers for different values of the variable, as shown in the next examples.

EXAMPLE 4
EVALUATING A POLYNOMIAL

Find the value of $3x^4 + 5x^3 - 4x - 4$ when $x = -2$ and when $x = 3$.

First, substitute -2 for x.

$$\begin{aligned} 3x^4 + 5x^3 - 4x - 4 &= 3(-2)^4 + 5(-2)^3 - 4(-2) - 4 \\ &= 3 \cdot 16 + 5 \cdot (-8) + 8 - 4 \\ &= 48 - 40 + 8 - 4 \\ &= 12 \end{aligned}$$

Next, replace x with 3.

$$\begin{aligned} 3x^4 + 5x^3 - 4x - 4 &= 3(3)^4 + 5(3)^3 - 4(3) - 4 \\ &= 3 \cdot 81 + 5 \cdot 27 - 12 - 4 \\ &= 362 \end{aligned}$$ ■

CAUTION Notice the use of parentheses around the numbers that are substituted for the variable in Example 4. This is particularly important when substituting a negative number for a variable that is raised to a power, so that the sign of the product is correct.

5 Polynomials may be added, subtracted, multiplied, and divided. Polynomial addition and subtraction are explained in the rest of this section.

ADDING POLYNOMIALS

To add two polynomials, add like terms.

EXAMPLE 5 **ADDING POLYNOMIALS VERTICALLY**

Add $6x^3 - 4x^2 + 3$ and $-2x^3 + 7x^2 - 5$.

Write like terms in columns.

$$\begin{array}{r} 6x^3 - 4x^2 + 3 \\ \underline{-2x^3 + 7x^2 - 5} \end{array}$$

Now add, column by column.

$$\begin{array}{rrr} 6x^3 & -4x^2 & 3 \\ \underline{-2x^3} & \underline{7x^2} & \underline{-5} \\ 4x^3 & 3x^2 & -2 \end{array}$$

Add the three sums together.

$$4x^3 + 3x^2 + (-2) = 4x^3 + 3x^2 - 2 \quad \blacksquare$$

The polynomials in Example 5 also could be added horizontally, as shown in the next example.

EXAMPLE 6 **ADDING POLYNOMIALS HORIZONTALLY**

Add $6x^3 - 4x^2 + 3$ and $-2x^3 + 7x^2 - 5$.

Write the sum as

$$(6x^3 - 4x^2 + 3) + (-2x^3 + 7x^2 - 5).$$

Use the associative and commutative properties to rewrite this sum with the parentheses removed and with the subtractions changed to additions of inverses.

$$6x^3 + (-4x^2) + 3 + (-2x^3) + 7x^2 + (-5)$$

Place like terms together.

$$6x^3 + (-2x^3) + (-4x^2) + 7x^2 + 3 + (-5)$$

Combine like terms to get

$$4x^3 + 3x^2 + (-2), \qquad \text{or simply} \qquad 4x^3 + 3x^2 - 2,$$

the same answer found in Example 5. ■

6 Earlier, the difference $x - y$ was defined as $x + (-y)$. (We find the difference $x - y$ by adding x and the opposite of y.) For example,

$$7 - 2 = 7 + (-2) = 5 \qquad \text{and} \qquad -8 - (-2) = -8 + 2 = -6.$$

A similar method is used to subtract polynomials.

SUBTRACTING POLYNOMIALS Subtract two polynomials by changing all the signs on the second polynomial and adding the result to the first polynomial.

EXAMPLE 7 **SUBTRACTING POLYNOMIALS**

Subtract: $(5x - 2) - (3x - 8)$.

By the definition of subtraction,

$$(5x - 2) - (3x - 8) = (5x - 2) + [-(3x - 8)].$$

As shown in Chapter 1, the distributive property gives

$$-(3x - 8) = -1(3x - 8) = -3x + 8,$$

so

$$(5x - 2) - (3x - 8) = (5x - 2) + (-3x + 8) = 2x + 6. \quad ■$$

EXAMPLE 8 **SUBTRACTING POLYNOMIALS**

Subtract $6x^3 - 4x^2 + 2$ from $11x^3 + 2x^2 - 8$.

Write the problem.

$$(11x^3 + 2x^2 - 8) - (6x^3 - 4x^2 + 2)$$

Change all the signs in the second polynomial and add the two polynomials.

$$(11x^3 + 2x^2 - 8) + (-6x^3 + 4x^2 - 2) = 5x^3 + 6x^2 - 10$$

We can check a subtraction problem by using the fact that if $a - b = c$, then $a = b + c$. For example, $6 - 2 = 4$. Check by writing $6 = 2 + 4$, which is correct. Check the polynomial subtraction above by adding $6x^3 - 4x^2 + 2$ and $5x^3 + 6x^2 - 10$. Since the sum is $11x^3 + 2x^2 - 8$, the subtraction was performed correctly. ■

Subtraction also can be done in columns.

EXAMPLE 9 **SUBTRACTING POLYNOMIALS VERTICALLY**

Use the method of subtracting by columns to find

$$(14y^3 - 6y^2 + 2y - 5) - (2y^3 - 7y^2 - 4y + 6).$$

Arrange like terms in columns.

$$\begin{array}{r} 14y^3 - 6y^2 + 2y - 5 \\ \underline{2y^3 - 7y^2 - 4y + 6} \end{array}$$

Change all signs in the second row, and then add.

$$\begin{array}{rl} 14y^3 - 6y^2 + 2y - 5 & \\ \underline{-2y^3 + 7y^2 + 4y - 6} & \text{Change all signs.} \\ 12y^3 + y^2 + 6y - 11 & \text{Add.} \end{array}$$ ■

Either the horizontal or the vertical method may be used for adding and subtracting polynomials.

3.1 EXERCISES

In each polynomial, add like terms where possible. See Example 2. In Exercises 11–16, write the results in descending powers of the variable.

1. $2r^5 + (-3r^5)$

2. $-19y^2 + 9y^2$

3. $3x^5 + 2x^5 - 4x^5$

4. $6x^3 - 8x^7 - 9x^3$

5. $-4p^7 + 8p^7 - 5p^7$

6. $-3a^8 + 4a^8 - 3a^8 + 2a^8$

7. $4y^2z + 3y^2z - 2y^2z + y^2z$

8. $3r^5t^2 - 8r^5t^2 + r^5t^2 - 2r^5t^2$

9. $-5p^5q^3 + 8p^5q^3 - 2p^5q^3 - p^5q^3$

10. $6k^3n - 9k^3n + 8k^3n - 2k^3n$

11. $y^4 + 8y^4 - 9y^2 + 6y^2 + 10y^2$

12. $11a^2 - 10a^2 + 2a^2 - a^6 + 2a^6$

13. $4z^5 - 9z^3 + 8z^2 + 10z^5$

14. $-9m^3 + 2m^3 - 11m^3 + 15m^2 - 9m$

15. $-.82q^2 + 1.72q - .37 + 1.99q^2 - .32q + .12$

16. $5.8r^3 - 2.7r^2 + 5.4r - 6.7r^3 + 1.4r - r^2$

For each polynomial, first simplify, if possible; then give the degree of the polynomial and tell whether it is (a) a monomial, (b) a binomial, (c) a trinomial, (d) none of these. See Example 3.

17. $5x^4 - 8x$

18. $4y - 8y$

19. $23x^9y - \frac{1}{2}x^2y^2 + xy^3$

20. $2m^7n - 3m^6n^2 + 2m^5n^3 + mn^2$

21. $x^8 + 3x^7 - 5x^4$

22. $2xw^2 - 2x^2w$

23. $\frac{3}{5}x^5z^3 + \frac{2}{5}x^5z^3$

24. $\frac{9}{11}x^2$

25. $2m^8 - 5m^9$

Tell whether each statement is true always, sometimes *or* never. *(See Objective 3.)*

26. A binomial is a polynomial.

27. A polynomial is a trinomial.

28. A trinomial is a binomial.

29. A monomial has no coefficient.

30. A binomial is a trinomial.

31. A polynomial of degree 4 has 4 terms.

Find the value of each polynomial when (a) $x = 2$ and when (b) $x = -1$. See Example 4.

32. $2x^2 - 4x$

33. $8x + 5x^2 + 2$

34. $2x^5 - 4x^4 + 5x^3 - x^2$

35. $2x^2 + 5x + 1$

36. $-3x^2 + 14x - 2$

37. $-2x^2 + 3$

38. $-5x^2 + 4x + 5$

39. $-x^2 - x$

40. $-x^2 - x + 2$

Add or subtract as indicated. See Examples 5 and 9.

41. Add.

$$\begin{array}{r} 4a^3 - 4a^2 \\ \underline{6a^3 + 5a^2} \end{array}$$

42. Add.

$$\begin{array}{r} 12x^4 - x^2 \\ \underline{8x^4 + 3x^2} \end{array}$$

43. Subtract.

$$\begin{array}{r} 2n^5 - 5n^3 + 6 \\ \underline{3n^5 + 7n^3 + 8} \end{array}$$

44. Subtract.

$$\begin{array}{r} 3r^2 - 4r + 2 \\ \underline{7r^2 + 2r - 3} \end{array}$$

45. Add.

$$\begin{array}{r} 9m^3 - 5m^2 + 4m - 8 \\ \underline{3m^3 + 6m^2 + 8m - 6} \end{array}$$

46. Add.

$$\begin{array}{r} 12r^5 + 11r^4 - 7r^3 - 2r^2 - 5r - 3 \\ \underline{-\ 8r^5 - 10r^4 + 3r^3 + 2r^2 - 5r + 7} \end{array}$$

47. Subtract.

$$\begin{array}{r} 5a^4 - 3a^3 + 2a^2 \qquad\qquad \\ \underline{a^3 - \ a^2 + a - 1} \end{array}$$

48. Add.

$$\begin{array}{r} 3w^2 - 5w + 2 \\ 4w^2 + 6w - 5 \\ \underline{8w^2 + 7w - 2} \end{array}$$

Perform the indicated operations. See Examples 6–8.

49. $(3r^2 + 5r - 6) + (2r - 5r^2)$

50. $(8m^2 - 7m) - (3m^2 + 7m)$

51. $(x^2 + x) - (3x^2 + 2x - 1)$

52. $(3x^2 + 2x + 5) + (8x^2 - 5x - 4)$

53. $(16x^3 - x^2 + 3x) + (-12x^3 + 3x^2 + 2x)$

54. $(-2b^6 + 3b^4 - b^2) - (-b^6 + 2b^4 + 2b^2)$

55. $(7y^4 + 3y^2 + 2y) - (-18y^4 - 5y^2 - y)$

56. $(3x^2 + 2x + 5) - (-7x^2 - 8x + 2) + (3x^2 - 4x + 7)$

Find the perimeter of the geometric figures with sides of the following lengths.

57. rectangle; length: $3x^2 + 5x + 13$; width: $2x^2 - 6x + 10$

58. rectangle; length: $5k^2 + 8x$; width: $2k^2 + 4x$

59. triangle; sides: $3r^2 + 2r$, $r^3 + r^2 + 4$, $9r + 6$

60. triangle; sides: $4p^3 - 2p^2 + 3p$, $p^3 + 2p$, $5p^2 - p - 4$

The sum of the three angles in any triangle is 180°. Use this fact to find the value of the variables in the following exercises.

61.

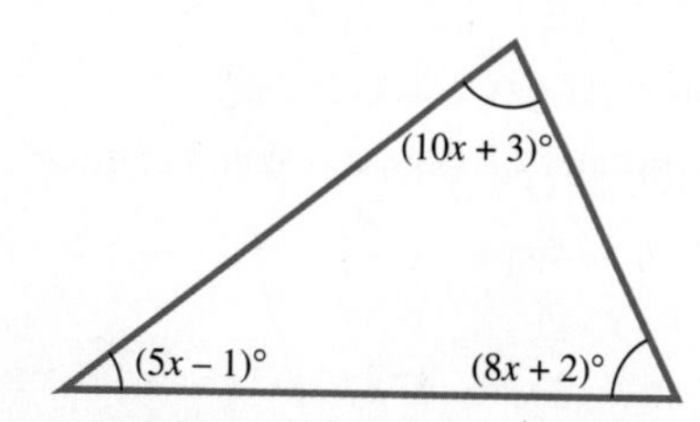

62.

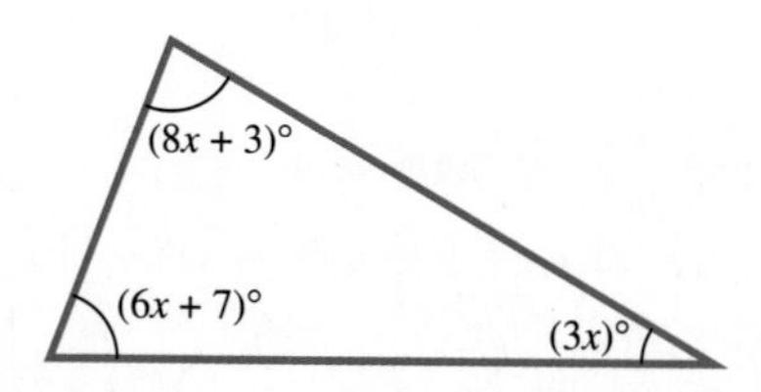

Perform the indicated operations in Exercises 63–66.

63. $[(8m^2 + 4m - 7) - (2m^2 - 5m + 2)] - (m^2 + m + 1)$

64. $[(9b^3 - 4b^2 + 3b + 2) - (-2b^3 - 3b^2 + b)] - (8b^3 + 6b + 4)$

65. Subtract $9x^2 - 6x + 5$ from $3x^2 - 2$.

66. Add $x^3 + 4x^2 - 2x + 3$ to $2x^3 - 8x^2 - 5$.

67. Find the sum of $10x^4 - 3x^3 + 2x + 1$ and $3x^3 - 6x^2 + 8x - 2$.

68. Find the difference when $9x^4 + 3x^2 + 5$ is subtracted from $8x^4 - 2x^3 + x - 1$.

69. Why is the term 3^4 *not* degree 4?

70. Can a polynomial in more than one variable be written in descending powers? Explain your answer and give examples.

PREVIEW EXERCISES

Evaluate each expression. See Section 1.3.

71. $2 \cdot 2 \cdot 2 \cdot 2 \cdot 2 \cdot 2$

72. $3 \cdot 3 \cdot 3$

73. $5 \cdot 5 \cdot 5 \cdot 5$

74. $4 \cdot 4 \cdot 4 \cdot 4 \cdot 4$

75. $\frac{2}{3} \cdot \frac{2}{3} \cdot \frac{2}{3}$

76. $\frac{5}{8} \cdot \frac{5}{8}$

77. $(2 \cdot 2 \cdot 2)(2 \cdot 2 \cdot 2 \cdot 2)$

78. $(3 \cdot 3)(3 \cdot 3 \cdot 3)$

3.2 EXPONENTS

OBJECTIVES

1. USE EXPONENTS.
2. USE THE PRODUCT RULE FOR EXPONENTS.
3. USE THE RULE $(a^m)^n = a^{mn}$.
4. USE THE RULE $(ab)^m = a^m b^m$.
5. USE THE RULE $\left(\frac{a}{b}\right)^m = \frac{a^m}{b^m}$.

1 Recall from Section 1.2 that in the expression 5^2, the number 5 is the *base* and 2 is the *exponent*. The expression 5^2 is called an *exponential expression*. Usually we do not write the exponent when it is 1; however, sometimes it is convenient to do so. In general, for any quantity a, $a^1 = a$.

EXAMPLE 1 DETERMINING THE BASE AND EXPONENT IN AN EXPONENTIAL EXPRESSION

Evaluate each exponential expression. Name the base and the exponent.

	BASE	EXPONENT
(a) $5^4 = 5 \cdot 5 \cdot 5 \cdot 5 = 625$	5	4
(b) $-5^4 = -1 \cdot 5^4 = -1 \cdot (5 \cdot 5 \cdot 5 \cdot 5) = -625$	5	4
(c) $(-5)^4 = (-5)(-5)(-5)(-5) = 625$	-5	4

■

CAUTION It is important to understand the differences between parts (b) and (c) of Example 1. In -5^4 the lack of parentheses shows that the exponent 4 refers only to the base 5, and not -5; in $(-5)^4$ the parentheses show that the exponent 4 refers to the base -5. In summary, $-a^n$ and $(-a)^n$ are not necessarily the same.

EXPRESSION	BASE	EXPONENT	EXAMPLE
$-a^n$	a	n	$-3^2 = -(3 \cdot 3) = -9$
$(-a)^n$	$-a$	n	$(-3)^2 = (-3)(-3) = 9$

2 By the definition of exponents,

$$\begin{aligned} 2^4 \cdot 2^3 &= (2 \cdot 2 \cdot 2 \cdot 2)(2 \cdot 2 \cdot 2) \\ &= 2 \cdot 2 \cdot 2 \cdot 2 \cdot 2 \cdot 2 \cdot 2 \\ &= 2^7. \end{aligned}$$

Also,

$$\begin{aligned} 6^2 \cdot 6^3 &= (6 \cdot 6)(6 \cdot 6 \cdot 6) \\ &= 6 \cdot 6 \cdot 6 \cdot 6 \cdot 6 \\ &= 6^5. \end{aligned}$$

Generalizing from these examples, $2^4 \cdot 2^3 = 2^{4+3} = 2^7$ and $6^2 \cdot 6^3 = 6^{2+3} = 6^5$, suggests the **product rule for exponents.**

PRODUCT RULE FOR EXPONENTS

For any positive integers m and n, $a^m \cdot a^n = a^{m+n}$.
Example: $6^2 \cdot 6^5 = 6^{2+5} = 6^7$

CAUTION The bases must be the same before the product rule for exponents can be applied.

EXAMPLE 2 USING THE PRODUCT RULE

Use the product rule for exponents to find each product.

(a) $6^3 \cdot 6^5 = 6^{3+5} = 6^8$ by the product rule.

(b) $(-4)^7(-4)^2 = (-4)^{7+2} = (-4)^9$

(c) $x^2 \cdot x = x^2 \cdot x^1 = x^{2+1} = x^3$

(d) $m^4m^3m^5 = m^{4+3+5} = m^{12}$

(e) The product rule does not apply to the product $2^3 \cdot 3^2$, since the bases are different.

(f) The product rule does not apply to $2^3 + 2^4$, since it is a *sum,* not a *product.* ■

EXAMPLE 3 USING THE PRODUCT RULE

Multiply $2x^3$ and $3x^7$.

Since $2x^3$ means $2 \cdot x^3$ and $3x^7$ means $3 \cdot x^7$, we can use the associative and commutative properties to get

$$2x^3 \cdot 3x^7 = (2 \cdot 3) \cdot (x^3 \cdot x^7) = 6x^{10}. \quad ■$$

CAUTION Be careful you understand the difference between *adding* and *multiplying* exponential expressions. For example,

$$8x^3 + 5x^3 = 13x^3, \qquad \text{but} \qquad (8x^3)(5x^3) = 8 \cdot 5x^{3+3} = 40x^6.$$

3 We can simplify an expression such as $(8^3)^2$ with the product rule for exponents as follows.

$$\begin{aligned}(8^3)^2 &= (8^3)(8^3)\\ &= 8^{3+3}\\ &= 8^6\end{aligned}$$

The exponents in $(8^3)^2$ are multiplied to give the exponent in 8^6. As another example,

$$\begin{aligned}(5^2)^3 &= 5^2 \cdot 5^2 \cdot 5^2\\ &= 5^{2+2+2}\\ &= 5^6,\end{aligned}$$

and $2 \cdot 3 = 6$. These examples suggest **power rule (a) for exponents.**

POWER RULE (A) FOR EXPONENTS

For any positive integers m and n, $(a^m)^n = a^{mn}$.
Example: $(3^2)^4 = 3^{2 \cdot 4} = 3^8$

EXAMPLE 4 USING POWER RULE (A)

Use power rule (a) for exponents to simplify each expression.

(a) $(2^5)^3 = 2^{5 \cdot 3} = 2^{15}$

(b) $(5^7)^2 = 5^{7(2)} = 5^{14}$

(c) $(x^2)^5 = x^{2(5)} = x^{10}$

(d) $(n^3)^2 = n^{3(2)} = n^6$ ■

4 The properties studied in Chapter 1 can be used to develop two more rules for exponents. Using the definition of an exponential expression and the commutative and associative properties, we can rewrite the expression $(4 \cdot 8)^3$ as follows.

$$\begin{aligned}(4 \cdot 8)^3 &= (4 \cdot 8)(4 \cdot 8)(4 \cdot 8) && \text{Definition of exponent}\\ &= (4 \cdot 4 \cdot 4) \cdot (8 \cdot 8 \cdot 8) && \text{Commutative and associative properties}\\ &= 4^3 \cdot 8^3 && \text{Definition of exponent}\end{aligned}$$

This example suggests **power rule (b) for exponents.**

POWER RULE (B) FOR EXPONENTS

For any positive integer m, $(ab)^m = a^m b^m$.
Example: $(2p)^5 = 2^5 p^5$

EXAMPLE 5 USING POWER RULE (B)

Use power rule (b) for exponents to simplify each expression.

(a) $(3xy)^2 = 3^2x^2y^2 = 9x^2y^2$

(b)
$$\begin{aligned}5(pq)^2 &= 5(p^2q^2) && \text{Power rule (b)}\\ &= 5p^2q^2 && \text{Multiply.}\end{aligned}$$

(c)
$$\begin{aligned}3(2m^2p^3)^4 &= 3[2^4(m^2)^4(p^3)^4] && \text{Power rule (b)}\\ &= 3 \cdot 2^4m^8p^{12} && \text{Power rule (a)}\\ &= 48m^8p^{12}\end{aligned}$$ ■

5 Since the quotient $\frac{a}{b}$ can be written as $a\left(\frac{1}{b}\right)$, we can use power rule (b), together with some of the properties of real numbers, to get **power rule (c) for exponents.**

POWER RULE (C) FOR EXPONENTS

For any positive integer m, $\left(\frac{a}{b}\right)^m = \frac{a^m}{b^m}$ $(b \neq 0)$.

Example: $\left(\frac{5}{3}\right)^2 = \frac{5^2}{3^2}$

EXAMPLE 6 USING POWER RULE (C)

Use power rule (c) for exponents to simplify each expression.

(a) $\left(\frac{2}{3}\right)^5 = \frac{2^5}{3^5}$ **(b)** $\left(\frac{m}{n}\right)^3 = \frac{m^3}{n^3}$, $(n \neq 0)$ ■

The rules for exponents discussed in this section are summarized below. These rules are basic to the study of algebra and should be *memorized.*

RULES FOR EXPONENTS

For positive integers m and n:

		EXAMPLES
Product rule	$a^m \cdot a^n = a^{m+n}$	$6^2 \cdot 6^5 = 6^{2+5} = 6^7$
Power rules (a)	$(a^m)^n = a^{mn}$	$(3^2)^4 = 3^{2 \cdot 4} = 3^8$
(b)	$(ab)^m = a^m b^m$	$(2p)^5 = 2^5 p^5$
(c)	$\left(\frac{a}{b}\right)^m = \frac{a^m}{b^m}$ $(b \neq 0)$	$\left(\frac{5}{3}\right)^2 = \frac{5^2}{3^2}$

As shown in the next example, more than one rule may be needed to simplify an expression with exponents.

EXAMPLE 7 USING COMBINATIONS OF RULES

Use the rules for exponents to simplify each expression.

(a) $\left(\frac{2}{3}\right)^2 \cdot 2^3 = \frac{2^2}{3^2} \cdot \frac{2^3}{1}$ Power rule (c)

$= \frac{2^2 \cdot 2^3}{3^2 \cdot 1}$ Multiply the fractions.

$= \frac{2^5}{3^2}$ Product rule

(b) $(5x)^3(5x)^4 = (5x)^7$ Product rule

$= 5^7 x^7$ Power rule (b)

(c) $(2x^2y^3)^4(3xy^2)^3 = 2^4(x^2)^4(y^3)^4 \cdot 3^3x^3(y^2)^3$ — Power rule (b)

$= 2^4 \cdot 3^3x^8y^{12}x^3y^6$ — Power rule (a)

$= 16 \cdot 27x^{11}y^{18}$ — Product rule

$= 432x^{11}y^{18}$ ■

3.2 EXERCISES

Identify the base and exponent for each exponential expression. See Example 1.

1. 5^{12} **2.** a^6 **3.** $(3m)^4$ **4.** $(5k)^3$

5. -2^4 and $(-2)^4$ **6.** -125^3 and $(-125)^3$ **7.** $(-24)^2$ and -24^2 **8.** $-(-3)^5$ and -3^5

9. $-r^6$ and $(-r)^6$ **10.** $5y^3$ and $(5y)^3$

Write each expression using exponents.

11. $3 \cdot 3 \cdot 3 \cdot 3 \cdot 3$ **12.** $4 \cdot 4 \cdot 4$ **13.** $(-2)(-2)(-2)(-2)(-2)$

14. $(-1)(-1)(-1)(-1)$ **15.** $\frac{1}{(-2)(-2)(-2)}$ **16.** $\frac{1}{2 \cdot 2 \cdot 2 \cdot 2 \cdot 2}$

17. $p \cdot p \cdot p \cdot p \cdot p$ **18.** $\frac{1}{a \cdot a \cdot a \cdot a \cdot a \cdot a}$ **19.** $(-2z)(-2z)(-2z)$

20. $(-3m)(-3m)(-3m)(-3m)$

Evaluate each expression. For example, $5^2 + 5^3 = 25 + 125 = 150$.

21. 4^3 **22.** -4^3 **23.** $(-4)^2$ **24.** -4^2

25. $2^2 + 2^5$ **26.** $4^2 + 4^1$ **27.** $(-4)^2 - (-2)^2$ **28.** $(-2)^3 - (-3)^2$

Use the product rule to simplify each expression. Write each answer in exponential form. See Examples 2 and 3.

29. $4^2 \cdot 4^3$ **30.** $3^5 \cdot 3^4$ **31.** $3^4 \cdot 3^7$ **32.** $2^5 \cdot 2^{15}$

33. $4^3 \cdot 4^5 \cdot 4^{10}$ **34.** $2^3 \cdot 2^4 \cdot 2^6$ **35.** $(-3)^3(-3)^2$ **36.** $(-4)^5(-4)^3$

37. $y^3 \cdot y^4 \cdot y^7$ **38.** $a^8 \cdot a^5 \cdot a^2$ **39.** $r \cdot r^5 \cdot r^4 \cdot r^7$ **40.** $m^9 \cdot m \cdot m^5 \cdot m^8$

41. $(-9r^3)(7r^6)$ **42.** $(8a^9)(-3a^{14})$ **43.** $(2p^4)(5p^9)$ **44.** $(3q^8)(7q^5)$

In each of the following exercises, first add *the given expressions (if possible); then start over and* multiply *them. See Example 3.*

45. $4m^3, 9m^3$ **46.** $8y^2, 7y^2$ **47.** $-12p, 11p$ **48.** $3q^4, 5q^4$

49. $7r, 3r, 5r$ **50.** $9a^3, 2a^3, 3a^3$ **51.** $-5a^2, 3a^3$ **52.** $6r^4, -8r^5$

Use the power rules for exponents to simplify each expression. Write each answer in exponential form. See Examples 4–6.

53. $(6^3)^2$ **54.** $(8^4)^6$ **55.** $(9^3)^2$ **56.** $(2^3)^4$

57. $(-5^2)^4$ **58.** $(-2^3)^2$ **59.** $(-4^2)^3$ **60.** $(-3^5)^3$

61. $(5m)^3$ **62.** $(2xy)^4$ **63.** $(-2pq)^4$ **64.** $(-3ab)^5$

65. $\left(\frac{-3x^5}{4}\right)^2$ **66.** $\left(\frac{4m^3n^2}{5}\right)^4$ **67.** $\left(\frac{5a^2b}{c^4}\right)^3$ **68.** $\left(\frac{2y^2z^4}{w^3}\right)^5$

Use the rules for exponents to simplify each expression. See Example 7.

69. $\left(\frac{4}{3}\right)^5 \cdot (4)^3$ **70.** $\left(\frac{5}{3}y\right)^2$ **71.** $(3m)^2(3m)^5$ **72.** $(-5p)^4(-5p)^2$

73. $(8z)^6(8z)^3$ **74.** $(2p)^4(2p)$ **75.** $(2m^2n)^3(mn^2)$ **76.** $(3p^2q^2)^3(q^4)$

77. $(5ab^2)^5(5a^3b)^2$ **78.** $(-r^3s)^4(r^2s)^3$

Simplify each expression. Assume that all variables represent positive integers.

79. $5^r \cdot 5^{7r}$ **80.** $6^{5p} \cdot 6^p$ **81.** $(2m)^p$ **82.** $(3k)^t$

83. $\left(\frac{4^2}{3^3}\right)^r$ **84.** $\left(\frac{5^3}{7^4}\right)^k$ **85.** $\left(\frac{2p^m}{q^r}\right)^n$ **86.** $\left(\frac{4r^a}{r^b}\right)^c$

Find the areas of the following figures. (Leave π in the answer for Exercise 90.)*

87.

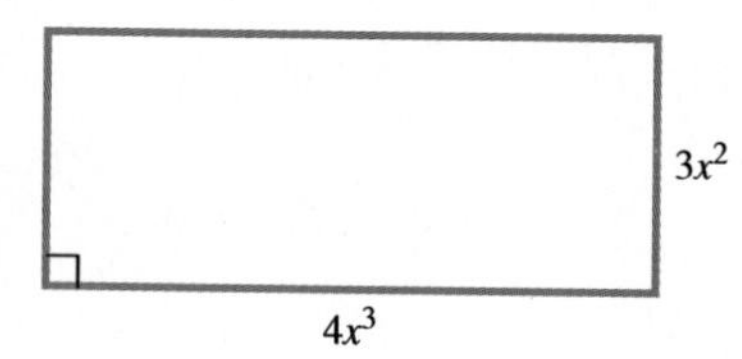

88.

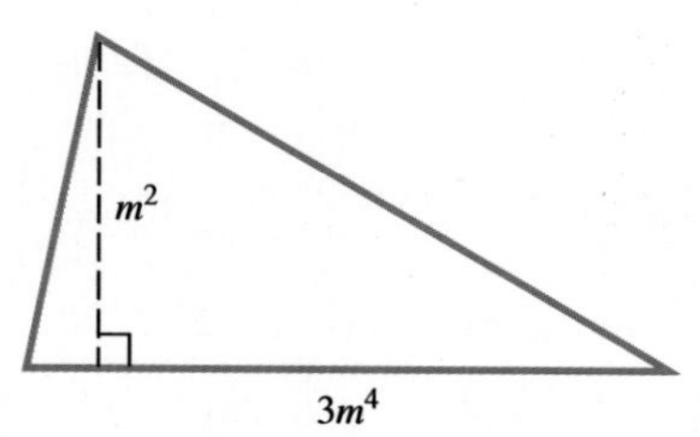

89.

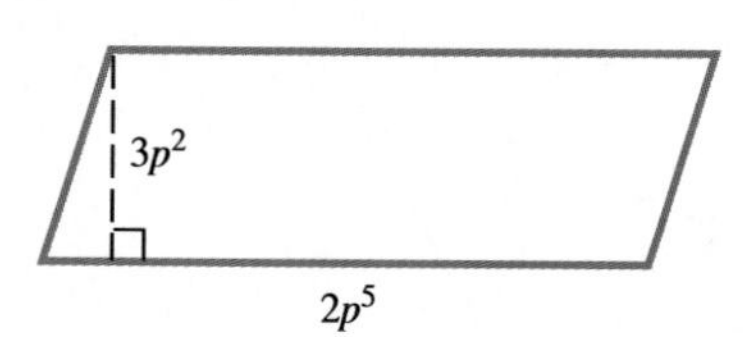

90.

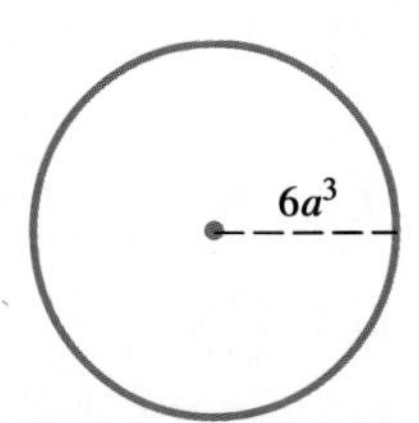

---

91. If $a^2 = 10$, what is a^6? (See Objective 3.)

92. Assume a is a positive number greater than 1. Arrange the following terms in order from smallest to largest: $-(-a)^3$, $-a^3$, $(-a)^4$, $-a^4$. Explain how you decided on the order. (See Objective 1.)

93. In your own words, describe a rule to tell whether an exponential expression with a negative base is positive or negative. (See Objective 1.)

PREVIEW EXERCISES

Multiply.

94. $p(2p)$ **95.** $3k(5k)$ **96.** $5x^2(2x)$ **97.** $9r^3(2r)$

98. $7m^3(8m^2)$ **99.** $4y^5(7y^2)$ **100.** $6p^5(5p^4)$ **101.** $2z^8(5z^3)$

---

*The small square in the figures for Exercises 87–89 indicates a right angle (90°).

3.3 MULTIPLICATION OF POLYNOMIALS

OBJECTIVES

1. MULTIPLY A MONOMIAL AND A POLYNOMIAL.
2. MULTIPLY TWO POLYNOMIALS.
3. MULTIPLY BINOMIALS BY THE FOIL METHOD.

1 As shown earlier, the product of two monomials is found by using the rules for exponents and the commutative and associative properties. For example,

$$(-8m^6)(-9m^4) = (-8)(-9)(m^6)(m^4) = 72m^{6+4} = 72m^{10}.$$

CAUTION It is important not to confuse the *addition* of terms with the *multiplication* of terms. For example,

$$7q^5 + 2q^5 = 9q^5, \quad \text{but} \quad (7q^5)(2q^5) = 7 \cdot 2q^{5+5} = 14q^{10}.$$

To find the product of a monomial and a polynomial with more than one term, we use the distributive property and then the method shown above.

EXAMPLE 1 **MULTIPLYING A MONOMIAL AND A POLYNOMIAL**

Use the distributive property to find each product.

(a) $4x^2(3x + 5)$

$$\begin{aligned} 4x^2(3x + 5) &= (4x^2)(3x) + (4x^2)(5) && \text{Distributive property} \\ &= 12x^3 + 20x^2 && \text{Multiply monomials.} \end{aligned}$$

(b) $-8m^3(4m^3 + 3m^2 + 2m - 1)$

$$\begin{aligned} &-8m^3(4m^3 + 3m^2 + 2m - 1) \\ &\quad = (-8m^3)(4m^3) + (-8m^3)(3m^2) && \text{Distributive property} \\ &\quad\quad + (-8m^3)(2m) + (-8m^3)(-1) \\ &\quad = -32m^6 - 24m^5 - 16m^4 + 8m^3 && \text{Multiply monomials.} \end{aligned}$$

■

2 The distributive property is used repeatedly to find the product of any two polynomials. For example, to find the product of the polynomials $x + 1$ and $x - 4$, think of $x - 4$ as a single quantity and use the distributive property as follows.

$$(x + 1)(x - 4) = x(x - 4) + 1(x - 4)$$

Now use the distributive property twice to find $x(x - 4)$ and $1(x - 4)$.

$$\begin{aligned} x(x - 4) + 1(x - 4) &= x(x) + x(-4) + 1(x) + 1(-4) \\ &= x^2 - 4x + x - 4 \\ &= x^2 - 3x - 4 \end{aligned}$$

A rule for multiplying any two polynomials is given below.

MULTIPLYING POLYNOMIALS Multiply two polynomials by multiplying each term of the second polynomial by each term of the first polynomial and adding the products.

EXAMPLE 2 **MULTIPLYING TWO POLYNOMIALS**

Find the product of $4m^3 - 2m^2 + 4m$ and $m^2 + 5$.

Multiply each term of the second polynomial by each term of the first. (Either polynomial can be written first in the product.)

$$\begin{aligned}(m^2 + 5)&(4m^3 - 2m^2 + 4m)\\ &= m^2(4m^3) - m^2(2m^2) + m^2(4m) + 5(4m^3) - 5(2m^2) + 5(4m)\\ &= 4m^5 - 2m^4 + 4m^3 + 20m^3 - 10m^2 + 20m\end{aligned}$$

Now combine like terms.

$$= 4m^5 - 2m^4 + 24m^3 - 10m^2 + 20m \quad ■$$

When at least one of the factors in a product of polynomials has three or more terms, the multiplication can be simplified by writing one polynomial above the other.

EXAMPLE 3 **MULTIPLYING VERTICALLY**

Multiply $x^3 + 2x^2 + 4x + 1$ by $3x + 5$.

Start by writing the polynomials as follows.

$$\begin{array}{r} x^3 + 2x^2 + 4x + 1 \\ 3x + 5 \\ \hline \end{array}$$

It is not necessary to line up terms in columns, because any terms may be multiplied (not just like terms). Begin by multiplying each of the terms in the top row by 5.

Step 1

$$\begin{array}{r} x^3 + 2x^2 + 4x + 1 \\ 3x + 5 \\ \hline 5x^3 + 10x^2 + 20x + 5 \end{array} \quad 5(x^3 + 2x^2 + 4x + 1)$$

Notice how this process is similar to multiplication of whole numbers. Now multiply each term in the top row by $3x$. Be careful to place the like terms in columns, since the final step will involve addition (as in multiplying two whole numbers).

Step 2

$$\begin{array}{rrrrr} & x^3 + & 2x^2 + & 4x + & 1 \\ & & & 3x + & 5 \\ \hline & 5x^3 + & 10x^2 + & 20x + & 5 \\ 3x^4 + & 6x^3 + & 12x^2 + & 3x & \end{array} \quad 3x(x^3 + 2x^2 + 4x + 1)$$

Step 3 Add like terms.

$$\begin{array}{rrrrr} & x^3 + & 2x^2 + & 4x + & 1 \\ & & & 3x + & 5 \\ \hline & 5x^3 + & 10x^2 + & 20x + & 5 \\ 3x^4 + & 6x^3 + & 12x^2 + & 3x & \\ \hline 3x^4 + & 11x^3 + & 22x^2 + & 23x + & 5 \end{array}$$

The product is $3x^4 + 11x^3 + 22x^2 + 23x + 5$. ■

3 In algebra, many of the polynomials to be multiplied are both binomials (with just two terms). For these products a shortcut that eliminates the need to write out all the steps is used. To develop this shortcut, let us first multiply $x + 3$ and $x + 5$ using the distributive property.

$$\begin{aligned}(x + 3)(x + 5) &= x(x + 5) + 3(x + 5) \\ &= x(x) + x(5) + 3(x) + 3(5) \\ &= x^2 + 5x + 3x + 15 \\ &= x^2 + 8x + 15\end{aligned}$$

The first term in the second line, $(x)(x)$, is the product of the first terms of the two binomials.

$(x + 3)(x + 5)$ Multiply the first terms: $(x)(x)$.

The term $(x)(5)$ is the product of the first term of the first binomial and the last term of the second binomial. This is the **outer product.**

$(x + 3)(x + 5)$ Multiply the outer terms: $(x)(5)$.

The term $(3)(x)$ is the product of the last term of the first binomial and the first term of the second binomial. The product of these middle terms is the **inner product.**

$(x + 3)(x + 5)$ Multiply the inner terms: $(3)(x)$.

Finally, $(3)(5)$ is the product of the last terms of the two binomials.

$(x + 3)(x + 5)$ Multiply the last terms: $(3)(5)$.

In the third step of the multiplication above, the inner product and the outer product are added. This step should be performed mentally, so that the three terms of the answer can be written without extra steps as

$$(x + 3)(x + 5) = x^2 + 8x + 15.$$

A summary of these steps is given below. This procedure is sometimes called the **FOIL method,** which comes from the abbreviation for *first, outer, inner, last*.

MULTIPLYING BINOMIALS BY THE FOIL METHOD

Step 1 **Multiply the first terms.** Multiply the two first terms of the binomials to get the first term of the answer.

Step 2 **Find the outer and inner products.** Find the outer product and the inner product and add them (mentally if possible) to get the middle term of the answer.

Step 3 **Multiply the last terms.** Multiply the two last terms of the binomials to get the last term of the answer.

EXAMPLE 4 USING THE FOIL METHOD

Find the product $(x + 8)(x - 6)$ by the FOIL method.

Step 1 F Multiply the *first* terms.

$$x(x) = x^2$$

Step 2 O Find the product of the *outer* terms.

$$x(-6) = -6x$$

I Find the product of the *inner* terms.

$$8(x) = 8x$$

Add the outer and inner products mentally.

$$-6x + 8x = 2x$$

Step 3 L Multiply the *last* terms.

$$8(-6) = -48$$

The product of $x + 8$ and $x - 6$ is found by adding the terms found in the three steps above, so

$$(x + 8)(x - 6) = x^2 - 6x + 8x - 48 = x^2 + 2x - 48.$$

As a shortcut, this product can be found in the following manner.

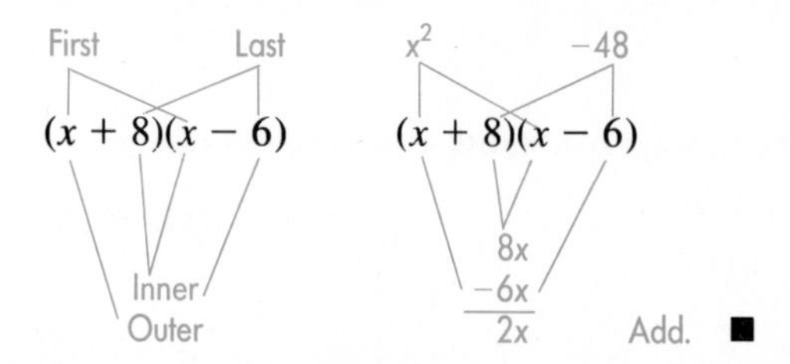

Sometimes it is not possible to add the inner and outer products of the FOIL method, as shown in the next example.

EXAMPLE 5
USING THE FOIL METHOD

Multiply $9x - 2$ and $3y + 1$.

First	$(9x - 2)(3y + 1)$	$27xy$
Outer	$(9x - 2)(3y + 1)$	$9x$
Inner	$(9x - 2)(3y + 1)$	$-6y$
Last	$(9x - 2)(3y + 1)$	-2

$$(9x - 2)(3y + 1) = \overset{\text{F}}{27xy} + \overset{\text{O}}{9x} - \overset{\text{I}}{6y} - \overset{\text{L}}{2} \quad \blacksquare$$

EXAMPLE 6
USING THE FOIL METHOD

Find the following products.

(a) $$(2k + 5y)(k + 3y) = \overset{\text{F}}{(2k)(k)} + \overset{\text{O}}{(2k)(3y)} + \overset{\text{I}}{(5y)(k)} + \overset{\text{L}}{(5y)(3y)}$$
$$= 2k^2 + 6ky + 5ky + 15y^2$$
$$= 2k^2 + 11ky + 15y^2$$

(b) $(7p + 2q)(3p - q) = 21p^2 - pq - 2q^2$ ■

NOTE The inner and outer products are often like terms that may be combined, as in Example 6.

3.3 EXERCISES

Find each product. See Example 1.

1. $(-4x^5)(8x^2)$
2. $(-3x^7)(2x^5)$
3. $(5y^4)(3y^7)$
4. $(10p^2)(5p^3)$
5. $(15a^4)(2a^5)$
6. $(-3m^6)(-5m^4)$
7. $(9r^3)(-2s^2)$
8. $2m(3m + 2)$
9. $-5p(6 - 3p)$
10. $3p(-2p^3 + 4p^2)$
11. $4x(3 + 2x + 5x^3)$
12. $-8z(2z + 3z^2 + 3z^3)$
13. $7y(3 + 5y^2 - 2y^3)$
14. $2y^3(3 + 2y + 5y^4)$
15. $-2m^4(3m^2 + 5m + 6)$
16. $-4r^3(-7r^2 + 8r - 9)$
17. $-9a^5(-3a^6 - 2a^4 + 8a^2)$
18. $3a^2(2a^2 - 4ab + 5b^2)$
19. $4z^3(8z^2 + 5zy - 3y^2)$
20. $7m^3n^2(3m^2 + 2mn - n^3)$
21. $2p^2q(3p^2q^2 - 5p + 2q^2)$

Find each product. See Examples 2 and 3.

22. $(6x + 1)(2x^2 + 4x + 1)$
23. $(9y - 2)(8y^2 - 6y + 1)$
24. $(9a + 2)(9a^2 + a + 1)$
25. $(2r - 1)(3r^2 + 4r - 4)$
26. $(4m + 3)(5m^3 - 4m^2 + m - 5)$
27. $(y + 4)(3y^4 - 2y^2 + 1)$
28. $(2x - 1)(3x^5 - 2x^3 + x^2 - 2x + 3)$
29. $(2a + 3)(a^4 - a^3 + a^2 - a + 1)$
30. $(5x^2 + 2x + 1)(x^2 - 3x + 5)$
31. $(2m^2 + m - 3)(m^2 - 4m + 5)$

Use FOIL to find each product. See Examples 4–6.

32. $(r-1)(r+3)$ **33.** $(x+2)(x-5)$ **34.** $(x-7)(x-3)$

35. $(r+3)(r+6)$ **36.** $(2x-1)(3x+2)$ **37.** $(4y-5)(2y+1)$

38. $(6z+5)(z-3)$ **39.** $(3x-1)(2x+3)$ **40.** $(2r-1)(4r+3)$

41. $(11m-10)(10m+11)$ **42.** $(4+5x)(5-4x)$ **43.** $(8+3x)(2-x)$

44. $(-3+2r)(4+r)$ **45.** $(-5+6z)(2-z)$ **46.** $(-3+a)(-5-2a)$

47. $(-6-3y)(1-4y)$ **48.** $(p+3q)(p+q)$ **49.** $(2r-3s)(3r+s)$

50. $(5y+z)(2y-z)$ **51.** $(9m+4k)(2m-3k)$ **52.** $(8y-9z)(y+5z)$

53. $(4r+9s)(-2r+5s)$ **54.** $(7m+11n)(3m-8n)$

Find the products in Exercises 55–64.

55. $\left(3p+\frac{5}{4}q\right)\left(2p-\frac{5}{3}q\right)$ **56.** $\left(-x+\frac{2}{3}y\right)\left(3x-\frac{3}{4}y\right)$

57. $(m^3-4)(2m^3+3)$ **58.** $(4a^2+b^2)(a^2-2b^2)$

59. $(2k^3+h^2)(k^2-3h^2)$ **60.** $(4x^3-5y^4)(x^2+y)$

61. $3p^3(2p^2+5p)(p^3+2p+1)$ **62.** $5k^2(k^2-k+4)(k^3-3)$

63. $-2x^5(3x^2+2x-5)(4x+2)$ **64.** $-4x^3(3x^4+2x^2-x)(-2x+1)$

---

65. Rephrase the distributive property in your own words.

66. Explain in words how to find the product of two monomials with numerical coefficients.

PREVIEW EXERCISES

Multiply out each of the following squares. See Section 3.2.

67. $(3m)^2$ **68.** $(5p)^2$ **69.** $(-2r)^2$

70. $(-5a)^2$ **71.** $(4x^2)^2$ **72.** $(8y^3)^2$

3.4 SPECIAL PRODUCTS

OBJECTIVES

1 SQUARE BINOMIALS.

2 FIND THE PRODUCT OF THE SUM AND DIFFERENCE OF TWO TERMS.

3 FIND HIGHER POWERS OF BINOMIALS.

In the previous section, we saw how the distributive property is used to multiply any two polynomials. In this section, we develop patterns for certain binomial products that occur frequently.

1 The square of a binomial can be found quickly by using the method shown in Example 1.

EXAMPLE 1 SQUARING A BINOMIAL

Find $(m + 3)^2$.

Squaring $m + 3$ by the FOIL method gives

$$(m + 3)(m + 3) = m^2 + 6m + 9. \quad \blacksquare$$

The result has the square of both the first and the last terms of the binomial:

$$m^2 = m^2 \qquad \text{and} \qquad 3^2 = 9.$$

The middle term is twice the product of the two terms of the binomial, since both the outer and inner products are $(m)(3)$ and

$$(m)(3) + (m)(3) = 2(m)(3) = 6m.$$

This example suggests the following rule.

SQUARE OF A BINOMIAL

The square of a binomial is a trinomial consisting of the square of the first term, plus twice the product of the two terms, plus the square of the last term of the binomial. For x and y,

$$(x + y)^2 = x^2 + 2xy + y^2$$

and

$$(x - y)^2 = x^2 - 2xy + y^2.$$

EXAMPLE 2 SQUARING BINOMIALS

Use the rule to square each binomial.

(a) $(5z - 1)^2 = (5z)^2 - 2(5z)(1) + (1)^2 = 25z^2 - 10z + 1$

Recall that $(5z)^2 = 5^2z^2 = 25z^2$.

(b) $(3b + 5r)^2 = (3b)^2 + 2(3b)(5r) + (5r)^2 = 9b^2 + 30br + 25r^2$

(c) $(2a - 9x)^2 = 4a^2 - 36ax + 81x^2$

(d) $\left(4m + \frac{1}{2}\right)^2 = (4m)^2 + 2(4m)\left(\frac{1}{2}\right) + \left(\frac{1}{2}\right)^2 = 16m^2 + 4m + \frac{1}{4} \quad \blacksquare$

CAUTION A common error in squaring a binomial is forgetting the middle term of the product. In general, $(a + b)^2 \neq a^2 + b^2$.

2 Binomial products of the form $(x + y)(x - y)$ also occur frequently. In these products, one binomial is the sum of two terms, and the other is the difference of the same two terms. As an example, the product of $a + 2$ and $a - 2$ is

$$(a + 2)(a - 2) = a^2 - 2a + 2a - 4 = a^2 - 4.$$

As we can show with the FOIL method, the product of $x + y$ and $x - y$ is the difference of two squares.

PRODUCT OF THE SUM AND DIFFERENCE OF TWO TERMS

The product of the sum and difference of the two terms x and y is

$$(x + y)(x - y) = x^2 - y^2.$$

EXAMPLE 3
FINDING THE PRODUCT OF THE SUM AND DIFFERENCE OF TWO TERMS

Find each product.

(a) $(x + 4)(x - 4)$
Use the pattern for the sum and difference of two terms.

$$(x + 4)(x - 4) = x^2 - 4^2 = x^2 - 16$$

(b) $(3 - w)(3 + w)$
By the commutative property this product is the same as $(3 + w)(3 - w)$.

$$(3 - w)(3 + w) = (3 + w)(3 - w) = 3^2 - w^2 = 9 - w^2$$

(c) $(a - b)(a + b) = a^2 - b^2$. ■

EXAMPLE 4
FINDING THE PRODUCT OF THE SUM AND DIFFERENCE OF TWO TERMS

Find each product.

(a) $(5m + 3)(5m - 3)$
Use the rule for the product of the sum and difference of two terms.

$$(5m + 3)(5m - 3) = (5m)^2 - 3^2 = 25m^2 - 9$$

(b) $(4x + y)(4x - y) = (4x)^2 - y^2 = 16x^2 - y^2$

(c) $\left(z - \frac{1}{4}\right)\left(z + \frac{1}{4}\right) = z^2 - \frac{1}{16}$ ■

The product formulas of this section will be very useful in later work, particularly in Chapter 4. Therefore, it is important to memorize these formulas and practice using them.

3 The methods used in the previous section and this section can be combined to find higher powers of binomials.

EXAMPLE 5
FINDING HIGHER POWERS OF BINOMIALS

Find each product.

(a) $(x + 5)^3$

$$\begin{aligned}
(x + 5)^3 &= (x + 5)^2(x + 5) && a^3 = a^2 \cdot a \\
&= (x^2 + 10x + 25)(x + 5) && \text{Square the binomial.} \\
&= x^3 + 10x^2 + 25x + 5x^2 + 50x + 125 && \text{Multiply polynomials.} \\
&= x^3 + 15x^2 + 75x + 125 && \text{Combine terms.}
\end{aligned}$$

(b) $(2y - 3)^4$

$$\begin{aligned}(2y-3)^4 &= (2y-3)^2(2y-3)^2 && a^4 = a^2 \cdot a^2\\ &= (4y^2 - 12y + 9)(4y^2 - 12y + 9) && \text{Square each binomial.}\\ &= 16y^4 - 48y^3 + 36y^2 - 48y^3 + 144y^2 && \text{Multiply the polynomials.}\\ &\quad - 108y + 36y^2 - 108y + 81 &&\\ &= 16y^4 - 96y^3 + 216y^2 - 216y + 81 && \text{Combine terms.} \blacksquare\end{aligned}$$

3.4 EXERCISES

Find each square. See Examples 1 and 2.

1. $(m + 2)^2$
2. $(x + 8)^2$
3. $(r - 3)^2$
4. $(z - 5)^2$
5. $(x + 2y)^2$
6. $(3m - p)^2$
7. $(5p + 2q)^2$
8. $(8a - 3b)^2$
9. $(4a + 5b)^2$
10. $(9y + z)^2$
11. $(3r + 2s)^2$
12. $(6m - 7k)^2$

Find the following products. See Examples 3 and 4.

13. $(a + 8)(a - 8)$
14. $(k + 5)(k - 5)$
15. $(2 + p)(2 - p)$
16. $(4 - 3t)(4 + 3t)$
17. $(2m + 5)(2m - 5)$
18. $(2b + 5)(2b - 5)$
19. $(3x + 4y)(3x - 4y)$
20. $(6a - p)(6a + p)$
21. $(5y + 3x)(5y - 3x)$
22. $(4c + 5d)(4c - 5d)$
23. $(3h - 2j)(3h + 2j)$
24. $(6m + 7n)(6m - 7n)$

Find each product. See Example 5.

25. $(m - 5)^3$
26. $(p + 3)^3$
27. $(2a + 1)^3$
28. $(3m - 1)^3$
29. $(y + 4)^4$
30. $(z - 2)^4$
31. $(3r - 2t)^4$
32. $(2z + 5y)^4$

Find the products in Exercises 33–40.

33. $\left(2z - \frac{5}{2}x\right)^2$
34. $\left(6a - \frac{3}{2}b\right)^2$
35. $\left(2m - \frac{5}{3}\right)\left(2m + \frac{5}{3}\right)$
36. $\left(3a - \frac{4}{5}\right)\left(3a + \frac{4}{5}\right)$
37. $(x^2 - 1)(x^2 + 1)$
38. $(2 + m^3)(2 - m^3)$
39. $(7y^2 + 10z)(7y^2 - 10z)$
40. $(6x + 5y^2)(6x - 5y^2)$

41. Let $a = 2$ and $b = 5$. Evaluate $(a + b)^2$ and $a^2 + b^2$. Does $(a + b)^2 = a^2 + b^2$? (See Objective 1.)

42. Let $p = 7$ and $q = 3$. Evaluate $(p - q)^2$ and $p^2 - q^2$. Does $(p - q)^2 = p^2 - q^2$? (See Objective 1.)

Write a mathematical expression for the area of each of the following geometric figures. (See Objectives 1 and 2.)

43.

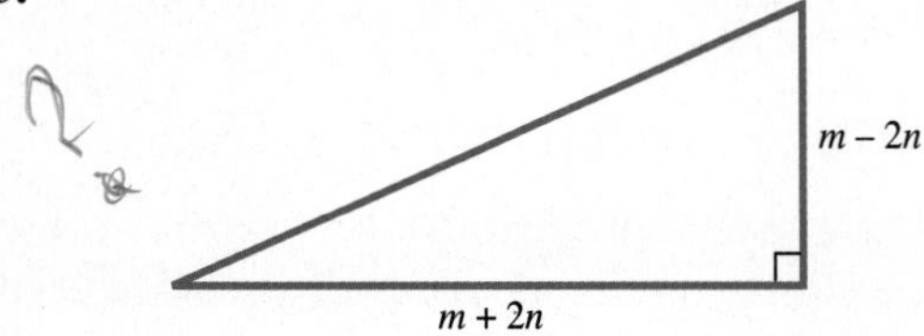

44.

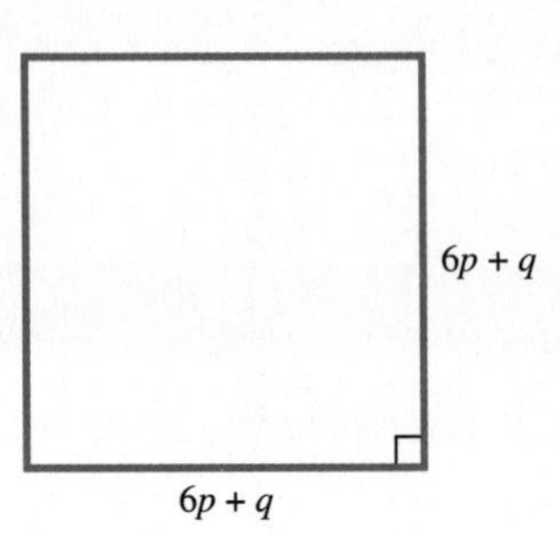

45.

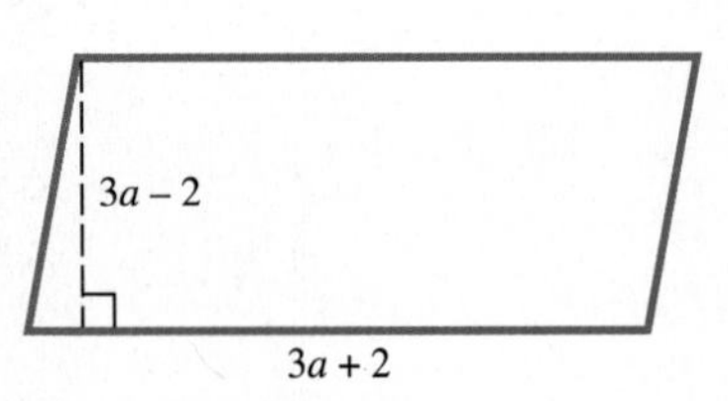

46.

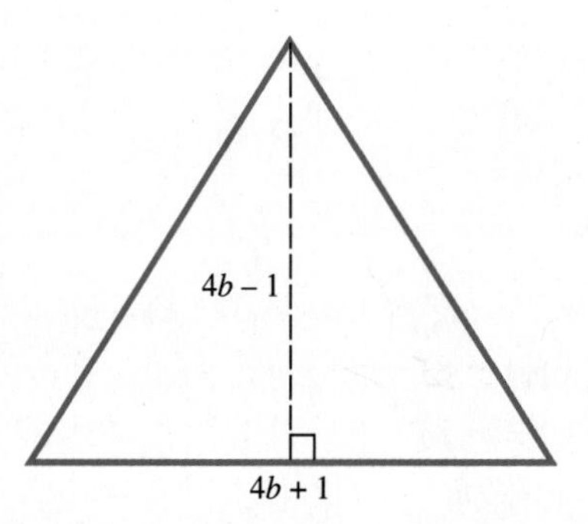

47. Explain how the expressions $x^2 + y^2$ and $(x + y)^2$ differ. (See Objective 1.)

48. Does $a^3 + b^3 = (a + b)^3$? Explain your answer. (See Objective 3.)

PREVIEW EXERCISES

Give the reciprocal of each of the following quantities. See Section 1.1.

49. 5

50. -2

51. $-\frac{1}{4}$

52. $\frac{1}{8}$

53. $\frac{3}{4}$

54. $-\frac{5}{2}$

55. $-\frac{3}{8}$

56. $\frac{7}{2}$

3.5 THE QUOTIENT RULE AND INTEGER EXPONENTS

OBJECTIVES

1 USE ZERO AS AN EXPONENT.
2 USE NEGATIVE NUMBERS AS EXPONENTS.
3 USE THE QUOTIENT RULE FOR EXPONENTS.
4 USE COMBINATIONS OF RULES.
5 USE VARIABLES AS EXPONENTS.

1 In Section 3.2 we studied the product rule for exponents. The rule for division with exponents is similar to the product rule for exponents. For example,

$$\frac{6^5}{6^2} = \frac{6 \cdot 6 \cdot 6 \cdot 6 \cdot 6}{6 \cdot 6} = 6 \cdot 6 \cdot 6 = 6^3.$$

The difference between the exponents, $5 - 2$, gives the exponent in the quotient, 3. Also,

$$\frac{m^4}{m^2} = \frac{m \cdot m \cdot m \cdot m}{m \cdot m} = m \cdot m = m^2,$$

and $4 - 2 = 2$. Generalizing from these examples, the difference between the exponents gives the exponent of the answer.

If the exponents in the numerator and denominator are equal, then, for example,

$$\frac{6^5}{6^5} = \frac{6 \cdot 6 \cdot 6 \cdot 6 \cdot 6}{6 \cdot 6 \cdot 6 \cdot 6 \cdot 6} = 1.$$

If, however, the exponents are subtracted as above,

$$\frac{6^5}{6^5} = 6^{5-5} = 6^0.$$

This means that 6^0 should equal 1. A zero exponent is defined as follows.

DEFINITION OF ZERO EXPONENT

For any nonzero real number a, $a^0 = 1$.
Example: $17^0 = 1$.

EXAMPLE 1 USING ZERO EXPONENTS

Evaluate each exponential expression.

(a) $60^0 = 1$

(b) $(-60)^0 = 1$

(c) $-(60^0) = -(1) = -1$

(d) $y^0 = 1$, if $y \neq 0$.

(e) $6y^0 = 6(1) = 6$, if $y \neq 0$

(f) $(6y)^0 = 1$, if $y \neq 0$ ■

CAUTION Notice the difference between parts (b) and (c) of Example 1. In Example 1(b) the base is -60 and the exponent is 0. Any nonzero base raised to a zero exponent is 1. But in Example 1(c), the base is 60. Then $60^0 = 1$, and $-60^0 = -1$.

2 The discussion at the beginning of this section showed that

$$\frac{6^5}{6^2} = 6^3,$$

where the exponent in the denominator was subtracted from the exponent in the numerator. If the denominator exponent were larger than the numerator exponent, subtracting would result in a negative exponent.

For example,

$$\frac{6^2}{6^5} = 6^{2-5} = 6^{-3}.$$

On the other hand,

$$\frac{6^2}{6^5} = \frac{6 \cdot 6}{6 \cdot 6 \cdot 6 \cdot 6 \cdot 6} = \frac{1}{6 \cdot 6 \cdot 6} = \frac{1}{6^3},$$

so that

$$6^{-3} = \frac{1}{6^3}.$$

This example suggests that **negative exponents** be defined as follows.

DEFINITION OF A NEGATIVE EXPONENT

For any nonzero real number a and any integer n, $\quad a^{-n} = \frac{1}{a^n}.$

Example: $3^{-2} = \frac{1}{3^2}$

By definition, a^{-n} and a^n are reciprocals, since

$$a^n \cdot a^{-n} = a^n \cdot \frac{1}{a^n} = 1.$$

The definition of a^{-n} also can be written as

$$a^{-n} = \frac{1}{a^n} = \left(\frac{1}{a}\right)^n.$$

For example, using the last result above,

$$6^{-3} = \left(\frac{1}{6}\right)^3 \quad \text{and} \quad \left(\frac{1}{3}\right)^{-2} = 3^2.$$

EXAMPLE 2 USING NEGATIVE EXPONENTS Simplify by using the definition of negative exponents.

(a) $3^{-2} = \frac{1}{3^2} = \frac{1}{9}$

(b) $5^{-3} = \frac{1}{5^3} = \frac{1}{125}$

(c) $\left(\frac{1}{2}\right)^{-3} = 2^3 = 8$

(d) $\left(\frac{2}{5}\right)^{-4} = \left(\frac{5}{2}\right)^4$ $\quad \frac{2}{5}$ and $\frac{5}{2}$ are reciprocals.

(e) $4^{-1} - 2^{-1} = \frac{1}{4} - \frac{1}{2} = \frac{1}{4} - \frac{2}{4} = -\frac{1}{4}$

(f) $p^{-2} = \frac{1}{p^2}, \quad p \neq 0$

(g) $\frac{1}{x^{-4}}, \quad x \neq 0$

$$\frac{1}{x^{-4}} = \frac{1^{-4}}{x^{-4}} \qquad 1^{-4} = 1$$

$$= \left(\frac{1}{x}\right)^{-4} \qquad \text{Power rule (c)}$$

$$= x^4 \qquad \frac{1}{x} \text{ and } x \text{ are reciprocals.}$$

(h) $\frac{2}{x^{-4}}, \quad x \neq 0$

Write $\frac{2}{x^{-4}}$ as $2 \cdot \frac{1}{x^{-4}}$. Then use the result from part (g).

$$\frac{2}{x^{-4}} = 2 \cdot \frac{1}{x^{-4}} = 2 \cdot x^4 \text{ or } 2x^4 \quad ■$$

CAUTION A negative exponent does not indicate a negative number; negative exponents lead to reciprocals.

EXPRESSION	EXAMPLE	
a^{-n}	$3^{-2} = \frac{1}{3^2} = \frac{1}{9}$	Not negative
$-a^{-n}$	$-3^{-2} = -\frac{1}{3^2} = -\frac{1}{9}$	Negative

Examples 2(g) and (h) suggest that the definition of negative exponent allows us to move factors in a fraction between the numerator and denominator if we also change the sign of the exponents. For example,

$$\frac{2^{-3}}{3^{-4}} = \frac{\frac{1}{2^3}}{\frac{1}{3^4}} \quad \text{Definition of negative exponent}$$

$$= \frac{1}{2^3} \cdot \frac{3^4}{1} \quad \text{Division by a fraction}$$

$$\frac{2^{-3}}{3^{-4}} = \frac{3^4}{2^3} \quad \text{Multiplication of fractions}$$

This fact is generalized below.

CHANGING FROM NEGATIVE TO POSITIVE EXPONENTS

For any nonzero numbers a and b, and any integers m and n,

$$\frac{a^{-m}}{b^{-n}} = \frac{b^n}{a^m}.$$

Example: $\frac{3^{-5}}{2^{-4}} = \frac{2^4}{3^5}$

EXAMPLE 3

CHANGING FROM NEGATIVE TO POSITIVE EXPONENTS

Write with only positive exponents. Assume all variables represent nonzero real numbers.

(a) $\frac{4^{-2}}{5^{-3}} = \frac{5^3}{4^2}$

(b) $\frac{m^{-5}}{p^{-1}} = \frac{p^1}{m^5} = \frac{p}{m^5}$

(c) $\frac{a^{-2}b}{3d^{-3}} = \frac{bd^3}{3a^2}$

(d) $x^3y^{-4} = \frac{x^3y^{-4}}{1} = \frac{x^3}{y^4}$ ■

3 Now that zero and negative exponents have been defined, the **quotient rule for exponents** can be given.

QUOTIENT RULE FOR EXPONENTS

For any nonzero real number a and any integers m and n,

$$\frac{a^m}{a^n} = a^{m-n}.$$

Example: $\frac{5^8}{5^4} = 5^{8-4} = 5^4$

EXAMPLE 4
USING THE QUOTIENT RULE FOR EXPONENTS

Simplify by using the quotient rule for exponents. Write answers with positive exponents.

(a) $\dfrac{5^8}{5^6} = 5^{8-6} = 5^2$

(b) $\dfrac{4^2}{4^9} = 4^{2-9} = 4^{-7} = \dfrac{1}{4^7}$

(c) $\dfrac{5^{-3}}{5^{-7}} = 5^{-3-(-7)} = 5^4$ or $\dfrac{5^{-3}}{5^{-7}} = \dfrac{5^7}{5^3} = 5^{7-3} = 5^4$

(d) $\dfrac{q^5}{q^{-3}} = q^{5-(-3)} = q^8, \quad q \neq 0$

(e) $\dfrac{3^2x^5}{3^4x^3} = \dfrac{3^2}{3^4} \cdot \dfrac{x^5}{x^3} = 3^{2-4} \cdot x^{5-3}$

$= 3^{-2}x^2 = \dfrac{x^2}{3^2} = \dfrac{x^2}{9}, \quad x \neq 0$ ■

The definitions and rules for exponents given in this section and Section 3.2 are summarized below.

DEFINITIONS AND RULES FOR EXPONENTS

If no denominators are zero, for any integers m and n:

			EXAMPLES
Product rule		$a^m \cdot a^n = a^{m+n}$	$7^4 \cdot 7^3 = 7^7$
Zero exponent		$a^0 = 1$	$(-3)^0 = 1$
Negative exponent		$a^{-n} = \dfrac{1}{a^n}$	$5^{-3} = \dfrac{1}{5^3}$
Quotient rule		$\dfrac{a^m}{a^n} = a^{m-n}$	$\dfrac{2^2}{2^5} = 2^{-3} = \dfrac{1}{2^3}$
Power rules	(a)	$(a^m)^n = a^{mn}$	$(4^2)^3 = 4^6$
	(b)	$(ab)^m = a^mb^m$	$(3k)^4 = 3^4k^4$
	(c)	$\left(\dfrac{a}{b}\right)^m = \dfrac{a^m}{b^m}$	$\left(\dfrac{2}{3}\right)^{10} = \dfrac{2^{10}}{3^{10}}$
	(d)	$\dfrac{a^{-m}}{b^{-n}} = \dfrac{b^n}{a^m}$	$\dfrac{5^{-3}}{3^{-5}} = \dfrac{3^5}{5^3}$
	(e)	$\left(\dfrac{a}{b}\right)^{-m} = \left(\dfrac{b}{a}\right)^m$	$\left(\dfrac{4}{7}\right)^{-2} = \left(\dfrac{7}{4}\right)^2$

4 As shown in the next example, sometimes we may need to use more than one rule to simplify an expression.

EXAMPLE 5 USING A COMBINATION OF RULES

Use a combination of the rules for exponents to simplify each expression.

(a) $\dfrac{(4^2)^3}{4^5}$

Use power rule (a) and then the quotient rule.

$$\frac{(4^2)^3}{4^5} = \frac{4^6}{4^5} = 4^{6-5} = 4^1 = 4$$

(b) $(2x)^3(2x)^2$

Use the product rule first. Then use power rule (b).

$$(2x)^3(2x)^2 = (2x)^5 = 2^5x^5 \quad \text{or} \quad 32x^5$$

(c) $\left(\dfrac{2x^3}{5}\right)^{-4}$

Use power rules (b), (c), and (e).

$$\left(\frac{2x^3}{5}\right)^{-4} = \left(\frac{5}{2x^3}\right)^4$$ Change the base to its reciprocal and change the sign of the exponent.

$$= \frac{5^4}{2^4(x^3)^4}$$ Power rules

$$= \frac{5^4}{16x^{12}}$$ Power rule (a)

(d) $$\left(\frac{3x^{-2}}{4^{-1}y^3}\right)^{-3} = \frac{3^{-3}x^6}{4^3y^{-9}}$$ Power rules

$$= \frac{x^6y^9}{3^3 \cdot 4^3}$$ Change the base to its reciprocal and change the sign of the exponent.

$$= \frac{x^6y^9}{27(64)}$$ Definition of exponent

$$= \frac{x^6y^9}{1728}$$ ■

NOTE Since the steps can be done in several different orders, there are many equally good ways to simplify a problem like Example 5(d).

5 All the rules given in the box also apply when variables are used as exponents, as long as the variables represent only integers, an assumption we shall make throughout this chapter.

EXAMPLE 6 USING VARIABLE EXPONENTS

Use the rules for exponents to simplify the following.

(a) $3x^k \cdot 2x^5 = 3 \cdot 2 \cdot x^k \cdot x^5 = 6x^{k+5}$

(b) $\dfrac{y^{7z}}{y^{3z}} = y^{7z-3z} = y^{4z}$

(c) $(2a^4)^r = 2^r \cdot a^{4r} = 2^r a^{4r}$

(d) $z^q \cdot z^{5q} \cdot z^{-4q} = z^{q+5q+(-4q)} = z^{2q}$

(e) $(2m)^a \cdot m^{-a} = 2^a m^a m^{-a} = 2^a m^{a-a} = 2^a m^0 = 2^a \cdot 1 = 2^a$ ■

3.5 EXERCISES

Evaluate or simplify each expression. Write with only positive exponents. Assume all variables represent nonzero numbers. See Examples 1–3.

1. $4^0 + 5^0$ **2.** $3^0 + 8^0$ **3.** $(-9)^0 + 9^0$ **4.** $(-8)^0 + (-8)^0$

5. 3^{-3} **6.** 2^{-5} **7.** $(-12)^{-1}$ **8.** $(-6)^{-2}$

9. $\left(\dfrac{1}{2}\right)^{-5}$ **10.** $\left(\dfrac{1}{5}\right)^{-2}$ **11.** $\left(\dfrac{1}{2}\right)^{-1}$ **12.** $\left(\dfrac{5}{4}\right)^{-2}$

13. $(3p)^0$ **14.** $3p^0$ **15.** $-3p^{-1}$ **16.** $-3(-p^{-2})$

17. $2^{-1} + 3^{-1}$ **18.** $3^{-1} - 4^{-1}$ **19.** $5^{-1} + 4^{-1}$ **20.** $3^{-1} + 6^{-1}$

21. $\dfrac{5^{-2}}{3^{-4}}$ **22.** $\dfrac{7^{-1}}{2^{-5}}$ **23.** $\dfrac{a^{-2}}{b^{-5}}$ **24.** $\dfrac{r^{-3}}{5^{-2}}$

Use the quotient rule to simplify each expression. Write each answer with positive exponents. Assume that all variables represent nonzero real numbers. See Example 4.

25. $\dfrac{4^7}{4^2}$ **26.** $\dfrac{11^5}{11^3}$ **27.** $\dfrac{8^3}{8^9}$ **28.** $\dfrac{5^4}{5^{10}}$

29. $\dfrac{6^{-4}}{6^2}$ **30.** $\dfrac{7^{-5}}{7^3}$ **31.** $\dfrac{14^{-2}}{14^{-5}}$ **32.** $\dfrac{6^{-3}}{6^{-8}}$

33. $\dfrac{x^6}{x^{-9}}$ **34.** $\dfrac{y^2}{y^{-5}}$ **35.** $\dfrac{z}{z^{-1}}$ **36.** $\dfrac{r^{-1}}{r}$

37. $\dfrac{1}{2^{-5}}$ **38.** $\dfrac{1}{4^{-3}}$ **39.** $\dfrac{2}{k^{-2}}$ **40.** $\dfrac{5}{p^{-5}}$

Simplify each expression. Give answers with only positive exponents. Assume that all variables represent nonzero numbers. See Examples 3 and 5.

41. $\dfrac{4^3 \cdot 4^{-5}}{4^7}$ **42.** $\dfrac{2^5 \cdot 2^{-4}}{2^{-1}}$ **43.** $\dfrac{5^4}{5^{-3} \cdot 5^{-2}}$ **44.** $\dfrac{8^6}{8^{-2} \cdot 8^5}$

45. $\dfrac{64^6}{32^6}$ **46.** $\dfrac{81^5}{9^5}$ **47.** $\dfrac{(3x)^5}{x^5}$ **48.** $\dfrac{m^8}{(2m)^8}$

49. $\left(\dfrac{5m^{-2}}{m^{-1}}\right)^2$ **50.** $\left(\dfrac{4x^3}{x^{-1}}\right)^{-1}$ **51.** $\dfrac{x^7(x^8)^{-2}}{(x^{-2})^3}$ **52.** $\dfrac{(m^3)^2(m^{-2})^4}{(m^{-1})^6}$

53. $\dfrac{3^{-1}a^{-2}}{3^2a^{-4}}$ **54.** $\dfrac{2^{-5}p^{-3}}{2^{-7}p^5}$ **55.** $\dfrac{4k^{-3}m^5}{4^{-1}k^{-7}m^{-3}}$ **56.** $\dfrac{6^{-1}y^{-2}z^5}{6^2y^{-1}z^{-2}}$

Simplify each expression. Assume that all variables represent nonzero integers. Give answers with positive exponents only. See Example 6.

57. $5^r \cdot 5^{7r} \cdot 5^{-2r}$ **58.** $6^{5p} \cdot 6^p \cdot 6^{-2p}$ **59.** $x^{-a} \cdot x^{-3a} \cdot x^{-7a}$ **60.** $\dfrac{a^{6y}}{a^{2y}}$

61. $\dfrac{q^{-3k}}{q^{-8k}}$ **62.** $\dfrac{z^{-7m}}{z^{-12m}}$ **63.** $(6 \cdot p^{-3})^{-y}$ **64.** $(2 \cdot a^{-p})^4$

Evaluate each expression.

65. $2 \cdot 3^{-1} + 4 \cdot 2^{-1}$ **66.** $5 \cdot 4^{-2} + 3 \cdot 2^{-3}$ **67.** $-4 \cdot 2^{-2} + 3 \cdot 2^{-3}$ **68.** $4 \cdot 3^{-1} - 2 \cdot 3^{-2}$

Simplify each expression. Give answers with only positive exponents. Assume all variables are nonzero real numbers.

69. $\dfrac{(4a^2b^3)^{-2}(2ab^{-1})^3}{(a^3b)^{-4}}$ **70.** $\dfrac{(m^6n)^{-2}(m^2n^{-2})^3}{m^{-1}n^{-2}}$

71. $\dfrac{(2y^{-1}z^2)^2(3y^{-2}z^{-3})^3}{(y^3z^2)^{-1}}$ **72.** $\dfrac{(3p^{-2}q^3)^2(5p^{-1}q^{-4})^{-1}}{(p^2q^{-2})^{-3}}$

---

73. Is the expression $2x^3 - 4x^2 + x^{-1}$ a polynomial? Why? (See Objective 2.)

74. What is the reciprocal of a^{-n}? Explain. (See Objective 2.)

75. If $a^{-3} = 10$, what is $\dfrac{1}{a^3}$? (See Objective 2.)

PREVIEW EXERCISES

Use the distributive property to write each product as a sum of terms. Write answers with positive exponents only. Simplify each term. See Section 1.9.

76. $\dfrac{1}{2p}(4p^2 + 2p + 8)$ **77.** $\dfrac{1}{5x}(5x^2 - 10x + 45)$

78. $\dfrac{1}{3m}(m^3 + 9m^2 - 6m)$ **79.** $\dfrac{1}{4y}(y^4 + 6y^2 + 8)$

3.6 THE QUOTIENT OF A POLYNOMIAL AND A MONOMIAL

OBJECTIVE

1 DIVIDE A POLYNOMIAL BY A MONOMIAL.

1 The quotient rule for exponents is used to divide a monomial by another monomial. For example,

$$\frac{12x^2}{6x} = 2x, \qquad \frac{25m^5}{5m^2} = 5m^3, \qquad \text{and} \qquad \frac{30a^2b^8}{15a^3b^3} = \frac{2b^5}{a}.$$

These examples suggest the following rule.

DIVIDING A POLYNOMIAL BY A MONOMIAL

To divide a polynomial by a monomial, divide each term of the polynomial by the monomial:

$$\frac{a+b}{c} = \frac{a}{c} + \frac{b}{c} \qquad (c \neq 0).$$

EXAMPLE 1 **DIVIDING A POLYNOMIAL BY A MONOMIAL**

Divide $5m^5 - 10m^3$ by $5m^2$.

Use the rule above, with + replaced by −.

$$\frac{5m^5 - 10m^3}{5m^2} = \frac{5m^5}{5m^2} - \frac{10m^3}{5m^2} = m^3 - 2m$$

Recall from arithmetic that division problems can be checked by multiplication:

$$\frac{63}{7} = 9 \quad \text{because} \quad 7 \cdot 9 = 63.$$

To check the polynomial quotient, multiply $m^3 - 2m$ by $5m^2$. Since

$$5m^2(m^3 - 2m) = 5m^5 - 10m^3,$$

the quotient is correct.

Since division by 0 is undefined, the quotient

$$\frac{5m^5 - 10m^3}{5m^2}$$

is undefined if $m = 0$. In the rest of the chapter, assume that no denominators are 0. ■

EXAMPLE 2 **DIVIDING A POLYNOMIAL BY A MONOMIAL**

Divide $\dfrac{16a^5 - 12a^4 + 8a^2}{4a^3}$.

Divide each term of $16a^5 - 12a^4 + 8a^2$ by $4a^3$.

$$\frac{16a^5 - 12a^4 + 8a^2}{4a^3} = \frac{16a^5}{4a^3} - \frac{12a^4}{4a^3} + \frac{8a^2}{4a^3}$$

$$= 4a^2 - 3a + \frac{2}{a}$$

The result is not a polynomial because of the expression $2/a$, which has a variable in the denominator. While the sum, difference, and product of two polynomials are always polynomials, the quotient of two polynomials may not be.

Again, check by multiplying.

$$4a^3\left(4a^2 - 3a + \frac{2}{a}\right) = 4a^3(4a^2) - 4a^3(3a) + 4a^3\left(\frac{2}{a}\right)$$

$$= 16a^5 - 12a^4 + 8a^2 \quad ■$$

EXAMPLE 3 **DIVIDING A POLYNOMIAL BY A MONOMIAL**

Divide $-7x^3 + 12x^4 + 4x$ by $4x$.

The polynomial should be written in descending order before dividing. Write it as $12x^4 - 7x^3 + 4x$; then divide by $4x$.

$$\frac{12x^4 - 7x^3 + 4x}{4x} = \frac{12x^4}{4x} - \frac{7x^3}{4x} + \frac{4x}{4x}$$

$$= 3x^3 - \frac{7x^2}{4} + 1 = 3x^3 - \frac{7}{4}x^2 + 1$$

Check by multiplication. ■

CAUTION In Example 3, notice that the quotient $\frac{4x}{4x} = 1$. It is a common error to leave that term out of the answer. Checking by multiplication will show that the answer $3x^3 - \frac{7}{4}x^2$ is not correct.

EXAMPLE 4 **DIVIDING A POLYNOMIAL BY A MONOMIAL**

Divide $180y^{10} - 150y^8 + 120y^6 - 90y^4 + 100y$ by $-30y^2$.

Using the methods of this section,

$$\frac{180y^{10} - 150y^8 + 120y^6 - 90y^4 + 100y}{-30y^2}$$

$$= \frac{180y^{10}}{-30y^2} - \frac{150y^8}{-30y^2} + \frac{120y^6}{-30y^2} - \frac{90y^4}{-30y^2} + \frac{100y}{-30y^2}$$

$$= -6y^8 + 5y^6 - 4y^4 + 3y^2 - \frac{10}{3y}.$$

To check, multiply this result by $-30y^2$. ■

3.6 EXERCISES

Divide.

1. $\frac{4x^2}{2x}$
2. $\frac{8m^7}{2m}$
3. $\frac{10a^3}{5a}$
4. $\frac{36p^8}{4p^3}$
5. $\frac{27k^4m^5}{3km^6}$
6. $\frac{18x^5y^6}{3x^2y^2}$
7. $\frac{-15m^3p^2}{5mp^4}$
8. $\frac{-32a^4b^5}{4a^5b}$

Divide each polynomial by $2m$. *See Examples 1–4.*

9. $60m^4 - 20m^2$
10. $120m^6 - 60m^3 + 80m^2$
11. $10m^5 - 16m^2 + 8m^3$
12. $6m^5 - 4m^3 + 2m^2$
13. $8m^5 - 4m^3 + 4m^2$
14. $8m^3 - 4m^2 + 6m$
15. $2m^5 - 4m^2 + 8m$
16. $m^2 + m + 1$
17. $2m^2 - 2m + 5$

Divide each polynomial by $-5q$. *See Examples 1–4.*

18. $-30q^4 + 15q^3 - 10q$
19. $25q^3 - 10q^2 + 5q$
20. $15q^6 - 30q^4 + 5q^2 - q$
21. $-40q^5 + 35q^3 - 20$
22. $q^5 - 10q^3 + 8q^2$
23. $3q^4 + 5q^2 - 6q$
24. $q^3 + q^2 - 9q + 1$
25. $5q^3 - 3q^2 + 2q - 7$
26. $3q^5 - 2q^4 + 10q^2$

Divide each polynomial by $3x^2$. *See Examples 1–4.*

27. $3x^4 + 9x^3$
28. $15x^2 - 9x^3$
29. $12x^4 + 3x^2 - 3x^3$
30. $45x^3 + 15x^2 - 9x^5$
31. $36x + 24x^2 + 3x^3$
32. $4x^4 - 3x^3 + 2x$
33. $x^3 + 6x^2 - x$
34. $-3x^4 + 6x^5 + 2 + 9x^2$

Perform each division in Exercises 35–42. See Examples 1–4.

35. $\frac{8k^4 - 12k^3 - 2k^2 + 7k - 3}{2k}$
36. $\frac{27r^4 - 36r^3 - 6r^2 + 26r - 2}{3r}$
37. $\frac{100p^5 - 50p^4 + 30p^3 - 30p}{-10p^2}$
38. $\frac{2m^5 - 6m^4 + 8m^2}{-2m^3}$
39. $(16y^5 - 8y^2 + 12y) \div (4y^2)$
40. $(20a^4 - 15a^5 + 25a^3) \div (15a^4)$
41. $(120x^{11} - 60x^{10} + 140x^9 - 100x^8) \div (10x^{12})$
42. $(5 + x + 6x^2 + 8x^3) \div (3x^4)$

---

43. Evaluate $\frac{5y + 6}{2}$ when $y = 2$. Evaluate $5y + 3$ when $y = 2$. Does $\frac{5y + 6}{2} = 5y + 3$? (See Objective 1.)

44. Evaluate $\frac{10r + 7}{5}$ when $r = 1$. Evaluate $2r + 7$ when $r = 1$. Does $\frac{10r + 7}{5} = 2r + 7$? (See Objective 1.)

Solve each problem.

45. What polynomial, when divided by $3x^2$, yields $4x^3 + 3x^2 - 4x + 2$ as a quotient?

46. What polynomial, when divided by $4m^3$, yields $-6m^2 + 4m$ as a quotient?

47. The quotient of a certain polynomial and $-7y^2$ is $9y^2 + 3y + 5 - 2/y$. Find the polynomial.

48. The quotient of a certain polynomial and a is $2a^2 + 3a + 5$. Find the polynomial.

49. If a polynomial in x of degree 4 is divided by a monomial in x of degree 2, what is the degree of the quotient? (Assume the quotient is a polynomial.) (See Objective 1.)

50. Suppose a polynomial of degree n is divided by a monomial of degree m to get a *polynomial* quotient. What is the degree of the quotient? Explain. (See Objective 1.)

PREVIEW EXERCISES

Find each product. See Section 3.2.

51. $x(2x^2 - 5x + 1)$

52. $m(3m^3 - 2m^2 + 5)$

53. $-4k(5k^3 - 3k^2 + 2k)$

54. $-7z(z^2 - 5z + 6)$

55. $3m^5(2m^5 - 4m^3 + m^2)$

56. $2p^3(8p^2 - 4p + 9)$

Subtract. See Section 3.1.

57. $\begin{array}{r} 8x - 5 \\ \underline{3x - 7} \end{array}$

58. $\begin{array}{r} 5a - 4 \\ \underline{7a - 2} \end{array}$

59. $\begin{array}{r} 6x^2 - 4 \\ \underline{8x^2 + 7} \end{array}$

60. $\begin{array}{r} 3k^2 - 4k + 1 \\ \underline{4k^2 + 7k - 5} \end{array}$

61. $\begin{array}{r} 9x^2 - 4x - 7 \\ \underline{2x^2 + 6x - 8} \end{array}$

62. $\begin{array}{r} 4m^2 - 5m - 1 \\ \underline{3m^2 + 2m - 5} \end{array}$

3.7 THE QUOTIENT OF TWO POLYNOMIALS

OBJECTIVE

1 DIVIDE A POLYNOMIAL BY A POLYNOMIAL.

1 A method of "long division" is used to divide a polynomial by a polynomial (other than a monomial). This method is similar to the method of long division used for two whole numbers. For comparison, the division of whole numbers is shown alongside the division of polynomials. Both polynomials should be written with descending powers before beginning the division process.

Step 1

Divide 27 into 6696.

$$27\overline{)6696}$$

Divide $2x + 3$ into $8x^3 - 4x^2 - 14x + 15$.

$$2x + 3\overline{)8x^3 - 4x^2 - 14x + 15}$$

Step 2

27 divides into 66 **2** times; $2 \cdot 27 =$ **54**.

$$\begin{array}{r} 2 \\ 27\overline{)6696} \\ \underline{54} \end{array}$$

$2x$ divides into $8x^3$ $\mathbf{4x^2}$ times; $4x^2(2x + 3) = \mathbf{8x^3 + 12x^2}$.

$$\begin{array}{r} 4x^2 \\ 2x + 3\overline{)8x^3 - 4x^2 - 14x + 15} \\ \underline{8x^3 + 12x^2} \end{array}$$

Step 3

Subtract: $66 - 54 = 12$; then bring down the next digit.

$$\begin{array}{r} 2 \\ 27\overline{)6696} \\ \underline{54} \\ 129 \end{array}$$

Subtract: $-4x^2 - 12x^2 = -16x^2$; then bring down the next term.

$$\begin{array}{r} 4x^2 \\ 2x + 3\overline{)8x^3 - 4x^2 - 14x + 15} \\ \underline{8x^3 + 12x^2} \\ -16x^2 - 14x \end{array}$$

(To subtract two polynomials, change the sign of the second and then add.)

Step 4

27 divides into 129 **4** times; $4 \cdot 27 =$ **108**.

$$\begin{array}{r} 24 \\ 27\overline{)6696} \\ \underline{54} \\ 129 \\ \underline{108} \end{array}$$

$2x$ divides into $-16x^2$ $\mathbf{-8x}$ times; $-8x(2x + 3) = \mathbf{-16x^2 - 24x}$.

$$\begin{array}{r} 4x^2 - 8x \\ 2x + 3\overline{)8x^3 - 4x^2 - 14x + 15} \\ \underline{8x^3 + 12x^2} \\ -16x^2 - 14x \\ \underline{-16x^2 - 24x} \end{array}$$

Step 5

Subtract: $129 - 108 = 21$; then bring down the next digit.

$$\begin{array}{r} 24 \\ 27\overline{)6696} \\ \underline{54} \\ 129 \\ \underline{108} \\ 216 \end{array}$$

Subtract: $-14x - (-24x) = 10x$; then bring down the next term.

$$\begin{array}{r} 4x^2 - 8x \\ 2x + 3\overline{)8x^3 - 4x^2 - 14x + 15} \\ \underline{8x^3 + 12x^2} \\ -16x^2 - 14x \\ \underline{-16x^2 - 24x} \\ 10x + 15 \end{array}$$

Step 6

27 divides into 216 **8** times; $8 \cdot 27 =$ **216**.

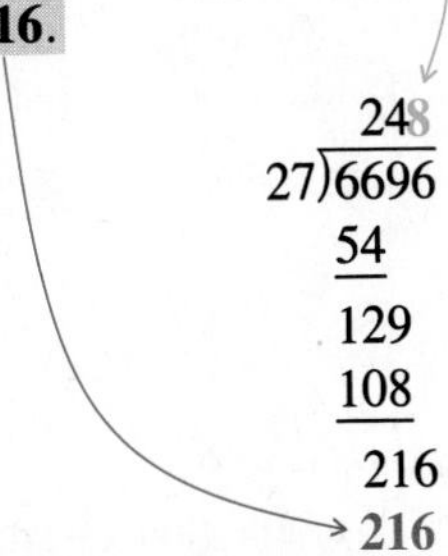

$2x$ divides into $10x$ **5** times; $5(2x + 3) =$ $\mathbf{10x + 15}$.

$$\begin{array}{r} 4x^2 - 8x + 5 \\ 2x + 3\overline{)8x^3 - 4x^2 - 14x + 15} \\ \underline{8x^3 + 12x^2} \\ -16x^2 - 14x \\ \underline{-16x^2 - 24x} \\ 10x + 15 \\ \underline{10x + 15} \end{array}$$

6696 divided by 27 is 248. There is no remainder.

$8x^3 - 4x^2 - 14x + 15$ divided by $2x + 3$ is $4x^2 - 8x + 5$. There is no remainder.

Step 7

Check by multiplication.
$27 \cdot 248 = 6696$

Check by multiplication.
$(2x + 3)(4x^2 - 8x + 5)$
$= 8x^3 - 4x^2 - 14x + 15$

Notice that at each step in the polynomial division process, the *first* term was divided into the *first* term.

EXAMPLE 1
DIVIDING A POLYNOMIAL BY A POLYNOMIAL

Divide $4x^3 - 4x^2 + 5x - 8$ by $2x - 1$.

$$\begin{array}{r} 2x^2 - x + 2 \\ 2x - 1\overline{)4x^3 - 4x^2 + 5x - 8} \\ \underline{4x^3 - 2x^2} \quad\quad\quad\quad \\ -2x^2 + 5x \quad\quad \\ \underline{-2x^2 + x} \quad\quad \\ 4x - 8 \\ \underline{4x - 2} \\ -6 \end{array}$$

Step 1 $2x$ divides into $4x^3$ $\mathbf{(2x^2)}$ times; $2x^2(2x - 1) = 4x^3 - 2x^2$.
Step 2 Subtract; bring down the next term.
Step 3 $2x$ divides into $-2x^2$ $\mathbf{(-x)}$ times; $-x(2x - 1) = -2x^2 + x$.
Step 4 Subtract; bring down the next term.
Step 5 $2x$ divides into $4x$ **2** times; $2(2x - 1) = 4x - 2$.
Step 6 Subtract. The remainder is -6.

Thus, $2x - 1$ divides into $4x^3 - 4x^2 + 5x - 8$ with a quotient of $2x^2 - x + 2$ and a remainder of -6. Write the remainder as a fraction with $2x - 1$ as the denominator. The result is not a polynomial because of the remainder.

$$\frac{4x^3 - 4x^2 + 5x - 8}{2x - 1} = 2x^2 - x + 2 + \frac{-6}{2x - 1}$$

Step 7

Check by multiplication.

$$(2x - 1)\left(2x^2 - x + 2 + \frac{-6}{2x - 1}\right)$$
$$= (2x - 1)(2x^2) + (2x - 1)(-x) + (2x - 1)(2) + (2x - 1)\left(\frac{-6}{2x - 1}\right)$$
$$= 4x^3 - 2x^2 - 2x^2 + x + 4x - 2 - 6$$
$$= 4x^3 - 4x^2 + 5x - 8 \quad \blacksquare$$

EXAMPLE 2
DIVIDING INTO A POLYNOMIAL WITH MISSING TERMS

Divide $x^3 + 2x - 3$ by $x - 1$.

Here the polynomial $x^3 + 2x - 3$ is missing the x^2 term. When terms are missing, use 0 as the coefficient for the missing terms.

$$x^3 + 2x - 3 = x^3 + 0x^2 + 2x - 3$$

Now divide.

$$
\begin{array}{r}
x^2 + x + 3 \\
x - 1\overline{)x^3 \quad 0x^2 + 2x - 3} \\
\underline{x^3 - x^2} \\
x^2 + 2x \\
\underline{x^2 - x} \\
3x - 3 \\
\underline{3x - 3}
\end{array}
$$

The remainder is 0. The quotient is $x^2 + x + 3$. Check by multiplication.

$$(x^2 + x + 3)(x - 1) = x^3 + 2x - 3 \quad \blacksquare$$

EXAMPLE 3

DIVIDING BY A POLYNOMIAL WITH MISSING TERMS

Divide $x^4 + 2x^3 + 2x^2 - x - 1$ by $x^2 + 1$.

Since $x^2 + 1$ has a missing x term, write it as $x^2 + 0x + 1$. Then proceed through the division process.

$$
\begin{array}{r}
x^2 + 2x + 1 \\
x^2 + 0x + 1\overline{)x^4 + 2x^3 + 2x^2 - x - 1} \\
\underline{x^4 + 0x^3 + x^2} \phantom{- x - 1} \\
2x^3 + x^2 - x \\
\underline{2x^3 + 0x^2 + 2x} \\
x^2 - 3x - 1 \\
\underline{x^2 + 0x + 1} \\
-3x - 2
\end{array}
$$

When the result of subtracting ($-3x - 2$, in this case) is a polynomial of smaller degree than the divisor ($x^2 + 0x + 1$), that polynomial is the remainder. Write the result as

$$x^2 + 2x + 1 + \frac{-3x - 2}{x^2 + 1}. \quad \blacksquare$$

3.7 EXERCISES

Perform each division. (Be sure each polynomial is written in descending powers of the variable.) See Examples 1 and 2.

1. $\dfrac{x^2 - x - 6}{x - 3}$

2. $\dfrac{m^2 - 2m - 24}{m + 4}$

3. $\dfrac{2y^2 + 9y - 35}{y + 7}$

4. $\dfrac{y^2 + 2y + 1}{y + 1}$

5. $\dfrac{p^2 + 2p - 20}{p + 6}$

6. $\dfrac{x^2 + 11x + 16}{x + 8}$

7. $\dfrac{r^2 - 8r + 15}{r - 3}$

8. $\dfrac{t^2 - 3t - 10}{t - 5}$

9. $\dfrac{12m^2 - 20m + 3}{2m - 3}$

10. $\dfrac{9w^2 + 6w + 10}{3w - 2}$

11. $\dfrac{2x^2 + 3}{2x + 4}$

12. $\dfrac{4m^2 + 5}{2m - 6}$

13. $\dfrac{12r^2 - 17r + 5}{3r - 5}$

14. $\dfrac{15y^2 + 11y - 60}{3y + 7}$

15. $\dfrac{14k^2 + 19k - 30}{7k - 8}$

16. $\dfrac{15m^2 + 34m + 28}{5m + 3}$

17. $\dfrac{2x^3 + 3x - x^2 + 2}{2x + 1}$

18. $\dfrac{-11t^2 + 12t^3 + 18 + 9t}{4t + 3}$

19. $\dfrac{8k^4 - 12k^3 - 2k^2 + 7k - 6}{2k - 3}$

20. $\dfrac{27r^4 - 36r^3 - 6r^2 + 26r - 24}{3r - 4}$

21. $\dfrac{3y^3 + y^2 + 4y + 1}{y + 1}$

22. $\dfrac{2r^3 - 5r^2 - 6r + 15}{r - 3}$

23. $\dfrac{3k^3 - 4k^2 - 6k + 10}{k - 2}$

24. $\dfrac{5z^3 - z^2 + 10z + 2}{z + 2}$

25. $\dfrac{6p^4 - 16p^3 + 15p^2 - 5p + 10}{3p + 1}$

26. $\dfrac{6r^4 - 11r^3 - r^2 + 16r - 8}{2r - 3}$

Perform each division. See Examples 2 and 3.

27. $\dfrac{-6x - x^2 + x^4}{x^2 - 2}$

28. $\dfrac{5 - 2r^2 + r^4}{r^2 - 1}$

29. $\dfrac{-m^3 - 9m^4 - 22m^2 + 4m^5 - 15}{4m^2 - m - 3}$

30. $\dfrac{6x^4 - x^3 + x^2 + 2x^5 + 5 - x}{2x^2 + 2x + 1}$

31. $\dfrac{y^3 + 1}{y + 1}$

32. $\dfrac{x^4 - 1}{x^2 - 1}$

33. $\dfrac{a^4 - 1}{a^2 + 1}$

34. $\dfrac{p^5 - 2}{p^2 - 1}$

35. $(2y^2 - 5y - 3) \div (2y + 4)$

36. $(3a^2 - 11a + 16) \div (2a + 6)$

37. $(9w^2 + 5w + 10) \div (3w - 2)$

38. $(4m^4 - 3m^3 + m^2 - m) \div (4m^2 + 3)$

39. $(6p^5 + 4p^4 - p^3 + 3p + 2) \div (3p^3 + 2p)$

40. $(2x^5 + 6x^4 - x^3 - x^2 + 5) \div (2x^2 + 2x + 1)$

---

41. In the division problem $(4x^4 + 2x^3 - 14x^2 + 19x + 10) \div (2x + 5) = 2x^3 - 4x^2 + 3x + 2$, what polynomial is the divisor? What polynomial is the quotient? (See Objective 1.)

42. When dividing one polynomial by another, how do you know when to stop dividing? (See Objective 1.)

PREVIEW EXERCISES

Evaluate each of the following.

43. 10(6427)

44. 100(72.69)

45. 1000(1.23)

46. 10,000(26.94)

47. $34 \div 10$

48. $6501 \div 100$

49. $237 \div 1000$

50. $42 \div 10{,}000$

3.8 AN APPLICATION OF EXPONENTS: SCIENTIFIC NOTATION

OBJECTIVES

1 EXPRESS NUMBERS IN SCIENTIFIC NOTATION.
2 CONVERT NUMBERS IN SCIENTIFIC NOTATION TO NUMBERS WITHOUT EXPONENTS.
3 USE SCIENTIFIC NOTATION IN CALCULATIONS.

FOCUS ON PROBLEM SOLVING

The distance to Earth from the planet Pluto is 4.58×10^9 kilometers. In April 1983, Pioneer 10 transmitted radio signals from Pluto to Earth at the speed of light, 3.00×10^5 kilometers per second. How long (in seconds) did it take for the signals to reach Earth?

As this problem indicates, the numbers used in science are often extremely large or extremely small. Because of the difficulty of working with many zeros, these numbers are expressed with exponents. The problem stated above is Exercise 49 in the exercises for this section. After working through this section, you should be able to solve this problem.

In **scientific notation,** a number is written in the form $a \times 10^n$, where n is an integer and $1 \leq |a| < 10$. When a number is multiplied by a power of 10, such as 10^1, $10^2 = 100$, $10^3 = 1000$, $10^{-1} = .1$, $10^{-2} = .01$, $10^{-3} = .001$, and so on, the net effect is to move the decimal point to the right if it is a positive power and to the left if it is a negative power. This is shown in the examples below. (In work with scientific notation, the times symbol, $\times$, is used.)

$23.19 \times 10^1 = 23.19 \times 10 = 231.9$ — Decimal point moves 1 place to the right.

$23.19 \times 10^2 = 23.19 \times 100 = 2319.$ — Decimal point moves 2 places to the right.

$23.19 \times 10^3 = 23.19 \times 1000 = 23190.$ — Decimal point moves 3 places to the right.

$23.19 \times 10^{-1} = 23.19 \times .1 = 2.319$ — Decimal point moves 1 place to the left.

$23.19 \times 10^{-2} = 23.19 \times .01 = .2319$ — Decimal point moves 2 places to the left.

$23.19 \times 10^{-3} = 23.19 \times .001 = .02319$ — Decimal point moves 3 places to the left.

1 A number in scientific notation is always written with the decimal point after the first nonzero digit and then multiplied by the appropriate power of 10. For example, 35 is written 3.5×10^1, or 3.5×10; 56,200 is written 5.62×10^4, since

$$56{,}200 = 5.62 \times 10{,}000 = 5.62 \times 10^4.$$

The steps involved in writing a number in scientific notation are given below.

WRITING A NUMBER IN SCIENTIFIC NOTATION

Step 1 Move the decimal point to the right of the first nonzero digit.
Step 2 Count the number of places you moved the decimal point.
Step 3 The number of places in Step 2 is the absolute value of the exponent on 10.
Step 4 The exponent on 10 is positive if you made the number smaller in Step 1. The exponent is negative if you made the number larger in Step 1. If the decimal point is not moved, the exponent is 0.

EXAMPLE 1 USING SCIENTIFIC NOTATION

Write each number in scientific notation.

(a) 93,000,000

The number will be written in scientific notation as 9.3×10^n. To find the value of n, first compare 9.3 with the original number, 93,000,000. Is the number larger or smaller? Here, 9.3 is *smaller* than 93,000,000. Therefore, we must multiply by a *positive* power of 10 so the product 9.3×10^n will equal the larger number.

Move the decimal point to follow the first nonzero digit (the 9). Count the number of places the decimal point was moved.

$$93{,}000{,}000 \quad \text{7 places}$$

Since the decimal point was moved 7 places, and since n is positive, $93{,}000{,}000 = 9.3 \times 10^7$.

(b) $63{,}200{,}000{,}000 = 6.3200000000 = 6.32 \times 10^{10}$

10 places

(c) .00462

Move the decimal point to the right of the first nonzero digit, and count the number of places the decimal point was moved.

$$.00462 \quad \text{3 places}$$

Since 4.62 is *larger* than .00462, the exponent must be *negative*.

$$.00462 = 4.62 \times 10^{-3}$$

(d) $.0000762 = 7.62 \times 10^{-5}$ ■

2 To convert a number written in scientific notation to a number without exponents, work in reverse. Multiplying by a positive power of 10 will make the number larger; multiplying by a negative power of 10 will make the number smaller.

EXAMPLE 2 WRITING NUMBERS WITHOUT EXPONENTS

Write each number without exponents.

(a) 6.2×10^3

Since the exponent is positive, make 6.2 larger by moving the decimal point 3 places to the right.

$$6.2 \times 10^3 = 6.200 = 6200.$$

(b) $4.283 \times 10^5 = 4.28300 = 428{,}300$ Move 5 places to the right.

(c) $7.04 \times 10^{-3} = .00704$ Move 3 places to the left.

As these examples show, the exponent tells the number of places that the decimal point is moved. ■

3 The next example shows how scientific notation can be used with products and quotients.

EXAMPLE 3
MULTIPLYING AND DIVIDING WITH SCIENTIFIC NOTATION

Write each product or quotient without exponents.

(a) $(6 \times 10^3)(5 \times 10^{-4})$

$$(6 \times 10^3)(5 \times 10^{-4}) = (6 \times 5)(10^3 \times 10^{-4})$$ Commutative and associative properties

$$= 30 \times 10^{-1}$$ Product rule for exponents

$$= 3.0$$ Write without exponents.

(b) $\dfrac{4 \times 10^{-5}}{2 \times 10^3} = \dfrac{4}{2} \times \dfrac{10^{-5}}{10^3} = 2 \times 10^{-8} = .00000002$ ■

NOTE Multiplying or dividing numbers written in scientific notation may produce an answer in the form $a \times 10^0$. Since $10^0 = 1$, this equals 1. For example,

$$(8 \times 10^{-4})(5 \times 10^4) = 40 \times 10^0 = 40.$$

Some calculators will accept data and display answers in scientific notation. For instance, 6.191736×10^6 is displayed on some calculators as 6.191736 E + 6. See your calculator manual for further information.

3.8 EXERCISES

Express each number in scientific notation. See Example 1.

1. 6,835,000,000 **2.** 321,000,000,000,000 **3.** 8,360,000,000,000

4. 6850 **5.** 215 **6.** 683

7. 25,000 **8.** 110,000,000 **9.** .035

10. .005 **11.** .0101 **12.** .0000006

13. .000012 **14.** .000000982

Write each number without exponents. See Example 2.

15. 8.1×10^9 **16.** 3.5×10^2

17. 9.132×10^6 **18.** 2.14×10^0

19. 3.24×10^8

20. 4.35×10^4

21. 3.2×10^{-4}

22. 5.76×10^{-5}

23. 4.1×10^{-2}

24. 1.79×10^{-3}

Perform the indicated operations with the numbers in scientific notation, and then write your answers without exponents. See Examples 2 and 3.

25. $(2 \times 10^8) \times (4 \times 10^{-3})$

26. $(5 \times 10^4) \times (3 \times 10^{-2})$

27. $(4 \times 10^{-1}) \times (1 \times 10^{-5})$

28. $(6 \times 10^{-5}) \times (2 \times 10^4)$

29. $(7 \times 10^3) \times (2 \times 10^2) \times (3 \times 10^{-4})$

30. $(3 \times 10^{-5}) \times (3 \times 10^2) \times (5 \times 10^{-2})$

31. $\dfrac{9 \times 10^5}{3 \times 10^{-1}}$

32. $\dfrac{12 \times 10^{-4}}{4 \times 10^4}$

33. $\dfrac{4 \times 10^{-3}}{2 \times 10^{-2}}$

34. $\dfrac{5 \times 10^{-1}}{1 \times 10^{-5}}$

35. $\dfrac{2.6 \times 10^5}{2 \times 10^2}$

36. $\dfrac{9.5 \times 10^{-1}}{5 \times 10^3}$

37. $(1.2 \times 10^2) \times (5 \times 10^{-3}) \div (2.4 \times 10^3)$

38. $(4.6 \times 10^{-3}) \times (2 \times 10^{-1}) \div (4 \times 10^5)$

39. $\dfrac{7.2 \times 10^{-3} \times 1.6 \times 10^5}{4 \times 10^{-2} \times 3.6 \times 10^9}$

40. $\dfrac{8.7 \times 10^{-2} \times 1.2 \times 10^{-6}}{3 \times 10^{-4} \times 2.9 \times 10^{11}}$

Write the numbers in each statement in scientific notation.

41. The atmospheric pressure on the surface of Mars is about .007 bar.

42. Phillip Morris Co., the leading advertiser in the United States, spent $1,558,000 on advertising in a recent year.

43. The distance from Earth to the sun is about 92,900,000 miles.

44. The planet Pluto is .0017 as big as Earth.

Write the numbers in each statement without exponents.

45. A nanosecond is 1×10^{-9} seconds.

46. The average life span of a human is 2×10^9 seconds.

47. The body of a 150-pound person contains 2.3×10^{-4} pounds of copper and 6×10^{-3} pounds of iron.

48. In the food chain that links the largest sea creature, the whale, to the smallest, the diatom, 4×10^{14} diatoms sustain a medium-sized whale for only a few hours.

49. The distance to Earth from the planet Pluto is 4.58×10^9 kilometers. In April 1983, Pioneer 10 transmitted radio signals from Pluto to Earth at the speed of light, 3.00×10^5 kilometers per second. How long (in seconds) did it take for the signals to reach Earth? (Refer to Objective 3. *Hint:* Use the formula $d = rt$ solved for time.)

50. In Exercise 49, how many hours did it take for the signals to reach Earth?

CHAPTER 3 GLOSSARY

KEY TERMS

3.1 polynomial A polynomial is a term or the sum of a finite number of terms.

descending powers A polynomial in x is written in descending powers if the exponents on x decrease left to right.

degree of a term The degree of a term is the sum of the exponents on the variables.

degree of a polynomial The degree of a polynomial in one variable is the highest exponent found in any term of the polynomial.

trinomial A trinomial is a polynomial with exactly three terms.

binomial A binomial is a polynomial with exactly two terms.

monomial A monomial is a polynomial with exactly one term.

3.2 exponential expression An exponential expression is a base with an exponent.

3.8 scientific notation A number written as $a \times 10^n$, where $1 \leq |a| < 10$ and n is an integer, is in scientific notation.

NEW SYMBOLS

x^{-n} x to the negative n power

CHAPTER 3 QUICK REVIEW

SECTION	CONCEPTS	EXAMPLES
3.1 POLYNOMIALS	**Addition:** Add like terms.	Add: $\begin{array}{r} 2x^2 + 5x - 3 \\ 5x^2 - 2x + 7 \\ \hline 7x^2 + 3x + 4 \end{array}$
	Subtraction: Change the signs of the terms in the second polynomial and add to the first polynomial.	Subtract: $(2x^2 + 5x - 3) - (5x^2 - 2x + 7)$ $= (2x^2 + 5x - 3) + (-5x^2 + 2x - 7)$ $= -3x^2 + 7x - 10$
3.2 EXPONENTS	For any integers m and n with no denominators zero:	
	Product rule $a^m \cdot a^n = a^{m+n}$	$2^4 \cdot 2^5 = 2^9$
	Power rules	
	(a) $(a^m)^n = a^{mn}$	$(3^4)^2 = 3^8$
	(b) $(ab)^m = a^m b^m$	$(6a)^5 = 6^5 a^5$
	(c) $\left(\frac{a}{b}\right)^m = \frac{a^m}{b^m}$ $(b \neq 0)$	$\left(\frac{2}{3}\right)^4 = \frac{2^4}{3^4}$
	(d) $\frac{a^{-m}}{b^{-n}} = \frac{b^n}{a^m}$	$\frac{4^{-2}}{3^{-5}} = \frac{3^5}{4^2}$
	(e) $\left(\frac{a}{b}\right)^{-m} = \left(\frac{b}{a}\right)^m$	$\left(\frac{6}{5}\right)^{-3} = \left(\frac{5}{6}\right)^3$

SECTION	CONCEPTS	EXAMPLES
3.3 MULTIPLICATION OF POLYNOMIALS	Multiply each term of the first polynomial by each term of the second polynomial. Then add like terms.	Multiply: $$\begin{array}{r} 3x^3 - 4x^2 + 2x - 7 \\ 4x + 3 \\ \hline 9x^3 - 12x^2 + 6x - 21 \\ 12x^4 - 16x^3 + 8x^2 - 28x \\ \hline 12x^4 - 7x^3 - 4x^2 - 22x - 21 \end{array}$$
	FOIL Method *Step 1* Multiply the two first terms to get the first term of the answer.	Find $(2x + 3)(5x - 4)$. $2x(5x) = 10x^2$
	Step 2 Find the outer product and the inner product and mentally add them, when possible, to get the middle term of the answer.	$2x(-4) + 3(5x) = 7x$
	Step 3 Multiply the two last terms to get the last term of the answer.	$3(-4) = -12$ The product of $(2x + 3)$ and $(5x - 4)$ is $10x^2 + 7x - 12$.
3.4 SPECIAL PRODUCTS	**Square of a Binomial** $(a + b)^2 = a^2 + 2ab + b^2$ $(a - b)^2 = a^2 - 2ab + b^2$	$(3x + 1)^2 = 9x^2 + 6x + 1$ $(2m - 5n)^2 = 4m^2 - 20mn + 25n^2$
	Product of the Sum and Difference of Two Terms $(a + b)(a - b) = a^2 - b^2$	$(4a + 3)(4a - 3) = 16a^2 - 9$
3.5 THE QUOTIENT RULE AND INTEGER EXPONENTS	If $a \neq 0$, for integers m and n:	
	Zero exponent $a^0 = 1$	$15^0 = 1$
	Negative exponent $a^{-n} = \frac{1}{a^n}$	$5^{-2} = \frac{1}{5^2} = \frac{1}{25}$
	Quotient rule $\frac{a^m}{a^n} = a^{m-n}$	$\frac{4^8}{4^3} = 4^5$
3.6 THE QUOTIENT OF A POLYNOMIAL AND A MONOMIAL	Divide each term of the polynomial by the monomial: $\frac{a + b}{c} = \frac{a}{c} + \frac{b}{c}$	Divide: $\frac{4x^3 - 2x^2 + 6x - 8}{2x} = 2x^2 - x + 3 - \frac{4}{x}$

SECTION	CONCEPTS	EXAMPLES
3.7 THE QUOTIENT OF TWO POLYNOMIALS	Use "long division."	Divide: $$\begin{array}{r} 2x - 5 + \dfrac{-1}{3x+4} \\ 3x + 4\overline{)6x^2 - 7x - 21} \\ \underline{6x^2 + 8x} \\ -15x - 21 \\ \underline{-15x - 20} \\ -1 \end{array}$$
3.8 AN APPLICATION OF EXPONENTS: SCIENTIFIC NOTATION	To write a number in scientific notation (as $a \times 10^n$), move the decimal point to follow the first nonzero digit. The number of digits the decimal point is moved is the absolute value of n. If moving the decimal point makes the number smaller, n is positive. Otherwise, n is negative. If the decimal point is not moved, n is 0.	$247 = 2.47 \times 10^2$ $0.0051 = 5.1 \times 10^{-3}$ $4.8 = 4.8 \times 10^0$

CHAPTER 3 REVIEW EXERCISES

[3.1] *Combine terms where possible in the following polynomials. Write the answers in descending powers of the variable. Give the degree of the answer. Identify each polynomial as a monomial, binomial, trinomial, or none of these.*

1. $9m^2 + 11m^2 + 2m^2$

2. $-4p + p^3 - p^2 + 8p + 2$

3. $12a^5 - 9a^4 + 8a^3 + 2a^2 - a + 3$

4. $-7y^5 - 8y^4 - y^5 + y^4 + 9y$

5. $-7x^5 - 8x - x^5 + x + 9x^3$

6. $-5z^3 + 7 - 6z^2 + 8z$

7. $(6p^2 - p - 8) - (-4p^2 + 2p - 3)$

8. $(12r^4 - 7r^3 + 2r^2) - (5r^4 - 3r^3 + 2r^2 - 1)$

Add or subtract as indicated.

9. Add.

$$\begin{array}{r} -2a^3 + 5a^2 \\ \underline{3a^3 - a^2} \end{array}$$

10. Subtract.

$$\begin{array}{r} 6y^2 - 8y + 2 \\ \underline{5y^2 + 2y - 7} \end{array}$$

11. Subtract.

$$\begin{array}{r} -12k^4 - 8k^2 + 7k \\ \underline{k^4 + 7k^2 - 11k} \end{array}$$

12. $(2m^3 - 8m^2 + 4) + (3m^3 + 2m^2 - 7)$

13. $(12r^4 - 7r^3 + 2r^2) - (5r^4 - 3r^3 + 2r^2)$

[3.2] *Simplify. Assume all variables are nonzero real numbers.*

14. $\left(\frac{3}{4}\right)^4$ **15.** $4^2 + 4^3$ **16.** $(4)^5(4)^3$

17. $5^4 \cdot 5^7$ **18.** $(2^4)^2$ **19.** $(3x^2)^3$

20. $(-5m^3)^2$ **21.** $(ab^0c^3)^4$ **22.** $\left(\frac{2m^3n}{p^2}\right)^2$

[3.3] *Find each product.*

23. $5x(2x - 11)$ **24.** $-3p^3(2p^2 - 5p)$

25. $2y^2(-11y^2 + 2y + 9)$ **26.** $-m^5(8m^2 - 10m + 6)$

27. $(a + 2)(a^2 - 4a + 1)$ **28.** $(3r - 2)(2r^2 + 4r - 3)$

29. $(5p^2 + 3p)(p^3 - p^2 + 5)$ **30.** $(m - 9)(m + 2)$

31. $(3k - 6)(2k + 1)$ **32.** $(a + 3b)(2a - b)$

33. $(6k + 5q)(2k - 7q)$ **34.** $(5p + 3)(p + 2)(p - 1)$

[3.4] *Find each product.*

35. $(a + 4)^2$ **36.** $(3p - 2)^2$ **37.** $(2r + 5t)^2$

38. $(8z - 3y)^2$ **39.** $(6m - 5)(6m + 5)$ **40.** $(2z + 7)(2z - 7)$

41. $(5a + 6b)(5a - 6b)$ **42.** $(9y + 8z)(9y - 8z)$ **43.** $(r + 2)^3$

[3.5] *Simplify. Write each answer with only positive exponents. Assume that all variables are nonzero real numbers.*

44. -7^{-2} **45.** $\left(\frac{5}{8}\right)^{-2}$ **46.** $2^{-1} + 4^{-1}$ **47.** $(5^{-2})^{-4}$

48. $9^3 \cdot 9^{-5}$ **49.** $(-3)^7(-3)^3$ **50.** $\frac{6^{-5}}{6^{-3}}$ **51.** $\frac{x^{-7}}{x^{-9}}$

52. $(9^3)^{-2}$ **53.** $\frac{y^4 \cdot y^{-2}}{y^{-5}}$ **54.** $(6r^{-2})^{-1}$ **55.** $(3p)^4(3p^{-7})$

56. $\frac{ab^{-3}}{a^4b^2}$ **57.** $\frac{(6r^{-1})^2(2r^{-4})}{r^{-5}(r^2)^{-3}}$

[3.6] *Perform each division.*

58. $\frac{-15y^4}{9y^2}$ **59.** $\frac{12x^3y^2}{6xy}$ **60.** $\frac{6y^4 - 12y^2 + 18y}{6y}$

61. $(2p^3 - 6p^2 + 5p) \div (2p^2)$ **62.** $(-10m^4n^2 + 5m^3n^2 + 6m^2n^4) \div (5m^2n)$

[3.7] *Perform each division.*

63. $\dfrac{2r^2 + 3r - 14}{r - 2}$

64. $\dfrac{10a^3 + 9a^2 - 14a + 9}{5a - 3}$

65. $\dfrac{2k^4 + 3k^3 + 9k^2 - 8}{2k^2 + k + 1}$

66. $\dfrac{m^4 + 4m^3 - 5m^2 - 12m + 6}{m^2 - 3}$

[3.8] *Write each number in scientific notation.*

67. 15,800,000 **68.** .0004251 **69.** .0000976 **70.** .784

Write each number without exponents.

71. 1.2×10^4

72. 4.253×10^{-4}

73. $(6 \times 10^4) \times (1.5 \times 10^3)$

74. $(2 \times 10^{-3}) \times (4 \times 10^5)$

75. $\dfrac{9 \times 10^{-2}}{3 \times 10^2}$

76. $\dfrac{8 \times 10^4}{2 \times 10^{-2}}$

MIXED REVIEW EXERCISES

Perform the indicated operations. Write with positive exponents only. Assume all variables represent nonzero real numbers.

77. $5^0 + 7^0$

78. $\left(\dfrac{6r^2p}{5}\right)^3$

79. $(12a + 1)(12a - 1)$

80. 2^{-4}

81. $(8^{-3})^4$

82. $\dfrac{12m^2 - 11m - 10}{3m - 5}$

83. $\dfrac{2p^3 - 6p^2 + 5p}{2p^2}$

84. $\dfrac{(2m^{-5})(3m^2)^{-1}}{m^{-2}(m^{-1})^2}$

85. $(3k - 6)(2k^2 + 4k + 1)$

86. $\dfrac{r^9 \cdot r^{-5}}{r^{-2} \cdot r^{-7}}$

87. $(2r + 5s)^2$

88. $(-5y^2 + 3y - 11) + (4y^2 - 7y + 15)$

89. $(2r + 5)(5r - 2)$

90. $(-5m^3)^2$

91. $\dfrac{2y^3 + 17y^2 + 37y + 7}{2y + 7}$

92. $(25x^2y^3 - 8xy^2 + 15x^3y) \div (10x^2y^3)$

93. $-8 - [6 - (3 + p)]$

94. $(6p^2 - p - 8) - (-4p^2 + 2p - 3)$

95. $\dfrac{5^8}{5^{19}}$

96. $(-7 + 2k)^2$

97. $(5^2)^4$

98. $(3k + 1)(2k - 3)$

99. $\dfrac{(m^2)^3}{(m^4)^2}$

100. Explain why the degree of 2^5 is *not* 5.

101. Explain why $(a + b)^2 \neq a^2 + b^2$.

CHAPTER 3 TEST

For each polynomial, combine terms; then give the degree of the polynomial. Finally, select the most specific description from this list: (a) trinomial, (b) binomial, (c) monomial, (d) none of these.

1. $3x^2 + 6x - 4x^2$

2. $11m^3 - m^2 + m^4 + m^4 - 7m^2$

Perform the indicated operation.

3. $(2x^5 - 4x + 7) - (x^5 + x^2 - 2x - 5)$

4. $(y^2 - 5y - 3) + (3y^2 + 2y) - (y^2 - y - 1)$

Evaluate each expression.

5. $\left(\frac{4}{3}\right)^3$

6. $(5)^{-2}$

Simplify. Write each answer using only positive exponents.

7. $5^2 \cdot 5^6 \cdot 5$

8. $\frac{8^4}{8^9}$

9. $\left(\frac{6q^{-2}}{q^{-3}}\right)^{-1} \quad (q \neq 0)$

10. $\frac{(2p^2)^3 p^4}{p^5} \quad (p \neq 0)$

11. Why is 2^{-4} positive?

In Exercises 12–15, perform the indicated operations.

12. $6m^2(m^3 + 2m^2 - 3m + 7)$

13. $(5m + 6)(3m - 11)$

14. $(2t + 5s)(7t - 4s)$

15. $(3k + 2)(2k^2 - 7k + 8)$

16. $(2k + 5m)^2$

17. $\left(2r + \frac{1}{2}t\right)^2$

18. $(6p^2 - 5r)(6p^2 + 5r)$

19. $\frac{-15y^5 - 8y^4 + 6y^3}{3y^2}$

20. $(10r^3 + 25r^2 - 15r + 8) \div (5r^3)$

21. $\frac{3x^3 - 2x^2 - 6x - 4}{x - 2}$

22. $\frac{10r^4 + 4r^3 - 25r^2 - 6r + 20}{2r^2 - 3}$

23. Write .000379 in scientific notation.

Write each number without exponents.

24. $(6 \times 10^{-4}) \times (1.5 \times 10^6)$

25. $\frac{5.6 \times 10^{-7}}{1.4 \times 10^{-3}}$

CHAPTER FOUR

FACTORING AND APPLICATIONS

An object shot straight up with an initial velocity of 64 feet per second will go up 64 feet, then fall back again. To find the number of seconds until it begins to fall requires solving the equation $t^2 - 4t + 4 = 0$. This equation can be solved by factoring.

Factoring a polynomial reverses the process of multiplication, which was discussed in Chapter 3. As shown in Sections 4.5 and 4.7, factoring is used to solve certain equations and inequalities, and therefore, is important in many applications of algebra. Factoring will also be used to reduce algebraic fractions to lowest terms in Chapter 5.

4.1 FACTORS; THE GREATEST COMMON FACTOR

OBJECTIVES

1 FIND THE GREATEST COMMON FACTOR OF A LIST OF TERMS.

2 FACTOR OUT THE GREATEST COMMON FACTOR.

3 FACTOR BY GROUPING.

Recall from Chapter 1, to **factor** means to write a quantity as a product. That is, factoring is the opposite of multiplication. For example,

Multiplication	*Factoring*
$6 \cdot 2 = 12,$	$12 = 6 \cdot 2.$
↑ ↑ (Factors) ↑ (Product)	↑ (Product) ↑ ↑ (Factors)

Other factored forms of 12 are $(-6)(-2)$, $3 \cdot 4$, $(-3)(-4)$, $12 \cdot 1$, and $(-12)(-1)$. More than two factors may be used, so another factored form of 12 is $2 \cdot 2 \cdot 3$. The positive integer factors of 12 are

$$1, 2, 3, 4, 6, 12.$$

1 An integer that is a factor of two or more integers is called a **common factor** of those integers. For example, 6 is a common factor of 18 and 24 since 6 is a factor of both 18 and 24. Other common factors of 18 and 24 are 1, 2, and 3. The **greatest common factor** of a list of integers is the largest common factor of those integers. Thus, 6 is the greatest common factor of 18 and 24, since it is the largest of the common factors of these numbers.

Recall from Chapter 1, a prime number has only itself and 1 as factors. In Section 1.1, we factored numbers into prime factors. This is the first step in finding the greatest common factor of a list of numbers.

Find the greatest common factor of a list of numbers as follows.

FINDING THE GREATEST COMMON FACTOR

Step 1 **Factor.** Write each number in prime factored form.

Step 2 **List common factors.** List each prime number that is a factor of every number in the list.

Step 3 **Choose smallest exponents.** Use as exponents on the prime factors the *smallest* exponent from the prime factored forms. (If a prime does not appear in one of the prime factored forms, it cannot appear in the greatest common factor.)

Step 4 **Multiply.** Multiply the primes from Step 3. If there are no primes left after Step 3, the greatest common factor is 1.

EXAMPLE 1
FINDING THE GREATEST COMMON FACTOR FOR NUMBERS

Find the greatest common factor for each list of numbers.

(a) 30, 45

First write each number in prime factored form.

$$30 = 2 \cdot 3 \cdot 5$$
$$45 = 3^2 \cdot 5$$

Now, take each prime the *least* number of times it appears in all the factored forms. There is no 2 in the prime factored form of 45, so there will be no 2 in the greatest common factor. The least number of times 3 appears in all the factored forms is 1, and the least number of times 5 appears is also 1. From this, the greatest common factor is

$$3^1 \cdot 5^1 = 15.$$

(b) 72, 120, 432

Find the prime factored form of each number.

$$72 = 2^3 \cdot 3^2$$
$$120 = 2^3 \cdot 3 \cdot 5$$
$$432 = 2^4 \cdot 3^3$$

The least number of times 2 appears in all the factored forms is 3, and the least number of times 3 appears is 1. There is no 5 in the prime factored form of either 72 or 432, so the greatest common factor is

$$2^3 \cdot 3 = 24.$$

(c) 10, 11, 14

Write the prime factored form of each number.

$$10 = 2 \cdot 5$$
$$11 = 11$$
$$14 = 2 \cdot 7$$

There are no primes common to all three numbers, so the greatest common factor is 1. ■

The greatest common factor also can be found for a list of variables. For example, the terms x^4, x^5, x^6, and x^7 have x^4 as the greatest common factor, because 4 is the smallest exponent on x.

NOTE The exponent on a variable in the greatest common factor is the *smallest* exponent that appears on the factors.

EXAMPLE 2
FINDING THE GREATEST COMMON FACTOR FOR VARIABLE TERMS

Find the greatest common factor for each list of terms.

(a) $21m^7,\ -18m^6,\ 45m^8,\ -24m^5$

$$21m^7 = 3 \cdot 7 \cdot m^7$$
$$-18m^6 = -1 \cdot 2 \cdot 3^2 \cdot m^6$$
$$45m^8 = 3^2 \cdot 5 \cdot m^8$$
$$-24m^5 = -1 \cdot 2^3 \cdot 3 \cdot m^5$$

First, 3 is the greatest common factor of the coefficients 21, −18, 45, and −24. The smallest exponent on m is 5, so the greatest common factor of the terms is $3m^5$.

(b) $x^4y^2,\ x^7y^5,\ x^3y^7,\ y^{15}$

$$x^4y^2 = x^4 \cdot y^2$$
$$x^7y^5 = x^7 \cdot y^5$$
$$x^3y^7 = x^3 \cdot y^7$$
$$y^{15} = y^{15}$$

There is no x in the last term, y^{15}, so x will not appear in the greatest common factor. There is a y in each term, however, and 2 is the smallest exponent on y. The greatest common factor is y^2. ■

2 The idea of a greatest common factor can be used to write a polynomial (a sum) in factored form as a product. For example, the polynomial

$$3m + 12$$

consists of the two terms $3m$ and 12. The greatest common factor for these two terms is 3. Write $3m + 12$ so that each term is a product with 3 as one factor.

$$3m + 12 = 3 \cdot m + 3 \cdot 4$$

Now use the distributive property.

$$3m + 12 = 3 \cdot m + 3 \cdot 4 = 3(m + 4)$$

The factored form of $3m + 12$ is $3(m + 4)$. This process is called **factoring out the greatest common factor.**

EXAMPLE 3
FACTORING OUT THE GREATEST COMMON FACTOR

Factor out the greatest common factor.

(a) $20m^5 + 10m^4 + 15m^3$

The greatest common factor for the terms of this polynomial is $5m^3$.

$$20m^5 + 10m^4 + 15m^3 = (5m^3)(4m^2) + (5m^3)(2m) + (5m^3)3$$
$$= 5m^3(4m^2 + 2m + 3)$$

Check this work by multiplying $5m^3$ and $4m^2 + 2m + 3$. You should get the original polynomial as your answer.

(b) $x^5 + x^3 = (x^3)x^2 + (x^3)1 = x^3(x^2 + 1)$

(c) $20m^7p^2 - 36m^3p^4 = 4m^3p^2(5m^4 - 9p^2)$

(d) $a(a + 3) + 4(a + 3)$

The binomial $a + 3$ is the greatest common factor here.

$$a(a + 3) + 4(a + 3) = (a + 3)(a + 4) \quad ■$$

CAUTION Be careful to avoid the common error of leaving out the 1 in a problem like Example 3(b). Always be sure that the factored form can be multiplied out to give the original polynomial.

3 Common factors are used in **factoring by grouping,** as explained in the next example.

EXAMPLE 4
FACTORING BY GROUPING

Factor by grouping.

(a) $2x + 6 + ax + 3a$

The first two terms have a common factor of 2, and the last two terms have a common factor of a.

$$2x + 6 + ax + 3a = 2(x + 3) + a(x + 3)$$

The expression is still not in factored form, because it is the *sum* of two terms. Now, however, $x + 3$ is a common factor and can be factored out.

$$2x + 6 + ax + 3a = 2(x + 3) + a(x + 3) = (x + 3)(2 + a)$$

The final result is in factored form because it is a *product*. Note that the goal in factoring by grouping is to get a common factor, $x + 3$ here, so that the last step is possible.

(b) $m^2 + 6m + 2m + 12 = m(m + 6) + 2(m + 6)$
$$= (m + 6)(m + 2)$$

(c) $6xy - 21x - 8y + 28 = 3x(2y - 7) - 4(2y - 7) = (2y - 7)(3x - 4)$

Must be same

Since the quantities in parentheses in the second step must be the same, it was necessary here to factor out -4 rather than 4. ■

CAUTION Note the careful use of signs in Example 4(c). Sign errors often occur when grouping with negative signs.

Use these steps when factoring four terms by grouping.

FACTORING BY GROUPING

Step 1 Write the four terms so that the first two have a common factor and the last two have a common factor.

Step 2 Use the distributive property to factor each group of two terms.

Step 3 If possible, factor a common binomial factor from the results of Step 2.

Step 4 If Step 2 does not result in a common binomial factor, try grouping the terms of the original polynomial in a different way. (If two terms still have no common binomial factor, the polynomial cannot be factored.)

4.1 EXERCISES

Find the greatest common factor for each list of terms. See Examples 1 and 2.

1. $12y, 24$
2. $72m, 12$
3. $30p^2, 20p^3, 40p^5$
4. $14r^5, 28r^2, 56r^8$
5. $18r, 32y, 11z$
6. $45m^2, 12n, 7p^2$
7. $18m^2n^2, 36m^4n^5, 12m^3n$
8. $50p^5r^2, 25p^4r^7, 30p^7r^8$
9. $32y^4x^5, 24y^7x, 36y^3x^5$

Complete the factoring.

10. $18 = 9(\quad)$
11. $3x^2 = 3x(\quad)$
12. $8x^3 = 8x(\quad)$
13. $9m^4 = 3m^2(\quad)$
14. $12p^5 = 6p^3(\quad)$
15. $-8z^9 = 4z^5(\quad)$
16. $-15k^{11} = -5k^8(\quad)$
17. $x^2y^3 = xy(\quad)$
18. $a^3b^2 = a^2b(\quad)$
19. $27a^3b^2 = 9a^2b(\quad)$
20. $14x^4y^3 = 2xy(\quad)$
21. $-16m^3n^3 = 4mn^2(\quad)$

Factor out the greatest common factor. See Example 3.

22. $10p + 20p^2$
23. $49a^2 + 14a$
24. $100a^4 + 16a^2$
25. $121p^5 - 33p^4$
26. $8p^2 - 4p^4$
27. $11z^2 - 100$
28. $12z^2 - 11y^4$
29. $19y^3p^2 + 38y^2p^3$
30. $100m^5 - 50m^3 + 100m^2$
31. $13y^6 + 26y^5 - 39y^3$
32. $5x^4 + 25x^3 - 20x^2$
33. $45q^4p^5 - 36qp^6 + 81q^2p^3$
34. $c(x + 2) - d(x + 2)$
35. $r(5 - x) + t(5 - x)$
36. $m(m + 2n) + n(m + 2n)$
37. $3p(1 - 4p) - 2q(1 - 4p)$

Factor by grouping. See Example 4.

38. $m^2 + 2m + nm + 2n$

39. $a^2 - 2a + ab - 2b$

40. $8k + 6 + 4kq + 3q$

41. $7z^2 + 3z + 14mz + 6m$

42. $5m + 15 - 2mp - 6p$

43. $2y^2 - 6y - 2xy + 6x$

44. $18r^2 + 12ry - 3ry - 2y^2$

45. $3a^3 + 3ab^2 + 2a^2b + 2b^3$

46. $a^5 + 2a^5b - 3 - 6b$

47. $a^2b - 4a - ab^4 + 4b^3$

Factor out the greatest common factor.

48. $m^6n^5 - 2m^5 + 5m^3n^5$

49. $125z^5a^3 - 60z^4a^5 + 85z^3a^4$

50. $30a^2m^2 + 60a^3m + 180a^3m^2$

51. $33y^8 - 44y^{12} + 77y^3 + 11y^4$

52. $26g^6h^4 + 13g^3h^4 - 39g^4h^3$

53. $36a^4b^3 + 32a^5b^2 - 48a^6b^3$

Decide whether each of the following expressions is factored. (See Objective 2.)

54. $x^2 + 5x$

55. $3m - 2m^2$

56. $y^2(y - 5)$

57. $5k(2k - p)$

58. $4(m - 1) + n(m - 1)$

59. $(2x + y)(x - y)$

60. $(3 + 4p)(2p - 1)$

61. $a(3 - b) + 4(3 - b)$

PREVIEW EXERCISES

Find each product. See Sections 3.3 and 3.4.

62. $(x + 2)(x + 5)$

63. $(m + 3)(m + 4)$

64. $(y + 7)(y - 3)$

65. $(r + 5)(r - 9)$

66. $(q - 5)(q - 7)$

67. $(y - 4)(y - 8)$

68. $(p - 10)(p + 10)$

69. $(a + 7)(a - 7)$

70. $(m - 12)(m + 12)$

4.2 FACTORING TRINOMIALS

OBJECTIVES

1 FACTOR TRINOMIALS WITH A COEFFICIENT OF 1 FOR THE SQUARED TERM.

2 FACTOR SUCH POLYNOMIALS AFTER FACTORING OUT THE GREATEST COMMON FACTOR.

1 Using FOIL, we can show that the product of the polynomials $k - 3$ and $k + 1$ is

$$(k - 3)(k + 1) = k^2 - 2k - 3.$$

Now suppose we are given the polynomial $k^2 - 2k - 3$ and want to rewrite it as the product $(k - 3)(k + 1)$. This product is called the **factored form** of $k^2 - 2k - 3$, and the process of finding the factored form is called **factoring.** The discussion of factoring in this section is limited to trinomials like $x^2 - 2x - 24$ or $y^2 + 2y - 15$, where the coefficient of the squared term is 1.

When factoring polynomials with only integer coefficients, we use only integers for the numerical factors. For example, $x^2 + 5x + 6$ can be factored by finding integers a and b such that

$$x^2 + 5x + 6 = (x + a)(x + b).$$

To find these integers a and b, we first multiply the two factors on the right-hand side of the equation:

$$(x + a)(x + b) = x^2 + ax + bx + ab.$$

By the distributive property,

$$x^2 + ax + bx + ab = x^2 + (a + b)x + ab.$$

Comparing this result with $x^2 + 5x + 6$ shows that we must find integers a and b having a sum of 5 and a product of 6.

$$x^2 + 5x + 6 = x^2 + (a + b)x + ab.$$

Sum of a and b is 5. Product of a and b is 6.

Since many pairs of integers have a sum of 5, it is best to begin by listing those pairs of integers whose product is 6. Both 5 and 6 are positive, so only pairs in which both integers are positive need be considered.

Product	*Sum*	
$1 \cdot 6 = 6$	$1 + 6 = 7$	
$2 \cdot 3 = 6$	$2 + 3 = 5$	Sum is 5.

Both pairs have a product of 6, but only the pair 2 and 3 has a sum of 5. So 2 and 3 are the required integers, and

$$x^2 + 5x + 6 = (x + 2)(x + 3).$$

We can check by multiplying the binomials using FOIL. Make sure that the sum of the outer and inner products produces the correct middle term.

$$(x + 2)(x + 3) = x^2 + 5x + 6$$

$$\begin{array}{r} 2x \\ 3x \\ \hline 5x \end{array} \quad \text{Add.}$$

This method of factoring can be used only for trinomials having the coefficient of the squared term equal to 1. Methods for factoring other trinomials will be given in the next section.

EXAMPLE 1 **FACTORING A TRINOMIAL WITH ALL TERMS POSITIVE**

Factor $m^2 + 9m + 14$.

Look for two integers whose product is 14 and whose sum is 9. List the pairs of integers whose product is 14. Then examine the sums. Again, only positive integers are needed because all signs are positive.

14, 1	$14 + 1 = 15$	
7, 2	$7 + 2 = 9$	Sum is 9.

From the list, 7 and 2 are the required integers, since $7 \cdot 2 = 14$ and $7 + 2 = 9$. Thus $m^2 + 9m + 14 = (m + 2)(m + 7)$. ■

NOTE In Example 1, the answer also could have been written $(m + 7)(m + 2)$. Because of the commutative property of multiplication, the order of the factors does not matter.

The trinomials in the examples so far had all positive terms. If a trinomial has one or more negative terms, both positive and negative factors must be considered.

EXAMPLE 2 **FACTORING A TRINOMIAL WITH TWO NEGATIVE TERMS**

Factor $p^2 - 2p - 15$.

Find two integers whose product is -15 and whose sum is -2. If these numbers do not come to mind right away, find them (if they exist) by listing all the pairs of integers whose product is -15. Because the last term, -15, is negative, we need pairs of integers with different signs.

15, −1	$15 + (-1) = 14$	
5, −3	$5 + (-3) = 2$	
−15, 1	$-15 + 1 = -14$	
−5, 3	$-5 + 3 = -2$	Sum is −2.

The necessary integers are -5 and 3, and

$$p^2 - 2p - 15 = (p - 5)(p + 3). \quad ■$$

EXAMPLE 3 **FACTORING A TRINOMIAL WITH ONE NEGATIVE TERM**

(a) Factor $x^2 - 5x + 12$.

List all pairs of integers whose product is 12. Since the middle term is negative and the last term is positive, we need pairs with both numbers negative. Then examine the sums.

−12, −1	$-12 + (-1) = -13$
−6, −2	$-6 + (-2) = -8$
−3, −4	$-3 + (-4) = -7$

None of the pairs of integers has a sum of -5. Because of this, the trinomial $x^2 - 5x + 12$ *cannot be factored using only integer factors,* showing that it is a **prime polynomial.**

(b) $k^2 - 8k + 11$

There is no pair of integers whose product is 11 and whose sum is -8, so $k^2 - 8k + 11$ is a prime polynomial. ■

We can now summarize the procedure for factoring a trinomial of the form $x^2 + bx + c$.

FACTORING $x^2 + bx + c$

Find two integers whose product is c and whose sum is b.

1. Both integers must be positive if b and c are positive.
2. Both integers must be negative if c is positive and b is negative.
3. One integer must be positive and one must be negative if c is negative.

EXAMPLE 4
FACTORING A TRINOMIAL WITH TWO VARIABLES

Factor $z^2 - 2bz - 3b^2$.

To factor $z^2 - 2bz - 3b^2$, look for two expressions whose product is $-3b^2$ and whose sum is $-2b$. The expressions are $-3b$ and b, with

$$z^2 - 2bz - 3b^2 = (z - 3b)(z + b). \quad \blacksquare$$

2 The trinomial in the next example does not have a coefficient of 1 for the squared term. (In fact, there is no squared term.) A preliminary step must be taken before using the steps discussed above.

EXAMPLE 5
FACTORING A TRINOMIAL WITH A COMMON FACTOR

Factor $4x^5 - 28x^4 + 40x^3$.

First, factor out the greatest common factor, $4x^3$.

$$4x^5 - 28x^4 + 40x^3 = 4x^3(x^2 - 7x + 10)$$

Now factor $x^2 - 7x + 10$. The integers -5 and -2 have a product of 10 and a sum of -7. The completely factored form is

$$4x^5 - 28x^4 + 40x^3 = 4x^3(x - 5)(x - 2). \quad \blacksquare$$

CAUTION When factoring, always remember to look for a common factor first. Do not forget to include the common factor as part of the answer. Multiplying out the factored form should always give the original polynomial.

4.2 EXERCISES

Complete the factoring.

1. $x^2 + 10x + 21 = (x + 7)(\qquad)$

2. $p^2 + 11p + 30 = (p + 5)(\qquad)$

3. $r^2 + 15r + 56 = (r + 7)(\qquad)$

4. $x^2 + 15x + 44 = (x + 4)(\qquad)$

5. $t^2 - 14t + 24 = (t - 2)(\qquad)$

6. $x^2 - 9x + 8 = (x - 1)(\qquad)$

7. $x^2 - 12x + 32 = (x - 4)(\qquad)$

8. $y^2 - 2y - 15 = (y + 3)(\qquad)$

9. $m^2 + 2m - 24 = (m - 4)(\quad)$

10. $x^2 + 9x - 22 = (x - 2)(\quad)$

11. $x^2 - 7xy + 10y^2 = (x - 2y)(\quad)$

12. $k^2 - 3kh - 28h^2 = (k - 7h)(\quad)$

Factor as completely as possible. If a polynomial cannot be factored, write prime. See Examples 1–3.

13. $x^2 + 6x + 5$
14. $y^2 + 9y + 8$
15. $a^2 + 9a + 20$
16. $b^2 + 8b + 15$
17. $x^2 - 8x + 7$
18. $m^2 + m - 20$
19. $p^2 + 4p + 5$
20. $n^2 + 4n + 12$
21. $y^2 - 6y + 8$
22. $r^2 - 11r + 30$
23. $s^2 + 2s - 35$
24. $h^2 + 11h + 12$
25. $n^2 - 12n - 35$
26. $a^2 - 2a - 99$
27. $b^2 - 11b + 24$
28. $x^2 - 9x + 20$
29. $k^2 - 10k + 25$
30. $z^2 - 14z + 49$

Factor completely in Exercises 31–48. See Examples 4 and 5.

31. $x^2 + 4ax + 3a^2$
32. $x^2 - mx - 6m^2$
33. $y^2 - by - 30b^2$
34. $z^2 + 2zx - 15x^2$
35. $x^2 + xy - 30y^2$
36. $a^2 - ay - 56y^2$
37. $r^2 - 2rs + s^2$
38. $m^2 - 2mn - 3n^2$
39. $p^2 - 3pq - 10q^2$
40. $c^2 - 5cd + 4d^2$
41. $3m^3 + 12m^2 + 9m$
42. $3y^5 - 18y^4 + 15y^3$
43. $6a^2 - 48a - 120$
44. $h^7 - 5h^6 - 14h^5$
45. $3r^3 - 30r^2 + 72r$
46. $2x^6 - 8x^5 - 42x^4$
47. $3x^4 - 3x^3 - 90x^2$
48. $2y^3 - 8y^2 - 10y$

Solve.

49. Use the FOIL method from Section 3.2 to show that $(2x + 4)(x - 3) = 2x^2 - 2x - 12$. If you are asked to completely factor $2x^2 - 2x - 12$, why would it be incorrect to give $(2x + 4)(x - 3)$ as your answer? (See Objective 2.)

50. If you are asked to completely factor the polynomial $3x^2 + 9x - 12$, why would it be incorrect to give $(x - 1)(3x + 12)$ as your answer? (See Objective 2.)

51. What polynomial can be factored to give $(y - 7)(y + 6)$?

52. What polynomial can be factored to give $(a + 9)(a + 6)$?

Factor completely.

53. $a^5 + 3a^4b - 4a^3b^2$
54. $m^3n - 2m^2n^2 - 3mn^3$
55. $y^3z + y^2z^2 - 6yz^3$
56. $k^7 - 2k^6m - 15k^5m^2$
57. $z^{10} - 4z^9y - 21z^8y^2$
58. $x^9 + 5x^8w - 24x^7w^2$
59. $(a + b)x^2 + (a + b)x - 12(a + b)$
60. $(x + y)n^2 + (x + y)n + 16(x + y)$
61. $(2p + q)r^2 - 12(2p + q)r + 27(2p + q)$
62. $(3m - n)k^2 - 13(3m - n)k + 40(3m - n)$

PREVIEW EXERCISES

Find each product. See Section 3.3.

63. $(2y - 7)(y + 4)$
64. $(3a + 2)(2a + 1)$
65. $(5z + 2)(3z - 2)$
66. $(4m - 3)(2m + 5)$
67. $(4p + 1)(2p - 3)$
68. $(6r - 2)(3r + 1)$

4.3 MORE ON FACTORING TRINOMIALS

OBJECTIVES

1 FACTOR TRINOMIALS BY GROUPING WHEN THE COEFFICIENT OF THE SQUARED TERM IS NOT 1.

2 FACTOR TRINOMIALS BY TRIAL AND ERROR.

Trinomials such as $2x^2 + 7x + 6$, in which the coefficient of the squared term is *not* 1, can be factored with an extension of the method presented in the last section.

1 Recall that a trinomial such as $m^2 + 3m + 2$ is factored by finding two numbers whose product is 2 and whose sum is 3. To factor $2x^2 + 7x + 6$, look for two integers whose product is $2 \cdot 6 = 12$ and whose sum is 7.

$$2x^2 + 7x + 6$$

Sum is 7.

Product is $2 \cdot 6 = 12$.

By considering the pairs of positive integers whose product is 12, the necessary integers are found to be 3 and 4. Use these integers to write the middle term, $7x$, as $7x = 3x + 4x$. With this, the trinomial $2x^2 + 7x + 6$ becomes

$$2x^2 + 7x + 6 = 2x^2 + \underbrace{3x + 4x}_{7x = 3x + 4x} + 6.$$

Factor the new polynomial by grouping as in Section 4.1.

$$\begin{aligned} 2x^2 + 3x + 4x + 6 &= x(2x + 3) + 2(2x + 3) \\ &= (2x + 3)(x + 2) \end{aligned}$$

The common factor of $2x + 3$ was factored out to get

$$2x^2 + 7x + 6 = (2x + 3)(x + 2).$$

Check by finding the product of $2x + 3$ and $x + 2$.

The middle term in the polynomial $2x^2 + 7x + 6$ could have been written as $7x = 4x + 3x$ to get

$$\begin{aligned} 2x^2 + 7x + 6 &= 2x^2 + 4x + 3x + 6 \\ &= 2x(x + 2) + 3(x + 2) \\ &= (x + 2)(2x + 3). \end{aligned}$$

Either result is correct.

EXAMPLE 1 **FACTORING TRINOMIALS BY GROUPING**

Factor each trinomial.

(a) $6r^2 + r - 1$

We must find two integers with a product of $6(-1) = -6$ and a sum of 1.

Sum is 1.

$$6r^2 + r - 1 = 6r^2 + 1r - 1$$

Product is $6(-1) = -6$.

The integers are -2 and 3. Write the middle term, $+r$, as $-2r + 3r$, so that

$$6r^2 + r - 1 = 6r^2 - 2r + 3r - 1.$$

Factor by grouping on the right-hand side.

$$\begin{aligned} 6r^2 + r - 1 &= 6r^2 - 2r + 3r - 1 \\ &= 2r(3r - 1) + 1(3r - 1) \qquad \text{The binomials must be the same.} \\ &= (3r - 1)(2r + 1) \end{aligned}$$

(b) $12z^2 - 5z - 2$

Look for two integers whose product is $12(-2) = -24$ and whose sum is -5. The required integers are 3 and -8, and

$$\begin{aligned} 12z^2 - 5z - 2 &= 12z^2 + 3z - 8z - 2 \qquad -5z = 3z - 8z \\ &= 3z(4z + 1) - 2(4z + 1) \qquad \text{Factor each group.} \\ &= (4z + 1)(3z - 2). \qquad \text{Factor out } 4z + 1. \end{aligned}$$

■

2 The rest of this section shows an alternative method of factoring trinomials in which the coefficient of the squared term is not 1. This method uses trial and error. In the next example, the alternative method is used to factor $2x^2 + 7x + 6$, the same trinomial factored at the beginning of this section.

To factor $2x^2 + 7x + 6$ by trial and error, we must use FOIL backwards. We want to write $2x^2 + 7x + 6$ as the product of two binomials.

$$2x^2 + 7x + 6 = (\qquad)(\qquad)$$

The product of the two first terms of the binomials is $2x^2$. The possible factors of $2x^2$ are $2x$ and x or $-2x$ and $-x$. Since all terms of the trinomial are positive, only positive factors should be considered. Thus, we have

$$2x^2 + 7x + 6 = (2x \qquad)(x \qquad).$$

The product of the two last terms, 6, can be factored as $6 \cdot 1$, $1 \cdot 6$, $2 \cdot 3$, or $3 \cdot 2$. Try each pair to find the pair that gives the correct middle term.

Since $2x + 6 = 2(x + 3)$, the binomial $2x + 6$ has a common factor of 2, while $2x^2 + 7x + 6$ has no common factor other than 1. The product $(2x + 6)(x + 1)$ cannot be correct.

NOTE If the original polynomial has no common factor, then none of its binomial factors will either.

Now try the numbers 2 and 3 as factors of 6. Because of the common factor of 2 in $2x + 2$, $(2x + 2)(x + 3)$ will not work. Try $(2x + 3)(x + 2)$.

$$(2x + 3)(x + 2) = 2x^2 + 7x + 6 \qquad \text{Correct}$$

$$\begin{array}{r} 3x \\ 4x \\ \hline 7x \end{array} \quad \text{Add.}$$

Finally, we see that $2x^2 + 7x + 6$ factors as

$$2x^2 + 7x + 6 = (2x + 3)(x + 2).$$

Check by multiplying $2x + 3$ and $x + 2$.

EXAMPLE 2
FACTORING A TRINOMIAL BY TRIAL AND ERROR

Factor $8p^2 + 14p + 5$.

The number 8 has several possible pairs of factors, but 5 has only 1 and 5 or -1 and -5. For this reason, it is easier to begin by considering the factors of 5. Ignore the negative factors since all coefficients in the trinomial are positive. If $8p^2 + 14p + 5$ can be factored, the factors will have the form

$$(\quad + 5)(\quad + 1).$$

The possible pairs of factors of $8p^2$ are $8p$ and p, or $4p$ and $2p$. Try various combinations, checking the middle term in each case.

$$(8p + 5)(p + 1) \qquad \text{Incorrect}$$

$$\begin{array}{r} 5p \\ 8p \\ \hline 13p \end{array} \quad \text{Add.}$$

$$(p + 5)(8p + 1) \qquad \text{Incorrect}$$

$$\begin{array}{r} 40p \\ p \\ \hline 41p \end{array} \quad \text{Add.}$$

$$(4p + 5)(2p + 1) \qquad \text{Correct}$$

$$\begin{array}{r} 10p \\ 4p \\ \hline 14p \end{array} \quad \text{Add.}$$

Since $14p$ is the correct middle term, $8p^2 + 14p + 5$ factors as $(4p + 5)(2p + 1)$. ■

EXAMPLE 3
FACTORING A TRINOMIAL BY TRIAL AND ERROR

Factor $6x^2 - 11x + 3$.

Since 3 has only 1 and 3 or -1 and -3 as factors, it is better here to begin by factoring 3. The last term of the trinomial $6x^2 - 11x + 3$ is positive and the middle term has a negative coefficient, so only negative factors should be considered. Try -3 and -1 as factors of 3:

$$(\quad - 3)(\quad - 1).$$

The factors of $6x^2$ may be either $6x$ and x, or $2x$ and $3x$. Try $2x$ and $3x$.

$(2x - 3)(3x - 1)$ Correct

$-9x$
$-2x$
$-11x$ Add.

These factors give the correct middle term, so

$$6x^2 - 11x + 3 = (2x - 3)(3x - 1). \blacksquare$$

EXAMPLE 4
FACTORING A TRINOMIAL BY TRIAL AND ERROR

Factor $8x^2 + 6x - 9$.

The integer 8 has several possible pairs of factors, as does -9. Since the last term is negative, one positive factor and one negative factor of -9 are needed. Since the coefficient of the middle term is small, it is wise to avoid large factors such as 8 or 9. Let us try 4 and 2 as factors of 8, and 3 and -3 as factors of -9, and check the middle term.

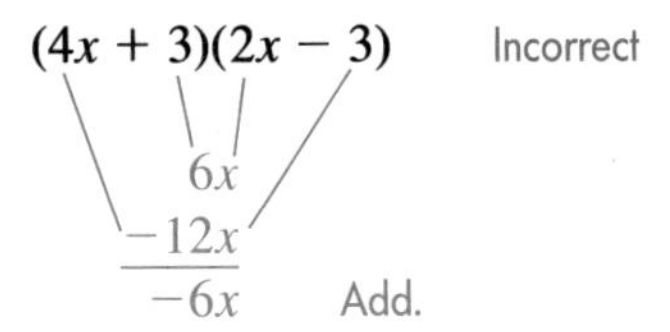

Let us try exchanging 3 and -3.

$(4x - 3)(2x + 3)$ Correct

$-6x$
$12x$
$6x$ Add.

This time we got the correct middle term, so

$$8x^2 + 6x - 9 = (4x - 3)(2x + 3). \blacksquare$$

EXAMPLE 5
FACTORING A TRINOMIAL WITH TWO VARIABLES

Factor $12a^2 - ab - 20b^2$.

There are several pairs of factors of $12a^2$, including $12a$ and a, $6a$ and $2a$, and $3a$ and $4a$, just as there are many pairs of factors of $-20b^2$, including $-20b$ and b, $10b$ and $-2b$, $-10b$ and $2b$, $4b$ and $-5b$, and $-4b$ and $5b$. Once again, since the desired middle term is small, we shall avoid the larger factors. Let us try as factors $6a$ and $2a$ and $4b$ and $-5b$.

$$(6a + 4b)(2a - 5b)$$

This cannot be correct, as mentioned before, since $6a + 4b$ has a common factor while the given trinomial has none. Let us try $3a$ and $4a$ with $4b$ and $-5b$.

$$(3a + 4b)(4a - 5b) = 12a^2 + ab - 20b^2 \quad \text{Incorrect}$$

Here the middle term has the wrong sign, so we change the signs in the factors.

$$(3a - 4b)(4a + 5b) = 12a^2 - ab - 20b^2 \quad \text{Correct} \quad ■$$

EXAMPLE 6 FACTORING A TRINOMIAL WITH A COMMON FACTOR

Factor $28x^5 - 58x^4 - 30x^3$.

First factor out the greatest common factor, $2x^3$.

$$28x^5 - 58x^4 - 30x^3 = 2x^3(14x^2 - 29x - 15)$$

Now try to factor $14x^2 - 29x - 15$. Try $7x$ and $2x$ as factors of $14x^2$ and -3 and 5 as factors of -15.

$$(7x - 3)(2x + 5) = 14x^2 + 29x - 15 \quad \text{Incorrect}$$

The middle term differs only in sign, so change the signs in the two factors.

$$(7x + 3)(2x - 5) = 14x^2 - 29x - 15 \quad \text{Correct}$$

Finally, the factored form of $28x^5 - 58x^4 - 30x^3$ is

$$28x^5 - 58x^4 - 30x^3 = 2x^3(7x + 3)(2x - 5). \quad ■$$

CAUTION Do not forget to include the common factor in the final result.

EXAMPLE 7 FACTORING A TRINOMIAL WITH A NEGATIVE COMMON FACTOR

Factor $-24a^3 - 42a^2 + 45a$

The common factor could be $3a$ or $-3a$. If we factor out $-3a$, the first term of the trinomial factor will be positive, which will make it easier to factor.

$$-24a^3 - 42a^2 + 45a = -3a(8a^2 + 14a - 15) \quad \text{Factor out the greatest common factor.}$$
$$= -3a(4a - 3)(2a + 5) \quad \text{Use trial and error.} \quad ■$$

4.3 EXERCISES

Complete the factoring.

1. $2x^2 - x - 1 = (2x + 1)(\quad)$

2. $3a^2 + 5a + 2 = (3a + 2)(\quad)$

3. $5b^2 - 16b + 3 = (5b - 1)(\quad)$

4. $2x^2 + 11x + 12 = (2x + 3)(\quad)$

5. $4y^2 + 17y - 15 = (y + 5)(\quad)$

6. $7z^2 + 10z - 8 = (z + 2)(\quad)$

7. $6a^2 + 7ab - 20b^2 = (2a \quad)(3a \quad)$

8. $9m^2 - 3mn - 2n^2 = (3m \quad)(3m \quad)$

9. $4x^3 - 10x^2 - 6x = 2x(\qquad) = 2x(2x \qquad)(x \qquad)$

10. $15r^3 - 39r^2 - 18r = 3r(\qquad) = 3r(5r \qquad)(r \qquad)$

11. $6m^6 + 7m^5 - 20m^4 = m^4(\qquad) = m^4(\qquad 4)(\qquad 5)$

12. $16y^5 - 4y^4 - 6y^3 = 2y^3(\qquad) = 2y^3(\qquad 3)(\qquad 1)$

Factor as completely as possible. Remember to look for a greatest common factor first. Use either method. See Examples 1–7.

13. $2x^2 + 7x + 3$

14. $3y^2 + 13y + 4$

15. $3a^2 + 10a + 7$

16. $7r^2 + 8r + 1$

17. $4r^2 + r - 3$

18. $3p^2 + 2p - 8$

19. $15m^2 + m - 2$

20. $6x^2 + x - 1$

21. $8m^2 - 10m - 3$

22. $2a^2 + 30 - 17a$

23. $5a^2 - 6 - 7a$

24. $11s + 12s^2 - 5$

25. $3r^2 + r - 10$

26. $20x^2 - 28x - 3$

27. $4y^2 + 69y + 17$

28. $21m^2 + 13m + 2$

29. $38x^2 + 23x + 2$

30. $20y^2 + 39y - 11$

31. $10x^2 + 11x - 6$

32. $2 + 7b + 6b^2$

33. $10 + 19w + 6w^2$

34. $20q^2 - 41q + 20$

35. $6q^2 + 23q + 21$

36. $8x^2 + 47x - 6$

37. $10m^2 - 23m + 12$

38. $4t^2 - 5t - 6$

39. $8k^2 + 2k - 15$

40. $15p^2 - p - 6$

41. $10m^2 - m - 24$

42. $16a^2 + 30a + 9$

43. $8x^2 - 14x + 3$

44. $-24b^2 + 37b + 5$

45. $-40m^2 - m + 6$

46. $15a^2 + 22a + 8$

47. $12p^2 + 7pq - 12q^2$

48. $6m^2 - 5mn - 6n^2$

49. $25a^2 + 25ab + 6b^2$

50. $6x^2 - 5xy - y^2$

51. $6a^2 - 7ab - 5b^2$

52. $2m^3 + 2m^2 - 40m$

53. $15n^4 - 39n^3 + 18n^2$

54. $24a^4 + 10a^3 - 4a^2$

55. $-32z^2w^4 + 20zw^4 + 12w^4$

56. $-15x^2y^2 + 7xy^2 + 4y^2$

57. $4k^4 - 2k^3w - 6k^2w^2$

58. $4a^4 - a^3b - 3a^2b^2$

59. $6m^6n + 7m^5n^2 + 2m^4n^3$

60. $12k^3q^4 - 4k^2q^5 - kq^6$

61. $18z^3y - 3z^2y^2 - 105zy^3$

62. $18x^2(y - 3)^2 + 5x(y - 3)^2 - 75(y - 3)^2$

63. $25q^2(m + 1)^3 - 5q(m + 1)^3 - 2(m + 1)^3$

Find all integers k so that the following trinomials can be factored using the methods of this section. (See Objectives 1 and 2.)

64. $5x^2 + kx - 1$

65. $2c^2 + kc - 3$

66. $2m^2 + km + 5$

67. $3y^2 + ky + 3$

---

68. Explain how the signs of the last terms of the two binomial factors of a trinomial are determined. (See Objectives 1 and 2.)

PREVIEW EXERCISES

Find each product. See Sections 3.3 and 3.4.

69. $(3r - 1)(3r + 1)$

70. $(5p - 3q)(5p + 3q)$

71. $(2m - 3)^2$

72. $(x - 1)(x^2 + x + 1)$

73. $(2z - 3)(4z^2 + 6z + 9)$

74. $(3k - 1)(9k^2 + 3k + 1)$

4.4 SPECIAL FACTORIZATIONS

OBJECTIVES

1. FACTOR THE DIFFERENCE OF TWO SQUARES.
2. FACTOR A PERFECT SQUARE TRINOMIAL.
3. FACTOR THE DIFFERENCE OF TWO CUBES.
4. FACTOR THE SUM OF TWO CUBES.

1 Recall from the last chapter that

$$(x + y)(x - y) = x^2 - y^2.$$

Based on this product, a **difference of two squares** can be factored as follows.

DIFFERENCE OF TWO SQUARES

$$x^2 - y^2 = (x + y)(x - y)$$

EXAMPLE 1 **FACTORING A DIFFERENCE OF SQUARES**

Factor each difference of two squares.

(a) $x^2 - 49 = x^2 - 7^2 = (x + 7)(x - 7)$

(b) $y^2 - m^2 = (y + m)(y - m)$

(c) $z^2 - \frac{9}{16} = z^2 - \left(\frac{3}{4}\right)^2 = \left(z + \frac{3}{4}\right)\left(z - \frac{3}{4}\right)$

(d) $p^2 + 16$

This polynomial is not the *difference* of two squares. Using FOIL,

$$(p + 4)(p - 4) = p^2 - 16$$

$$(p - 4)(p - 4) = p^2 - 8p + 16$$

and

$$(p + 4)(p + 4) = p^2 + 8p + 16$$

so $p^2 + 16$ is a prime polynomial. ■

CAUTION As Example 1(d) suggests, the sum of two squares usually cannot be factored.

EXAMPLE 2 **FACTORING A DIFFERENCE OF SQUARES**

Factor $25m^2 - 16$.

This is the difference of two squares, since

$$25m^2 - 16 = (5m)^2 - 4^2.$$

Now factor $(5m)^2 - 4^2$ as

$$(5m + 4)(5m - 4).$$ ■

EXAMPLE 3 **FACTORING MORE COMPLEX DIFFERENCES OF SQUARES**

Factor completely.

(a) $9a^2 - 4b^2 = (3a)^2 - (2b)^2 = (3a + 2b)(3a - 2b)$

(b) $81y^2 - 36$

First factor out the common factor of 9.

$$81y^2 - 36 = 9(9y^2 - 4)$$
$$= 9(3y + 2)(3y - 2)$$

(c) $p^4 - 36 = (p^2)^2 - 6^2 = (p^2 + 6)(p^2 - 6)$

Neither $p^2 + 6$ nor $p^2 - 6$ can be factored further.

(d) $m^4 - 16 = (m^2)^2 - 4^2$

$$= (m^2 + 4)(m^2 - 4) \qquad \text{Difference of squares}$$
$$= (m^2 + 4)(m + 2)(m - 2). \qquad \text{Difference of squares}$$ ■

CAUTION A common error is to forget to factor the difference of two squares a second time when several steps are required, as in Example 3(d).

2 The expressions 144, $4x^2$, and $81m^6$ are called *perfect squares,* since

$$144 = 12^2, \qquad 4x^2 = (2x)^2, \qquad \text{and} \qquad 81m^6 = (9m^3)^2.$$

A **perfect square trinomial** is a trinomial that is the square of a binomial. For example, $x^2 + 8x + 16$ is a perfect square trinomial, since it is the square of the binomial $x + 4$:

$$x^2 + 8x + 16 = (x + 4)^2.$$

For a trinomial to be a perfect square, two of its terms must be perfect squares. For this reason, $16x^2 + 4x + 15$ cannot be a perfect square trinomial since only the term $16x^2$ is a perfect square.

On the other hand, even though two of the terms are perfect squares, the trinomial may not be a perfect square trinomial. For example, $x^2 + 6x + 36$ has two perfect square terms, but it is not a perfect square trinomial. (Try to find a binomial that can be squared to give $x^2 + 6x + 36$.)

Multiply to see that the square of a binomial gives the following perfect square trinomials.

PERFECT SQUARE TRINOMIAL

$$x^2 + 2xy + y^2 = (x + y)^2$$
$$x^2 - 2xy + y^2 = (x - y)^2$$

The middle term of a perfect square trinomial is always twice the product of the two terms in the squared binomial. (This was shown in Section 3.3.) Use this to check any attempt to factor a trinomial that appears to be a perfect square.

EXAMPLE 4
FACTORING A PERFECT SQUARE TRINOMIAL

Factor $x^2 + 10x + 25$.

The term x^2 is a perfect square, and so is 25. Try to factor the trinomial as

$$x^2 + 10x + 25 = (\mathbf{x} + \mathbf{5})^2.$$

To check, take twice the product of the two terms in the squared binomial.

$$\text{Twice} \rightarrow \mathbf{2} \cdot \mathbf{x} \cdot \mathbf{5} = 10x$$

First term of binomial (x); Last term of binomial (5)

Since $10x$ is the middle term of the trinomial, the trinomial is a perfect square and can be factored as $(x + 5)^2$. ■

EXAMPLE 5
FACTORING PERFECT SQUARE TRINOMIALS

Factor each perfect square trinomial.

(a) $x^2 - 22x + 121$

The first and last terms are perfect squares ($121 = 11^2$). Check to see whether the middle term of $x^2 - 22x + 121$ is twice the product of the first and last terms of the binomial $x - 11$.

$$\text{Twice} \rightarrow \mathbf{2} \cdot \mathbf{x} \cdot \mathbf{11} = 22x$$

First term of binomial (x); Last term of binomial (11)

Since twice the product of the first and last terms of the binomial is the middle term, $x^2 - 22x + 121$ is a perfect square trinomial and

$$x^2 - 22x + 121 = (x - 11)^2.$$

(b) $9m^2 - 24m + 16 = (3m)^2 - 2(3m)(4) + 4^2 = (3m - 4)^2$

Twice (2); First term ($3m$); Last term (4)

(c) $25y^2 + 20y + 16$

The first and last terms are perfect squares.

$$25y^2 = (5y)^2 \qquad \text{and} \qquad 16 = 4^2$$

Twice the product of the first and last terms of the binomial $5y + 4$ is

$$2 \cdot 5y \cdot 4 = 40y,$$

which is not the middle term of $25y^2 + 20y + 16$. This polynomial is not a perfect square. In fact, the polynomial cannot be factored even with the methods of Section 4.3; it is a prime polynomial. ■

NOTE The sign of the second term in the squared binomial is always the same as the sign of the middle term in the trinomial. Also, the first and last terms of a perfect square trinomial must be positive, since they are squares. For example, the polynomial $x^2 - 2x - 1$ cannot be a perfect square because the last term is negative.

3 The difference of two squares was factored above; it is possible also to factor the **difference of two cubes.** Use the following pattern.

DIFFERENCE OF TWO CUBES

$$x^3 - y^3 = (x - y)(x^2 + xy + y^2)$$

This pattern *should be memorized*. Multiply on the right to see that the pattern gives the correct factors.

$$\begin{array}{r} x^2 + xy + y^2 \\ x - y \\ \hline -x^2y - xy^2 - y^3 \\ x^3 + x^2y + xy^2 \quad\quad \\ \hline x^3 \quad\quad\quad\quad\quad - y^3 \end{array}$$

CAUTION The polynomial $x^3 - y^3$ is not equivalent to $(x - y)^3$, because $(x - y)^3$ is factored as

$$(x - y)^3 = (x - y)(x - y)(x - y)$$
$$= (x - y)(x^2 - 2xy + y^2)$$

but

$$x^3 - y^3 = (x - y)(x^2 + xy + y^2).$$

EXAMPLE 6 **FACTORING DIFFERENCES OF CUBES**

Factor the following.

(a) $m^3 - 125$

Let $x = m$ and $y = 5$ in the pattern for the difference of two cubes.

$$x^3 - y^3 = (x - y)(x^2 + xy + y^2)$$
$$m^3 - 125 = m^3 - 5^3 = (m - 5)(m^2 + 5m + 5^2) \quad \text{Let } x = m, y = 5.$$
$$= (m - 5)(m^2 + 5m + 25)$$

(b) $8p^3 - 27$

Substitute into the rule using $2p$ for x and 3 for y.

$$8p^3 - 27 = (2p)^3 - 3^3$$
$$= (2p - 3)[(2p)^2 + (2p)3 + 3^2]$$
$$= (2p - 3)(4p^2 + 6p + 9)$$

(c) $4m^3 - 32 = 4(m^3 - 8)$

$$= 4(m^3 - 2^3)$$
$$= 4(m - 2)(m^2 + 2m + 4)$$

(d) $125t^3 - 216s^6 = (5t)^3 - (6s^2)^3$

$$= (5t - 6s^2)[(5t)^2 + (5t)(6s^2) + (6s^2)^2]$$
$$= (5t - 6s^2)(25t^2 + 30ts^2 + 36s^4) \quad ■$$

CAUTION A common error in factoring the difference of two cubes, $x^3 - y^3 = (x - y)(x^2 + xy + y^2)$, is to try to factor $x^2 + xy + y^2$. It is easy to confuse this factor with a perfect square trinomial, $x^2 + 2xy + y^2$. Because of the lack of a 2 in $x^2 + xy + y^2$, it is very unusual to be able to further factor an expression of the form $x^2 + xy + y^2$.

4 A sum of two squares, such as $m^2 + 25$, cannot be factored, but the **sum of two cubes** can be factored by the following pattern, *which should be memorized.*

SUM OF TWO CUBES

$$x^3 + y^3 = (x + y)(x^2 - xy + y^2)$$

Compare the pattern for the *sum* of two cubes with the pattern for the *difference* of two cubes. The only difference between them is the positive and negative signs.

$$a^3 - b^3 = (a - b)(a^2 + ab + b^2)$$

$$a^3 + b^3 = (a + b)(a^2 - ab + b^2)$$

Observing these relationships should help you to remember these patterns.

EXAMPLE 7 FACTORING SUMS OF CUBES

Factor.

(a) $k^3 + 27 = k^3 + 3^3$

$= (k + 3)(k^2 - 3k + 3^2)$

$= (k + 3)(k^2 - 3k + 9)$

(b) $8m^3 + 125 = (2m)^3 + 5^3$

$= (2m + 5)[(2m)^2 - (2m)(5) + 5^2]$

$= (2m + 5)(4m^2 - 10m + 25)$

(c) $1000a^6 + 27b^3 = (10a^2)^3 + (3b)^3$

$= (10a^2 + 3b)[(10a^2)^2 - (10a^2)(3b) + (3b)^2]$

$= (10a^2 + 3b)(100a^4 - 30a^2b + 9b^2)$ ■

The methods of factoring discussed in this section are summarized here. All these rules should be memorized.

SPECIAL FACTORIZATIONS

Difference of two squares	$x^2 - y^2 = (x + y)(x - y)$
Perfect square trinomials	$x^2 + 2xy + y^2 = (x + y)^2$
	$x^2 - 2xy + y^2 = (x - y)^2$
Difference of two cubes	$x^3 - y^3 = (x - y)(x^2 + xy + y^2)$
Sum of two cubes	$x^3 + y^3 = (x + y)(x^2 - xy + y^2)$

CAUTION Remember the *sum* of two *squares* usually cannot be factored.

4.4 EXERCISES

Factor each binomial completely. If a polynomial cannot be factored, write prime. *See Examples 1–3.*

1. $x^2 - 16$ **2.** $m^2 - 25$ **3.** $a^2 - b^2$ **4.** $r^2 - t^2$

5. $m^2 - 1$ **6.** $y^2 - 9$ **7.** $25m^2 - 16$ **8.** $144y^2 - 25$

9. $36t^2 - 16$ **10.** $9 - 36a^2$ **11.** $25a^2 - 16r^2$ **12.** $100k^2 - 49m^2$

13. $x^2 + 16$ **14.** $m^2 + 100$ **15.** $p^4 - 49$ **16.** $r^4 - 9$

17. $a^4 - 1$ **18.** $x^4 - 16$ **19.** $m^4 - 81$ **20.** $p^4 - 256$

Factor any perfect square trinomials. See Examples 4 and 5.

21. $a^2 + 4a + 4$ **22.** $p^2 + 2p + 1$ **23.** $x^2 - 10x + 25$

24. $y^2 - 8y + 16$ **25.** $49 + 14a + a^2$ **26.** $100 - 20m + m^2$

27. $k^2 + 121 + 22k$ **28.** $r^2 + 144 + 24r$ **29.** $y^2 - 10y + 100$

30. $4c^2 + 12c + 9$ **31.** $16t^2 - 40t + 25$ **32.** $25h^2 - 20h + 4$

33. $9x^2 + 24x + 16$ **34.** $100a^2 - 140ab + 49b^2$ **35.** $49x^2 + 28xy + 4y^2$

36. $64y^2 - 48ya + 9a^2$ **37.** $x^3y + 6x^2y^2 + 9xy^3$ **38.** $4k^3w + 20k^2w^2 + 25kw^3$

Factor each sum or difference of cubes. See Examples 6 and 7.

39. $a^3 + 1$ **40.** $a^3 - 1$ **41.** $x^3 - 8$ **42.** $m^3 + 8$

43. $p^3 + q^3$ **44.** $k^3 - h^3$ **45.** $27x^3 - 1$ **46.** $64p^3 + n^3$

47. $8p^3 + q^3$ **48.** $y^3 - 8x^3$ **49.** $27a^3 - 64b^3$ **50.** $125t^3 + 8s^3$

51. $64x^3 + 125y^3$ **52.** $216z^3 - w^3$ **53.** $125m^3 - 8p^3$ **54.** $27r^3 + 1000s^3$

Factor completely in Exercises 55–64.

55. $64y^6 + 1$ **56.** $m^6 - 8$ **57.** $8k^6 - 27q^3$ **58.** $125z^3 + 64r^6$

59. $(m + n)^2 - (m - n)^2$ **60.** $(a - b)^3 - (a + b)^3$

61. $(x^2 + 2x + 1) - 4$

62. $(a + 1)^2 - 4$

63. $m^2 - p^2 + 2m + 2p$

64. $3r - 3k + 3r^2 - 3k^2$

Find the value of the indicated variables. (See Objective 2.)

65. Find a value of b so that $x^2 + bx + 25 = (x + 5)^2$.

66. For what value of c is $4m^2 - 12m + c = (2m - 3)^2$?

67. Find a so that $ay^2 - 12y + 4 = (3y - 2)^2$.

68. Find b so that $100a^2 + ba + 9 = (10a + 3)^2$.

---

69. In your own words, explain why the pattern $x^2 + y^2$ is prime. (See Objective 1.)

70. Why is the sum of squares $4x^2 + 64y^2$ factorable?

PREVIEW EXERCISES

Solve each equation. See Section 2.2.

71. $m - 2 = 0$

72. $r + 1 = 0$

73. $3k - 2 = 0$

74. $4z + 5 = 0$

75. $7a + 9 = 0$

76. $3x + 7 = 0$

77. $8y + 5 = 0$

78. $12k - 11 = 0$

SUMMARY EXERCISES ON FACTORING*

As you factor a polynomial, these questions will help you decide on a suitable factoring technique.

FACTORING A POLYNOMIAL

Step 1 Is there a common factor?

Step 2 How many terms are in the polynomial?

Two terms Check to see whether it is either the difference of two squares or the sum or difference of two cubes.

Three terms Is it a perfect square trinomial? If the trinomial is not a perfect square, check to see whether the coefficient of the squared term is 1. If so, use the method of Section 4.2. If the coefficient of the squared term of the trinomial is not 1, use the general factoring methods of Section 4.3.

Four terms Can the polynomial be factored by grouping?

Step 3 Can any factors be factored further?

---

*This exercise set includes all kinds of factoring methods. The exercises are randomly mixed to give you practice at deciding which method should be used.

EXERCISES

Factor as completely as possible.

1. $a^2 - 4a - 12$
2. $a^2 + 17a + 72$
3. $6y^2 - 6y - 12$
4. $7y^6 + 14y^5 - 168y^4$
5. $6a + 12b + 18c$
6. $m^2 - 3mn - 4n^2$
7. $p^2 - 17p + 66$
8. $z^2 - 6z + 7z - 42$
9. $10z^2 - 7z - 6$
10. $2m^2 - 10m - 48$
11. $m^2 - n^2 + 5m - 5n$
12. $15y + 5$
13. $8a^5 - 8a^4 - 48a^3$
14. $8k^2 - 10k - 3$
15. $z^2 - 3za - 10a^2$
16. $50z^2 - 100$
17. $x^2 - 4x - 5x + 20$
18. $100n^2r^2 + 30nr^3 - 50n^2r$
19. $6n^2 - 19n + 10$
20. $9y^2 + 12y - 5$
21. $16x + 20$
22. $m^2 + 2m - 15$
23. $6y^2 - 5y - 4$
24. $m^2 - 81$
25. $6z^2 + 31z + 5$
26. $5z^2 + 24z - 5 + 3z + 15$
27. $4k^2 - 12k + 9$
28. $8p^2 + 23p - 3$
29. $54m^2 - 24z^2$
30. $8m^2 - 2m - 3$
31. $3k^2 + 4k - 4$
32. $45a^3b^5 - 60a^4b^2 + 75a^6b^4$
33. $14k^3 + 7k^2 - 70k$
34. $5 + r - 5s - rs$
35. $y^4 - 16$
36. $20y^5 - 30y^4$
37. $8m - 16m^2$
38. $k^2 - 16$
39. $z^3 - 8$
40. $y^2 - y - 56$
41. $k^2 + 9$
42. $27p^{10} - 45p^9 - 252p^8$
43. $32m^9 + 16m^5 + 24m^3$
44. $8m^3 + 125$
45. $16r^2 + 24rm + 9m^2$
46. $z^2 - 12z + 36$
47. $15h^2 + 11hg - 14g^2$
48. $5z^3 - 45z^2 + 70z$
49. $k^2 - 11k + 30$
50. $64p^2 - 100m^2$
51. $3k^3 - 12k^2 - 15k$
52. $y^2 - 4yk - 12k^2$
53. $1000p^3 + 27$
54. $64r^3 - 343$
55. $6 + 3m + 2p + mp$
56. $2m^2 + 7mn - 15n^2$
57. $16z^2 - 8z + 1$
58. $125m^4 - 400m^3n + 195m^2n^2$
59. $108m^2 - 36m + 3$
60. $100a^2 - 81y^2$
61. $64m^2 - 40mn + 25n^2$
62. $4y^2 - 25$
63. $32z^3 + 56z^2 - 16z$
64. $10m^2 + 25m - 60$
65. $20 + 5m + 12n + 3mn$
66. $4 - 2q - 6p + 3pq$
67. $6a^2 + 10a - 4$
68. $36y^6 - 42y^5 - 120y^4$
69. $a^3 - b^3 + 2a - 2b$
70. $16k^2 - 48k + 36$
71. $64m^2 - 80mn + 25n^2$
72. $72y^3z^2 + 12y^2 - 24y^4z^2$
73. $8k^2 - 2kh - 3h^2$
74. $2a^2 - 7a - 30$
75. $(m + 1)^3 + 1$
76. $8a^3 - 27$
77. $10y^2 - 7yz - 6z^2$
78. $m^2 - 4m + 4$
79. $8a^2 + 23ab - 3b^2$
80. $a^4 - 625$
81. $9m^2 - 64$
82. $24k^4p + 60k^3p^2 + 150k^2p^3$
83. $9z^2 + 64$
84. $15t - 15 - t^2 + 1$
85. $a^2 + 8a + 16$
86. $z^2 - z^3 + m - mz$

4.5 SOLVING QUADRATIC EQUATIONS BY FACTORING

OBJECTIVES

1 SOLVE QUADRATIC EQUATIONS BY FACTORING.

2 SOLVE OTHER EQUATIONS BY FACTORING.

In this section we introduce *quadratic equations,* equations that contain a squared term and no terms of higher degree.

QUADRATIC EQUATION

An equation that can be put in the form

$$ax^2 + bx + c = 0,$$

where a, b, and c are real numbers, with $a \neq 0$, is a **quadratic equation.**

The form $ax^2 + bx + c = 0$ is the **standard form** of a quadratic equation. For example,

$$x^2 + 5x + 6 = 0, \qquad 2a^2 - 5a = 3, \qquad \text{and} \qquad y^2 = 4$$

are all quadratic equations but only $x^2 + 5x + 6 = 0$ is in standard form.

1 Some quadratic equations can be solved by factoring. A more general method for solving those equations that cannot be solved by factoring is given in Chapter 9. We use the **zero-factor property** to solve a quadratic equation by factoring.

ZERO-FACTOR PROPERTY

If a and b are real numbers and if $ab = 0$, then $a = 0$ or $b = 0$.

In other words, if the product of two numbers is zero, then at least one of the numbers must be zero. This means that one number *must* be 0, but both *may be* 0.

EXAMPLE 1 USING THE ZERO-FACTOR PROPERTY

Solve the equation $(x + 3)(2x - 1) = 0$.

The product $(x + 3)(2x - 1)$ is equal to zero. By the zero-factor property, the only way that the product of these two factors can be zero is if at least one of the factors is zero. Therefore, either $x + 3 = 0$ or $2x - 1 = 0$. Solve each of these two linear equations as in Chapter 2.

$$\begin{aligned} x + 3 &= 0 \quad \text{or} \quad & 2x - 1 &= 0 & & \\ x &= -3 \quad \text{or} \quad & 2x &= 1 & & \text{Add to both sides.} \\ & & x &= \frac{1}{2} & & \text{Divide by 2.} \end{aligned}$$

Since both of these equations have a solution, the given equation $(x + 3)(2x - 1) = 0$ has two solutions, -3 and $1/2$. Check these answers by substituting -3 for x in the original equation. Then start over and substitute $1/2$ for x.

If $x = -3$, then

$$(-3 + 3)[2(-3) - 1] = 0 \quad ?$$
$$0(-7) = 0 \quad ?$$
$$0 = 0. \quad \text{True}$$

If $x = 1/2$, then

$$\left(\frac{1}{2} + 3\right)\left(2 \cdot \frac{1}{2} - 1\right) = 0 \quad ?$$
$$\frac{7}{2}(1 - 1) = 0 \quad ?$$
$$\frac{7}{2} \cdot 0 = 0 \quad ?$$
$$0 = 0. \quad \text{True}$$

Both -3 and $1/2$ produce true statements, so they are solutions to the original equation. ■

NOTE The word "or" as used in Example 1 means "one or the other or both."

In Example 1 the equation to be solved was presented with the polynomial in factored form. If the polynomial in an equation is not already factored, first make sure that the equation is in standard form. Then factor the polynomial.

EXAMPLE 2 **SOLVING A QUADRATIC EQUATION NOT IN STANDARD FORM**

Solve the equation $x^2 - 5x = -6$.

First, rewrite the equation with all terms on one side by adding 6 on both sides.

$$x^2 - 5x = -6$$
$$x^2 - 5x + 6 = 0 \qquad \text{Add 6.}$$

Now factor $x^2 - 5x + 6$. Find two numbers whose product is 6 and whose sum is -5. These two numbers are -2 and -3, so the equation becomes

$$(x - 2)(x - 3) = 0.$$

Proceed as in Example 1. Set each factor equal to 0.

$$x - 2 = 0 \quad \text{or} \quad x - 3 = 0$$

Solve the equation on the left by adding 2 on both sides. In the equation on the right, add 3 on both sides. Doing this gives

$$x = 2 \quad \text{or} \quad x = 3.$$

Check both solutions by substituting first 2 and then 3 for x in the original equation. ■

In summary, go through the following steps to solve quadratic equations by factoring.

SOLVING A QUADRATIC EQUATION BY FACTORING

Step 1 **Write in standard form.** Write the equation in standard form: all terms on one side of the equals sign, with 0 on the other side.

Step 2 **Factor.** Factor completely.

Step 3 **Use the zero-factor property.** Set each factor with a variable equal to 0, and solve the resulting equations.

Step 4 **Check.** Check each solution in the original equation.

EXAMPLE 3 **SOLVING A QUADRATIC EQUATION WITH A COMMON FACTOR**

Solve the equation $4p^2 + 40 = 26p$.

Subtract $26p$ from each side and write in descending powers to get

$$4p^2 - 26p + 40 = 0.$$

Factor out the common factor of 2:

$$2(2p^2 - 13p + 20) = 0.$$

Factor $2p^2 - 13p + 20$ as $(2p - 5)(p - 4)$, giving

$$2(2p - 5)(p - 4) = 0.$$

Set each of these three factors equal to 0.

$$2 = 0 \quad \text{or} \quad 2p - 5 = 0 \quad \text{or} \quad p - 4 = 0$$

The equation $2 = 0$ has no solution. Solve the equation in the middle by first adding 5 on both sides of the equation. Then divide both sides by 2. Solve the equation on the right by adding 4 to both sides.

$$\begin{aligned} 2p - 5 &= 0 \quad \text{or} \quad p - 4 = 0 \\ 2p &= 5 \quad \text{or} \quad p = 4 \\ p &= \frac{5}{2} \end{aligned}$$

The solutions of $4p^2 + 40 = 26p$ are 5/2 and 4; check them by substituting in the original equation. ■

CAUTION A common error is to include 2 as a solution in Example 3.

EXAMPLE 4 **SOLVING QUADRATIC EQUATIONS**

Solve each equation.

(a) $16m^2 - 25 = 0$

Factor the left-hand side of the equation as the difference of two squares.

$$(4m + 5)(4m - 5) = 0$$

Set each factor equal to 0.

$$4m + 5 = 0 \quad \text{or} \quad 4m - 5 = 0$$

Solve each equation.

$$4m = -5 \quad \text{or} \quad 4m = 5$$

$$m = -\frac{5}{4} \quad \text{or} \quad m = \frac{5}{4}$$

The two solutions, $-5/4$ and $5/4$, should be checked in the original equation.

(b) $y^2 = 2y$

First write the equation in standard form.

$$y^2 - 2y = 0 \qquad \text{Standard form}$$

$$y(y - 2) = 0 \qquad \text{Factor.}$$

$$y = 0 \quad \text{or} \quad y - 2 = 0 \qquad \text{Set each factor equal to 0.}$$

$$y = 2 \qquad \text{Solve.}$$

The solutions are 0 and 2.

(c) $k(2k + 1) = 3$

Multiply on the left-hand side and then get all terms on one side.

$$k(2k + 1) = 3$$

$$2k^2 + k = 3$$

$$2k^2 + k - 3 = 0$$

Now factor.

$$(k - 1)(2k + 3) = 0$$

Set each factor equal to 0 and solve the equations.

$$k - 1 = 0 \quad \text{or} \quad 2k + 3 = 0$$

$$k = 1 \quad \text{or} \quad 2k = -3$$

$$k = -\frac{3}{2}$$

The two solutions are 1 and $-3/2$. ■

CAUTION In Example 4(b) it is tempting to begin by dividing both sides of the equation by y to get $y = 2$. Note, however, that the other solution, 0, is not found by this method.

In Example 4(c) the zero-factor property could not be used to solve the original equation because of the 3 on the right. Remember that the zero-factor property applies only to a product that equals 0.

2 The zero-factor property also can be used to solve equations that result in more than two factors with variables, as shown in Example 5. (These equations are *not* quadratic equations. Why not?)

EXAMPLE 5
SOLVING EQUATIONS WITH MORE THAN TWO VARIABLE FACTORS

Solve the equation $6z^3 - 6z = 0$.

First, factor out the greatest common factor in $6z^3 - 6z$.

$$6z^3 - 6z = 0$$
$$6z(z^2 - 1) = 0$$

Now factor $z^2 - 1$ as $(z + 1)(z - 1)$ to get

$$6z(z + 1)(z - 1) = 0.$$

By an extension of the zero-factor property, this product can equal zero only if at least one of the factors is zero. Write three equations, one for each factor with a variable.

$$6z = 0 \quad \text{or} \quad z + 1 = 0 \quad \text{or} \quad z - 1 = 0$$

Solving these three equations gives three solutions,

$$z = 0 \quad \text{or} \quad z = -1 \quad \text{or} \quad z = 1.$$

Check by substituting, in turn, 0, -1, and 1 in the original equation. ■

EXAMPLE 6
SOLVING EQUATIONS WITH MORE THAN TWO VARIABLE FACTORS

Solve the equation $(3x - 1)(x^2 - 9x + 20) = 0$.

Factor $x^2 - 9x + 20$ as $(x - 5)(x - 4)$. Then rewrite the original equation as

$$(3x - 1)(x - 5)(x - 4) = 0.$$

Set each of these three factors equal to 0.

$$3x - 1 = 0 \quad \text{or} \quad x - 5 = 0 \quad \text{or} \quad x - 4 = 0$$

Solving these three equations gives

$$x = \frac{1}{3} \quad \text{or} \quad x = 5 \quad \text{or} \quad x = 4$$

as the solutions of the original equation. Check each solution. ■

CAUTION In Example 6, it would be unproductive to begin by multiplying the two factors together. Keep in mind the zero-factor property requires the product of two or more factors equal to zero. Always consider first whether an equation is given in the appropriate form for the zero-factor property.

4.5 EXERCISES

Solve each equation. See Example 1.

1. $(x - 2)(x + 4) = 0$
2. $(y - 3)(y + 5) = 0$
3. $(3x + 5)(2x - 1) = 0$
4. $(2a + 3)(a - 2) = 0$
5. $(5p + 1)(2p - 1) = 0$
6. $(3k - 8)(k + 7) = 0$
7. $(2m + 9)(3m - 1) = 0$
8. $(9a - 2)(3a + 1) = 0$
9. $(x - 1)(3x + 5) = 0$
10. $(k - 3)(k + 5) = 0$
11. $(3r - 7)(2r + 8) = 0$
12. $(5a + 2)(3a - 1) = 0$

Solve each equation. See Examples 2–4.

13. $x^2 + 5x + 6 = 0$
14. $y^2 - 3y + 2 = 0$
15. $r^2 - 5r - 6 = 0$
16. $y^2 - y - 12 = 0$
17. $m^2 + 3m - 28 = 0$
18. $p^2 - p - 6 = 0$
19. $a^2 = 24 - 5a$
20. $m^2 = 3m + 4$
21. $z^2 = -2 - 3z$
22. $p^2 = 2p + 3$
23. $3a^2 + 5a - 2 = 0$
24. $6r^2 - r - 2 = 0$
25. $2k^2 - k - 10 = 0$
26. $6x^2 - 7x - 5 = 0$
27. $18a^2 = 15 - 39a$
28. $18s^2 + 24s = -8$
29. $25p^2 + 20p + 4 = 0$
30. $10b^2 + 15b - 45 = 0$
31. $20b^2 = 32b + 16$
32. $16r^2 - 25 = 0$
33. $4k^2 - 9 = 0$
34. $9m^2 - 36 = 0$
35. $16x^2 - 64 = 0$
36. $6b^2 - 4b = 0$
37. $2c^2 + 3c = 0$
38. $5x^2 = 5x$
39. $6z^2 = 10z$
40. $m(m - 7) = -10$
41. $z(2z + 7) = 4$
42. $2(x^2 - 66) = -13x$
43. $3(m^2 + 4) = 20m$
44. $3r(r + 1) = (2r + 3)(r + 1)$
45. $(3k + 1)(k + 1) = 2k(k + 3)$
46. $12k(k - 4) = 3(k - 4)$
47. $y^2 = 4(y - 1)$

Solve each equation. See Examples 5 and 6.

48. $(2r - 5)(3r^2 - 16r + 5) = 0$
49. $(3m - 4)(6m^2 + m - 2) = 0$
50. $(2x + 7)(x^2 - 2x - 3) = 0$
51. $(x - 1)(6x^2 + x - 12) = 0$
52. $x^3 - 25x = 0$
53. $m^3 - 4m = 0$
54. $9y^3 - 49y = 0$
55. $16r^3 - 9r = 0$
56. $r^3 - 2r^2 - 8r = 0$
57. $x^3 - x^2 - 6x = 0$
58. $a^3 + a^2 - 20a = 0$
59. $y^3 - 6y^2 + 8y = 0$
60. $r^4 = 2r^3 + 15r^2$
61. $x^3 = 3x + 2x^2$
62. $6p^2(p + 1) = 4(p + 1) - 5p(p + 1)$
63. $6x^2(2x + 3) - 5x(2x + 3) = 4(2x + 3)$
64. $(k + 3)^2 - (2k - 1)^2 = 0$
65. $(4y - 3)^3 - 9(4y - 3) = 0$

Solve each problem.

66. Explain why the solutions of $(x - 3)(x + 2) = 1$ are not found from the two equations

$$x - 3 = 1 \qquad \text{and} \qquad x + 2 = 1.$$

(See Objective 1.)

67. What is wrong with the following solution?

$$4x^2 = 4x$$
$$x = 1 \qquad \text{Divide both sides by } 4x.$$

(See Objective 1.)

PREVIEW EXERCISES

Solve each problem. See Sections 2.4 and 2.5.

68. An animal lover has 3 more cats than dogs. She has a total of 7 cats and dogs. How many cats does she have?

69. A small motorboat engine requires 1/5 as much oil as gasoline. How much oil would be used in 6 gallons of the mixture?

70. The length of a rectangle is 3 meters more than the width. The perimeter of the rectangle is 34 meters. Find the width of the rectangle.

71. A rectangle has a length 4 meters less than twice the width. The perimeter of the rectangle is 4 meters more than five times the width. Find the width of the rectangle.

4.6 APPLICATIONS OF QUADRATIC EQUATIONS

OBJECTIVES

1 SOLVE PROBLEMS ABOUT GEOMETRIC FIGURES.

2 SOLVE PROBLEMS USING THE PYTHAGOREAN FORMULA.

FOCUS ON PROBLEM SOLVING

Wei-Jen works due north of home. Her husband Alan works due east. They leave for work at the same time. By the time Wei-Jen is 5 miles from home, the distance between them is 1 mile more than Alan's distance from home. How far from home is Alan?

Many applied problems require the solution of a quadratic equation in order to solve the original problem. The text and examples in this section should prepare you to solve the problem given above, which is Exercise 19 in the exercises for this section. After working through this section, you should be able to solve this problem.

PROBLEM SOLVING

We are now ready to use factoring to solve quadratic equations that arise from applied problems. Most problems in this section will require one of the formulas given in Appendix A. The general approach is the same as in Chapter 2. We still follow the six steps listed in Section 2.4 and continue the work with formulas and geometric problems begun in Section 2.5. ■

1 We begin with a problem about the area of a rectangle.

EXAMPLE 1 SOLVING AN AREA PROBLEM

The width of a rectangular floor is 4 meters less than the length. The area of the floor is 96 square meters. Find the length and width of the floor.

Let x = the length of the floor;

$x - 4$ = the width (the width is 4 less than the length).

See Figure 4.1. The area of a rectangle is given by the formula

$$\text{area} = LW = \text{length} \times \text{width}.$$

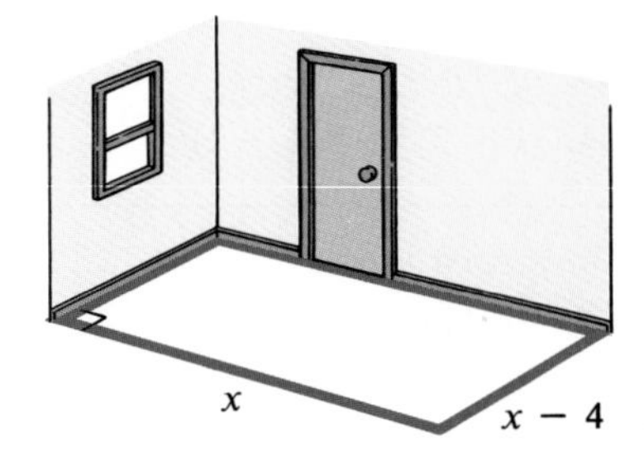

FIGURE 4.1

Substitute 96 for the area, x for the length, and $x - 4$ for the width into the formula.

$$A = LW$$

$$96 = x(x - 4) \quad \text{Let } A = 96,\ L = x,\ W = x - 4$$

Multiply on the right.

$$96 = x^2 - 4x \qquad \text{Distributive property}$$

$$0 = x^2 - 4x - 96 \qquad \text{Subtract 96 from both sides.}$$

$$0 = (x - 12)(x + 8) \qquad \text{Factor.}$$

Set each factor equal to 0.

$$x - 12 = 0 \quad \text{or} \quad x + 8 = 0$$

Solve each equation.

$$x = 12 \quad \text{or} \quad x = -8$$

The solutions of the equations are 12 and -8. Since a rectangle cannot have a negative length, discard the solution -8. Then 12 meters is the length of the floor, and $12 - 4 = 8$ meters is the width. As a check, note that the width is 4 less than the length and the area is $8 \cdot 12 = 96$ square meters as required. ■

CAUTION In an applied problem, always be careful to check solutions against physical facts.

The next application involves *perimeter,* the distance around a figure, as well as area.

EXAMPLE 2

SOLVING AN AREA AND PERIMETER PROBLEM

The length of a rectangular rug is 4 feet more than the width. The area of the rug is numerically 1 more than the perimeter. See Figure 4.2. Find the length and width of the rug.

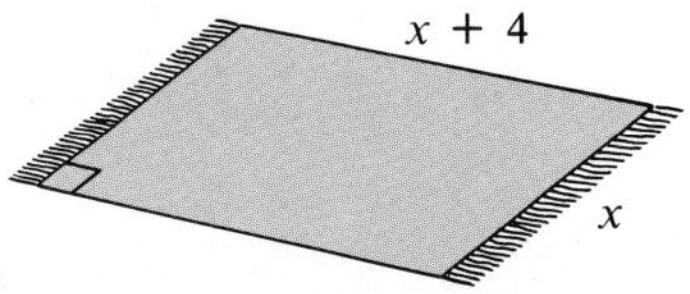

Let x = the width of the rug;

$x + 4$ = the length of the rug.

The area is the product of the length and width, so

$$A = LW.$$

$x + 4$

x

FIGURE 4.2

Substituting $x + 4$ for the length and x for the width gives

$$A = (x + 4)x.$$

Now substitute into the formula for perimeter.

$$P = 2L + 2W$$

$$P = 2(x + 4) + 2x$$

According to the information given in the problem, the area is numerically 1 more than the perimeter.

The area	is	1	more than	the perimeter.
↓	↓	↓	↓	↓
$(x + 4)x$	$=$	1	$+$	$2(x + 4) + 2x$

Simplify and solve this equation.

$$x^2 + 4x = 1 + 2x + 8 + 2x \qquad \text{Distributive property}$$
$$x^2 + 4x = 9 + 4x \qquad \text{Combine terms.}$$
$$x^2 = 9 \qquad \text{Subtract } 4x \text{ from both sides.}$$
$$x^2 - 9 = 0 \qquad \text{Subtract 9 from both sides.}$$
$$(x + 3)(x - 3) = 0 \qquad \text{Factor.}$$
$$x + 3 = 0 \quad \text{or} \quad x - 3 = 0 \qquad \text{Zero-factor property}$$
$$x = -3 \quad \text{or} \quad x = 3$$

A rectangle cannot have a negative width, so ignore -3. The only valid solution is 3, so the width is 3 feet and the length is $3 + 4 = 7$ feet. Check to see that the area is numerically 1 more than the perimeter. The rug is 3 feet wide and 7 feet long. ■

2 The next example requires the **Pythagorean formula** from geometry.

PYTHAGOREAN FORMULA

If a right triangle (a triangle with a 90° angle) has longest side of length c and two other sides of length a and b, then

$$a^2 + b^2 = c^2.$$

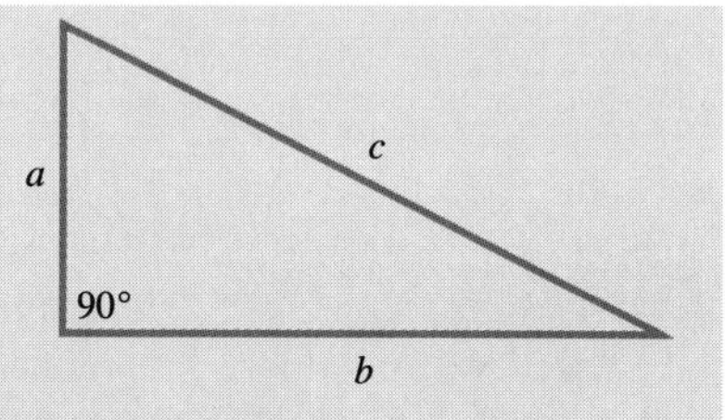

(See the figure.) The longest side, the **hypotenuse,** is opposite the right angle. The two shorter sides are the **legs** of the triangle.

EXAMPLE 3 USING THE PYTHAGOREAN FORMULA

The hypotenuse of a right triangle is 2 feet more than the shorter leg. The longer leg is 1 foot more than the shorter leg. Find the lengths of the sides of the triangle.

Let x be the length of the shorter leg. Then

$$x = \text{shorter leg},$$
$$x + 1 = \text{longer leg},$$
$$x + 2 = \text{hypotenuse}.$$

Place these on a right triangle, as in Figure 4.3.

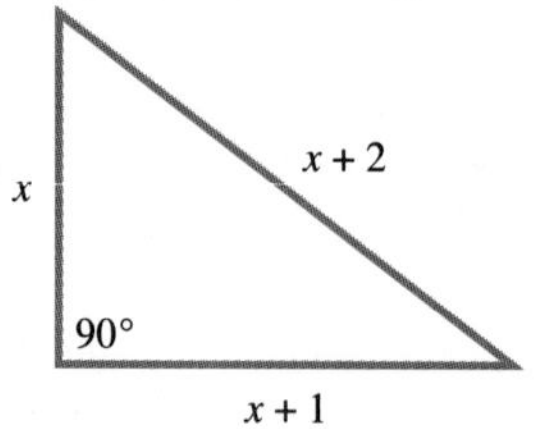

FIGURE 4.3

Substitute into the Pythagorean formula,

$$a^2 + b^2 = c^2$$
$$x^2 + (x + 1)^2 = (x + 2)^2.$$

Since $(x + 1)^2 = x^2 + 2x + 1$, and since $(x + 2)^2 = x^2 + 4x + 4$, the equation becomes

$$x^2 + x^2 + 2x + 1 = x^2 + 4x + 4.$$

$$x^2 - 2x - 3 = 0 \qquad \text{Get 0 on one side.}$$

$$(x - 3)(x + 1) = 0 \qquad \text{Factor.}$$

$$x - 3 = 0 \quad \text{or} \quad x + 1 = 0 \qquad \text{Set each factor equal to 0 and solve.}$$

$$x = 3 \quad \text{or} \quad x = -1$$

Since -1 cannot be the length of a side of a triangle, 3 is the only possible answer. The triangle has a shorter leg of length 3 feet, a longer leg of length $3 + 1 = 4$ feet, and a hypotenuse of length $3 + 2 = 5$ feet. Check that $3^2 + 4^2 = 5^2$. ■

CAUTION When solving a problem involving the Pythagorean formula, be sure that the expressions for the sides are properly placed.

$$\text{leg}^2 + \text{leg}^2 = \text{hypotenuse}^2$$

4.6 EXERCISES

In Exercises 1–4,
(a) *choose the appropriate formula (see Appendix A);*
(b) *write an equation using the formula and the given information;*
(c) *solve the equation;*
(d) *use the solution to find the indicated dimensions of each figure.*

1. Area = 64 square centimeters

$x - 6$

$x + 6$

2. Area = 2 square meters

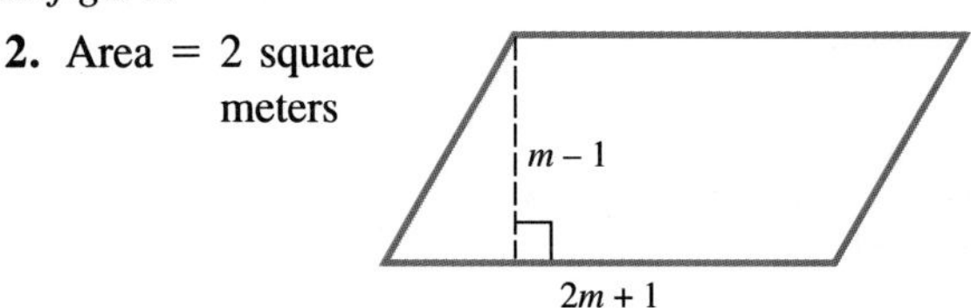

3. Area = 10 square feet

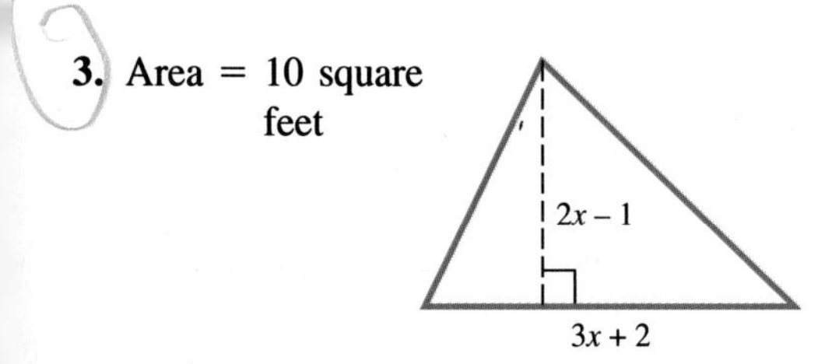

4. Area = 16 square inches

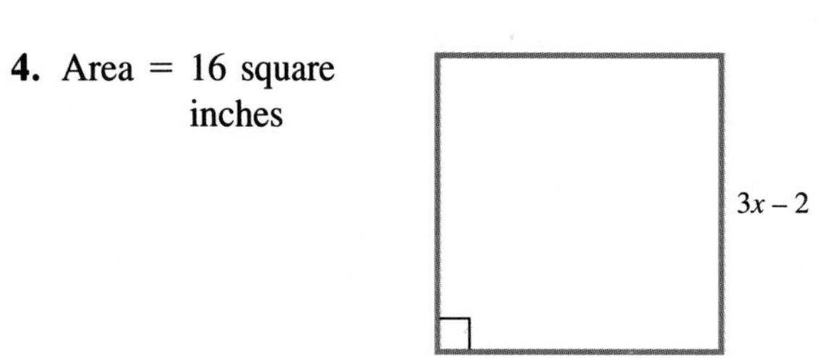

Solve each of the following problems. See Examples 1 and 2.

5. The length of a rectangular picture is 5 centimeters more than the width. The area is 66 square centimeters. Find the width of the picture.

6. The length of the floor of a rectangular closet is 1 foot more than the width. The area is 56 square feet. Find the width of the closet.

7. A manufacturer wants to make a rectangular box with a width 1/2 the length. The area of the bottom of the box is to be 162 square inches. What dimensions should the bottom of the box have?

8. The length of the cover of a book is to be 1.5 times the width. The area is to be 37.5 square inches. What dimensions should be used for the cover?

9. The length of a rectangle is 3 inches more than the width. The area is numerically 4 less than the perimeter. Find the width of the rectangle.

10. The width of a rectangle is 5 meters less than the length. The area is numerically 10 more than the perimeter. Find the length of the rectangle.

Exercises 11 and 12 require the formula for the area of a triangle.

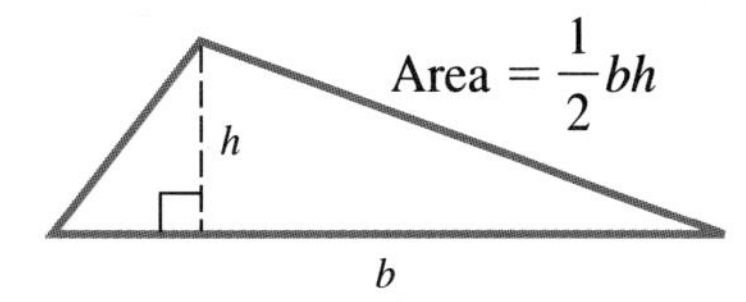

11. The area of a triangle is 25 square centimeters. The base is twice the height. Find the height of the triangle.

12. The height of a triangle is 3 inches more than the base. The area of the triangle is 27 square inches. Find the base of the triangle.

Work the following problems.

13. One square has sides 1 foot less than the length of the sides of a second square. If the difference of the areas of the two squares is 37 square feet, find the length of the side of the second square.

14. The sides of one square have a length 2 meters more than the sides of a second square. If the area of the larger square is subtracted from three times the area of the smaller square, the answer is 12 square meters. Find the length of the side of the second square.

15. Thuy wishes to build a box to hold his tools. The box is to be 4 feet high, and the width of the box is to be 1 foot less than the length. The volume of the box will be 120 cubic feet. Find the length of the box. (*Hint:* The formula for the volume of a box is $V = LWH$.)

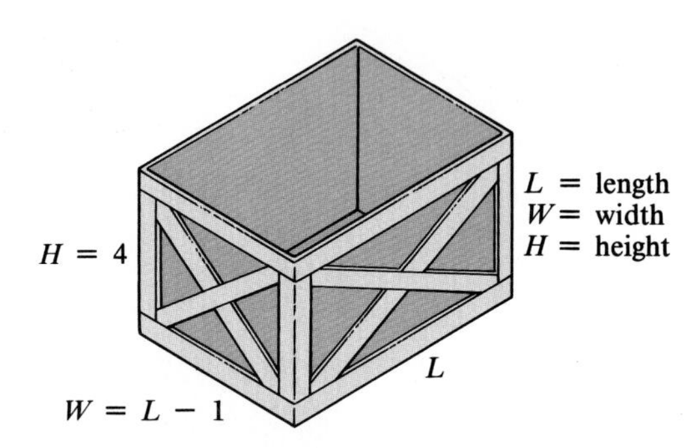

16. The volume of a box must be 315 cubic meters. The length of the box is to be 7 meters, and the height is to be 4 meters more than the width. Find the width of the box.

The following exercises require the Pythagorean formula. See Example 3.

17. The hypotenuse of a right triangle is 1 centimeter longer than the longer leg. The shorter leg is 7 centimeters shorter than the longer leg. Find the length of the longer leg of the triangle.

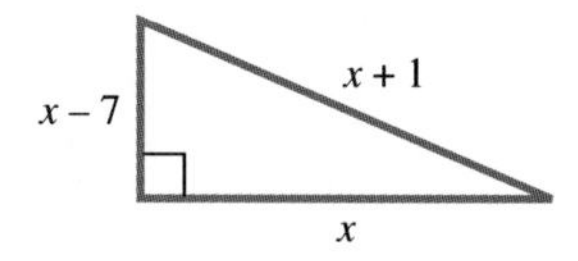

18. The longer leg of a right triangle is 1 meter longer than the shorter leg. The hypotenuse is 1 meter shorter than twice the shorter leg. Find the length of the shorter leg of the triangle.

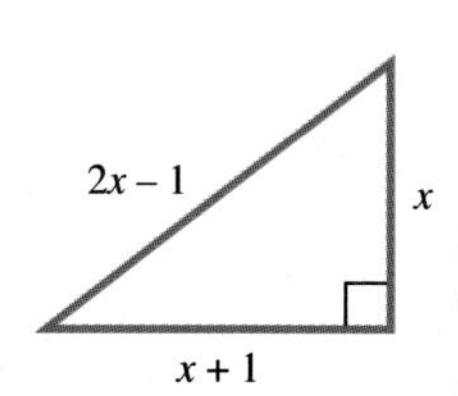

4.7 SOLVING QUADRATIC INEQUALITIES

OBJECTIVE

1 SOLVE QUADRATIC INEQUALITIES AND GRAPH THEIR SOLUTIONS.

1 A **quadratic inequality** is an inequality that involves a second-degree polynomial. Examples of quadratic inequalities include

$$2x^2 + 3x - 5 < 0, \qquad x^2 \le 4, \qquad \text{and} \qquad x^2 + 5x + 6 > 0.$$

Examples 1 and 2 show how to solve such inequalities.

EXAMPLE 1 **SOLVING A QUADRATIC INEQUALITY**

Solve $x^2 - 3x - 10 \le 0$.

To begin, find the solution of the corresponding quadratic equation,

$$x^2 - 3x - 10 = 0.$$

Factor to get $\qquad (x - 5)(x + 2) = 0$

from which $\qquad x - 5 = 0 \quad \text{or} \quad x + 2 = 0$

$$x = 5 \quad \text{or} \quad x = -2.$$

Since 5 and -2 are the only values that satisfy $x^2 - 3x - 10 = 0$, all other values of x will make $x^2 - 3x - 10$ either less than 0 (< 0) or greater than 0 (> 0). The values $x = 5$ and $x = -2$ determine three regions on the number line, as shown in Figure 4.4. Region A includes all numbers less than -2, region B includes the numbers between -2 and 5, and region C includes all numbers greater than 5.

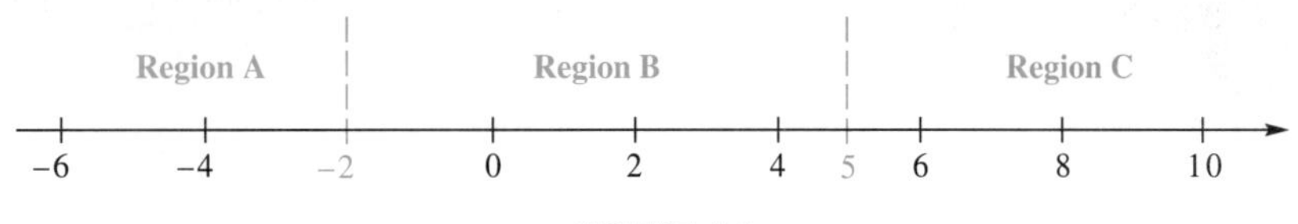

FIGURE 4.4

All values of x in a given region will cause $x^2 - 3x - 10$ to have the same sign (either positive or negative). Test one value of x from each region to see which regions satisfy $x^2 - 3x - 10 \le 0$. First, are the points in region A part of the solution? As a trial value, choose any number less than -2, say -6.

$$x^2 - 3x - 10 \le 0 \qquad \text{Original inequality}$$

$$(-6)^2 - 3(-6) - 10 \le 0 \quad ? \qquad \text{Let } x = -6.$$

$$36 + 18 - 10 \le 0 \quad ? \qquad \text{Simplify.}$$

$$44 \le 0 \qquad \text{False}$$

Since $44 \le 0$ is false, the points in Region A do not belong to the solution.

What about Region B? Try the value $x = 0$.

$$0^2 - 3(0) - 10 \le 0 \quad ? \quad \text{Let } x = 0.$$
$$-10 \le 0 \quad \text{True}$$

Since $-10 \le 0$ is true, the points in Region B do belong to the solution.

Try $x = 6$ to check Region C.

$$6^2 - 3(6) - 10 \le 0 \quad ? \quad \text{Let } x = 6.$$
$$36 - 18 - 10 \le 0 \quad ? \quad \text{Simplify.}$$
$$8 \le 0 \quad \text{False}$$

Since $8 \le 0$ is false, the points in Region C do not belong to the solution.

The points in Region B are the only ones that satisfy $x^2 - 3x - 10 \le 0$. As shown in Figure 4.5, the solution includes the points in Region B together with the endpoints -2 and 5. The solution is written

$$-2 \le x \le 5. \quad ■$$

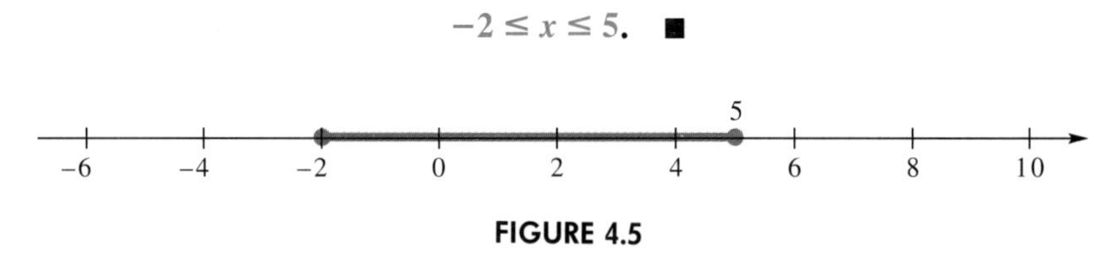

FIGURE 4.5

To summarize, we use the following steps to solve a quadratic inequality.

SOLVING A QUADRATIC INEQUALITY

Step 1 **Write an equation.** Change the inequality to an equation.
Step 2 **Solve.** Use the zero-factor property to solve the equation from Step 1.
Step 3 **Determine the regions.** Use the solutions of the equation in Step 2 to determine regions on the number line.
Step 4 **Test each region.** Choose a number from each region. Substitute the number into the original inequality. If the number satisfies the inequality, all numbers in that region satisfy the inequality.
Step 5 **Write the solutions.** Write the solutions as inequalities.

EXAMPLE 2
SOLVING A QUADRATIC INEQUALITY

Solve $x^2 + 5x + 6 > 0$.

Begin by factoring the corresponding equation $x^2 + 5x + 6 = 0$, to get $(x + 2)(x + 3) = 0$. The solutions of the equation are -2 and -3. These points determine three regions on the number line. See Figure 4.6. This time, these points will not belong to the solution because only values of x that make $x^2 + 5x + 6$ *greater than* 0 are solutions.

Region A Region B Region C

−4 −3 −2 −1 0

FIGURE 4.6

Do the points in Region A belong to the solution? Decide by selecting any point in Region A, such as −4. Does −4 satisfy the original inequality?

$$x^2 + 5x + 6 > 0 \qquad \text{Original inequality}$$
$$(-4)^2 + 5(-4) + 6 > 0 \quad ? \qquad \text{Let } x = -4.$$
$$16 - 20 + 6 > 0 \quad ? \qquad \text{Simplify.}$$
$$2 > 0 \qquad \text{True}$$

Since $2 > 0$ is true, all the points in Region A belong to the solution of the given inequality.

For Region B, choose a value between −3 and −2, say −2 1/2, or −5/2.

$$\left(-\frac{5}{2}\right)^2 + 5\left(-\frac{5}{2}\right) + 6 > 0 \quad ? \qquad \text{Let } x = -\frac{5}{2}.$$
$$\frac{25}{4} + \left(-\frac{25}{2}\right) + 6 > 0 \quad ? \qquad \text{Simplify.}$$
$$-\frac{1}{4} > 0 \qquad \text{False}$$

Since $-1/4 > 0$ is false, no point in Region B belongs to the solution.

For Region C, try the number 0.

$$0^2 + 5(0) + 6 > 0 \quad ? \qquad \text{Let } x = 0.$$
$$6 > 0 \qquad \text{True}$$

Since $6 > 0$ is true, the points in Region C belong to the solution.

The solution is shown in Figure 4.7. The graph of $x^2 + 5x + 6 > 0$ includes all values of x less than −3, and all values of x greater than −2. Write the solution as

$$x < -3 \quad \text{or} \quad x > -2. \blacksquare$$

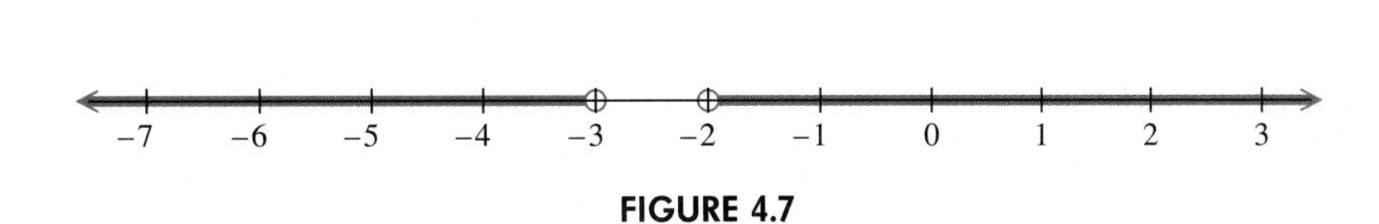

FIGURE 4.7

CAUTION There is no shortcut way to write the solution $x < -3$ or $x > -2$.

4.7 EXERCISES

Solve each inequality and graph the solution. See Examples 1 and 2.

1. $(m + 2)(m - 5) < 0$
2. $(k - 1)(k + 3) > 0$
3. $(t + 6)(t + 5) \geq 0$
4. $(g - 2)(g - 4) \leq 0$
5. $(a + 3)(a - 3) < 0$
6. $(b - 2)(b + 2) > 0$
7. $(a + 6)(a - 7) \geq 0$
8. $(z - 5)(z - 4) \leq 0$
9. $m^2 + 5m + 6 > 0$
10. $y^2 - 3y + 2 < 0$
11. $z^2 - 4z - 5 \leq 0$
12. $3p^2 - 5p - 2 \leq 0$
13. $5m^2 + 3m - 2 < 0$
14. $2k^2 + 7k - 4 > 0$
15. $6r^2 - 5r - 4 < 0$
16. $6r^2 + 7r - 3 > 0$
17. $q^2 - 7q + 6 < 0$
18. $2k^2 - 7k - 15 \leq 0$
19. $6m^2 + m - 1 > 0$
20. $30r^2 + 3r - 6 \leq 0$
21. $12p^2 + 11p + 2 < 0$
22. $a^2 - 16 < 0$
23. $9m^2 - 36 > 0$
24. $r^2 - 100 \geq 0$
25. $r^2 > 16$
26. $m^2 \geq 25$

The following inequalities are not quadratic inequalities, but they may be solved in a similar manner.

27. $(a + 2)(3a - 1)(a - 4) \geq 0$
28. $(2p - 7)(p - 1)(p + 3) \leq 0$
29. $(r - 2)(r^2 - 3r - 4) < 0$
30. $(m + 5)(m^2 - m - 6) > 0$

A sign graph *for an algebraic expression in x is a number line labeled with plus and minus signs to indicate where the expression is positive and where it is negative. Match each of the following expressions with its sign graph below.*

31. $x - 2$
32. $-(x - 2)$
33. $(x - 2)^2$
34. $-(x - 2)^2$

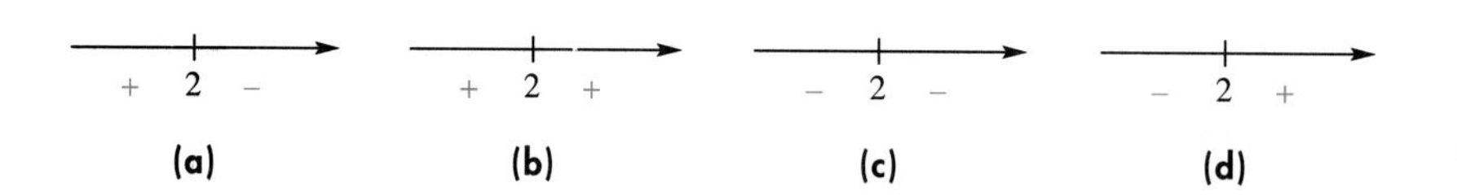

(a) (b) (c) (d)

PREVIEW EXERCISES

Write in lowest terms. See Section 1.1.

35. $\frac{8}{12}$
36. $\frac{5}{10}$
37. $\frac{14}{42}$
38. $\frac{18}{32}$
39. $\frac{50}{72}$
40. $\frac{25}{60}$
41. $\frac{26}{156}$
42. $\frac{34}{136}$

CHAPTER 4 GLOSSARY

KEY TERMS

4.1 factor An expression A is a factor of an expression B if B can be divided by A with zero remainder.

common factor An expression that is a factor of two or more expressions is a common factor of those expressions.

greatest common factor The greatest common factor is the largest quantity that is a factor of each of a group of quantities.

4.2 factored form The indicated product of two polynomials is called the factored form of the product polynomial.

factoring The process of finding the factored form of a polynomial is called factoring.

prime polynomial A prime polynomial is a polynomial that cannot be factored.

4.4 perfect square trinomial A perfect square trinomial is a trinomial that can be factored as the square of a binomial.

4.5 quadratic equation A quadratic equation is an equation that can be written in the form $ax^2 + bx + c = 0$, with $a \neq 0$.

standard form The form $ax^2 + bx + c = 0$ is the standard form of a quadratic equation.

4.6 hypotenuse The longest side of a right triangle, opposite the right angle, is called the hypotenuse.

legs The two shorter sides of a right triangle are called the legs of the triangle.

4.7 quadratic inequality A quadratic inequality is an inequality that involves a second-degree polynomial.

CHAPTER 4 QUICK REVIEW

SECTION	CONCEPTS	EXAMPLES
4.1 FACTORS; THE GREATEST COMMON FACTOR	**Finding the Greatest Common Factor** *Step 1* Write each expression in prime factored form. *Step 2* List each prime that is a factor of every expression in the list. *Step 3* Use as exponents on the common prime factors the *smallest* exponent from the prime factored forms. (If a prime does not appear in one of the prime factored forms, it cannot appear in the greatest common factor.)	Find the greatest common factor of $4x^2y, \quad -6x^2y^3, \quad 2xy^2.$ $4x^2y = 2^2 \cdot x^2 \cdot y$ $-6x^2y^3 = -1 \cdot 2 \cdot 3 \cdot x^2 \cdot y^3$ $2xy^2 = 2 \cdot x \cdot y^2$ The greatest common factor is $2xy$.

SECTION	CONCEPTS	EXAMPLES
	Step 4 Multiply the primes of Step 3. If there are no primes left after Step 3, the greatest common factor is 1.	
	Factoring by Grouping	Factor: $3x^2 + 5x - 24xy - 40y$.
	Step 1 Write the four terms so that the first two have a common factor and the last two have a common factor.	**1.** $(3x^2 + 5x) + (-24xy - 40y)$
	Step 2 Use the distributive property to factor each group of two terms.	**2.** $x(3x + 5) - 8y(3x + 5)$
	Step 3 If possible, factor a common binomial factor from the results of Step 2.	**3.** $(3x + 5)(x - 8y)$
	Step 4 If Step 2 does not result in a common binomial factor, try grouping the terms of the original polynomial in a different way.	
4.2 FACTORING TRINOMIALS	To factor $x^2 + mx + n$, find integers a and b so that $a + b = m$ and $ab = n$. Then $x^2 + mx + n = (x + a)(x + b)$.	Factor: $x^2 + 6x + 8$ $a + b = 6$ and $ab = 8$ $a = 2$ and $b = 4$ $x^2 + 6x + 8 = (x + 2)(x + 4)$
4.3 MORE ON FACTORING TRINOMIALS	To factor $px^2 + mx + n$, find integers a, b, c, and d so that $ac = p$, $bd = n$, and $ad + bc = m$. Then $px^2 + mx + n = (ax + b)(cx + d)$.	Factor: $3x^2 + 14x - 5$ $ac = 3, \quad bd = -5$ $ad + bc = 14$ By trial and error or by grouping, $3x^2 + 14x - 5 = (3x - 1)(x + 5)$.
4.4 SPECIAL FACTORIZATIONS	$x^2 - y^2 = (x + y)(x - y)$ $x^2 + 2xy + y^2 = (x + y)^2$ $x^2 - 2xy + y^2 = (x - y)^2$ $x^3 - y^3 = (x - y)(x^2 + xy + y^2)$ $x^3 + y^3 = (x + y)(x^2 - xy + y^2)$	$4x^2 - 9 = (2x + 3)(2x - 3)$ $9x^2 + 6x + 1 = (3x + 1)^2$ $4x^2 - 20x + 25 = (2x - 5)^2$ $m^3 - 8 = m^3 - 2^3$ $= (m - 2)(m^2 + 2m + 4)$ $27 + z^3 = 3^3 + z^3$ $= (3 + z)(9 - 3z + z^2)$

SECTION	CONCEPTS	EXAMPLES
4.5 SOLVING QUADRATIC EQUATIONS BY FACTORING	**Zero-Factor Property** If a and b are real numbers and if $ab = 0$, then $a = 0$ or $b = 0$.	If $(x - 2)(x + 3) = 0$, then $x - 2 = 0$ or $x + 3 = 0$.
	Solving a Quadratic Equation by Factoring	Solve $2x^2 = 7x + 15$.
	Step 1 Write the equation in standard form: all terms on one side of the equals sign, with 0 on the other side.	**1.** $2x^2 - 7x - 15 = 0$
	Step 2 Factor completely.	**2.** $(2x + 3)(x - 5) = 0$
	Step 3 Set each factor with a variable equal to 0, and solve the resulting equations.	**3.** $2x + 3 = 0$ or $x - 5 = 0$ $2x = -3$ $\quad$ $x = 5$ $x = -\frac{3}{2}$
	Step 4 Check each solution in the original equation.	**4.** Check that both solutions satisfy the original equation.
	Remember: Not all quadratic equations can be solved by factoring.	
4.6 APPLICATIONS OF QUADRATIC EQUATIONS	**Pythagorean Formula** In a right triangle, the square of the hypotenuse equals the sum of the squares of the legs. $a^2 + b^2 = c^2$	In a right triangle one leg measures 2 feet longer than the other. The hypotenuse measures 4 feet longer than the shorter leg. Find the lengths of the three sides of the triangle. Let x = the length of the shorter leg. Then $x^2 + (x + 2)^2 = (x + 4)^2$ $x^2 + x^2 + 4x + 4 = x^2 + 8x + 16$ $2x^2 + 4x + 4 = x^2 + 8x + 16$ $x^2 - 4x - 12 = 0$ $(x + 2)(x - 6) = 0$ $x + 2 = 0$ or $x - 6 = 0$ $x = -2$ or $x = 6$. Discard -2 as a solution. Check that the sides are 6, $6 + 2 = 8$, and $6 + 4 = 10$ feet in length.

SECTION	CONCEPTS	EXAMPLES
4.7 SOLVING QUADRATIC INEQUALITIES	**Solving a Quadratic Inequality**	Solve: $2x^2 + 5x - 3 < 0$.
	Step 1 Change the inequality to an equation.	**1.** $2x^2 + 5x - 3 = 0$
	Step 2 Use the zero-factor property to solve the equation from Step 1.	**2.** $(2x - 1)(x + 3) = 0$ $2x - 1 = 0$ or $x + 3 = 0$ $2x = 1$ or $x = -3$ $x = \frac{1}{2}$
	Step 3 Use the solutions of the equation in Step 2 to determine regions on the number line.	**3.** 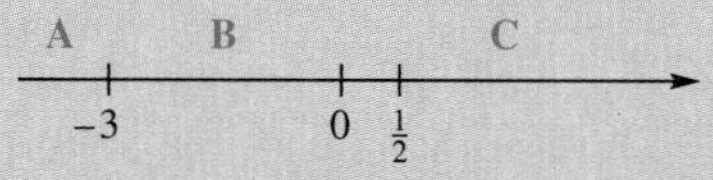
	Step 4 Choose a number from each region. Substitute the number into the original inequality. If the number satisfies the inequality, all numbers in that region satisfy the inequality.	**4.** Choose −4 from A: $2(-4)^2 + 5(-4) - 3 < 0$? $2(16) - 20 - 3 < 0$? $9 < 0$ False Choose 0 from B: $2(0)^2 + 5(0) - 3 < 0$? $0 + 0 - 3 < 0$? $-3 < 0$ True Choose 1 from C: $2(1)^2 + 5(1) - 3 < 0$? $2 + 5 - 3 < 0$? $4 < 0$ False Only the numbers from region B satisfy the inequality.
	Step 5 Write the solutions as inequalities.	**5.** The solution is written $-3 < x < 1/2$.

CHAPTER 4 REVIEW EXERCISES

[4.1] *Factor out the greatest common factor.*

1. $6 - 18r^5 + 12r^3$

2. $32y^4r^3 - 48y^5r^2 + 24y^7r^5$

Factor by grouping.

3. $6p^2 + 9p + 2pq + 3q$

4. $12r^2 + 18rq - 10rt - 15qt$

[4.2] *Factor completely.*

5. $r^2 - 6r - 27$

6. $p^2 + p - 30$

7. $z^2 - 7z - 44$

8. $z^2 - 11zx + 10x^2$

9. $p^7 - p^6q - 2p^5q^2$

10. $5p^6 - 45p^5 + 70p^4$

[4.3] *Factor completely.*

11. $2k^2 - 5k + 2$

12. $3z^2 + 11z - 4$

13. $6r^2 - 5r - 6$

14. $3p^2 - 2pq - 8q^2$

15. $7m^2 + 19mn - 6n^2$

16. $10r^3s + 17r^2s^2 + 6rs^3$

[4.4] *Factor completely.*

17. $100a^2 - 9$

18. $49y^2 - 25z^2$

19. $144p^2 - 36q^2$

20. $y^4 - 625$

21. $16m^2 + 40mn + 25n^2$

22. $25a^2 + 15ab + 9b^2$

23. $54x^3 - 72x^2 + 24x$

24. $125r^3 - 216s^3$

25. $343x^3 + 64$

[4.5] *Solve each equation. Check each solution.*

26. $r^2 + 3r = 10$

27. $m^2 - 5m + 4 = 0$

28. $k^2 = 8k - 15$

29. $p^2 = 12(p - 3)$

30. $(2p + 3)(p^2 - 4p + 3) = 0$

31. $x^3 - 9x = 0$

[4.6] *Solve each applied problem.*

32. The floor plan for a house is a rectangle with length 7 meters more than its width. The area is 170 square meters. Find the width and length of the house.

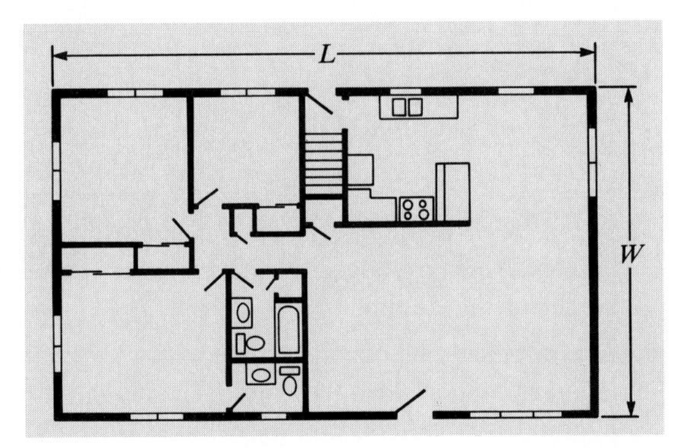

33. The triangular sail of a schooner has an area of 30 square meters. The height of the sail is 4 meters more than the base. Find the base of the sail.

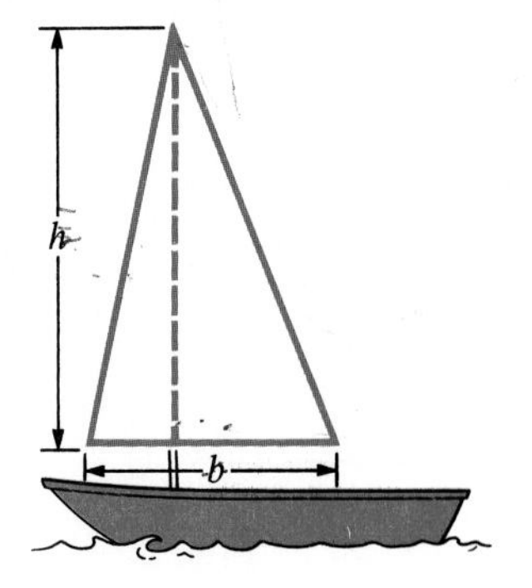

34. Two cars left an intersection at the same time. One traveled north. The other traveled 14 miles farther, but to the east. How far apart were they then, if the distance between them was 4 miles more than the distance traveled east?

35. A ladder is leaning against a building. The distance from the bottom of the ladder to the building is 4 feet less than the length of the ladder. How high up the side of the building is the top of the ladder if that distance is 2 feet less than the length of the ladder?

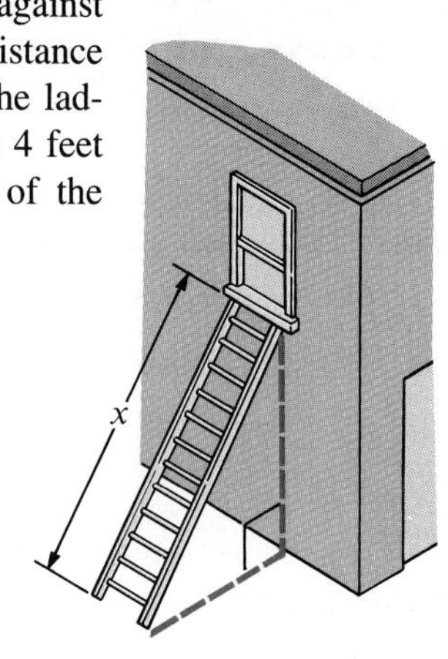

[4.7] *Solve each inequality.*

36. $(q + 5)(q - 3) > 0$

37. $(2r - 1)(r + 4) \geq 0$

38. $m^2 - 5m + 6 \leq 0$

39. $z^2 - 8z + 15 < 0$

40. $2p^2 + 5p - 12 \geq 0$

41. $(m + 3)(m - 1)(2m + 5) \geq 0$

MIXED REVIEW EXERCISES

Factor completely.

42. $r^2 - 4rs - 96s^2$

43. $24k^5 - 20k^4 + 4k^3$

44. $15m^3n^4 - 20m^2n^5 + 50m^3n^6$

45. $1000t^3 - 27$

46. $m^2 + 9$

47. $4y^2 + 3y + 8y + 6$

48. $6m^3 - 21m^2 - 45m$

49. $100y^6 - 50y^3 + 300y^4$

50. $y^2 - 8yz + 15z^2$

51. $9r^2 - 42r + 49$

52. $2z^2 + 9zy - 5y^2$

53. $p^2 + 2pq - 120q^2$

54. $15m^2 + 20mp - 12mn - 16np$

Solve the following.

55. $3k^2 - 11k - 20 = 0$

56. $(3z - 1)(z^2 + 3z + 2) = 0$

57. $100b^2 - 49 = 0$

58. Find the weight of the spherical ball bearing shown in the figure. The weight in ounces is given approximately by $W = 4.8V$, where V is the volume in cubic inches. See Appendix A for the formula for the volume of a sphere. Use 3.14 for π and give the answer to the nearest hundredth.

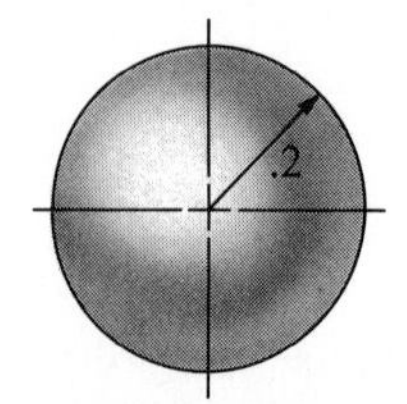

59. The area of a triangle is 12 square meters. The base is 2 meters longer than the height. Find the height of the triangle.

60. A pyramid has a rectangular base with a length that is 2 meters more than the width. The height of the pyramid is 6 meters, and the volume is 48 cubic meters. Find the width of the base.

61. A lot is shaped like a right triangle. The hypotenuse is 3 meters longer than the longer leg. The longer leg is 6 meters more than twice the length of the shorter leg. Find the lengths of the sides of the lot.

62. A bicyclist heading east and a motorist traveling south left an intersection at the same time. When the motorist had gone 17 miles farther than the bicyclist, the distance between them was 1 mile more than the distance traveled by the motorist. How far apart were they then? (*Hint:* Draw a sketch.)

63. Although $(2x + 8)(3x - 4) = 6x^2 + 16x - 32$, the polynomial is not factored completely. Explain why and give the completely factored form.

64. Explain why $(3x - 15)(x + 2)$ is not the completely factored form of $3x^2 - 9x - 30$ and factor the polynomial completely.

CHAPTER 4 TEST

Factor completely.

1. $6ab - 36ab^2$

2. $15k^2t + 25kt^2$

3. $16m^2 - 24m^3 + 32m^4$

4. $28pq + 14p + 56p^2q^3$

5. $m^2 - 9m + mp - 3p$

6. $12 - 6a + 2b - ab$

7. $x^2 - 4x - 45$

8. $3p^4q + 18p^3q - 21p^2q$

9. $3a^2 + 13a - 10$

10. $30z^5 - 69z^4 - 15z^3$

11. $12r^2p^2 + 19rp^2 + 5p^2$

12. $6t^2 - tx - 2x^2$

13. $50m^2 - 98$

14. $144a^2 - 169b^2$

15. $a^4 - 625$

16. $4p^2 + 12p + 9$

17. $25z^2 - 10z + 1$

18. $4y^3 + 16y^2 + 16y$

19. $8p^3 - 125$

20. $27r^3 + 64t^6$

21. $m^2 - n^2 - 4m - 4n$

Solve each equation.

22. $3x^2 + 5x = 2$

23. $p(2p + 3) = 20$

24. $(m - 3)(6m^2 - 11m - 10) = 0$

25. $z^3 = 16z$

Solve each applied problem.

26. The length of a rectangular flower bed is 1 foot less than twice its width. The area is 15 square feet. Find the dimensions of the flower bed.

27. A carpenter needs to cut a brace to support a wall stud. See the figure. The brace should be 7 feet less than three times the length of the stud. The brace will be fastened on the floor 1 foot less than twice the length of the stud away from the stud. How long should the brace be?

28. Explain why the expression $2(x + 4) + y(x + 4)$ is not completely factored and give the completely factored form.

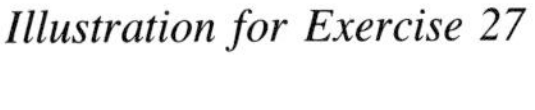

Illustration for Exercise 27

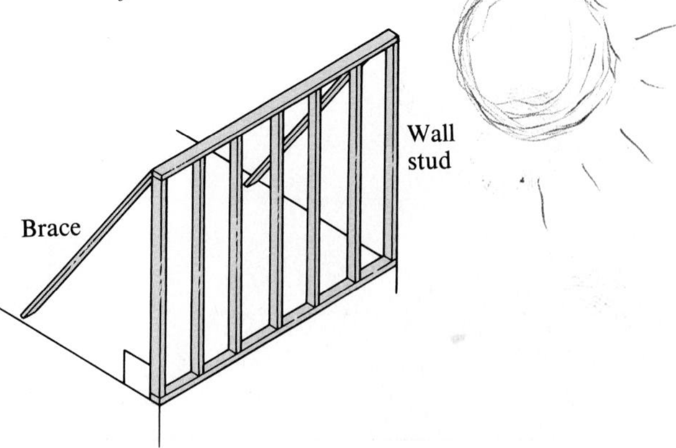

Graph the solution of each inequality.

29. $2p^2 + 5p - 3 \le 0$

30. $m^2 + 2m - 24 > 0$

CHAPTER FIVE

RATIONAL EXPRESSIONS

1988	Carl Lewis, United States	28 feet, 7 1/4 inches
1984	Carl Lewis, United States	28 feet, 1/4 inch
1980	Lutz Dombrowski, East Germany	28 feet, 1/4 inch
1976	Arnie Robinson, United States	27 feet, 4 1/2 inches
1972	Randy Williams, United States	27 feet, 1/2 inch
1968	Bob Beamon, United States	29 feet, 2 1/2 inches

Shown here are the winners and their winning jumps in six recent Olympic Games Long Jump competitions. Notice that each jump required a fractional part of an inch when it was measured. Whole numbers are not sufficiently accurate to measure with enough precision in many instances in everyday life. Similarly, polynomials are not sufficient in describing certain situations that occur in algebra, and so quotients of polynomials, known as *rational expressions,* are studied. Rational expressions are the algebraic equivalent of fractions in arithmetic. The techniques of factoring, studied in Chapter 4, are essential in working with rational expressions.

5.1 THE FUNDAMENTAL PROPERTY OF RATIONAL EXPRESSIONS

OBJECTIVES

1 FIND THE VALUES FOR WHICH A RATIONAL EXPRESSION IS UNDEFINED.

2 FIND THE NUMERICAL VALUE OF A RATIONAL EXPRESSION.

3 WRITE RATIONAL EXPRESSIONS IN LOWEST TERMS.

The quotient of two integers (with divisor not zero) is called a rational number. In the same way, the quotient of two polynomials with divisor not equal to zero is called a *rational expression.*

RATIONAL EXPRESSION

A **rational expression** is an expression of the form

$$\frac{P}{Q}$$

where P and Q are polynomials, with $Q \neq 0$.

Examples of rational expressions include $\frac{-6x}{x^3 + 8}$, $\frac{9x}{y + 3}$, and $\frac{2m^3}{9}$.

1 A fraction with a zero denominator is *not* a rational expression, since division by zero is not possible. For that reason, be careful when substituting a number for a variable in the denominator of a rational expression. For example, in

$$\frac{8x^2}{x - 3}$$

x can take on any value except 3. When $x = 3$, the denominator becomes $3 - 3 = 0$, making the expression undefined.

In order to determine the values for which a rational expression is undefined, use the following procedure.

DETERMINING WHEN A RATIONAL EXPRESSION IS UNDEFINED

Step 1 Set the denominator of the rational expression equal to 0.
Step 2 Solve this equation.
Step 3 The solutions of the equation are the values which make the rational expression undefined.

This procedure is illustrated in Example 1.

NOTE The numerator of a rational expression may be *any* number.

EXAMPLE 1
FINDING VALUES THAT MAKE RATIONAL EXPRESSIONS UNDEFINED

Find any values for which the following rational expressions are undefined.

(a) $\dfrac{p + 5}{3p + 2}$

Remember that the *numerator* may be any number; we must find any value of p that makes the *denominator* equal to 0. Find these values according to the procedure described above.

Step 1 Set the denominator equal to 0.

$$3p + 2 = 0$$

Step 2 Solve this equation.

$$3p = -2$$

$$p = -\frac{2}{3}$$

Step 3 Since $p = -2/3$ will make the denominator zero, the given expression is undefined for $-2/3$.

(b) $\dfrac{9m^2}{m^2 - 5m + 6}$

Find the numbers that make the denominator zero by solving the equation

$$m^2 - 5m + 6 = 0.$$

Factor the polynomial and set each factor equal to zero.

$$(m - 2)(m - 3) = 0$$

$$m - 2 = 0 \quad \text{or} \quad m - 3 = 0$$

$$m = 2 \quad \text{or} \quad m = 3$$

The original expression is undefined for 2 and for 3.

(c) $\dfrac{2r}{r^2 + 1}$

For this rational expression to be undefined, $r^2 + 1$ must equal 0.

$$r^2 + 1 = 0$$
$$r^2 = -1 \qquad \text{Subtract 1.}$$

This equation, $r^2 = -1$, has no real number solutions, since the square of a real number can never be negative. Thus, there are no values for which this rational expression is undefined. (Equivalently, the rational expression is always defined.) ■

2 The following example shows how to find the numerical value of a rational expression for a given value of the variable.

EXAMPLE 2
EVALUATING A RATIONAL EXPRESSION

Find the numerical value of $\dfrac{3x + 6}{2x - 4}$ for the given values of x.

(a) $x = 1$

Find the value of the rational expression by substituting 1 for x.

$$\frac{3x + 6}{2x - 4} = \frac{3(1) + 6}{2(1) - 4} \qquad \text{Let } x = 1.$$
$$= \frac{9}{-2}$$
$$= -\frac{9}{2}$$

(b) $x = 2$

Substituting 2 for x makes the denominator zero, so the rational expression is undefined when $x = 2$. ■

3 A fraction such as 2/3 is said to be in lowest terms. How can "lowest terms" be defined? We use the idea of greatest common factor to give this definition, which applies to all rational expressions.

LOWEST TERMS

A rational expression P/Q $(Q \neq 0)$ is in **lowest terms** if the greatest common factor of its numerator and denominator is 1.

Because a rational expression represents a number for each value of the variable that does not make the denominator zero, the properties of rational numbers also apply to rational expressions. For example, the fundamental property of rational expressions permits rational expressions to be written in lowest terms.

FUNDAMENTAL PROPERTY OF RATIONAL EXPRESSIONS

If P/Q is a rational expression and if K represents any rational expression, where $K \neq 0$, then

$$\frac{PK}{QK} = \frac{P}{Q}.$$

This property is based on the identity property of multiplication, since

$$\frac{PK}{QK} = \frac{P}{Q} \cdot \frac{K}{K} = \frac{P}{Q} \cdot 1 = \frac{P}{Q}.$$

The next example shows how to write both a rational number and a rational expression in lowest terms. Notice the similarity in the procedures.

EXAMPLE 3 **WRITING IN LOWEST TERMS**

Write each expression in lowest terms.

(a) $\dfrac{30}{72}$ **(b)** $\dfrac{14k^2}{2k^3}$

Begin by factoring.

$$\frac{30}{72} = \frac{2 \cdot 3 \cdot 5}{2 \cdot 2 \cdot 2 \cdot 3 \cdot 3} \qquad \frac{14k^2}{2k^3} = \frac{2 \cdot 7 \cdot k \cdot k}{2 \cdot k \cdot k \cdot k}$$

Group any factors common to the numerator and denominator.

$$\frac{30}{72} = \frac{5(2 \cdot 3)}{2 \cdot 2 \cdot 3(2 \cdot 3)} \qquad \frac{14k^2}{2k^3} = \frac{7(2 \cdot k \cdot k)}{k(2 \cdot k \cdot k)}$$

Use the fundamental property.

$$\frac{30}{72} = \frac{5}{2 \cdot 2 \cdot 3} = \frac{5}{12} \qquad \frac{14k^2}{2k^3} = \frac{7}{k}$$ ■

EXAMPLE 4 **WRITING IN LOWEST TERMS**

Write each rational expression in lowest terms.

(a) $\dfrac{3x-12}{5x-20}$

Begin by factoring both numerator and denominator. Then use the fundamental property.

$$\frac{3x-12}{5x-20} = \frac{3(x-4)}{5(x-4)}$$
$$= \frac{3}{5}$$

The rational expression $\dfrac{3x-12}{5x-20}$ is equal to $\dfrac{3}{5}$ for all values of $x \neq 4$ (since the denominator of the original rational expression is 0 when x is 4).

(b) $\dfrac{m^2 + 2m - 8}{2m^2 - m - 6}$

Always begin by factoring both numerator and denominator, if possible. Then use the fundamental property.

$$\frac{m^2 + 2m - 8}{2m^2 - m - 6} = \frac{(m + 4)(m - 2)}{(2m + 3)(m - 2)}$$
$$= \frac{m + 4}{2m + 3}$$

Thus, $\dfrac{m^2 + 2m - 8}{2m^2 - m - 6} = \dfrac{m + 4}{2m + 3}$ for $m \neq -\dfrac{3}{2}$ or 2, since the denominator of the original expression is 0 for these values of x. ■

From now on, statements of equality of rational expressions will be written with the understanding that they apply only to those real numbers that make neither denominator equal to zero.

CAUTION One of the most common errors in algebra occurs when students attempt to write rational expressions in lowest terms before factoring. The fundamental property is applied only *after* the numerator and denominator are expressed in factored form. For example, in Example 4(a), it would be *wrong* to eliminate the x in the numerator and denominator; x is not a factor of either expression.

EXAMPLE 5
WRITING IN LOWEST TERMS (COMMON FACTOR OF −1)

Write $\dfrac{x - y}{y - x}$ in lowest terms.

At first glance, there does not seem to be any way in which $x - y$ and $y - x$ can be factored to get a common factor. However, one way to approach this is to notice that -1 is a factor of the numerator, since

$$x - y = -1(-x + y) = -1(y - x).$$

Now the fundamental property can be used to simplify the rational expression.

$$\frac{x - y}{y - x} = \frac{-1(y - x)}{1(y - x)} = \frac{-1}{1} = -1$$

Notice also that the -1 could be factored from the denominator instead, obtaining the same result. ■

Example 5 suggests the following rule.

The quotient of two nonzero expressions that differ only in sign is -1.

This rule will be used often in working with rational expressions.

EXAMPLE 6 **WRITING IN LOWEST TERMS**

Write each rational expression in lowest terms.

(a) $\dfrac{2 - m}{m - 2}$

Since $2 - m$ and $m - 2$ (or $-2 + m$) differ only in sign, use the rule above.

$$\frac{2 - m}{m - 2} = -1$$

(b) $\dfrac{4x^2 - 9}{6 - 4x}$

Factor the numerator and denominator, and use the rule above.

$$\frac{4x^2 - 9}{6 - 4x} = \frac{(2x + 3)(2x - 3)}{2(3 - 2x)}$$

$$= \frac{2x + 3}{2}(-1)$$

$$= -\frac{2x + 3}{2}$$

(c) $\dfrac{3 + r}{3 - r}$

The quantities $3 + r$ and $3 - r$ do *not* differ only in sign. This rational expression cannot be written in simpler form. ■

NOTE The form of the answer given in Example 6(b) is only one of several acceptable forms. The $-$ sign representing the -1 factor is in front of the fraction, on the same line as the fraction bar. The -1 factor may be placed in front of the fraction, in the numerator, or in the denominator. Some other acceptable forms of the answer are

$$\frac{-(2x + 3)}{2}, \quad \frac{-2x - 3}{2}, \quad \text{and} \quad \frac{2x + 3}{-2}.$$

However, can you see why $\dfrac{-2x + 3}{2}$ is *not* an acceptable form?

5.1 EXERCISES

Find any values for which the following expressions are undefined. See Example 1.

1. $\dfrac{3}{4x}$

2. $\dfrac{5}{2x}$

3. $\dfrac{x^2}{x + 5}$

4. $\dfrac{3x^2}{2x - 1}$

5. $\dfrac{a + 4}{a^2 - 8a + 15}$

6. $\dfrac{p + 6}{p^2 - p - 12}$

7. $\dfrac{9y}{y^2 + 16}$

8. $\dfrac{12z}{z^2 + 100}$

Find the numerical value of each expression when **(a)** $x = 2$ *and* **(b)** $x = -3$. *See Example 2.*

9. $\frac{4x - 2}{3x}$ **10.** $\frac{-5x + 1}{2x}$ **11.** $\frac{4x^2 - 2x}{3x}$ **12.** $\frac{x^2 - 1}{x}$

13. $\frac{(-8x)^2}{3x + 9}$ **14.** $\frac{2x^2 + 5}{3 + x}$ **15.** $\frac{x + 8}{x^2 - 4x + 2}$ **16.** $\frac{2x - 1}{x^2 - 7x + 3}$

17. $\frac{5x^2}{6 - 3x - x^2}$ **18.** $\frac{-2x^2}{8 + x - x^2}$ **19.** $\frac{2x + 5}{x^2 + 3x - 10}$ **20.** $\frac{3x - 7}{2x^2 - 3x - 2}$

---

21. If 2 is substituted for x in the rational expression $\frac{x - 2}{x^2 - 4}$, the result is $\frac{0}{0}$. A commonly heard statement is "Any number divided by itself is 1." Does this mean that this expression is equal to 1 for $x = 2$? If not, explain. (See Objective 1.)

22. For $x \neq 2$, the rational expression $\frac{2(x - 2)}{x - 2}$ is equal to 2. Can the same be said for $\frac{2x - 2}{x - 2}$? Explain. (See Objective 3.)

Write each expression in lowest terms. See Examples 3–6.

23. $\frac{12k^2}{6k}$ **24.** $\frac{9m^3}{3m}$ **25.** $\frac{-12m^2p}{9mp^2}$

26. $\frac{6a^2b^3}{-24a^3b^2}$ **27.** $\frac{6(y + 2)}{8(y + 2)}$ **28.** $\frac{9(m + 2)}{5(m + 2)}$

29. $\frac{(x + 1)(x - 1)}{(x + 1)^2}$ **30.** $\frac{3(t + 5)}{(t + 5)(t - 1)}$ **31.** $\frac{12m^2 - 9}{3}$

32. $\frac{15p^2 - 10}{5}$ **33.** $\frac{32y + 20}{24}$ **34.** $\frac{40q - 25}{20}$

35. $\frac{2q - 6}{5q - 10}$ **36.** $\frac{9p + 12}{21p + 28}$ **37.** $\frac{m^2 - n^2}{m + n}$

38. $\frac{a^2 - b^2}{a - b}$ **39.** $\frac{5m^2 - 5m}{10m - 10}$ **40.** $\frac{3y^2 - 3y}{2(y - 1)}$

41. $\frac{16r^2 - 4s^2}{4r - 2s}$ **42.** $\frac{11s^2 - 22s^3}{6 - 12s}$ **43.** $\frac{m^2 - 4m + 4}{m^2 + m - 6}$

44. $\frac{a^2 - a - 6}{a^2 + a - 12}$ **45.** $\frac{x^2 + 3x - 4}{1 - x^2}$ **46.** $\frac{8m^2 + 6m - 9}{9 - 16m^2}$

Write each expression in lowest terms. See Examples 5 and 6.

47. $\frac{m - 5}{5 - m}$ **48.** $\frac{3 - p}{p - 3}$ **49.** $\frac{-a + b}{b - a}$ **50.** $\frac{b - a}{a - b}$

51. $\frac{x^2 - 1}{1 - x}$ **52.** $\frac{p^2 - q^2}{q - p}$ **53.** $\frac{m^2 - 4m}{4m - m^2}$ **54.** $\frac{s^2 - r^2}{r^2 - s^2}$

55. Which one of the following rational expressions is *not* equivalent to $\frac{4-3x}{7}$? (See Objective 3.)

(a) $-\frac{-4+3x}{7}$ (b) $-\frac{4-3x}{-7}$

(c) $\frac{-4+3x}{-7}$ (d) $\frac{-(3x+4)}{7}$

56. One of the following is equal to 1 for *all* real numbers. Which one is it? (See Objectives 1 and 2.)

(a) $\frac{k^2+2}{k^2+2}$ (b) $\frac{4-m}{4-m}$

(c) $\frac{2x+9}{2x+9}$ (d) $\frac{x^2-1}{x^2-1}$

Write each expression in lowest terms.

57. $\frac{a+b+a^2+ba}{ab+b^2}$

58. $\frac{2p+pq-8-4q}{8+4q}$

59. $\frac{m^2-n^2-4m-4n}{2m-2n-8}$

60. $\frac{x^2y+y+x^2z+z}{xy+xz}$

61. $\frac{b^3-a^3}{a^2-b^2}$

62. $\frac{k^3+8}{k^2-4}$

63. $\frac{z^3+27}{z^3-3z^2+9z}$

64. $\frac{1-8r^3}{8r^2+4r+2}$

PREVIEW EXERCISES

Multiply or divide as indicated. See Section 1.1.

65. $\frac{3}{4}\cdot\frac{5}{8}$

66. $\frac{7}{10}\cdot\frac{3}{5}$

67. $\frac{8}{15}\cdot\frac{20}{3}$

68. $\frac{15}{4}\cdot\frac{12}{5}$

69. $\frac{6}{5}\div\frac{3}{10}$

70. $\frac{21}{8}\div\frac{7}{4}$

71. $\frac{27}{8}\div\frac{5}{12}$

72. $\frac{2}{3}\div\frac{4}{9}$

5.2 MULTIPLICATION AND DIVISION OF RATIONAL EXPRESSIONS

OBJECTIVES

1 MULTIPLY RATIONAL EXPRESSIONS.

2 DIVIDE RATIONAL EXPRESSIONS.

1 The product of two fractions is found by multiplying the numerators and multiplying the denominators. Rational expressions are multiplied in the same way.

MULTIPLYING RATIONAL EXPRESSIONS

The product of the rational expressions P/Q and R/S is

$$\frac{P}{Q}\cdot\frac{R}{S}=\frac{PR}{QS}.$$

The next example shows the multiplication of both two rational numbers and two rational expressions. This parallel discussion lets you compare the steps.

EXAMPLE 1
MULTIPLYING RATIONAL EXPRESSIONS

Multiply. Write answers in lowest terms.

(a) $\frac{3}{10} \cdot \frac{5}{9}$ **(b)** $\frac{6}{x} \cdot \frac{x^2}{12}$

Find the product of the numerators and the product of the denominators.

$$\frac{3}{10} \cdot \frac{5}{9} = \frac{3 \cdot 5}{10 \cdot 9} \qquad \frac{6}{x} \cdot \frac{x^2}{12} = \frac{6 \cdot x^2}{x \cdot 12}$$

Use the fundamental property to write each product in lowest terms.

$$\frac{3}{10} \cdot \frac{5}{9} = \frac{1 \cdot 3 \cdot 5}{2 \cdot 5 \cdot 3 \cdot 3} = \frac{1}{6} \qquad \frac{6}{x} \cdot \frac{x^2}{12} = \frac{6 \cdot x \cdot x}{2 \cdot 6 \cdot x} = \frac{x}{2}$$

Notice in the second step above that the products were left in factored form since common factors must be identified to write the product in lowest terms. ■

NOTE It is also possible to divide out common factors in the numerator and denominator *before* multiplying the rational expressions. Many people use this method with success. For example, to multiply 6/5 and 35/22, the following method can be used.

$$\frac{6}{5} \cdot \frac{35}{22} = \frac{2 \cdot 3}{5} \cdot \frac{5 \cdot 7}{2 \cdot 11} \qquad \text{Identify common factors.}$$

$$= \frac{3 \cdot 7}{11} \qquad \text{Lowest terms}$$

$$= \frac{21}{11} \qquad \text{Multiply in numerator.}$$

EXAMPLE 2
MULTIPLYING RATIONAL EXPRESSIONS

Multiply and express the product in lowest terms: $\frac{x+y}{2x} \cdot \frac{x^2}{(x+y)^2}$.

Use the definition of multiplication.

$$\frac{x+y}{2x} \cdot \frac{x^2}{(x+y)^2} = \frac{(x+y)x^2}{2x(x+y)^2} \qquad \text{Multiply numerators; multiply denominators.}$$

$$= \frac{(x+y)x \cdot x}{2x(x+y)(x+y)} \qquad \text{Meaning of 2 as an exponent}$$

$$= \frac{x}{2(x+y)} \cdot \frac{x(x+y)}{x(x+y)} \qquad \text{Definition of multiplication}$$

$$= \frac{x}{2(x+y)} \qquad \text{Lowest terms}$$

Notice how in the third line, the factor $\frac{x(x+y)}{x(x+y)}$ appears. Since it is equal to 1, the final product is $\frac{x}{2(x+y)}$. ■

EXAMPLE 3
MULTIPLYING RATIONAL EXPRESSIONS

Multiply and express the product in lowest terms:

$$\frac{x^2 + 3x}{x^2 - 3x - 4} \cdot \frac{x^2 - 5x + 4}{x^2 + 2x - 3}.$$

First factor the numerators and denominators whenever possible. Then use the fundamental property to write the product in lowest terms.

$$\frac{x^2 + 3x}{x^2 - 3x - 4} \cdot \frac{x^2 - 5x + 4}{x^2 + 2x - 3}$$

$$= \frac{x(x + 3)}{(x - 4)(x + 1)} \cdot \frac{(x - 4)(x - 1)}{(x + 3)(x - 1)} \qquad \text{Factor.}$$

$$= \frac{x(x + 3)(x - 4)(x - 1)}{(x - 4)(x + 1)(x + 3)(x - 1)} \qquad \text{Multiply numerators; multiply denominators.}$$

$$= \frac{x}{x + 1} \qquad \text{Lowest terms}$$

The quotients $\frac{x + 3}{x + 3}$, $\frac{x - 4}{x - 4}$, and $\frac{x - 1}{x - 1}$ are all equal to 1, justifying the final product $\frac{x}{x + 1}$. ■

2 To develop a method for dividing rational numbers and rational expressions, consider the following problem. Suppose that you have 7/8 of a gallon of milk and you wish to find how many quarts you have. Since a quart is 1/4 of a gallon, you must ask yourself, "How many 1/4s are there in 7/8?" This would be interpreted as

$$\frac{7}{8} \div \frac{1}{4} \qquad \text{or} \qquad \frac{\frac{7}{8}}{\frac{1}{4}}$$

since the fraction bar means division.

The fundamental property of rational expressions discussed earlier can be applied to rational number values of P, Q, and K. With $P = 7/8$, $Q = 1/4$, and $K = 4$,

$$\frac{P}{Q} = \frac{P \cdot K}{Q \cdot K} = \frac{\frac{7}{8} \cdot 4}{\frac{1}{4} \cdot 4} = \frac{\frac{7}{8} \cdot 4}{1} = \frac{7}{8} \cdot \frac{4}{1}.$$

So, to divide 7/8 by 1/4, we must multiply 7/8 by the reciprocal of 1/4, namely 4. Since $(7/8)(4) = 7/2$, there are 7/2 or 3 1/2 quarts in 7/8 gallon.

The discussion above justifies the rule for dividing rational numbers: to divide a/b by c/d, multiply a/b by the reciprocal of c/d. Division of rational expressions is defined in the same way.

DIVIDING RATIONAL EXPRESSIONS

If P/Q and R/S are any two rational expressions, with $R/S \neq 0$, then

$$\frac{P}{Q} \div \frac{R}{S} = \frac{P}{Q} \cdot \frac{S}{R} = \frac{PS}{QR}$$

The next example shows the division of two rational numbers and the division of two rational expressions.

EXAMPLE 4 **DIVIDING RATIONAL EXPRESSIONS**

Divide the following fractions. Write answers in lowest terms.

(a) $\dfrac{5}{8} \div \dfrac{7}{16}$ **(b)** $\dfrac{y}{y+3} \div \dfrac{4y}{y+5}$

Multiply the first expression and the reciprocal of the second.

$$\frac{5}{8} \div \frac{7}{16} = \frac{5}{8} \cdot \frac{16}{7} \quad \text{Reciprocal of } \frac{7}{16}$$

$$= \frac{5 \cdot 16}{8 \cdot 7}$$

$$= \frac{5 \cdot 8 \cdot 2}{8 \cdot 7}$$

$$= \frac{5 \cdot 2}{7}$$

$$= \frac{10}{7}$$

$$\frac{y}{y+3} \div \frac{4y}{y+5}$$

$$= \frac{y}{y+3} \cdot \frac{y+5}{4y} \quad \text{Reciprocal of } \frac{4y}{y+5}$$

$$= \frac{y(y+5)}{(y+3)(4y)}$$

$$= \frac{y+5}{4(y+3)} \quad \text{Fundamental property} \blacksquare$$

EXAMPLE 5 **DIVIDING RATIONAL EXPRESSIONS**

Divide $\dfrac{(3m)^2}{(2p)^3} \div \dfrac{6m^3}{16p^2}$.

Use the properties of exponents as necessary.

$$\frac{(3m)^2}{(2p)^3} \div \frac{6m^3}{16p^2} = \frac{9m^2}{8p^3} \div \frac{6m^3}{16p^2}$$

$$= \frac{9m^2}{8p^3} \cdot \frac{16p^2}{6m^3} \quad \text{Multiply by the reciprocal.}$$

$$= \frac{9 \cdot 16m^2p^2}{8 \cdot 6p^3m^3} \quad \text{Definition of multiplication; commutative property}$$

$$= \frac{3(48m^2p^2)}{mp(48m^2p^2)} \quad \text{Factor.}$$

$$= \frac{3}{mp} \quad \text{Fundamental property} \blacksquare$$

EXAMPLE 6
DIVIDING RATIONAL EXPRESSIONS

Divide and write the quotient in lowest terms: $\dfrac{x^2-4}{(x+3)(x-2)} \div \dfrac{(x+2)(x+3)}{2x}$.

$$\frac{x^2-4}{(x+3)(x-2)} \div \frac{(x+2)(x+3)}{2x}$$

$$= \frac{x^2-4}{(x+3)(x-2)} \cdot \frac{2x}{(x+2)(x+3)}$$ Use the definition of division.

$$= \frac{(x+2)(x-2)}{(x+3)(x-2)} \cdot \frac{2x}{(x+2)(x+3)}$$ Be sure numerators and denominators are factored.

$$= \frac{(x+2)(x-2)(2x)}{(x+3)(x-2)(x+2)(x+3)}$$ Multiply numerators and multiply denominators.

$$= \frac{2x}{(x+3)^2}$$ Use the fundamental property to write in lowest terms. ■

In Example 6, only the numerator had to be factored. Remember that *all* numerators and denominators must be factored before the fundamental property can be applied. The next example requires more factoring than was necessary in Example 6.

EXAMPLE 7
DIVIDING RATIONAL EXPRESSIONS

Divide and write the quotient in lowest terms: $\dfrac{m^2-4}{m^2-1} \div \dfrac{2m^2+4m}{1-m}$.

$$\frac{m^2-4}{m^2-1} \div \frac{2m^2+4m}{1-m}$$

$$= \frac{m^2-4}{m^2-1} \cdot \frac{1-m}{2m^2+4m}$$ Use the definition of division.

$$= \frac{(m+2)(m-2)}{(m+1)(m-1)} \cdot \frac{1-m}{2m(m+2)}$$ Factor.

$$= \frac{(m+2)(m-2)}{(m+1)(m-1)} \cdot \frac{1-m}{2m(m+2)}$$ $1-m$ and $m-1$ differ only in sign.

$$= \frac{-1(m-2)}{2m(m+1)}$$ From Section 5.1, $\dfrac{1-m}{m-1} = -1$.

$$= \frac{2-m}{2m(m+1)}$$ Use the distributive property in the numerator.

As mentioned in Section 5.1, there are equivalent ways to write the quotient $\dfrac{2-m}{2m(m+1)}$. Some of them are $-\dfrac{m-2}{2m(m+1)}$, $\dfrac{-m+2}{2m(m+1)}$, and $\dfrac{m-2}{-2m(m+1)}$. ■

The procedure for multiplying or dividing rational expressions is summarized below.

MULTIPLYING OR DIVIDING RATIONAL EXPRESSIONS

Step 1 If the operation is division, use the definition of division to rewrite as multiplication.
Step 2 Factor numerators and denominators completely.
Step 3 Multiply numerators and multiply denominators.
Step 4 Write the answer in lowest terms.

5.2 EXERCISES

Multiply or divide. Write each answer in lowest terms. See Examples 1 and 5.

1. $\frac{9m^2}{16} \cdot \frac{4}{3m}$ **2.** $\frac{21z^4}{8} \cdot \frac{12}{7z^3}$ **3.** $\frac{4p^2}{8p} \cdot \frac{3p^3}{16p^4}$ **4.** $\frac{6x^3}{9x} \cdot \frac{12x}{x^2}$

5. $\frac{8a^4}{12a^3} \cdot \frac{9a^5}{3a^2}$ **6.** $\frac{14p^5}{2p^2} \cdot \frac{8p^6}{28p^3}$ **7.** $\frac{3r^2}{9r^3} \div \frac{8r^4}{6r^5}$ **8.** $\frac{25m^{10}}{9m^5} \div \frac{15m^6}{10m^4}$

9. $\frac{3m^2}{(4m)^3} \div \frac{9m^3}{32m^4}$ **10.** $\frac{5x^3}{(4x)^2} \div \frac{15x^2}{8x^4}$ **11.** $\frac{-6r^4}{3r^5} \div \frac{(2r^2)^2}{-4}$ **12.** $\frac{-10a^6}{3a^2} \div \frac{(3a)^3}{81a}$

---

13. Explain in your own words how to multiply rational expressions. (See Objective 1.)

14. Explain in your own words how to divide rational expressions. (See Objective 2.)

Multiply or divide. Write each answer in lowest terms. See Examples 2–4, 6, and 7.

15. $\frac{a+b}{2} \cdot \frac{12}{(a+b)^2}$ **16.** $\frac{3(x-1)}{y} \cdot \frac{2y}{5(x-1)^2}$ **17.** $\frac{2k+8}{6} \div \frac{3k+12}{2}$

18. $\frac{5m+25}{10} \cdot \frac{12}{6m+30}$ **19.** $\frac{9y-18}{6y+12} \cdot \frac{3y+6}{15y-30}$ **20.** $\frac{12p+24}{36p-36} \div \frac{6p+12}{8p-8}$

21. $\frac{3r+12}{8} \cdot \frac{16r}{9r+36}$ **22.** $\frac{2r+2p}{8z} \div \frac{r^2+rp}{72}$ **23.** $\frac{y^2-16}{y+3} \div \frac{y-4}{y^2-9}$

24. $\frac{9(y-4)^2}{8(z+3)^2} \cdot \frac{16(z+3)}{3(y-4)}$ **25.** $\frac{6(m+2)}{3(m-1)^2} \div \frac{(m+2)^2}{9(m-1)}$ **26.** $\frac{4y+12}{2y-10} \div \frac{y^2-9}{y^2-y-20}$

27. $\frac{2-y}{8} \cdot \frac{7}{y-2}$ **28.** $\frac{9-2z}{3} \cdot \frac{9}{2z-9}$ **29.** $\frac{8-r}{8+r} \div \frac{r-8}{r+8}$

30. $\frac{6-y}{6+2y} \div \frac{6+y}{3+y}$ **31.** $\frac{m^2-16}{4-m} \cdot \frac{-4+m}{-4-m}$ **32.** $\frac{6r-18}{3r^2+2r-8} \cdot \frac{12r-16}{4r-12}$

33. $\frac{k^2-k-6}{k^2+k-12} \div \frac{k^2+2k-3}{k^2+3k-4}$ **34.** $\frac{m^2+3m+2}{m^2+5m+4} \cdot \frac{m^2+10m+24}{m^2+5m+6}$

35. $\dfrac{z^2 - z - 6}{z^2 - 2z - 8} \cdot \dfrac{z^2 + 7z + 12}{z^2 - 9}$

36. $\dfrac{y^2 + y - 2}{y^2 + 3y - 4} \div \dfrac{y + 2}{y + 3}$

37. $\dfrac{2m^2 - 5m - 12}{m^2 - 10m + 24} \div \dfrac{4m^2 - 9}{m^2 - 9m + 18}$

38. $\dfrac{2m^2 + 7m + 3}{m^2 - 9} \cdot \dfrac{m^2 - 3m}{2m^2 + 11m + 5}$

39. $\dfrac{m^2 + 2mp - 3p^2}{m^2 - 3mp + 2p^2} \div \dfrac{m^2 + 4mp + 3p^2}{m^2 + 2mp - 8p^2}$

40. $\dfrac{r^2 + rs - 12s^2}{r^2 - rs - 20s^2} \div \dfrac{r^2 - 2rs - 3s^2}{r^2 + rs - 30s^2}$

41. $\dfrac{(x + 1)^3(x + 4)}{x^2 + 5x + 4} \div \dfrac{x^2 + 2x + 1}{x^2 + 3x + 2}$

42. $\dfrac{(q - 3)^4(q + 2)}{q^2 + 3q + 2} \div \dfrac{q^2 - 6q + 9}{q^2 + 4q + 4}$

In working the following exercises, remember the use of grouping symbols (Section 1.2), how to factor sums and differences of cubes (Section 4.4), and how to factor by grouping (Section 4.3).

43. $\left(\dfrac{x^2 + 10x + 25}{x^2 + 10x} \cdot \dfrac{10x}{x^2 + 15x + 50}\right) \div \dfrac{x + 5}{x + 10}$

44. $\left(\dfrac{m^2 - 12m + 32}{8m} \cdot \dfrac{m^2 - 8m}{m^2 - 8m + 16}\right) \div \dfrac{m - 8}{m - 4}$

45. $\dfrac{3a - 3b - a^2 + b^2}{4a^2 - 4ab + b^2} \cdot \dfrac{4a^2 - b^2}{2a^2 - ab - b^2}$

46. $\dfrac{4r^2 - t^2 + 10r - 5t}{2r^2 + rt + 5r} \cdot \dfrac{4r^3 + 4r^2t + rt^2}{2r + t}$

47. $\dfrac{-x^3 - y^3}{x^2 - 2xy + y^2} \div \dfrac{3y^2 - 3xy}{x^2 - y^2}$

48. $\dfrac{b^3 - 8a^3}{4a^3 + 4a^2b + ab^2} \div \dfrac{4a^2 + 2ab + b^2}{-a^3 - ab^3}$

PREVIEW EXERCISES

Write the prime factored form of each number. See Section 4.1.

49. 12 **50.** 50 **51.** 72 **52.** 105

Find the greatest common factor for each group of terms. See Section 4.1.

53. $12m, \quad 15m^2, \quad 6$

54. $45a^3, \quad 85a^2, \quad 105a^4$

55. $72p^2q^5, \quad 28p^5q^3r$

56. $36s^4t^3u^6, \quad 54s^5t^2u$

5.3 THE LEAST COMMON DENOMINATOR

OBJECTIVES

1 FIND LEAST COMMON DENOMINATORS.

2 REWRITE RATIONAL EXPRESSIONS WITH GIVEN DENOMINATORS.

1 In this section, we demonstrate a preliminary step needed to add or subtract rational expressions with different denominators. Just as with rational numbers, adding or subtracting rational expressions (to be discussed in the next section) often requires a **least common denominator,** the least expression that all denominators divide into without a remainder. For example, the least common denominator for 2/9 and 5/12 is 36, since 36 is the smallest number that both 9 and 12 divide into evenly.

Least common denominators often can be found by inspection. For example, the least common denominator for $1/6$ and $2/(3m)$ is $6m$. In other cases, a least common denominator can be found by a procedure similar to that used in Chapter 4 for finding the greatest common factor.

FINDING THE LEAST COMMON DENOMINATOR

Step 1 Factor each denominator into prime factors.

Step 2 List each different denominator factor the *greatest* number of times it appears in any denominator.

Step 3 Multiply the denominator factors from Step 2 to get the least common denominator.

When each denominator is factored into prime factors, every prime factor must divide evenly into the least common denominator. The least common denominator is often abbreviated LCD.

In Example 1, the least common denominator is found both for numerical denominators and algebraic denominators.

EXAMPLE 1
FINDING THE LEAST COMMON DENOMINATOR

Find the least common denominator for each pair of fractions.

(a) $\frac{1}{24}, \frac{7}{15}$ **(b)** $\frac{1}{8x}, \frac{3}{10x}$

Write each denominator in factored form, with numerical coefficients in prime factored form.

(a)
$$24 = 2 \cdot 2 \cdot 2 \cdot 3 = 2^3 \cdot 3$$
$$15 = 3 \cdot 5$$

(b)
$$8x = 2 \cdot 2 \cdot 2 \cdot x = 2^3 \cdot x$$
$$10x = 2 \cdot 5 \cdot x$$

The LCD is found by taking each different factor the greatest number of times it appears as a factor in any of the denominators.

(a) The factor 2 appears three times in one product and not at all in the other, so the greatest number of times 2 appears is three. The greatest number of times both 3 and 5 appear is one.

$$\text{LCD} = 2 \cdot 2 \cdot 2 \cdot 3 \cdot 5 = 2^3 \cdot 3 \cdot 5 = 120$$

(b) Here 2 appears three times in one product and once in the other, so the greatest number of times the 2 appears is three. The greatest number of times the 5 appears is one, and the greatest number of times x appears in either product is one.

$$\text{LCD} = 2 \cdot 2 \cdot 2 \cdot 5 \cdot x = 2^3 \cdot 5 \cdot x = 40x$$ ■

EXAMPLE 2
FINDING THE LCD

Find the least common denominator for $\frac{5}{6r^2}$ and $\frac{3}{4r^3}$.

Factor each denominator.

$$6r^2 = 2 \cdot 3 \cdot r^2$$
$$4r^3 = 2 \cdot 2 \cdot r^3 = 2^2 \cdot r^3$$

The greatest number of times 2 appears is two, the greatest number of times 3 appears is one, and the greatest number of times r appears is three; therefore,

$$\text{LCD} = 2^2 \cdot 3 \cdot r^3$$
$$= 12r^3. \quad \blacksquare$$

EXAMPLE 3
FINDING THE LCD

Find the LCD.

(a) $\frac{6}{5m}, \quad \frac{4}{m^2 - 3m}$

Factor each denominator.

$$5m = 5 \cdot m$$
$$m^2 - 3m = m(m - 3)$$

Take each different factor the greatest number of times it appears as a factor.

$$\text{LCD} = 5 \cdot m \cdot (m - 3)$$
$$= 5m(m - 3)$$

Since m is not a *factor* of $m - 3$, both factors, m and $m - 3$, must appear in the LCD.

(b) $\frac{1}{r^2 - 4r - 5}, \quad \frac{1}{r^2 - r - 20}, \quad \frac{1}{r^2 - 10r + 25}$

Factor each denominator.

$$r^2 - 4r - 5 = (r - 5)(r + 1)$$
$$r^2 - r - 20 = (r - 5)(r + 4)$$
$$r^2 - 10r + 25 = (r - 5)^2$$

The LCD is

$$(r - 5)^2(r + 1)(r + 4).$$

(c) $\frac{1}{q - 5}, \frac{3}{5 - q}$

The expression $5 - q$ can be written as $-1(q - 5)$, since

$$-1(q - 5) = -q + 5 = 5 - q.$$

Because of this, either $q - 5$ or $5 - q$ can be used as the LCD. $\blacksquare$

2 Once the least common denominator has been found, we can use the fundamental property to write equivalent rational expressions with this LCD. The next example shows how to do this with both numerical and algebraic fractions.

EXAMPLE 4
WRITING A FRACTION WITH A GIVEN DENOMINATOR

Rewrite each expression with the indicated denominator.

(a) $\dfrac{3}{8} = \dfrac{\quad}{40}$ **(b)** $\dfrac{9k}{25} = \dfrac{\quad}{50k}$

For each example, decide what quantity the denominator on the left must be multiplied by to get the denominator on the right. (Find this quantity by dividing.)

Multiply by $\dfrac{5}{5}$ to get a denominator of 40.

$$\frac{3}{8} = \frac{3}{8} \cdot \frac{5}{5} = \frac{15}{40}$$

$$\frac{5}{5} = 1$$

Get a denominator of $50k$ by multiplying by $\dfrac{2k}{2k}$.

$$\frac{9k}{25} = \frac{9k}{25} \cdot \frac{2k}{2k} = \frac{18k^2}{50k}$$

$$\frac{2k}{2k} = 1$$

Notice the use of the multiplicative identity property in each part of this example. ■

EXAMPLE 5
WRITING A FRACTION WITH A GIVEN DENOMINATOR

Rewrite the following rational expression with the indicated denominator.

$$\frac{12p}{p^2 + 8p} = \frac{\qquad}{p^3 + 4p^2 - 32p}$$

Begin by factoring each denominator.

$$p^2 + 8p = p(p + 8)$$
$$p^3 + 4p^2 - 32p = p(p^2 + 4p - 32) = p(p + 8)(p - 4)$$

The fractions may now be written as follows.

$$\frac{12p}{p(p + 8)} = \frac{\qquad}{p(p + 8)(p - 4)}$$

Comparing the two factored forms, we see that the denominator of the fraction on the left must be multiplied by $p - 4$; the numerator must also be multiplied by $p - 4$.

$$\frac{12p}{p^2 + 8p} = \frac{12p}{p(p + 8)} \cdot \frac{p - 4}{p - 4}$$ Multiplicative identity property

$$= \frac{12p(p - 4)}{p(p + 8)(p - 4)}$$ Multiplication of rational expressions

$$= \frac{12p^2 - 48p}{p^3 + 4p^2 - 32p}$$ Multiply the factors. ■

NOTE In the next section we will learn to add and subtract rational expressions, and this will often require the skill illustrated in Example 5. While it will often be beneficial to leave the denominator in factored form, we have multiplied the factors in the denominator in Example 5 to give the answer in the form the original problem was presented.

5.3 EXERCISES

Find the least common denominator for the following fractions. See Examples 1–3.

1. $\frac{5}{12}, \frac{7}{10}$

2. $\frac{1}{4}, \frac{5}{6}$

3. $\frac{7}{15}, \frac{11}{20}, \frac{5}{24}$

4. $\frac{9}{10}, \frac{12}{25}, \frac{11}{35}$

5. $\frac{17}{100}, \frac{13}{120}, \frac{29}{180}$

6. $\frac{17}{250}, \frac{1}{300}, \frac{127}{360}$

7. $\frac{9}{x^2}, \frac{8}{x^5}$

8. $\frac{2}{m^7}, \frac{3}{m^8}$

9. $\frac{2}{5p}, \frac{5}{6p}$

10. $\frac{4}{15k}, \frac{3}{4k}$

11. $\frac{7}{15y^2}, \frac{5}{36y^4}$

12. $\frac{4}{25m^3}, \frac{7}{10m^4}$

13. $\frac{1}{5a^2b^3}, \frac{2}{15a^5b}$

14. $\frac{7}{3r^4s^5}, \frac{2}{9r^6s^8}$

15. $\frac{7}{6p}, \frac{3}{4p-8}$

16. $\frac{7}{8k}, \frac{13}{12k-24}$

---

17. If two denominators have greatest common factor equal to 1, how can you easily find their least common denominator? (See Objective 1.)

18. Suppose that two fractions have as denominators a^k and a^r, where k and r are natural numbers, with $k > r$. What is their least common denominator? (See Objective 1.)

Find the least common denominators for the following fractions. See Examples 1–3.

19. $\frac{5}{32r^2}, \frac{9}{16r-32}$

20. $\frac{13}{18m^3}, \frac{8}{9m-36}$

21. $\frac{7}{6r-12}, \frac{4}{9r-18}$

22. $\frac{4}{5p-30}, \frac{5}{6p-36}$

23. $\frac{5}{12p+60}, \frac{1}{p^2+5p}, \frac{6}{p^2+10p+25}$

24. $\frac{1}{r^2+7r}, \frac{3}{5r+35}, \frac{7}{r^2+14r+49}$

25. $\frac{3}{8y+16}, \frac{2}{y^2+3y+2}$

26. $\frac{2}{9m-18}, \frac{9}{m^2-7m+10}$

27. $\frac{2}{m-3}, \frac{4}{3-m}$

28. $\frac{7}{8-a}, \frac{2}{a-8}$

29. $\frac{9}{p-q}, \frac{3}{q-p}$

30. $\frac{4}{z-x}, \frac{8}{x-z}$

31. $\frac{6}{a^2+6a}, \frac{4}{a^2+3a-18}$

32. $\frac{3}{y^2-5y}, \frac{2}{y^2-2y-15}$

33. $\frac{5}{k^2+2k-35}, \frac{3}{k^2+3k-40}, \frac{9}{k^2-2k-15}$

34. $\frac{9}{z^2+4z-12}, \frac{1}{z^2+z-30}, \frac{6}{z^2+2z-24}$

35. $\frac{3}{2y^2+7y-4}, \frac{7}{2y^2-7y+3}$

36. $\frac{2}{5a^2+13a-6}, \frac{6}{5a^2-22a+8}$

37. $\frac{1}{6r^2-r-15}, \frac{4}{3r^2-8r+5}$

38. $\frac{8}{2m^2-11m+14}, \frac{2}{2m^2-m-21}$

Rewrite each expression with the given denominator. See Examples 4 and 5.

39. $\frac{7}{11} = \frac{}{66}$

40. $\frac{5}{8} = \frac{}{56}$

41. $\frac{-11}{m} = \frac{}{8m}$

42. $\frac{-5}{z} = \frac{}{6z}$

43. $\frac{12}{35y} = \frac{}{70y^3}$

44. $\frac{17}{9r} = \frac{}{36r^2}$

45. $\frac{15m^2}{8k} = \frac{}{32k^4}$

46. $\frac{5t^2}{3y} = \frac{}{9y^2}$

47. $\frac{19z}{2z-6} = \frac{}{6z-18}$

48. $\frac{2r}{5r-5} = \frac{}{15r-15}$

49. $\frac{-2a}{9a-18} = \frac{}{18a-36}$

50. $\frac{-5y}{6y+18} = \frac{}{24y+72}$

51. $\frac{6}{k^2-4k} = \frac{}{k(k-4)(k+1)}$

52. $\frac{15}{m^2-9m} = \frac{}{m(m-9)(m+8)}$

53. $\frac{36r}{r^2-r-6} = \frac{}{(r-3)(r+2)(r+1)}$

54. $\frac{4m}{m^2-8m+15} = \frac{}{(m-5)(m-3)(m+2)}$

55. $\frac{a+2b}{2a^2+ab-b^2} = \frac{}{2a^3b+a^2b^2-ab^3}$

56. $\frac{m-4}{6m^2+7m-3} = \frac{}{12m^3+14m^2-6m}$

57. $\frac{4r-t}{r^2+rt+t^2} = \frac{}{t^3-r^3}$

58. $\frac{3x-1}{x^2+2x+4} = \frac{}{x^3-8}$

59. $\frac{2(z-y)}{y^2+yz+z^2} = \frac{}{y^4-z^3y}$

60. $\frac{2p+3q}{p^2+2pq+q^2} = \frac{}{(p+q)(p^3+q^3)}$

---

61. Write an explanation of how to find the least common denominator in a group of denominators. (See Objective 1.)

62. Write an explanation of how to write a rational expression as an equivalent rational expression with a given denominator. (See Objective 2.)

PREVIEW EXERCISES

Add or subtract as indicated. See Section 1.1.

63. $\frac{2}{3} + \frac{5}{3}$

64. $\frac{9}{5} + \frac{2}{5}$

65. $\frac{1}{2} + \frac{2}{5}$

66. $\frac{5}{4} + \frac{1}{3}$

67. $\frac{8}{10} - \frac{2}{15}$

68. $\frac{7}{12} - \frac{6}{15}$

69. $\frac{5}{24} + \frac{1}{18}$

70. $\frac{11}{36} + \frac{7}{45}$

5.4 ADDITION AND SUBTRACTION OF RATIONAL EXPRESSIONS

OBJECTIVES

1 ADD RATIONAL EXPRESSIONS HAVING THE SAME DENOMINATOR.

2 ADD RATIONAL EXPRESSIONS HAVING DIFFERENT DENOMINATORS.

3 SUBTRACT RATIONAL EXPRESSIONS.

1 The sum of two rational expressions is found with a procedure similar to that for adding two fractions.

ADDING RATIONAL EXPRESSIONS

If P/Q and R/Q are rational expressions, then

$$\frac{P}{Q}+\frac{R}{Q}=\frac{P+R}{Q}.$$

Again, the first example shows how the addition of rational expressions compares with that of rational numbers.

EXAMPLE 1 ADDING RATIONAL EXPRESSIONS WITH THE SAME DENOMINATOR

Add.

(a) $\frac{4}{7}+\frac{2}{7}$ **(b)** $\frac{3x}{x+1}+\frac{2x}{x+1}$

Since the denominators are the same, the sum is found by adding the two numerators and keeping the same (common) denominator.

$$\frac{4}{7}+\frac{2}{7}=\frac{4+2}{7}=\frac{6}{7}$$

$$\frac{3x}{x+1}+\frac{2x}{x+1}=\frac{3x+2x}{x+1}=\frac{5x}{x+1}$$ ■

2 Use the steps given below to add two rational expressions with different denominators. These are the same steps used to add fractions with different denominators.

ADDING WITH DIFFERENT DENOMINATORS

Step 1 Find the least common denominator (LCD).

Step 2 Rewrite each rational expression as a fraction with the least common denominator as the denominator.

Step 3 Add the numerators to get the numerator of the sum. The LCD is the denominator of the sum.

Step 4 Write the answer in lowest terms.

EXAMPLE 2 ADDING RATIONAL EXPRESSIONS WITH DIFFERENT DENOMINATORS

Add.

(a) $\frac{1}{12}+\frac{7}{15}$ **(b)** $\frac{2}{3y}+\frac{1}{4y}$

First find the LCD using the methods of the last section.

$$\text{LCD} = 2^2 \cdot 3 \cdot 5 = 60 \qquad \text{LCD} = 2^2 \cdot 3 \cdot y = 12y$$

Now rewrite each rational expression as a fraction with the LCD (either 60 or $12y$) as the denominator.

$$\frac{1}{12}+\frac{7}{15}=\frac{1(5)}{12(5)}+\frac{7(4)}{15(4)}=\frac{5}{60}+\frac{28}{60}$$

$$\frac{2}{3y}+\frac{1}{4y}=\frac{2(4)}{3y(4)}+\frac{1(3)}{4y(3)}=\frac{8}{12y}+\frac{3}{12y}$$

Since the fractions now have common denominators, add the numerators. (Write in lowest terms if necessary.)

$$\frac{5}{60} + \frac{28}{60} = \frac{5 + 28}{60} = \frac{33}{60} = \frac{11}{20}$$

$$\frac{8}{12y} + \frac{3}{12y} = \frac{8 + 3}{12y} = \frac{11}{12y}$$ ■

In the next example, we show the four steps listed earlier in the section.

EXAMPLE 3 ADDING RATIONAL EXPRESSIONS

Add, and write the sum in lowest terms.

$$\frac{2x}{x^2 - 1} + \frac{-1}{x + 1}$$

Step 1 Since the denominators are different, find the LCD.

$$x^2 - 1 = (x + 1)(x - 1)$$

$$x + 1 \text{ is prime.}$$

The LCD is $(x + 1)(x - 1)$.

Step 2 Rewrite each rational expression as a fraction with denominator $(x + 1)(x - 1)$.

$$\frac{2x}{x^2 - 1} + \frac{-1}{x + 1} = \frac{2x}{(x + 1)(x - 1)} + \frac{-1(x - 1)}{(x + 1)(x - 1)}$$ Multiply second fraction by $\frac{x - 1}{x - 1}$.

$$= \frac{2x}{(x + 1)(x - 1)} + \frac{-x + 1}{(x + 1)(x - 1)}$$ Distributive property

Step 3 Add the numerators to get the numerator of the sum. The denominator is the LCD, $(x + 1)(x - 1)$.

$$= \frac{2x - x + 1}{(x + 1)(x - 1)}$$ Add numerators; keep the same denominator.

$$= \frac{x + 1}{(x + 1)(x - 1)}$$ Combine like terms in the numerator.

Step 4 Write the answer in lowest terms.

$$= \frac{1(x + 1)}{(x + 1)(x - 1)}$$ Identity property for multiplication

$$= \frac{1}{x - 1}$$ Fundamental property of rational expressions ■

In the rest of the examples in this section, the steps will not be numbered.

EXAMPLE 4
ADDING RATIONAL EXPRESSIONS

Add, and express the answer in lowest terms.

$$\frac{2x}{x^2 + 5x + 6} + \frac{x + 1}{x^2 + 2x - 3}$$

Begin by factoring the denominators completely.

$$\frac{2x}{(x + 2)(x + 3)} + \frac{x + 1}{(x + 3)(x - 1)}$$

The least common denominator is $(x + 2)(x + 3)(x - 1)$. By the fundamental property of rational expressions,

$$\frac{2x}{(x + 2)(x + 3)} + \frac{x + 1}{(x + 3)(x - 1)}$$

$$= \frac{2x(x - 1)}{(x + 2)(x + 3)(x - 1)} + \frac{(x + 1)(x + 2)}{(x + 3)(x - 1)(x + 2)}$$

$$= \frac{2x(x - 1) + (x + 1)(x + 2)}{(x + 2)(x + 3)(x - 1)}$$ Add numerators; keep the same denominator.

$$= \frac{2x^2 - 2x + x^2 + 3x + 2}{(x + 2)(x + 3)(x - 1)}$$ Distributive property

$$= \frac{3x^2 + x + 2}{(x + 2)(x + 3)(x - 1)}$$ Combine terms.

Since $3x^2 + x + 2$ cannot be factored, the rational expression cannot be reduced. It is usually best to leave the denominator in factored form since it is then easier to identify common factors in the numerator and denominator. ■

In some problems, rational expressions to be added or subtracted have denominators that are negatives of each other. The next example illustrates how to proceed in such a problem.

EXAMPLE 5
ADDING RATIONAL EXPRESSIONS WITH DENOMINATORS THAT ARE NEGATIVES

Add, and express the answer in lowest terms.

$$\frac{y}{y - 2} + \frac{8}{2 - y}$$

To get a common denominator of $y - 2$, multiply the second expression by -1 in both the numerator and the denominator.

$$\frac{y}{y - 2} + \frac{8}{2 - y} = \frac{y}{y - 2} + \frac{8(-1)}{(2 - y)(-1)}$$

$$= \frac{y}{y - 2} + \frac{-8}{y - 2}$$ Distributive property

$$= \frac{y - 8}{y - 2}$$ Add numerators; keep the same denominator.

If we had chosen to use $2 - y$ as the common denominator, the final answer would be in the form $\frac{8 - y}{2 - y}$, which is equivalent to $\frac{y - 8}{y - 2}$. ■

3 To *subtract* rational expressions, use the following rule.

SUBTRACTING RATIONAL EXPRESSIONS

If P/Q and R/Q are rational expressions, then

$$\frac{P}{Q} - \frac{R}{Q} = \frac{P - R}{Q}.$$

EXAMPLE 6 SUBTRACTING RATIONAL EXPRESSIONS

Subtract: $\frac{2m}{m - 1} - \frac{2}{m - 1}$.

By the definition of subtraction,

$$\frac{2m}{m - 1} - \frac{2}{m - 1} = \frac{2m - 2}{m - 1}$$ Subtract numerators; keep the same denominator.

$$= \frac{2(m - 1)}{m - 1}$$ Factor the numerator.

$$= 2.$$ Write in lowest terms. ■

EXAMPLE 7 SUBTRACTING RATIONAL EXPRESSIONS

Subtract: $\frac{9}{x - 2} - \frac{3}{x}$.

The least common denominator is $x(x - 2)$.

$$\frac{9}{x - 2} - \frac{3}{x} = \frac{9x}{x(x - 2)} - \frac{3(x - 2)}{x(x - 2)}$$ Get the least common denominator.

$$= \frac{9x - 3(x - 2)}{x(x - 2)}$$ Subtract numerators; keep the same denominator.

$$= \frac{9x - 3x + 6}{x(x - 2)}$$ Distributive property

$$= \frac{6x + 6}{x(x - 2)}$$ Combine like terms in the numerator. ■

NOTE It would not be wrong to factor the final numerator in Example 7 to get an answer in the form $\frac{6(x + 1)}{x(x - 2)}$; however, the fundamental property would not apply, since there are no common factors that would allow us to write the answer in lower terms. In general, we will leave numerators in polynomial form, and denominators in factored form.

CAUTION Sign errors often occur in subtraction problems like the ones in Examples 7 and 8. Remember that the numerator of the fraction being subtracted must be treated as a single quantity. Parentheses can be used to avoid this common error.

EXAMPLE 8 SUBTRACTING RATIONAL EXPRESSIONS

Subtract, and express the answer in lowest terms.

$$\frac{3x}{x+5} - \frac{2x-5}{x+5}$$

The denominators are the same, so keep the denominator and subtract the numerators. Use parentheses around $2x - 5$.

$$\frac{3x}{x+5} - \frac{2x-5}{x+5} = \frac{3x-(2x-5)}{x+5} \quad \text{Subtract numerators; keep the same denominator.}$$

$$= \frac{3x-2x+5}{x+5} \quad \text{Distributive property}$$

$$= \frac{x+5}{x+5} \quad \text{Combine like terms.}$$

$$= 1 \quad \text{Lowest terms}$$

■

EXAMPLE 9 SUBTRACTING RATIONAL EXPRESSIONS

Subtract: $\dfrac{6x}{x^2-2x+1} - \dfrac{1}{x^2-1}$.

Begin by factoring.

$$\frac{6x}{x^2-2x+1} - \frac{1}{x^2-1} = \frac{6x}{(x-1)(x-1)} - \frac{1}{(x-1)(x+1)}$$

From the factored denominators, we can identify the common denominator, $(x-1)(x-1)(x+1)$. Use the factor $x - 1$ twice, since it appears twice in the first denominator.

$$\frac{6x}{(x-1)(x-1)} - \frac{1}{(x-1)(x+1)}$$

$$= \frac{6x(x+1)}{(x-1)(x-1)(x+1)} - \frac{1(x-1)}{(x-1)(x-1)(x+1)} \quad \text{Fundamental property}$$

$$= \frac{6x(x+1)-1(x-1)}{(x-1)(x-1)(x+1)} \quad \text{Subtract.}$$

$$= \frac{6x^2+6x-x+1}{(x-1)(x-1)(x+1)} \quad \text{Distributive property}$$

$$= \frac{6x^2+5x+1}{(x-1)(x-1)(x+1)} \quad \text{Combine like terms.}$$

The result may be written as $\dfrac{6x^2+5x+1}{(x-1)^2(x+1)}$. ■

EXAMPLE 10 SUBTRACTING RATIONAL EXPRESSIONS

Subtract: $\dfrac{q}{q^2 - 4q - 5} - \dfrac{3}{2q^2 - 13q + 15}$.

Start by factoring each denominator.

$$\frac{q}{q^2 - 4q - 5} - \frac{3}{2q^2 - 13q + 15} = \frac{q}{(q + 1)(q - 5)} - \frac{3}{(q - 5)(2q - 3)}$$

Now rewrite each of the two rational expressions with the least common denominator, $(q + 1)(q - 5)(2q - 3)$. Then subtract numerators.

$$\frac{q}{(q + 1)(q - 5)} - \frac{3}{(q - 5)(2q - 3)}$$

$$= \frac{q(2q - 3)}{(q + 1)(q - 5)(2q - 3)} - \frac{3(q + 1)}{(q + 1)(q - 5)(2q - 3)} \quad \text{Get a common denominator.}$$

$$= \frac{q(2q - 3) - 3(q + 1)}{(q + 1)(q - 5)(2q - 3)} \quad \text{Subtract.}$$

$$= \frac{2q^2 - 3q - 3q - 3}{(q + 1)(q - 5)(2q - 3)} \quad \text{Distributive property}$$

$$= \frac{2q^2 - 6q - 3}{(q + 1)(q - 5)(2q - 3)} \quad \text{Combine like terms.}$$ ■

5.4 EXERCISES

Note: In many problems involving sums and differences of rational expressions, several different equivalent forms of the answer exist. If your answer does not look exactly like the one given in the back of the book, check to see if your answer is an equivalent form.

Find the sums or differences. Write the answers in lowest terms. See Examples 1 and 8.

1. $\dfrac{2}{p} + \dfrac{5}{p}$
2. $\dfrac{3}{r} + \dfrac{6}{r}$
3. $\dfrac{9}{k} - \dfrac{12}{k}$
4. $\dfrac{15}{z} - \dfrac{25}{z}$
5. $\dfrac{y}{y + 1} + \dfrac{1}{y + 1}$
6. $\dfrac{3m}{m - 4} + \dfrac{-12}{m - 4}$
7. $\dfrac{p - q}{3} - \dfrac{p + q}{3}$
8. $\dfrac{a + b}{2} - \dfrac{a - b}{2}$
9. $\dfrac{m^2}{m + 6} + \dfrac{6m}{m + 6}$
10. $\dfrac{y^2}{y - 1} + \dfrac{-y}{y - 1}$
11. $\dfrac{q^2 - 4q}{q - 2} + \dfrac{4}{q - 2}$
12. $\dfrac{z^2 - 10z}{z - 5} + \dfrac{25}{z - 5}$
13. $\dfrac{3r + 1}{r - 4} - \dfrac{2r + 5}{r - 4}$
14. $\dfrac{7t + 1}{t + 3} - \dfrac{6t - 2}{t + 3}$

15. Explain the procedure used to add rational expressions that have the same denominator. (See Objective 1.)

16. Explain the procedure used to subtract rational expressions that have the same denominator. (See Objective 3.)

Find the sums or differences. See Examples 2, 5, and 7.

17. $\frac{m}{3} + \frac{1}{2}$

18. $\frac{p}{6} - \frac{2}{3}$

19. $\frac{4}{3} - \frac{1}{y}$

20. $\frac{8}{5} - \frac{2}{a}$

21. $\frac{5m}{6} - \frac{2m}{3}$

22. $\frac{3}{x} + \frac{5}{2x}$

23. $\frac{4 + 2k}{5} + \frac{2 + k}{10}$

24. $\frac{5 - 4r}{8} - \frac{2 - 3r}{6}$

25. $\frac{m + 2}{m} + \frac{m + 3}{4m}$

26. $\frac{2x - 5}{x} + \frac{x - 1}{2x}$

27. $\frac{6}{y^2} - \frac{2}{y}$

28. $\frac{3}{p} + \frac{5}{p^2}$

29. $\frac{-1}{x^2} + \frac{3}{xy}$

30. $\frac{9}{p^2} + \frac{4}{px}$

31. $\frac{8}{x - 2} - \frac{4}{x + 2}$

Find the sums or differences. See Examples 3–7, 9, and 10.

32. $\frac{6}{m - 5} - \frac{2}{m + 5}$

33. $\frac{2x}{x + y} - \frac{3x}{2x + 2y}$

34. $\frac{1}{a - b} - \frac{a}{4a - 4b}$

35. $\frac{1}{m^2 - 9} + \frac{1}{3m + 9}$

36. $\frac{-6}{y^2 - 4} - \frac{3}{2y + 4}$

37. $\frac{1}{m^2 - 1} - \frac{1}{m^2 + 3m + 2}$

38. $\frac{1}{y^2 - 4} + \frac{3}{y^2 + 5y + 6}$

39. $\frac{4}{2 - q} + \frac{7}{q - 2}$

40. $\frac{9}{8 - y} + \frac{6}{y - 8}$

41. $\frac{3}{4p - 5} + \frac{9}{5 - 4p}$

42. $\frac{8}{3 - 7y} - \frac{2}{7y - 3}$

---

43. Explain the procedure used to add rational expressions with different denominators. (See Objective 2.)

44. Explain the procedure used to subtract rational expressions with different denominators. (See Objectives 2 and 3.)

Perform the indicated operations.

45. $\frac{8}{m - 2} + \frac{3}{5m} + \frac{7}{5m(m - 2)}$

46. $\frac{-1}{7z} + \frac{3}{z + 2} + \frac{4}{7z(z + 2)}$

47. $\frac{4}{r^2 - r} + \frac{6}{r^2 + 2r} - \frac{1}{r^2 + r - 2}$

48. $\frac{6}{k^2 + 3k} - \frac{1}{k^2 - k} + \frac{2}{k^2 + 2k - 3}$

49. $\frac{4y - 1}{2y^2 + 5y - 3} - \frac{y + 3}{6y^2 + y - 2}$

50. $\frac{2q + 1}{3q^2 + 10q - 8} - \frac{3q + 5}{2q^2 + 5q - 12}$

51. $\frac{x + 3y}{x^2 + 2xy + y^2} + \frac{x - y}{x^2 + 4xy + 3y^2}$

52. $\frac{m}{m^2 - 1} + \frac{m - 1}{m^2 + 2m + 1}$

53. $\frac{r + y}{18r^2 + 12ry - 3ry - 2y^2} + \frac{3r - y}{36r^2 - y^2}$

54. $\frac{2x - z}{2x^2 - 4xz + 5xz - 10z^2} - \frac{x + z}{x^2 - 4z^2}$

55. $\dfrac{3p-2}{1-4q^4} - \dfrac{p^2+2}{2pq^2-8q^2+p-4}$

56. $\dfrac{t+5}{64t^3+1} + \dfrac{1-4t}{16t^2-4t+1}$

Perform the indicated operations in Exercises 57–60. Remember the order of operations.

57. $\left(\dfrac{-k}{2k^2-5k-3} + \dfrac{3k-2}{2k^2-k-1}\right)\dfrac{2k+1}{k-1}$

58. $\left(\dfrac{3p+1}{2p^2+p-6} - \dfrac{5p}{3p^2-p}\right)\dfrac{2p-3}{p+2}$

59. $\dfrac{k^2+4k+16}{k+4}\left(\dfrac{-5}{16-k^2} + \dfrac{2k+3}{k^3-64}\right)$

60. $\dfrac{m-5}{2m+5}\left(\dfrac{-3m}{m^2-25} - \dfrac{m+4}{125-m^3}\right)$

---

61. A rectangle has a width of $\dfrac{y+1}{3}$ and a length of $\dfrac{y-2}{4}$. **(a)** Find the perimeter. **(b)** Find the area.

62. The dimensions (length and width) of a rectangle are $\dfrac{9}{2p}$ and $\dfrac{4}{p^2}$. **(a)** Find the perimeter. **(b)** Find the area.

PREVIEW EXERCISES

Perform the indicated operations. See Sections 1.1 and 1.5–1.8.

63. $-1\frac{2}{3} + 4\frac{1}{2}$

64. $2\frac{5}{8} - 3\frac{1}{4}$

65. $-4\frac{1}{4} \div 1\frac{3}{8}$

66. $-6\frac{1}{2} \div -3\frac{1}{2}$

5.5 COMPLEX FRACTIONS

OBJECTIVES

1. SIMPLIFY A COMPLEX FRACTION BY WRITING IT AS A DIVISION PROBLEM (METHOD 1).
2. SIMPLIFY A COMPLEX FRACTION BY MULTIPLYING THE NUMERATOR AND THE DENOMINATOR BY THE LCD OF ALL THE FRACTIONS WITHIN THE COMPLEX FRACTION (METHOD 2).

The quotient of two mixed numbers in arithmetic, such as $2\frac{1}{2} \div 3\frac{1}{4}$, can be written as a fraction:

$$2\frac{1}{2} \div 3\frac{1}{4} = \frac{2\frac{1}{2}}{3\frac{1}{4}} = \frac{2+\frac{1}{2}}{3+\frac{1}{4}}.$$

The last expression is the quotient of expressions that involve fractions. In algebra, some rational expressions also have fractions in the numerator, or denominator, or both.

COMPLEX FRACTION A rational expression with fractions in the numerator, denominator, or both, is called a **complex fraction**.

Examples of complex fractions include

$$\frac{2+\frac{1}{2}}{3+\frac{1}{4}}, \qquad \frac{\frac{3x^2-5x}{6x^2}}{2x-\frac{1}{x}}, \qquad \text{and} \qquad \frac{3+x}{5-\frac{2}{x}}.$$

The parts of a complex fraction are named as follows.

$$\frac{\frac{2}{p}-\frac{1}{q}}{\frac{3}{p}+\frac{5}{q}}$$

← Numerator of complex fraction
← Main fraction bar
← Denominator of complex fraction

1 Since the main fraction bar represents division in a complex fraction, one method of simplifying a complex fraction involves rewriting it as a division problem.

METHOD 1

To simplify a complex fraction:
Step 1 Write both the numerator and denominator as single fractions.
Step 2 Change the complex fraction to a division problem.
Step 3 Perform the indicated division.

Once again, in this section the first example shows complex fractions from both arithmetic and algebra. You should notice the similarities as you read through it.

EXAMPLE 1 SIMPLIFYING COMPLEX FRACTIONS BY METHOD 1

Simplify each complex fraction.

(a) $\dfrac{\frac{2}{3}+\frac{5}{9}}{\frac{1}{4}+\frac{1}{12}}$ **(b)** $\dfrac{6+\frac{3}{x}}{\frac{x}{4}+\frac{1}{8}}$

First, write each numerator as a single fraction.

$$\frac{2}{3}+\frac{5}{9}=\frac{2(3)}{3(3)}+\frac{5}{9}=\frac{6}{9}+\frac{5}{9}=\frac{11}{9}$$

$$6+\frac{3}{x}=\frac{6}{1}+\frac{3}{x}=\frac{6x}{x}+\frac{3}{x}=\frac{6x+3}{x}$$

Do the same thing with each denominator.

$$\frac{1}{4}+\frac{1}{12}=\frac{1(3)}{4(3)}+\frac{1}{12}=\frac{3}{12}+\frac{1}{12}=\frac{4}{12}$$

$$\frac{x}{4}+\frac{1}{8}=\frac{x(2)}{4(2)}+\frac{1}{8}=\frac{2x}{8}+\frac{1}{8}=\frac{2x+1}{8}$$

The original complex fraction can now be written as follows.

$$\frac{\frac{11}{9}}{\frac{4}{12}} \qquad \Bigg| \qquad \frac{\frac{6x+3}{x}}{\frac{2x+1}{8}}$$

Now use the rule for division and the fundamental property.

$$\frac{11}{9} \div \frac{4}{12} = \frac{11}{9} \cdot \frac{12}{4} = \frac{11 \cdot 3 \cdot 4}{3 \cdot 3 \cdot 4} = \frac{11}{3}$$

$$\frac{6x+3}{x} \div \frac{2x+1}{8} = \frac{6x+3}{x} \cdot \frac{8}{2x+1} = \frac{3(2x+1)}{x} \cdot \frac{8}{2x+1} = \frac{24}{x} \quad \blacksquare$$

EXAMPLE 2 SIMPLIFYING A COMPLEX FRACTION BY METHOD 1

Simplify the complex fraction $\dfrac{\frac{xp}{q^3}}{\frac{p^2}{qx^2}}$.

Here, the numerator and denominator are already single fractions, so use the division rule and then the fundamental property: $\dfrac{xp}{q^3} \div \dfrac{p^2}{qx^2} = \dfrac{xp}{q^3} \cdot \dfrac{qx^2}{p^2} = \dfrac{x^3}{q^2p}$. ■

EXAMPLE 3 SIMPLIFYING A COMPLEX FRACTION BY METHOD 1

Simplify the complex fraction $\dfrac{\frac{x}{x+y}}{\frac{1}{x}+\frac{1}{y}}$.

Since the numerator is already a single fraction, it is only necessary to simplify the denominator. Do this by adding $1/x$ and $1/y$.

$$\frac{1}{x} + \frac{1}{y} = \frac{1}{x} \cdot \frac{y}{y} + \frac{1}{y} \cdot \frac{x}{x} \qquad \text{Multiply each fraction by 1 to get a common denominator, } xy.$$

$$= \frac{y}{xy} + \frac{x}{xy} \qquad \text{Multiply fractions.}$$

$$= \frac{x+y}{xy} \qquad \text{Add fractions.}$$

The numerator is the single fraction $x/(x + y)$, and now the denominator is the single fraction $(x + y)/(xy)$. Write as a division problem and perform the division.

$$\frac{x}{x+y} \div \frac{x+y}{xy} = \frac{x}{x+y} \cdot \frac{xy}{x+y} \qquad \text{Definition of division}$$

$$= \frac{x^2y}{(x+y)^2} \qquad \text{Multiply fractions.} \quad \blacksquare$$

2 As an alternative method, a complex fraction may be simplified by a method that uses the identity property of multiplication. Since any expression can be multiplied by a form of 1 to get an equivalent expression, we may multiply both the numerator and the denominator of a complex fraction by the same nonzero expression to get an equivalent complex fraction. If we choose the expression to be the LCD of all the fractions within the complex fraction, the complex fraction will be simplified. This is Method 2.

METHOD 2

To simplify a complex fraction:

Step 1 Find the LCD of all fractions within the complex fraction.

Step 2 Multiply both the numerator and the denominator of the complex fraction by this LCD, using the distributive property as necessary. Reduce to lowest terms.

In the next example, Method 2 is used to simplify the same complex fractions as in Example 1.

EXAMPLE 4 SIMPLIFYING COMPLEX FRACTIONS BY METHOD 2

Simplify each complex fraction.

(a) $\dfrac{\frac{2}{3}+\frac{5}{9}}{\frac{1}{4}+\frac{1}{12}}$ **(b)** $\dfrac{6+\frac{3}{x}}{\frac{x}{4}+\frac{1}{8}}$

Find the LCD for all denominators in the complex fraction.

The LCD for 3, 9, 4, and 12 is 36. | The LCD for x, 4, and 8 is $8x$.

Multiply numerator and denominator of the complex fraction by the LCD.

$$\frac{\frac{2}{3}+\frac{5}{9}}{\frac{1}{4}+\frac{1}{12}} = \frac{36\left(\frac{2}{3}+\frac{5}{9}\right)}{36\left(\frac{1}{4}+\frac{1}{12}\right)}$$

$$= \frac{36\left(\frac{2}{3}\right)+36\left(\frac{5}{9}\right)}{36\left(\frac{1}{4}\right)+36\left(\frac{1}{12}\right)}$$

$$= \frac{24+20}{9+3}$$

$$= \frac{44}{12} = \frac{4\cdot 11}{4\cdot 3}$$

$$= \frac{11}{3}$$

$$\frac{6+\frac{3}{x}}{\frac{x}{4}+\frac{1}{8}} = \frac{8x\left(6+\frac{3}{x}\right)}{8x\left(\frac{x}{4}+\frac{1}{8}\right)}$$

$$= \frac{8x(6)+8x\left(\frac{3}{x}\right)}{8x\left(\frac{x}{4}\right)+8x\left(\frac{1}{8}\right)} \qquad \text{Distributive property}$$

$$= \frac{48x+24}{2x^2+x}$$

$$= \frac{24(2x+1)}{x(2x+1)} \qquad \text{Factor.}$$

$$= \frac{24}{x} \qquad \text{Lowest terms}$$

■

EXAMPLE 5 SIMPLIFYING A COMPLEX FRACTION BY METHOD 2

Simplify the complex fraction $\dfrac{\dfrac{3}{5m} - \dfrac{2}{m^2}}{\dfrac{9}{2m} + \dfrac{3}{4m^2}}$.

The least common denominator for $5m$, m^2, $2m$, and $4m^2$ is $20m^2$. Multiply numerator and denominator by $20m^2$.

$$\frac{\dfrac{3}{5m} - \dfrac{2}{m^2}}{\dfrac{9}{2m} + \dfrac{3}{4m^2}} = \frac{20m^2\left(\dfrac{3}{5m} - \dfrac{2}{m^2}\right)}{20m^2\left(\dfrac{9}{2m} + \dfrac{3}{4m^2}\right)}$$

$$= \frac{20m^2\left(\dfrac{3}{5m}\right) - 20m^2\left(\dfrac{2}{m^2}\right)}{20m^2\left(\dfrac{9}{2m}\right) + 20m^2\left(\dfrac{3}{4m^2}\right)} \qquad \text{Distributive property}$$

$$= \frac{12m - 40}{90m + 15}$$

Factoring the numerator and denominator of this result leads to

$$\frac{4(3m - 10)}{15(6m + 1)}.$$

Either of these answers is acceptable. There are other acceptable forms as well. ∎

The two methods shown in this section can be applied to a complex fraction to simplify it. You may want to choose one method and stick to it in order to eliminate confusion. However, some students prefer to use Method 1 for problems like Example 2, which is the quotient of two fractions. They prefer Method 2 for problems like Examples 1, 3, 4, and 5, which have sums or differences in the numerators or denominators or both.

5.5 EXERCISES

Note: In many problems involving complex fractions, several different equivalent forms of the answer exist. If your answer does not look exactly like the one given in the back of the book, check to see if your answer is an equivalent form.

Simplify each complex fraction. Use either method. See Examples 1–5.

1. $\dfrac{\dfrac{3}{7}}{\dfrac{2}{3}}$ **2.** $\dfrac{\dfrac{9}{10}}{\dfrac{5}{6}}$ **3.** $\dfrac{-\dfrac{4}{3}}{\dfrac{2}{9}}$ **4.** $\dfrac{-\dfrac{5}{6}}{\dfrac{5}{4}}$ **5.** $\dfrac{\dfrac{p}{q^2}}{\dfrac{p^2}{q}}$ **6.** $\dfrac{\dfrac{a}{x}}{\dfrac{a^2}{2x}}$

7. $\dfrac{\frac{x}{y}}{\frac{x^2}{y}}$

8. $\dfrac{\frac{p^4}{r}}{\frac{p^2}{r^2}}$

9. $\dfrac{\frac{m^3p^4}{5m}}{\frac{8mp^5}{p^2}}$

10. $\dfrac{\frac{9z^5x^3}{2x}}{\frac{8z^2x^5}{3z}}$

11. $\dfrac{\frac{x+1}{4}}{\frac{x-3}{x}}$

12. $\dfrac{\frac{m+6}{m}}{\frac{m-6}{2}}$

13. $\dfrac{\frac{3}{q}}{\frac{1-q}{6q^2}}$

14. $\dfrac{\frac{6}{x}}{\frac{1+x}{8x^5}}$

15. $\dfrac{\frac{2x+3}{12}}{\frac{x-1}{15}}$

16. $\dfrac{\frac{15}{k+1}}{\frac{10}{3(k+1)}}$

17. $\dfrac{\frac{5}{8}+\frac{2}{3}}{\frac{7}{3}-\frac{1}{4}}$

18. $\dfrac{\frac{6}{5}-\frac{1}{9}}{\frac{2}{5}+\frac{5}{3}}$

19. $\dfrac{1-\frac{3}{8}}{2+\frac{1}{4}}$

20. $\dfrac{3+\frac{5}{4}}{1-\frac{7}{8}}$

21. $\dfrac{\frac{3}{y}+1}{\frac{3+y}{2}}$

22. $\dfrac{6+\frac{2}{r}}{\frac{r+2}{3}}$

23. $\dfrac{\frac{1}{x}+x}{\frac{x^2+1}{8}}$

24. $\dfrac{\frac{3}{m}-m}{\frac{3-m^2}{4}}$

25. $\dfrac{x+\frac{1}{x}}{\frac{4}{x}+y}$

26. $\dfrac{y-\frac{6}{y}}{y+\frac{2}{y}}$

27. $\dfrac{\frac{p+3}{p}}{\frac{1}{p}+\frac{1}{5}}$

28. $\dfrac{r+\frac{1}{r}}{\frac{1}{r}-r}$

29. $\dfrac{\frac{2}{p^2}-\frac{3}{5p}}{\frac{4}{p}+\frac{1}{4p}}$

30. $\dfrac{\frac{2}{m^2}-\frac{3}{m}}{\frac{2}{5m^2}+\frac{1}{3m}}$

31. $\dfrac{\frac{5}{x^2y}-\frac{2}{xy^2}}{\frac{3}{x^2y^2}+\frac{4}{xy}}$

32. $\dfrac{\frac{1}{m^3p}+\frac{2}{mp^2}}{\frac{4}{mp}+\frac{1}{m^2p}}$

33. $\dfrac{\frac{1}{4}-\frac{1}{a^2}}{\frac{1}{2}+\frac{1}{a}}$

34. $\dfrac{\frac{1}{9}-\frac{1}{m^2}}{\frac{1}{3}+\frac{1}{m}}$

35. $\dfrac{\frac{1}{z+5}}{\frac{4}{z^2-25}}$

36. $\dfrac{\frac{a}{a+1}}{\frac{2}{a^2-1}}$

37. $\dfrac{\frac{1}{m+1}-1}{\frac{1}{m+1}+1}$

38. $\dfrac{\frac{2}{x-1}+2}{\frac{2}{x-1}-2}$

---

39. In a fraction, what operation does the fraction bar represent? (See Objective 1.)

40. What property of real numbers justifies Method 2 of simplifying complex fractions? (See Objective 2.)

41. Write an explanation of Method 1 for simplifying complex fractions. (See Objective 1.)

42. Write an explanation of Method 2 for simplifying complex fractions. (See Objective 2.)

Simplify the complex fractions.

43. $\dfrac{\dfrac{1}{m-1}+\dfrac{2}{m+2}}{\dfrac{2}{m+2}-\dfrac{1}{m-3}}$

44. $\dfrac{\dfrac{5}{r+3}-\dfrac{1}{r-1}}{\dfrac{2}{r+2}+\dfrac{3}{r+3}}$

45. $1-\dfrac{1}{1+\dfrac{1}{1+1}}$

46. $3-\dfrac{2}{4+\dfrac{2}{4-2}}$

47. $r+\dfrac{r}{4-\dfrac{2}{6+2}}$

48. $\dfrac{2q}{7}-\dfrac{q}{6+\dfrac{8}{4+4}}$

PREVIEW EXERCISES

Use the distributive property to simplify. See Section 2.1.

49. $9\left(\dfrac{4x}{3}+\dfrac{2}{9}\right)$

50. $8\left(\dfrac{3r}{4}-\dfrac{9}{8}\right)$

51. $12\left(\dfrac{11p^2}{3}-\dfrac{9p}{4}\right)$

52. $6\left(\dfrac{5z^2}{2}+\dfrac{8z}{3}\right)$

Solve each equation. See Sections 2.3 and 4.5.

53. $3x-5=7x+3$

54. $9z+2=7z-6$

55. $8(2q+5)-1=7q$

56. $6(z-3)+5=8z$

57. $k^2+3k-4=0$

58. $p^2-6p-7=0$

5.6 EQUATIONS INVOLVING RATIONAL EXPRESSIONS

OBJECTIVES

1. DISTINGUISH BETWEEN EXPRESSIONS WITH RATIONAL COEFFICIENTS AND EQUATIONS WITH TERMS THAT ARE RATIONAL EXPRESSIONS.
2. SOLVE EQUATIONS WITH RATIONAL EXPRESSIONS.
3. SOLVE A FORMULA FOR A SPECIFIED VARIABLE.

In Section 2.3 we learned how to solve equations with fractions as coefficients. By using the multiplication property of equality, we can clear the fractions by multiplying through by the least common denominator (LCD). In this section we will examine this procedure in more detail.

1 Before solving equations with rational expressions, it is necessary to understand the difference between *sums* and *differences* of terms with rational coefficients, and *equations* with terms that are rational expressions. Sums and differences are *simplified,* while equations are *solved.*

EXAMPLE 1
DISTINGUISHING BETWEEN EXPRESSIONS AND EQUATIONS

Identify each of the following as an expression or an equation. If it is an expression, simplify it. If it is an equation, solve it.

(a) $\frac{3}{4}x - \frac{2}{3}x$

This is a difference of two terms, so it is an expression. (There is no equals sign.) Simplify by finding the LCD, writing each coefficient with this LCD, and combining like terms.

$$\frac{3}{4}x - \frac{2}{3}x = \frac{9}{12}x - \frac{8}{12}x \qquad \text{Get a common denominator.}$$

$$= \frac{1}{12}x \qquad \text{Combine like terms.}$$

(b) $\frac{3}{4}x - \frac{2}{3}x = \frac{1}{2}$

Because of the equals sign, this is an equation to be solved. Use the multiplication property of equality to clear fractions. The LCD is 12.

$$\frac{3}{4}x - \frac{2}{3}x = \frac{1}{2}$$

$$12\left(\frac{3}{4}x - \frac{2}{3}x\right) = 12\left(\frac{1}{2}\right) \qquad \text{Multiply by 12.}$$

$$12\left(\frac{3}{4}x\right) - 12\left(\frac{2}{3}x\right) = 12\left(\frac{1}{2}\right) \qquad \text{Distributive property}$$

$$9x - 8x = 6 \qquad \text{Multiply.}$$

$$x = 6 \qquad \text{Combine like terms.}$$

The solution, 6, should be checked in the original equation. ■

The ideas of Example 1 can be summarized as follows.

When adding or subtracting, the LCD must be kept throughout the simplification. When solving an equation, the LCD is used to multiply both sides so that denominators are eliminated.

2 The next few examples illustrate the method of clearing an equation of fractions in order to solve it.

EXAMPLE 2
SOLVING AN EQUATION INVOLVING RATIONAL EXPRESSIONS

Solve $\frac{p}{2} - \frac{p-1}{3} = 1$.

Multiply both sides by the least common denominator, 6.

$$6\left(\frac{p}{2} - \frac{p-1}{3}\right) = 6 \cdot 1$$

$$6\left(\frac{p}{2}\right) - 6\left(\frac{p-1}{3}\right) = 6 \qquad \text{Distributive property}$$

$$3p - 2(p-1) = 6$$

Be careful to put parentheses around $p - 1$; otherwise an incorrect solution may be found. Continue simplifying and solve the equation.

$$3p - 2p + 2 = 6 \quad \text{Distributive property}$$
$$p + 2 = 6 \quad \text{Combine like terms.}$$
$$p = 4 \quad \text{Subtract 2.}$$

Check to see that 4 is correct by replacing p with 4 in the original equation. ■

CAUTION The most common error in equations like the one found in Example 2 occurs when parentheses are not used for the numerator $p - 1$ in the second fraction. If parentheses are not used, the sign of the last term on the left side will be incorrect.

The equations in Example 1(b) and Example 2 did not have variables in denominators. When solving equations that have a variable in a denominator, remember that the number 0 cannot be used as a denominator. Therefore, the solution cannot be a number that will make the denominator equal 0.

EXAMPLE 3
SOLVING AN EQUATION INVOLVING RATIONAL EXPRESSIONS

Solve $\frac{x}{x-2} = \frac{2}{x-2} + 2.$

Multiply both sides by the least common denominator, $x - 2$.

$$(x-2)\left(\frac{x}{x-2}\right) = (x-2)\left(\frac{2}{x-2}\right) + (x-2)(2)$$

$$x = 2 + 2x - 4 \quad \text{Distributive property}$$
$$x = -2 + 2x \quad \text{Combine terms.}$$
$$-x = -2 \quad \text{Subtract } 2x.$$
$$x = 2 \quad \text{Multiply by } -1.$$

The proposed solution is 2. If we substitute 2 into the original equation, we get

$$\frac{2}{2-2} = \frac{2}{2-2} + 2 \quad ?$$
$$\frac{2}{0} = \frac{2}{0} + 2. \quad ?$$

Notice that 2 makes both denominators equal to 0. Because 0 cannot be the denominator of a fraction, this equation has no solution. ■

While it is always a good idea to check solutions to guard against arithmetic and algebraic errors, it is *essential* to check proposed solutions when variables appear in denominators in the original equation. Some students like to determine which numbers cannot be solutions *before* solving the equation.

SOLVING EQUATIONS WITH RATIONAL EXPRESSIONS

Step 1 Multiply both sides of the equation by the least common denominator. (This clears the equation of fractions.)
Step 2 Solve the resulting equation.
Step 3 Check each proposed solution by substituting it in the original equation. Reject any that cause a denominator to equal 0.

EXAMPLE 4 SOLVING AN EQUATION INVOLVING RATIONAL EXPRESSIONS

Solve $\dfrac{2m}{m^2-4}+\dfrac{1}{m-2}=\dfrac{2}{m+2}$.

Notice that 2 and -2 cannot be solutions of this equation. (Why?) Since $m^2-4=(m+2)(m-2)$, use $(m+2)(m-2)$ as the least common denominator. Multiply both sides by $(m+2)(m-2)$.

$$(m+2)(m-2)\left(\frac{2m}{m^2-4}+\frac{1}{m-2}\right)=(m+2)(m-2)\frac{2}{m+2}$$

$$(m+2)(m-2)\frac{2m}{(m+2)(m-2)}+(m+2)(m-2)\frac{1}{(m-2)}$$
$$=(m+2)(m-2)\frac{2}{m+2} \quad \text{Distributive property}$$

$$\begin{aligned} 2m+m+2 &= 2(m-2) && \text{Combine terms and use the}\\ 3m+2 &= 2m-4 && \text{distributive property.}\\ 3m-2m &= -4-2 && \text{Subtract } 2m\text{; subtract 2.}\\ m &= -6 \end{aligned}$$

-6 does not make an original denominator equal to 0. It can be verified by substitution that it is a solution for the given equation. ■

EXAMPLE 5 SOLVING AN EQUATION INVOLVING RATIONAL EXPRESSIONS

Solve $\dfrac{2}{x^2-x}=\dfrac{1}{x^2-1}$.

Begin by finding a least common denominator. Since x^2-x can be factored as $x(x-1)$, and x^2-1 can be factored as $(x+1)(x-1)$, the least common denominator is $x(x+1)(x-1)$.

$$\frac{2}{x(x-1)}=\frac{1}{(x+1)(x-1)} \quad \text{Factor the denominators.}$$

Notice that 0, -1, and 1 cannot be solutions of this equation. Multiply both sides of the equation by $x(x+1)(x-1)$.

$$x(x+1)(x-1)\frac{2}{x(x-1)} = x(x+1)(x-1)\frac{1}{(x+1)(x-1)}.$$

$$2(x+1) = x$$

$$2x + 2 = x \qquad \text{Distributive property}$$

$$2 = -x \qquad \text{Subtract } 2x.$$

$$x = -2 \qquad \text{Multiply by } -1.$$

The proposed solution is -2, which does not make any denominator equal 0. A check will verify that no arithmetic or algebraic errors have been made. Since -2 satisfies the equation, it is the solution. ■

EXAMPLE 6
SOLVING AN EQUATION INVOLVING RATIONAL EXPRESSIONS

Solve $\dfrac{1}{x-1} + \dfrac{1}{2} = \dfrac{2}{x^2-1}$.

Factor the denominator on the right.

$$\frac{1}{x-1} + \frac{1}{2} = \frac{2}{(x+1)(x-1)}$$

Notice that 1 and -1 cannot be solutions of this equation. The least common denominator is $2(x+1)(x-1)$. Multiply both sides of the equation by this common denominator.

$$2(x+1)(x-1)\left(\frac{1}{x-1} + \frac{1}{2}\right) = 2(x+1)(x-1)\frac{2}{(x+1)(x-1)}$$

$$2(x+1)(x-1)\frac{1}{x-1} + 2(x+1)(x-1)\frac{1}{2}$$

$$= 2(x+1)(x-1)\frac{2}{(x+1)(x-1)}$$

$$2(x+1) + (x+1)(x-1) = 4$$

$$2x + 2 + x^2 - 1 = 4 \qquad \text{Distributive property}$$

$$x^2 + 2x + 1 = 4$$

$$x^2 + 2x - 3 = 0 \qquad \text{Get 0 on the right side.}$$

Factoring gives

$$(x+3)(x-1) = 0.$$

$$x + 3 = 0 \quad \text{or} \quad x - 1 = 0 \qquad \text{Zero-factor property}$$

$$x = -3 \quad \text{or} \quad x = 1$$

-3 and 1 are proposed solutions. However, as noted above, 1 makes an original denominator equal to 0, so 1 is not a solution. However, -3 is a solution, as can be shown by substituting -3 for x in the original equation. ■

EXAMPLE 7
SOLVING AN EQUATION INVOLVING RATIONAL EXPRESSIONS

Solve $\dfrac{1}{k^2 + 4k + 3} + \dfrac{1}{2k + 2} = \dfrac{3}{4k + 12}$.

Factoring each denominator gives the equation

$$\frac{1}{(k + 1)(k + 3)} + \frac{1}{2(k + 1)} = \frac{3}{4(k + 3)}.$$

The LCD is $4(k + 1)(k + 3)$, indicating that -1 and -3 cannot be solutions of the equation. Multiply both sides by this LCD.

$$4(k + 1)(k + 3)\left(\frac{1}{(k + 1)(k + 3)} + \frac{1}{2(k + 1)}\right) = 4(k + 1)(k + 3)\frac{3}{4(k + 3)}$$

$$4(k + 1)(k + 3)\frac{1}{(k + 1)(k + 3)} + 2 \cdot 2(k + 1)(k + 3)\frac{1}{2(k + 1)}$$

$$= 4(k + 1)(k + 3)\frac{3}{4(k + 3)}$$

$$4 + 2(k + 3) = 3(k + 1)$$

$$4 + 2k + 6 = 3k + 3 \qquad \text{Distributive property}$$

$$2k + 10 = 3k + 3$$

$$10 - 3 = 3k - 2k$$

$$7 = k$$

The proposed solution, 7, does not make an original denominator equal to zero. A check shows that the algebra is correct, so 7 is the solution of the equation. ■

3 Solving a formula for a specified variable was discussed in Chapter 2. In the next example, this procedure is applied to a formula involving fractions.

EXAMPLE 8
SOLVING FOR A SPECIFIED VARIABLE

Solve the formula $\dfrac{1}{a} = \dfrac{1}{b} + \dfrac{1}{c}$ for c.

The LCD of all the fractions in the equation is abc, so multiply both sides by abc.

$$abc\left(\frac{1}{a}\right) = abc\left(\frac{1}{b} + \frac{1}{c}\right)$$

$$abc\left(\frac{1}{a}\right) = abc\left(\frac{1}{b}\right) + abc\left(\frac{1}{c}\right) \qquad \text{Distributive property}$$

$$bc = ac + ab$$

Since we are solving for c, get all terms with c on one side of the equation. Do this here by subtracting ac from both sides.

$$bc - ac = ab \qquad \text{Subtract } ac.$$

Factor out the common factor c on the left.

$$c(b - a) = ab \qquad \text{Factor out } c.$$

Finally, divide both sides by the coefficient of c, which is $b - a$.

$$c = \frac{ab}{b - a} \quad \blacksquare$$

CAUTION Students often have trouble in the step that involves factoring out the variable for which they are solving. In Example 8, we had to factor out c on the left side so that we could divide both sides by $b - a$.

When solving an equation for a specified variable, be sure that the specified variable appears on only one side of the equals sign in the final equation.

5.6 EXERCISES

Identify each of the following as an expression or an equation. If it is an expression, simplify it. If it is an equation, solve it. See Example 1.

1. $\frac{4}{7}x + \frac{3}{5}x$

2. $\frac{7}{8}x + \frac{1}{5}x$

3. $\frac{4}{7}x + \frac{3}{5}x = 1$

4. $\frac{7}{8}x + \frac{1}{5}x = 1$

5. $\frac{2}{3}y - \frac{7}{4}y$

6. $\frac{3}{5}y - \frac{7}{10}y$

7. $\frac{2}{3}y - \frac{7}{4}y = -13$

8. $\frac{3}{5}y - \frac{7}{10}y = 1$

---

9. Suppose that you were asked to solve $\frac{4}{9}x - \frac{2}{3}x$. How would you respond? (See Objective 1.)

10. Explain how the LCD is used in a different way when adding and subtracting rational expressions as compared to solving equations with rational expressions. (See Objective 1.)

Solve each equation and check your answers. See Examples 1(b) and 2.

11. $\frac{1}{4} = \frac{x}{2}$

12. $\frac{2}{m} = \frac{5}{12}$

13. $\frac{9}{k} = \frac{3}{4}$

14. $\frac{15}{f} = \frac{30}{8}$

15. $\frac{7}{y} = \frac{8}{3}$

16. $\frac{2}{z} = \frac{11}{5}$

17. $\frac{6}{x} - \frac{4}{x} = 5$

18. $\frac{3}{x} + \frac{2}{x} = 5$

19. $\frac{x}{2} - \frac{x}{4} = 6$

20. $\frac{4}{y} + \frac{2}{3} = 1$

21. $\frac{9}{m} = 5 - \frac{1}{m}$

22. $\frac{3x}{5} + 2 = \frac{1}{4}$

23. $\frac{2t}{7} - 5 = t$

24. $\frac{1}{2} + \frac{2}{m} = 1$

25. $\frac{x + 1}{2} = \frac{x + 2}{3}$

26. $\frac{t-4}{3} = t + 2$

27. $\frac{3m}{2} + m = 5$

28. $\frac{2k+3}{k} = \frac{3}{2}$

29. $\frac{5-y}{y} + \frac{3}{4} = \frac{7}{y}$

30. $\frac{x}{x-4} = \frac{2}{x-4} + 5$

31. $\frac{m-2}{5} = \frac{m+8}{10}$

32. $\frac{2p+8}{9} = \frac{10p+4}{27}$

33. $\frac{5r-3}{7} = \frac{15r-2}{28}$

34. $\frac{2y-1}{y} + 2 = \frac{1}{2}$

35. $\frac{3r+1}{3} + \frac{r}{6} = \frac{23}{6}$

36. $\frac{m-2}{4} + \frac{m+1}{3} = \frac{10}{3}$

37. $\frac{y+2}{5} + \frac{y-5}{3} = \frac{7}{5}$

Solve the following equations. Be very careful with signs. See Example 2.

38. $\frac{a+7}{8} - \frac{a-2}{3} = \frac{4}{3}$

39. $\frac{m+2}{5} - \frac{m-6}{7} = 2$

40. $\frac{p}{2} - \frac{p-1}{4} = \frac{5}{4}$

41. $\frac{r}{6} - \frac{r-2}{3} = \frac{-4}{3}$

42. $\frac{5y}{3} - \frac{2y-1}{4} = \frac{1}{4}$

43. $\frac{8k}{5} - \frac{3k-4}{2} = \frac{5}{2}$

---

44. For the equation $\frac{1}{x+2} + \frac{2}{x+3} = \frac{4}{x-5}$, what values of x could not possibly be solutions, since they would cause a denominator to equal 0? (See Objective 2.)

45. If we multiply both sides of the equation $\frac{6}{x+5} = \frac{6}{x+5}$ by $x + 5$, we get $6 = 6$. Are all real numbers solutions of this equation? Explain. (See Objective 2.)

Solve each equation and check your answers. See Examples 3–5.

46. $\frac{8}{2k-4} + \frac{3}{5k-10} = \frac{23}{5}$

47. $\frac{1}{3p+15} + \frac{5}{4p+20} = \frac{19}{24}$

48. $\frac{m}{2m+2} = \frac{-2m}{4m+4} + \frac{2m-3}{m+1}$

49. $\frac{5p+1}{3p+3} = \frac{5p-5}{5p+5} + \frac{3p-1}{p+1}$

50. $\frac{x+1}{x-3} = \frac{4}{x-3} + 6$

51. $\frac{p}{p-2} + 4 = \frac{2}{p-2}$

52. $\frac{2}{k-3} - \frac{3}{k+3} = \frac{12}{k^2-9}$

53. $\frac{1}{r+5} - \frac{3}{r-5} = \frac{-10}{r^2-25}$

54. $\frac{4}{p} - \frac{2}{p+1} = 3$

55. $\frac{6}{r} + \frac{1}{r-2} = 3$

56. $\frac{2}{m-1} + \frac{1}{m+1} = \frac{5}{4}$

57. $\frac{5}{z-2} + \frac{10}{z+2} = 7$

---

58. If you are solving a formula for the letter k, and your steps lead you to the equation

$$kr - mr = km,$$

what would be your next step? (See Objective 3.)

59. If you are solving a formula for the letter k, and your steps lead you to the equation

$$kr - km = mr,$$

what would be your next step? (See Objective 3.)

Solve for the specified variable. See Example 8.

60. $P = \frac{kT}{V}$; for T

61. $I = \frac{kE}{R}$; for R

62. $N = \frac{kF}{d}$; for d

63. $F = \frac{k}{r}$; for r

64. $F = \frac{k}{d - D}$; for D

65. $I = \frac{E}{R + r}$; for R

66. $I = \frac{E}{R + r}$; for r

67. $S = \frac{a}{1 - r}$; for r

68. $h = \frac{2A}{B + b}$; for b

69. $\frac{1}{x} = \frac{1}{y} - \frac{1}{z}$; for y

70. $\frac{3}{k} = \frac{1}{p} + \frac{1}{q}$; for q

71. $9x + \frac{3}{z} = \frac{5}{y}$; for z

72. $2a - \frac{5}{b} + \frac{1}{c} = 0$; for b

73. $\frac{1}{a} = \frac{1}{b} + \frac{1}{c}$; for a

Solve each equation and check the answers. See Examples 6 and 7.

74. $\frac{2}{y} = \frac{y}{5y - 12}$

75. $\frac{8x + 3}{x} = 3x$

76. $\frac{x + 4}{x^2 - 3x + 2} - \frac{5}{x^2 - 4x + 3} = \frac{x - 4}{x^2 - 5x + 6}$

77. $\frac{3y}{y^2 + 5y + 6} = \frac{5y}{y^2 + 2y - 3} - \frac{2}{y^2 + y - 2}$

78. $\frac{3}{r^2 + r - 2} - \frac{1}{r^2 - 1} = \frac{7}{2(r^2 + 3r + 2)}$

79. $\frac{m}{m^2 + m - 2} + \frac{m}{m^2 - 1} = \frac{m}{m^2 + 3m + 2}$

PREVIEW EXERCISES

Write a mathematical expression for each exercise. See Section 2.8.

80. Eryn drives from Philadelphia to Pittsburgh, a distance of 288 miles, in t hours. Find her rate in miles per hour.

81. Mark drives for 10 hours, traveling from City A to City B, a distance of d kilometers. Find his rate in kilometers per hour.

82. Natalie flies her small plane from St. Louis to Chicago, a distance of 289 miles, at z miles per hour. Find her time in hours.

83. Joshua can do a job in x hours. What portion of the job is done in 1 hour?

84. Jennifer needs b hours to tune up her car. How much of the job does she do in 1 hour?

SUMMARY ON RATIONAL EXPRESSIONS

We have seen how to perform the four operations of arithmetic with rational expressions, and how to solve equations with rational expressions. The exercises in this summary provide an opportunity for the student to work a mixed variety of problems of these types. In so doing, recall the procedures explained in the earlier sections of this chapter. They are summarized here.

Multiplication of Rational Expressions	Multiply numerators and multiply denominators. Use the fundamental principle to express in lowest terms.
Division of Rational Expressions	First, change the second fraction to its reciprocal; then multiply as described above.
Addition of Rational Expressions	Find the least common denominator (LCD) if necessary. Add numerators, and keep the same denominator. Express in lowest terms.
Subtraction of Rational Expressions	Find the LCD if necessary. Subtract numerators (use parentheses as required), and keep the same denominator. Express in lowest terms.
Solving Equations with Rational Expressions	Multiply both sides of the equation by the LCD of all the rational expressions in the equation. Solve, using methods described in earlier chapters. Be sure to check all proposed solutions and reject any that cause a denominator to equal 0.

A common error that was mentioned in Section 5.6 bears repeating here. Students often confuse *operations* on rational expressions with the *solution of equations* with rational expressions. For example, the four possibie operations on the rational expressions $\frac{1}{x}$ and $\frac{1}{x-2}$ can be performed as follows.

Add: $\frac{1}{x} + \frac{1}{x-2} = \frac{x-2}{x(x-2)} + \frac{x}{x(x-2)}$ Write with a common denominator.

$= \frac{x-2+x}{x(x-2)}$ Add numerators; keep the same denominator.

$= \frac{2x-2}{x(x-2)}$ Combine like terms.

Subtract: $\frac{1}{x} - \frac{1}{x-2} = \frac{x-2}{x(x-2)} - \frac{x}{x(x-2)}$ Write with a common denominator.

$= \frac{x-2-x}{x(x-2)}$ Subtract numerators; keep the same denominator.

$= \frac{-2}{x(x-2)}$ Combine like terms.

Multiply: $\frac{1}{x} \cdot \frac{1}{x-2} = \frac{1}{x(x-2)}$ Multiply numerators and multiply denominators.

Divide: $\frac{1}{x} \div \frac{1}{x-2} = \frac{1}{x} \cdot \frac{x-2}{1} = \frac{x-2}{x}$ Change to multiplication by the reciprocal of the second fraction.

On the other hand, consider the *equation*

$$\frac{1}{x} + \frac{1}{x-2} = \frac{3}{4}.$$

First notice that neither 0 nor 2 could possibly be solutions of this equation, since each will cause a denominator to equal 0. We can use the multiplication property of equality by multiplying both sides by the LCD, $4x(x - 2)$, giving an equation with no denominators.

$$4x(x-2)\frac{1}{x} + 4x(x-2)\frac{1}{x-2} = 4x(x-2)\frac{3}{4}$$

$$4x - 8 + 4x = 3x^2 - 6x \quad \text{Distributive property}$$

$$0 = 3x^2 - 14x + 8 \quad \text{Get 0 on one side.}$$

$$0 = (3x - 2)(x - 4) \quad \text{Factor.}$$

$$3x - 2 = 0 \quad \text{or} \quad x - 4 = 0 \quad \text{Zero-factor property}$$

$$x = \frac{2}{3} \quad \text{or} \quad x = 4$$

Both 2/3 and 4 are solutions, since neither make a denominator equal to zero.

In conclusion, remember the following points when working exercises involving rational expressions:

1. The fundamental principle is applied only after numerators and denominators have been *factored*.
2. When adding and subtracting rational expressions, the common denominator must be kept throughout the problem.
3. Always look to see if the answer is in lowest terms; if it is not, use the fundamental principle.
4. When solving equations with rational expressions, reject any proposed solution that causes an original denominator to equal 0.

EXERCISES

In each of the following exercises, decide whether the given rational expressions should be added, subtracted, multiplied, or divided, and perform the operation, or else solve the given equation.

1. $\frac{6}{m} + \frac{2}{m}$

2. $\frac{b^2c^3}{b^5c^4} \cdot \frac{c^5}{b^7}$

3. $\frac{2}{x^2 + 2x - 3} \div \frac{8x^2}{3x - 3}$

4. $\frac{4}{m - 2} = 1$

5. $\frac{2}{x + 1} + \frac{5}{x - 1} = \frac{10}{x^2 - 1}$

6. $\frac{3}{x + 3} + \frac{4}{x + 6} = \frac{9}{x^2 + 9x + 18}$

7. $\dfrac{2r^2 - 3r - 9}{2r^2 - r - 6} \cdot \dfrac{r^2 + 2r - 8}{r^2 - 2r - 3}$

8. $\dfrac{1}{m^2 - 3m} + \dfrac{4}{m^2 - 9}$

9. $\dfrac{p + 3}{8} = \dfrac{p - 2}{9}$

10. $\dfrac{4t^2 - t}{6t^2 + 10t} \div \dfrac{8t^2 + 2t - 1}{3t^2 + 11t + 10}$

11. $\dfrac{5}{y - 1} + \dfrac{2}{3y - 3}$

12. $\dfrac{1}{z} + \dfrac{1}{z + 2} = \dfrac{8}{15}$

13. $\dfrac{2}{r - 1} + \dfrac{1}{r} = \dfrac{5}{2}$

14. $\dfrac{2}{y} - \dfrac{7}{5y}$

15. $\dfrac{4}{9z} - \dfrac{3}{2z}$

16. $\dfrac{r - 3}{2} = \dfrac{2r - 5}{5}$

17. $\dfrac{x}{x - 2} + \dfrac{3}{x + 2} = \dfrac{8}{x^2 - 4}$

18. $\dfrac{2z}{z + 3} = \dfrac{2}{z - 3} - \dfrac{12}{z^2 - 9}$

19. $\dfrac{1}{m^2 + 5m + 6} + \dfrac{2}{m^2 + 4m + 3}$

20. $\dfrac{2k^2 - 3k}{20k^2 - 5k} \div \dfrac{2k^2 - 5k + 3}{4k^2 + 11k - 3}$

5.7 APPLICATIONS OF RATIONAL EXPRESSIONS

OBJECTIVES

1. SOLVE APPLICATIONS WITH RATIONAL EXPRESSIONS INVOLVING UNKNOWN NUMBERS.
2. SOLVE APPLICATIONS WITH RATIONAL EXPRESSIONS INVOLVING DISTANCE.
3. SOLVE APPLICATIONS WITH RATIONAL EXPRESSIONS INVOLVING WORK.
4. SOLVE PROBLEMS ABOUT VARIATION.

FOCUS ON PROBLEM SOLVING

A quantity, its $\frac{2}{3}$, its $\frac{1}{2}$, and its $\frac{1}{7}$, added together, become 33. What is the quantity?

If you think that applying mathematics to problem solving is a recent development, think again! This problem is taken from the Rhind papyrus, an Egyptian artifact from which we have learned much about the mathematics of early Egypt. It dates back to approximately 1650 B.C. The suggested method of solving this problem, and others like it, was to guess at the solution and then, if the guess was not correct, to use a correction factor to find the actual answer. This process was called "the rule of false position." Modern methods of solving problems of this type involve writing equations with rational expressions, and then using the methods described in Section 5.6.

This problem is Exercise 5 in the exercises for this section. After working through this section, you should be able to solve this problem.

PROBLEM SOLVING

Every time we learn to solve a new type of equation, we are then able to apply our knowledge to solving new types of applications. In Section 5.6 we learned how to solve equations involving rational expressions, and now we can solve applications that involve this type of equation. The problem solving techniques of the earlier chapters still apply. The main difference between the problems of this section and those of earlier sections is that the equations for these problems involve rational expressions. ■

1 In order to prepare for more meaningful applications, we begin with an example about an unknown number.

EXAMPLE 1
SOLVING A PROBLEM ABOUT AN UNKNOWN NUMBER

If the same number is added to both the numerator and the denominator of the fraction 3/4, the result is 5/6. Find the number.

Let $x =$ the number added to the numerator and the denominator. Then

$$\frac{3+x}{4+x}$$

represents the result of adding the same number to both the numerator and denominator. Since this result is 5/6,

$$\frac{3+x}{4+x} = \frac{5}{6}.$$

Solve this equation by multiplying both sides by the least common denominator, $6(4 + x)$.

$$\begin{aligned} 6(4+x)\frac{3+x}{4+x} &= 6(4+x)\frac{5}{6} \\ 6(3+x) &= 5(4+x) \\ 18+6x &= 20+5x \\ x &= 2 \end{aligned}$$

Check the solution in the words of the original problem: if 2 is added to both the numerator and denominator of 3/4, the result is 5/6, as required. ■

2 We now look at more meaningful applications that lead to equations with rational expressions.

PROBLEM SOLVING

In Section 2.8 we first saw applications involving distance, rate, and time. Recall the importance of setting up a chart in these problems, so that the information is summarized in an organized fashion. The next example shows how to solve this type of problem, where a rational expression appears in the equation. ■

EXAMPLE 2
SOLVING A PROBLEM ABOUT DISTANCE, RATE, AND TIME

The Big Muddy River has a current of 3 miles per hour. A motorboat takes the same amount of time to go 12 miles downstream as it takes to go 8 miles upstream. What is the speed of the boat in still water?

This problem requires the distance formula,

$$d = rt \text{ (distance = rate} \times \text{time).}$$

Let x = the speed of the boat in still water. Since the current pushes the boat when the boat is going downstream, the speed of the boat downstream will be the sum of the speed of the boat and the speed of the current, or $x + 3$ miles per hour. Also, the boat's speed going upstream is $x - 3$ miles per hour. The information given in the problem is summarized in this chart.

	d	r	t
Downstream	12	$x + 3$	
Upstream	8	$x - 3$	

Fill in the last column, representing time by solving the formula $d = rt$ for t.

$$d = rt$$

$$\frac{d}{r} = t \qquad \text{Divide by } r.$$

Then the time upstream is the distance divided by the rate, or

$$\frac{8}{x - 3},$$

and the time downstream is also the distance divided by the rate, or

$$\frac{12}{x + 3}.$$

Now complete the chart.

	d	r	t
Downstream	12	$x + 3$	$\frac{12}{x + 3}$
Upstream	8	$x - 3$	$\frac{8}{x - 3}$

Times are equal.

According to the original problem, the time upstream equals the time downstream. The two times from the chart must therefore be equal, giving the equation

$$\frac{12}{x + 3} = \frac{8}{x - 3}.$$

Solve this equation by multiplying both sides by $(x + 3)(x - 3)$.

$$(x + 3)(x - 3)\frac{12}{x + 3} = (x + 3)(x - 3)\frac{8}{x - 3}.$$

$$12(x - 3) = 8(x + 3)$$

$$12x - 36 = 8x + 24 \qquad \text{Distributive property}$$

$$4x = 60 \qquad \text{Subtract } 8x\text{; add 36.}$$

$$x = 15 \qquad \text{Divide by 4.}$$

The speed of the boat in still water is 15 miles per hour. Check this solution by first finding the speed of the boat downstream, which is $15 + 3 = 18$ miles per hour. Traveling 12 miles would take

$$d = rt$$

$$12 = 18t$$

$$t = \frac{2}{3} \text{ hour.}$$

On the other hand, the speed of the boat upstream is $15 - 3 = 12$ miles per hour, and traveling 8 miles would take

$$d = rt$$

$$8 = 12t$$

$$t = \frac{2}{3} \text{ hour.}$$

The time upstream equals the time downstream, as required. ■

3 Suppose that you can mow your lawn in 4 hours. Then after 1 hour, you will have mowed 1/4 of the lawn. After 2 hours, you will have mowed 2/4 or 1/2 of the lawn, and so on. This idea is generalized as follows.

RATE OF WORK

If a job can be completed in t units of time, then the rate of work is

$$\frac{1}{t} \text{ job per unit of time.}$$

PROBLEM SOLVING

The relationship between problems involving work and problems involving distance is a very close one. Recall that the formula $d = rt$ says that distance traveled is equal to rate of travel multiplied by time traveled. Similarly, the fractional part of a job accomplished is equal to the rate of the work multiplied by the time worked. In the lawn mowing example, after 3 hours, the fractional part of the job done is

$$\underbrace{\frac{1}{4}}_{\text{Rate of work}} \cdot \underbrace{3}_{\text{Time worked}} = \underbrace{\frac{3}{4}}_{\text{Fractional part of job done}}.$$

After 4 hours, $(1/4)(4) = 1$ whole job has been done.

These ideas are used in solving problems about the length of time needed to do a job. These problems are often called work problems. ■

EXAMPLE 3
SOLVING A PROBLEM ABOUT WORK

With a riding lawn mower, John, the groundskeeper in a large park, can cut the lawn in 8 hours. With a small mower, his assistant Walt needs 14 hours to cut the same lawn. If both John and Walt work on the lawn, how long will it take to cut it?

Let x = the number of hours it will take for John and Walt to mow the lawn, working together.

Certainly, x will be less than 8, since John alone can mow the lawn in 8 hours. Begin by making a chart as shown. Remember that based on the previous discussion, John's rate alone is 1/8 job per hour, and Walt's rate is 1/14 job per hour.

	Rate	Time working together	Fractional part of the job done when working together
John	$\frac{1}{8}$	x	$\frac{1}{8}x$
Walt	$\frac{1}{14}$	x	$\frac{1}{14}x$

Sum is 1 whole job.

Since together John and Walt complete 1 whole job, we must add their individual fractional parts and set the sum equal to 1.

$$\underbrace{\frac{1}{8}x}_{\text{Fractional part done by John}} + \underbrace{\frac{1}{14}x}_{\text{Fractional part done by Walt}} = \underbrace{1}_{\text{1 whole job}}$$

$$56\left(\frac{1}{8}x + \frac{1}{14}x\right) = 56(1) \quad \text{Multiply by LCD, 56.}$$

$$56\left(\frac{1}{8}x\right) + 56\left(\frac{1}{14}x\right) = 56(1) \quad \text{Distributive property}$$

$$7x + 4x = 56$$

$$11x = 56 \quad \text{Combine like terms.}$$

$$x = \frac{56}{11} \quad \text{Divide by 11.}$$

Working together, John and Walt can mow the lawn in 56/11 hours, or 5 1/11 hours. ■

4 Suppose that gasoline costs \$1.50 per gallon. Then 1 gallon costs \$1.50, 2 gallons cost 2(\$1.50) = \$3.00, 3 gallons cost 3(\$1.50) = \$4.50, and so on. Each time, the total cost is obtained by multiplying the number of gallons by the price per gallon. In general, if k equals the price per gallon and x equals the number of gallons, then the total cost y is equal to kx. Notice that as number of gallons increases, total cost increases.

The preceding discussion is an example of variation. Equations with fractions often result when discussing variation. As in the gasoline example, two variables **vary directly** if one is a constant multiple of the other.

DIRECT VARIATION

y **varies directly** as x if there exists a constant k such that

$$y = kx.$$

EXAMPLE 4 **USING DIRECT VARIATION**

Suppose y varies directly as x, and $y = 20$ when $x = 4$. Find y when $x = 9$.

Since y varies directly as x, there is a constant k such that $y = kx$. Also, $y = 20$ when $x = 4$. Substituting these values into $y = kx$ gives

$$y = kx$$
$$20 = k \cdot 4, \qquad \text{Let } y = 20,\ x = 4.$$

from which $k = 5$. Since $y = kx$ and $k = 5$, $y = 5x$. When $x = 9$,

$$y = 5x = 5 \cdot 9 = 45.$$

Thus, $y = 45$ when $x = 9$. ■

There are other types of variation. Suppose that a rectangle has area 48 square units and length 24 units. Then its width is 2 units. However, if the area stays the same (48 square units) and the length decreases to 12 units, the width increases to 4 units. As length decreases, width increases. This is an example of **inverse variation:** $W = \frac{A}{L}$.

Some other types of variation are defined below, where k represents a constant.

TYPES OF VARIATION

y varies directly as the square of x	$y = kx^2$
m varies inversely as p	$m = \frac{k}{p}$
r varies inversely as the square of s	$r = \frac{k}{s^2}$

EXAMPLE 5 **USING INVERSE VARIATION**

Suppose z varies inversely as t, and $z = 8$ when $t = 5$. Find z when $t = 20$.

Since z varies inversely as t, there is a constant k such that $z = \frac{k}{t}$. Find k by replacing z with 8 and t with 5.

$$8 = \frac{k}{5} \qquad \text{Let } z = 8 \text{ and } t = 5.$$

Multiply both sides by 5 to get $k = 40$, so that $z = \frac{40}{t}$. When $t = 20$,

$$z = \frac{40}{20} = 2. \quad ■$$

5.7 EXERCISES

Solve each problem. See Example 1.

1. One-third of a number is 2 more than one-sixth of the same number. What is the number?

2. The numerator of the fraction 13/15 is increased by an amount so that the value of the resulting fraction is 7/5. By what amount was the numerator increased?

3. In a certain fraction, the denominator is 4 less than the numerator. If 3 is added to both the numerator and the denominator, the result is equal to 3/2. Find the original fraction.

4. The denominator of a certain fraction is three times the numerator. If 2 is added to the numerator and subtracted from the denominator, the resulting fraction is equal to 1. Find the original fraction.

5. A quantity, its 2/3, its 1/2, and its 1/7, added together, become 33. What is the quantity? (From the *Rhind Mathematical Papyrus*.)

6. A quantity, its 3/4, its 1/2, and its 1/3, added together, become 93. What is the quantity?

7. The profits from a carnival are to be given to two scholarships so that one scholarship receives 3/2 as much money as the other. If the total amount given to the two scholarships is $780, how much goes to the scholarship that receives the lesser amount?

8. A mother and her daughter worked four days at a job. The daughter earned 2/5 as much as her mother. They earned a total of $1344. How much did each earn per day?

9. A child takes 5/8 the number of pills that an adult takes for the same illness. Together the child and the adult use 26 pills. Find the number used by the adult.

10. An apprentice is paid 3/4 the salary of an experienced journeyman. If the total wages paid an apprentice and a journeyman are $56,000, find the amount paid to the journeyman.

11. Suppose that Audrey walks D miles at R miles per hour in the same time that Kenneth walks d miles at r miles per hour. Give an equation relating D, R, d, and r. (See Objective 2.)

12. If a boat travels m miles per hour in still water, what is its rate when it travels upstream in a river with current 5 miles per hour? What is its rate downstream in the river? (See Objective 2.)

Solve each problem. See Example 2.

13. Reynaldo flew his airplane 500 miles against the wind in the same time it took him to fly it 600 miles with the wind. If the speed of the wind was 10 miles per hour, what was the average speed of his plane? (Let x = speed of the plane in still air.)

	d	r	t
Against the wind	500	$x - 10$	
With the wind	600	$x + 10$	

14. Sam can row 4 miles per hour in still water. It takes as long to row 8 miles upstream as 24 miles downstream. How fast is the current? (Let x = speed of the current.)

	d	r	t
Upstream	8	$4 - x$	
Downstream	24	$4 + x$	

15. A boat goes 210 miles downriver in the same time it can go 140 miles upriver. The speed of the current is 5 miles per hour. Find the speed of the boat in still water.

16. A boat can go 20 miles against a current in the same time that it can go 60 miles with the current. The current is 4 miles per hour. Find the speed of the boat with no current.

17. Elena flew from Philadelphia to Des Moines at 180 miles per hour and from Des Moines to Philadelphia at 150 miles per hour. The trip at the slower speed took 1 hour longer than the trip at the higher speed. Find the distance between the two cities. (Assume there was no wind in either direction.)

18. The distance from Seattle, Washington, to Victoria, British Columbia, is about 148 miles by ferry. It takes about 4 hours less to travel by ferry from Victoria to Vancouver, British Columbia, a distance of about 74 miles. What is the average speed of the ferry?

19. If it takes Elayn 10 hours to do a job, what is her rate? (See Objective 3.)

20. If it takes Clay 12 hours to do a job, how much of the job does he do in 8 hours? (See Objective 3.)

Solve each problem. See Example 3.

21. Jorge can paint a room, working alone, in 8 hours. Caterina can paint the same room, working alone, in 6 hours. How long will it take them if they work together?

22. Roberto can tune up his Bronco in 2 hours working alone. His brother, Marco, can do the job in 3 hours working alone. How long would it take them if they worked together?

23. A certain copier can do a printing job in 7 hours working alone. It takes a smaller copier 12 hours to do the same job working alone. How long would it take them to do the job if they work together?

24. Wing can fix a meal of barbecued chicken in 3 hours, but Tammie can prepare the same meal in only 1 hour. How long would it take them to prepare the meal working together?

25. One pipe can fill a swimming pool in 6 hours, and another pipe can do it in 9 hours. How long will it take the two pipes working together to fill the pool 3/4 full?

26. Dennis can do a job in 4 days. When Dennis and Sue work together, the job takes 2 1/3 days. How long would the job take Sue working alone?

27. An inlet pipe can fill a swimming pool in 9 hours, and an outlet pipe can empty the pool in 12 hours. Through an error, both pipes are left open. How long will it take to fill the pool?

28. A cold water faucet can fill a sink in 12 minutes, and a hot water faucet can fill it in 15. The drain can empty the sink in 25 minutes. If both faucets are on and the drain is open, how long will it take to fill the sink?

29. Refer to Exercise 27. Assume the error was discovered after both pipes had been running for 3 hours, and the outlet pipe was then closed. How much more time would then be required to fill the pool? (*Hint:* How much of the job had been done when the error was discovered?)

Solve the following problems about variation. See Examples 4 and 5.

30. If x varies directly as y, and $x = 9$ when $y = 2$, find x when y is 6.

31. If z varies directly as x, and $z = 15$ when $x = 4$, find z when x is 8.

32. If m varies directly as p^2, and $m = 20$ when $p = 2$, find m when p is 5.

33. If a varies directly as b^2, and $a = 48$ when $b = 4$, find a when $b = 7$.

34. If p varies inversely as q^2, and $p = 4$ when $q = 1/2$, find p when $q = 3/2$.

35. If z varies inversely as x^2, and $z = 9$ when $x = 2/3$, find z when $x = 5/4$.

36. The circumference of a circle varies directly as the radius. A circle with a radius of 7 centimeters has a circumference of 43.96 centimeters. Find the circumference if the radius changes to 11 centimeters.

37. The pressure exerted by a certain liquid at a given point varies directly as the depth of the point beneath the surface of the liquid. The pressure at 10 feet is 50 pounds per square inch. What is the pressure at 20 feet?

38. The force required to compress a spring varies directly as the change in the length of the spring. If a force of 12 pounds is required to compress a certain spring 3 inches, how much force is required to compress the spring 5 inches?

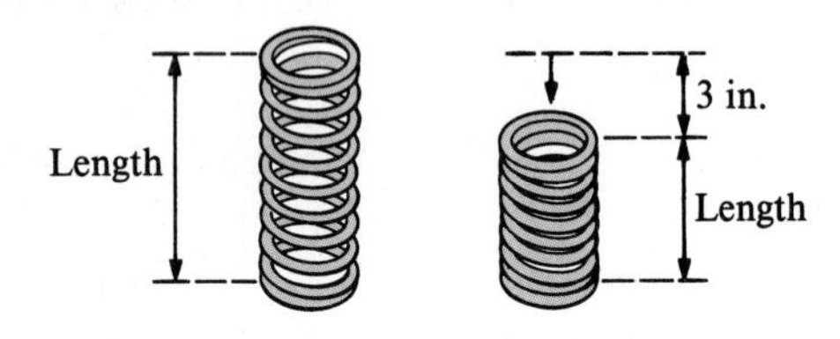

39. The illumination produced by a light source varies inversely as the square of the distance from the source. If the illumination produced 4 feet from a light source is 75 footcandles, find the illumination produced 9 feet from the same source.

Use the ideas presented in this section to solve each of the following. This is a mixed group of exercises. See Examples 1–5.

40. An experienced employee can enter tax data from standard returns into a computer twice as fast as a new employee. Working together, it takes the employees 2 hours to enter a fixed amount of data. How long would it take the experienced employee working alone to enter the same amount of data?

41. If r varies inversely as s, and $r = 7$ when $s = 8$, find r when $s = 12$.

42. If three times a number is added to twice its reciprocal, the answer is 5. Find the number.

43. An airplane, maintaining a constant airspeed, takes as long to go 450 miles with the wind as it does to go 375 miles against the wind. If the wind is blowing at 15 miles per hour, what is the speed of the plane in still air?

44. If y varies inversely as x, and $y = 10$ when $x = 3$, find y when $x = 12$.

45. The current in a simple electrical circuit varies inversely as the resistance. If the current is 50 amps (an *amp* is a unit for measuring current) when the resistance is 10 ohms (an *ohm* is a unit for measuring resistance), find the current if the resistance is 5 ohms.

46. Becky can go 30 miles downstream in the same time that she can go 10 miles upstream. If the speed of the current is 5 miles per hour, find the speed of her boat in still water.

47. If twice a number is subtracted from three times its reciprocal, the result is 1. Find the number.

48. For a body falling freely from rest (disregarding air resistance), the distance the body falls varies directly as the square of the time. If an object is dropped from the top of a tower 400 feet high and hits the ground in 5 seconds, how far did it fall in the first 3 seconds?

49. One painter can paint a house three times faster than another. Working together, they can paint a house in 4 days. How long would it take the faster painter working alone?

Choose the correct responses. Assume that the constant of variation, k, is positive. (See Objective 4.)

50. **(a)** If y varies directly as x, then as y increases, x ___?___ (decreases/increases).

(b) If y varies inversely as x, then as y increases, x ___?___ (decreases/increases).

PREVIEW EXERCISES

Find the value of y when **(a)** $x = 2$ *and* **(b)** $x = -4$. *See Sections 1.3 and 2.5.*

51. $y = 4x - 7$ **52.** $y = 3 - 2x$ **53.** $4x - y = 1$ **54.** $y + 3x = 2$

55. $3x + 7y = 10$ **56.** $2x - 3y = 5$ **57.** $x = -3y$ **58.** $y = 2x$

CHAPTER 5 GLOSSARY

KEY TERMS

5.1 rational expression The quotient of two polynomials with denominator not equal to zero is called a rational expression.

lowest terms A rational expression is written in lowest terms if the greatest common factor of its numerator and denominator is 1.

5.3 least common denominator (LCD) The smallest expression that all denominators divide into evenly without remainder is called the least common denominator.

5.5 complex fraction A rational expression with fractions in the numerator, denominator, or both, is called a complex fraction.

5.7 direct variation y varies directly as x if there is a constant k such that $y = kx$.

inverse variation y varies inversely as x if there is a constant k such that $y = k/x$.

CHAPTER 5 QUICK REVIEW

SECTION	CONCEPTS	EXAMPLES
5.1 THE FUNDAMENTAL PROPERTY OF RATIONAL EXPRESSIONS	To find the values for which a rational expression is not defined, set the denominator equal to zero and solve the equation.	Find the values for which the expression $$\frac{x-4}{x^2-16}$$ is not defined. $$x^2 - 16 = 0$$ $$(x-4)(x+4) = 0$$ $$x - 4 = 0 \quad \text{or} \quad x + 4 = 0$$ $$x = 4 \quad \text{or} \quad x = -4$$ The rational expression is not defined for 4 or -4.
	To write a rational expression in lowest terms, (1) factor; and (2) use the fundamental property to remove common factors from the numerator and denominator.	Express $\frac{x^2-1}{(x-1)^2}$ in lowest terms. $$\frac{x^2-1}{(x-1)^2} = \frac{(x-1)(x+1)}{(x-1)(x-1)}$$ $$= \frac{x+1}{x-1}$$

SECTION	CONCEPTS	EXAMPLES
5.2 MULTIPLICATION AND DIVISION OF RATIONAL EXPRESSIONS	**Multiplication** **1.** Factor. **2.** Multiply numerators and multiply denominators. **3.** Write in lowest terms.	Multiply: $\frac{3x+9}{x-5} \cdot \frac{x^2-3x-10}{x^2-9}$. $= \frac{3(x+3)}{x-5} \cdot \frac{(x-5)(x+2)}{(x+3)(x-3)}$ $= \frac{3(x+3)(x-5)(x+2)}{(x-5)(x+3)(x-3)}$ $= \frac{3(x+2)}{x-3}$
	Division **1.** Factor. **2.** Multiply the first rational expression by the reciprocal of the second rational expression. **3.** Write in lowest terms.	Divide: $\frac{2x+1}{x+5} \div \frac{6x^2-x-2}{x^2-25}$. $= \frac{2x+1}{x+5} \div \frac{(2x+1)(3x-2)}{(x+5)(x-5)}$ $= \frac{2x+1}{x+5} \cdot \frac{(x+5)(x-5)}{(2x+1)(3x-2)}$ $= \frac{x-5}{3x-2}$
5.3 THE LEAST COMMON DENOMINATOR	**Finding the LCD** **1.** Factor each denominator into prime factors. **2.** List each different denominator factor the greatest number of times it appears in any denominator. **3.** Multiply the denominators from Step 2 to get the LCD.	Find the LCD for $\frac{3}{k^2-8k+16}$ and $\frac{1}{4k^2-16k}$. $k^2-8k+16=(k-4)^2$ $4k^2-16k=4k(k-4)$ $\text{LCD} = (k-4)^2 \cdot 4 \cdot k$ $= 4k(k-4)^2$
	Writing a Rational Expression with the LCD as Denominator **1.** Factor both denominators. **2.** Decide what factors the denominator must be multiplied by to equal the LCD. **3.** Multiply the rational expression by that factor over itself (multiply by 1).	Find the numerator: $\frac{5}{2z^2-6z} = \frac{\quad}{4z^3-12z^2}$. $\frac{5}{2z(z-3)} = \frac{\quad}{4z^2(z-3)}$ $2z(z-3)$ must be multiplied by $2z$. $\frac{5}{2z(z-3)} \cdot \frac{2z}{2z} = \frac{10z}{4z^2(z-3)}$ $= \frac{10z}{4z^3-12z^2}$

SECTION	CONCEPTS	EXAMPLES
5.4 ADDITION AND SUBTRACTION OF RATIONAL EXPRESSIONS	**Adding Rational Expressions** **1.** Find the LCD. **2.** Rewrite each rational expression as a fraction with the LCD as denominator. **3.** Add the numerators to get the numerator of the sum. The LCD is the denominator of the sum. **4.** Write in lowest terms.	Add: $\frac{2}{3m+6} + \frac{m}{m^2-4}$. $\frac{2}{3(m+2)} + \frac{m}{(m+2)(m-2)}$ The LCD is $3(m+2)(m-2)$. $= \frac{2(m-2)}{3(m+2)(m-2)} + \frac{3m}{3(m+2)(m-2)}$ $= \frac{2m-4+3m}{3(m+2)(m-2)}$ $= \frac{5m-4}{3(m+2)(m-2)}$
	Subtracting Rational Expressions Follow the same steps as for addition, but subtract in Step 3.	Subtract: $\frac{6}{k+4} - \frac{2}{k}$. The LCD is $k(k+4)$. $\frac{6k}{(k+4)k} - \frac{2(k+4)}{k(k+4)}$ $= \frac{6k-2(k+4)}{k(k+4)}$ $= \frac{6k-2k-8}{k(k+4)} = \frac{4k-8}{k(k+4)}$
5.5 COMPLEX FRACTIONS	**Simplifying Complex Fractions** *Method 1* Simplify the numerator and denominator separately. Then divide the simplified numerator by the simplified denominator.	Simplify. **(1)** $\frac{\frac{1}{a}-a}{1-a} = \frac{\frac{1}{a}-\frac{a^2}{a}}{1-a} = \frac{\frac{1-a^2}{a}}{1-a}$ $= \frac{1-a^2}{a} \cdot \frac{1}{1-a}$ $= \frac{(1-a)(1+a)}{a(1-a)} = \frac{1+a}{a}$
	Method 2 Multiply numerator and denominator of the complex fraction by the LCD of all the denominators in the complex fraction.	**(2)** $\frac{\frac{1}{a}-a}{1-a} = \frac{\frac{1}{a}-a}{1-a} \cdot \frac{a}{a} = \frac{\frac{a}{a}-a^2}{(1-a)a}$ $= \frac{1-a^2}{(1-a)a} = \frac{(1+a)(1-a)}{(1-a)a}$ $= \frac{1+a}{a}$

<table>
<tr><th>SECTION</th><th>CONCEPTS</th><th>EXAMPLES</th></tr>
<tr><td>5.6 EQUATIONS INVOLVING RATIONAL EXPRESSIONS</td><td>Solving Equations with Rational Expressions
1. Using the multiplication property of equality, multiply each side of the equation by the LCD of all denominators in the equation.
2. Solve the resulting equation which should have no fractions.
3. Check each proposed solution. Reject any that make a denominator equal to 0.</td><td>Solve $\frac{2}{x-1} + \frac{3}{4} = \frac{5}{x-1}$.
The LCD is $4(x-1)$. Note that 1 cannot be a solution.
$4(x-1)\left(\frac{2}{x-1} + \frac{3}{4}\right) = 4(x-1)\left(\frac{5}{x-1}\right)$
$4(x-1)\left(\frac{2}{x-1}\right) + 4(x-1)\left(\frac{3}{4}\right)$
$= 4(x-1)\left(\frac{5}{x-1}\right)$
$8 + 3(x-1) = 20$
$8 + 3x - 3 = 20$
$3x = 15$
$x = 5$
The proposed solution, 5, checks.</td></tr>
<tr><td>5.7 APPLICATIONS OF RATIONAL EXPRESSIONS</td><td>Solving Problems About Distance 1. State what the variable represents. 2. Use a chart to identify distance, rate, and time. 3. Solve $d = rt$ for the unknown quantity. 4. From the wording in the problem, decide the relationship between the quantities. Use those expressions to write an equation. 5. Solve the equation. 6. Check the solution.</td><td>On a trip from Sacramento to Monterey, Marge traveled at an average speed of 60 miles per hour. The return trip, at an average speed of 64 miles per hour, took 1/4 hour less. How far did she travel between the two cities? |	d	r	$t = \frac{d}{r}$
Going	x	60	$\frac{x}{60}$
Returning	x	64	$\frac{x}{64}$ | Since the time for the return trip was 1/4 hour less, the time going equals the time returning plus 1/4. $\frac{x}{60} = \frac{x}{64} + \frac{1}{4}$ The solution of this equation is $x = 240$. She traveled 240 miles.</td></tr>
<tr><td></td><td>Solving Problems About Work
1. State what the variable represents.
2. Put the information from the problem in a chart. If a job is done in t units of time, the rate is $1/t$.</td><td>It takes the regular mail carrier 6 hours to cover her route. A substitute takes 8 hours to cover the same route. How long would it take them to cover the route together?
Let x = the number of hours it would take them working together.</td></tr>
</table>

SECTION	CONCEPTS	EXAMPLES

CONCEPTS

3. Write the equation. The sum of the fractional parts should equal 1 (whole job).
4. Solve the equation.
5. Check the solution.

EXAMPLES

The rate of the regular carrier is 1/6 job per hour; the rate of the substitute is 1/8 job per hour.

Multiply rate times time to get the fractional part of the job done.

	Rate	Time	Part of the job done
Regular	$\frac{1}{6}$	x	$\frac{1}{6}x$
Substitute	$\frac{1}{8}$	x	$\frac{1}{8}x$

The equation is

$$\frac{1}{6}x + \frac{1}{8}x = 1.$$

Multiply by the LCD, 24.

$$24\left(\frac{1}{6}x + \frac{1}{8}x\right) = 24(1)$$

$$4x + 3x = 24$$

$$7x = 24$$

$$x = \frac{24}{7}$$

It would take them 24/7 or 3 3/7 hours to cover the route together.

CONCEPTS

Solving Variation Problems

1. Write the variation equation using $y = kx$ or $y = k/x$.
2. Find k by substituting the given values of x and y into the equation.
3. Write the equation with the value of k from Step 2 and the given value of x or y. Solve for the remaining variable.

EXAMPLES

If a varies inversely as b, and $a = 4$ when $b = 4$, find a when $b = 6$.

The equation for inverse variation is

$$a = \frac{k}{b}.$$

Substitute $a = 4$ and $b = 4$.

$$4 = \frac{k}{4}.$$

The solution is $k = 16$. Let $k = 16$ and $b = 6$ in the variation equation.

$$a = \frac{16}{6} = \frac{8}{3}.$$

CHAPTER 5 REVIEW EXERCISES

[5.1] *Find any values for which the following expressions are undefined.*

1. $\frac{2}{7x}$
2. $\frac{3}{m-5}$
3. $\frac{r-3}{r^2-2r-8}$
4. $\frac{3z+5}{2z^2+5z-3}$

Find the numerical value of each expression when **(a)** $x = 3$ *and* **(b)** $x = -1$.

5. $\frac{x^2}{x+2}$
6. $\frac{5x+3}{2x-1}$
7. $\frac{8x}{x^2-2}$
8. $\frac{x-5}{x-3}$

Write each expression in lowest terms.

9. $\frac{15p^2}{5p}$
10. $\frac{6y^2z^3}{9y^4z^2}$
11. $\frac{9x^2-16}{6x+8}$
12. $\frac{m-5}{5-m}$

[5.2] *Find each product or quotient. Write each answer in lowest terms.*

13. $\frac{10p^5}{5} \div \frac{3p^7}{20}$
14. $\frac{8z^2}{(4z)^3} \div \frac{4z^5}{32z}$
15. $\frac{7y+14}{8y-5} \div \frac{4y+8}{16y-10}$
16. $\frac{3k+5}{k+2} \cdot \frac{k^2-4}{18k^2-50}$
17. $\frac{2p^2+3p-2}{p^2+5p+6} \cdot \frac{p^2-2p-15}{2p^2-7p-15}$
18. $\frac{8r^2+23r-3}{64r^2-1} \div \frac{r^2-4r-21}{64r^2+16r+1}$

[5.3] *Find the least common denominator for the following fractions.*

19. $\frac{1}{15}, \frac{7}{30}, \frac{4}{45}$
20. $\frac{3}{8y}, \frac{7}{12y^2}, \frac{1}{16y^3}$
21. $\frac{1}{y^2+2y}, \frac{4}{y^2+6y+8}$
22. $\frac{3}{z^2+z-6}, \frac{2}{z^2+4z+3}$

Rewrite each rational expression with the given denominator.

23. $\frac{4}{9} = \frac{}{45}$
24. $\frac{12}{m} = \frac{}{5m}$
25. $\frac{3}{8m^2} = \frac{}{24m^3}$
26. $\frac{12}{y-4} = \frac{}{8y-32}$
27. $\frac{-2k}{3k+15} = \frac{}{15k+75}$
28. $\frac{12y}{y^2-y-2} = \frac{}{(y-2)(y+1)(y-4)}$

[5.4] *Add or subtract as indicated. Write all answers in lowest terms.*

29. $\frac{11}{3r} - \frac{8}{3r}$
30. $\frac{b}{b+5} + \frac{5}{b+5}$
31. $\frac{7}{k} - \frac{2}{5}$
32. $\frac{3+5m}{2} - \frac{m}{4}$
33. $\frac{2}{y+1} - \frac{3}{y-1}$
34. $\frac{2}{r-5} + \frac{3}{5-r}$
35. $\frac{10}{p^2-2p} - \frac{2}{p^2-5p+6}$

[5.5] *Simplify each complex fraction.*

36. $\dfrac{\dfrac{5k-1}{k}}{\dfrac{4k+3}{8k}}$

37. $\dfrac{\dfrac{6}{r}-1}{\dfrac{6-r}{4r}}$

38. $\dfrac{\dfrac{1}{a+b}-1}{\dfrac{1}{a+b}+1}$

[5.6] *Solve each equation. Check your answers.*

39. $\dfrac{2}{p}=\dfrac{5}{8}$

40. $\dfrac{3}{k}-\dfrac{2}{k}=7$

41. $\dfrac{y}{2}-\dfrac{y}{5}=6$

42. $\dfrac{3}{4}r-1=-4$

43. $\dfrac{z+3}{8}=\dfrac{z-2}{3}$

44. $\dfrac{2}{z}=\dfrac{z+1}{z+3}$

45. $\dfrac{3y-1}{y-2}=\dfrac{5}{y-2}+1$

46. $\dfrac{3}{m-2}+\dfrac{1}{m-1}=\dfrac{7}{m^2-3m+2}$

47. $\dfrac{p+2}{p^2-1}+\dfrac{2p}{p^2+2p+1}=\dfrac{2}{p-1}$

Solve for the specified variable.

48. $m=\dfrac{Rv}{t}$; for t

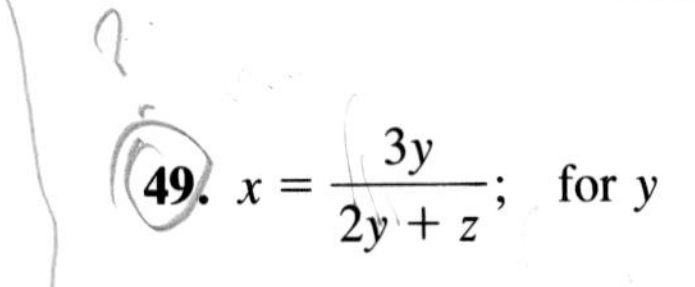

49. $x=\dfrac{3y}{2y+z}$; for y

[5.7] *Solve each problem.*

50. When half a number is subtracted from two-thirds of a number, the result is 2. Find the number.

51. A certain fraction has a numerator that is 4 more than the denominator. If 6 is subtracted from the denominator, the result is 3. Find the original fraction.

52. The commission received by a salesperson for selling a small car is 2/3 that received for selling a large car. On a recent day, Linda sold one of each, earning a commission of \$300. Find the commission for each type of car.

53. Eva flew her plane 400 kilometers with the wind in the same time it took her to go 200 kilometers against the wind. The speed of the wind is 50 kilometers per hour. Find the speed of the plane in still air.

54. A man can plant his garden in 5 hours, working alone. His daughter can do the same job in 8 hours. How long would it take them if they worked together?

55. One roofer can roof a house twice as fast as another. Working together, they can roof the house in 1 1/3 days. How long would it take the faster roofer working alone?

56. If m varies directly as q^2, and $m = 8$ when $q = 4$, find m when $q = 6$.

57. If r varies inversely as s, and $r = 9$ when $s = 1/2$, find r when $s = 2$.

58. If z varies inversely as y^2, and $z = 5$ when $y = 2$, find z when $y = 3/4$.

MIXED REVIEW EXERCISES

Perform the indicated operations.

59. $\dfrac{3p + q}{5} - \dfrac{p - q}{3}$

60. $\dfrac{6 - y}{6 + y} \div \dfrac{y - 6}{y + 6}$

61. $\dfrac{z + \dfrac{1}{x}}{z - \dfrac{1}{x}}$

62. $\dfrac{z^2 + z - 2}{z^2 + 7z + 10} \div \dfrac{z - 3}{z + 5}$

63. $\dfrac{8}{r^2} - \dfrac{3}{2r}$

Solve the following.

64. $\dfrac{1}{k} + \dfrac{3}{r} = \dfrac{5}{z}$; for r

65. $\dfrac{5 + m}{m} + \dfrac{3}{4} = \dfrac{-2}{m}$

---

66. When Mario and Luigi work together on a job, they can do it in 3 3/7 days. Mario can do the job working alone in 8 days. How long would it take Luigi working alone?

67. $\dfrac{y}{3} - \dfrac{y - 2}{8} = -1$

68. If x varies directly as y, and $x = 12$ when $y = 5$, find x when $y = 3$.

69. Five times a number is added to three times the reciprocal of the number, giving 17/2. Find the number.

70. A boat goes 7 miles per hour in still water. It takes as long to go 20 miles upstream as 50 miles downstream. Find the speed of the current.

CHAPTER 5 TEST

1. Find any values for which $\dfrac{8k + 1}{k^2 - 4k + 3}$ is undefined.

2. Find the numerical value of $\dfrac{6r + 1}{2r^2 - 3r - 20}$ when **(a)** $r = -1$ and **(b)** $r = 4$.

Write each rational expression in lowest terms.

3. $\dfrac{8m^2p^2}{6m^3p^5}$

4. $\dfrac{5y^3 - 5y}{2y + 2}$

Multiply or divide, as indicated. Write all answers in lowest terms.

5. $\dfrac{a^6b}{a^3} \cdot \dfrac{b^2}{a^2b^3}$

6. $\dfrac{8y - 16}{9} \div \dfrac{6 - 3y}{5}$

7. $\dfrac{6m^2 - m - 2}{8m^2 + 10m + 3} \cdot \dfrac{4m^2 + 7m + 3}{3m^2 + 5m + 2}$

8. $\dfrac{5a^2 + 7a - 6}{2a^2 + 3a - 2} \div \dfrac{5a^2 + 17a - 12}{2a^2 + 5a - 3}$

Find the least common denominator for the following fractions.

9. $\frac{3}{10p^2}, \frac{1}{25p^3}, \frac{-7}{30p^5}$

10. $\frac{r-1}{2r^2+7r+6}, \frac{2r+1}{2r^2-7r-15}$

Rewrite each rational expression with the given denominator.

11. $\frac{11}{7r} = \frac{}{49r^2}$

12. $\frac{5}{8m-16} = \frac{}{24m^2-48m}$

Add or subtract as indicated. Write all answers in lowest terms.

13. $\frac{5}{x} - \frac{6}{x}$

14. $\frac{-3}{a+1} + \frac{5}{6a+6}$

15. $\frac{m^2}{m-3} + \frac{m+1}{3-m}$

16. $\frac{3}{2k^2+3k-2} - \frac{k}{k^2+3k+2}$

Simplify each complex fraction.

17. $\dfrac{\frac{2p}{k^2}}{\frac{3p^2}{k^3}}$

18. $\dfrac{\frac{1}{p+4} - 2}{\frac{1}{p+4} + 2}$

---

19. What values of x could not be solutions of the equation

$$\frac{2}{x+1} - \frac{3}{x-4} = 6,$$

since they would cause a denominator to equal zero?

Solve each equation and check your answers.

20. $\frac{p}{p-2} = \frac{2}{p-2} + 3$

21. $\frac{2}{z^2-2z-3} = \frac{3}{z-3} + \frac{2}{z+1}$

Solve each problem.

22. If four times a number is added to half the number, the result is 3. Find the number.

23. The current in a river is 5 miles per hour. A boat can go 125 miles downstream in the same time as it can go 75 miles upstream. Find the speed of the boat in still water.

24. A man can paint a room in his house, working alone, in 5 hours. His wife can do the job in 4 hours. How long will it take them to paint the room if they work together?

25. If x varies directly as y, and $x = 8$ when $y = 12$, find x when $y = 28$.

CHAPTER SIX

GRAPHING LINEAR EQUATIONS

It is important in many situations (in business or in science, for example) to be able to make predictions based on known data. An executive may wish to predict next year's costs, profits, or sales, for instance. Scientists are currently trying to predict whether or not the earth will continue to get increasingly warmer. One way to make such predictions is to construct an equation that the known data satisfies and use that equation to predict future data.

In this chapter we discuss how the relationship between two variables can be presented pictorially with a graph or algebraically with an equation. These ideas extend the work in earlier chapters where we graphed the solutions of equations or inequalities with one variable on number lines.

6.1 LINEAR EQUATIONS IN TWO VARIABLES

OBJECTIVES

1. WRITE A SOLUTION AS AN ORDERED PAIR.
2. DECIDE WHETHER A GIVEN ORDERED PAIR IS A SOLUTION OF A GIVEN EQUATION.
3. COMPLETE ORDERED PAIRS FOR A GIVEN EQUATION.
4. COMPLETE A TABLE OF VALUES.
5. PLOT ORDERED PAIRS.

Most of the equations discussed so far, such as

$$3x + 5 = 12 \qquad \text{or} \qquad 2x^2 + x + 5 = 0,$$

have had only one variable. Equations in two variables, like

$$y = 4x + 5 \qquad \text{or} \qquad 2x + 3y = 6,$$

are discussed in this chapter. Both of these equations are examples of *linear equations*.

LINEAR EQUATION

A **linear equation** in two variables is an equation that can be put in the form

$$Ax + By = C,$$

where A, B, and C are real numbers and A and B are not both 0.

1 A solution of a linear equation in two variables requires two numbers, one for each variable. For example, the equation $y = 4x + 5$ is satisfied if x is replaced with 2 and y is replaced with 13, since

$$13 = 4(2) + 5. \qquad \text{Let } x = 2;\ y = 13.$$

The pair of numbers $x = 2$ and $y = 13$ gives a solution of the equation $y = 4x + 5$. The phrase "$x = 2$ and $y = 13$" is abbreviated

$$(2, 13).$$

The x-value, 2, and the y-value, 13, are given as a pair of numbers, written inside parentheses. The x-value is always given first. A pair of numbers such as (2, 13) is called an **ordered pair.** As the name indicates, the order in which the numbers are written is important. The ordered pairs (2, 13) and (13, 2) are not the same. The second pair indicates that $x = 13$ and $y = 2$.

Of course, letters other than x and y may be used in the equation with the numbers in the ordered pair placed so the letters are in alphabetical order. For example, one solution to the equation $3p - q = -11$ is $p = -2$ and $q = 5$. This solution is written as the ordered pair $(-2, 5)$.

2 The next example shows how to decide whether an ordered pair is a solution of an equation. An ordered pair that is a solution of an equation is said to *satisfy* the equation.

EXAMPLE 1 DECIDING WHETHER AN ORDERED PAIR SATISFIES AN EQUATION

Decide whether the given ordered pair is a solution of the given equation.

(a) (3, 2); $2x + 3y = 12$

To see whether (3, 2) is a solution of the equation $2x + 3y = 12$, substitute 3 for x and 2 for y in the given equation.

$$\begin{aligned} 2x + 3y &= 12 \\ 2(3) + 3(2) &= 12 \quad ? \qquad \text{Let } x = 3; \text{ let } y = 2. \\ 6 + 6 &= 12 \quad ? \\ 12 &= 12 \qquad \text{True} \end{aligned}$$

This result is true, so (3, 2) satisfies $2x + 3y = 12$.

(b) $(-2, -7)$; $m + 5n = 33$

$$\begin{aligned} (-2) + 5(-7) &= 33 \quad ? \qquad \text{Let } m = -2; \text{ let } n = -7. \\ -2 + (-35) &= 33 \quad ? \\ -37 &= 33 \qquad \text{False} \end{aligned}$$

This result is false, so $(-2, -7)$ is *not* a solution of $m + 5n = 33$. ■

3 Choosing a number for one variable in a linear equation makes it possible to find the value of the other variable, as shown in the next example.

EXAMPLE 2 COMPLETING AN ORDERED PAIR

Complete the given ordered pairs for the equation $y = 4x + 5$.

(a) (7,   )

In this ordered pair, $x = 7$. (Remember that x always comes first.) Find the corresponding value of y by replacing x with 7 in the equation $y = 4x + 5$.

$$y = 4(7) + 5 = 28 + 5 = 33$$

The ordered pair is (7, 33).

(b) (, 13)

In this ordered pair, $y = 13$. Find the value of x by replacing y with 13 in the equation and then solving for x.

$$y = 4x + 5$$
$$13 = 4x + 5 \quad \text{Let } y = 13.$$
$$8 = 4x \quad \text{Subtract 5 from both sides.}$$
$$2 = x \quad \text{Divide both sides by 4.}$$

The ordered pair is (2, 13). ■

4 Ordered pairs of an equation often are displayed in a **table of values** as in the next example. The table may be written either vertically or horizontally. We will write these tables horizontally in this section and vertically in Section 6.2.

EXAMPLE 3 COMPLETING A TABLE OF VALUES

Complete the given table of values for the equation $x - 2y = 8$.

x	2	10		
y			0	−2

Complete the first two ordered pairs by letting $x = 2$ and $x = 10$, respectively.

If $x = 2$, then $x - 2y = 8$ becomes

$$2 - 2y = 8$$
$$-2y = 6$$
$$y = -3.$$

If $x = 10$, then $x - 2y = 8$ becomes

$$10 - 2y = 8$$
$$-2y = -2$$
$$y = 1.$$

Now complete the last two ordered pairs by letting $y = 0$ and $y = -2$, respectively.

If $y = 0$, then $x - 2y = 8$ becomes

$$x - 2(0) = 8$$
$$x - 0 = 8$$
$$x = 8.$$

If $y = -2$, then $x - 2y = 8$ becomes

$$x - 2(-2) = 8$$
$$x + 4 = 8$$
$$x = 4.$$

The completed table of values is as follows.

x	2	10	8	4
y	−3	1	0	−2

■

EXAMPLE 4 COMPLETING A TABLE OF VALUES

Complete the given table of values for the equation $x = 5$.

x			
y	−2	6	3

The given equation is $x = 5$. No matter which value of y might be chosen, the value of x is always the same, 5.

x	5	5	5
y	−2	6	3

■

NOTE When an equation such as $x = 5$ is discussed along with equations in two variables, think of $x = 5$ as an equation in two variables by rewriting $x = 5$ as $x + 0y = 5$. This form of the equation shows that for any value of y, the value of x is 5.

Each of the equations discussed in this section has many ordered pairs as solutions. Each choice of a real number for one variable will lead to a particular real number for the other variable. This is true of linear equations in general: linear equations in two variables have an infinite number of solutions.

Earlier we used a number line to graph solutions of equations in one variable. Techniques for graphing the solutions of an equation in *two* variables will be shown in this section. The solutions of such an equation are *ordered pairs* of numbers in the form (x, y), so *two* number lines are needed, one for x and one for y. These two number lines are drawn as shown in Figure 6.1. The horizontal number line is called the ***x*-axis.** The vertical line is called the ***y*-axis.** Together, the x-axis and y-axis form a **coordinate system.**

The coordinate system is divided into four regions, called **quadrants.** These quadrants are numbered counterclockwise, as shown in Figure 6.1. Points on the axes themselves are not in any quadrant. The point at which the x-axis and y-axis meet is called the **origin.** The origin is labeled 0 in Figure 6.1.

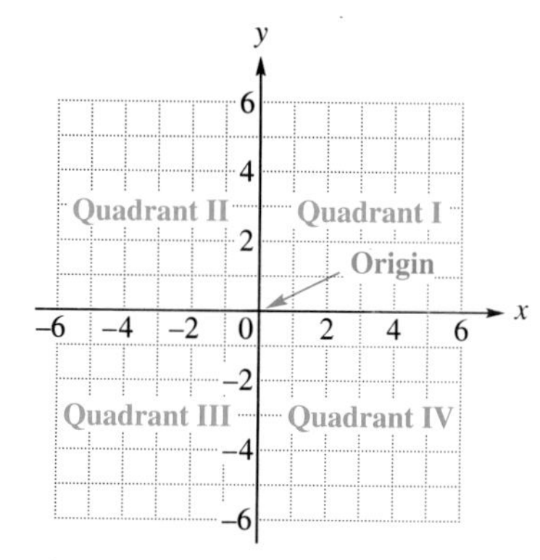

FIGURE 6.1

5 By referring to the two axes, every point on the plane can be associated with an ordered pair. The numbers in the ordered pair are called the **coordinates** of the point. For example, locate the point associated with the ordered pair (2, 3), by starting at the origin. Since the x-coordinate is 2, go 2 units to the right along the x-axis. Then since the y-coordinate is 3, turn and go up 3 units on a line parallel to the y-axis. This is called **plotting** the point (2, 3). (See Figure 6.2.) From now on we will refer to the point with x-coordinate 2 and y-coordinate 3 as the point (2, 3).

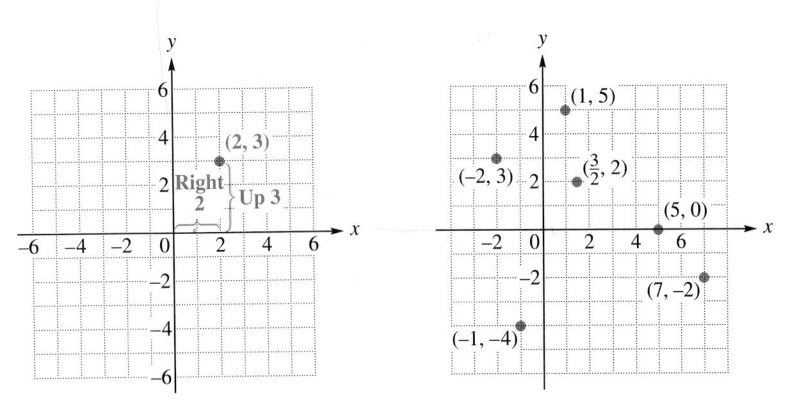

FIGURE 6.2

FIGURE 6.3

EXAMPLE 5
PLOTTING ORDERED PAIRS

Plot the given points on a coordinate system.

(a) (1, 5) **(b)** (−2, 3) **(c)** (−1, −4)

(d) (7, −2) **(e)** $\left(\frac{3}{2}, 2\right)$ **(f)** (5, 0)

Locate the point (−1, −4), for example, by first going 1 unit to the left along the x-axis. Then turn and go 4 units down, parallel to the y-axis. Plot the point (3/2, 2), by going 3/2 (or 1 1/2) units to the right along the x-axis. Then turn and go 2 units up, parallel to the y-axis. Figure 6.3 shows the graphs of the points in this example. ■

EXAMPLE 6
COMPLETING ORDERED PAIRS FOR AN APPLICATION

A company has found that the cost to produce x calculators is $y = 25x + 250$, where y represents the cost in cents. Complete the following table of values.

x	1	2	3
y			

To complete the ordered pair (1,), let $x = 1$.

$$y = 25x + 250$$
$$y = 25(1) + 250 \qquad \text{Let } x = 1.$$
$$y = 25 + 250$$
$$y = 275$$

This gives the ordered pair (1, 275), which says that the cost to produce 1 calculator is 275 cents or \$2.75. Complete the ordered pairs (2,) and (3,) as follows.

$$y = 25x + 250 \qquad y = 25x + 250$$
$$y = 25(2) + 250 \qquad y = 25(3) + 250$$
$$y = 50 + 250 \qquad y = 75 + 250$$
$$y = 300 \qquad y = 325$$

This gives the ordered pairs (2, 300) and (3, 325). ■

The ordered pairs (2, 300) and (3, 325), along with several other ordered pairs that satisfy $y = 25x + 250$, are graphed in Figure 6.4. Notice how the axes are labeled. In this application x represents the number of calculators and y represents the corresponding cost in cents. Different scales are used on the two axes, since the y-values in the ordered pairs are much larger than the x-values. Here, each square represents 50 units in the vertical direction and 1/2 unit in the horizontal direction.

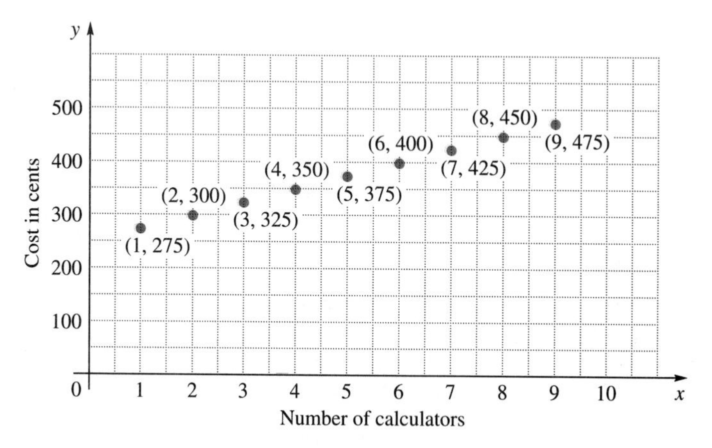

FIGURE 6.4

6.1 EXERCISES

Decide whether the given ordered pair is a solution of the given equation. See Example 1.

1. $x + y = 9$; (2, 7)

2. $3x + y = 8$; (0, 8)

3. $2x - y = 6$; (2, −2)

4. $2x + y = 5$; (2, 1)

5. $y = 3x$; (1, 3)

6. $x = -4y$; (8, −2)

7. $x = -6$; (−6, 8)

8. $x + 4 = 0$; (−5, 4)

9. $y = 2$; (9, 2)

---

10. Explain why there are an infinite number of solutions for a linear equation in two variables. (See Objective 4.)

11. In your own words, tell how to plot the ordered pair (−2, 3). (See Objective 5.)

Complete the given ordered pairs for the equation $y = 3x + 5$. *See Example 2.*

12. (2,)

13. (5,)

14. (8,)

15. (0,)

16. (−3,)

17. (−4,)

18. (, 14)

19. (, −10)

Complete the table of values for each equation. See Example 3.

20. $y = 2x + 1$

x	3	0	−1
y			

21. $y = 3x - 5$

x	2	0	−3
y			

22. $y = 8 - 3x$

x	2	0	−3
y			

23. $y = -2 - 5x$

x	4	0	−4
y			

24. $2a + b = 9$

a	0	3	12
b			

25. $-3m + n = 4$

m	1	0	−2
n			

Complete the given table of values for each equation. See Example 4.

26. $x = -4$

x			
y	6	2	−3

27. $y = -8$

x	4	0	−4
y			

28. $x + 9 = 0$

x			
y	8	3	0

29. $y - 4 = 0$

x	9	−5	0
y			

30. $y = -3x$

x	−2	0	
y			−6

31. $x = 4y$

x	0		−8
y		3	

Give the ordered pairs that correspond to the points labeled in the figure.

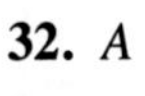

32. *A*

33. *B*

34. *C*

35. *D*

36. *E*

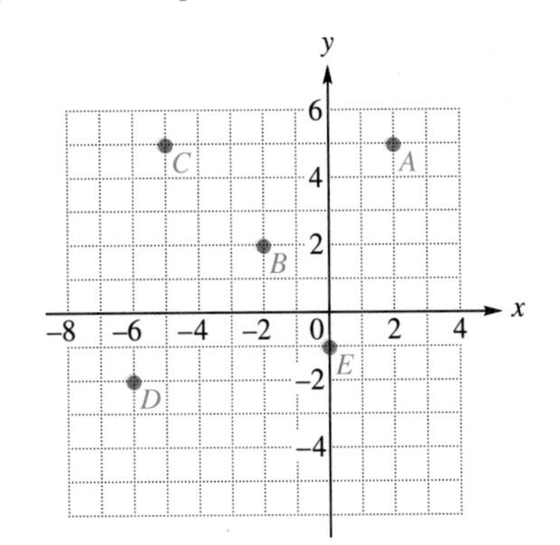

---

37. Do (4, −1) and (−1, 4) represent the same ordered pair? Explain why. (See Objective 1.)

38. Do the ordered pairs (3, 4) and (4, 3) correspond to the same point on the plane? Explain why. (See Objective 5.)

Plot the following ordered pairs on a coordinate system. See Example 5.

39. (6, 1) **40.** (4, −2) **41.** (−4, −5) **42.** (−2, 4)

43. (−5, −1) **44.** (−3, 5) **45.** (3, −5) **46.** (4, 0)

47. (−2, 0) **48.** (0, 6) **49.** (0, −5) **50.** (0, 0)

Without plotting the given point, state the quadrant in which each point lies.

51. (2, 3) **52.** (2, −3) **53.** (−2, 3) **54.** (−2, −3)

55. (−1, −1) **56.** (4, 7) **57.** (−3, 6) **58.** (1, −5)

59. (5, −4) **60.** (9, −1) **61.** (0, 0) **62.** (−2, 0)

Complete the table of values using the given equation. Then plot the ordered pairs. See Examples 2–5.

63. $y = 2x + 6$

x	0	2		
y			0	2

64. $y = 8 - 4x$

x	0	3		
y			0	16

65. $3x + 5y = 15$

x	0	10		
y			0	6

66. $2x - 5y = 10$

x	0	10		
y			0	-6

67. $y = 3x$

x	0	-2	4	
y				-3

68. $x + 2y = 0$

x	0		4	
y		3		-1

69. $y + 2 = 0$

x	5	0	-3	-2
y				

70. $x - 4 = 0$

x				
y	7	0	-4	4

In statistics, ordered pairs are used to decide whether two quantities (such as the height and weight of an individual) are related in such a way that one can be predicted when given the other. Ordered pairs that give these quantities for a number of individuals are plotted on a graph (called a scatter diagram*). If the points lie approximately on a line, the variables have a linear relationship.*

71. Make a scatter diagram by plotting the following ordered pairs of heights and weights for six women on the given axes: (62, 105), (65, 130), (67, 142), (63, 115), (66, 120), (60, 98). As shown in the figure provided, the horizontal axis is used to represent the heights, given as the first coordinates, and the vertical axis represents the weights, given as the second coordinates. (We could have assigned the first coordinates to weights and the second coordinates to heights.)

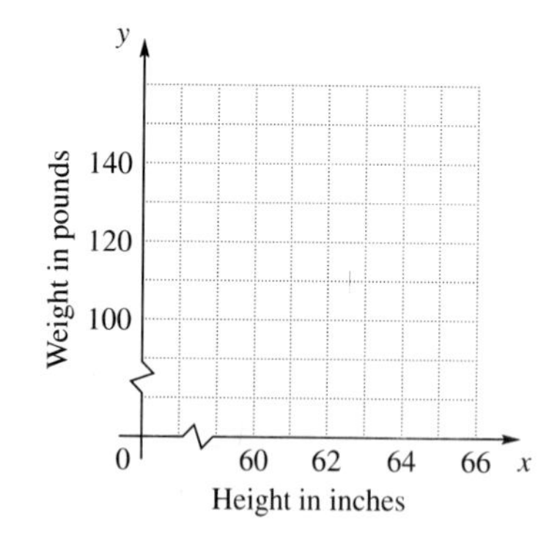

72. In Exercise 71, is there a linear relationship between height and weight? (Do the points lie approximately on a straight line?)

73. Make a scatter diagram by plotting the following points, which give the annual cost (in thousands of dollars) to attend a certain private college for selected years: (1988, 9.2), (1989, 9.8), (1990, 10.5), (1991, 11.0). Put years on the horizontal axis. Is this a linear relationship?

74. Repeat Exercise 73 using the following points which give the annual cost (in thousands of dollars) to attend a certain public college for selected years: (1988, 4.6), (1989, 5.0), (1990, 5.3), (1991, 5.8).

(a) Is this a linear relationship?

(b) What do the numbers on the x-axis represent?

(c) What do the numbers on the y-axis represent?

PREVIEW EXERCISES

Solve each equation. See Section 2.2.

75. $3m + 5 = 13$

76. $2k - 7 = 9$

77. $3 - y = 12$

78. $-4 - x = 6$

79. $7 + 5p = 12$

80. $9 - 3r = 15$

81. $8 - q = -7$

82. $-3 + 8z = -19$

6.2 GRAPHING LINEAR EQUATIONS IN TWO VARIABLES

OBJECTIVES

1 GRAPH LINEAR EQUATIONS BY COMPLETING AND PLOTTING ORDERED PAIRS.
2 FIND INTERCEPTS.
3 GRAPH LINEAR EQUATIONS OF THE FORM $Ax + By = 0$.
4 GRAPH LINEAR EQUATIONS OF THE FORM $y = k$ OR $x = k$.

There are an infinite number of ordered pairs that satisfy an equation in two variables. For example, we find ordered pairs that are solutions of the equation $x + 2y = 7$ by choosing as many values of x (or y) as we wish, and then completing each ordered pair. For instance, if we choose $x = 1$, then

$$\begin{aligned} x + 2y &= 7 \\ 1 + 2y &= 7 \quad \text{Let } x = 1. \\ 2y &= 6 \\ y &= 3, \end{aligned}$$

so the ordered pair (1, 3) is a solution of the equation. In the same way, $(-3, 5)$, $(3, 2)$, $(0, 7/2)$, $(-2, 9/2)$, $(-1, 4)$, $(6, 1/2)$, and $(7, 0)$ are all solutions of the equation $x + 2y = 7$. These ordered pairs have been plotted in Figure 6.5.

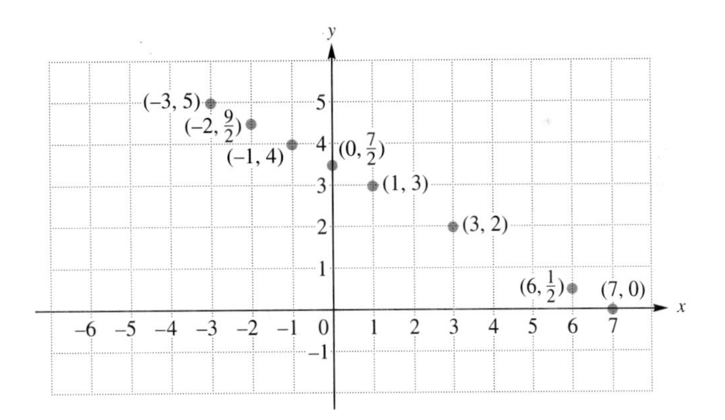

FIGURE 6.5

1 Notice that the points plotted in this figure all appear on a straight line. The line that goes through these points is shown in Figure 6.6. In fact, all ordered pairs satisfying the equation $x + 2y = 7$ correspond to points that lie on this same straight line. This line gives a "picture" of all the solutions of the equation $x + 2y = 7$. Only a portion of the line is shown here, but it extends indefinitely in both directions, as suggested by the arrowhead on each end of the line. The line is called the **graph** of the equation and the process of plotting the ordered pairs and drawing the line through the corresponding points is called **graphing.**

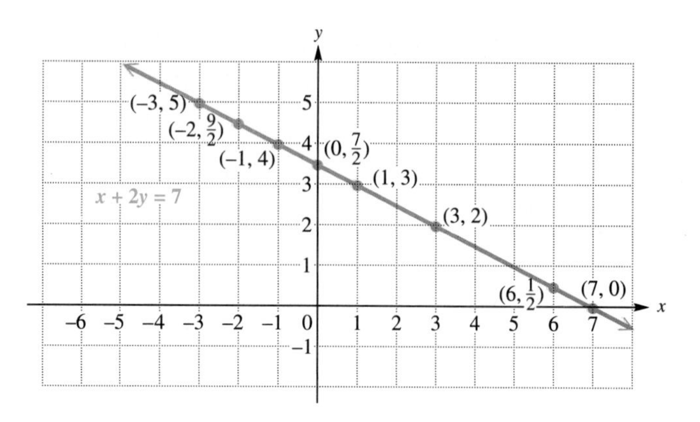

FIGURE 6.6

The preceding discussion can be generalized.

GRAPH OF A LINEAR EQUATION The graph of any linear equation in two variables is a straight line.

Notice that the word *line* appears in the name "*linear* equation."

Since two distinct points determine a line, a straight line can be graphed by finding any two different points on the line. However, it is a good idea to plot a third point as a check.

EXAMPLE 1 **GRAPHING A LINEAR EQUATION**

Graph the linear equation $3x + 2y = 6$.

For most linear equations, two different points on the graph can be found by first letting $x = 0$, and then letting $y = 0$.

If $x = 0$:	If $y = 0$:
$3x + 2y = 6$	$3x + 2y = 6$
$3(0) + 2y = 6$	$3x + 2(0) = 6$
$0 + 2y = 6$	$3x + 0 = 6$
$2y = 6$	$3x = 6$
$y = 3.$	$x = 2.$

These results give the ordered pairs (0, 3) and (2, 0). Get a third ordered pair (as a check) by letting x or y equal some other number. For example, let $x = -2$. Replace x with -2 in the given equation.

$$3x + 2y = 6$$
$$3(-2) + 2y = 6 \qquad \text{Let } x = -2.$$
$$-6 + 2y = 6$$
$$2y = 12$$
$$y = 6$$

The ordered pair is $(-2, 6)$. The three ordered pairs are shown in the table of values with Figure 6.7. Plot the corresponding points, then draw a line through them. This line, shown in Figure 6.7, is the graph of $3x + 2y = 6$. ■

x	y
0	3
2	0
−2	6

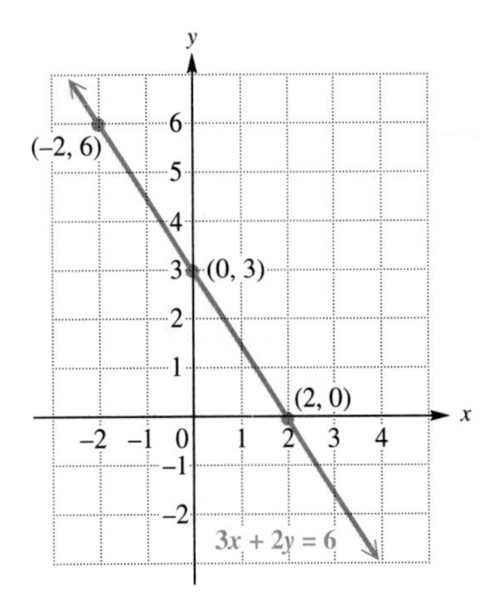

FIGURE 6.7

EXAMPLE 2
GRAPHING A LINEAR EQUATION

Graph the linear equation $4x = 5y + 20$.

Although this equation is not in the form $Ax + By = C$, it could be put in that form and so is a linear equation. Replace x by 0, then replace y by 0, to find two ordered pairs.

$$\begin{aligned} 4x &= 5y + 20 \\ 4(0) &= 5y + 20 \quad && x = 0 \\ 0 &= 5y + 20 \\ -5y &= 20 \quad && \text{Subtract } 5y. \\ y &= -4 \end{aligned}$$

$$\begin{aligned} 4x &= 5y + 20 \\ 4x &= 5(0) + 20 \quad && y = 0 \\ 4x &= 20 \\ x &= 5 \end{aligned}$$

The ordered pairs are $(0, -4)$ and $(5, 0)$. Get a third ordered pair (as a check), by choosing some number other than 0 for x or y; for example, choose $y = 2$. Replacing y with 2 in the equation $4x = 5y + 20$ leads to the ordered pair $(15/2, 2)$ or $(7\ 1/2, 2)$. Plot the three ordered pairs and draw a line through the corresponding points. This line, shown in Figure 6.8, is the graph of $4x = 5y + 20$. ■

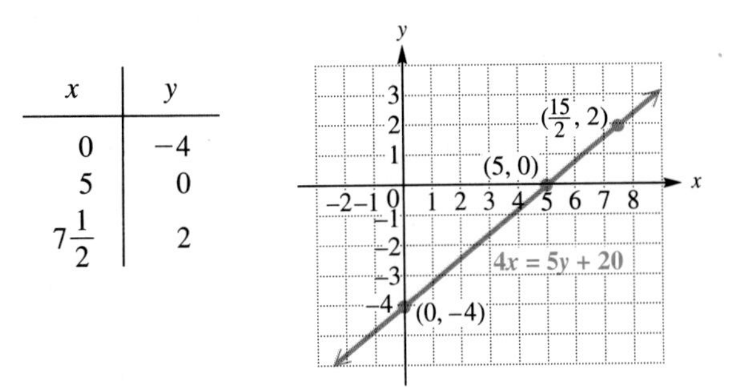

x	y
0	−4
5	0
$7\frac{1}{2}$	2

FIGURE 6.8

2 In Figure 6.8 the graph crosses the y-axis at $(0, -4)$ and the x-axis at $(5, 0)$. For this reason, $(0, -4)$ is called the ***y*-intercept** and $(5, 0)$ is called the ***x*-intercept** of the graph. The intercepts are particularly useful for graphing linear equations, as in Examples 1 and 2. The intercepts are found by replacing each variable, in turn, with 0 in the equation and solving for the value of the other variable.

FINDING INTERCEPTS

Find the x-intercept by letting $y = 0$ in the given equation and solving for x.
Find the y-intercept by letting $x = 0$ in the given equation and solving for y.

EXAMPLE 3 **FINDING INTERCEPTS**

Find the intercepts for the graph of $2x + y = 4$. Draw the graph.

Find the y-intercept by letting $x = 0$; find the x-intercept by letting $y = 0$.

$$\begin{aligned} 2x + y &= 4 \\ 2(0) + y &= 4 \\ 0 + y &= 4 \\ y &= 4 \end{aligned} \qquad \begin{aligned} 2x + y &= 4 \\ 2x + 0 &= 4 \\ 2x &= 4 \\ x &= 2 \end{aligned}$$

The y-intercept is $(0, 4)$. The x-intercept is $(2, 0)$. The graph with the two intercepts shown in color is given in Figure 6.9. Get a third point as a check. For example, choosing $x = 1$ gives $y = 2$. These three ordered pairs are shown in the table with Figure 6.9. Plot $(0, 4)$, $(2, 0)$, and $(1, 2)$ and draw a line through them. This line, shown in Figure 6.9, is the graph of $2x + y = 4$. ■

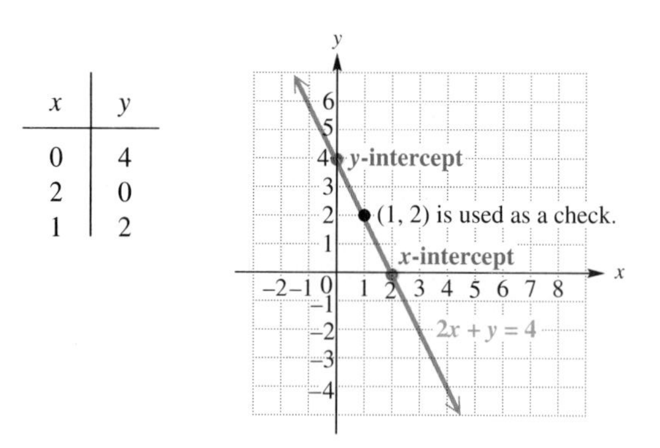

x	y
0	4
2	0
1	2

FIGURE 6.9

CAUTION When choosing x- or y-values to find ordered pairs to plot, be careful to choose so that the resulting points are not too close together. For example, using $(-1, -1)$, $(0, 0)$, and $(1, 1)$ may result in an inaccurate line. It is better to choose points where the x-values differ by at least 2.

3 In the earlier examples, the x- and y-intercepts were used to help draw the graphs. This is not always possible, as the following examples show. Example 4 shows what to do when the x- and y-intercepts are the same point.

EXAMPLE 4 **GRAPHING AN EQUATION OF THE FORM $Ax + By = 0$**

Graph the linear equation $x - 3y = 0$.

If we let $x = 0$, then $y = 0$, giving the ordered pair $(0, 0)$. Letting $y = 0$ also gives $(0, 0)$. This is the same ordered pair, so choose two additional values for x or y. Choosing 2 for y gives $x - 3 \cdot 2 = 0$, giving the ordered pair $(6, 2)$. For a check point, choose -6 for x getting -2 for y. This ordered pair, $(-6, -2)$, along with $(0, 0)$ and $(6, 2)$, was used to get the graph shown in Figure 6.10. ■

x	y
0	0
6	2
−6	−2

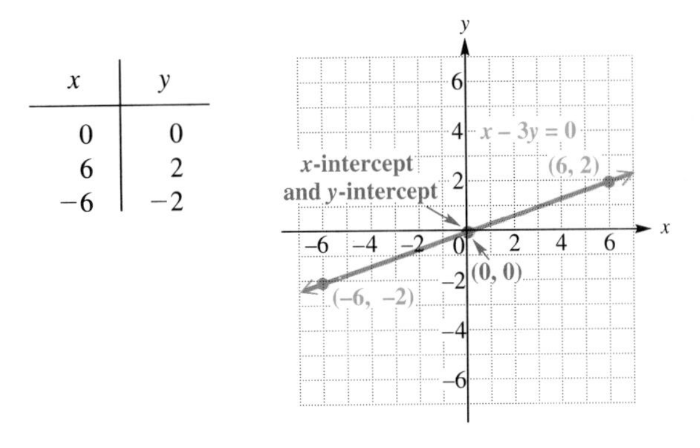

FIGURE 6.10

Example 4 can be generalized as follows.

LINE THROUGH THE ORIGIN

If A and B are real numbers, the graph of a linear equation of the form

$$Ax + By = 0$$

goes through the origin $(0, 0)$.

4 The equation $y = -4$ is a linear equation in which the coefficient of x is 0. (Write $y = -4$ as $0x + y = -4$ to see this.) Also, $x = 3$ is a linear equation in which the coefficient of y is 0. These equations lead to horizontal or vertical straight lines, as the next examples show.

EXAMPLE 5
GRAPHING AN EQUATION OF THE FORM $y = k$

Graph the linear equation $y = -4$.

As the equation states, for any value of x, y is always equal to -4. To get ordered pairs that are solutions of this equation, choose any numbers for x but always let y equal -4. Three ordered pairs that work are shown in the table of values with Figure 6.11. Drawing a line through these points gives the horizontal line shown in Figure 6.11. ■

x	y
-2	-4
0	-4
3	-4

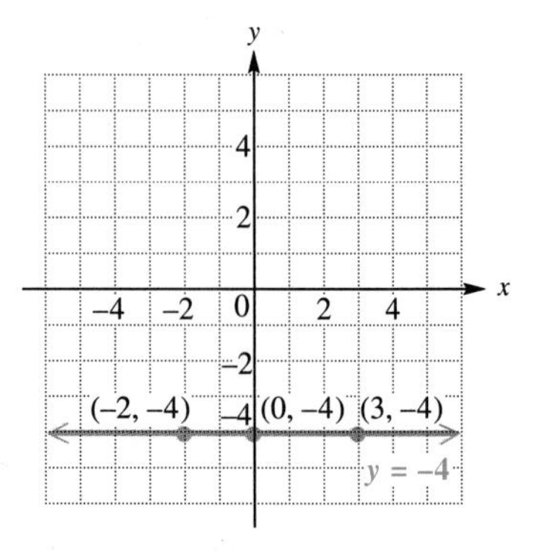

FIGURE 6.11

HORIZONTAL LINE

The graph of the linear equation $y = k$, where k is a real number, is the horizontal line going through the point $(0, k)$.

EXAMPLE 6
GRAPHING AN EQUATION OF THE FORM $x = k$

Graph the linear equation $x - 3 = 0$.

First add 3 to each side of the equation $x - 3 = 0$ to get $x = 3$. All the ordered pairs that are solutions of this equation have an x-value of 3. Any number can be used for y. Three ordered pairs that satisfy the equation are shown in the table of values with Figure 6.12. Drawing a line through these points gives the vertical line shown in Figure 6.12. ■

x	y
3	3
3	0
3	-2

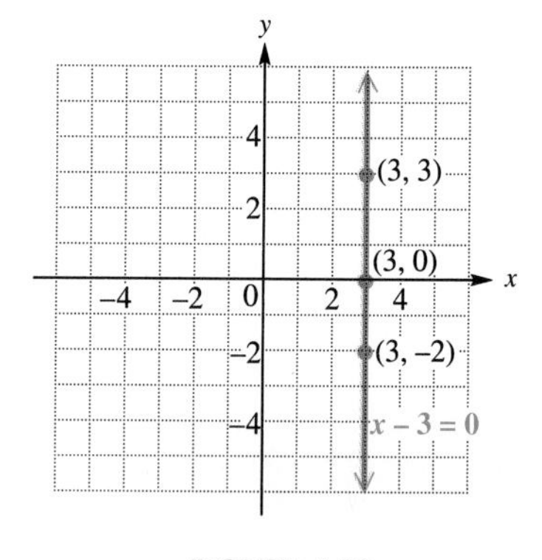

FIGURE 6.12

VERTICAL LINE

The graph of the linear equation $x = k$, where k is a real number, is the vertical line going through the point $(k, 0)$.

In particular, notice that the horizontal line $y = 0$ is the x-axis and the vertical line $x = 0$ is the y-axis.

The different forms of straight-line equations and the methods of graphing them are summarized on the next page.

GRAPHING STRAIGHT LINES

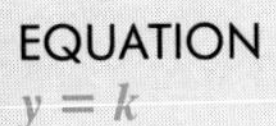

EQUATION	TO GRAPH	EXAMPLE
$y = k$	Draw a horizontal line, through $(0, k)$.	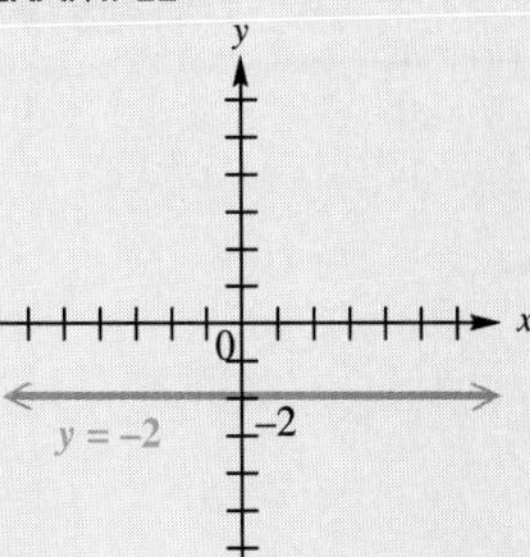
$x = k$	Draw a vertical line, through $(k, 0)$.	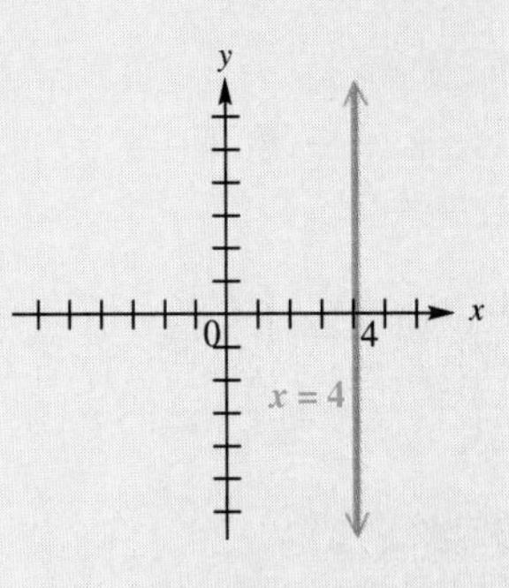
$Ax + By = 0$	Graph goes through $(0, 0)$. Get additional points that lie on the graph by choosing any value of x, or y, except 0.	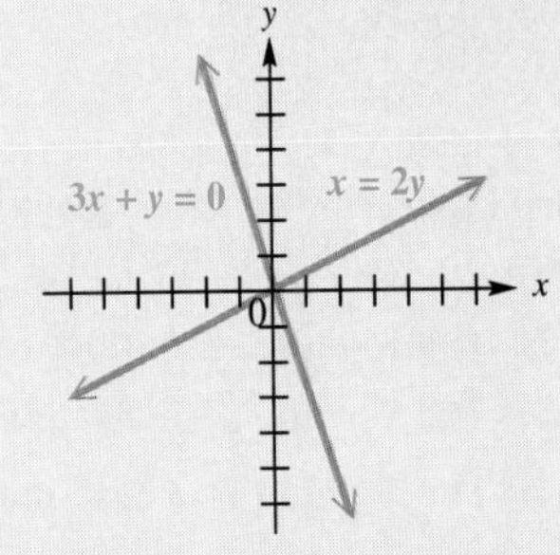
$Ax + By = C$ but not of the types above	Find any two points the line goes through. A good choice is to find the intercepts: let $x = 0$, and find the corresponding value of y; then let $y = 0$, and find x. As a check, get a third point by choosing a value of x or y that has not yet been used.	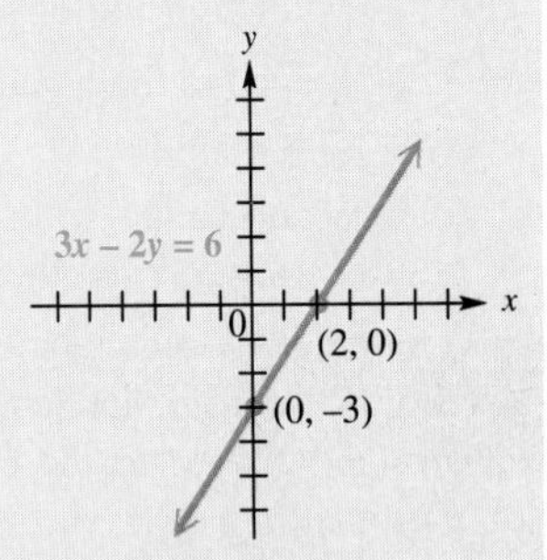

6.2 EXERCISES

Complete the ordered pairs for each equation. Then graph the equation by plotting the points and drawing a line through them. See Examples 1 and 2.

1. $x + y = 5$ (0,) (, 0) (2,)

2. $y = x - 3$ (0,) (, 0) (5,)

3. $y = x + 4$ (0,) (, 0) (−2,)

4. $y + 5 = x$ (0,) (, 0) (6,)

5. $y = 3x - 6$ (0,) (, 0) (3,)

6. $x = 2y + 1$ (0,) (, 0) (3,)

7. $2x + 5y = 20$ (0,) (, 0) (5,)

8. $3x - 4y = 12$ (0,) (, 0) (8,)

9. $x + 5 = 0$ (, 2) (, 0) (, −3)

10. $y - 4 = 0$ (3,) (0,) (−2,)

Find the intercepts for each equation. See Example 3.

11. $2x + 3y = 6$

12. $7x + 2y = 14$

13. $3x - 5y = 9$

14. $6x - 5y = 12$

15. $2y = 5x$

16. $x = -3y$

17. $-2x = 8$

18. $6y = 12$

Graph each linear equation. See Examples 1–6.

19. $x - y = 2$

20. $x + y = 6$

21. $x + 2y = 6$

22. $3x - y = 6$

23. $4x = 3y - 12$

24. $5x = 2y - 10$

25. $3x = 6 - 2y$

26. $2x + 3y = 12$

27. $2x - 7y = 14$

28. $3x + 5y = 15$

29. $3x + 7y = 14$

30. $6x - 5y = 30$

31. $y = 2x$

32. $y = -3x$

33. $y + 6x = 0$

34. $x - 4y = 0$

35. $x + 2 = 0$

36. $y - 3 = 0$

37. $y = -1$

38. $x = 2$

Translate each of the statements of Exercises 39–44 into an equation. Then graph the equation.

39. The x-value is 2 more than the y-value.

40. The y-value is 3 less than the x-value.

41. The y-value is 3 less than twice the x-value.

42. The x-value is 4 more than three times the y-value.

43. If 3 is added to the y-value, the result is 4 less than twice the x-value.

44. If 6 is subtracted from 4 times the y-value, the result is three times the x-value.

45. The height h of a woman (in centimeters) can be estimated from the length of her radius bone r (from the wrist to the elbow) with the following formula: $h = 73.5 + 3.9r$. Estimate the heights of women with radius bones of the following lengths.
(a) 23 centimeters
(b) 25 centimeters
(c) 20 centimeters
(d) Graph $h = 73.5 + 3.9r$.

46. As a rough estimate, the weight of a man taller than about 60 inches is approximated by $y = 5.5x - 220$, where x is the height of the person in inches, and y is the weight in pounds. Estimate the weights of men whose heights are as follows.
(a) 62 inches **(b)** 64 inches
(c) 68 inches **(d)** 72 inches
(e) Graph $y = 5.5x - 220$.

47. The graph shows the value of a certain automobile over its first five years. Use the graph to estimate

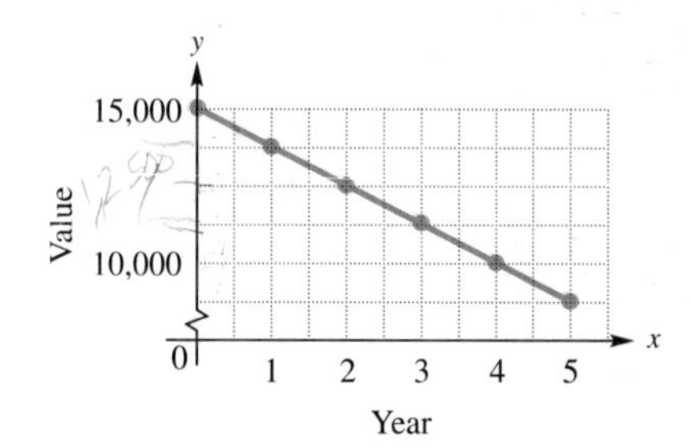

the depreciation (loss in value) during the following years.

(a) First **(b)** Second **(c)** Fifth

(d) What is the total depreciation over the 5-year period?

48. The demand for an item is closely related to its price. As price goes up, demand goes down. On the other hand, when price goes down, demand goes up. Suppose the demand for a certain fashionable watch is 1000 when its price is \$30 and 8000 when it costs \$15.

(a) Let x be the price and y be the demand for the watch. Graph the two given pairs of prices and demands.

(b) Assume the relationship is linear. Draw a line through the two points from part (a). From your graph estimate the demand if the price drops to \$10.

(c) Use the graph to estimate the price if the demand is 4000.

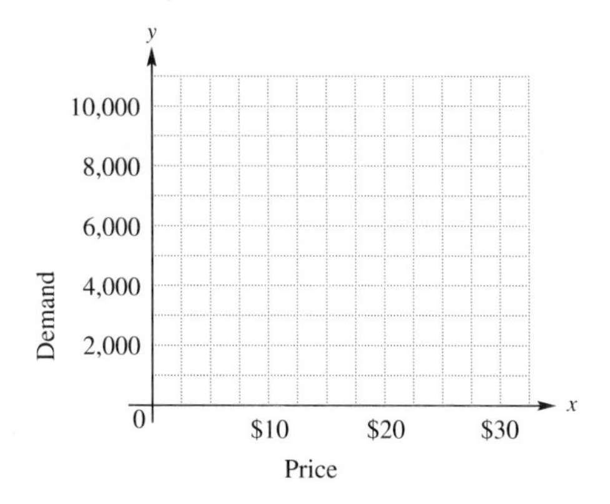

49. Why are equations of the form $Ax + By = C$ called linear equations? (See Objective 1.)

50. Compare and contrast the linear equations of the form $Ax + By = C$ with the linear equations $ax + b = c$ discussed in Chapter 2. How are they similar? How do they differ?

PREVIEW EXERCISES

Find each quotient. See Section 1.8.

51. $\dfrac{4-2}{8-5}$ **52.** $\dfrac{-3-5}{2-7}$ **53.** $\dfrac{-2-(-4)}{3-(-1)}$

54. $\dfrac{5-(-7)}{-4-(-1)}$ **55.** $\dfrac{-2-(-5)}{-9-12}$ **56.** $\dfrac{-6-3}{4-(-5)}$

57. $\dfrac{-9-4}{-2-3}$ **58.** $\dfrac{12-(-4)}{-3-5}$ **59.** $\dfrac{-2-(-7)}{3-3}$

6.3 THE SLOPE OF A LINE

OBJECTIVES

1. FIND THE SLOPE OF A LINE GIVEN TWO POINTS.
2. FIND THE SLOPE FROM THE EQUATION OF A LINE.
3. USE THE SLOPE TO DETERMINE WHETHER TWO LINES ARE PARALLEL, PERPENDICULAR, OR NEITHER.

We can graph a straight line if at least two different points on the line are known. A line can also be graphed by using just one point on the line, along with the ''steepness'' of the line.

1 One way to measure the steepness of a line is to compare the vertical change in the line (the rise) to the horizontal change (the run) while moving along the line from one fixed point to another. This measure of steepness is called the *slope* of the line.

Figure 6.13 shows a line with the points (x_1, y_1) and (x_2, y_2). (Read x_1 as "x-sub-one" and x_2 as "x-sub-two.") Moving along the line from the point (x_1, y_1) to the point (x_2, y_2) causes y to change by $y_2 - y_1$ units. This is the vertical change. Similarly, x changes by $x_2 - x_1$ units, the horizontal change. The ratio of the change in y to the change in x gives the slope of the line. We usually denote slope with the letter m. The slope of a line is defined as follows.

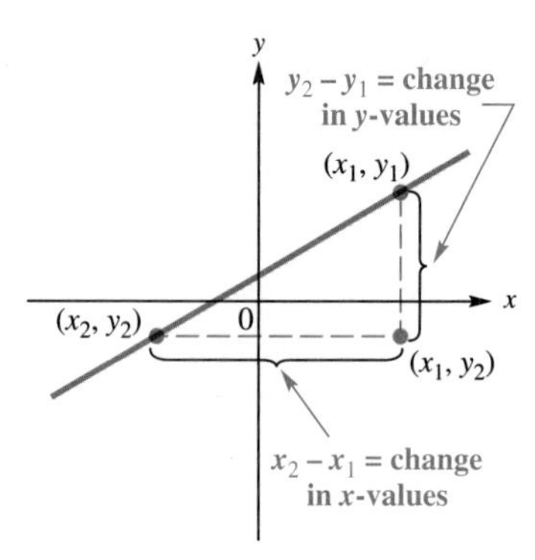

FIGURE 6.13

SLOPE FORMULA

The **slope** of the line through the points (x_1, y_1) and (x_2, y_2) is

$$m = \frac{\text{change in } y}{\text{change in } x} = \frac{y_2 - y_1}{x_2 - x_1} \quad \text{if } x_1 \neq x_2.$$

The slope of a line tells how fast y changes for each unit of change in x; that is, the slope gives the rate of change in y for each unit of change in x.

The idea of slope is useful in many everyday situations. For example, a highway with a 10% grade (or slope) rises 1 meter for every 10 meters horizontally. Architects specify the pitch of a roof by indicating the slope: a 5/12 roof means that the roof rises 5 feet for every 12 feet in the horizontal direction. The slope of a stairwell also indicates the ratio of the vertical rise to the horizontal run. See Figure 6.14.

1 meter

10 meters

A 10% grade

(a)

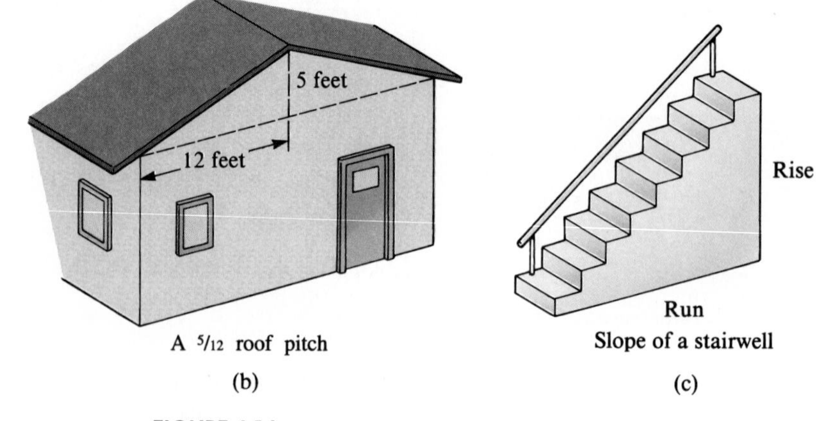

A 5/12 roof pitch

(b)

Slope of a stairwell

(c)

FIGURE 6.14

EXAMPLE 1
FINDING THE SLOPE OF A LINE

Find the slope of each of the following lines.

(a) The line through $(-4, 7)$ and $(1, -2)$
Use the definition of slope. Let $(-4, 7) = (x_2, y_2)$ and $(1, -2) = (x_1, y_1)$. Then

$$\text{slope} = \frac{\text{change in } y}{\text{change in } x}$$

$$m = \frac{y_2 - y_1}{x_2 - x_1}$$

$$= \frac{7 - (-2)}{-4 - 1} = \frac{9}{-5} = -\frac{9}{5}.$$

See Figure 6.15.

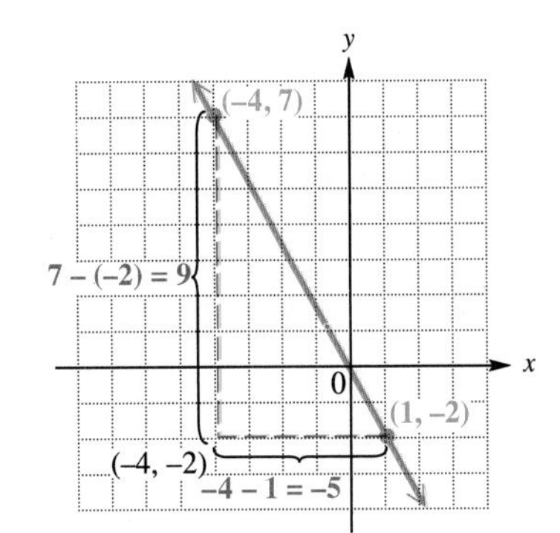

FIGURE 6.15

(b) The line through $(12, -5)$ and $(-9, -2)$

$$m = \frac{-5 - (-2)}{12 - (-9)} = \frac{-3}{21} = -\frac{1}{7}$$

The same slope is found by subtracting in reverse order.

$$\frac{-2 - (-5)}{-9 - 12} = \frac{3}{-21} = -\frac{1}{7} \quad \blacksquare$$

CAUTION It makes no difference which point is (x_1, y_1) or (x_2, y_2); however, it is important to be consistent. Start with the x- and y-values of one point (either one) and subtract the corresponding values of the other point.

In Example 1(a) the slope is negative and the corresponding line in Figure 6.15 falls from left to right. As Figure 6.16(a) shows, this is generally true of lines with negative slopes. Lines with positive slopes go up (rise) from left to right, as shown in Figure 6.16(b).

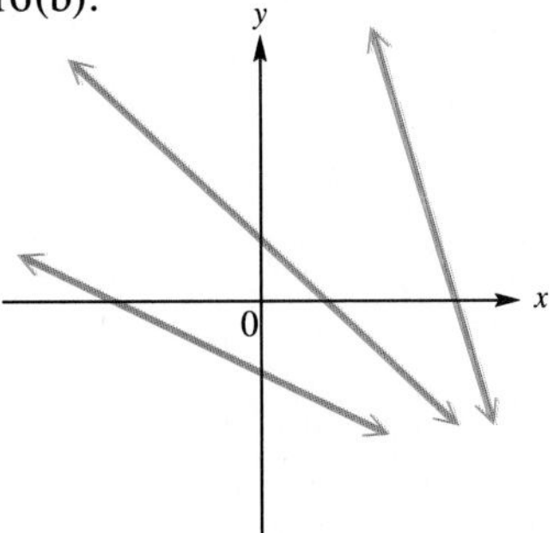

Lines with negative slopes

(a)

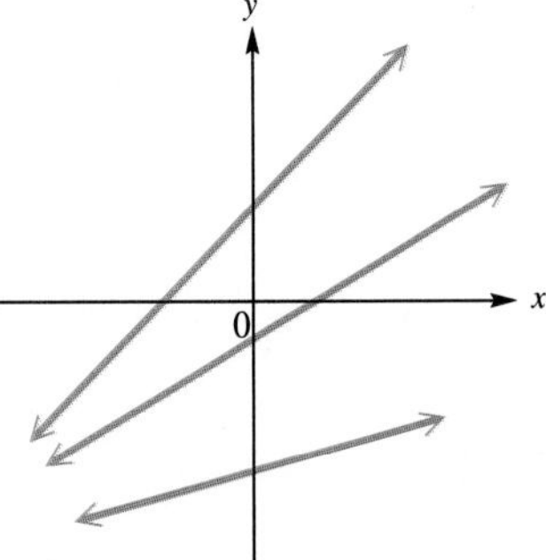

Lines with positive slopes

(b)

FIGURE 6.16

POSITIVE AND NEGATIVE SLOPES

A line with positive slope rises from left to right.

A line with negative slope falls from left to right.

The next examples illustrate slopes of horizontal and vertical lines.

EXAMPLE 2

FINDING THE SLOPE OF A HORIZONTAL LINE

Find the slope of the line through $(-8, 4)$ and $(2, 4)$.

Use the definition of slope.

$$m = \frac{4 - 4}{-8 - 2} = \frac{0}{-10} = 0$$

As shown in Figure 6.17 by a sketch of the line through these two points, the line through the points is horizontal, with equation $y = 4$. *All horizontal lines have a slope of* 0, since the difference in y-values is always 0. ■

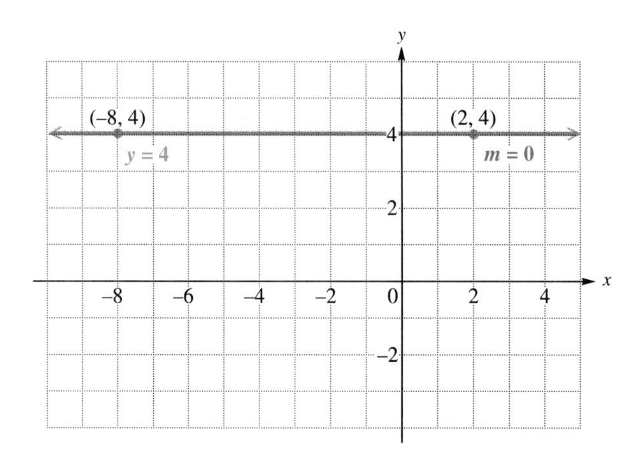

FIGURE 6.17

EXAMPLE 3

FINDING THE SLOPE OF A VERTICAL LINE

Find the slope of the line through $(6, 2)$ and $(6, -9)$.

$$m = \frac{2 - (-9)}{6 - 6}$$

$$= \frac{11}{0} \qquad \text{Undefined}$$

Since division by 0 is undefined, the slope is undefined. The graph in Figure 6.18 shows that the line through these two points is vertical, with equation $x = 6$. Since all points on a vertical line have the same x-value, *the slope of any vertical line is undefined*. ■

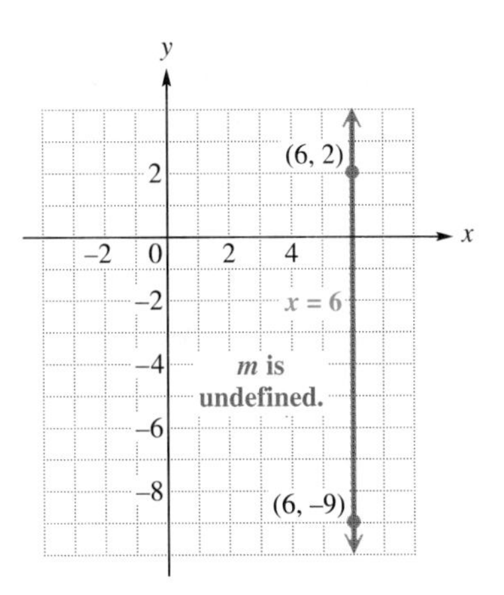

FIGURE 6.18

SLOPES OF HORIZONTAL AND VERTICAL LINES

Horizontal lines, with equations of the form $y = k$, **have a slope of 0.**
Vertical lines, with equations of the form $x = k$, **have undefined slope.**

2 The slope of a line also can be found directly from its equation. For example, the slope of the line

$$y = -3x + 5$$

can be found from two points on the line. Get these two points by choosing two different values of x, say -2 and 4, and finding the corresponding y-values.

If $x = -2$:	If $x = 4$:
$y = -3(-2) + 5$	$y = -3(4) + 5$
$y = 6 + 5$	$y = -12 + 5$
$y = 11.$	$y = -7.$

The ordered pairs are $(-2, 11)$ and $(4, -7)$. Now find the slope.

$$m = \frac{11 - (-7)}{-2 - 4}$$

$$= \frac{18}{-6} = -3$$

The slope, -3, is the same number as the coefficient of x in the equation $y = -3x + 5$. It can be shown that this always happens, *as long as the equation is solved for y*. This fact is used to find the slope of a line from its equation.

SLOPE OF A LINE FROM ITS EQUATION

Step 1 Solve the equation for y.
Step 2 The slope is given by the coefficient of x.

EXAMPLE 4 FINDING SLOPE FROM AN EQUATION

Find the slope of each of the following lines.

(a) $2x - 5y = 4$
Solve the equation for y.

$$2x - 5y = 4$$
$$-5y = -2x + 4 \quad \text{Subtract } 2x \text{ from each side.}$$
$$y = \frac{2}{5}x - \frac{4}{5} \quad \text{Divide each side by } -5.$$

The slope is given by the coefficient of x, so the slope is

$$m = \frac{2}{5}.$$

(b) $8x + 4y = 1$
Solve the equation for y.

$$8x + 4y = 1$$
$$4y = -8x + 1 \quad \text{Subtract } 8x \text{ from each side.}$$
$$y = -2x + \frac{1}{4} \quad \text{Divide each side by } 4.$$

The slope of this line is given by the coefficient of x, -2. ■

3 Two lines in a plane that never intersect are **parallel.** Slopes can be used to tell whether two lines are parallel. For example, Figure 6.19 shows the graph of $x + 2y = 4$ and the graph of $x + 2y = -6$. These lines appear to be parallel. Solve for y to find that both $x + 2y = 4$ and $x + 2y = -6$ have a slope of $-1/2$. Nonvertical parallel lines always have equal slopes.

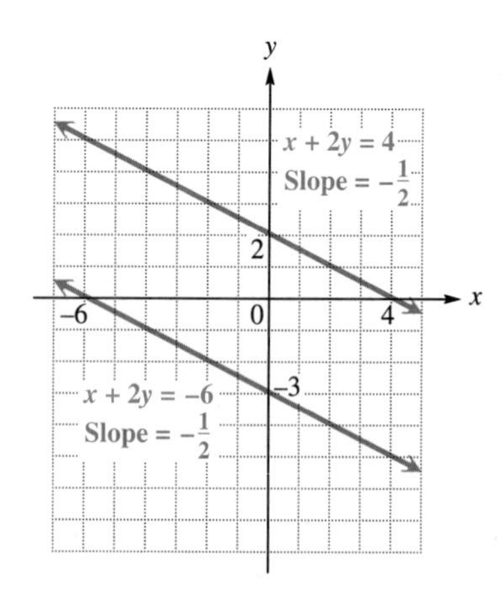

FIGURE 6.19

Figure 6.20 shows the graph of $x + 2y = 4$ and the graph of $2x - y = 6$. These lines appear to be **perpendicular** (meet at a 90° angle). Solving for y shows that the slope of $x + 2y = 4$ is $-1/2$, while the slope of $2x - y = 6$ is 2. The product of $-1/2$ and 2 is

$$\left(-\frac{1}{2}\right)(2) = -1.$$

This is true in general; the product of the slopes of two perpendicular lines is always -1. This means that the slopes of perpendicular lines are negative reciprocals: if one slope is the nonzero number a, the other is $-1/a$.

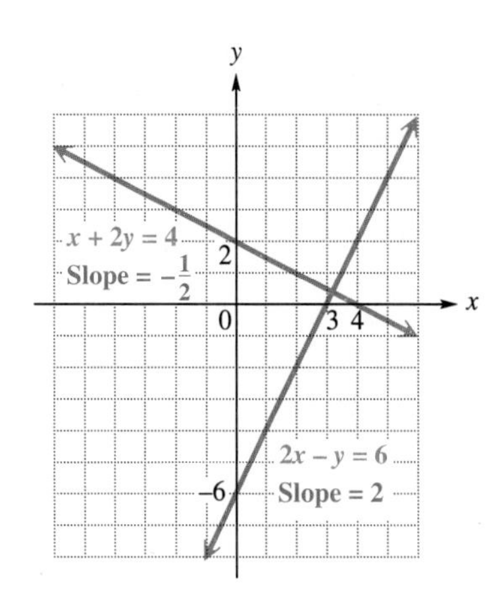

FIGURE 6.20

PARALLEL AND PERPENDICULAR LINES

Two nonvertical lines with the same slope are parallel; two perpendicular lines have slopes that are negative reciprocals.

EXAMPLE 5
DECIDING WHETHER TWO LINES ARE PARALLEL OR PERPENDICULAR

Decide whether the lines are *parallel, perpendicular,* or *neither*.

(a) $x + 2y = 7$
$-2x + y = 3$

Find the slope of each line by first solving each equation for y.

$$x + 2y = 7 \qquad\qquad -2x + y = 3$$

$$2y = -x + 7 \qquad\qquad y = 2x + 3$$

$$y = -\frac{1}{2}x + \frac{7}{2}$$

Slope: $-\dfrac{1}{2}$ Slope: 2

Since the slopes are not equal, the lines are not parallel. Check the product of the slopes: $(-1/2)(2) = -1$. The two lines are perpendicular because the product of their slopes is -1, indicating that the slopes are negative reciprocals.

(b) $3x - y = 4$
$6x - 2y = 9$

Find the slopes. Both lines have a slope of 3, so the lines are parallel.

(c) $4x + 3y = 6$
$2x - y = 5$

Here the slopes are $-4/3$ and 2. These lines are neither parallel nor perpendicular. ■

6.3 EXERCISES

Use the coordinates of the indicated points to find the slope of each of the following lines. See Example 1.

1.

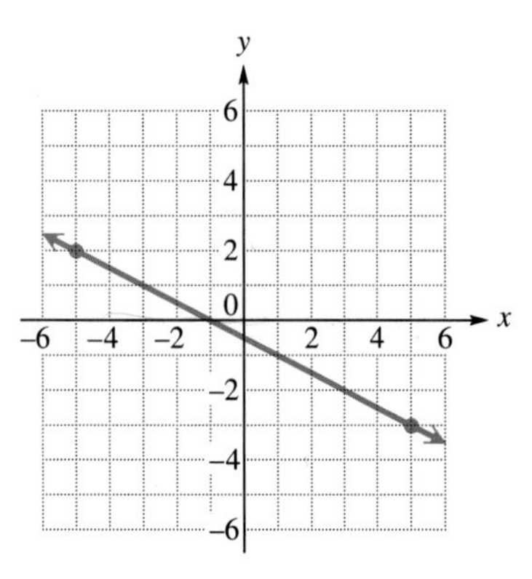

2.

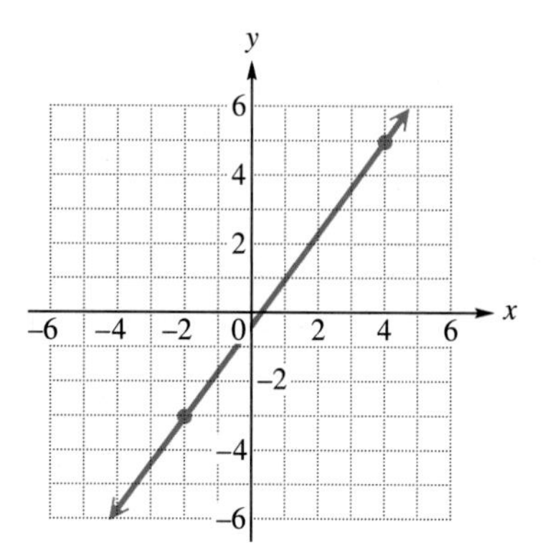

3.

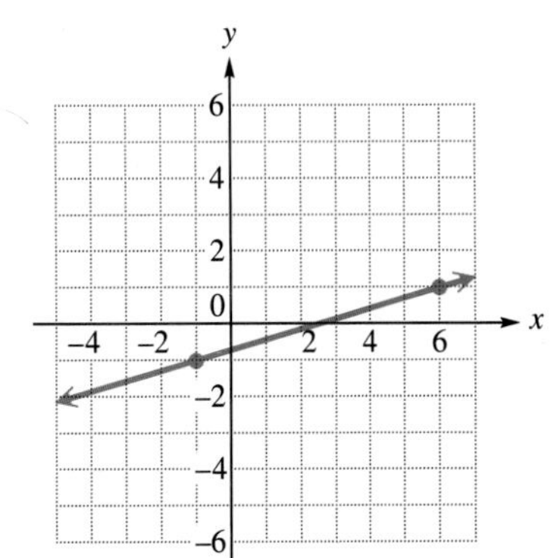

4.

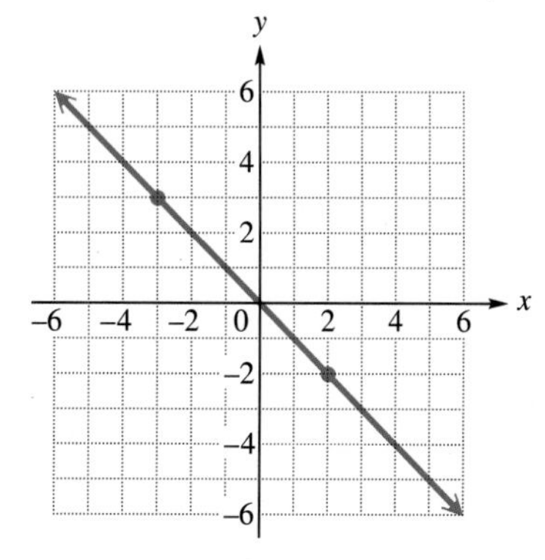

5.

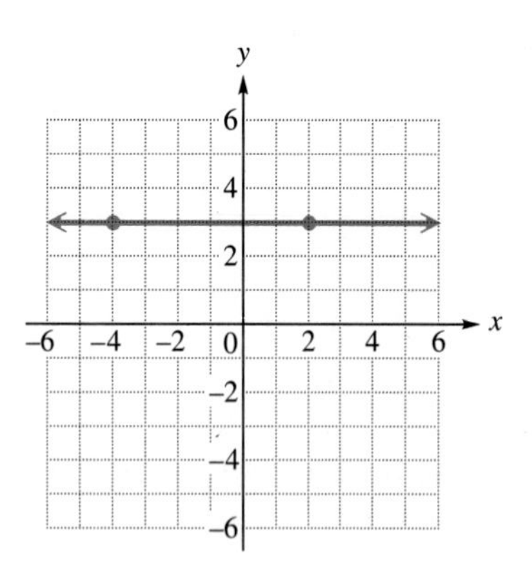

6.

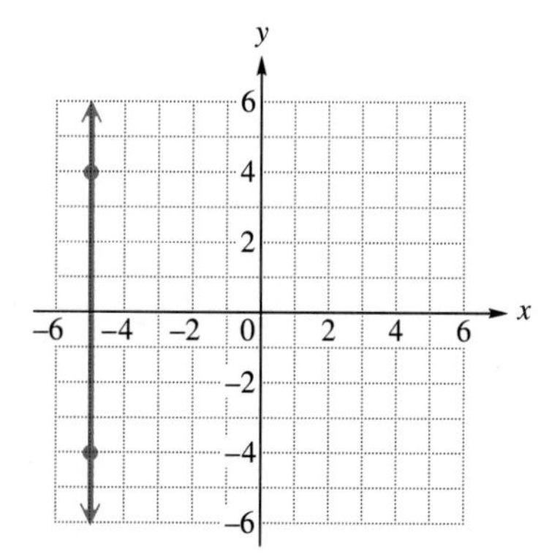

Find the slope of the line going through each pair of points. See Examples 1–3.

7. $(-4, 1)$, $(2, 8)$ **8.** $(3, 7)$, $(5, 2)$ **9.** $(-1, 2)$, $(-3, -7)$

10. $(5, -4)$, $(-5, -9)$ **11.** $(8, 0)$, $(0, 5)$ **12.** $(0, -3)$, $(2, 0)$

13. $(-1, 6)$, $(4, 6)$ **14.** $(5, 3)$, $(5, -2)$ **15.** $(-9, 1)$, $(-9, 0)$

16. Give an example of slope as used in your own experience. (See Objective 1.)

Find the slope of each of the following lines. See Example 4.

17. $y = 5x + 2$ **18.** $y = -x + 4$ **19.** $y = x + 1$

20. $y = 6 - 5x$ **21.** $y = 3 + 9x$ **22.** $2x + y = 5$

23. $4x - y = 8$ **24.** $-6x + 4y = 1$ **25.** $3x - 2y = 5$

26. $2x + 5y = 4$ **27.** $9x + 7y = 5$ **28.** $y + 4 = 0$

Match the following slopes to lines (a)–(d) *below. (See Objective 1.)*

29. $-\frac{1}{2}$ **30.** 0 **31.** undefined **32.** $\frac{1}{2}$

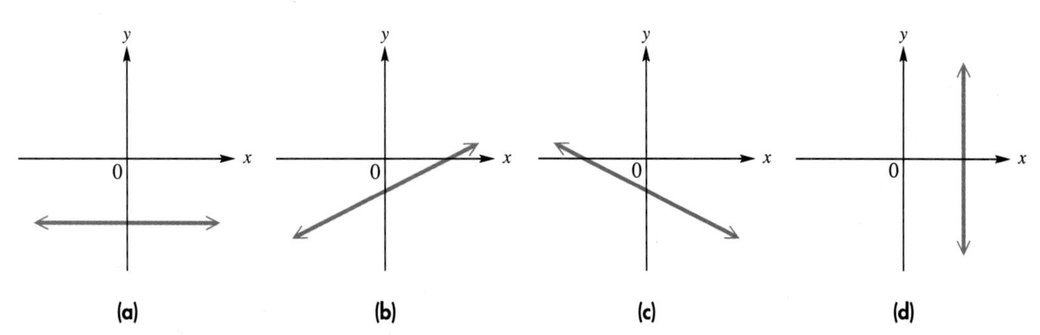

Decide whether the lines in each exercise are parallel, perpendicular, or neither. See Example 5.

33. $x + y = 5$, $x - y = 1$ **34.** $y - x = 3$, $y - x = 5$ **35.** $y - x = 4$, $y + x = 3$

36. $2x - 5y = 4$, $4x - 10y = 1$ **37.** $3x - 2y = 4$, $2x + 3y = 1$ **38.** $3x - 5y = 2$, $5x + 3y = -1$

39. $4x - 3y = 4$, $8x - 6y = 0$ **40.** $x - 4y = 2$, $2x + 4y = 1$ **41.** $8x - 9y = 2$, $3x + 6y = 1$

42. $5x - 3y = 8$, $3x - 5y = 10$ **43.** $6x + y = 12$, $x - 6y = 12$ **44.** $2x - 5y = 11$, $4x + 5y = 2$

---

45. Give an example of parallel lines in your daily surroundings.

46. Give an example of perpendicular lines in your daily surroundings.

Find the slope of each of the following lines.

47. $\frac{2}{3}y = \frac{5}{4}x - 3$

48. $\frac{3}{4}y - \frac{2}{5}x = 6$

49. $\frac{y}{2} + \frac{x}{4} = 12$

50. $\frac{5}{4}x + \frac{1}{4}y = -3$

Find the slope (or pitch) of the roofs shown in the figures. Measurements are given in feet.

51.

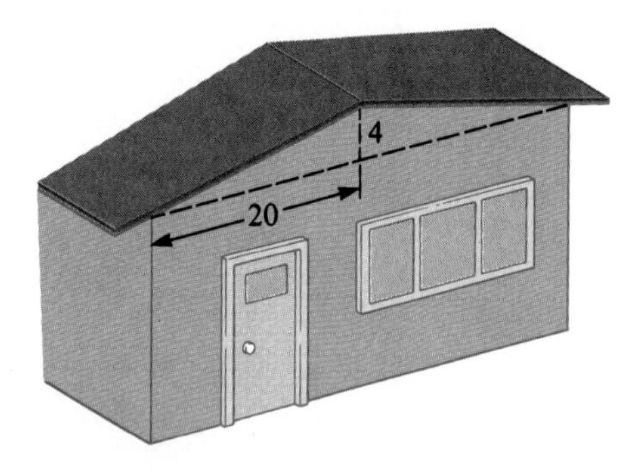

52.

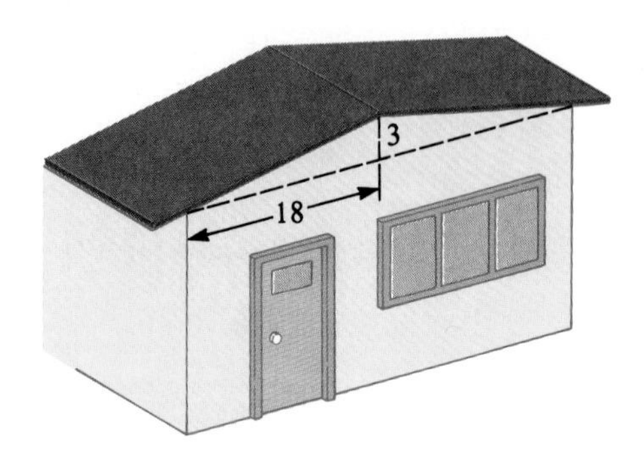

What is the slope (or grade) of the hills shown in the figures? Measurements are given in meters.

53.

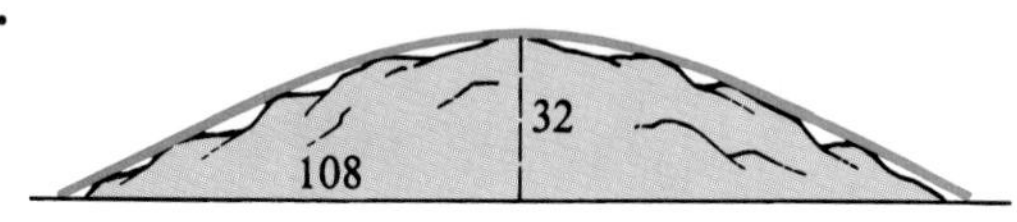

54.

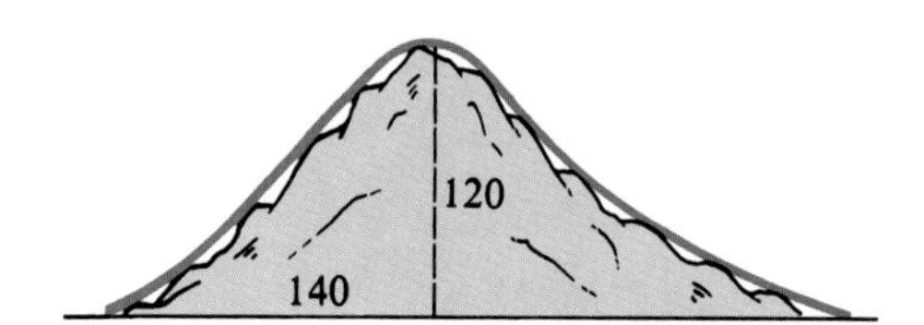

PREVIEW EXERCISES

Solve for y. Simplify your answers. See Section 2.5.

55. $y - 2 = 3(x - 4)$

56. $y + 1 = 2(x - 5)$

57. $y - (-3) = -2(x - 1)$

58. $y - (-4) = -(x + 1)$

59. $y - (-5) = 4[x - (-1)]$

60. $y - 2 = -3[x - (-1)]$

61. $y - \left(-\frac{3}{5}\right) = -\frac{1}{2}[x - (-3)]$

62. $y - \left(-\frac{5}{8}\right) = \frac{3}{8}(x - 5)$

6.4 EQUATIONS OF A LINE

OBJECTIVES

1. WRITE AN EQUATION OF A LINE GIVEN ITS SLOPE AND y-INTERCEPT.
2. GRAPH A LINE GIVEN ITS SLOPE AND A POINT ON THE LINE.
3. WRITE AN EQUATION OF A LINE GIVEN ITS SLOPE AND ANY POINT ON THE LINE.
4. WRITE AN EQUATION OF A LINE GIVEN TWO POINTS ON THE LINE.

The last section showed how to find the slope of a line from the equation of the line. For example, the slope of the line having the equation $y = 2x + 3$ is 2, the coefficient of x. What does the number 3 represent? If $x = 0$, the equation becomes

$$y = 2(0) + 3 = 0 + 3 = 3.$$

Since $y = 3$ corresponds to $x = 0$, $(0, 3)$ is the y-intercept of the graph of $y = 2x + 3$. An equation like $y = 2x + 3$ that is solved for y is said to be in **slope-intercept form** because both the slope and the y-intercept of the line can be found from the equation.

SLOPE-INTERCEPT FORM

The slope-intercept form of the equation of a line with slope m and y-intercept $(0, b)$ is

$$y = mx + b.$$

1 Given the slope and y-intercept of a line, we can use the slope-intercept form to find an equation of the line.

EXAMPLE 1 FINDING AN EQUATION OF A LINE

Find an equation of each of the following lines.

(a) With slope 5 and y-intercept $(0, 3)$
Use the slope-intercept form. Let $m = 5$ and $b = 3$.

$$y = mx + b$$
$$y = 5x + 3$$

(b) With slope $\frac{2}{3}$ and y-intercept $(0, -1)$

Here $m = \frac{2}{3}$ and $b = -1$.

$$y = mx + b$$
$$y = \frac{2}{3}x - 1$$ ■

2 The slope and y-intercept can be used to graph a line. For example, to graph $y = \frac{2}{3}x - 1$, first locate the y-intercept, $(0, -1)$, on the y-axis. From the definition of slope and the fact that the slope of this line is 2/3,

$$m = \frac{\text{difference in } y\text{-values}}{\text{difference in } x\text{-values}} = \frac{2}{3}.$$

Another point P on the graph of the line can be found by counting from the y-intercept 2 units up and then counting 3 units to the right. The line is then drawn through point P and the y-intercept, as shown in Figure 6.21. This method can be extended to graph a line given its slope and any point on the line.

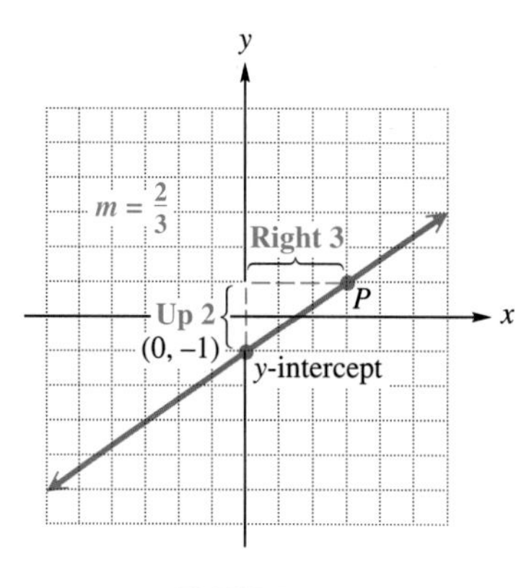

FIGURE 6.21

FIGURE 6.22

EXAMPLE 2
GRAPHING A LINE GIVEN A POINT AND THE SLOPE

Graph the line through $(-2, 3)$ with slope -4.

First, locate the point $(-2, 3)$. Write the slope as

$$m = \frac{\text{difference in } y\text{-values}}{\text{difference in } x\text{-values}} = -4 = \frac{-4}{1}.$$

(We could have used $4/-1$ instead.) Another point on the line is located by counting 4 units down (because of the negative sign) and then 1 unit to the right. Finally, draw the line through this new point P and the given point $(-2, 3)$. See Figure 6.22. ■

3 An equation of a line also can be found from any point on the line and the slope of the line. Let m represent the slope of the line and let (x_1, y_1) represent the given point on the line. Let (x, y) represent any other point on the line. Then by the definition of slope,

$$\frac{y - y_1}{x - x_1} = m$$

or

$$y - y_1 = m(x - x_1).$$

This result is the **point-slope form** of the equation of a line.

POINT-SLOPE FORM

The point-slope form of the equation of a line with slope m going thro[ugh]
(x_1, y_1) is

$$y - y_1 = m(x - x_1).$$

This very important result should be memorized.

EXAMPLE 3
USING THE POINT-SLOPE FORM TO WRITE AN EQUATION

Find an equation of each of the following lines. Write the equation in the form $Ax + By = C$.

(a) Through $(-2, 4)$, with slope -3

The given point is $(-2, 4)$ so $x_1 = -2$ and $y_1 = 4$. Also, $m = -3$. Substitute these values into the point-slope form.

$$\begin{aligned}
y - y_1 &= m(x - x_1) \\
y - 4 &= -3[x - (-2)] \\
y - 4 &= -3(x + 2) \\
y - 4 &= -3x - 6 && \text{Distributive property} \\
y &= -3x - 2 && \text{Add 4.} \\
3x + y &= -2 && \text{Add } 3x.
\end{aligned}$$

The last equation is in the form $Ax + By = C$.

(b) Through $(4, 2)$, with slope $3/5$

Use $x_1 = 4$, $y_1 = 2$, and $m = 3/5$ in the point-slope form.

$$\begin{aligned}
y - y_1 &= m(x - x_1) \\
y - 2 &= \frac{3}{5}(x - 4)
\end{aligned}$$

Multiply both sides by 5 to clear of fractions.

$$\begin{aligned}
5(y - 2) &= 5 \cdot \frac{3}{5}(x - 4) \\
5(y - 2) &= 3(x - 4) \\
5y - 10 &= 3x - 12 && \text{Distributive property} \\
5y &= 3x - 2 && \text{Add 10.} \\
-3x + 5y &= -2 && \text{Subtract } 3x. \quad \blacksquare
\end{aligned}$$

4 The point-slope form also can be used to find an equation of a line when two points on the line are known.

EXAMPLE 4
FINDING THE EQUATION OF A LINE FROM TWO POINTS

Find an equation of the line through the points $(-2, 5)$ and $(3, 4)$. Write the equation in the form $Ax + By = C$.

First find the slope of the line, using the definition of slope.

$$\text{slope} = \frac{5 - 4}{-2 - 3} = \frac{1}{-5} = -\frac{1}{5}$$

...v use either $(-2, 5)$ or $(3, 4)$ and the point-slope form. Using $(3, 4)$ gives

$$y - y_1 = m(x - x_1)$$

$$y - 4 = -\frac{1}{5}(x - 3)$$

$$5(y - 4) = -1(x - 3) \qquad \text{Multiply by 5.}$$

$$5y - 20 = -x + 3 \qquad \text{Distributive property}$$

$$5y = -x + 23 \qquad \text{Add 20 on each side.}$$

$$x + 5y = 23. \qquad \text{Add } x \text{ on each side.}$$

The same result would be found by using $(-2, 5)$ for (x_1, y_1). ■

A summary of the types of linear equations is given here.

LINEAR EQUATIONS

Equation	Description
$Ax + By = C$	**Standard form (A, B, and C integers, $A > 0$, $B \neq 0$)** Slope is $-A/B$. x-intercept is $(C/A, 0)$. y-intercept is $(0, C/B)$.
$x = k$	**Vertical line** Slope is undefined. x-intercept is $(k, 0)$.
$y = k$	**Horizontal line** Slope is 0. y-intercept is $(0, k)$.
$y = mx + b$	**Slope-intercept form** Slope is m. y-intercept is $(0, b)$.
$y - y_1 = m(x - x_1)$	**Point-slope form** Slope is m. Line goes through (x_1, y_1).

CAUTION The above definition of "standard form" is not the same in all texts. Also, a linear equation can be written as $Ax + By = C$ in many different (equally correct) ways. For example, $3x + 4y = 12$, $6x + 8y = 24$, and $9x + 12y = 36$ all represent the same set of ordered pairs. Let us agree that $3x + 4y = 12$ is preferable to the other forms because the greatest common factor of 3, 4, and 12 is 1.

6.4 EXERCISES

Write an equation for each line given its slope and y-intercept. See Example 1.

1. $m = 3$, y-intercept $(0, 5)$

2. $m = -2$, y-intercept $(0, 4)$

3. $m = -1$, y-intercept $(0, -6)$

4. $m = \frac{5}{3}$, y-intercept $\left(0, \frac{1}{2}\right)$

5. $m = \frac{2}{5}$, y-intercept $\left(0, -\frac{1}{4}\right)$

6. $m = 0$, y-intercept $(0, -5)$

Graph the line going through the given point and having the given slope. (In Exercises 15–18, recall the type of lines having 0 slope and undefined slope.) See Example 2.

7. $(2, 5)$, $m = \frac{1}{2}$

8. $(-4, -3)$, $m = -\frac{2}{5}$

9. $(-1, -1)$, $m = -\frac{3}{8}$

10. $(0, 2)$, $m = \frac{7}{4}$

11. $(-3, 0)$, $m = \frac{5}{9}$

12. $(2, 9)$, $m = -4$

13. $(4, -1)$, $m = 2$

14. $(3, -5)$, $m = 1$

15. $(1, 2)$, $m = 0$

16. $(-4, -8)$, $m = 0$

17. $(3, 5)$, undefined slope

18. $(2, 3)$, undefined slope

Write each equation in slope intercept form (by solving for y) and graph it using the slope and y-intercept.

19. $2x - y = 8$

20. $x - y = -7$

21. $3x + y = -5$

22. $-x + 2y = 4$

For each linear equation, identify its graph from (a)–(d) *below.*

23. $y = 1 - x$

24. $y = \frac{1}{2}x$

25. $y = 2$

26. $y = x - 1$

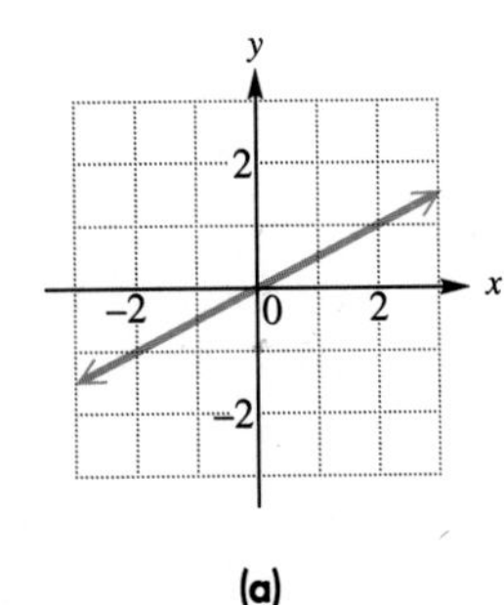

(a)

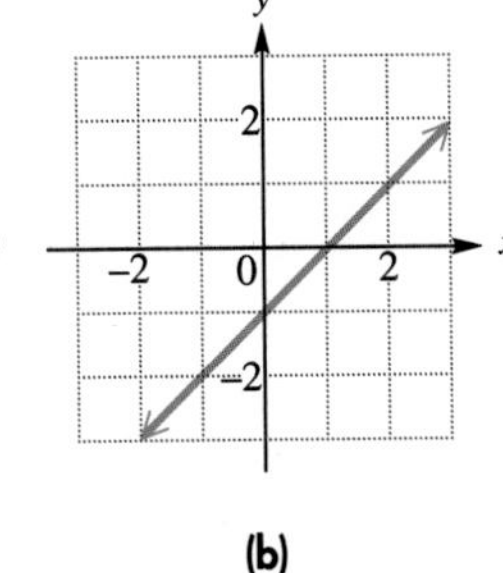

(b)

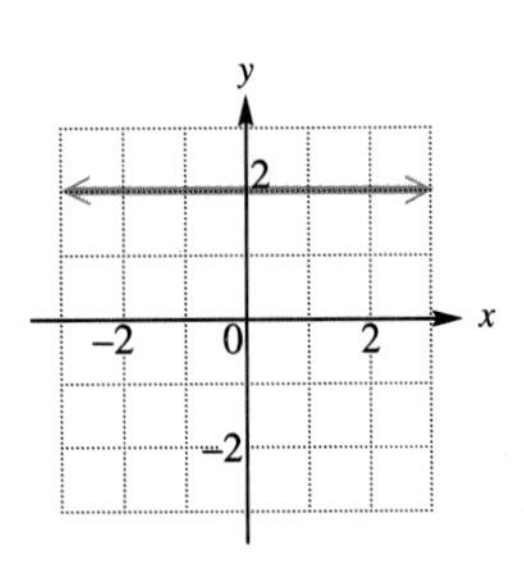

(c)

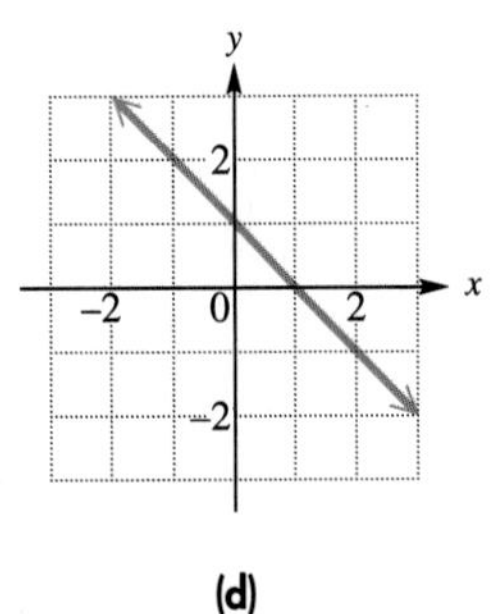

(d)

Write an equation for the line passing through the given point and having the given slope. Write the equation in the form $Ax + By = C$. See Example 3.

27. $(5, 3)$, $m = 2$

28. $(1, 4)$, $m = 3$

29. $(2, -8)$, $m = -2$

30. $(-1, 7)$, $m = -4$

31. $(3, 5)$, $m = \frac{2}{3}$

32. $(2, -4)$, $m = \frac{4}{5}$

33. $(-3, -2)$, $m = -\frac{3}{4}$

34. x-intercept $(-8, 0)$, $m = -\frac{5}{9}$

35. x-intercept $(6, 0)$, $m = -\frac{8}{11}$

Write an equation of the line passing through each pair of points. Write the equations in the form $Ax + By = C$. See Example 4.

36. $(7, 4)$, $(8, 5)$

37. $(-2, 1)$, $(3, 4)$

38. $(-8, -2)$, $(-1, -7)$

39. $(3, -4)$, $(-2, -1)$

40. $(-7, -5)$, $(-9, -2)$

41. $(0, 2)$, $(3, 0)$

42. $(4, 0)$, $(0, -2)$

43. $(2, -5)$, $(-4, 7)$

44. $(3, -7)$, $(-5, 0)$

45. $\left(\frac{1}{2}, \frac{3}{2}\right)$, $\left(-\frac{1}{4}, \frac{5}{4}\right)$

46. $\left(-\frac{2}{3}, \frac{8}{3}\right)$, $\left(\frac{1}{3}, \frac{7}{3}\right)$

47. $\left(-1, \frac{5}{8}\right)$, $\left(\frac{1}{8}, 2\right)$

---

48. Write an equation for the line through $(-1, 2)$ and
(a) parallel to the line $3x - 2y = 6$;
(b) perpendicular to the line $3x - 2y = 6$.

49. Write an equation for the line through $(5, 4)$ and
(a) perpendicular to the line $x + 3y = 4$;
(b) parallel to the line $x + 3y = 4$.

The cost y to produce x items is often expressed as $y = mx + b$. The number b gives the fixed cost (the cost that is the same no matter how many items are produced), and the number m is the variable cost (the cost to produce an additional item). Write the cost equation for each of the following. All costs are in dollars.

50. Fixed cost 50, variable cost 9

51. Fixed cost 100, variable cost 12

52. Fixed cost 70.50, variable cost 3.50

53. Refer to Exercise 51 and find the total cost to make **(a)** 50 items and **(b)** 125 items.

54. Refer to Exercise 52 and find the total cost to make **(a)** 25 items and **(b)** 110 items.

The sales of a company for a given year can be written as an ordered pair in which the first number gives the year (perhaps since the company started business) and the second number gives the sales for that year. If the sales increase at a steady rate, a linear equation for sales can be found. Sales in millions of dollars for two years are given for each of two companies below.

Company A		Company B	
Year	Sales	Year	Sales
x	y	x	y
1	24	1	18
5	48	4	27

55. (a) Write two ordered pairs of the form (year, sales) for Company A.
(b) Write an equation of the line through the two pairs in part (a). This is a sales equation for Company A.

56. (a) Write two ordered pairs of the form (year, sales) for Company B.
(b) Write an equation of the line through the two pairs in part (a). This is a sales equation for Company B.

57. In your own words, explain how to graph a line when the slope of the line and a point on the line are given. (See Objective 2.)

58. Compare the advantages and disadvantages of the point-slope form and the slope-intercept form. (See Objectives 1 and 3.)

PREVIEW EXERCISES

Solve and graph each inequality. See Section 2.9.

59. $m + 3 < 10$ **60.** $k - 5 \geq 10$ **61.** $2p + 7 \leq 8$ **62.** $3p - 9 \geq 4$

63. $6 - 2p < 12$ **64.** $5 - 3p < -8$ **65.** $-4 - p > 7$ **66.** $-8 - 3p < 4$

6.5 GRAPHING LINEAR INEQUALITIES IN TWO VARIABLES

OBJECTIVES

1 GRAPH $\leq$ OR $\geq$ LINEAR INEQUALITIES.

2 GRAPH $<$ OR $>$ LINEAR INEQUALITIES.

3 GRAPH INEQUALITIES WITH A BOUNDARY THROUGH THE ORIGIN.

In Section 6.2 we discussed methods for graphing linear equations, such as the equation $2x + 3y = 6$. Now this discussion is extended to **linear inequalities in two variables,** such as

$$2x + 3y \leq 6.$$

(Recall that $\leq$ is read "is less than or equal to.")

1 The inequality $2x + 3y \leq 6$ means that

$$2x + 3y < 6 \quad \text{or} \quad 2x + 3y = 6.$$

As we found at the beginning of this chapter, the graph of $2x + 3y = 6$ is a line. This line divides the plane into two regions. The graph of the solutions of the inequality $2x + 3y < 6$ will include only *one* of these regions. We find the required region by solving the given inequality for y.

$$2x + 3y \leq 6$$

$$3y \leq -2x + 6 \qquad \text{Subtract } 2x.$$

$$y \leq -\frac{2}{3}x + 2 \qquad \text{Divide by 3.}$$

By this last statement, ordered pairs in which y *is less than or equal to* $(-2/3)x + 2$ will be solutions to the inequality. The ordered pairs in which y is equal to $(-2/3)x + 2$ are on a line, so the pairs in which y *is less than* $(-2/3)x + 2$ will be *below* that line. To indicate the solution, shade the region below the line, as in Figure 6.23. The shaded region, along with the line, is the desired graph.

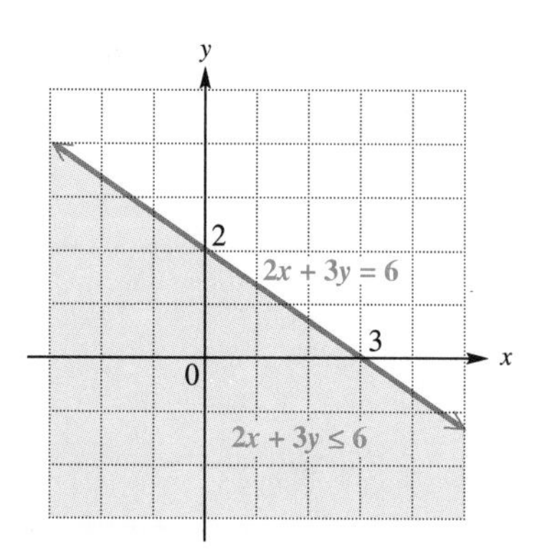

FIGURE 6.23

As an alternative method, a test point gives a quick way to find the correct region to shade. Choose any point *not* on the line. Because (0, 0) is easy to substitute into an inequality, it is often a good choice, and we will use it here.

Substitute 0 for x and 0 for y in the given inequality to see whether the resulting statement is true or false. In the example above,

$$2x + 3y \leq 6 \qquad \text{Original inequality}$$
$$2(0) + 3(0) \leq 6 \quad ? \qquad \text{Let } x = 0 \text{ and } y = 0.$$
$$0 + 0 \leq 6 \quad ?$$
$$0 \leq 6. \qquad \text{True}$$

Since the last statement is true, shade the region that includes the test point (0, 0). This agrees with the result shown in Figure 6.23.

2 Inequalities that do not include the equals sign are graphed in a similar way.

EXAMPLE 1
GRAPHING A LINEAR INEQUALITY

Graph the inequality $x - y > 5$.

This inequality does not include the equals sign. Therefore, the points on the line $x - y = 5$ do not belong to the graph. However the line still serves as a boundary for two regions, one of which satisfies the inequality. To graph the inequality, first graph the equation $x - y = 5$. Use a dashed line to show that the points on the line are *not* solutions of the inequality $x - y > 5$. Choose a test point to see which side of the line satisfies the inequality. Let us choose (1, −2) this time.

$$x - y > 5 \qquad \text{Original inequality}$$
$$1 - (-2) > 5 \quad ? \qquad \text{Let } x = 1 \text{ and } y = -2.$$
$$3 > 5 \qquad \text{False}$$

Since $3 > 5$ is false, the graph of the inequality is *not* the region that contains (1, −2). Shade the other region, as shown in Figure 6.24. This shaded region is the required graph. Check that the correct region is shaded by selecting a test point in the shaded region and substituting for x and y in the inequality $x - y > 5$. For example, use $(4, -3)$ from the shaded region, as follows.

$$x - y > 5$$
$$4 - (-3) > 5 \quad ?$$
$$7 > 5 \qquad \text{True}$$

This verifies that the correct region was shaded in Figure 6.24. ■

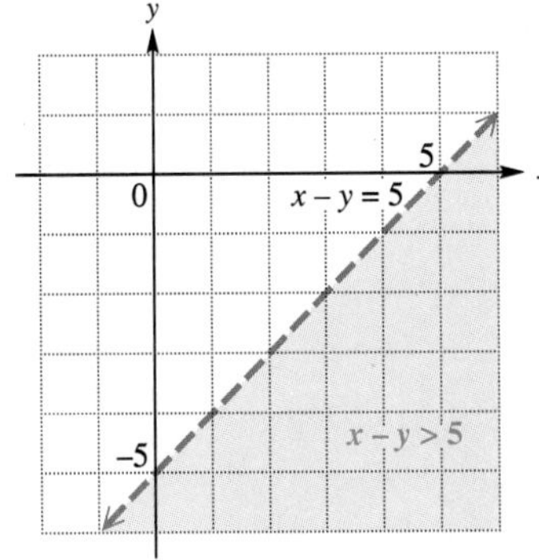

FIGURE 6.24

A summary of the steps used to graph a linear inequality in two variables follows.

GRAPHING A LINEAR INEQUALITY

Step 1 **Graph the boundary.** Graph the line that is the boundary of the region. Use the methods of Section 6.2. Draw a solid line if the inequality has $\leq$ or $\geq$; draw a dashed line if the inequality has $<$ or $>$.

Step 2 **Shade the appropriate side.** Use any point not on the line as a test point. Substitute for x and y in the *inequality*. If a true statement results, shade the side containing the test point. If a false statement results, shade the other side.

EXAMPLE 2
GRAPHING A LINEAR INEQUALITY

Graph the inequality $2x - 5y \geq 10$.

First graph the equation $2x - 5y = 10$. Use a solid line to show that the points on the line are solutions of the inequality $2x - 5y \geq 10$. Choose any test point not on the line. Again, we choose $(0, 0)$.

$$2x - 5y \geq 10 \qquad \text{Original inequality}$$
$$2(0) - 5(0) \geq 10 \quad ? \qquad \text{Let } x = 0 \text{ and } y = 0.$$
$$0 - 0 \geq 10 \quad ?$$
$$0 \geq 10 \qquad \text{False}$$

Since $0 \geq 10$ is false, shade the region *not* containing the test point $(0, 0)$. (See Figure 6.25.) ■

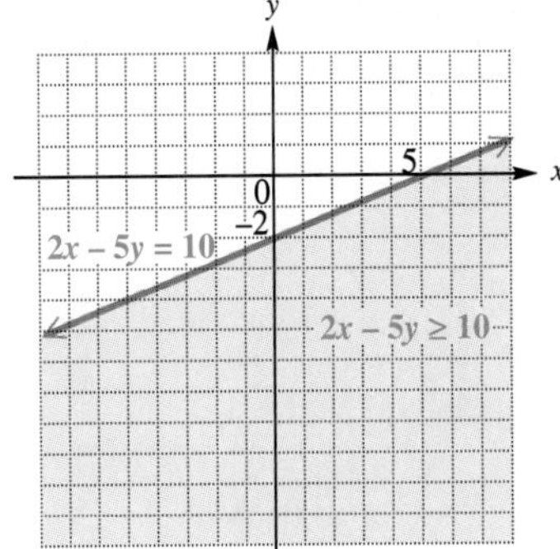

FIGURE 6.25

EXAMPLE 3
GRAPHING A LINEAR INEQUALITY

Graph the inequality $x \leq 3$.

First, graph $x = 3$, a vertical line going through the point (3, 0). Use a solid line. (Why?) Choose (0, 0) as a test point.

$x \leq 3$		Original inequality
$0 \leq 3$	?	Let $x = 0$.
$0 \leq 3$		True

Since $0 \leq 3$ is true, shade the region containing (0, 0), as in Figure 6.26. ■

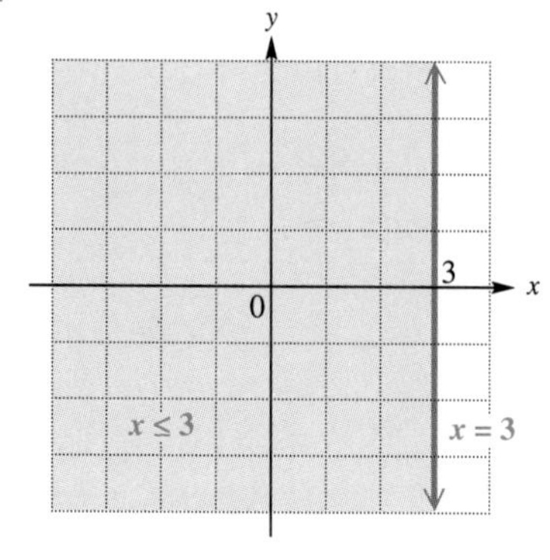

FIGURE 6.26

3 The next example shows how to graph an inequality having a boundary line going through the origin, an inequality in which (0, 0) cannot be used as a test point.

EXAMPLE 4
GRAPHING A LINEAR INEQUALITY

Graph the inequality $x \leq 2y$.

Begin by graphing $x = 2y$. Some ordered pairs that can be used to graph this line are (0, 0), (6, 3), and (4, 2). Use a solid line. The point (0, 0) cannot be used as a test point since (0, 0) is on the line $x = 2y$. Instead, choose a test point off the line $x = 2y$. For example, let us choose (1, 3), which is not on the line.

$x \leq 2y$		Original inequality
$1 \leq 2(3)$	?	Let $x = 1$ and $y = 3$.
$1 \leq 6$		True

Since $1 \leq 6$ is true, shade the side of the graph containing the test point (1, 3). (See Figure 6.27.) ■

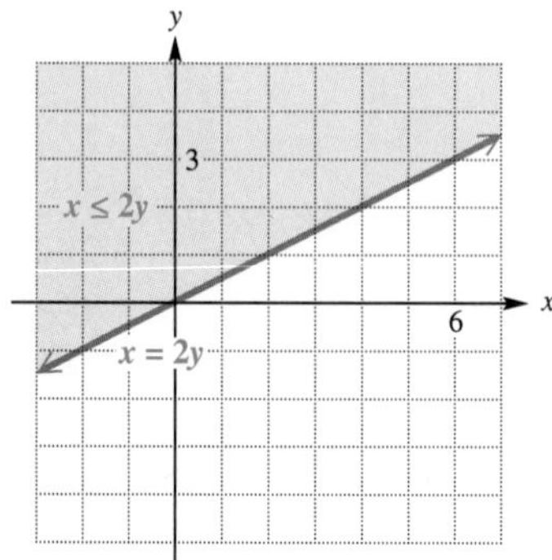

FIGURE 6.27

6.5 EXERCISES

In Exercises 1–12, the straight line for each inequality has been drawn. Complete each graph by shading the correct region. See Example 1.

1. $x + y \leq 4$

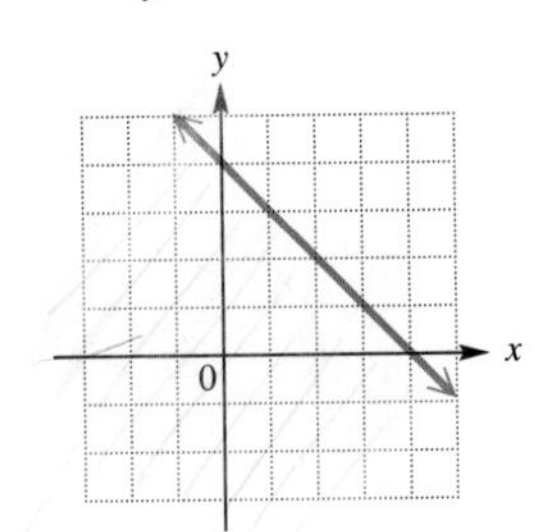

2. $x + y \geq 2$

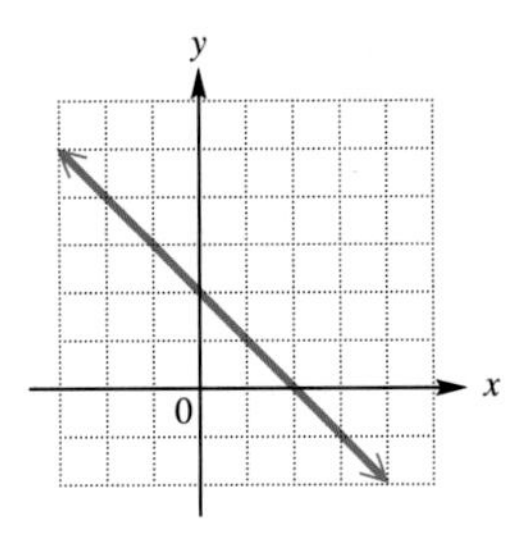

3. $x + 2y \leq 7$

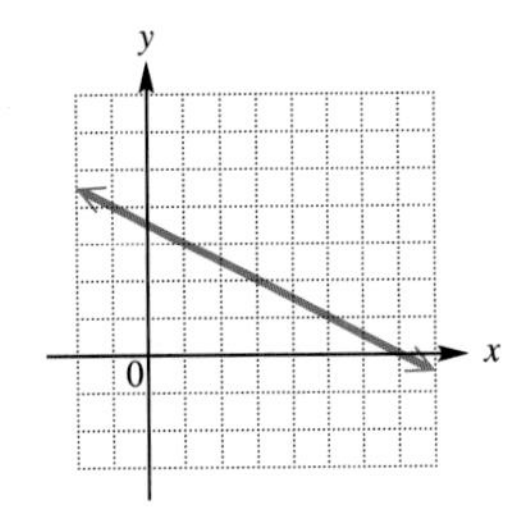

4. $2x + y \leq 5$

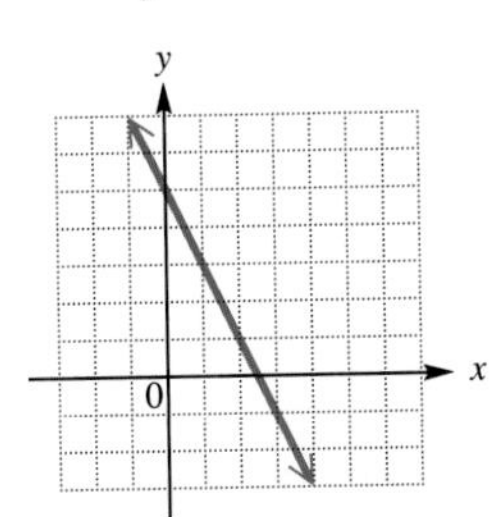

5. $-3x + 4y < 12$

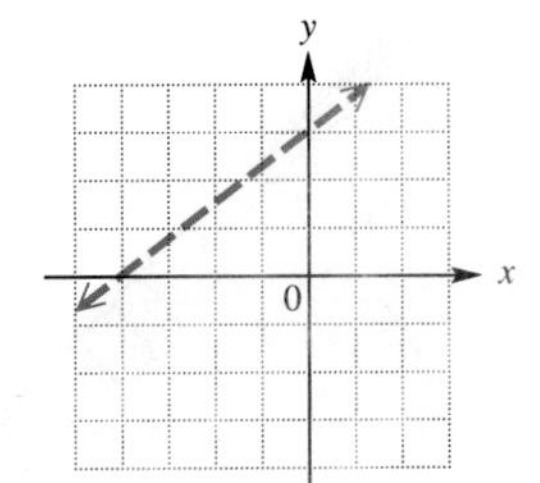

6. $4x - 5y > 20$

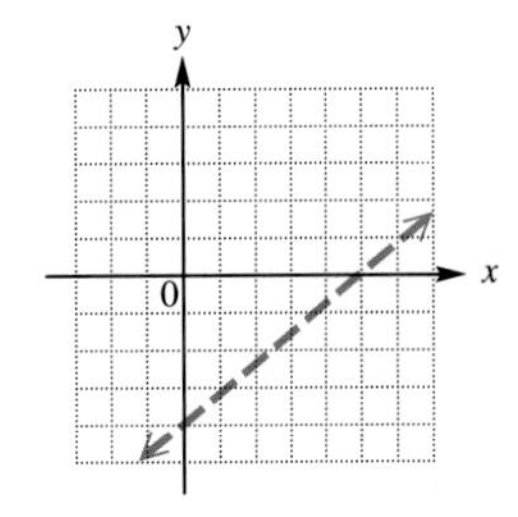

7. $5x + 3y > 15$

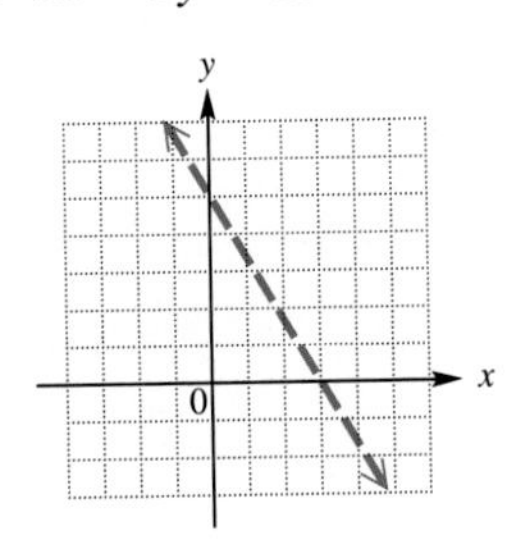

8. $6x - 5y < 30$

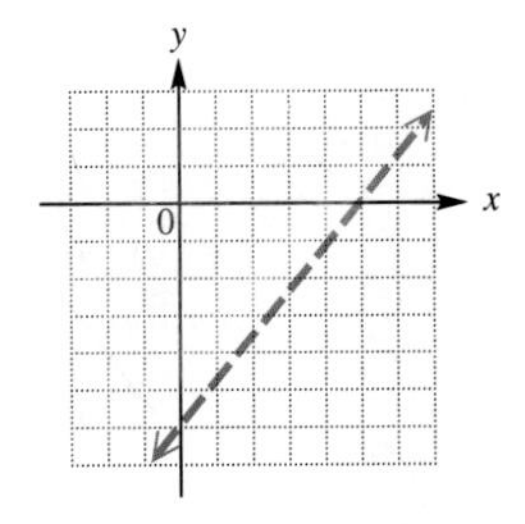

9. $x < 4$

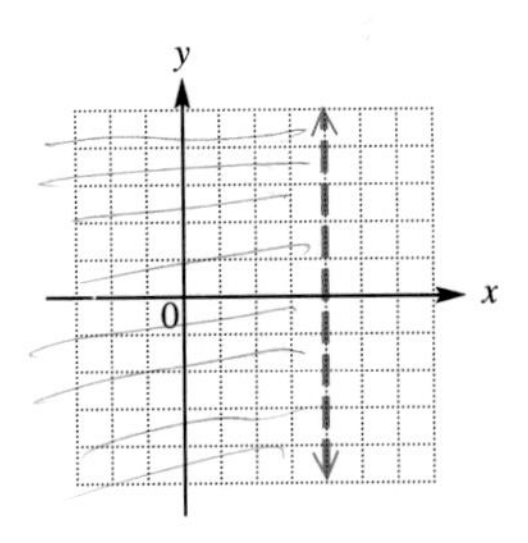

10. $y > -1$

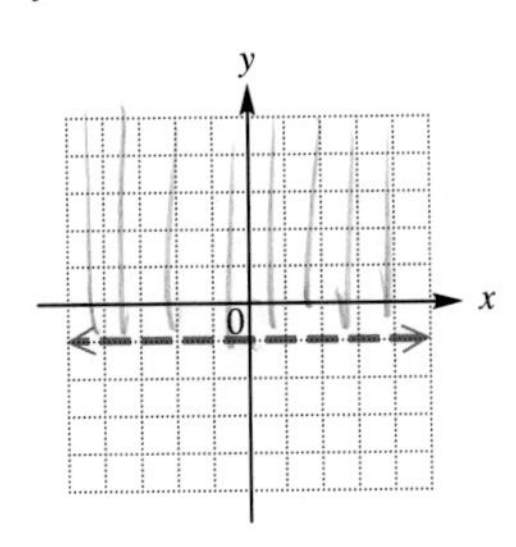

11. $x \leq 4y$

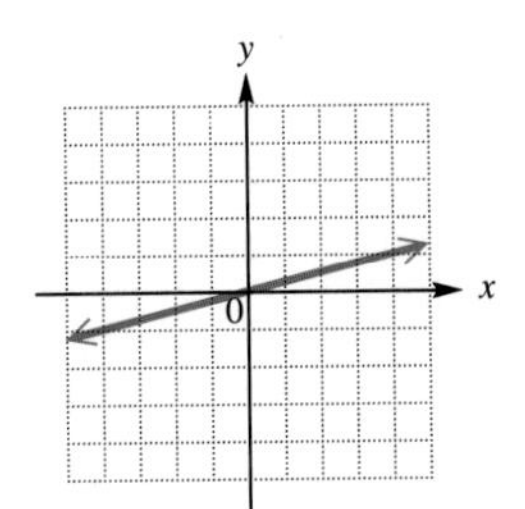

12. $-2x > y$

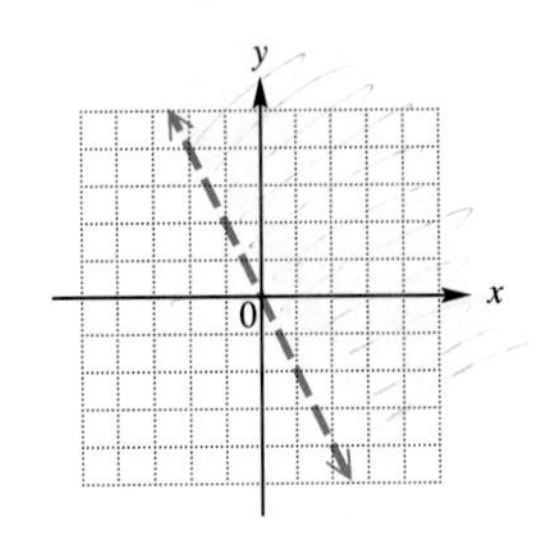

Graph each linear inequality. See Examples 1–4.

13. $x + y \leq 8$ **14.** $x + y \geq 4$ **15.** $x + 2y \geq 4$ **16.** $x + 3y \leq 6$

17. $2x + 3y > 6$ **18.** $3x + 4y > 12$ **19.** $3x + 7y \geq 21$ **20.** $2x + 5y \geq 10$

21. $x < 3$ **22.** $x < -2$ **23.** $y \leq 2$ **24.** $y \leq -3$

25. $x \geq -2$ **26.** $x \leq 3y$ **27.** $x \leq 5y$ **28.** $x \geq -2y$

29. $-4x \leq y$ **30.** $2x + 3y \geq 0$ **31.** $3x + 4y \leq 0$ **32.** $x + y < 0$

Match each of the following inequalities with its graph in (a)–(d) *below.*

33. $y < 2x - 1$ **34.** $x \geq 2$ **35.** $x + y \geq 3$ **36.** $y < x$

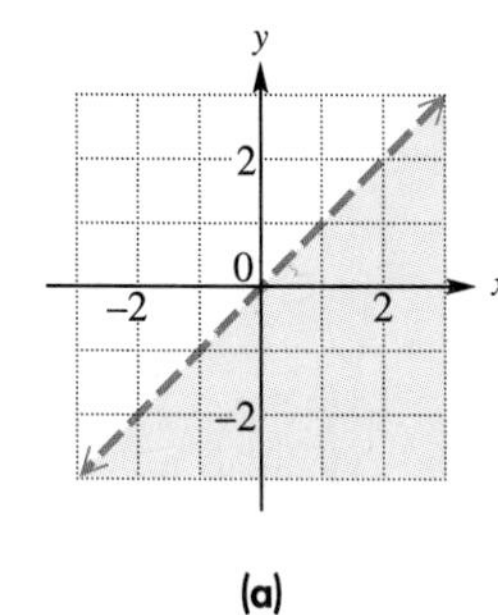

(a)

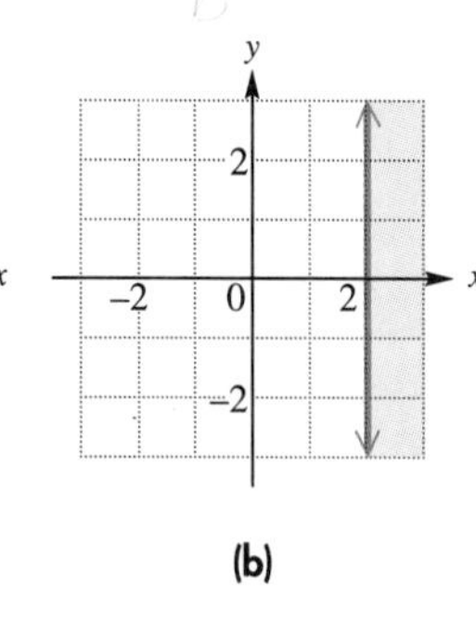

(b)

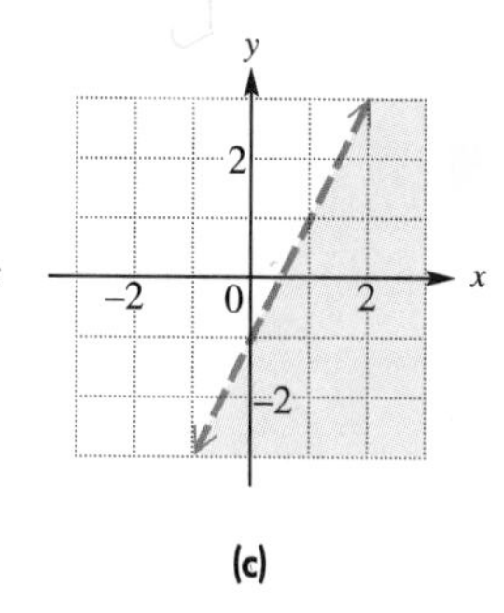

(c)

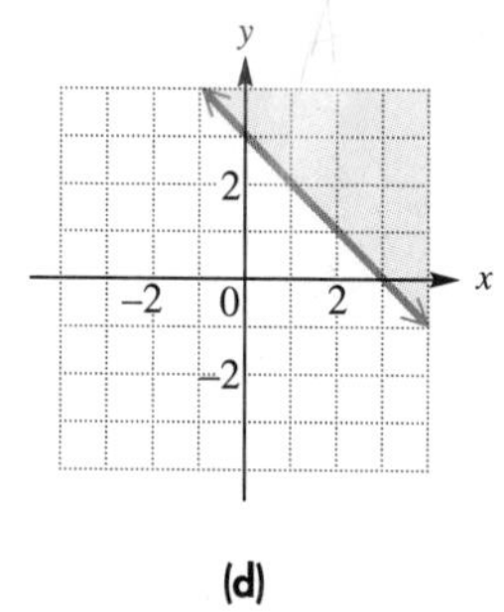

(d)

For each of the following problems: **(a)** *Graph the inequality. Here $x \geq 0$ and $y \geq 0$, so graph only the part of the inequality in quadrant I.* **(b)** *Give some ordered pairs that satisfy the inequality.*

37. The Sweet Tooth Candy Company uses x pounds of chocolate for chocolate cookies and y pounds of chocolate for fudge. The company has 200 pounds of chocolate available, so that

$$x + y \leq 200.$$

38. A company will ship x units of merchandise to outlet I and y units of merchandise to outlet II. The company must ship a total of at least 500 units to these two outlets. This can be expressed by writing

$$x + y \geq 500.$$

39. A toy manufacturer makes stuffed bears and geese. It takes 20 minutes to sew a bear and 30 minutes to sew a goose. There is a total of 480 minutes of sewing time available to make x bears and y geese. These restrictions lead to the inequality

$$20x + 30y \leq 480.$$

40. Ms. Branson takes x vitamin C tablets each day at a cost of 10¢ each and y multivitamins each day at a cost of 15¢ each. She wants the total cost to be no more than 50¢ a day. This can be expressed by writing

$$10x + 15y \leq 50.$$

41. Compare the linear inequalities $2x + 3 < 6$ and $2x + 3y < 6$. How do they differ? How do their graphs differ? How are they similar?

42. Describe the graph of the inequality $y > 2x$. (See Objective 2.)

PREVIEW EXERCISES

Find the value of $3x^2 - 2x + 5$ *for each value of* x. *See Section 3.1.*

43. 2 **44.** -4 **45.** 3 **46.** 0

Find the value of $2x^4 - x^2$ *for each value of* x.

47. 3 **48.** -2 **49.** -1 **50.** -3

6.6 FUNCTIONS

OBJECTIVES

1 UNDERSTAND THE DEFINITION OF A RELATION.
2 UNDERSTAND THE DEFINITION OF A FUNCTION.
3 DECIDE WHETHER AN EQUATION DEFINES A FUNCTION.
4 FIND DOMAINS AND RANGES.
5 USE $f(x)$ NOTATION.

In Section 6.1, the equation $y = 25x + 250$ was used to find the cost y in cents to produce x calculators. Choosing values for x and using the equation to find the corresponding values of y led to a set of ordered pairs (x, y). In each ordered pair, y (the cost) was *related* to x (the number of calculators produced) by the equation $y = 25x + 250$.

1 In an ordered pair (x, y), x and y are called the **components** of the ordered pair. Any set of ordered pairs is called a **relation.** The set of all first components in the ordered pairs of a relation is the **domain** of the relation, and the set of all second components in the ordered pairs is the **range** of the relation. Recall from Chapter 1 that sets are written with set braces, { }.

EXAMPLE 1 DEFINING A RELATION

(a) The relation {(0, 1), (2, 5), (3, 8), (4, 2)} has domain {0, 2, 3, 4} and range {1, 2, 5, 8}. The correspondence between the elements of the domain and the elements of the range is shown in Figure 6.28.

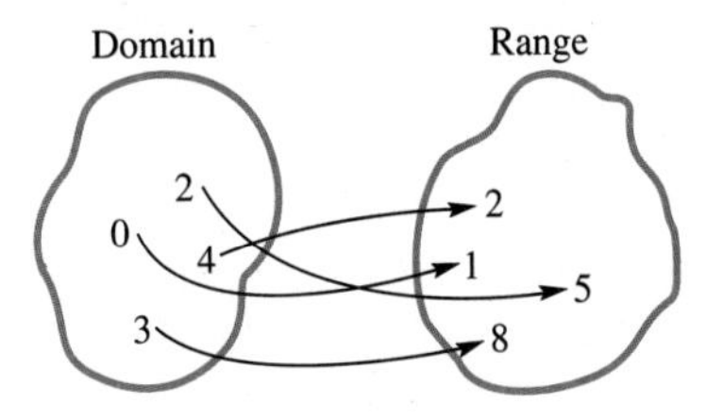

FIGURE 6.28

(b) Figure 6.29 shows a relation where the domain elements represent four mothers and the range elements represent eight children. The figure shows the correspondence between each mother and her children. The relation also could be written as the set of ordered pairs {(A, 8), (B, 1), (B, 2), (C, 3), (C, 4), (C, 7), (D, 5), (D, 6)}. ■

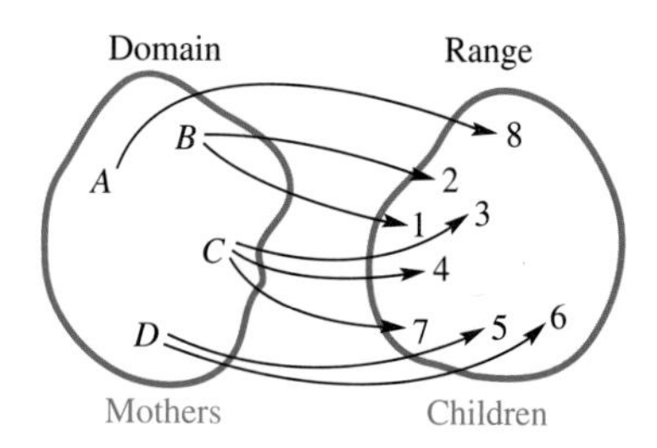

FIGURE 6.29

2 A special type of relation, called a *function*, is particularly useful in applications.

FUNCTION

A **function** is a set of ordered pairs in which each first component corresponds to exactly one second component.

By this definition, the relation in Example 1(a) is a function. However, the relation in Example 1(b) is *not* a function, because at least one first component (mother) corresponds to more than one second component (child). Notice that if the ordered pairs in Example 1(b) are reversed, with the child as the first component and the mother as the second component, the result *is* a function.

The simple relations given here were defined by listing the ordered pairs or by showing the correspondence with a figure. Most useful functions have an infinite number of ordered pairs and must be defined with an equation that tells how to get the second component given the first component. It is customary to use an equation with x and y as the variables, where x represents the first component and y the second component in the ordered pairs.

EXAMPLE 2
DEFINING A FUNCTION

Some everyday examples of functions are given here.

(a) The **cost** y in dollars charged by an express mail company is a function of the **weight in pounds** x determined by the equation $y = 1.5(x - 1) + 9$.

(b) In one state, the sales tax is 6% of the price of an item. The **tax** y on a particular item is a function of the **price** x, so that

$$y = .06x. \quad ■$$

3 Given a graph of an equation, the definition of a function can be used to decide whether the graph represents a function or not. By the definition of a function, each value of x must lead to exactly one value of y. In Figure 6.30, the indicated value of x leads to two values of y, so this graph is not the graph of a function. A vertical line can be drawn that cuts this graph in more than one point.

On the other hand, in Figure 6.31 any vertical line will cut the graph in no more than one point. Because of this, the graph in Figure 6.31 is the graph of a function.

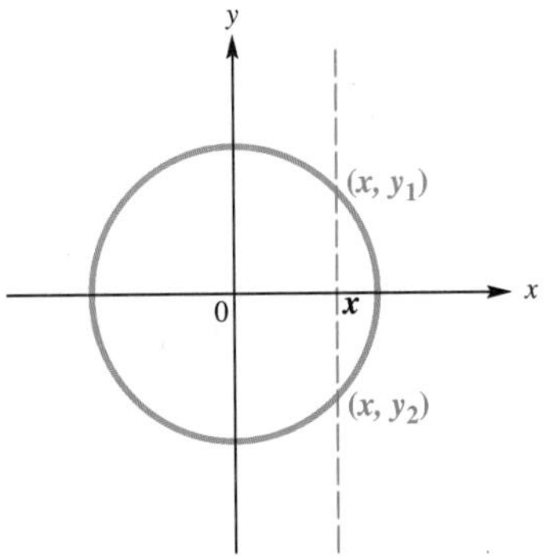

FIGURE 6.30

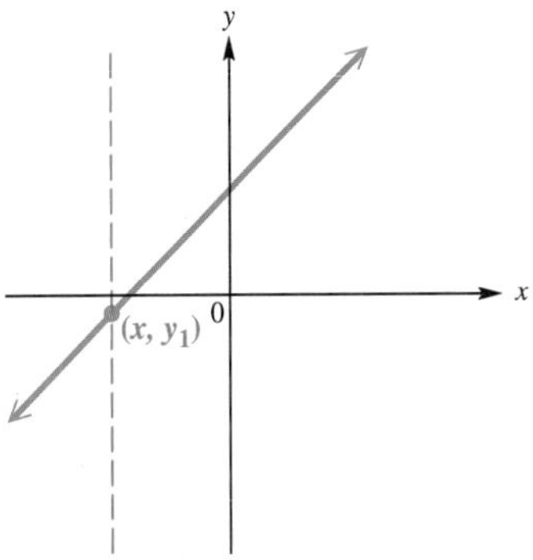

FIGURE 6.31

This idea gives the **vertical line test** for a function.

VERTICAL LINE TEST If a vertical line cuts a graph in more than one point, the graph is not the graph of a function.

As Figure 6.31 suggests, any nonvertical line is the graph of a function. For this reason, any linear equation of the form $Ax + By = C$, where $B \neq 0$, defines a function.

EXAMPLE 3

DECIDING WHETHER A RELATION DEFINES A FUNCTION

Decide whether or not the following relations are functions.

(a) $y = 2x - 9$

This linear equation can be written as $2x - 1y = 9$. Since the graph of this equation is a line that is not vertical, the equation defines a function.

(b)

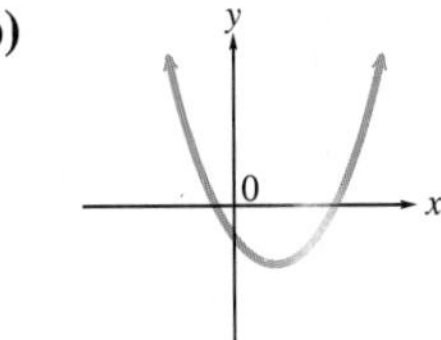

FIGURE 6.32

Use the vertical line test. Any vertical line will cross the graph in Figure 6.32 just once, so this is the graph of a function.

(c)

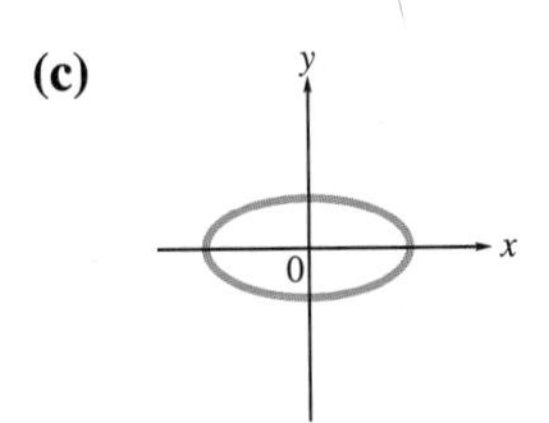

FIGURE 6.33

The vertical line test shows that the graph in Figure 6.33 is not the graph of a function; a vertical line can cross the graph twice.

(d) $x = 4$

The graph of $x = 4$ is a vertical line, so the equation does not define a function.

(e) $x = y^2$

We need to decide whether any x-value corresponds to more than one y-value. Suppose $x = 36$. Then

$$x = y^2 \qquad \text{becomes} \qquad y^2 = 36.$$

The equation $y^2 = 36$ has *two* solutions: $y = 6$ and $y = -6$. Because the *one* x-value, 36, leads to *two* y-values, the equation $x = y^2$ does not define a function.

(f) $2x + y < 4$

Since this is an inequality, any choice of a value for x will lead to an *infinite* number of y-values. For example, if $x = 3$,

$$\begin{aligned} 2x + y &< 4 \\ 2(3) + y &< 4 \qquad \text{Let } x = 3. \\ 6 + y &< 4 \\ y &< -2. \end{aligned}$$

If $x = 3$, y can be *any* value less than -2. Since $x = 3$ leads to more than one y-value, the inequality $2x + y < 4$ does not define a function. ■

Some generalizations are suggested by Example 3. Inequalities never define functions, since each x-value leads to an infinite number of y-values. An equation in which y is squared cannot define a function because most x-values will lead to two y-values. This is true for any *even* power of y, such as y^4, y^6, and so on. Similarly, an equation involving $|y|$ does not define a function, because an x-value may lead to more than one y-value.

4 By the definitions of domain and range given for relations, the set of all numbers that can be used as replacements for x in a function is the domain of the function, and the set of all possible values of y is the range of the function.

EXAMPLE 4

FINDING THE DOMAIN AND RANGE OF A FUNCTION

Find the domain and range for the following functions.

(a) $y = 6x - 9$

Any number at all may be used for x, so the domain is the set of all real numbers. Also, any number may be used for y, so the range is also the set of all real numbers.

(b) $y = x^2$

Any number can be squared, so the domain is the set of all real numbers. However, since the square of a real number cannot be negative and since $y = x^2$, the values of y cannot be negative, making the range the set of all nonnegative numbers, written as $y \geq 0$. ■

5 It is common to use the letters f, g, and h to name functions. For example, the function $y = 3x + 5$ is often written

$$f(x) = 3x + 5,$$

where $f(x)$ is read "f of x." The notation $f(x)$ is another way of writing y in a function. For the function $f(x) = 3x + 5$, if $x = 7$ then

$$\begin{aligned} f(7) &= 3 \cdot 7 + 5 && \text{Let } x = 7. \\ &= 21 + 5 = 26. \end{aligned}$$

Read this result, $f(7) = 26$, as "f of 7 equals 26." The notation $f(7)$ means the value of y when x is 7. The statement $f(7) = 26$ says that the value of y is 26 when x is 7.

To find $f(-3)$, replace x with -3.

$$\begin{aligned} f(x) &= 3x + 5 \\ f(-3) &= 3(-3) + 5 && \text{Let } x = -3. \\ f(-3) &= -9 + 5 = -4 \end{aligned}$$

CAUTION The symbol $f(x)$ does *not* mean f times x. It represents the y-value that corresponds to x.

f(x) NOTATION

In the notation $f(x)$,
- f is the name of the function,
- x is the domain value,
- $f(x)$ is the range value y for the domain value x.

EXAMPLE 5 USING FUNCTION NOTATION

For the function with $f(x) = x^2 - 3$, find the following.

(a) $f(2)$

Replace x with 2.

$$\begin{aligned} f(x) &= x^2 - 3 \\ f(2) &= 2^2 - 3 && \text{Let } x = 2. \\ f(2) &= 4 - 3 = 1 \end{aligned}$$

(b) $f(0) = 0^2 - 3 = 0 - 3 = -3$

(c) $f(-3) = (-3)^2 - 3 = 9 - 3 = 6$ ■

EXAMPLE 6 USING FUNCTION NOTATION

Let $P(x) = 5x^2 - 4x + 3$. Find the following. (This function is named P instead of f or g to show that it is a polynomial.)

(a) $P(0)$

Replace x with 0.

$$\begin{aligned} P(x) &= 5x^2 - 4x + 3 \\ P(0) &= 5 \cdot 0^2 - 4 \cdot 0 + 3 \\ P(0) &= 3 \end{aligned}$$

(b) $P(-2) = 5 \cdot (-2)^2 - 4 \cdot (-2) + 3 = 20 + 8 + 3 = 31$

(c) $P(3) = 5 \cdot 3^2 - 4 \cdot 3 + 3 = 36$ ■

6.6 EXERCISES

Decide which of the following relations are functions. Give the domain and range in Exercises 1–6. See Examples 1 and 2.

1. $\{(-4, 3), (-2, 1), (0, 5), (-2, -4)\}$

2. $\{(3, 7), (1, 4), (0, -2), (-1, -1), (-2, 4)\}$

3. $\{(-2, 3), (-1, 2), (0, 0), (1, 2), (2, 3)\}$

4. $\{(1, 5), (5, 7), (5, 9), (7, 11)\}$

5.

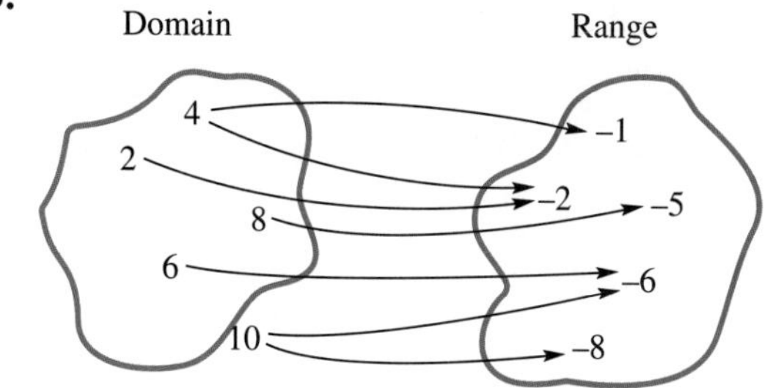

6.

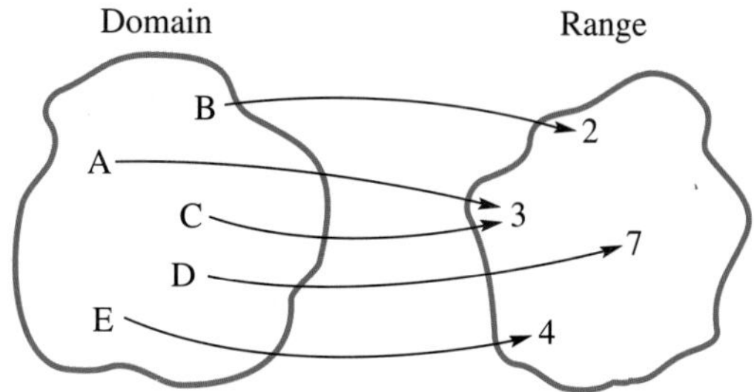

7.

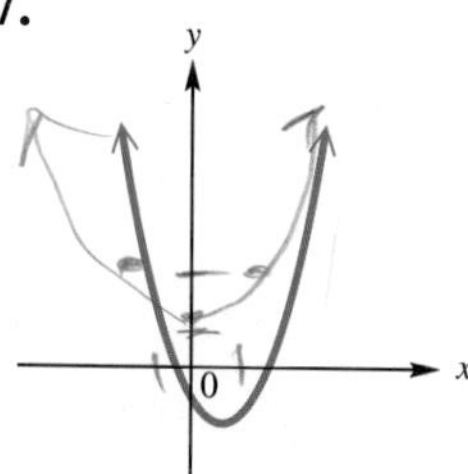

8.

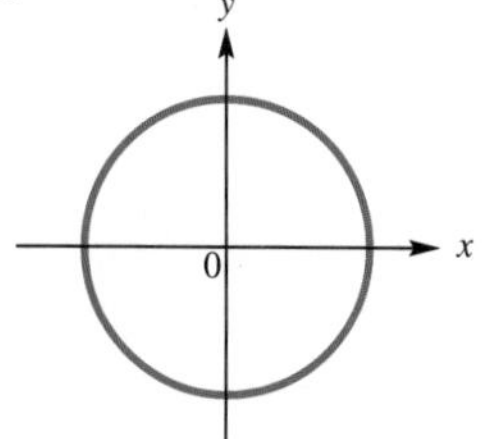

9.

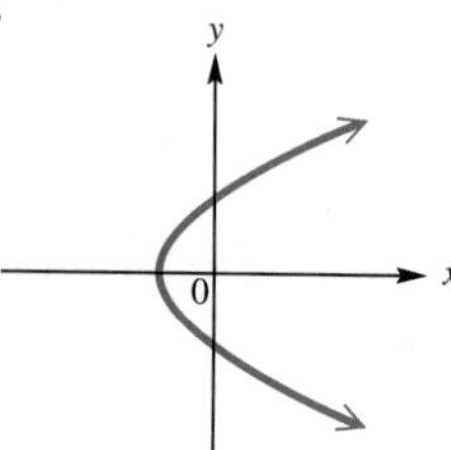

10.

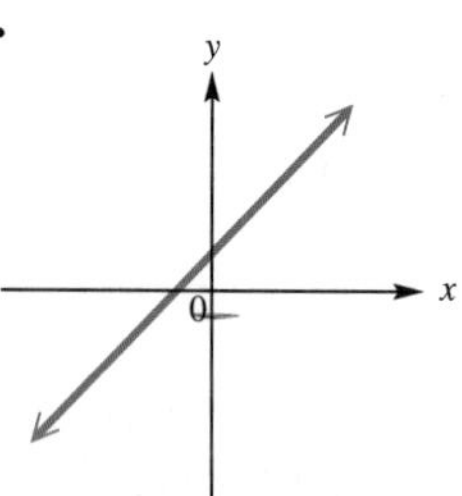

11.

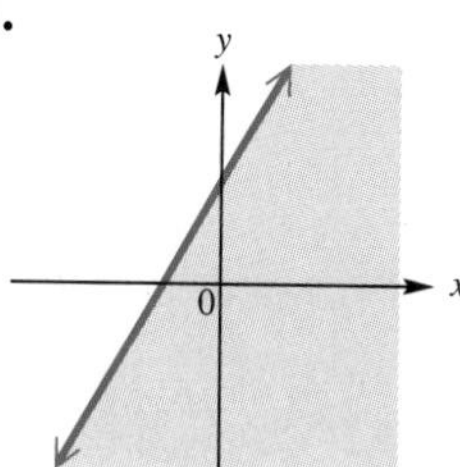

12.

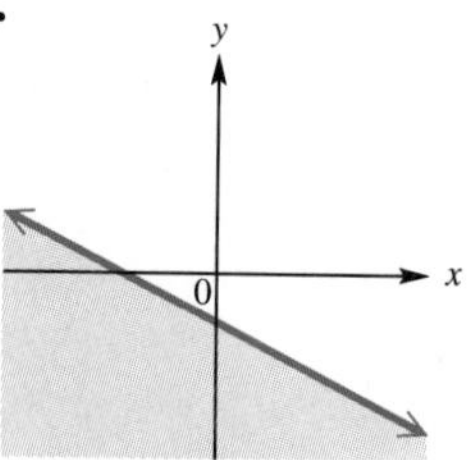

Decide which of the following are functions. See Example 3.

13. $y = 5x - 1$

14. $y = 4x + 5$

15. $2x + 3y = 6$

16. $4x - 3y = 12$

17. $y = x^2 + 3$

18. $y = 5 - x^2$

19. $x = y^2 - 4$

20. $y = x^2 + 6$

21. $2x + y < 6$

22. $3x - 4y > 2$

23. $y = \dfrac{1}{x}$

24. $y = \dfrac{1}{x + 2}$

Find the domain and range of the following functions. See Example 4.

25. $y = 2x + 5$

26. $y = 5x - 6$

27. $2x - y = 6$

28. $2x + 3y = 12$

29. $y = x^2 - 3$

30. $y = x^2 + 4$

31. $y = -x^2 + 6$

32. $y = -x^2 - 8$

For the following, find **(a)** $f(2)$, **(b)** $f(0)$, *and* **(c)** $f(-3)$. *See Example 5.*

33. $f(x) = 4x - 1$
34. $f(x) = -5 - 6x$
35. $f(x) = x^2 + 2$
36. $f(x) = x^2 - 5$
37. $f(x) = (x - 3)^2$
38. $f(x) = -(x - 4)^2$
39. $f(x) = -|x + 2|$
40. $f(x) = -|x - 3|$

For the following, find **(a)** $P(0)$, **(b)** $P(-3)$, *and* **(c)** $P(2)$. *See Example 6.*

41. $P(x) = x^2 + 2x$
42. $P(x) = 2x^2 + 3x - 6$
43. $P(x) = -x^2 - 8x + 9$
44. $P(x) = -3x^2 + 4x - 2$
45. $P(x) = x^3 - 4x^2 + 1$
46. $P(x) = x^3 + 5x^2 - 11x + 2$

Recall that $|x|$ *represents the absolute value of* x. *For example,* $|3| = 3$, $|-5| = 5$, *and so on. Decide which of the following are functions.*

47. $y = |x|$
48. $x = |y|$
49. $x = |y| - 3$
50. $y = |x| + 8$

---

51. Many scientists believe that rising carbon dioxide (CO_2) levels will lead to higher global temperatures and significant climate change. One study showed that the average CO_2 concentration (in appropriate units) at Mauna Loa, Hawaii increased according to the linear function $f(x) = 1.7x + 200$, where x represents the number of years since 1900. For example, $x = 10$ represents 1910, $x = 50$ represents 1950, and so on.

(a) Find $f(60)$. **(b)** Find $f(90)$.

(c) Estimate the CO_2 concentration, that is, $f(x)$, in the year 2000.

52. Compare the definitions of a relation and a function. How are they alike? How are they different? (See Objectives 1 and 2.)

53. In your own words, explain the meaning of domain of a function. (See Objective 4.)

PREVIEW EXERCISES

Add the polynomials. See Section 3.1.

54. $\begin{array}{r} x - 2y \\ \underline{3x + 2y} \end{array}$

55. $\begin{array}{r} 5x - 7y \\ \underline{12x + 7y} \end{array}$

56. $\begin{array}{r} 9a - 5b \\ \underline{-9a + 7b} \end{array}$

57. $\begin{array}{r} -11p + 8q \\ \underline{11p - 9q} \end{array}$

CHAPTER 6 GLOSSARY

KEY TERMS

6.1 linear equation An equation that can be written in the form $Ax + By = C$ is a linear equation in two variables.

ordered pair A pair of numbers written between parentheses in which the order is important is called an ordered pair.

table of values A table showing ordered pairs of numbers is called a table of values.

coordinate system An x-axis and y-axis at right angles form a coordinate system.

quadrants A coordinate system divides the plane into four regions called quadrants.

origin The point at which the x-axis and y-axis intersect is called the origin.

coordinates The numbers in an ordered pair are called the coordinates of the corresponding point.

plot To plot an ordered pair is to find the corresponding point on a coordinate system.

6.2 **graph** The graph of an equation is the set of points that correspond to all ordered pairs that satisfy the equation.

y-intercept If a graph crosses the y-axis at k, then the y-intercept is $(0, k)$.

x-intercept If a graph crosses the x-axis at k, then the x-intercept is $(k, 0)$.

6.3 **slope** The slope of a line is the ratio of the change in y compared to the change in x when moving along the line.

parallel lines Two lines in a plane that never intersect are parallel.

perpendicular lines Perpendicular lines intersect at a 90° angle.

6.5 **linear inequality** An inequality that can be written in the form $Ax + By < C$ or $Ax + By > C$ is a linear inequality in two variables.

6.6 **components** In an ordered pair (x, y), x and y are called the components of the ordered pair.

relation A relation is a set of ordered pairs.

domain The domain of a relation is the set of all first components of the ordered pairs.

range The range of a relation is the set of all second components of the ordered pairs.

function A function is a set of ordered pairs in which each first component corresponds to exactly one second component.

NEW SYMBOLS

(a, b) an ordered pair

(x_1, y_1) x-sub-one, y-sub-one

m slope

$f(x)$ function of x

CHAPTER 6 QUICK REVIEW

SECTION	CONCEPTS	EXAMPLES
6.1 LINEAR EQUATIONS IN TWO VARIABLES	An ordered pair is a solution of an equation if it satisfies the equation.	Is $(2, -5)$ or $(0, -6)$ a solution of $4x - 3y = 18$? $4(2) - 3(-5) = 23 \neq 18$ $(2, -5)$ is not a solution. $4(0) - 3(-6) = 18$ $(0, -6)$ is a solution.
	If a value of either variable in an equation is given, the other variable can be found by substitution.	Complete the ordered pair (0,) for $3x = y + 4$. $3(0) = y + 4$ $0 = y + 4$ $-4 = y$ The ordered pair is $(0, -4)$.
	Plot the ordered pair $(-2, 4)$ by starting at the origin, going 2 units to the left, then going 4 units up.	

SECTION	CONCEPTS	EXAMPLES
6.2 GRAPHING LINEAR EQUATIONS IN TWO VARIABLES	The graph of $y = k$ is a horizontal line through $(0, k)$.	y; $y = 2$; x; 0
	The graph of $x = k$ is a vertical line through $(k, 0)$.	y; $x = -3$; x; 0
	The graph of $Ax + By = 0$ goes through the origin. Find and plot another point that satisfies the equation. Then draw the line through the two points.	y; $2x + 3y = 0$; x; 0; $(3, -2)$
	To graph a linear equation: **1.** Find at least two ordered pairs that satisfy the equation. **2.** Plot the corresponding points. **3.** Draw a straight line through the points.	y; $x - 2y = 4$; $(4, 0)$; x; 0; $(0, -2)$
6.3 THE SLOPE OF A LINE	The slope of the line through (x_1, y_1) and (x_2, y_2) is $$m = \frac{y_2 - y_1}{x_2 - x_1}, \ (x_1 \neq x_2).$$	The line through $(-2, 3)$ and $(4, -5)$ has slope $$m = \frac{-5 - 3}{4 - (-2)} = \frac{-8}{6} = -\frac{4}{3}.$$
	Horizontal lines have slope 0.	The line $y = -2$ has slope 0.
	Vertical lines have undefined slope.	The line $x = 4$ has **undefined slope**.

SECTION	CONCEPTS	EXAMPLES
	To find the slope of a line from its equation, solve for y. The slope is the coefficient of x.	Find the slope of $3x - 4y = 12$. $-4y = -3x + 12$ $y = \frac{3}{4}x - 3$ Slope is $\frac{3}{4}$.
6.4 EQUATIONS OF A LINE	**Slope-Intercept Form** $y = mx + b$ m is the slope. $(0, b)$ is the y-intercept.	Find an equation of the line with slope 2 and y-intercept $(0, -5)$. The equation is $y = 2x - 5$.
	Point-Slope Form $y - y_1 = m(x - x_1)$ m is the slope. (x_1, y_1) is a point on the line.	Find an equation of the line with slope $-\frac{1}{2}$ through $(-4, 5)$. $y - 5 = -\frac{1}{2}(x - (-4))$ $2(y - 5) = -(x + 4)$ $2y - 10 = -x - 4$ $x + 2y = 6$
6.5 GRAPHING LINEAR INEQUALITIES IN TWO VARIABLES	**1.** Graph the line that is the boundary of the region. Make it solid if the inequality is $\leq$ or $\geq$; make it dashed if the inequality is $<$ or $>$. **2.** Use any point not on the line as a test point. Substitute for x and y in the inequality. If the result is true, shade the side of the line containing the test point; if the result is false, shade the other side.	Graph $2x + y \leq 5$. Graph the line $2x + y = 5$. Make it solid because of $\leq$. Use $(1, 0)$ as a test point. $2(1) + 0 \leq 5$? $2 \leq 5$ True Shade the side of the line containing $(1, 0)$. 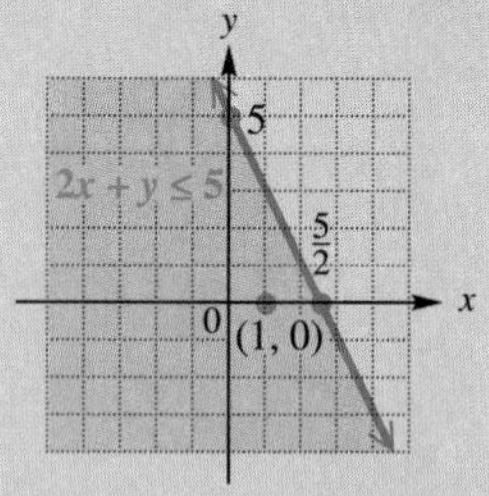
6.6 FUNCTIONS	**Vertical Line Test** If a vertical line cuts a graph in more than one point, the graph is not the graph of a function.	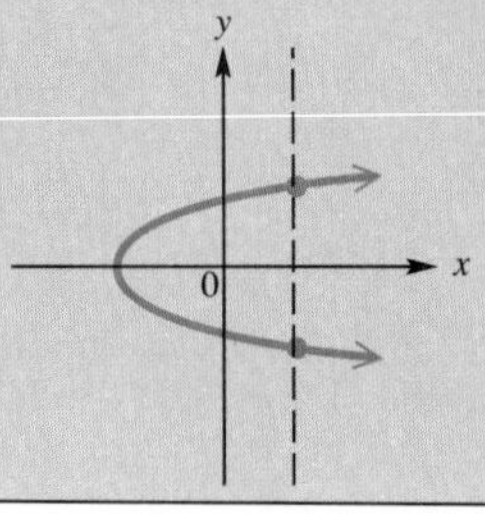 The graph shown is not the graph of a function.

CHAPTER 6 REVIEW EXERCISES

[6.1] *Decide whether the given ordered pair is a solution of the given equation.*

1. $x + y = 7$; $(3, 4)$
2. $2x + y = 5$; $(1, 4)$
3. $x + 3y = 9$; $(1, 3)$
4. $2x + 5y = 7$; $(1, 1)$
5. $3x - y = 4$; $(1, -1)$
6. $5x - 3y = 16$; $(2, -2)$

Complete the given ordered pairs or tables of values for each equation.

7. $y = 3x - 2$ $\quad(-1,\ \ )\ (0,\ \ )\ (\ \ , 5)$
8. $2y = 4x + 1$ $\quad(0,\ \ )\ (\ \ , 0)\ (\ \ , 2)$
9. $x + 4 = 0$

x			
y	-3	0	5

10. $y - 5 = 0$

x	-2	0	8
y			

[6.2] *Graph each linear equation.*

11. $y = 2x + 3$
12. $x + y = 5$
13. $2x - y = 5$
14. $x + 2y = 0$
15. $y + 3 = 0$
16. $x - y = 0$

Find the intercepts for the graph of each equation.

17. $y = 2x - 5$
18. $2x + y = 7$
19. $5x - 2y = 0$

[6.3] *Find the slope of each line.*

20. Through $(2, 3)$ and $(-1, 1)$
21. Through $(0, 0)$ and $(-1, -2)$
22. Through $(2, 5)$ and $(2, -2)$
23. $y = 3x - 1$
24. $y = 8$
25. $x = 2$
26. $2x - 5y = 3$

Decide whether the lines are parallel, perpendicular, or neither.

27. $3x + 2y = 5$; $6x + 4y = 12$
28. $x - 3y = 8$; $3x + y = 6$
29. $4x + 3y = 10$; $3x - 4y = 12$
30. $x - 2y = 3$; $x + 2y = 3$

[6.4] *Write an equation for each line in the form $Ax + By = C$.*

31. $m = 3$, y-intercept $(0, -2)$
32. $m = -1$, y-intercept $\left(0, \frac{3}{4}\right)$
33. $m = \frac{2}{3}$, y-intercept $(0, 5)$
34. Through $(5, -2)$, $m = 1$
35. Through $(-1, 4)$, $m = \frac{2}{3}$
36. Through $(1, -1)$, $m = -\frac{3}{4}$
37. Through $(2, 1)$ and $(-2, 2)$
38. Through $(-2, 6)$ and $(3, 6)$

[6.5] *Complete the graph of each linear inequality by shading the correct region.*

39. $x - y \leq 3$

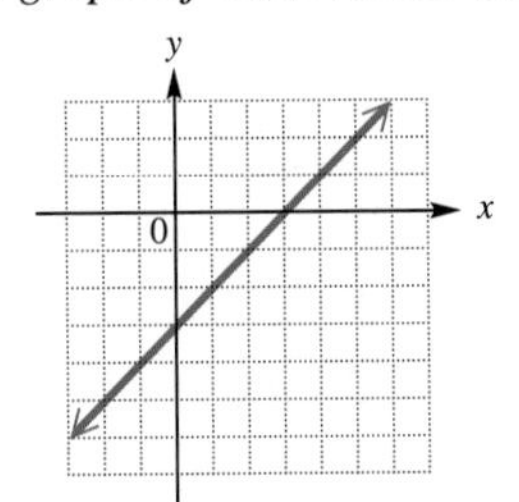

40. $3x - y \geq 5$

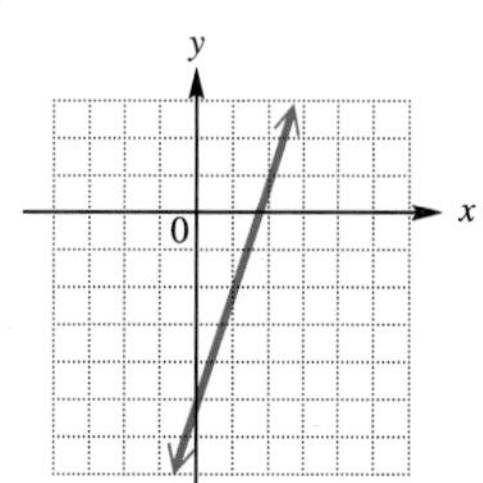

Graph each linear inequality.

41. $x + 2y \leq 6$

42. $3x + 5y \geq 9$

[6.6] *Decide which of the following are functions.*

43. $\{(-2, 4), (0, 8), (2, 5), (2, 3)\}$

44. $\{(8, 3), (7, 4), (6, 5), (5, 6), (4, 7)\}$

45.

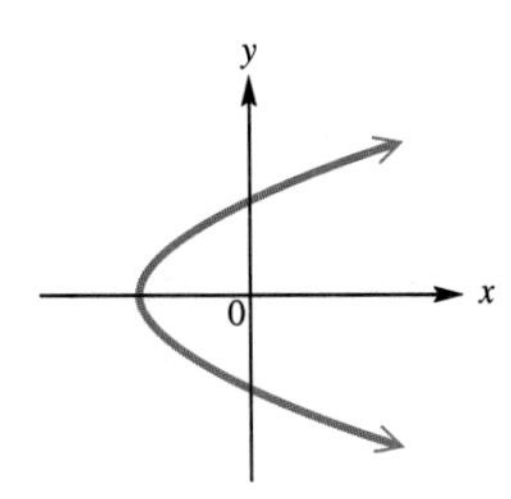

46.

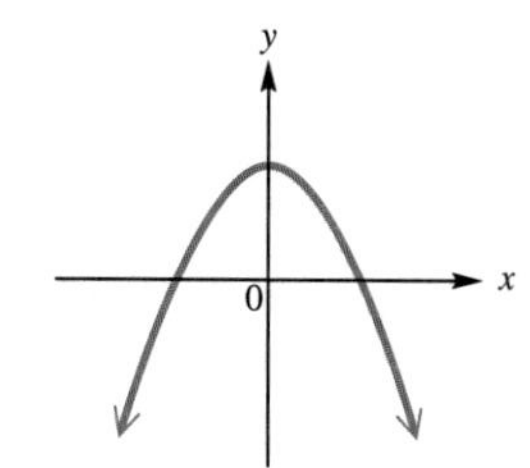

47. $2x + 3y = 12$

48. $y = x^2$

49. $x - 5y < 10$

Find the domain and range of the following functions.

50. $4x - 3y = 12$

51. $y = x^2 + 1$

52. $y = |x - 1|$

Find **(a)** $f(2)$ *and* **(b)** $f(-1)$.

53. $f(x) = 3x + 2$

54. $f(x) = 2x^2 - 1$

55. $f(x) = |x + 3|$

56. Let $P(x) = -x^2 - 4x + 2$. Find $P(-1)$ and $P(-3)$.

MIXED REVIEW EXERCISES

Find the slope and the intercepts of each line.

57. Through $(4, -1)$ and $(-2, -3)$

58. $y = \frac{2}{3}x + 5$

Which of the following are functions?

59.

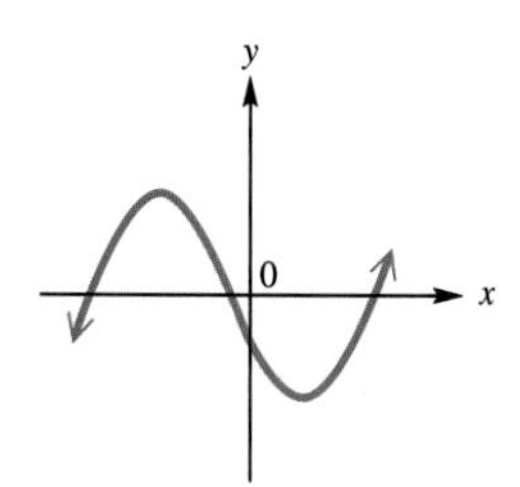

60. $y = |x|$

Complete the given ordered pairs for each equation.

61. $4x + 3y = 9$; $(0,\ )$, $(\ , 0)$, $(-2,\ )$

62. $x = 2y$; $(0,\ )$, $(8,\ )$, $(\ , -3)$

Write an equation for each line in the form $Ax + By = C$.

63. Through $(5, 0)$ and $(5, -1)$

64. $m = -\frac{1}{4}$, y-intercept $\left(0, -\frac{5}{4}\right)$

65. Through $(8, 6)$, $m = -3$

66. Through $(3, -5)$ and $(-4, -1)$

67. Find the domain and range of $y = 2x + 1$.

In Exercises 68–71, graph each relation.

68. $3y = x$ **69.** $y = 4x - 1$ **70.** $y \leq -2$ **71.** $x - 2y > 0$

---

72. Find **(a)** $f(2)$ and **(b)** $f(-1)$ for $f(x) = -(x + 1)^2$.

73. What kind of inequality has a dashed line as a boundary? Explain why.

74. Since two points determine a line, why do we usually find three ordered pairs to graph a line?

CHAPTER 6 TEST

Complete the ordered pairs using the given equations.

1. $2x + 7y = 21$; (0,) (, 0) (3,) (, 2)

2. $x = 3y$; (0,) (, 2) (8,) (−12,)

Graph each linear equation. Give the x- and y-intercepts.

3. $x + y = 4$ **4.** $3x - 4y = 18$ **5.** $2x + y = 0$ **6.** $x + 5 = 0$

Find the slope of each line.

7. Through $(-2, 4)$ and $(5, 1)$ **8.** $4x + 7y = 10$ **9.** $x - 5 = 0$

Write an equation for each line in the form $Ax + By = C$.

10. Through $(1, -3)$, $m = -4$

11. $m = -\frac{3}{4}$, y-intercept $(0, 2)$

12. Through $(-2, -6)$ and $(-1, 3)$

Graph each linear inequality.

13. $2x + y \leq 8$ **14.** $y < -3$

Decide which are functions.

15.

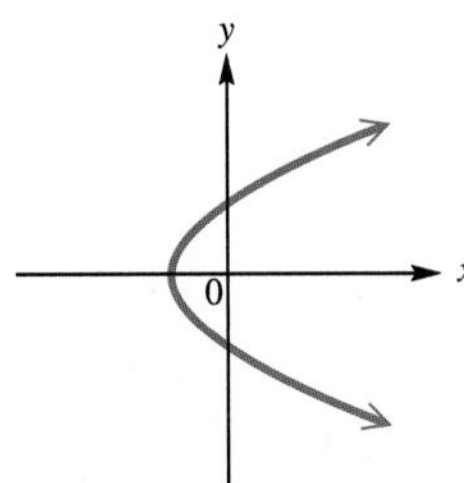

16. $y = 3x + 5$ **17.** $x = y^2 - 1$

18. Find the domain and range of $y = 2x^2 + 1$.

19. For $f(x) = 6x - 2$ find $f(-3)$ and $f(5)$.

20. Explain the reason for the names slope-intercept form and point-slope form of the equation of a line.

CHAPTER SEVEN

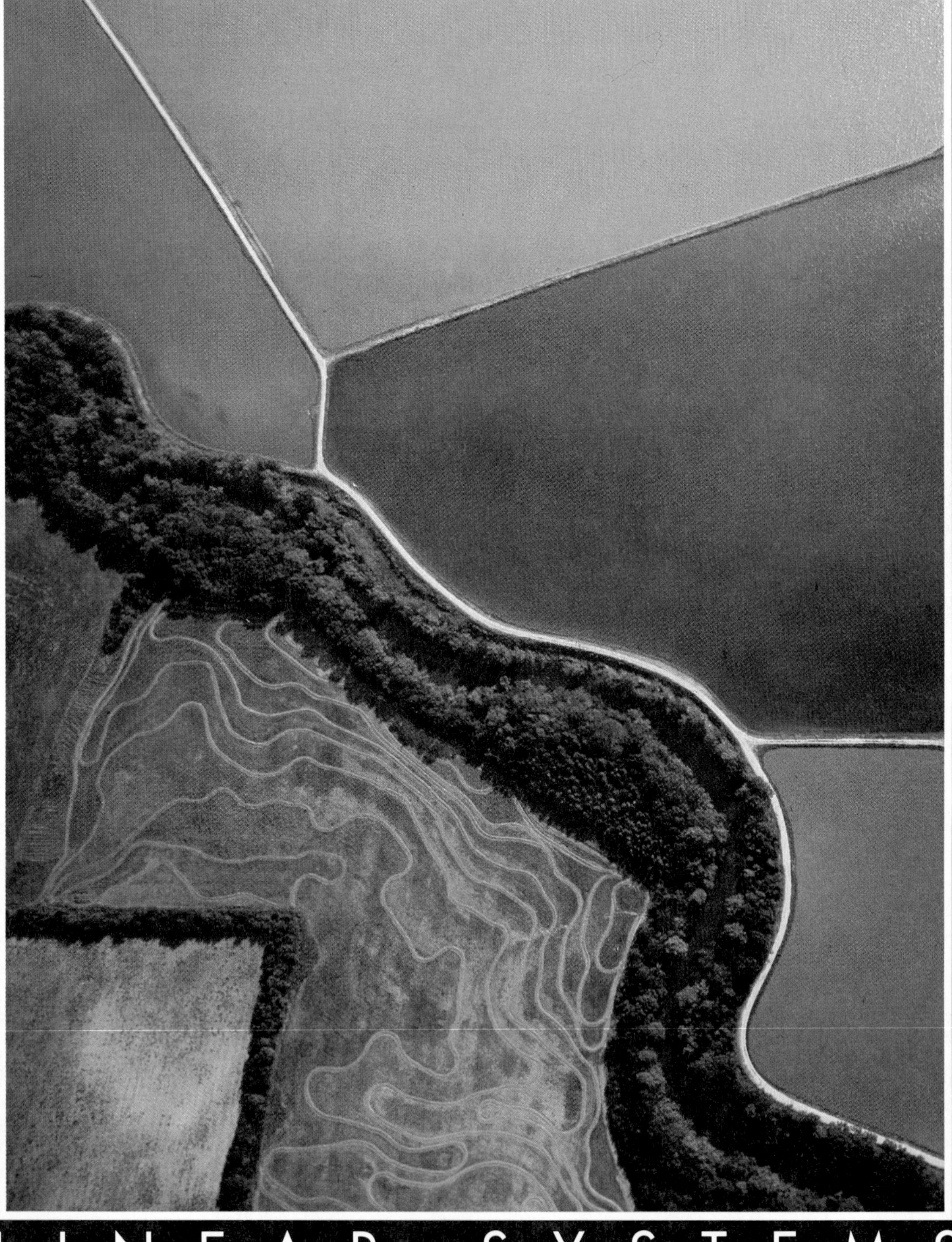

LINEAR SYSTEMS

When a number of equations in several variables are considered simultaneously, we have what is known as a system of equations. In this chapter we study linear systems of equations and inequalities, concentrating on systems in two variables.

In more advanced mathematics courses, systems of equations and inequalities in more than two variables are studied. Such systems have many useful applications. One such application occurred with the study of data sent back to earth from Mars from the U.S. Viking spacecrafts. On July 20, 1976, Viking I landed on Mars and was joined by Viking II several months later. Teams of scientists obtained banks of useful information by studying photos sent back to Earth via radio beams. For example, they found the exact location of the two spacecrafts, the size of Mars, the orientation of its axis, and its rate of spin. All of these measurements were obtained using systems of equations.

7.1 SOLVING SYSTEMS OF LINEAR EQUATIONS BY GRAPHING

OBJECTIVES

1. DECIDE WHETHER A GIVEN ORDERED PAIR IS A SOLUTION OF A SYSTEM.
2. SOLVE LINEAR SYSTEMS BY GRAPHING.
3. IDENTIFY SYSTEMS WITH NO SOLUTIONS OR WITH AN INFINITE NUMBER OF SOLUTIONS.
4. IDENTIFY INCONSISTENT SYSTEMS OR SYSTEMS WITH DEPENDENT EQUATIONS WITHOUT GRAPHING.

A **system of linear equations** consists of two or more linear equations with the same variables. Examples of systems of linear equations include

$$\begin{aligned} 2x + 3y &= 4 \\ 3x - y &= -5 \end{aligned} \qquad \begin{aligned} x + 3y &= 1 \\ -y &= 4 - 2x \end{aligned} \qquad \text{and} \qquad \begin{aligned} x - y &= 1 \\ y &= 3. \end{aligned}$$

In the system on the right, think of $y = 3$ as an equation in two variables by writing it as $0x + y = 3$.

1 Applications often require solving a system of equations. The **solution of a system** of linear equations includes all the ordered pairs that make both equations true at the same time.

EXAMPLE 1 DETERMINING WHETHER AN ORDERED PAIR IS A SOLUTION

Is $(4, -3)$ a solution of the following systems?

(a) $x + 4y = -8$
$3x + 2y = 6$

To decide whether $(4, -3)$ is a solution of the system, substitute 4 for x and -3 for y in each equation.

$x + 4y = -8$			$3x + 2y = 6$	
$4 + 4(-3) = -8$	?		$3(4) + 2(-3) = 6$	?
$4 + (-12) = -8$	?		$12 + (-6) = 6$	?
$-8 = -8$		True	$6 = 6$	True

Since $(4, -3)$ satisfies both equations, it is a solution.

(b) $2x + 5y = -7$
$3x + 4y = 2$
Again, substitute 4 for x and -3 for y in both equations.

$2x + 5y = -7$			$3x + 4y = 2$	
$2(4) + 5(-3) = -7$	?		$3(4) + 4(-3) = 2$	?
$8 + (-15) = -7$	?		$12 + (-12) = 2$	?
$-7 = -7$		True	$0 = 2$	False

Here $(4, -3)$ is not a solution since it does not satisfy the second equation. ■

Several methods of solving a system of two linear equations in two variables are discussed in this chapter.

2 One way to find the solution of a system of two linear equations is to graph both equations on the same axes. The graph of each line shows points whose coordinates satisfy the equation of that line. The coordinates of any point where the lines intersect give a solution of the system. Since two different straight lines can intersect at no more than one point, there can never be more than one solution for such a system.

EXAMPLE 2
SOLVING SYSTEMS BY GRAPHING

Solve each system of equations by graphing both equations on the same axes.

(a) $2x + 3y = 4$
$3x - y = -5$
As shown in Chapter 6, these two equations can be graphed by plotting several points for each line. Some ordered pairs that satisfy each equation are shown below.

$2x + 3y = 4$

x	y
0	$\frac{4}{3}$
2	0
-2	$\frac{8}{3}$

$3x - y = -5$

x	y
0	5
$-\frac{5}{3}$	0
-2	-1

The lines in Figure 7.1 suggest that the graphs intersect at the point $(-1, 2)$. Check this by substituting -1 for x and 2 for y in both equations. Since $(-1, 2)$ satisfies both equations, the solution of this system is $(-1, 2)$.

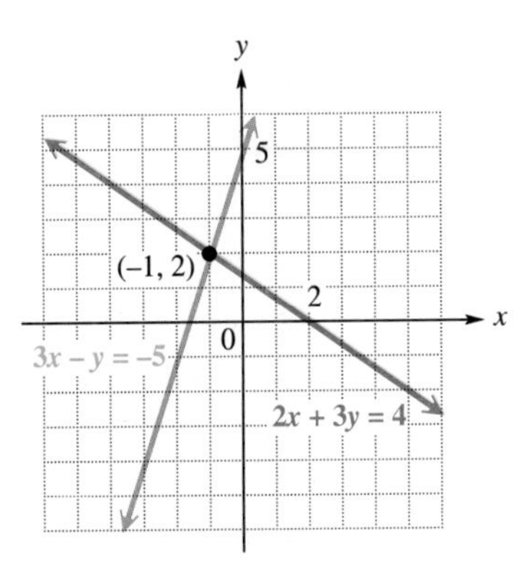

FIGURE 7.1

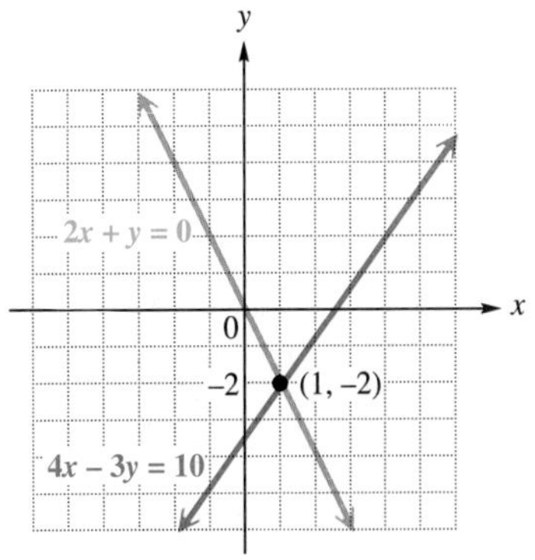

FIGURE 7.2

(b) $2x + y = 0$
$4x - 3y = 10$

Find the solution of the system by graphing the two lines on the same axes. As suggested by Figure 7.2, the solution is $(1, -2)$, the point at which the graphs of the two lines intersect. Check by substituting 1 for x and -2 for y in both equations of the system. ■

CAUTION A difficulty with the graphing method of solution is that it may not be possible to determine from the graph the exact coordinates of the point that represents the solution. For this reason, algebraic methods of solution are explained later in this chapter. The graphing method does, however, show geometrically how solutions are found.

3 Sometimes the graphs of the two equations in a system either do not intersect at all or are the same line, as in the systems of Example 3.

EXAMPLE 3
SOLVING SPECIAL SYSTEMS

Solve each system by graphing.

(a) $2x + y = 2$
$2x + y = 8$

The graphs of these lines are shown in Figure 7.3. The two lines are parallel and have no points in common. For a system whose equations lead to graphs with no points in common, we write "no solution."

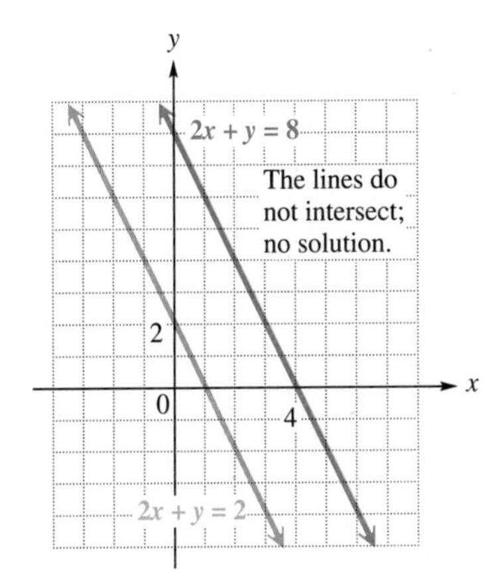

FIGURE 7.3

(b) $2x + 5y = 1$
$6x + 15y = 3$

The graphs of these two equations are the same line. See Figure 7.4. The second equation can be obtained by multiplying both sides of the first equation by 3. In this case, every point on the line is a solution of the system, and the solution is an infinite number of ordered pairs. We write ''infinite number of solutions'' to indicate this case. ■

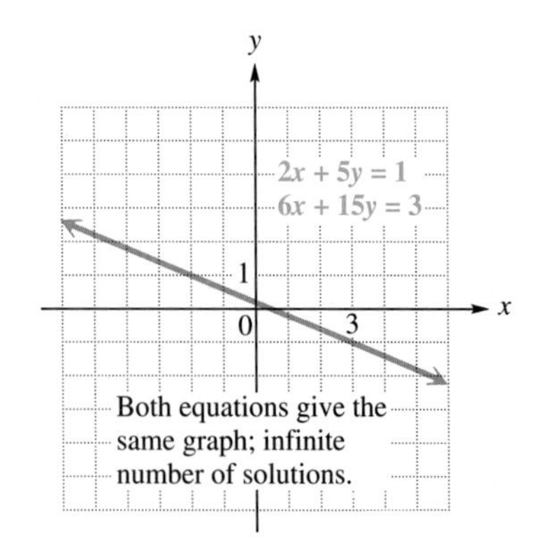

FIGURE 7.4

Each system in Example 2 has a solution. A system with a solution is called a **consistent system.** A system of equations with no solution, such as the one in Example 3(a), is called an **inconsistent system.** The equations in Example 2 are independent equations. **Independent equations** have different graphs. The equations of the system in Example 3(b) have the same graph. Because they are different forms of the same equation, these equations are called **dependent equations.** Examples 2 and 3 show the three cases that may occur in a system of equations with two unknowns.

POSSIBLE TYPES OF SOLUTIONS

1. The graphs cross at exactly one point, which gives the (single) solution of the system. The system is consistent and the equations are independent.

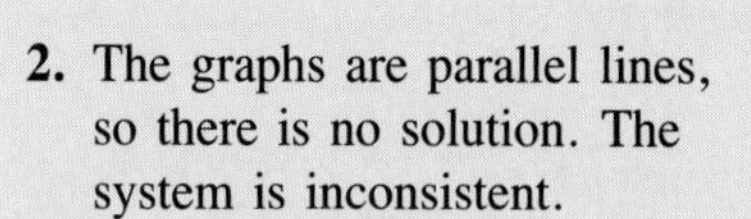

2. The graphs are parallel lines, so there is no solution. The system is inconsistent.

3. The graphs are the same line. There are an infinite number of solutions. The equations are dependent.

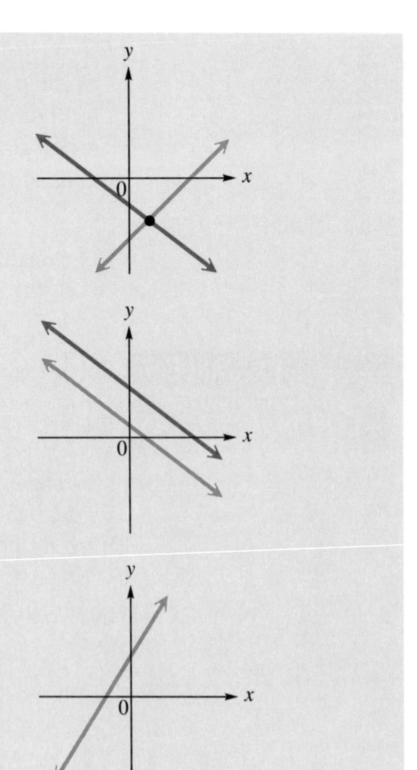

4 Example 3 showed that the graphs of an inconsistent system are parallel lines and the graphs of a system of dependent equations are the same line. These special kinds of systems can be recognized without graphing by using slopes.

EXAMPLE 4
IDENTIFYING THE THREE CASES USING SLOPES

Describe each system without graphing.

(a) $3x + 2y = 6$
$-2y = 3x - 5$

Write each equation in slope-intercept form by solving for y.

$$3x + 2y = 6 \qquad -2y = 3x - 5$$
$$2y = -3x + 6 \qquad y = -\frac{3}{2}x + \frac{5}{2}$$
$$y = -\frac{3}{2}x + 3$$

Both equations have a slope of $-3/2$ but they have different y-intercepts. The previous chapter showed that lines with the same slope are parallel, so these equations have graphs that are parallel lines. The system has no solution.

(b) $2x - y = 4$
$x = \frac{y}{2} + 2$

Again, write the equations in slope-intercept form.

$$2x - y = 4 \qquad x = \frac{y}{2} + 2$$
$$-y = -2x + 4 \qquad \frac{y}{2} + 2 = x$$
$$y = 2x - 4$$
$$\frac{y}{2} = x - 2$$
$$y = 2x - 4$$

The equations are exactly the same; their graphs are the same line. The system has an infinite number of solutions.

(c) $x - 3y = 5$
$2x + y = 8$

In slope-intercept form, the equations are as follows.

$$x - 3y = 5 \qquad 2x + y = 8$$
$$-3y = -x + 5 \qquad y = -2x + 8$$
$$y = \frac{1}{3}x - \frac{5}{3}$$

The graphs of these equations are neither parallel lines nor the same line since the slopes are different. There will be exactly one solution to this system. ■

7.1 EXERCISES

Decide whether the given ordered pair is the solution of the given system. See Example 1.

1. $(2, -5)$
$3x + y = 1$
$2x + 3y = -11$

2. $(-1, 6)$
$2x + y = 4$
$3x + 2y = 9$

3. $(4, -2)$
$x + y = 2$
$2x + 5y = 2$

4. $(-6, 3)$
$x + 2y = 0$
$3x + 5y = 3$

5. $(2, 0)$
$3x + 5y = 6$
$4x + 2y = 5$

6. $(0, -4)$
$2x - 5y = 20$
$3x + 6y = -20$

7. $(5, 2)$
$4x = 26 - 3y$
$3x = 29 - 7y$

8. $(9, 1)$
$2x = 23 - 5y$
$3x = 24 - 2y$

9. $(6, -8)$
$2y = -x - 10$
$3y = 2x + 30$

10. $(-5, 2)$
$5y = 3x + 20$
$-3y = 2x + 4$

11. $(-3, -6)$
$2x - y = 0$
$2y = x - 9$

12. $(-2, 5)$
$3x + 4y = 8$
$3y = -x + 13$

13. Explain why a system of two linear equations cannot have exactly two solutions. (See Objective 3.)

14. Explain the difficulties that may be encountered when solving a system of two linear equations using the graphing method. (See Objective 2.)

Solve each system of equations by graphing both equations on the same axes. See Examples 2 and 3. If the system is inconsistent or the equations are dependent, say so.

15. $x + y = 6$
$x - y = 2$

16. $x + y = -1$
$x - y = 3$

17. $x + y = 4$
$y - x = 4$

18. $y - x = -5$
$x + y = 1$

19. $x + 2y = 2$
$x - 2y = 6$

20. $4x + y = 2$
$2x - y = 4$

21. $x + 2y = 4$
$2x + 4y = 12$

22. $2x - y = 6$
$4x - 2y = 8$

23. $2x - y = 4$
$4x = 2y + 8$

24. $3x = 5 - y$
$6x + 2y = 10$

25. $3x - y = 3$
$-x + y = -3$

26. $-x + 3y = -4$
$x + 2y = 4$

27. $3x + 2y = 6$
$3x - 4y = 24$

28. $3x + 2y = -10$
$x - 2y = -6$

29. $4x + y = 5$
$3x - 2y = 12$

30. $3x - 4y = -24$
$5x + 2y = -14$

31. $y = x + 1$
$y = 3x - 1$

32. $y = x + 6$
$y = -x - 2$

33. $3x = y + 5$
$6x - 2y = 5$

34. $4y + 10 = x$
$2x - 3 = 8y$

35. $x + y = 4$
$y = -2$

36. $2x + y = 1$
$y = 3$

37. $x = 2$
$3x - y = -1$

38. $x = 5$
$2x - y = 4$

39. $\frac{3}{5}x + \frac{2}{5}y = 2$
$2x - \frac{3}{2}y = -\frac{15}{2}$

40. $\frac{2}{3}x + \frac{5}{3}y = 4$
$\frac{1}{4}x - \frac{1}{2}y = -\frac{3}{4}$

Without graphing, answer the following questions for each linear system. See Example 4.

(a) *Is the system inconsistent, are the equations dependent, or neither?*

(b) *Is the graph a pair of intersecting lines, a pair of parallel lines, or one line?*

(c) *Does the system have one solution, no solution, or an infinite number of solutions?*

41. $y - x = -5$
$x + y = 1$

42. $2x + y = 6$
$x - 3y = -4$

43. $x + 2y = 0$
$4y = -2x$

44. $4x + y = 2$
$2x - y = 4$

45. $3x - 2y = -3$
$3x + y = 6$

46. $y = 3x$
$y + 3 = 3x$

47. $5x + 4y = 7$
$10x + 8y = 4$

48. $2x + 3y = 12$
$2x - y = 4$

An application of mathematics in the study of economics deals with **supply and demand.** *Typically, as the price of an item increases, the demand for the item decreases, while the supply increases. (There are exceptions to this, however.) If supply and demand can be described by straight line equations, the point at which the lines intersect determines the equilibrium supply and equilibrium demand. Suppose that an economist has studied the supply and demand for aluminum siding and has come up with the conclusion that the price per unit, p, and the demand, x, are related by the demand equation* $p = 60 - \frac{3}{4}x$, *while the supply is given by the equation* $p = \frac{3}{4}x$. *The graphs of these two equations are shown at the right.*

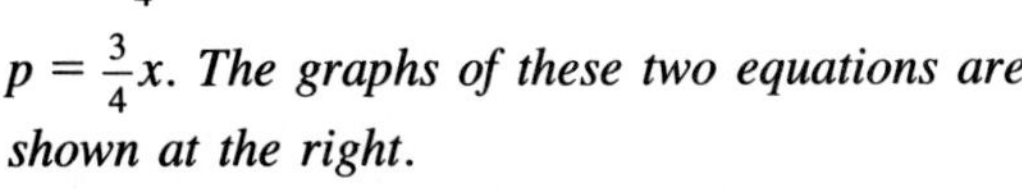

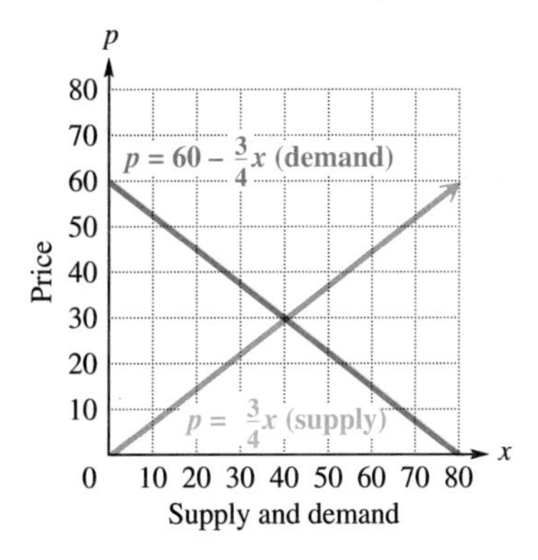

Answer the following questions about this graph.

49. At what value of x does supply equal demand?

50. At what value of p does supply equal demand?

51. What are the coordinates of the point of intersection of the two lines?

52. When $x > 40$, does demand exceed supply or does supply exceed demand?

PREVIEW EXERCISES

Add. See Section 3.1.

53. $2m + 3n$
$\underline{5m - n}$

54. $6p + 2q$
$\underline{4p + 5q}$

55. $x - 2y$
$\underline{-x + 3y}$

56. $3r - 2t$
$\underline{5r + 2t}$

57. $-2x + y$ and $-y + 5x$

58. $4a + 3b$ and $-3b - 2a$

59. $6z + 5w$ and $-5w - 6z$

60. $-3m + 2n$ and $-2n + 3m$

7.2 SOLVING SYSTEMS OF LINEAR EQUATIONS BY ADDITION

OBJECTIVES

1. SOLVE LINEAR SYSTEMS BY ADDITION.
2. MULTIPLY ONE OR BOTH EQUATIONS OF A SYSTEM SO THAT THE ADDITION METHOD CAN BE USED.
3. USE AN ALTERNATIVE METHOD TO FIND THE SECOND VALUE IN A SYSTEM.
4. USE THE ADDITION METHOD TO SOLVE AN INCONSISTENT SYSTEM.
5. USE THE ADDITION METHOD TO SOLVE A SYSTEM OF DEPENDENT EQUATIONS.

Graphing to solve a system of equations has a serious drawback: It is difficult to estimate a solution such as $(1/3, -5/6)$ accurately from a graph.

1 An algebraic method that depends on the addition property of equality can be used to solve systems. As mentioned earlier, adding the same quantity to each side of an equation results in equal sums.

$$\text{If} \quad A = B, \qquad \text{then} \quad A + C = B + C.$$

This addition can be taken a step further. Adding *equal* quantities, rather than the *same* quantity, to both sides of an equation also results in equal sums.

$$\text{If} \quad A = B \quad \text{and} \quad C = D, \qquad \text{then} \quad A + C = B + D.$$

The use of the addition property to solve systems is called the **addition method** for solving systems of equations. For most systems, this method is more efficient than graphing.

NOTE When using the addition method, the idea is to eliminate one of the variables. To do this, one of the variables must have coefficients that are negatives in the two equations. Keep this in mind throughout the examples in this section.

EXAMPLE 1 USING THE ADDITION METHOD

Use the addition method to solve the system

$$x + y = 5$$
$$x - y = 3.$$

Each equation in this system is a statement of equality, so, as discussed above, the sum of the right sides equals the sum of the left sides. Adding in this way gives

$$(x + y) + (x - y) = 5 + 3.$$

Combine terms and simplify to get

$$2x = 8$$
$$x = 4. \qquad \text{Divide by 2.}$$

This result, $x = 4$, gives the x-value of the solution of the given system. To find the y-value of the solution, substitute 4 for x in either of the two equations of the system. Choosing the first equation, $x + y = 5$, gives

$$x + y = 5$$
$$4 + y = 5 \qquad \text{Let } x = 4.$$
$$y = 1.$$

The solution, (4, 1), can be checked by substituting 4 for x and 1 for y in both equations of the given system.

$$x + y = 5 \qquad\qquad x - y = 3$$
$$4 + 1 = 5 \quad ? \qquad\qquad 4 - 1 = 3 \quad ?$$
$$5 = 5 \quad \text{True} \qquad\qquad 3 = 3 \quad \text{True}$$

Since both results are true, the solution of the given system is (4, 1). ■

CAUTION A system is not completely solved until values for *both* x and y are found. Do not make the common mistake of only finding the value of one variable.

In general, we use the following steps to solve a linear system of equations by the addition method.

SOLVING LINEAR SYSTEMS BY ADDITION

Step 1 Write both equations of the system in the form $Ax + By = C$.

Step 2 Multiply one or both equations by appropriate numbers so that the coefficients of x (or y) are negatives of each other.

Step 3 Add the two equations to get an equation with only one variable.

Step 4 Solve the equation from Step 3.

Step 5 Substitute the solution from Step 4 into either of the original equations to find the value of the remaining variable.

Step 6 Write the solution as an ordered pair and check the answer.

It does not matter which variable is eliminated first. Usually we choose the one that is more convenient to work with.

EXAMPLE 2
USING THE ADDITION METHOD

Solve the system

$$y = 2x - 11$$
$$5x - y = 26.$$

Step 1 Rewrite the first equation in the form $Ax + By = C$.

$$-2x + y = -11 \qquad \text{Subtract } 2x.$$
$$5x - y = 26$$

Step 2 Because the coefficients of y are 1 and -1, it is not necessary to multiply either equation by a number.

Step 3 Add the two equations.

$$\begin{array}{rl} -2x + y = & -11 \\ 5x - y = & 26 \\ \hline 3x \quad\quad = & 15 \end{array} \qquad \text{Add in columns.}$$

Step 4 Solve the equation.

$$3x = 15$$
$$x = 5 \qquad \text{Divide by 3.}$$

Step 5 Find the value of y by substituting 5 for x in either of the original equations. Choosing the first gives

$$y = 2x - 11$$
$$y = 2(5) - 11 \qquad \text{Let } x = 5.$$
$$y = 10 - 11$$
$$y = -1.$$

Step 6 The solution is $(5, -1)$. Check the solution by substitution into both of the original equations. Let $x = 5$ and $y = -1$.

$$\begin{array}{rlr} y = 2x - 11 & & \\ -1 = 2(5) - 11 & ? & \\ -1 = -1 & & \text{True} \end{array} \qquad \begin{array}{rlr} 5x - y = 26 & & \\ 5(5) - (-1) = 26 & ? & \\ 26 = 26 & & \text{True} \end{array}$$

Since $(5, -1)$ is a solution of *both* equations, it is the solution of the system. ■

In the remaining examples, we will not specifically number the steps.

2 In both examples above, a variable was eliminated by the addition step. Sometimes it is necessary to multiply both sides of one or both equations in a system by some number before the addition step will eliminate a variable.

NOTE The method of identifying equations by numbers in parentheses, as used in Example 3, will occasionally be used in explanation of examples.

EXAMPLE 3 **MULTIPLYING WHILE USING THE ADDITION METHOD**

Solve the system

$$x + 3y = 7 \qquad (1)$$
$$2x + 5y = 12. \qquad (2)$$

Adding the two equations gives $3x + 8y = 19$, which does not help to solve the system. However, if both sides of equation (1) are first multiplied by -2, the terms with the variable x will drop out after adding.

$$-2(x + 3y) = -2(7)$$
$$-2x - 6y = -14 \qquad (3)$$

Now add equations (3) and (2).

$$\begin{aligned} -2x - 6y &= -14 \qquad (3) \\ 2x + 5y &= 12 \qquad (2) \\ \hline -y &= -2 \\ y &= 2 \end{aligned}$$

Substituting $y = 2$ into equation (1) gives

$$\begin{aligned} x + 3y &= 7 \\ x + 3(2) &= 7 \qquad \text{Let } y = 2. \\ x + 6 &= 7 \\ x &= 1. \end{aligned}$$

The solution of this system is (1, 2). Check that this ordered pair satisfies both of the original equations. ■

EXAMPLE 4
MULTIPLYING BOTH EQUATIONS WHEN USING THE ADDITION METHOD

Solve the system

$$\begin{aligned} 2x + 3y &= -15 \qquad (1) \\ 5x + 2y &= 1. \qquad (2) \end{aligned}$$

Here the multiplication property of equality will be used with both equations instead of just one, as in Example 3. Multiply by numbers that will cause the coefficients of x (or of y) in the two equations to be additive inverses of each other. For example, multiply both sides of equation (1) by 5, and both sides of equation (2) by -2.

$$\begin{aligned} 10x + 15y &= -75 \qquad \text{Multiply equation (1) by 5.} \\ -10x - 4y &= -2 \qquad \text{Multiply equation (2) by } -2. \\ \hline 11y &= -77 \qquad \text{Add.} \\ y &= -7 \end{aligned}$$

Substituting -7 for y in either (1) or (2) gives $x = 3$. Check that the solution of the system is $(3, -7)$. ■

3 In some cases it is easier to find the value of the second variable in a system by using the addition method twice. The next example shows this method.

EXAMPLE 5
FINDING THE SECOND VALUE USING AN ALTERNATIVE METHOD

Solve the system

$$\begin{aligned} 4x &= 9 - 3y \qquad (1) \\ 5x - 2y &= 8. \qquad (2) \end{aligned}$$

Rearrange the terms in equation (1) so that the like terms can be aligned in columns. Add $3y$ to both sides to get the system

$$4x + 3y = 9 \qquad (3)$$
$$5x - 2y = 8.$$

One way to proceed is to eliminate y by multiplying both sides of equation (3) by 2 and both sides of equation (2) by 3, and then adding.

$$\begin{aligned} 8x + 6y &= 18 \\ 15x - 6y &= 24 \\ \hline 23x \qquad &= 42 \\ x &= \frac{42}{23} \end{aligned}$$

Substituting 42/23 for x in one of the given equations would give y, but the arithmetic involved would be messy. Instead, solve for y by starting again with the original equations and eliminating x. Multiply both sides of equation (3) by 5 and both sides of equation (2) by -4, and then add.

$$\begin{aligned} 20x + 15y &= 45 \\ -20x + 8y &= -32 \\ \hline 23y &= 13 \\ y &= \frac{13}{23} \end{aligned}$$

Check that the solution is (42/23, 13/23). ■

When the value of the first variable is a fraction, the method used in Example 5 avoids errors that often occur when working with fractions. Of course, this method could be used in solving any system of equations.

4 In the previous section some of the systems had equations with graphs that were two parallel lines, while the equations of other systems had graphs that were the same line. These systems can also be solved with the addition method. The next example illustrates the solution of an inconsistent system, where the graphs of the equations are parallel lines.

EXAMPLE 6
USING THE ADDITION METHOD FOR AN INCONSISTENT SYSTEM

Solve the following system by the addition method.

$$2x + 4y = 5$$
$$4x + 8y = -9$$

Multiply both sides of $2x + 4y = 5$ by -2 and then add to $4x + 8y = -9$.

$$\begin{aligned} -4x - 8y &= -10 \\ 4x + 8y &= -9 \\ \hline 0 &= -19 \end{aligned} \quad \text{False}$$

The false statement $0 = -19$ shows that the given system is self-contradictory. *It has no solution.* This means that the graphs of the equations of this system are parallel lines, as shown in Figure 7.5. Since this system has no solution, it is inconsistent. ■

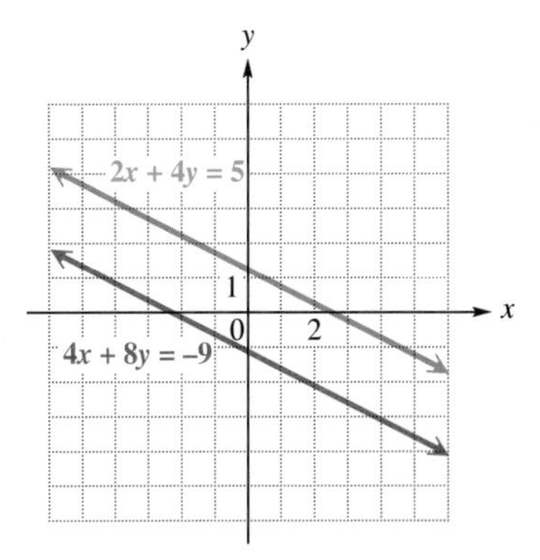

FIGURE 7.5

5 The next example shows the result of using the addition method when the equations of the system are dependent, with graphs that are the same line.

EXAMPLE 7 **USING ADDITION TO SOLVE A SYSTEM WITH DEPENDENT EQUATIONS**

Solve the following system by the addition method.

$$\begin{aligned} 3x - y &= 4 \\ -9x + 3y &= -12 \end{aligned}$$

Multiply both sides of the first equation by 3 and then add the two equations to get

$$\begin{aligned} 9x - 3y &= 12 \\ -9x + 3y &= -12 \\ \hline 0 &= 0. \end{aligned} \quad \text{True}$$

This result means that every solution of one equation is also a solution of the other, so the system has an infinite number of solutions: all the ordered pairs corresponding to points that lie on the common graph. As mentioned earlier, the equations of this system are dependent. In the answers at the back of this book, a solution of such a system of dependent equations is indicated by *infinite number of solutions*. A graph of the equations of this system is shown in Figure 7.6. ■

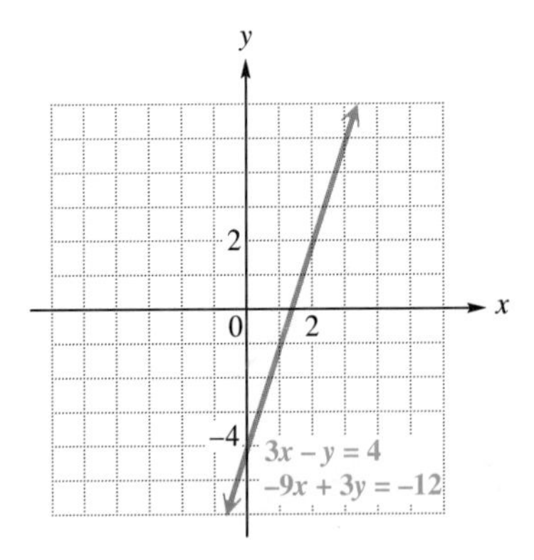

FIGURE 7.6

The possible situations that may occur when using the addition method are described in the chart on the following page.

SUMMARY OF SITUATIONS THAT MAY OCCUR

One of three situations may occur when the addition method is used to solve a linear system of equations.

1. The result of the addition step is a statement such as $x = 2$ or $y = -3$. The solution will be exactly one ordered pair. The graphs of the equations of the system will intersect at exactly one point.

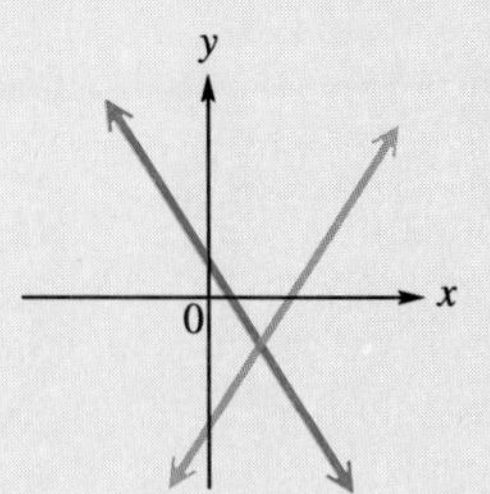

2. The result of the addition step is a false statement, such as $0 = 4$. In this case, the graphs are parallel lines and there is no solution for the system.

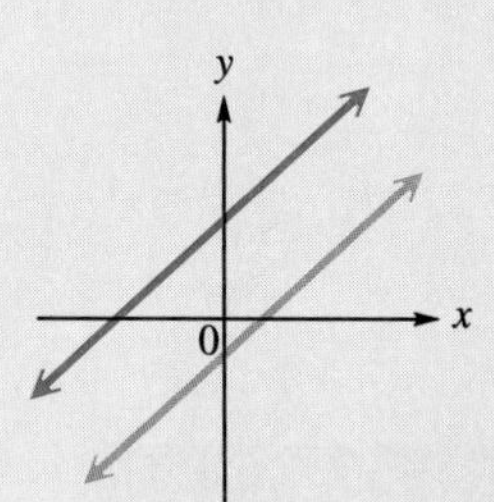

3. The result of the addition step is a true statement, such as $0 = 0$. The graphs of the equations of the system are the same line, and an infinite number of ordered pairs are solutions.

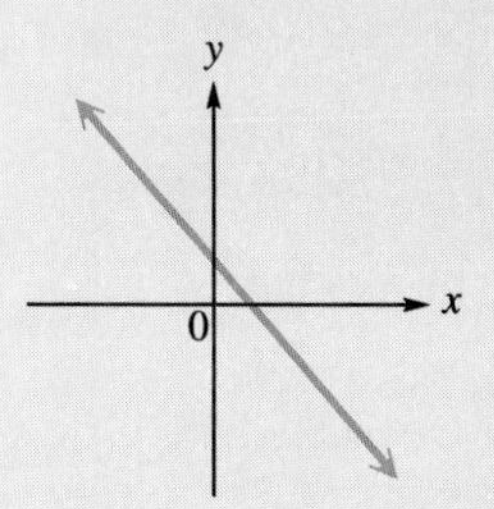

7.2 EXERCISES

Solve each system by the addition method. Check your answers. See Examples 1 and 2.

1. $\begin{aligned} x - y &= 3 \\ x + y &= -1 \end{aligned}$

2. $\begin{aligned} x + y &= 7 \\ x - y &= -3 \end{aligned}$

3. $\begin{aligned} x + y &= 2 \\ 2x - y &= 4 \end{aligned}$

4. $\begin{aligned} 3x - y &= 8 \\ x + y &= 4 \end{aligned}$

5. $\begin{aligned} 2x + y &= 14 \\ x - y &= 4 \end{aligned}$

6. $\begin{aligned} 2x + y &= 2 \\ -x - y &= 1 \end{aligned}$

7. $\begin{aligned} 3x + 2y &= 6 \\ -3x - y &= 0 \end{aligned}$

8. $\begin{aligned} 5x - y &= 9 \\ -5x + 2y &= -8 \end{aligned}$

9. $\begin{aligned} 6x - y &= 1 \\ -6x + 5y &= 7 \end{aligned}$

10. $\begin{aligned} 6x + y &= -2 \\ -6x + 3y &= -14 \end{aligned}$

11. $\begin{aligned} 2x - y &= 5 \\ 4x + y &= 4 \end{aligned}$

12. $\begin{aligned} x - 4y &= 13 \\ -x + 6y &= -18 \end{aligned}$

13. A student solves the system

$$5x - y = 15$$
$$7x + y = 21$$

and finds that $x = 3$, which is the correct value for x. The student gives the solution as "$x = 3$." Is this correct? Explain. (See Objective 1.)

14. A student solves the system

$$x + y = 4$$
$$2x + 2y = 8$$

and obtains the equation $0 = 0$. The student gives the solution as $(0, 0)$. Is this correct? Explain. (See Objective 5.)

Solve each system by the addition method. Check your answers. See Example 3.

15. $2x - y = 7$; $3x + 2y = 0$

16. $x + y = 7$; $-3x + 3y = -9$

17. $x + 3y = 16$; $2x - y = 4$

18. $4x - 3y = 8$; $2x + y = 14$

19. $x + 4y = -18$; $3x + 5y = -19$

20. $2x + y = 3$; $5x - 2y = -15$

21. $3x - 2y = -6$; $-5x + 4y = 16$

22. $-4x + 3y = 0$; $5x - 6y = 9$

23. $2x - y = -8$; $5x + 2y = -20$

24. $5x + 3y = -9$; $7x + y = -3$

25. $2x + y = 5$; $5x + 3y = 11$

26. $2x + 7y = -53$; $4x + 3y = -7$

Solve each system by the addition method. Check your answers. See Examples 2, 4, and 5.

27. $5x - 4y = -1$; $-7x + 5y = 8$

28. $3x + 2y = 12$; $5x - 3y = 1$

29. $3x + 5y = 33$; $4x - 3y = 15$

30. $2x + 5y = 3$; $5x = 3y + 23$

31. $3x + 7 = -5y$; $5x + 4y = 10$

32. $3y + 11 = -2x$; $5x + 2y = 22$

33. $2x + 3y = -12$; $5x = 7y - 30$

34. $2x + 9y = 16$; $5x = 6y + 40$

35. $4x + 7 = 3y$; $6x + 5y = 18$

Use the addition method to solve each system. Check your answers. See Examples 6 and 7.

36. $2x - y = 1$; $2x - y = 4$

37. $5x - 2y = 6$; $10x - 4y = 10$

38. $3x - 5y = 2$; $6x - 10y = 8$

39. $x + 3y = 5$; $2x + 6y = 10$

40. $6x - 2y = 12$; $-3x + y = -6$

41. $2x + 3y = 8$; $4x + 6y = 12$

42. $4x + y = 6$; $-8x - 2y = 21$

43. $3x + 5y = -19$; $6x + 7y = -23$

44. $3x - 2y = -13$; $9x + 5y = 16$

45. $24x + 12y = 19$; $16x - 18y = -9$

46. $9x + 4y = -4$; $6x + 6y = -11$

47. $9x - 5y = 7$; $21x + 20y = 10$

One way to solve a system that involves fractions as coefficients is to multiply each equation by the least common denominator of all the fractions in the equation to get integer coefficients. Then use the addition method as explained in this section. Solve each of the following systems using this procedure.

48. $2x + \frac{3}{4}y = -1$; $4x + \frac{7}{3}y = \frac{4}{3}$

49. $\frac{6}{5}x + \frac{1}{5}y = \frac{33}{5}$; $-x + \frac{3}{2}y = \frac{19}{2}$

50. $\frac{5}{3}x - y = -1$; $\frac{7}{2}x + \frac{5}{2}y = -\frac{41}{2}$

51. $\frac{5}{2}x + y = -\frac{11}{2}$
$\frac{3}{4}x - \frac{3}{4}y = -\frac{15}{4}$

52. $\frac{3}{2}x - \frac{3}{4}y = \frac{15}{4}$
$\frac{4}{3}x - \frac{5}{3}y = \frac{19}{3}$

53. $\frac{4}{3}x + \frac{1}{3}y = \frac{4}{3}$
$-\frac{8}{5}x - \frac{2}{5}y = -\frac{8}{5}$

54. Suppose that a system is in the form

$$Ax + By = C$$
$$Dx + Ey = F.$$

(a) Choose six consecutive integers for A, B, C, D, E, and F, and solve the system.

(b) Repeat part (a) with six different consecutive integers.

(c) Repeat part (a), again using six different consecutive integers (not the ones you used in part (b)).

(d) From your results in parts (a), (b), and (c), what do you think is true about a system in this form with consecutive integer values for A, B, C, D, E, and F? (While this result has not been proved here, it can be shown that it is true.)

PREVIEW EXERCISES

Solve each equation for y. See Section 2.5.

55. $3x + y = 4$ **56.** $-2x + y = 9$ **57.** $9x - 2y = 4$ **58.** $5x - 3y = 12$

Solve each equation. Check the solutions. See Sections 2.2, 2.3, and 5.6.

59. $-2(y - 2) + 5y = 10$ **60.** $2m - 3(4 - m) = 8$ **61.** $p + 4(6 - 2p) = 3$

62. $4\left(\frac{3 - 2k}{2}\right) + 3k = -3$ **63.** $4x - 2\left(\frac{1 - 3x}{2}\right) = 6$ **64.** $a + 3\left(\frac{1 - a}{2}\right) = 5$

7.3 SOLVING SYSTEMS OF LINEAR EQUATIONS BY SUBSTITUTION

OBJECTIVES

1 SOLVE LINEAR SYSTEMS BY SUBSTITUTION.

2 SOLVE LINEAR SYSTEMS WITH FRACTIONS.

1 The graphical method and the addition method for solving systems of linear equations were discussed in the previous sections. A third method, the substitution method, is particularly useful for solving systems where one equation is solved, or can be solved easily, for one of the variables.

EXAMPLE 1
SOLVING A SYSTEM BY SUBSTITUTION

Solve the system

$$3x + 5y = 26$$
$$y = 2x.$$

The second of these two equations says that $y = 2x$. Substituting $2x$ for y in the first equation gives

$$
\begin{aligned}
3x + 5y &= 26 \\
3x + 5(2x) &= 26 \qquad \text{Let } y = 2x. \\
3x + 10x &= 26 \\
13x &= 26 \\
x &= 2.
\end{aligned}
$$

Since $y = 2x$ and $x = 2$, $y = 2(2) = 4$. Check that the solution of the given system is $(2, 4)$. ■

A summary of the steps used in solving a system by substitution is given here.

SOLVING LINEAR SYSTEMS BY SUBSTITUTION

Step 1 Solve one of the equations for either variable. (If one of the variables has coefficient 1 or -1, choose it, since the substitution method is usually easier this way.)

Step 2 Substitute for that variable in the other equation. The result should be an equation with just one variable.

Step 3 Solve the equation from Step 2.

Step 4 Substitute the result from Step 3 into the equation from Step 1 to find the value of the other variable.

Step 5 Check the solution in both of the given equations.

Notice how the steps are used in the next example.

EXAMPLE 2
SOLVING A SYSTEM BY SUBSTITUTION

Use substitution to solve the system

$$2x + 5y = 7$$
$$x + y = -1.$$

Step 1 There are several ways that we can begin. Let us start by solving the second equation for x.

$$
\begin{aligned}
x + y &= -1 \\
x &= -1 - y \qquad \text{Subtract } y.
\end{aligned}
$$

Step 2 Substitute $-1 - y$ for x in the first equation. We must use parentheses around $-1 - y$.

$$2x + 5y = 7$$
$$2(-1 - y) + 5y = 7 \qquad \text{Let } x = -1 - y.$$

Step 3 Solve this equation.

$$-2 - 2y + 5y = 7 \qquad \text{Distributive property}$$
$$-2 + 3y = 7 \qquad \text{Combine like terms.}$$
$$3y = 9 \qquad \text{Add 2.}$$
$$y = 3 \qquad \text{Divide by 3.}$$

Step 4 Find x by letting $y = 3$ in $x = -1 - y$.

$$x = -1 - y$$
$$x = -1 - 3 \qquad \text{Let } y = 3.$$
$$x = -4$$

Step 5 The solution is $(-4, 3)$. Check by substituting into both of the original equations.

$$2x + 5y = 7 \qquad\qquad x + y = -1$$
$$2(-4) + 5(3) = 7 \quad ? \qquad\qquad -4 + 3 = -1 \quad ?$$
$$7 = 7 \quad \text{True} \qquad\qquad -1 = -1 \quad \text{True}$$

Since $(-4, 3)$ is a solution of *both* equations, it is the solution of the system. ■

In the rest of the examples, the steps are not numbered.

EXAMPLE 3
SOLVING A SYSTEM BY SUBSTITUTION

Use substitution to solve the system

$$x = 5 - 2y$$
$$2x + 4y = 6.$$

Substitute $5 - 2y$ for x in the second equation.

$$2x + 4y = 6$$
$$2(5 - 2y) + 4y = 6 \qquad \text{Let } x = 5 - 2y.$$
$$10 - 4y + 4y = 6 \qquad \text{Distributive property}$$
$$10 = 6 \qquad \text{False}$$

As shown in the last section, this false result means that the equations of the system have graphs that are parallel lines. The system is inconsistent and has no solution. ■

An important reason for learning the method of substitution is that some systems that involve higher degree equations *must* be solved by substitution. The next example shows a system that could more easily be solved by addition, but substitution is used in order to illustrate that it indeed can be used to solve any system in this chapter.

EXAMPLE 4
SOLVING A SYSTEM BY SUBSTITUTION

Use substitution to solve the system

$$2x + 3y = 8$$
$$4x + 3y = 4.$$

The substitution method requires an equation solved for one of the variables. Choose the first equation of the system, $2x + 3y = 8$, and solve for x. Start by subtracting $3y$ on both sides.

$$2x + 3y = 8$$
$$2x = 8 - 3y$$

Now divide both sides of this equation by 2.

$$x = \frac{8 - 3y}{2}$$

Finally, substitute this value for x in the second equation of the system.

$$4x + 3y = 4$$
$$4\left(\frac{8 - 3y}{2}\right) + 3y = 4 \qquad \text{Let } x = \frac{8 - 3y}{2}.$$
$$2(8 - 3y) + 3y = 4 \qquad \text{Divide 4 by 2.}$$
$$16 - 6y + 3y = 4 \qquad \text{Distributive property}$$
$$-3y = -12 \qquad \text{Combine terms; subtract 16.}$$
$$y = 4 \qquad \text{Divide by } -3.$$

Find x by letting $y = 4$ in $x = \dfrac{8 - 3y}{2}$.

$$x = \frac{8 - 3 \cdot 4}{2} = \frac{8 - 12}{2} = \frac{-4}{2} = -2$$

The solution of the given system is $(-2, 4)$. Check this solution in both equations. ■

EXAMPLE 5
SOLVING A SYSTEM BY SUBSTITUTION

Use substitution to solve the system

$$2x = 4 - y \qquad (1)$$
$$6 + 3y + 4x = 16 - x. \qquad (2)$$

Begin by simplifying equation (2) by adding x and subtracting 6 on both sides. This gives the simplified system

$$2x = 4 - y \qquad (1)$$
$$5x + 3y = 10. \qquad (3)$$

For the substitution method, one of the equations must be solved for either x or y. Since the coefficient of y in equation (1) is -1, avoid fractions by solving this equation for y.

$$2x = 4 - y \qquad (1)$$
$$2x - 4 = -y \qquad \text{Subtract 4.}$$
$$-2x + 4 = y \qquad \text{Multiply by } -1.$$

Now substitute $-2x + 4$ for y in equation (3).

$$5x + 3y = 10$$
$$5x + 3(-2x + 4) = 10 \qquad \text{Let } y = -2x + 4.$$
$$5x - 6x + 12 = 10 \qquad \text{Distributive property}$$
$$-x + 12 = 10 \qquad \text{Combine like terms.}$$
$$-x = -2 \qquad \text{Subtract 12.}$$
$$x = 2 \qquad \text{Multiply by } -1.$$

Since $y = -2x + 4$ and $x = 2$,

$$y = -2(2) + 4 = 0,$$

and the solution is (2, 0). Check the solution by substitution in both equations of the given system. ■

2 When a system includes equations with fractions as coefficients, eliminate the fractions by multiplying both sides by a common denominator. Then solve the resulting system.

EXAMPLE 6
SOLVING BY SUBSTITUTION WITH FRACTIONS AS COEFFICIENTS

Solve the system

$$3x + \frac{1}{4}y = 2 \qquad (1)$$
$$\frac{1}{2}x + \frac{3}{4}y = -\frac{5}{2}. \qquad (2)$$

Begin by eliminating fractions. Clear equation (1) of fractions by multiplying both sides by 4.

$$4\left(3x + \frac{1}{4}y\right) = 4(2) \qquad \text{Multiply by 4.}$$
$$4(3x) + 4\left(\frac{1}{4}y\right) = 4(2) \qquad \text{Distributive property}$$
$$12x + y = 8 \qquad (3)$$

Now clear equation (2) of fractions by multiplying both sides by the common denominator 4.

$$4\left(\frac{1}{2}x + \frac{3}{4}y\right) = 4\left(-\frac{5}{2}\right) \qquad \text{Multiply by 4.}$$

$$4\left(\frac{1}{2}x\right) + 4\left(\frac{3}{4}y\right) = 4\left(-\frac{5}{2}\right) \qquad \text{Distributive property}$$

$$2x + 3y = -10 \qquad (4)$$

The given system of equations has been simplified to

$$12x + y = 8 \qquad (3)$$
$$2x + 3y = -10. \qquad (4)$$

Solve the system by the substitution method. Equation (3) can be solved for y by subtracting $12x$ on each side.

$$12x + y = 8 \qquad (3)$$
$$y = -12x + 8$$

Now substitute the result for y in equation (4).

$$2x + 3(-12x + 8) = -10 \qquad \text{Let } y = -12x + 8.$$
$$2x - 36x + 24 = -10 \qquad \text{Distributive property}$$
$$-34x = -34 \qquad \text{Combine terms; subtract 24.}$$
$$x = 1 \qquad \text{Divide by } -34.$$

Using $x = 1$ in $y = -12x + 8$ gives $y = -4$. The solution is $(1, -4)$. Check by substituting 1 for x and -4 for y in the original equations. Verify that the same solution is found if the addition method is used to solve the system of equations (3) and (4). ■

7.3 EXERCISES

Solve each system by the substitution method. Check each solution. See Examples 1–4.

1. $x + y = 6$; $y = 2x$

2. $x + 3y = -11$; $y = -4x$

3. $3x + 2y = 26$; $x = y + 2$

4. $4x + 3y = -14$; $x = y - 7$

5. $x + 5y = 3$; $x = 2y + 10$

6. $5x + 2y = 14$; $y = 2x - 11$

7. $3x - y = 6$; $y = 3x - 5$

8. $4x - y = 4$; $y = 4x + 3$

9. $3x - 2y = 14$; $2x + y = 0$

10. $2x - 5 = -y$; $x + 3y = 0$

11. $x + y = 6$; $x - y = 4$

12. $3x - 2y = 13$; $x + y = 6$

13. $6x - 8y = 4$; $3x = 4y + 2$

14. $12x + 18y = 12$; $2x = 2 - 3y$

15. $2x + 3y = 11$; $4y - x = 0$

16. $4x + 5y = 5$; $2x + 3y = 1$

17. $6x + 5y = 13$; $3x + 2y = 4$

18. $5x - 2y = 0$; $3y + 2x = 19$

19. $3x + 4y = -10$; $x + 6 = 0$

20. $4x + y = 5$; $x - 2 = 0$

21. $5x + 2y = -19$; $y - 3 = 0$

22. Professor Brandsma gave the following item on a test in introductory algebra:

Solve the system

$$3x - y = 13$$
$$2x + 5y = 20$$

by the substitution method.

One student worked the problem by solving first for y in the first equation. Another student worked it by solving first for x in the second equation. Both students got the correct solution, (5, 2). Which student, do you think, had less work to do? Explain. (See Objective 1.)

23. Which one of the following systems would require the least amount of work if solved by the substitution method? Explain why. (See Objective 1.)

(a) $3x + 2y = 5$, $2x - 8y = 4$ **(b)** $9x + 7y = 4$, $3x + 2y = 1$

(c) $-6x + 5y = 8$, $4x + 8y = 3$ **(d)** $17x - y = 1$, $3x + 6y = 4$

Solve each system by either the addition method or the substitution method. First simplify equations where necessary. Check each solution. See Example 5.

24. $5x - 4y - 8x - 2 = 6x + 3y - 3$
$4x - y = -2y - 8$

25. $2x - 8y + 3y + 2 = 5y + 16$
$8x - 2y = 4x + 28$

26. $7x - 9 + 2y - 8 = -3y + 4x + 13$
$4y - 8x = -8 + 9x + 32$

27. $-2x + 3y = 12 + 2y$
$2x - 5y + 4 = -8 - 4y$

28. $2x + 5y = 7 + 4y - x$
$5x + 3y + 8 = 22 - x + y$

29. $y + 9 = 3x - 2y + 6$
$5 - 3x + 24 = -2x + 4y + 3$

30. $5x - 2y = 16 + 4x - 10$
$4x + 3y = 60 + 2x + y$

31. $4 + 4x - 3y = 34 + x$
$5y + 4x = 4y - 2 + 3x$

Solve each system by either the addition method or the substitution method. First clear all fractions. Check each solution. See Example 6.

32. $\frac{5}{3}x + 2y = \frac{1}{3} + y$
$2x - 3 + \frac{y}{3} = -2 + x$

33. $\frac{x}{6} + \frac{y}{6} = 1$
$-\frac{1}{2}x - \frac{1}{3}y = -5$

34. $\frac{x}{2} - \frac{y}{3} = \frac{5}{6}$
$\frac{x}{5} - \frac{y}{4} = \frac{1}{10}$

35. $\frac{x}{3} - \frac{3y}{4} = -\frac{1}{2}$
$\frac{2x}{3} + \frac{y}{2} = 3$

36. $\frac{x}{5} + 2y = \frac{8}{5}$
$\frac{3x}{5} + \frac{y}{2} = -\frac{7}{10}$

37. $\frac{x}{2} + \frac{y}{3} = \frac{7}{6}$
$\frac{x}{4} - \frac{3y}{2} = \frac{9}{4}$

38. $\frac{5x}{2} - \frac{y}{3} = \frac{5}{6}$
$\frac{4x}{3} + y = \frac{19}{3}$

39. $\frac{2x}{3} + \frac{y}{6} = 9$
$\frac{x}{4} - \frac{3y}{2} = -6$

40. $\frac{3x}{2} + \frac{y}{5} = 2$
$\frac{5x}{2} - \frac{3y}{10} = \frac{13}{2}$

PREVIEW EXERCISES

Solve each problem. See Sections 2.4, 2.5, 2.7, and 2.8.

41. If three times a number is added to 6, the result is 69. Find the number.

42. The product of 5 and 1 more than a number is 35. Find the number.

43. The perimeter of a rectangle is 46 feet. The width is 7 feet less than the length. Find the width.

44. The area of a rectangle is numerically 20 more than the width, and the length is 6 centimeters. What is the width?

45. A cashier has ten-dollar bills and twenty-dollar bills. There are 6 more tens than twenties. If there are 32 bills altogether, how many of them are twenties?

46. Jennifer Marshall traveled for 2 hours at a constant speed. Because of road work, she reduced her speed by 7 miles per hour for the next 2 hours. If she traveled 206 miles, what was her speed on the first part of the trip?

Photo for Exercise 45

7.4 APPLICATIONS OF LINEAR SYSTEMS

OBJECTIVES

1. USE LINEAR SYSTEMS TO SOLVE PROBLEMS ABOUT NUMBERS.
2. USE LINEAR SYSTEMS TO SOLVE PROBLEMS ABOUT QUANTITIES AND THEIR COSTS.
3. USE LINEAR SYSTEMS TO SOLVE PROBLEMS ABOUT MIXTURES.
4. USE LINEAR SYSTEMS TO SOLVE PROBLEMS ABOUT RATE OR SPEED USING THE DISTANCE FORMULA.

FOCUS ON PROBLEM SOLVING

A textbook author receives a $2 royalty for each of her algebra books sold, and a $3 royalty for each of her calculus books sold. During one royalty period, the two books together sold 13,000 copies and she received a total of $29,000 in royalties. How many of each kind of book were sold during that period?

In Chapter 2 we first saw the type of problem given here. At that time, our problem solving strategy was limited to the use of a single variable. Now that we are studying systems, many problems can be solved by using more than one variable. This problem is Exercise 19 in the exercises for this section. It can be solved using only one variable, but many students prefer to use a system to solve problems of this type. After working through this section, you should be able to solve this problem using a system of equations.

PROBLEM SOLVING

Many practical problems are more easily translated into equations if two variables are used. With two variables, a system of two equations is needed to find the desired solution. The examples in this section illustrate the method of solving applied problems using two equations and two variables.

Recall from Chapter 2 the steps used in solving applied problems. The steps presented there can be modified as follows to allow for two variables and two equations. ■

SOLVING APPLIED PROBLEMS WITH TWO VARIABLES

Step 1 **Choose the variables.**
Choose a variable to represent each of the two unknown values that must be found. Write down what each variable is to represent.

Step 2 **Draw figures or diagrams.**
If figures or diagrams can help, draw them. (In some cases, they will not apply.)

Step 3 **Write two equations.**
Translate the problem into a system of two equations using both variables.

Step 4 **Solve the system.**
Solve the system of equations, using either the addition or substitution method.

Step 5 **Answer the question(s).**
Answer the question or questions asked in the problem.

Step 6 **Check.**
Check your answer by using the original words of the problem. Be sure that your answer makes sense.

1 The first example shows how to use two variables to solve a problem about two unknown numbers.

EXAMPLE 1
SOLVING A PROBLEM ABOUT TWO NUMBERS

The sum of two numbers is 63. Their difference is 19. Find the two numbers.

Step 1 Let x = one number;
y = other number.

Step 2 A figure or diagram will not help in this problem, so go on to Step 3.
Step 3 From the information in the problem, set up a system of equations.

$$x + y = 63 \quad \text{The sum is 63.}$$
$$x - y = 19 \quad \text{The difference is 19.}$$

Step 4 Solve the system from Step 3. Here the addition method is used. Adding gives

$$\begin{aligned} x + y &= 63 \\ x - y &= 19 \\ \hline 2x \quad &= 82. \end{aligned}$$

From this last equation, $x = 41$. Substitute 41 for x in either equation to find $y = 22$.

Step 5 The numbers required in the problem are 41 and 22.

Step 6 The sum of 41 and 22 is 63, and their difference is $41 - 22 = 19$. The solution satisfies the conditions of the problem. ■

CAUTION If an applied problem asks for *two* values (as in Example 1), be sure to give both of them in your answer. Avoid the common error of giving only one of the values.

2 The next example shows how to solve a common type of applied problem involving two quantities and their costs.

PROBLEM SOLVING

Just as in Chapter 2, we can use a table or a box diagram to organize the information in order to solve an applied problem with two unknowns. We use a table in this example. ■

EXAMPLE 2
SOLVING A PROBLEM ABOUT QUANTITIES AND COSTS

Admission prices at a football game were \$6 for adults and \$2 for children. The total value of the tickets sold was \$2528, and 454 tickets were sold. How many adults and how many children attended the game?

Step 1 Let a = the number of adults' tickets sold;

c = the number of childrens' tickets sold.

Step 2 The information given in the problem is summarized in the table. The entries in the "total value" column were found by multiplying the number of tickets sold by the price per ticket.

KIND OF TICKET	NUMBER SOLD	COST OF EACH (IN DOLLARS)	TOTAL VALUE (IN DOLLARS)
Adult	a	6	$6a$
Child	c	2	$2c$
Total	454	—	2528

The total number of tickets sold was 454, so

$$a + c = 454.$$

Since the total value was \$2528, the final column leads to

$$6a + 2c = 2528.$$

Step 3 These two equations give the following system.

$$a + c = 454 \qquad (1)$$
$$6a + 2c = 2528 \qquad (2)$$

Step 4 We solve the system of equations with the addition method. First, multiply both sides of equation (1) by -2 to get

$$-2a - 2c = -908.$$

Then add this result to equation (2).

$$\begin{aligned} -2a - 2c &= -908 && \text{Multiply (1) by } -2.\\ 6a + 2c &= 2528 \\ \hline 4a \qquad &= 1620 && \text{Add.}\\ a &= 405 && \text{Divide by 4.} \end{aligned}$$

Substitute 405 for a in equation (1) to get

$$\begin{aligned} a + c &= 454 && (1)\\ 405 + c &= 454 && \text{Let } a = 405.\\ c &= 49. && \text{Subtract 405.} \end{aligned}$$

Step 5 There were 405 adults and 49 children at the game.

Step 6 Since 405 adults paid \$6 each and 49 children paid \$2 each, the value of tickets sold should be $405(6) + 49(2) = 2528$, or \$2528. This result agrees with the given information. ■

3 In Section 2.7 we learned how to solve mixture problems using one variable.

PROBLEM SOLVING

Many mixture problems can also be solved using two variables. In the next example we show how to solve a mixture problem using a system of equations. The ''box diagram'' method first introduced in Section 2.7 is used here once again. ■

EXAMPLE 3 SOLVING A MIXTURE PROBLEM (INVOLVING PERCENT)

A pharmacist needs 100 liters of 50% alcohol solution. She has on hand 30% alcohol solution and 80% alcohol solution, which she can mix. How many liters of each will be required to make the 100 liters of 50% alcohol solution?

Step 1 Let x = the number of liters of 30% alcohol needed;

y = the number of liters of 80% alcohol needed.

Step 2 Summarize the information using a box diagram. See Figure 7.7.

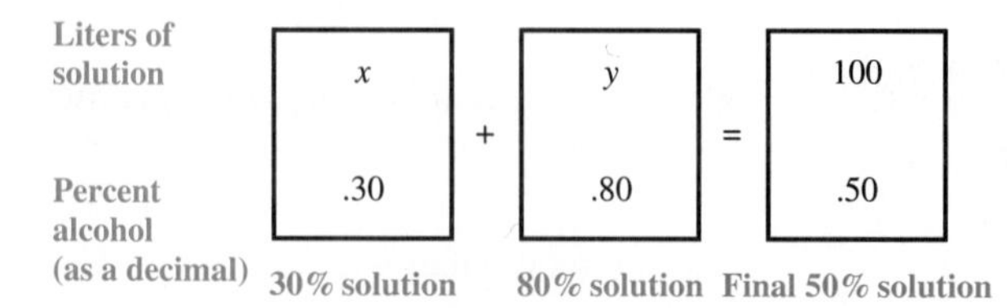

FIGURE 7.7

Step 3 We must write two equations. Since the total number of liters in the final mixture will be 100, the first equation is

$$x + y = 100.$$

To find the amount of pure alcohol in each mixture, multiply the number of liters by the concentration. The amount of pure alcohol in the 30% solution added to the amount of pure alcohol in the 80% solution will equal the amount of pure alcohol in the final 50% solution. This gives the second equation,

$$.30x + .80y = .50(100).$$

These two equations give the system

$$x + y = 100$$
$$.30x + .80y = 50. \qquad .50(100) = 50$$

Step 4 Solve this system by the substitution method. Solving the first equation of the system for x gives $x = 100 - y$. Substitute $100 - y$ for x in the second equation to get

$$.30(\mathbf{100 - y}) + .80y = 50 \qquad \text{Let } x = 100 - y.$$
$$30 - .30y + .80y = 50 \qquad \text{Distributive property}$$
$$.50y = 20 \qquad \text{Combine terms; subtract 30.}$$
$$y = 40. \qquad \text{Divide by .50.}$$

Then $x = 100 - y = 100 - 40 = 60$.

Step 5 The pharmacist should use 60 liters of the 30% solution and 40 liters of the 80% solution.

Step 6 Since $60 + 40 = 100$ and $.30(60) + .80(40) = 50$, this mixture will give the 100 liters of 50% solution, as required in the original problem.

The system in this problem could have been solved by the addition method. Also, we could have cleared decimals by multiplying both sides of the second equation by 100. ■

4 Problems that use the distance formula relating distance, rate, and time, were first introduced in Section 2.8.

PROBLEM SOLVING

In some cases, these problems can be solved by using a system of two linear equations. Keep in mind that setting up a chart and drawing a sketch will help in solving such problems. ■

EXAMPLE 4
SOLVING A PROBLEM ABOUT DISTANCE, RATE, AND TIME

Two executives in cities 400 miles apart drive to a business meeting at a location on the line between their cities. They meet after 4 hours. Find the speed of each car if one car travels 20 miles per hour faster than the other.

Step 1 Let x = the speed of the faster car;

y = the speed of the slower car.

Step 2 Use the formula that relates distance, rate, and time, $d = rt$. Since each car travels for 4 hours, the time, t, for each car is 4. This information is shown in the chart. The distance is found by using the formula $d = rt$ and the expressions already entered in the chart.

	r	t	d
Faster car	x	4	$4x$
Slower car	y	4	$4y$

Find d from $d = rt$.

Draw a sketch showing what is happening in the problem. See Figure 7.8.

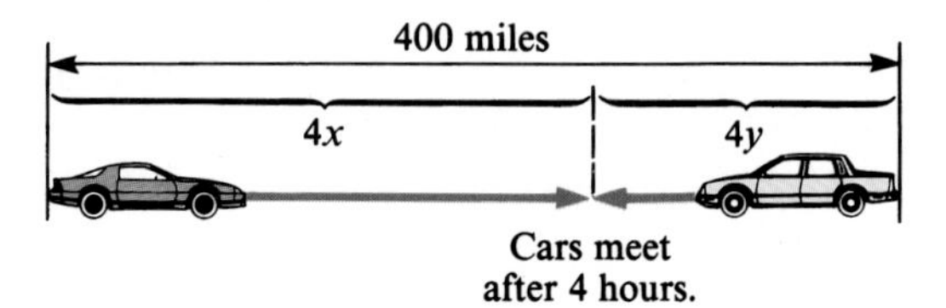

FIGURE 7.8

Step 3 As shown in the figure, since the total distance traveled by both cars is 400 miles, one equation is

$$4x + 4y = 400.$$

Because the faster car goes 20 miles per hour faster than the slower car, the second equation is

$$x = 20 + y.$$

Step 4 This system of equations,

$$4x + 4y = 400$$
$$x = 20 + y,$$

can be solved by substitution. Replace x with $20 + y$ in the first equation of the system and solve for y.

$$4(20 + y) + 4y = 400 \quad \text{Let } x = 20 + y.$$
$$80 + 4y + 4y = 400 \quad \text{Distributive property}$$
$$80 + 8y = 400 \quad \text{Combine like terms.}$$
$$8y = 320 \quad \text{Subtract 80.}$$
$$y = 40 \quad \text{Divide by 8.}$$

Step 5 Since $x = 20 + y$, and $y = 40$,

$$x = 20 + 40 = 60.$$

The speeds of the two cars are 40 miles per hour and 60 miles per hour.

Step 6 Check the answers. Since each car travels for 4 hours, the total distance traveled is

$$4(40) + 4(60) = 160 + 240 = 400$$

miles, as required. ■

The problems in this section also could be solved using only one variable, but for many of them the solution is simpler with two variables.

CAUTION Be careful! When you use two variables to solve a problem, you must write two equations.

7.4 EXERCISES

Write a system of equations for each problem. Then solve the system. Formulas are given in Appendices A and B. See Example 1.

1. The sum of two numbers is 52, and their difference is 34. Find the numbers.
2. Find two numbers whose sum is 56 and whose difference is 18.
3. Two angles are supplementary. The larger angle measures 20° more than three times the smaller. Find the measures of the angles.
4. Two angles are complementary. Their difference is 54°. Find the measures of the angles.
5. A certain number is three times as large as a second number. Their sum is 96. What are the two numbers?
6. One number is five times as large as another. The difference between the numbers is 48. Find the numbers.
7. In 1988, London's Heathrow Airport and Tokyo's Haneda Airport serviced a total of 69,702,340 passengers. Heathrow serviced 5,348,260 more passengers than did Haneda. How many passengers did each airport service?
8. In 1987–88, the number of learning disabled children served in programs for the handicapped was 2,518,000 more than the number served in programs for speech impaired children. Together, a total of 6,374,000 children with these two handicaps were served. How many learning disabled and how many speech impaired children were served?
9. Write an explanation of the method of solving an applied problem using a system of linear equations. (See Objective 1.)
10. Suppose that you are given a choice of methods to solve a problem, and you find that the problem can either be worked using one variable and one equation, or using two variables and two equations. Which method would you probably prefer? Why? (*Hint:* There is no right or wrong answer here. Explain your preference.)

Write a system of equations for each problem and then solve the system. See Example 2.

11. Mai Ling counts her \$1 and \$5 bills at the end of the day. She has a total of 81 bills, and the total value of the money is \$193. How many of each type of bill does she have?

TYPE OF BILL	NUMBER OF BILLS	TOTAL VALUE
\$1	x	
\$5	y	
Totals	81	\$193

12. Beverly David is a bank teller. At the end of a day she has a total of 86 ten-dollar bills and twenty-dollar bills. The total value of the money is \$1220. How many of each type of bill does she have?

TYPE OF BILL	NUMBER OF BILLS	TOTAL VALUE
\$10	x	\$10x
\$20	y	
Totals		

13. 193 tickets were sold for a soccer game. Student tickets cost \$.50 each and non-student tickets cost \$1.50 each. A total of \$206.50 was collected. How many of each type of ticket were sold?

14. Wally Smart bought some cassettes at \$5 each and some compact discs at \$9 each. He spent \$123 and got a total of 15 cassettes and discs. How many of each type did he buy?

15. Joanna Collins has twice as much money invested at 10% annual simple interest as she has at 8%. If her yearly income from these investments is \$700, how much does she have invested at each rate?

16. Jairo Santanilla has three times as much money invested at 9% interest as he does at 7%. If his yearly income from the investments is \$1360, how much does he have invested at each rate?

17. An artist bought some large canvases at \$7 each and some small ones at \$4 each, paying \$219 in total. Altogether, he bought 39 canvases. How many of each size did he buy?

18. A hospital bought a total of 146 bottles of glucose solution. Small bottles cost \$2 each, and large ones cost \$3 each. The total cost was \$336. How many of each size bottle were bought?

19. A textbook author receives a \$2 royalty for each of her algebra books sold, and a \$3 royalty for each of her calculus books sold. During one royalty period, the two books together sold 13,000 copies and she received a total of \$29,000 in royalties. How many of each kind of book were sold during that period?

20. The Earl K. Long Library bought 54 books. Some cost \$32 each and some cost \$44 each. The total cost of the books was \$1968. How many of each type of book did the library buy?

Write a system of equations for each problem and then solve the system. See Example 3.

21. A 40% dye solution is to be mixed with a 70% dye solution to get 60 liters of a 50% solution. How many liters of the 40% and 70% solutions will be needed?

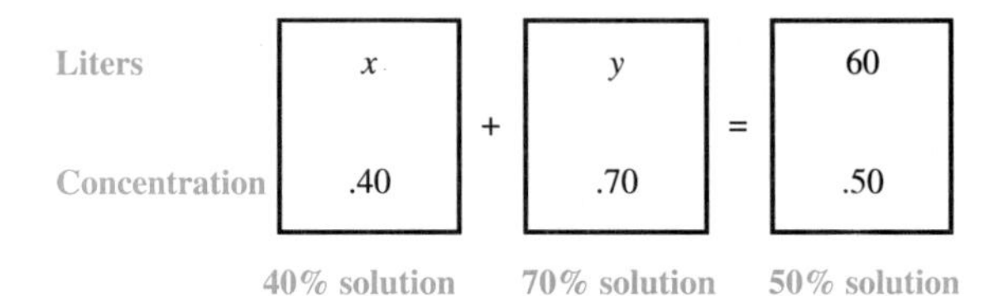

22. A 90% antifreeze solution is to be mixed with a 75% solution to make 40 liters of a 78% solution. How many liters of the 90% and 75% solutions will be used?

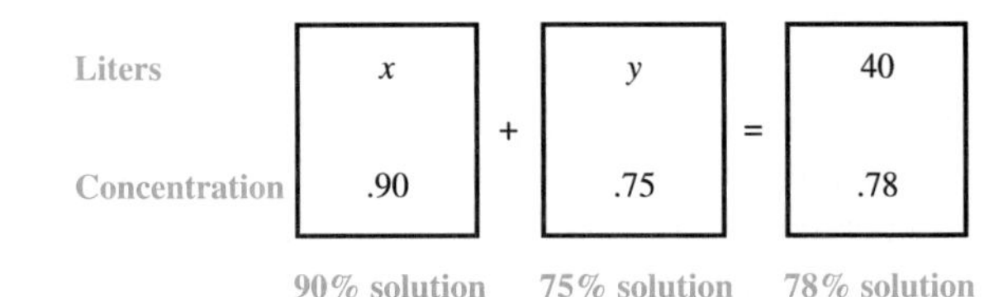

23. An alloy that is 60% copper is to be combined with an alloy that is 80% copper to obtain 40 pounds of a 65% alloy. How many pounds of the 60% and the 80% alloys will be needed?

24. How many liters of a 25% indicator solution and a 55% indicator solution should be mixed to obtain 24 liters of a 45% solution?

25. A grocer wishes to blend candy selling for \$1.20 a pound with candy selling for \$1.80 a pound to get a mixture that will be sold for \$1.40 a pound. How many pounds of the \$1.20 and the \$1.80 candy should be used to get 90 pounds of the mixture?

26. A merchant wishes to mix coffee worth \$6 per pound with coffee worth \$3 per pound to get 180 pounds of a mixture worth \$4 per pound. How many pounds of the \$6 and the \$3 coffee will be needed?

27. The owner of a nursery wants to mix some fertilizer worth \$7.00 per bag with some worth \$8.00 per bag to obtain 80 bags of mixture worth \$7.75 per bag. How many bags of each type should she use?

28. How many barrels of pickles worth \$40 per barrel and pickles worth \$60 per barrel must be mixed to obtain 100 barrels of a mixture worth \$48 per barrel?

Write a system of equations for each problem and then solve the system. See Example 4.

29. Two cars start from towns 300 miles apart and travel toward each other. They meet after 3 hours. Find the rate of each car if one travels 30 miles per hour slower than the other.

	r	t	d
Faster car	x	3	$3x$
Slower car	y	3	$3y$

30. Two trains start from stations 1000 miles apart and travel toward each other. They meet after 5 hours. Find the rate of each train if one travels 20 miles per hour faster than the other.

	r	t	d
Faster train	x	5	$5x$
Slower train	y	5	$5y$

31. It takes a boat 1 1/2 hours to go 12 miles downstream, and 6 hours to return. Find the speed of the boat in still water and the speed of the current. Let x = the speed of the boat in still water and y = the speed of the current.

	d	r	t
Downstream	12	$x + y$	$\frac{3}{2}$
Upstream	12	$x - y$	6

32. A boat takes 3 hours to go 24 miles upstream. It can go 36 miles downstream in the same time. Find the speed of the current and the speed of the boat in still water if x = the speed of the boat in still water and y = the speed of the current.

	d	r	t
Downstream	36	$x + y$	3
Upstream	24	$x - y$	3

33. If a plane can travel 400 miles per hour into the wind and 540 miles per hour with the wind, find the speed of the wind and the speed of the plane in still air.

34. A small plane travels 100 miles per hour with the wind and 60 miles per hour against it. Find the speed of the wind and the speed of the plane in still air.

Write a system of equations for each problem. Then solve the system.

35. Mr. Dawkins has \$10,000 to invest, part at 10% and part at 14% annual simple interest. He wants the interest income from the two investments to total \$1100 yearly. How much should he invest at each rate?

36. Garrett's Candy Shop wants to prepare 200 kilograms of a candy mixture for an Easter promotion. How much \$4.00 per kilogram candy should be mixed with \$5.20 per kilogram candy to get a mixture that can be sold for \$4.84 per kilogram?

37. A rectangle is twice as long as it is wide. Its perimeter is 60 inches. Find the dimensions of the rectangle.

38. Mrs. Gutierrez has twice as much money invested at 14% as she has at 16%. If her yearly income from investments is \$880, how much does she have invested at each rate?

39. A merchant wishes to mix nuts worth \$3 per pound with nuts worth \$1.50 per pound to get 45 pounds of a mixture worth \$2 per pound. How many pounds of the two kinds of nuts will be needed?

40. The perimeter of a triangle is 21 inches. If two sides are of equal length, and the third side is 3 inches longer than one of the equal sides, find the lengths of the three sides.

41. At the beginning of a walk for charity, John and Harriet are 30 miles apart. If they leave at the same time and walk in the same direction, John overtakes Harriet in 60 hours. If they walk toward each other, they meet in 5 hours. What are their speeds?

42. Mr. Anderson left Farmersville in a plane at noon to travel to Exeter. Mr. Bentley left Exeter in his automobile at 2 P.M. to travel to Farmersville. It is 400 miles from Exeter to Farmersville. If the sum of their speeds is 120 miles per hour, and if they met at 4 P.M., find the speed of each.

43. The Smith family is coming to visit, and no one knows how many children they have. Janet, one of the girls, says she has as many brothers as sisters; her brother Steve says he has twice as many sisters as brothers. How many boys and how many girls are in the family?

44. In the Lopez family, the number of boys is one more than half the number of girls. One of the Lopez boys, Rico, says that he has one more sister than brothers. How many boys and how many girls are in the family?

PREVIEW EXERCISES

Graph each linear inequality. See Section 6.5.

45. $x + y \leq 5$ **46.** $2x - y \geq 7$ **47.** $3x + 2y > 6$

48. $x + 3y < 9$ **49.** $5x + 4y > 0$ **50.** $3x - 4y < 0$

7.5 SOLVING SYSTEMS OF LINEAR INEQUALITIES

OBJECTIVE

1 SOLVE SYSTEMS OF LINEAR INEQUALITIES BY GRAPHING.

Graphing the solution of a linear inequality was discussed in Section 6.5. For example, to graph the solution of $x + 3y > 12$, first graph the line $x + 3y = 12$ by finding a few ordered pairs that satisfy the equation. Because the points on the line do not satisfy the inequality, use a dashed line. To decide which side of the line should be shaded, choose any test point not on the line, such as (0, 0). Substitute 0 for x and 0 for y in the given inequality.

$$x + 3y > 12$$

$$0 + 3(0) > 12 \quad ? \quad \text{Let } x = 0, y = 0.$$

$$0 > 12 \quad \text{False}$$

Since the test point does not satisfy the inequality, shade the region on the side of the line that does not include (0, 0), as in Figure 7.9.

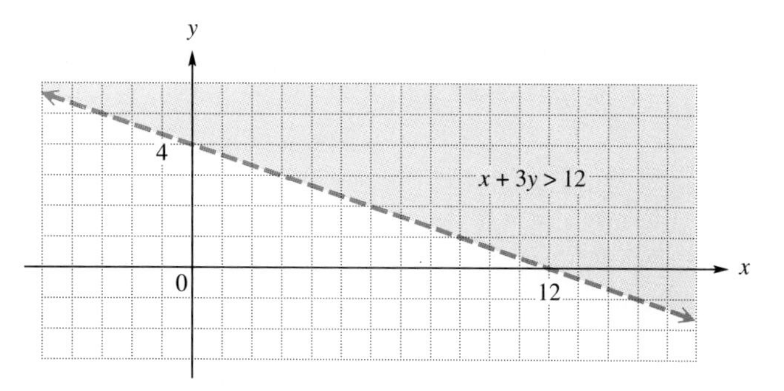

FIGURE 7.9

1 The same method is used to find the solution of a system of two linear inequalities, as shown in Examples 1–3. A **system of linear inequalities** consists of two or more linear inequalities. The **solution of a system of linear inequalities** includes all points that make all inequalities of the system true at the same time.

To solve a system of linear inequalities, use the following steps.

GRAPHING A SYSTEM OF LINEAR INEQUALITIES

Step 1 Graph each inequality using the method of Section 6.5.
Step 2 Indicate the solution of the system by using dark shading on the intersection of the two graphs (the region that the two graphs overlap).

EXAMPLE 1
SOLVING A SYSTEM OF LINEAR INEQUALITIES

Graph the solution of the linear system
$$3x + 2y \le 6$$
$$2x - 5y \ge 10.$$

Begin by graphing $3x + 2y \le 6$. To do this, graph $3x + 2y = 6$ as a solid line. Since (0, 0) makes the inequality true, shade the region containing (0, 0), as shown in Figure 7.10.

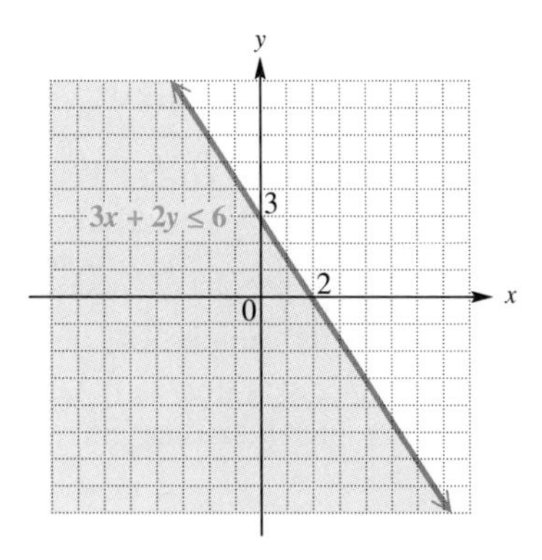

FIGURE 7.10

Now graph $2x - 5y \geq 10$. The solid line boundary is the graph of $2x - 5y = 10$. Since (0, 0) makes the inequality false, shade the region that does not contain (0, 0), as shown in Figure 7.11.

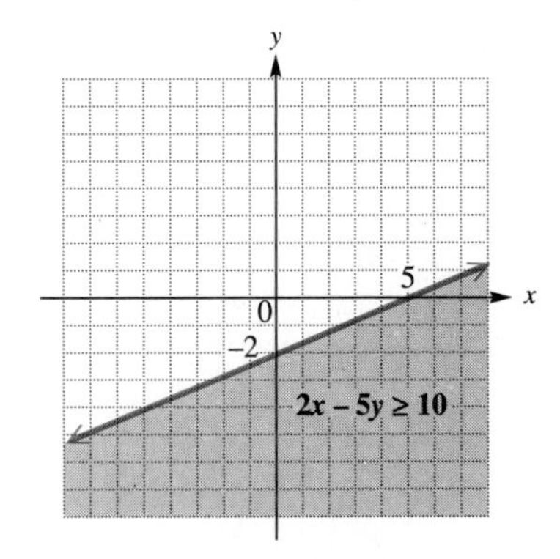

FIGURE 7.11

The solution of the system is given by the intersection (overlap) of the regions of the graphs in Figures 7.10 and 7.11. The solution is the darkest shaded region in Figure 7.12, and includes portions of the two boundary lines. ■

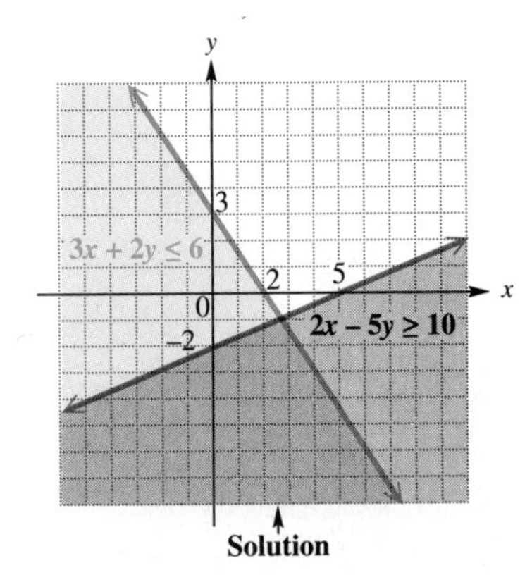

FIGURE 7.12

NOTE In practice, we usually do all the work on one set of axes at the same time. In the following examples, only one graph is shown. Be sure that the desired region of the solution is clearly indicated.

EXAMPLE 2
SOLVING A SYSTEM OF LINEAR INEQUALITIES

Graph the solution of the system

$$x - y > 5$$
$$2x + y < 2.$$

Figure 7.13 shows the graphs of both $x - y > 5$ and $2x + y < 2$. Dashed lines show that the graphs of the inequalities do not include their boundary lines. The solution of the system is the darkest shaded region in the figure. The solution does not include either boundary line. ■

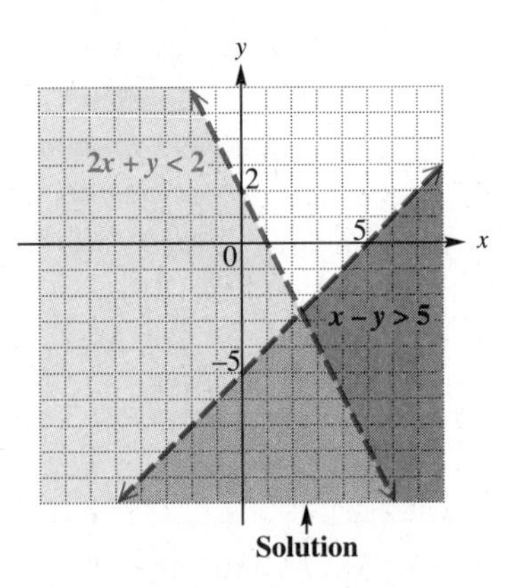

FIGURE 7.13

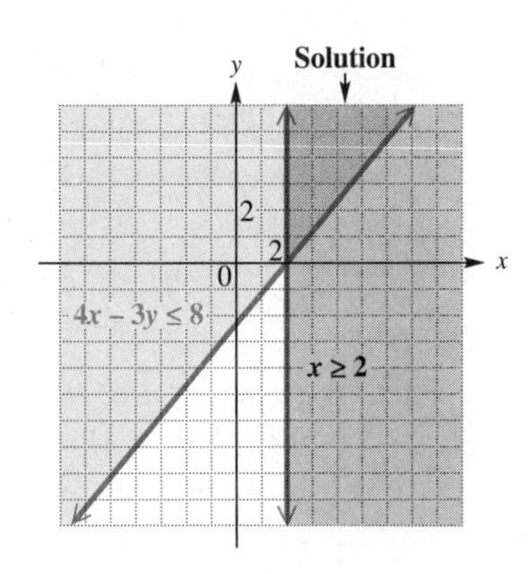

FIGURE 7.14

EXAMPLE 3
SOLVING A SYSTEM OF LINEAR INEQUALITIES

Graph the solution of the system

$$4x - 3y \leq 8$$
$$x \geq 2.$$

Recall that $x = 2$ is a vertical line through the point (2, 0). The graph of the solution is the darkest shaded region in Figure 7.14. ■

7.5 EXERCISES

Graph the solution of each system of linear inequalities. See Examples 1–3.

1. $x + y \leq 6$
$x - y \leq 1$

2. $x + y \geq 2$
$x - y \leq 3$

3. $2x - 3y \leq 6$
$x + y \geq -1$

4. $4x + 5y \leq 20$
$x - 2y \leq 5$

5. $x + 4y \leq 8$
$2x - y \leq 4$

6. $3x + y \leq 6$
$2x - y \leq 8$

7. $x - 4y \leq 3$
$x \geq 2y$

8. $2x + 3y \leq 6$
$x - y \geq 5$

9. $x + 2y \leq 4$
$x + 1 \geq y$

10. $y \leq 2x - 5$
$x - 3y \leq 2$

11. $4x + 3y \leq 6$
$x - 2y \geq 4$

12. $3x - y \leq 4$
$-6x + 2y \leq -10$

13. $x - 2y > 6$
$2x + y > 4$

14. $3x + y < 4$
$x + 2y > 2$

15. $x < 2y + 3$
$x + y > 0$

16. $2x + 3y < 6$
$4x + 6y > 18$

17. $x - 3y \leq 6$
$x \geq -1$

18. $2x + 5y \geq 10$
$x \leq 4$

19. $3x - 2y \geq 9$
$y \leq 3$

20. $x \geq -1$
$y \geq 4$

21. $x \geq 2$
$y \leq 3$

22. Which one of the following systems of linear inequalities is graphed in the figure? (See Objective 1.) Sect. 7.5 Ex. 22-gd.

(a) $x \leq 3$, $y \leq 1$ **(b)** $x \leq 3$, $y \geq 1$

(c) $x \geq 3$, $y \leq 1$ **(d)** $x \geq 3$, $y \geq 1$

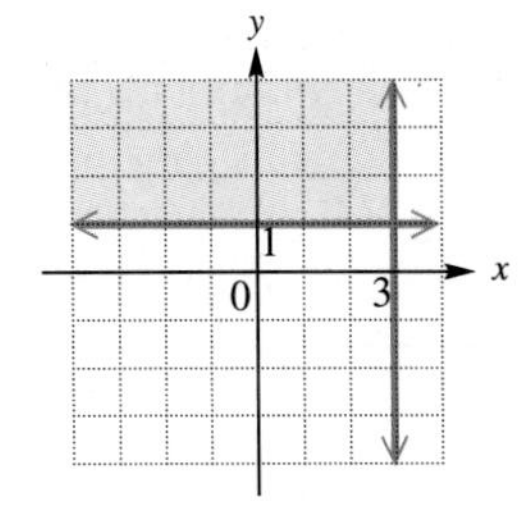

23. Suppose that your friend was absent from class (because of a bad cold) on the day that systems of linear inequalities were covered, and she asked you to write a short explanation of the process. Do this in a brief, concise explanation. (See Objective 1.)

The method of solving systems of linear inequalities can be extended to systems with more than two inequalities. For example, if there are three, graph each on the same set of axes and indicate the intersection (overlap) of all three regions. Use this method in Exercises 24–26.

24. $4x + 5y < 8$, $y > -2$, $x > -4$

25. $x + y \geq -3$, $x - y \leq 3$, $y \leq 3$

26. $3x - 2y \geq 6$, $x + y \leq 4$, $x \geq 0$, $y \geq -4$

PREVIEW EXERCISES

Find each value. See Section 1.2.

27. 5^2 **28.** 7^2 **29.** 11^2 **30.** 15^2

31. 6^3 **32.** 3^3 **33.** 2^4 **34.** 3^5

CHAPTER 7 GLOSSARY

KEY TERMS

7.1 **system of linear equations** A system of linear equations consists of two or more linear equations with the same variables.

solution of a system The solution of a system of linear equations includes all the ordered pairs that make all the equations of the system true at the same time.

consistent system A system of equations with a solution is a consistent system.

inconsistent system An inconsistent system of equations is a system with no solution.

independent equations Equations of a system that have different graphs are called independent equations.

dependent equations Equations of a system that have the same graph (because they are different forms of the same equation) are called dependent equations.

7.5 **system of linear inequalities** A system of linear inequalities contains two or more linear inequalities (and no other kinds of inequalities).

solution of a system of linear inequalities The solution of a system of linear inequalities consists of all points that make all inequalities of the system true at the same time.

CHAPTER 7 QUICK REVIEW

SECTION	CONCEPTS	EXAMPLES
7.1 SOLVING SYSTEMS OF LINEAR EQUATIONS BY GRAPHING	An ordered pair is a solution of a system if it makes all equations of the system true at the same time.	Is $(4, -1)$ a solution of the following system? $$x + y = 3$$ $$2x - y = 9$$ Yes, because $4 + (-1) = 3$, and $2(4) - (-1) = 9$ are both true.
	If the graphs of the equations of a system are both sketched on the same axes, the points of intersection, if any, are solutions of the system.	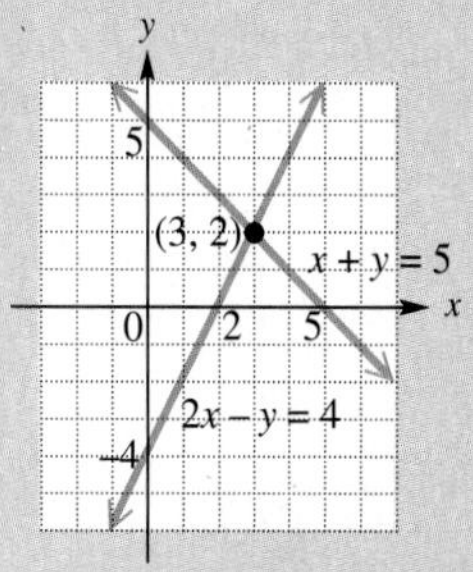 y, 5, (3, 2), $x + y = 5$, 0, 2, 5, x, $2x - y = 4$, −4 $(3, 2)$ is the solution of the system $$x + y = 5$$ $$2x - y = 4.$$
7.2 SOLVING SYSTEMS OF LINEAR EQUATIONS BY ADDITION	*Step 1* Write both equations in the form $Ax + By = C$.	Solve by addition. $$x + 3y = 7$$ $$3x - y = 1$$
	Step 2 Multiply one or both equations by appropriate numbers so that the coefficients of x (or y) are negatives of each other.	Multiply the top equation by -3 to eliminate the x terms by addition. $$-3x - 9y = -21$$ $$3x - y = 1$$
	Step 3 Add the equations to get an equation with only one variable.	$-10y = -20$ Add.
	Step 4 Solve the equation from Step 3.	$y = 2$ Divide by −10.

SECTION	CONCEPTS	EXAMPLES
	Step 5 Substitute the solution from Step 4 into either of the original equations to find the value of the remaining variable.	Substitute to get the value of x. $x + 3y = 7$ $x + 3(2) = 7$ Let $y = 2$. $x + 6 = 7$ Multiply. $x = 1$ Subtract 6.
	Step 6 Write the solution as an ordered pair, and check the answer.	The solution, (1, 2), checks.
	If the result of the addition step is a false statement, such as $0 = 4$, the graphs are parallel lines and there is *no solution* for the system.	$x - 2y = 6$ $-x + 2y = -2$ $0 = 4$ No solution
	If the result is a true statement, such as $0 = 0$, the graphs are the same line, and an *infinite number of ordered pairs are solutions*.	$x - 2y = 6$ $-x + 2y = -6$ $0 = 0$ Infinite number of solutions
7.3 SOLVING SYSTEMS OF LINEAR EQUATIONS BY SUBSTITUTION	Solve one equation for one variable, and substitute the expression into the other equation to get an equation in one variable. Solve the equation, and then substitute the solution into either of the original equations to obtain the value of the other variable. Check the answer.	Solve by substitution. $x + 2y = -5$ (1) $y = -2x - 1$ (2) Substitute $-2x - 1$ for y in equation (1). $x + 2(-2x - 1) = -5$ Solve to get $x = 1$. To find y, let $x = 1$ in equation (2): $y = -2(1) - 1 = -3$. The solution is $(1, -3)$.
7.4 APPLICATIONS OF LINEAR SYSTEMS		The sum of two numbers is 30. Their difference is 6. Find the numbers.
	Step 1 Choose a variable to represent each unknown value.	Let x represent one number. Let y represent the other number.

SECTION	CONCEPTS	EXAMPLES
	Step 2 Draw a figure or a diagram if it will help. *Step 3* Translate the problem into a system of two equations using both variables. *Step 4* Solve the system.	$x + y = 30$ $x - y = 6$ $2x = 36$ Add. $x = 18$ Divide by 2. Let $x = 18$ in the top equation: $18 + y = 30$. Solve to get $y = 12$.
	Step 5 Answer the question or questions asked.	The two numbers are 18 and 12.
	Step 6 Check the solution in the words of the problem.	$18 + 12 = 30$ and $18 - 12 = 6$, so the solution checks.
7.5 SOLVING SYSTEMS OF LINEAR INEQUALITIES	To solve a system of linear inequalities, graph each inequality on the same axes. (This was explained in Section 6.5.) The solution of the system is the overlap of the regions of the two graphs.	The shaded region is the solution of the system $2x + 4y \geq 5$ $x \geq 1$. 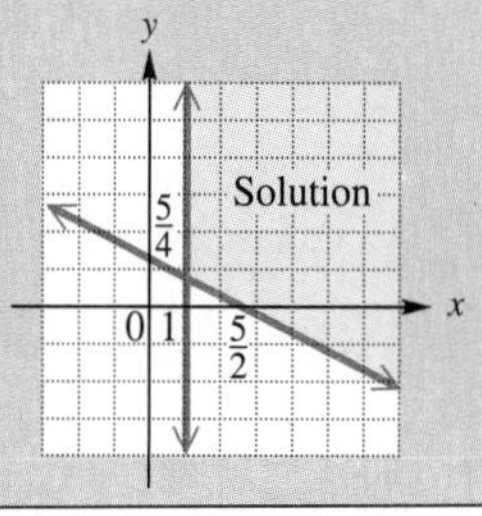

CHAPTER 7 REVIEW EXERCISES

[7.1] *Decide whether the given ordered pair is a solution of the given system.*

1. (3, 4)
$4x - 2y = 4$
$5x + y = 17$

2. (−2, 1)
$5x + 3y = -7$
$2x - 3y = -7$

3. (−5, 2)
$x - 4y = -10$
$2x + 3y = -4$

4. (6, 3)
$3x + 8y = 42$
$4x - 3y = 15$

5. (−1, −3)
$x + 2y = -7$
$-x + 3y = -8$

6. (2, 6)
$3x - y = 0$
$4x + 2y = 20$

7. Explain why (0, 0) must always be a solution of a system in the form

$$Ax + By = 0$$
$$Cx + Dy = 0.$$

8. Can a system of two linear equations in two unknowns have exactly three solutions? Explain.

Solve each system by graphing.

9. $x + y = 3$
$2x - y = 3$

10. $x - 2y = -6$
$2x + y = -2$

11. $2x + 3y = 1$
$4x - y = -5$

12. $x = 2y + 2$
$2x - 4y = 4$

13. $y + 2 = 0$
$3y = 6x$

14. $2x + 5y = 10$
$2x + 5y = -5$

[7.2] *Solve each system by the addition method. Identify any inconsistent systems or systems with dependent equations.*

15. $x + y = 6$
$2x + y = 8$

16. $3x - y = 13$
$x - 2y = 1$

17. $5x + 4y = 7$
$3x - 4y = 17$

18. $-4x + 3y = -7$
$6x - 5y = 11$

19. $3x - 4y = 7$
$6x - 8y = 14$

20. $2x + y = 5$
$2x + y = 8$

21. $3x - 2y = 14$
$5x - 4y = 24$

22. $2x + 6y = 18$
$3x + 5y = 19$

23. $3x - 4y = -1$
$-6x + 8y = 2$

24. $\frac{1}{3}x - \frac{1}{5}y = 3$
$\frac{1}{2}x + \frac{2}{5}y = 1$

25. $\frac{1}{6}x + \frac{1}{2}y = \frac{4}{3}$
$\frac{1}{4}x - \frac{1}{6}y = \frac{1}{6}$

[7.3] *Solve each system by the substitution method.*

26. $2x - 5y = 5$
$y = x + 2$

27. $3x + y = 14$
$x = 2y$

28. $5x + y = -6$
$x = 3y + 2$

29. $4x + 5y = 35$
$x = 2y - 1$

30. $6x + 5y = 9$
$2x - 3y = 17$

31. $2x + 3y = 5$
$3x + 4y = 8$

32. $\frac{1}{5}x + \frac{1}{8}y = 0$
$\frac{1}{2}x + \frac{1}{4}y = \frac{1}{2}$

33. $\frac{1}{3}x + \frac{1}{7}y = \frac{52}{21}$
$\frac{1}{2}x - \frac{1}{3}y = \frac{19}{6}$

[7.4] *Solve each problem by writing a system of equations and then solving it.*

34. One number is 2 more than twice as large as another. Their sum is 17. Find the numbers.

35. The perimeter of a rectangle is 40 meters. Its length is 1 1/2 times its width. Find the length and width of the rectangle.

36. A cashier has 20 bills, all of which are ten-dollar or twenty-dollar bills. The total value of the money is $250. How many of each type does he have?

37. Ms. Branson has $18,000 to invest. She wants the total annual income from the money to be $2600. She can invest part of it at 12% and the rest at 16%. How much should she invest at each rate?

38. A certain plane flying with the wind travels 540 miles in 2 hours. Later, flying against the same wind, the plane travels 690 miles in 3 hours. Find the speed of the plane in still air and the speed of the wind.

39. Two cars leave at the same time from towns 400 miles apart and travel toward each other. They meet after 4 hours. Find the average speed of each car if one car travels 10 miles per hour slower than the other.

[7.5] *Graph each system of linear inequalities.*

40. $x + y \geq 4$
$x - y \leq 2$

41. $x + y \leq 3$
$2x \geq y$

42. $y + 2 < 2x$
$x > 3$

43. $y \geq 3x$
$2x + 3y \leq 4$

MIXED REVIEW EXERCISES

Solve by any method.

44. $\frac{3x}{4} - \frac{y}{3} = \frac{7}{6}$
$\frac{x}{2} + \frac{2y}{3} = \frac{5}{3}$

45. $2x + y - x = 3y + 5$
$y + 2 = x - 5$

46. Candy that sells for \$1.30 a pound is to be mixed with candy selling for \$.90 a pound to get 50 pounds of a mix that will sell for \$1 per pound. How much of each type should be used?

47. $5x + 4y = 3$
$7x + 5y = 3$

48. $\frac{5x}{3} - \frac{y}{2} = -\frac{31}{3}$
$2x + \frac{y}{3} = \frac{26}{3}$

49. $2x + y < 6$
$y - 2x < 6$

50. $3x + 5y = -1$
$5x + 4y = 7$

51. $\frac{2x}{5} - \frac{y}{2} = \frac{7}{10}$
$\frac{x}{3} + y = 2$

52. $4x + y = 5$
$-12x - 3y = -15$

53. The sum of two numbers is 42, and their difference is 14. Find the numbers.

54. $x \geq 2$
$y \leq 5$

55. $\frac{x}{2} + \frac{y}{3} = -\frac{8}{3}$
$\frac{x}{4} + 2y = \frac{1}{2}$

56. $5x - 3 + y = 4y + 8$
$2y + 1 = x - 3$

CHAPTER 7 TEST

Solve each system by graphing.

1. $2x + y = 5$
$3x - y = 15$

2. $x + 2y = 6$
$2x - y = 7$

Solve each system by the addition method.

3. $2x - 5y = -13$
$3x + 5y = 43$

4. $4x + 3y = 25$
$5x + 4y = 32$

5. $6x + 5y = -13$
$-3x - \frac{5y}{2} = \frac{13}{2}$

6. $2x - y = 5$
$4x + 3y = 0$

7. $4x + 5y = 8$
$-8x - 10y = -6$

8. $\frac{6}{5}x - y = \frac{1}{5}$
$-\frac{2}{3}x + \frac{1}{6}y = \frac{1}{3}$

Solve each system by substitution.

9. $2x + y = 1$
$x = 8 + y$

10. $4x + 3y = 0$
$2x + y = -4$

Solve each system by any method.

11. $8 + 3x - 4y = 14 - 3y$
$3x + y + 12 = 9x - y$

12. $\frac{x}{2} - \frac{y}{4} = -4$
$\frac{2x}{3} + \frac{5y}{4} = 1$

---

13. Suppose that the graph of a system of two linear equations consists of lines that have the same slope but different y-intercepts. How many solutions does the system have?

Solve each problem by writing a system of equations and then solving it.

14. The sum of two numbers is 44. If one is doubled, the result is 1 more than the other. Find the numbers.

15. A local music store is having a sale. Some compact discs cost $9.00 and some cost $7.00. B. P. Gainey spent $86.00 and bought a total of 10 compact discs. How many did he buy at each price?

16. A 40% solution of acid is to be mixed with a 60% solution to get 100 liters of a 45% solution. How many liters of each solution should be used?

17. Two cars leave from Perham, Minnesota, and travel in the same direction. One car travels one and one third times as fast as the other. After 3 hours, they are 45 miles apart. What is the speed of each car?

Graph the solution of each system of inequalities.

18. $2x + 7y \leq 14$
$x - y \geq 1$

19. $2x - y \leq 6$
$4y + 12 \geq -3x$

20. $3x - 5y < 15$
$y < 2$

CHAPTER EIGHT

ROOTS AND RADICALS

The distance a person can see to the horizon from an airplane flying at 30,000 feet is approximately $1.22\sqrt{30{,}000}$ feet. The number $\sqrt{30{,}000}$ is a *square root* or *radical expression*. In this chapter we shall see how to evaluate a number like $\sqrt{30{,}000}$ and how to perform arithmetic operations with radical expressions.

Most roots and radicals represent real numbers and the properties of real numbers studied earlier apply to them. In particular, the rules we used in Chapter 3 in our work with polynomials will be used with radical expressions as well.

8.1 FINDING ROOTS

OBJECTIVES

1. FIND SQUARE ROOTS.
2. DECIDE WHETHER A GIVEN ROOT IS RATIONAL, IRRATIONAL, OR NOT A REAL NUMBER.
3. FIND DECIMAL APPROXIMATIONS FOR IRRATIONAL SQUARE ROOTS.
4. USE THE PYTHAGOREAN FORMULA.
5. FIND HIGHER ROOTS.

FOCUS ON PROBLEM SOLVING

Two cars leave Edmonton, Alberta, at the same time. One travels south at 25 miles per hour and the other travels east at 60 miles per hour. How far apart are they after 2 hours?

This problem can be solved by completing a right triangle and using the Pythagorean formula. The distance is given by a square root. Understanding how to find and express square roots as decimal numbers is necessary to solve a variety of practical problems. This problem is Exercise 79 in the exercises for this section. After working through this section, you should be able to solve the problem.

In Section 1.2 we discussed the idea of a *square* of a number. Recall that squaring a number means multiplying the number by itself.

$$\text{If } a = 7, \text{ then } a^2 = 7 \cdot 7 = 49.$$

$$\text{If } a = 10, \text{ then } a^2 = 10 \cdot 10 = 100.$$

$$\text{If } a = -5, \text{ then } a^2 = (-5) \cdot (-5) = 25.$$

$$\text{If } a = -\frac{1}{2}, \text{ then } a^2 = \left(-\frac{1}{2}\right) \cdot \left(-\frac{1}{2}\right) = \frac{1}{4}.$$

In this chapter, the opposite problem is considered.

$$\text{If } a^2 = 49, \text{ then } a = ?$$

$$\text{If } a^2 = 100, \text{ then } a = ?$$

$$\text{If } a^2 = 25, \text{ then } a = ?$$

$$\text{If } a^2 = \frac{1}{4}, \text{ then } a = ?$$

1 Finding a in the four statements on the previous page requires finding a number that can be multiplied by itself to result in the given number. The number a is called a **square root** of the number a^2.

EXAMPLE 1
FINDING THE SQUARE ROOTS OF A NUMBER

Find all square roots of 49.

Find a square root of 49 by thinking of a number that multiplied by itself gives 49. One square root is 7, since $7 \cdot 7 = 49$. Another square root of 49 is -7, since $(-7)(-7) = 49$. The number 49 has two square roots, 7 and -7. One is positive and one is negative. ■

The positive square root of a number is written with the symbol $\sqrt{}$. For example, the positive square root of 121 is 11, written

$$\sqrt{121} = 11.$$

The symbol $-\sqrt{}$ is used for the negative square root of a number. For example, the negative square root of 121 is -11, written

$$-\sqrt{121} = -11.$$

The symbol $\sqrt{}$ is called a **radical sign** and always represents the nonnegative square root. The number inside the radical sign is called the **radicand** and the entire expression, radical sign and radicand, is called a **radical.** An algebraic expression containing a radical is called a **radical expression.**

SQUARE ROOTS OF a

If a is a nonnegative real number,

$$\sqrt{a} \text{ is the positive square root of } a,$$
$$-\sqrt{a} \text{ is the negative square root of } a.$$

Also, for nonnegative a,

$$\sqrt{a} \cdot \sqrt{a} = (\sqrt{a})^2 = a \text{ and } -\sqrt{a} \cdot -\sqrt{a} = (-\sqrt{a})^2 = a.$$

When the square root of a positive real number is squared, the result is that positive real number. (Also, $(\sqrt{0})^2 = 0$.) This is illustrated in the next example.

EXAMPLE 2
SQUARING RADICAL EXPRESSIONS

Find the *square* of each radical expression.

(a) $\sqrt{13}$

$(\sqrt{13})^2 = 13$, by the definition of square root.

(b) $-\sqrt{29}$

$(-\sqrt{29})^2 = 29$ The square of a *negative* number is positive.

(c) $\sqrt{p^2 + 1}$

$(\sqrt{p^2 + 1})^2 = p^2 + 1$ ■

EXAMPLE 3
FINDING SQUARE ROOTS

Find each square root.

(a) $\sqrt{144}$

The radical $\sqrt{144}$ represents the positive square root of 144. Think of a positive number whose square is 144. The number is 12, so

$$\sqrt{144} = 12.$$

(b) $-\sqrt{1024}$

This symbol represents the negative square root of 1024. A calculator with a square root key can be used to find $\sqrt{1024} = 32$. Then,

$$-\sqrt{1024} = -32.$$

(c) $\sqrt{256} = 16$

(d) $-\sqrt{900} = -30$

(e) $\sqrt{\frac{4}{9}} = \frac{2}{3}$

(f) $-\sqrt{\frac{16}{49}} = -\frac{4}{7}$ ■

2 All numbers that have rational number square roots are called **perfect squares.** A number that is not a perfect square has a square root that is not a rational number. For example, $\sqrt{5}$ is not a rational number, because it cannot be written as the ratio of two integers. However, $\sqrt{5}$ is a real number and corresponds to a point on the number line. As mentioned earlier, a real number that is not rational is called an **irrational number.** The number $\sqrt{5}$ is irrational. Many square roots of integers are irrational.

If a is a positive number that is not a perfect square, then $\sqrt{a}$ is irrational.

Not every number has a *real number* square root. For example, there is no real number that can be squared to get -36. (The square of a real number can never be negative.) Because of this $\sqrt{-36}$ is not a real number.

If a is a negative number, then $\sqrt{a}$ is not a real number.

EXAMPLE 4
IDENTIFYING TYPES OF SQUARE ROOTS

Tell whether each square root is rational, irrational, or not a real number.

(a) $\sqrt{17}$

Since 17 is not a perfect square, $\sqrt{17}$ is irrational.

(b) $\sqrt{64}$

The number 64 is a perfect square, 8^2, so $\sqrt{64} = 8$, a rational number.

(c) $\sqrt{85}$ is irrational.

(d) $\sqrt{81}$ is rational ($\sqrt{81} = 9$).

(e) $\sqrt{-25}$

There is no real number whose square is -25. Therefore, $\sqrt{-25}$ is not a real number. ■

NOTE Not all irrational numbers are square roots of integers. For example, π (approximately 3.14159) is an irrational number that is not a square root of any integer.

3 Even if a number is irrational, a decimal that approximates the number can be found with a table or by using certain calculators. Square roots can be approximated as shown in Example 5.

EXAMPLE 5 APPROXIMATING IRRATIONAL SQUARE ROOTS

Find a decimal approximation for each square root. Round answers to the nearest thousandth.

(a) $\sqrt{11}$

Using the square root key of a calculator gives $\sqrt{11} \approx 3.31662479 \approx 3.317$ (rounded to the nearest thousandth), where $\approx$ means "is approximately equal to."

(b) $\sqrt{39} \approx 6.245$ **(c)** $\sqrt{745} \approx 27.295$ **(d)** $\sqrt{180} \approx 13.416$ ■

4 One application of square roots comes from using the Pythagorean formula. Recall from Section 4.6 that by this formula, if c is the length of the hypotenuse of a right triangle, and a and b are the lengths of the two legs, then

$$c^2 = a^2 + b^2.$$

EXAMPLE 6 USING THE PYTHAGOREAN FORMULA

Find the third side of each right triangle with sides a, b, and c, where c is the hypotenuse.

(a) $a = 3$, $b = 4$

Use the formula to find c^2 first.

$$\begin{aligned} c^2 &= a^2 + b^2 \\ &= 3^2 + 4^2 && \text{Let } a = 3 \text{ and } b = 4. \\ &= 9 + 16 = 25 && \text{Square and add.} \end{aligned}$$

Now find the positive square root of 25 to get c.

$$c = \sqrt{25} = 5$$

(Although -5 is also a square root of 25, the length of a side of a triangle must be a positive number.)

(b) $c = 9$, $b = 5$

Substitute the given values in the formula $c^2 = a^2 + b^2$. Then solve for a^2.

$$\begin{aligned} 9^2 &= a^2 + 5^2 && \text{Let } c = 9 \text{ and } b = 5. \\ 81 &= a^2 + 25 && \text{Square.} \\ 56 &= a^2 && \text{Subtract 25.} \end{aligned}$$

Again, we want only the positive root $a = \sqrt{56} \approx 7.483$. ■

CAUTION Be careful not to make the common mistake of thinking that $\sqrt{a^2 + b^2}$ equals $a + b$. As Example 6(a) shows,

$$\sqrt{9 + 16} = \sqrt{25} = 5 \neq \sqrt{9} + \sqrt{16} = 3 + 4,$$

so that, in general,

$$\sqrt{a^2 + b^2} \neq a + b.$$

The Pythagorean formula can be used to solve applied problems that involve right triangles.

PROBLEM SOLVING

When an applied problem involves lengths that form a right triangle, a good way to begin the solution is to sketch the triangle and label the three sides appropriately, using a variable as needed. Then use the Pythagorean formula to write an equation. ■

EXAMPLE 7 USING THE PYTHAGOREAN FORMULA

A ladder 10 feet long leans against a wall. The foot of the ladder is 6 feet from the base of the wall. How high up the wall does the top of the ladder rest?

As shown in Figure 8.1, a right triangle is formed with the ladder as the hypotenuse.

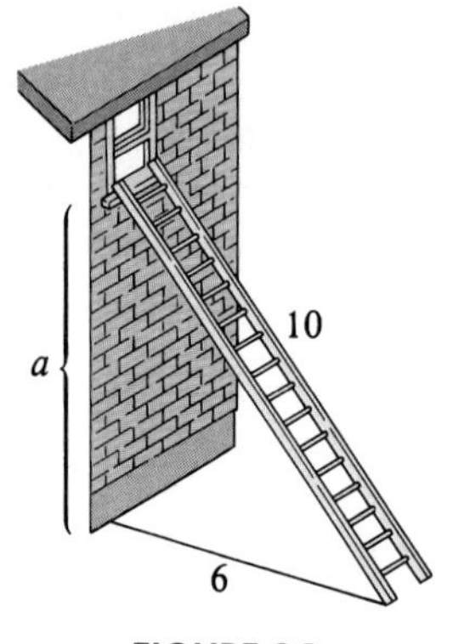

FIGURE 8.1

Let a represent the height of the top of the ladder. By the Pythagorean formula,

$$c^2 = a^2 + b^2$$

$$10^2 = a^2 + 6^2 \quad \text{Let } c = 10 \text{ and } b = 6.$$

$$100 = a^2 + 36 \quad \text{Square.}$$

$$64 = a^2 \quad \text{Subtract 36.}$$

$$\sqrt{64} = a$$

$$a = 8. \quad \sqrt{64} = 8$$

Choose the positive square root of 64 since a represents a length. The top of the ladder rests 8 feet up the wall. ■

5 Finding the square root of a number is the inverse of squaring a number. In a similar way, there are inverses to finding the cube of a number, or finding the fourth or higher power of a number. These inverses are called finding the **cube root,** written $\sqrt[3]{a}$, the **fourth root,** written $\sqrt[4]{a}$, and so on. In $\sqrt[n]{a}$, the number n is the **index** or **order** of the radical. It would be possible to write $\sqrt[2]{a}$ instead of $\sqrt{a}$, but the simpler symbol $\sqrt{a}$ is customary since the square root is the most commonly used root. A calculator that has a key marked $\sqrt[x]{}$ or x^y can be used to find these roots. When working with cube roots or fourth roots, it is helpful to memorize the first few *perfect cubes* ($2^3 = 8$, $3^3 = 27$, and so on), and the first few perfect fourth powers.

EXAMPLE 8
FINDING CUBE ROOTS

Find each cube root.

(a) $\sqrt[3]{8}$

Look for a number that can be cubed to give 8. Since $2^3 = 8$, then $\sqrt[3]{8} = 2$.

(b) $\sqrt[3]{-8}$

$$\sqrt[3]{-8} = -2 \quad \text{because} \quad (-2)^3 = -8. \quad \blacksquare$$

As these examples suggest, the cube root of a positive number is positive, and the cube root of a negative number is negative. *There is only one real number cube root for each real number.*

When the index of the radical is even (square root, fourth root, and so on), the radicand must be nonnegative to get a real number root. Also, for even indexes the symbols $\sqrt{\ }$, $\sqrt[4]{\ }$, $\sqrt[6]{\ }$, and so on are used for the *nonnegative* roots, which are called **principal roots.**

EXAMPLE 9
FINDING HIGHER ROOTS

Find each root.

(a) $\sqrt[4]{16}$

$\sqrt[4]{16} = 2$ because 2 is positive and $2^4 = 16$.

(b) $-\sqrt[4]{16} = -2$

(c) $\sqrt[4]{-16}$

There is no real number that equals $\sqrt[4]{-16}$ because a fourth power of a real number must be positive.

(d) $\sqrt[3]{64} = 4$ since $4^3 = 64$.

(e) $-\sqrt[5]{32}$

First find $\sqrt[5]{32}$. Since 2 is the number whose fifth power is 32, $\sqrt[5]{32} = 2$. If $\sqrt[5]{32} = 2$, then

$$-\sqrt[5]{32} = -2. \quad \blacksquare$$

8.1 EXERCISES

Find all square roots of each number. See Example 1.

1. 9 **2.** 16 **3.** 121 **4.** $\frac{196}{25}$

5. $\frac{400}{81}$ **6.** $\frac{900}{49}$ **7.** 625 **8.** 961

Find the square of each radical expression. See Example 2.

9. $\sqrt{16}$ **10.** $\sqrt{25}$ **11.** $-\sqrt{57}$

12. $-\sqrt{97}$ **13.** $\sqrt{4x^2 + 3}$ **14.** $\sqrt{7y^2 + 2}$

Find each root that is a real number. See Example 3.

15. $\sqrt{25}$ **16.** $\sqrt{36}$ **17.** $-\sqrt{64}$ **18.** $-\sqrt{100}$ **19.** $\sqrt{\frac{169}{49}}$

20. $\sqrt{-\frac{196}{25}}$ **21.** $-\sqrt{\frac{49}{9}}$ **22.** $-\sqrt{\frac{81}{16}}$ **23.** $\sqrt{-25}$

Write rational, irrational, *or* not a real number *for each radical. If a number is rational, give its exact value. If a number is irrational, use a calculator to give a decimal approximation to the nearest thousandth. See Examples 4 and 5.*

24. $\sqrt{16}$ **25.** $\sqrt{81}$ **26.** $\sqrt{15}$ **27.** $\sqrt{31}$

28. $\sqrt{-47}$ **29.** $\sqrt{-53}$ **30.** $\sqrt{68}$ **31.** $\sqrt{72}$

32. $-\sqrt{121}$ **33.** $-\sqrt{144}$ **34.** $\sqrt{110}$ **35.** $\sqrt{170}$

36. $\sqrt{-400}$ **37.** $\sqrt{-900}$ **38.** $-\sqrt{200}$ **39.** $-\sqrt{260}$

Use a calculator with a square root key to find the following. Round to the nearest thousandth.

40. $\sqrt{571}$ **41.** $\sqrt{693}$ **42.** $\sqrt{798}$ **43.** $\sqrt{453}$

44. $\sqrt{3.94}$ **45.** $\sqrt{1.03}$ **46.** $\sqrt{.00895}$ **47.** $\sqrt{.000402}$

48. Find a number whose square is 3. How many real numbers have a square of 3? (See Objective 1.)

49. Find a number whose square is 17. How many real numbers have a square of 17? (See Objective 1.)

Find the missing side of each right triangle with sides a, b, and c, where c is the longest side. Approximate answers to the nearest thousandth in Exercises 54–57. See Example 6.

50. $a = 6$, $b = 8$ **51.** $a = 5$, $b = 12$ **52.** $c = 17$, $a = 8$ **53.** $c = 26$, $b = 10$

54. $a = 10$, $b = 8$ **55.** $a = 9$, $b = 7$ **56.** $c = 12$, $b = 7$ **57.** $c = 8$, $a = 3$

Find each of the following roots that are real numbers. See Examples 8 and 9.

58. $\sqrt[3]{1000}$ **59.** $\sqrt[3]{8}$ **60.** $\sqrt[3]{125}$ **61.** $\sqrt[3]{216}$

62. $-\sqrt[3]{8}$ **63.** $-\sqrt[3]{64}$ **64.** $\sqrt[3]{-8}$ **65.** $\sqrt[3]{-27}$

66. $\sqrt[4]{1}$ **67.** $-\sqrt[4]{16}$ **68.** $\sqrt[4]{-16}$ **69.** $-\sqrt[4]{-625}$

70. $-\sqrt[4]{81}$ **71.** $-\sqrt[4]{256}$ **72.** $\sqrt[5]{1}$ **73.** $\sqrt[5]{-32}$

Work each problem in Exercises 74–79. See Example 7.

74. The hypotenuse of a triangle measures 9 inches and one leg measures 3 inches. What is the measure of the other leg? Give the answer to the nearest thousandth.

75. Two sides of a rectangle measure 8 centimeters and 15 centimeters. How long is the diagonal in the figure?

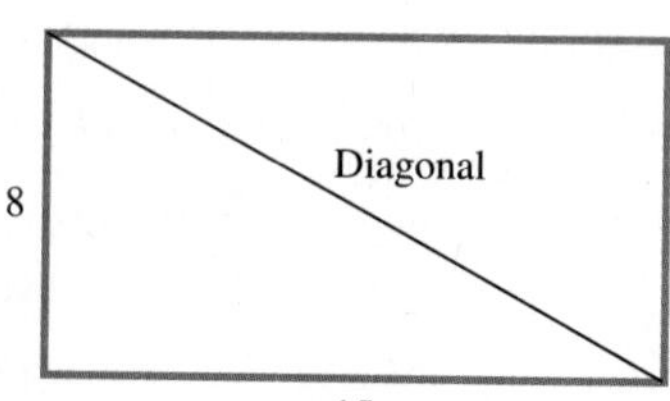

76. A guy wire is attached to the mast of a television antenna. It is attached 48 feet above ground level. If the wire is staked to the ground 36 feet from the base of the mast, how long is the wire?

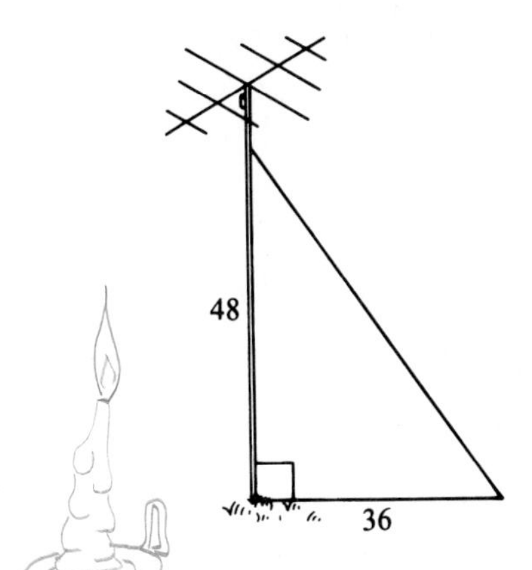

77. Margaret is flying a kite on 50 feet of string. How high is it above her hand if the horizontal distance between Margaret and the kite is 30 feet?

78. A boat is being pulled into a dock with a rope attached to the front of the boat. (See the figure.) When the boat is 4 feet from the dock, 5 feet of rope is extended. Find the height of the end of the rope above the dock.

79. Two cars leave Edmonton, Alberta, at the same time. One travels south at 25 miles per hour and the other travels east at 60 miles per hour. How far apart are they after 2 hours?

---

80. In your own words, define an irrational number. (See Objective 2.)

81. Complete the following: **(a)** $\sqrt{81} =$ _____. **(b)** If $a^2 = 81$, $a =$ _____. (See Objective 1.)

82. Are the answers in Exercise 81 the same or different? Why?

PREVIEW EXERCISES

Find all positive integer factors of each number. See Section 4.1.

83. 24 **84.** 70 **85.** 300 **86.** 48

87. 72 **88.** 75 **89.** 150 **90.** 120

8.2 MULTIPLICATION AND DIVISION OF RADICALS

OBJECTIVES

1 MULTIPLY RADICALS.

2 SIMPLIFY RADICALS USING THE PRODUCT RULE.

3 SIMPLIFY RADICAL QUOTIENTS USING THE QUOTIENT RULE.

4 USE THE PRODUCT AND QUOTIENT RULES TO SIMPLIFY HIGHER ROOTS.

1 Several useful rules for finding products and quotients of radicals are developed in this section. To illustrate the rule for products, notice that

$$\sqrt{4} \cdot \sqrt{9} = 2 \cdot 3 = 6 \qquad \text{and} \qquad \sqrt{4 \cdot 9} = \sqrt{36} = 6,$$

showing that

$$\sqrt{4} \cdot \sqrt{9} = \sqrt{4 \cdot 9}.$$

This result is a particular case of the more general product rule for radicals.

PRODUCT RULE FOR RADICALS

For nonnegative real numbers x and y,

$$\sqrt{x} \cdot \sqrt{y} = \sqrt{x \cdot y} \qquad \text{and} \qquad \sqrt{x \cdot y} = \sqrt{x} \cdot \sqrt{y}.$$

That is, the product of two radicals is the radical of the product.

EXAMPLE 1 USING THE PRODUCT RULE TO MULTIPLY RADICALS

Use the product rule for radicals to find each product.

(a) $\sqrt{2} \cdot \sqrt{3} = \sqrt{2 \cdot 3} = \sqrt{6}$

(b) $\sqrt{7} \cdot \sqrt{5} = \sqrt{35}$

(c) $\sqrt{11} \cdot \sqrt{a} = \sqrt{11a}$ Assume $a \geq 0$. ■

2 An important use of the product rule is in simplifying radicals. A radical is **simplified** when no perfect square factor remains under the radical sign. This is accomplished by using the product rule in the form $\sqrt{x \cdot y} = \sqrt{x} \cdot \sqrt{y}$. Example 2 shows how a radical may be simplified in this way.

EXAMPLE 2 USING THE PRODUCT RULE TO SIMPLIFY RADICALS

Simplify each radical.

(a) $\sqrt{20}$

Since 20 has a perfect square factor of 4,

$$\begin{aligned} \sqrt{20} &= \sqrt{4 \cdot 5} && \text{4 is a perfect square.} \\ &= \sqrt{4} \cdot \sqrt{5} && \text{Product rule} \\ &= 2\sqrt{5}. && \sqrt{4} = 2 \end{aligned}$$

Thus, $\sqrt{20} = 2\sqrt{5}$. Since 5 has no perfect square factor, other than 1, $2\sqrt{5}$ is called the **simplified form** of $\sqrt{20}$.

(b) $\sqrt{72}$

Begin by looking for the largest perfect square that is a factor of 72. This number is 36, so

$$\begin{aligned} \sqrt{72} &= \sqrt{36 \cdot 2} && \text{36 is a perfect square.} \\ &= \sqrt{36} \cdot \sqrt{2} && \text{Product rule} \\ &= 6\sqrt{2}. && \sqrt{36} = 6 \end{aligned}$$

Notice that 9 is a perfect square factor of 72. We could begin by factoring 72 as $9 \cdot 8$, to get

$$\sqrt{72} = \sqrt{9 \cdot 8} = 3\sqrt{8},$$

but then we would have to factor 8 as $4 \cdot 2$ in order to complete the simplification.

$$\sqrt{72} = 3\sqrt{8} = 3\sqrt{4 \cdot 2} = 3\sqrt{4} \cdot \sqrt{2} = 3 \cdot 2\sqrt{2} = 6\sqrt{2}$$

In either case, we obtain $6\sqrt{2}$ as the simplified form of $\sqrt{72}$; however, our work is simpler if we begin with the largest perfect square factor.

(c) $$\begin{aligned} \sqrt{300} &= \sqrt{100 \cdot 3} && \text{100 is a perfect square.} \\ &= \sqrt{100} \cdot \sqrt{3} && \text{Product rule} \\ &= 10\sqrt{3} && \sqrt{100} = 10 \end{aligned}$$

(d) $\sqrt{15}$

The number 15 has no perfect square factors (except 1), so $\sqrt{15}$ cannot be simplified further. ■

Sometimes the product rule can be used to simplify an answer, as Example 3 shows.

EXAMPLE 3 MULTIPLYING AND SIMPLIFYING RADICALS

Find each product and simplify.

(a) $$\begin{aligned} \sqrt{9} \cdot \sqrt{75} &= 3\sqrt{75} && \sqrt{9} = 3 \\ &= 3\sqrt{25 \cdot 3} && \text{25 is a perfect square.} \\ &= 3\sqrt{25} \cdot \sqrt{3} && \text{Product rule} \\ &= 3 \cdot 5\sqrt{3} && \sqrt{25} = 5 \\ &= 15\sqrt{3} && \text{Multiply.} \end{aligned}$$

(b) $$\begin{aligned} \sqrt{8} \cdot \sqrt{12} &= \sqrt{8 \cdot 12} && \text{Product rule} \\ &= \sqrt{4 \cdot 2 \cdot 4 \cdot 3} && \text{Factor; 4 is a perfect square.} \\ &= \sqrt{4} \cdot \sqrt{4} \cdot \sqrt{2 \cdot 3} && \text{Product rule} \\ &= 2 \cdot 2 \cdot \sqrt{6} && \sqrt{4} = 2 \\ &= 4\sqrt{6} && \text{Multiply.} \end{aligned}$$ ■

3 The *quotient rule for radicals* is very similar to the product rule. It, too, can be used either way.

QUOTIENT RULE FOR RADICALS

If x and y are nonnegative real numbers and $y \neq 0$,

$$\frac{\sqrt{x}}{\sqrt{y}} = \sqrt{\frac{x}{y}} \quad \text{and} \quad \sqrt{\frac{x}{y}} = \frac{\sqrt{x}}{\sqrt{y}}.$$

The quotient of the radicals is the radical of the quotient.

EXAMPLE 4 USING THE QUOTIENT RULE TO SIMPLIFY RADICALS

Simplify each radical.

(a) $\sqrt{\frac{25}{9}} = \frac{\sqrt{25}}{\sqrt{9}} = \frac{5}{3}$ Quotient rule

(b) $\frac{\sqrt{288}}{\sqrt{2}} = \sqrt{\frac{288}{2}} = \sqrt{144} = 12$ Quotient rule

(c) $\sqrt{\frac{3}{4}} = \frac{\sqrt{3}}{\sqrt{4}} = \frac{\sqrt{3}}{2}$ Quotient rule ■

EXAMPLE 5 USING THE QUOTIENT RULE TO DIVIDE RADICALS

Divide $27\sqrt{15}$ by $9\sqrt{3}$.

Use the quotient rule as follows.

$$\frac{27\sqrt{15}}{9\sqrt{3}} = \frac{27}{9} \cdot \frac{\sqrt{15}}{\sqrt{3}} = 3\sqrt{\frac{15}{3}} = 3\sqrt{5}$$ ■

Some problems require both the product and the quotient rules, as Example 6 shows.

EXAMPLE 6 USING BOTH THE PRODUCT AND QUOTIENT RULES

Simplify $\sqrt{\frac{3}{5}} \cdot \sqrt{\frac{4}{5}}$.

$$\sqrt{\frac{3}{5}} \cdot \sqrt{\frac{4}{5}} = \frac{\sqrt{3}}{\sqrt{5}} \cdot \frac{\sqrt{4}}{\sqrt{5}}$$ Quotient rule

$$= \frac{\sqrt{3} \cdot \sqrt{4}}{\sqrt{5} \cdot \sqrt{5}}$$ Multiply fractions.

$$= \frac{\sqrt{3} \cdot 2}{\sqrt{25}}$$ Product rule; $\sqrt{4} = 2$

$$= \frac{2\sqrt{3}}{5}$$ $\sqrt{25} = 5$ ■

The properties stated earlier in this section also apply when variables appear under the radical sign, as long as all the variables represent only nonnegative numbers. For example, $\sqrt{5^2} = 5$, but $\sqrt{(-5)^2} \neq -5$.

For a real number a, $\sqrt{a^2} = a$ only if a is nonnegative.

EXAMPLE 7
SIMPLIFYING RADICALS INVOLVING VARIABLES

Simplify each radical. Assume all variables represent positive real numbers.

(a) $\sqrt{25m^4} = \sqrt{25} \cdot \sqrt{m^4}$ Product rule
$= 5m^2$

(b) $\sqrt{64p^{10}} = 8p^5$ Product rule

(c) $\sqrt{r^9} = \sqrt{r^8 \cdot r}$
$= \sqrt{r^8} \cdot \sqrt{r} = r^4\sqrt{r}$ Product rule

(d) $\sqrt{\frac{5}{x^2}} = \frac{\sqrt{5}}{\sqrt{x^2}} = \frac{\sqrt{5}}{x}$ Quotient rule ■

4 The product rule and the quotient rule for radicals also work for other roots, as shown in Example 8. To simplify cube roots, look for factors that are *perfect cubes*. A **perfect cube** is a number with a rational cube root. For example, $\sqrt[3]{64} = 4$, and since 4 is a rational number, 64 is a perfect cube. Higher roots are handled in a similar manner.

PROPERTIES OF RADICALS

For all real numbers where the indicated roots exist,

$$\sqrt[n]{x} \cdot \sqrt[n]{y} = \sqrt[n]{xy} \quad \text{and} \quad \frac{\sqrt[n]{x}}{\sqrt[n]{y}} = \sqrt[n]{\frac{x}{y}}.$$

EXAMPLE 8
SIMPLIFYING HIGHER ROOTS

Simplify each radical.

(a) $\sqrt[3]{32} = \sqrt[3]{8 \cdot 4}$ 8 is a perfect cube.
$= \sqrt[3]{8} \cdot \sqrt[3]{4} = 2\sqrt[3]{4}$

(b) $\sqrt[3]{108} = \sqrt[3]{27 \cdot 4}$ 27 is a perfect cube.
$= \sqrt[3]{27} \cdot \sqrt[3]{4} = 3\sqrt[3]{4}$

(c) $\sqrt[4]{32} = \sqrt[4]{16} \cdot \sqrt[4]{2} = 2\sqrt[4]{2}$ 16 is a perfect fourth power.

(d) $\sqrt[3]{\frac{8}{125}} = \frac{\sqrt[3]{8}}{\sqrt[3]{125}} = \frac{2}{5}$

(e) $\sqrt[4]{\frac{16}{625}} = \frac{\sqrt[4]{16}}{\sqrt[4]{625}} = \frac{2}{5}$ ■

8.2 EXERCISES

Use the product rule to simplify each expression. See Examples 1–3.

1. $\sqrt{8} \cdot \sqrt{2}$ **2.** $\sqrt{27} \cdot \sqrt{3}$ **3.** $\sqrt{6} \cdot \sqrt{6}$ **4.** $\sqrt{11} \cdot \sqrt{11}$

5. $\sqrt{21} \cdot \sqrt{21}$ **6.** $\sqrt{17} \cdot \sqrt{17}$ **7.** $\sqrt{3} \cdot \sqrt{7}$ **8.** $\sqrt{2} \cdot \sqrt{5}$

9. $\sqrt{27}$ **10.** $\sqrt{45}$ **11.** $\sqrt{18}$ **12.** $\sqrt{75}$

13. $\sqrt{48}$ **14.** $\sqrt{80}$ **15.** $\sqrt{150}$ **16.** $\sqrt{700}$

17. $10\sqrt{27}$ **18.** $4\sqrt{8}$ **19.** $2\sqrt{20}$ **20.** $5\sqrt{80}$

21. $\sqrt{27} \cdot \sqrt{48}$ **22.** $\sqrt{75} \cdot \sqrt{27}$ **23.** $\sqrt{50} \cdot \sqrt{72}$ **24.** $\sqrt{98} \cdot \sqrt{8}$

25. $\sqrt{80} \cdot \sqrt{15}$ **26.** $\sqrt{60} \cdot \sqrt{12}$ **27.** $\sqrt{50} \cdot \sqrt{20}$ **28.** $\sqrt{72} \cdot \sqrt{12}$

Use the quotient rule and the product rule, as necessary, to simplify each expression. See Examples 4–6.

29. $\sqrt{\frac{100}{9}}$ **30.** $\sqrt{\frac{225}{16}}$ **31.** $\sqrt{\frac{30}{49}}$ **32.** $\sqrt{\frac{10}{121}}$

33. $\sqrt{\frac{1}{5}} \cdot \sqrt{\frac{4}{5}}$ **34.** $\sqrt{\frac{2}{3}} \cdot \sqrt{\frac{2}{27}}$ **35.** $\sqrt{\frac{2}{5}} \cdot \sqrt{\frac{8}{125}}$ **36.** $\sqrt{\frac{3}{8}} \cdot \sqrt{\frac{3}{2}}$

37. $\frac{\sqrt{75}}{\sqrt{3}}$ **38.** $\frac{\sqrt{200}}{\sqrt{2}}$ **39.** $\frac{\sqrt{48}}{\sqrt{3}}$ **40.** $\frac{\sqrt{72}}{\sqrt{8}}$

41. $\frac{15\sqrt{10}}{5\sqrt{2}}$ **42.** $\frac{18\sqrt{20}}{2\sqrt{10}}$ **43.** $\frac{25\sqrt{50}}{5\sqrt{5}}$ **44.** $\frac{26\sqrt{10}}{13\sqrt{5}}$

Simplify each expression. Assume that all variables represent positive real numbers. See Example 7.

45. $\sqrt{y} \cdot \sqrt{y}$ **46.** $\sqrt{m} \cdot \sqrt{m}$ **47.** $\sqrt{x^2}$ **48.** $\sqrt{y^2}$

49. $\sqrt{x^4}$ **50.** $\sqrt{y^4}$ **51.** $\sqrt{x^2y^4}$ **52.** $\sqrt{x^4y^8}$

53. $\sqrt{x^3}$ **54.** $\sqrt{y^5}$ **55.** $\sqrt{\frac{16}{x^2}}$ **56.** $\sqrt{\frac{100}{m^4}}$

57. $\sqrt{\frac{11}{r^4}}$ **58.** $\sqrt{\frac{23}{y^6}}$ **59.** $\sqrt{28m^3}$ **60.** $\sqrt{40p^4}$

Simplify each radical expression. Assume variables in denominators are not zero. See Example 8.

61. $\sqrt[3]{40}$ **62.** $\sqrt[3]{48}$ **63.** $\sqrt[3]{54}$ **64.** $\sqrt[3]{135}$

65. $\sqrt[3]{128}$ **66.** $\sqrt[3]{192}$ **67.** $\sqrt[4]{80}$ **68.** $\sqrt[4]{243}$

69. $\sqrt[3]{\frac{8}{27}}$ **70.** $\sqrt[4]{\frac{1}{256}}$ **71.** $\sqrt[4]{\frac{10{,}000}{81}}$ **72.** $\sqrt[3]{\frac{216}{125}}$

73. $\sqrt[3]{2} \cdot \sqrt[3]{4}$ **74.** $\sqrt[3]{9} \cdot \sqrt[3]{3}$ **75.** $\sqrt[4]{4} \cdot \sqrt[4]{4}$ **76.** $\sqrt[4]{3} \cdot \sqrt[4]{27}$

77. $\sqrt[3]{4x} \cdot \sqrt[3]{8x^2}$ **78.** $\sqrt[3]{25m} \cdot \sqrt[3]{125m^3}$ **79.** $\sqrt[3]{\frac{3m}{8n^3}}$ **80.** $\sqrt[4]{\frac{4k^2}{81p^4}}$

The volume of a cube is found with the formula $V = s^3$, where s is the length of an edge of the cube.

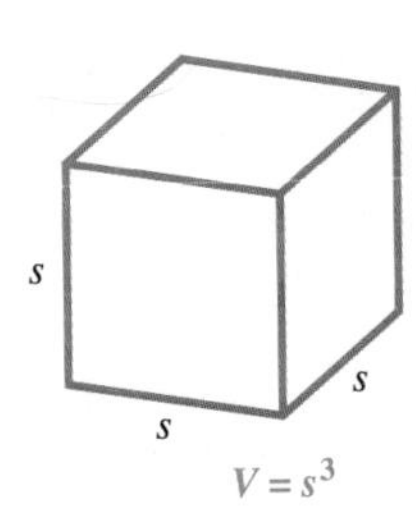

81. A container in the shape of a cube has a volume of 216 cubic centimeters. What is the depth of the container?

82. A cube-shaped box must be constructed to contain 128 cubic feet. What should the dimensions (height, width, and length) of the box be?

83. When we multiply two radicals with variables under the radical sign, such as $\sqrt{a} \cdot \sqrt{b} = \sqrt{ab}$, why is it important to know that both a and b represent nonnegative numbers? (See Objective 1.)

84. Why is it true that $\sqrt{m} \cdot \sqrt{m} = m$ only if m is not negative? (See Objective 1.)

85. Is it necessary to restrict k to a nonnegative number to say that $\sqrt[3]{k} \cdot \sqrt[3]{k} \cdot \sqrt[3]{k} = k$? Why? (See Objective 4.)

PREVIEW EXERCISES

Rewrite each fraction with the given denominator. See Section 5.3.

86. $\frac{9}{13} = \frac{}{39}$ **87.** $\frac{5}{11} = \frac{}{66}$ **88.** $\frac{2}{3} = \frac{}{15}$ **89.** $\frac{3}{14} = \frac{}{56}$

90. $\frac{2}{x} = \frac{}{3x}$ **91.** $\frac{5}{2y} = \frac{}{2y^2}$ **92.** $\frac{5}{\sqrt{3}} = \frac{}{2\sqrt{3}}$ **93.** $\frac{8}{\sqrt{5}} = \frac{}{3\sqrt{5}}$

8.3 ADDITION AND SUBTRACTION OF RADICALS

OBJECTIVES

1. ADD AND SUBTRACT RADICALS.
2. SIMPLIFY RADICAL SUMS AND DIFFERENCES.
3. SIMPLIFY RADICAL EXPRESSIONS INVOLVING MULTIPLICATION.

1 We add or subtract radicals by using the distributive property. For example,

$$8\sqrt{3} + 6\sqrt{3} = (8 + 6)\sqrt{3} \quad \text{Distributive property}$$
$$= 14\sqrt{3}.$$

Also,

$$2\sqrt{11} - 7\sqrt{11} = -5\sqrt{11}.$$

Only **like radicals,** those that are multiples of the *same root* of the *same number,* can be combined this way.

EXAMPLE 1 ADDING AND SUBTRACTING LIKE RADICALS

Add or subtract, as indicated.

(a) $3\sqrt{6} + 5\sqrt{6} = (3 + 5)\sqrt{6} = 8\sqrt{6}$ Distributive property

(b) $5\sqrt{10} - 7\sqrt{10} = (5 - 7)\sqrt{10} = -2\sqrt{10}$

(c) $\sqrt[3]{5} + \sqrt[3]{5} = 1\sqrt[3]{5} + 1\sqrt[3]{5} = (1 + 1)\sqrt[3]{5} = 2\sqrt[3]{5}$

(d) $\sqrt[4]{7} + 2\sqrt[4]{7} = 1\sqrt[4]{7} + 2\sqrt[4]{7} = 3\sqrt[4]{7}$

(e) $\sqrt{3} + \sqrt{7}$ cannot be simplified further. ■

2 Sometimes each radical in a sum or difference must be simplified first. Doing this may result in like radicals, which then can be added or subtracted.

EXAMPLE 2 **ADDING AND SUBTRACTING RADICALS THAT REQUIRE SIMPLIFICATION**

Simplify as much as possible.

(a)
$$\begin{aligned} 3\sqrt{2} + \sqrt{8} &= 3\sqrt{2} + \sqrt{4 \cdot 2} && \text{Factor.} \\ &= 3\sqrt{2} + \sqrt{4} \cdot \sqrt{2} && \text{Product rule} \\ &= 3\sqrt{2} + 2\sqrt{2} && \sqrt{4} = 2 \\ &= 5\sqrt{2} && \text{Add like radicals.} \end{aligned}$$

(b)
$$\begin{aligned} \sqrt{18} - \sqrt{27} &= \sqrt{9 \cdot 2} - \sqrt{9 \cdot 3} && \text{Factor.} \\ &= \sqrt{9} \cdot \sqrt{2} - \sqrt{9} \cdot \sqrt{3} && \text{Product rule} \\ &= 3\sqrt{2} - 3\sqrt{3} && \sqrt{9} = 3 \end{aligned}$$

Since $\sqrt{2}$ and $\sqrt{3}$ are unlike radicals, this difference cannot be simplified further.

(c)
$$\begin{aligned} 2\sqrt{12} + 3\sqrt{75} &= 2(\sqrt{4} \cdot \sqrt{3}) + 3(\sqrt{25} \cdot \sqrt{3}) && \text{Product rule} \\ &= 2(2\sqrt{3}) + 3(5\sqrt{3}) && \sqrt{4} = 2 \text{ and } \sqrt{25} = 5 \\ &= 4\sqrt{3} + 15\sqrt{3} && \text{Multiply.} \\ &= 19\sqrt{3} && \text{Add like radicals.} \end{aligned}$$

(d)
$$\begin{aligned} 3\sqrt[3]{16} + 5\sqrt[3]{2} &= 3(\sqrt[3]{8} \cdot \sqrt[3]{2}) + 5\sqrt[3]{2} && \text{Product rule} \\ &= 3(2\sqrt[3]{2}) + 5\sqrt[3]{2} && \sqrt[3]{8} = 2 \\ &= 6\sqrt[3]{2} + 5\sqrt[3]{2} && \text{Multiply.} \\ &= 11\sqrt[3]{2} && \text{Add like radicals.} \end{aligned}$$
■

3 Some radical expressions require both multiplication and addition (or subtraction). The order of operations presented earlier still applies.

EXAMPLE 3 **MULTIPLYING AND COMBINING TERMS IN RADICAL EXPRESSIONS**

Simplify each expression. Assume that all variables represent nonnegative real numbers.

(a)
$$\begin{aligned} \sqrt{5} \cdot \sqrt{15} + 4\sqrt{3} &= \sqrt{5 \cdot 15} + 4\sqrt{3} && \text{Product rule} \\ &= \sqrt{75} + 4\sqrt{3} && \text{Multiply.} \\ &= \sqrt{25 \cdot 3} + 4\sqrt{3} && \text{25 is a perfect square.} \\ &= \sqrt{25} \cdot \sqrt{3} + 4\sqrt{3} && \text{Product rule} \\ &= 5\sqrt{3} + 4\sqrt{3} && \sqrt{25} = 5 \\ &= 9\sqrt{3} && \text{Add like radicals.} \end{aligned}$$

(b) $\sqrt{2} \cdot \sqrt{6k} + \sqrt{27k} = \sqrt{12k} + \sqrt{27k}$ Product rule
$= \sqrt{4 \cdot 3k} + \sqrt{9 \cdot 3k}$ Factor.
$= \sqrt{4} \cdot \sqrt{3k} + \sqrt{9} \cdot \sqrt{3k}$ Product rule
$= 2\sqrt{3k} + 3\sqrt{3k}$ $\sqrt{4} = 2$ and $\sqrt{9} = 3$
$= 5\sqrt{3k}$ Add like radicals.

(c) $\sqrt[3]{2} \cdot \sqrt[3]{16m^3} - \sqrt[3]{108m^3} = \sqrt[3]{32m^3} - \sqrt[3]{108m^3}$ Product rule
$= \sqrt[3]{(8m^3)4} - \sqrt[3]{(27m^3)4}$ Factor.
$= 2m\sqrt[3]{4} - 3m\sqrt[3]{4}$ $\sqrt[3]{8m^3} = 2m$ and $\sqrt[3]{27m^3} = 3m$
$= -m\sqrt[3]{4}$ Subtract like radicals. ■

CAUTION Remember that a sum or difference of radicals can be simplified only if the radicals are *like radicals*. For example, $\sqrt{5} + 3\sqrt{5} = 4\sqrt{5}$, but $\sqrt{5} + 5\sqrt{3}$ cannot be simplified further. Also, $2\sqrt{3} + 5\sqrt[3]{3}$ cannot be simplified further.

8.3 EXERCISES

Simplify and add or subtract terms wherever possible. Assume all variables represent nonnegative numbers. See Examples 1 and 3.

1. $2\sqrt{3} + 5\sqrt{3}$
2. $6\sqrt{5} + 8\sqrt{5}$
3. $4\sqrt{7} - 9\sqrt{7}$
4. $6\sqrt{2} - 8\sqrt{2}$
5. $\sqrt{6} + \sqrt{6}$
6. $\sqrt{11} + \sqrt{11}$
7. $\sqrt{17} + 2\sqrt{17}$
8. $3\sqrt{19} + \sqrt{19}$
9. $5\sqrt{7} - \sqrt{7}$
10. $3\sqrt{27} - \sqrt{27}$
11. $3\sqrt{18} + \sqrt{2}$
12. $2\sqrt{48} - \sqrt{3}$
13. $-\sqrt{12} + \sqrt{75}$
14. $2\sqrt{27} - \sqrt{300}$
15. $5\sqrt{72} + 2\sqrt{50}$
16. $6\sqrt{18} - 4\sqrt{32}$
17. $5\sqrt{7} - 2\sqrt{28} + 6\sqrt{63}$
18. $3\sqrt{11} + 5\sqrt{44} - 3\sqrt{99}$
19. $9\sqrt{24} - 2\sqrt{54} + 3\sqrt{20}$
20. $2\sqrt{8} - 5\sqrt{32} + 2\sqrt{48}$
21. $5\sqrt{72} - 3\sqrt{48} - 4\sqrt{128}$
22. $4\sqrt{50} + 3\sqrt{12} + 5\sqrt{45}$
23. $\frac{1}{4}\sqrt{288} - \frac{1}{6}\sqrt{72}$
24. $\frac{2}{3}\sqrt{27} - \frac{3}{4}\sqrt{48}$
25. $\frac{3}{5}\sqrt{75} - \frac{2}{3}\sqrt{45}$
26. $\frac{5}{8}\sqrt{128} - \frac{3}{4}\sqrt{160}$
27. $\sqrt{9x} + \sqrt{49x} - \sqrt{16x}$
28. $\sqrt{4a} - \sqrt{16a} + \sqrt{9a}$
29. $\sqrt{3} \cdot \sqrt{7} + 2\sqrt{21}$
30. $\sqrt{13} \cdot \sqrt{2} + 3\sqrt{26}$
31. $\sqrt{6} \cdot \sqrt{2} + 3\sqrt{3}$
32. $4\sqrt{15} \cdot \sqrt{3} - 2\sqrt{5}$
33. $4\sqrt[3]{16} - 3\sqrt[3]{54}$
34. $5\sqrt[3]{128} + 3\sqrt[3]{250}$
35. $3\sqrt[3]{24} + 6\sqrt[3]{81}$
36. $2\sqrt[4]{48} - \sqrt[4]{243}$
37. $5\sqrt[4]{32} + 2\sqrt[4]{32} \cdot \sqrt[4]{4}$
38. $8\sqrt[3]{48} + 10\sqrt[3]{3} \cdot \sqrt[3]{18}$

Perform the indicated operations in Exercises 39–53. Assume that all variables represent nonnegative real numbers. See Example 3.

39. $\sqrt{6x^2} + x\sqrt{54}$
40. $\sqrt{75x^2} + x\sqrt{300}$
41. $\sqrt{20y^2} - 3y\sqrt{5}$
42. $3\sqrt{8x^2} - 4x\sqrt{2}$
43. $\sqrt{2b^2} + 3b\sqrt{18}$
44. $5\sqrt{75p^2} - 4\sqrt{27p^2}$

45. $-3\sqrt{32k} + 6\sqrt{8k}$
46. $2\sqrt{125x^2z} + 8x\sqrt{80z}$
47. $4p\sqrt{14m} - 6\sqrt{28mp^2}$
48. $3k\sqrt{24k^2h^2} + 9h\sqrt{54k^3}$
49. $6r\sqrt{27r^2s} + 3r^2\sqrt{3s}$
50. $6\sqrt[3]{8p^2} - 2\sqrt[3]{27p^2}$
51. $5\sqrt[4]{m^3} + 8\sqrt[4]{16m^3}$
52. $5\sqrt[4]{m^4} + 3\sqrt[4]{81m^4}$
53. $8k\sqrt[3]{54k} + 6\sqrt[3]{16k^4}$

---

54. A rectangular room has a width of $\sqrt{125}$ feet and a length of $\sqrt{245}$ feet. What is the perimeter?

55. Find the perimeter of a triangular lot with sides measuring $3\sqrt{27}$, $5\sqrt{12}$, and $2\sqrt{48}$ meters.

56. Explain what is wrong with the example $2\sqrt{5} + 2\sqrt{7} = 4\sqrt{12}$. (See Objective 1.)

57. What is wrong with the example $5\sqrt[3]{2} - 3\sqrt{2} = 2\sqrt[3]{2}$? (See Objective 1.)

58. Can the expression $4\sqrt{3} - 2\sqrt{45}$ be further simplified? Explain. (See Objective 3.)

59. Compare "adding like terms" as in the sum $4x + 3x = 7x$ and "adding like radicals" as in $4\sqrt{x} + 3\sqrt{x} = 7\sqrt{x}$ $(x > 0)$.

60. Compare the way we "simplify" the expressions $2m \cdot 3m^2 + m^3$ and $2\sqrt{5} \cdot 3\sqrt{2} + \sqrt{10}$.

PREVIEW EXERCISES

Find each product. See Sections 3.3 and 3.4.

61. $(m + 3)(m + 5)$
62. $(p - 2)(p + 7)$
63. $(2k - 3)(3k - 4)$
64. $(4z + 2)(5z - 8)$
65. $(3x - 1)^2$
66. $(2y + 5)^2$
67. $(a + 3)(a - 3)$
68. $(2n + 3)(2n - 3)$
69. $(4k + 5)(4k - 5)$

8.4 RATIONALIZING THE DENOMINATOR

OBJECTIVES

1. RATIONALIZE DENOMINATORS WITH SQUARE ROOTS.
2. WRITE RADICALS IN SIMPLIFIED FORM.
3. RATIONALIZE DENOMINATORS WITH CUBE ROOTS.

1 Decimal approximations for radicals were found in the first section of this chapter. For more complicated radical expressions it is easier to find these decimals if the denominators do not contain radicals. For example, the radical in the denominator of

$$\frac{\sqrt{3}}{\sqrt{2}}$$

can be eliminated by multiplying the numerator and the denominator by $\sqrt{2}$.

$$\frac{\sqrt{3}}{\sqrt{2}} = \frac{\sqrt{3} \cdot \sqrt{2}}{\sqrt{2} \cdot \sqrt{2}} = \frac{\sqrt{6}}{2} \qquad \text{Since } \sqrt{2} \cdot \sqrt{2} = 2$$

This process of changing the denominator from a radical (irrational number) to a rational number is called **rationalizing the denominator.** The value of the number is not changed; only the form of the number is changed, because the expression has been multiplied by 1 in the form $\sqrt{2}/\sqrt{2}$.

EXAMPLE 1 RATIONALIZING DENOMINATORS

Rationalize each denominator.

(a) $\dfrac{9}{\sqrt{6}}$

Multiply both numerator and denominator by $\sqrt{6}$.

$$\frac{9}{\sqrt{6}} = \frac{9 \cdot \sqrt{6}}{\sqrt{6} \cdot \sqrt{6}}$$

$$= \frac{9\sqrt{6}}{6} \qquad \sqrt{6} \cdot \sqrt{6} = 6$$

$$= \frac{3\sqrt{6}}{2} \qquad \text{Lowest terms}$$

(b) $\dfrac{12}{\sqrt{8}}$

The denominator here could be rationalized by multiplying by $\sqrt{8}$. However, the result can be found more directly by first simplifying the denominator.

$$\sqrt{8} = \sqrt{4} \cdot \sqrt{2} = 2\sqrt{2}$$

Then multiply numerator and denominator by $\sqrt{2}$.

$$\frac{12}{\sqrt{8}} = \frac{12}{2\sqrt{2}}$$

$$= \frac{12 \cdot \sqrt{2}}{2\sqrt{2} \cdot \sqrt{2}} \qquad \text{Multiply by } \frac{\sqrt{2}}{\sqrt{2}}.$$

$$= \frac{12\sqrt{2}}{2 \cdot 2} \qquad \sqrt{2} \cdot \sqrt{2} = 2$$

$$= \frac{12\sqrt{2}}{4} \qquad \text{Multiply.}$$

$$= 3\sqrt{2} \qquad \text{Lowest terms} \quad ■$$

2 A radical is considered to be in simplified form if the following three conditions are met.

SIMPLIFIED FORM OF A SQUARE ROOT RADICAL

1. The radicand contains no factor (except 1) that is a perfect square.
2. The radicand has no fractions.
3. No denominator contains a radical.

In the following examples, radicals are simplified according to these conditions. Radicals are considered simplified only if any denominators are rationalized, as shown in Examples 2–4.

EXAMPLE 2
SIMPLIFYING A RADICAL WITH A FRACTION

Simplify $\sqrt{\frac{27}{5}}$ by rationalizing the denominator.

First, use the quotient rule for radicals.

$$\sqrt{\frac{27}{5}} = \frac{\sqrt{27}}{\sqrt{5}}$$

Now multiply both numerator and denominator by $\sqrt{5}$.

$$\frac{\sqrt{27}}{\sqrt{5}} = \frac{\sqrt{27} \cdot \sqrt{5}}{\sqrt{5} \cdot \sqrt{5}}$$

$$= \frac{\sqrt{9 \cdot 3} \cdot \sqrt{5}}{5} \qquad \sqrt{5} \cdot \sqrt{5} = 5$$

$$= \frac{\sqrt{9} \cdot \sqrt{3} \cdot \sqrt{5}}{5} \qquad \text{Product rule}$$

$$= \frac{3 \cdot \sqrt{3 \cdot 5}}{5} = \frac{3\sqrt{15}}{5} \qquad \text{Product rule}$$

■

EXAMPLE 3
SIMPLIFYING A PRODUCT OF RADICALS

Simplify $\sqrt{\frac{5}{8}} \cdot \sqrt{\frac{1}{6}}$.

Use both the product rule and the quotient rule.

$$\sqrt{\frac{5}{8}} \cdot \sqrt{\frac{1}{6}} = \sqrt{\frac{5}{8} \cdot \frac{1}{6}} \qquad \text{Product rule}$$

$$= \sqrt{\frac{5}{48}} \qquad \text{Multiply.}$$

$$= \frac{\sqrt{5}}{\sqrt{48}} \qquad \text{Quotient rule}$$

To rationalize the denominator, first simplify, then multiply the numerator and denominator by $\sqrt{3}$ as follows.

$$\frac{\sqrt{5}}{\sqrt{48}} = \frac{\sqrt{5}}{\sqrt{16} \cdot \sqrt{3}} \qquad \text{Product rule}$$

$$= \frac{\sqrt{5}}{4\sqrt{3}} \qquad \sqrt{16} = 4$$

$$= \frac{\sqrt{5} \cdot \sqrt{3}}{4\sqrt{3} \cdot \sqrt{3}} \qquad \text{Rationalize the denominator.}$$

$$= \frac{\sqrt{15}}{4 \cdot 3} \qquad \text{Product rule}$$

$$= \frac{\sqrt{15}}{12} \qquad \text{Multiply.}$$

■

EXAMPLE 4 SIMPLIFYING A QUOTIENT OF RADICALS

Rationalize the denominator of $\frac{\sqrt{4x}}{\sqrt{y}}$. Assume that x and y represent positive real numbers.

Multiply numerator and denominator by $\sqrt{y}$.

$$\frac{\sqrt{4x}}{\sqrt{y}} = \frac{\sqrt{4x} \cdot \sqrt{y}}{\sqrt{y} \cdot \sqrt{y}} = \frac{\sqrt{4xy}}{y} = \frac{2\sqrt{xy}}{y} \quad \blacksquare$$

3 A denominator with a cube root is rationalized by changing the radicand in the denominator to a perfect cube, as shown in the next example.

EXAMPLE 5 RATIONALIZING A DENOMINATOR WITH A CUBE ROOT

Rationalize each denominator.

(a) $\sqrt[3]{\frac{3}{2}}$

Multiply the numerator and the denominator by enough factors of 2 to make the denominator a perfect cube. This will eliminate the radical in the denominator. Here, multiply by $\sqrt[3]{2^2}$.

$$\sqrt[3]{\frac{3}{2}} = \frac{\sqrt[3]{3}}{\sqrt[3]{2}} = \frac{\sqrt[3]{3} \cdot \sqrt[3]{2^2}}{\sqrt[3]{2} \cdot \sqrt[3]{2^2}} = \frac{\sqrt[3]{3 \cdot 2^2}}{\sqrt[3]{2^3}} = \frac{\sqrt[3]{12}}{2}$$

Since $\sqrt[3]{2^3} = \sqrt[3]{8} = 2$

(b) $\frac{\sqrt[3]{3}}{\sqrt[3]{4}}$

Since $4 \cdot 2 = 2^2 \cdot 2 = 2^3$, multiply numerator and denominator by $\sqrt[3]{2}$.

$$\frac{\sqrt[3]{3}}{\sqrt[3]{4}} = \frac{\sqrt[3]{3} \cdot \sqrt[3]{2}}{\sqrt[3]{4} \cdot \sqrt[3]{2}} = \frac{\sqrt[3]{6}}{\sqrt[3]{8}} = \frac{\sqrt[3]{6}}{2} \quad \blacksquare$$

CAUTION A common error in Example 5(a) is to multiply by $\sqrt[3]{2}$ instead of $\sqrt[3]{4}$. Notice that this would give a denominator of $\sqrt[3]{2} \cdot \sqrt[3]{2} = \sqrt[3]{4}$. Since 4 is not a perfect cube, the denominator is still not rationalized.

8.4 EXERCISES

Perform the indicated operations, and write all answers in simplest form. See Examples 1–3.

1. $\frac{6}{\sqrt{5}}$ **2.** $\frac{15}{\sqrt{15}}$ **3.** $\frac{3}{\sqrt{7}}$ **4.** $\frac{12}{\sqrt{10}}$ **5.** $\frac{8\sqrt{3}}{\sqrt{5}}$

6. $\frac{9\sqrt{6}}{\sqrt{5}}$ **7.** $\frac{12\sqrt{10}}{8\sqrt{3}}$ **8.** $\frac{9\sqrt{15}}{6\sqrt{2}}$ **9.** $\frac{8}{\sqrt{27}}$ **10.** $\frac{12}{\sqrt{18}}$

11. $\frac{12}{\sqrt{72}}$ **12.** $\frac{21}{\sqrt{45}}$ **13.** $\frac{\sqrt{10}}{\sqrt{5}}$ **14.** $\frac{\sqrt{6}}{\sqrt{3}}$ **15.** $\frac{\sqrt{40}}{\sqrt{3}}$

16. $\frac{\sqrt{5}}{\sqrt{8}}$ **17.** $\sqrt{\frac{1}{2}}$ **18.** $\sqrt{\frac{1}{8}}$ **19.** $\sqrt{\frac{9}{5}}$ **20.** $\sqrt{\frac{16}{7}}$

21. $\sqrt{\frac{3}{4}} \cdot \sqrt{\frac{1}{5}}$ **22.** $\sqrt{\frac{1}{10}} \cdot \sqrt{\frac{10}{3}}$ **23.** $\sqrt{\frac{21}{7}} \cdot \sqrt{\frac{21}{8}}$ **24.** $\sqrt{\frac{1}{11}} \cdot \sqrt{\frac{33}{16}}$

25. $\sqrt{\frac{2}{5}} \cdot \sqrt{\frac{3}{10}}$ **26.** $\sqrt{\frac{9}{8}} \cdot \sqrt{\frac{7}{16}}$ **27.** $\sqrt{\frac{16}{27}} \cdot \sqrt{\frac{1}{9}}$ **28.** $\sqrt{\frac{5}{8}} \cdot \sqrt{\frac{5}{6}}$

Write all answers in simplest form. Rationalize all denominators. Assume that all variables represent positive real numbers. See Example 4.

29. $\sqrt{\frac{6}{p}}$ **30.** $\sqrt{\frac{5}{x}}$ **31.** $\sqrt{\frac{3}{y}}$ **32.** $\sqrt{\frac{9}{k}}$ **33.** $\sqrt{\frac{2z^2}{x}}$ **34.** $\sqrt{\frac{3p^2}{q}}$

35. $\sqrt{\frac{5a^3}{b}}$ **36.** $\sqrt{\frac{7x^3}{y}}$ **37.** $\sqrt{\frac{6p^3}{3m}}$ **38.** $\sqrt{\frac{x^2}{4y}}$ **39.** $\sqrt{\frac{9a^2r}{5}}$ **40.** $\sqrt{\frac{2x^2z^4}{3y}}$

Rationalize the denominator of each cube root. Assume variables in denominators are not zero. See Example 5.

41. $\sqrt[3]{\frac{1}{2}}$ **42.** $\sqrt[3]{\frac{1}{4}}$ **43.** $\sqrt[3]{\frac{1}{32}}$ **44.** $\sqrt[3]{\frac{1}{5}}$ **45.** $\sqrt[3]{\frac{1}{11}}$ **46.** $\sqrt[3]{\frac{3}{2}}$

47. $\sqrt[3]{\frac{2}{5}}$ **48.** $\sqrt[3]{\frac{4}{9}}$ **49.** $\sqrt[3]{\frac{3}{4y^2}}$ **50.** $\sqrt[3]{\frac{3}{25x^2}}$ **51.** $\sqrt[3]{\frac{7m}{36n}}$ **52.** $\sqrt[3]{\frac{11p}{49q}}$

---

53. What should the fraction $\frac{1}{\sqrt[4]{2}}$ be multiplied by to rationalize the denominator?

54. To rationalize the denominator of $\frac{1}{\sqrt[5]{2}}$, by what number would you multiply the numerator and denominator?

In Exercises 55 and 56, leave answers as simplified radical expressions.

55. The period p of a pendulum is the time it takes for it to swing from one extreme to the other and back again. The value of p in seconds is given by

$$p = k \cdot \sqrt{\frac{L}{g}},$$

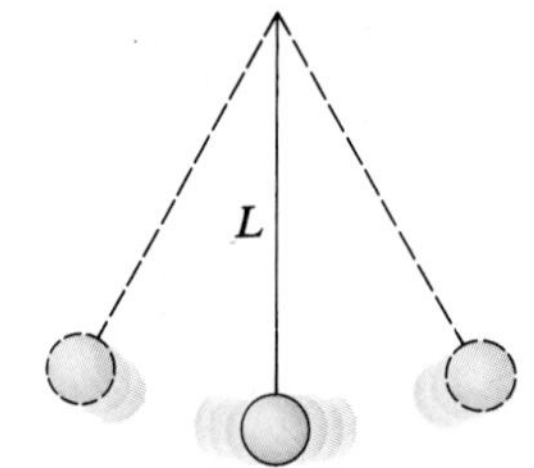

where L is the length of the pendulum, g is the acceleration due to gravity, and k is a constant. Find the period when $k = 6$, $L = 9$ feet, and $g = 32$ feet per second per second.

56. The velocity v of a meteorite approaching the earth is given by

$$v = \frac{k}{\sqrt{d}}$$

kilometers per second, where d is its distance from the center of the earth and k is a constant. What is the velocity of a meteorite that is 6000 kilometers away from the center of the earth, if $k = 450$?

57. When a denominator is rationalized, why is the value of the fraction not changed? (See Objective 1.)

58. What is the advantage of simplifying a denominator before rationalizing? (See Objective 1.)

PREVIEW EXERCISES

Combine like terms. See Section 2.1.

59. $2x + 3x - x$ **60.** $-5a + 2a + 7a$ **61.** $m + 2 - 3m - 5$

62. $-2p + 4p + 7 - 9 - 3p$ **63.** $4y + 3z + 5z - y$ **64.** $5w + 2t - w + 3t$

8.5 SIMPLIFYING RADICAL EXPRESSIONS

OBJECTIVES

1. SIMPLIFY RADICAL EXPRESSIONS WITH SUMS.
2. SIMPLIFY RADICAL EXPRESSIONS WITH PRODUCTS.
3. SIMPLIFY RADICAL EXPRESSIONS WITH QUOTIENTS.
4. WRITE RADICAL EXPRESSIONS WITH QUOTIENTS IN LOWEST TERMS.

It can be difficult to decide on the "simplest" form of a radical. The conditions for which a square root radical is in simplest form were listed in the previous section. Below is a set of guidelines that should be followed when you are simplifying radical expressions. Although the conditions are illustrated with square roots, they apply to higher roots as well.

SIMPLIFYING RADICAL EXPRESSIONS

1. If a radical represents a rational number, then that rational number should be used in place of the radical.

For example, $\sqrt{49}$ is simplified by writing 7; $\sqrt{64}$ by writing 8; $\sqrt{\frac{169}{9}}$ by writing $\frac{13}{3}$.

2. If a radical expression contains products of radicals, the product rule for radicals, $\sqrt{x} \cdot \sqrt{y} = \sqrt{xy}$, should be used to get a single radical.

For example, $\sqrt{3} \cdot \sqrt{2}$ is simplified to $\sqrt{6}$; $\sqrt{5} \cdot \sqrt{x}$ to $\sqrt{5x}$.

3. If a radicand has a factor that is a perfect square, the radical should be expressed as the product of the positive square root of the perfect square and the remaining radical factor. A similar statement applies to higher roots.

For example, $\sqrt{20}$ is simplified to $\sqrt{20} = \sqrt{4 \cdot 5} = \sqrt{4} \cdot \sqrt{5} = 2\sqrt{5}$; $\sqrt{75}$ to $5\sqrt{3}$; $\sqrt[3]{16} = \sqrt[3]{8 \cdot 2} = \sqrt[3]{8} \cdot \sqrt[3]{2} = 2\sqrt[3]{2}$.

4. If a radical expression contains sums or differences of radicals, the distributive property should be used to combine like radicals.

For example, $3\sqrt{2} + 4\sqrt{2}$ is combined as $7\sqrt{2}$, but $3\sqrt{2} + 4\sqrt{3}$ cannot be further combined.

5. Any denominator containing a radical should be rationalized.

For example, $\frac{5}{\sqrt{3}}$ is rationalized as $\frac{5}{\sqrt{3}} = \frac{5\sqrt{3}}{\sqrt{3} \cdot \sqrt{3}} = \frac{5\sqrt{3}}{3}$;

$\sqrt{\frac{3}{2}}$ is rationalized as $\sqrt{\frac{3}{2}} = \frac{\sqrt{3}}{\sqrt{2}} = \frac{\sqrt{3}}{\sqrt{2}} \cdot \frac{\sqrt{2}}{\sqrt{2}} = \frac{\sqrt{6}}{2}$.

1 The first example shows how radical expressions involving sums may be simplified.

EXAMPLE 1
ADDING RADICAL EXPRESSIONS

Simplify each of the following.

(a) $\sqrt{16} + \sqrt{9}$

Here $\sqrt{16} + \sqrt{9} = 4 + 3 = 7$.

(b) $5\sqrt{2} + 2\sqrt{18}$

First simplify $\sqrt{18}$.

$$\begin{aligned} 5\sqrt{2} + 2\sqrt{18} &= 5\sqrt{2} + 2(\sqrt{9} \cdot \sqrt{2}) && \text{9 is a perfect square.} \\ &= 5\sqrt{2} + 2(3\sqrt{2}) && \sqrt{9} = 3 \\ &= 5\sqrt{2} + 6\sqrt{2} && \text{Multiply.} \\ &= 11\sqrt{2} && \text{Add like radicals.} \end{aligned}$$

■

2 The next examples show how to simplify radical expressions involving products.

EXAMPLE 2
MULTIPLYING RADICAL EXPRESSIONS

Simplify $\sqrt{5}(\sqrt{8} - \sqrt{32})$.

Start by simplifying $\sqrt{8}$ and $\sqrt{32}$.

$$\sqrt{8} = 2\sqrt{2} \quad \text{and} \quad \sqrt{32} = 4\sqrt{2}$$

Now simplify inside the parentheses.

$$\begin{aligned} \sqrt{5}(\sqrt{8} - \sqrt{32}) &= \sqrt{5}(2\sqrt{2} - 4\sqrt{2}) \\ &= \sqrt{5}(-2\sqrt{2}) && \text{Subtract like radicals.} \\ &= -2\sqrt{5 \cdot 2} && \text{Product rule; commutative property} \\ &= -2\sqrt{10} && \text{Multiply.} \end{aligned}$$

■

EXAMPLE 3
MULTIPLYING RADICAL EXPRESSIONS

Find each product and simplify the answers.

(a) $(\sqrt{3} + 2\sqrt{5})(\sqrt{3} - 4\sqrt{5})$

The products of these sums of radicals can be found in the same way that we found the product of binomials in Chapter 3. The pattern of multiplication is the same, using the FOIL method.

$$\begin{aligned} &(\sqrt{3} + 2\sqrt{5})(\sqrt{3} - 4\sqrt{5}) \\ &= \overset{\text{F}}{\sqrt{3} \cdot \sqrt{3}} + \overset{\text{O}}{\sqrt{3}(-4\sqrt{5})} + \overset{\text{I}}{2\sqrt{5} \cdot \sqrt{3}} + \overset{\text{L}}{2\sqrt{5}(-4\sqrt{5})} \\ &= 3 - 4\sqrt{15} + 2\sqrt{15} - 8 \cdot 5 && \text{Product rule} \\ &= 3 - 2\sqrt{15} - 40 && \text{Add like radicals.} \\ &= -37 - 2\sqrt{15} && \text{Combine terms.} \end{aligned}$$

(b) $(\sqrt{3} + \sqrt{21})(\sqrt{3} - \sqrt{7})$

$$
\begin{aligned}
(\sqrt{3} + \sqrt{21})(\sqrt{3} - \sqrt{7}) \\
&= \sqrt{3}(\sqrt{3}) + \sqrt{3}(-\sqrt{7}) + \sqrt{21}(\sqrt{3}) + \sqrt{21}(-\sqrt{7}) && \text{FOIL} \\
&= 3 - \sqrt{21} + \sqrt{63} - \sqrt{147} && \text{Product rule} \\
&= 3 - \sqrt{21} + \sqrt{9} \cdot \sqrt{7} - \sqrt{49} \cdot \sqrt{3} && \text{Simplify radicals.} \\
&= 3 - \sqrt{21} + 3\sqrt{7} - 7\sqrt{3} && \sqrt{9} = 3 \text{ and } \sqrt{49} = 7
\end{aligned}
$$

Since there are no like radicals, no terms may be combined. ■

Since radicals represent real numbers, the special products of binomials discussed in Chapter 3 can be used to find products of radicals. Example 4 uses the rule for the product that gives the difference of two squares,

$$(a + b)(a - b) = a^2 - b^2.$$

EXAMPLE 4

USING A SPECIAL PRODUCT WITH RADICALS

Find each product.

(a) $(4 + \sqrt{3})(4 - \sqrt{3})$

Follow the pattern given above. Let $a = 4$ and $b = \sqrt{3}$.

$$
\begin{aligned}
(4 + \sqrt{3})(4 - \sqrt{3}) &= (4)^2 - (\sqrt{3})^2 \\
&= 16 - 3 = 13 && 4^2 = 16 \text{ and } (\sqrt{3})^2 = 3
\end{aligned}
$$

(b)
$$
\begin{aligned}
(\sqrt{12} - \sqrt{6})(\sqrt{12} + \sqrt{6}) &= (\sqrt{12})^2 - (\sqrt{6})^2 \\
&= 12 - 6 && (\sqrt{12})^2 = 12 \text{ and } (\sqrt{6})^2 = 6 \\
&= 6
\end{aligned}
$$
■

Both products in Example 4 resulted in rational numbers. The pairs of expressions in those products, $4 + \sqrt{3}$ and $4 - \sqrt{3}$, and $\sqrt{12} - \sqrt{6}$ and $\sqrt{12} + \sqrt{6}$, are called **conjugates** of each other.

3 Products of radicals similar to those in Example 4 can be used to rationalize the denominators in quotients with binomial denominators, such as

$$\frac{2}{4 - \sqrt{3}}.$$

By Example 4(a), if this denominator, $4 - \sqrt{3}$, is multiplied by $4 + \sqrt{3}$, then the product $(4 - \sqrt{3})(4 + \sqrt{3})$ is the rational number 13. Multiplying numerator and denominator by $4 + \sqrt{3}$ gives

$$\frac{2}{4 - \sqrt{3}} = \frac{2(4 + \sqrt{3})}{(4 - \sqrt{3})(4 + \sqrt{3})} = \frac{2(4 + \sqrt{3})}{13}.$$

The denominator now has been rationalized; it contains no radical signs.

USING CONJUGATES TO SIMPLIFY A RADICAL EXPRESSION

To simplify a radical expression with two terms in the denominator, where at least one of those terms is a radical, multiply both the numerator and the denominator by the conjugate of the denominator.

EXAMPLE 5 USING CONJUGATES TO RATIONALIZE A DENOMINATOR

Rationalize the denominator in the quotient $\dfrac{7}{3+\sqrt{5}}$.

The radical in the denominator can be eliminated by multiplying both numerator and denominator by $3-\sqrt{5}$, the conjugate of $3+\sqrt{5}$.

$$\frac{7}{3+\sqrt{5}} = \frac{7(3-\sqrt{5})}{(3+\sqrt{5})(3-\sqrt{5})} \qquad \text{Multiply by the conjugate.}$$

$$= \frac{7(3-\sqrt{5})}{3^2-(\sqrt{5})^2} \qquad (a+b)(a-b)=a^2-b^2$$

$$= \frac{7(3-\sqrt{5})}{9-5} \qquad 3^2=9 \text{ and } (\sqrt{5})^2=5$$

$$= \frac{7(3-\sqrt{5})}{4} \qquad \text{Subtract.}$$

■

EXAMPLE 6 USING CONJUGATES TO RATIONALIZE A DENOMINATOR

Simplify $\dfrac{6+\sqrt{2}}{\sqrt{2}-5}$.

Multiply numerator and denominator by $\sqrt{2}+5$.

$$\frac{6+\sqrt{2}}{\sqrt{2}-5} = \frac{(6+\sqrt{2})(\sqrt{2}+5)}{(\sqrt{2}-5)(\sqrt{2}+5)}$$

$$= \frac{6\sqrt{2}+30+2+5\sqrt{2}}{2-25} \qquad \text{FOIL}$$

$$= \frac{11\sqrt{2}+32}{-23} \qquad \text{Combine terms.}$$

$$= -\frac{11\sqrt{2}+32}{23} \qquad \frac{a}{-b}=-\frac{a}{b}$$

■

4 The final example shows how to write certain quotients with radicals in lowest terms.

EXAMPLE 7 WRITING A RADICAL QUOTIENT IN LOWEST TERMS

Write $\dfrac{3\sqrt{3}+15}{12}$ in lowest terms.

Factor the numerator, and then divide numerator and denominator by any common factors.

$$\frac{3\sqrt{3}+15}{12} = \frac{3(\sqrt{3}+5)}{3\cdot 4} = \frac{\sqrt{3}+5}{4}$$

■

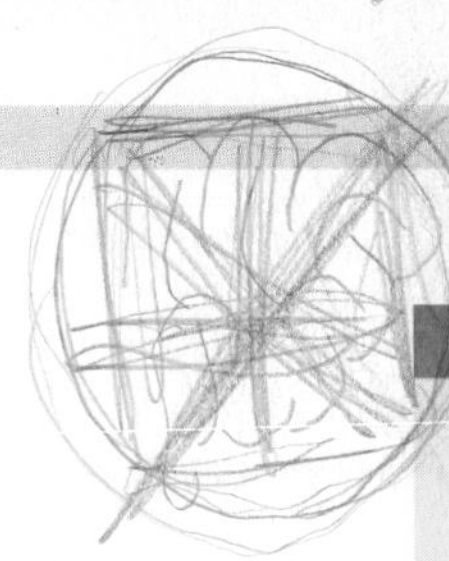
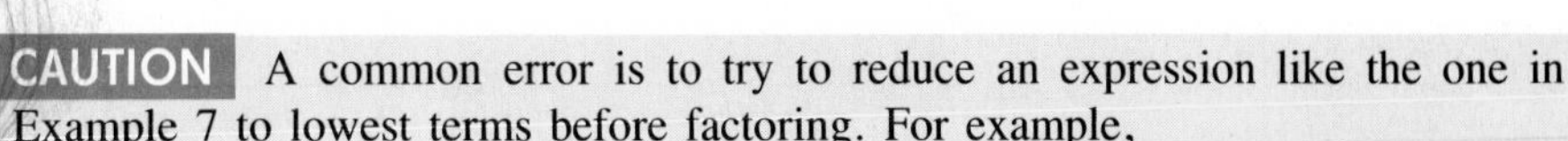

CAUTION A common error is to try to reduce an expression like the one in Example 7 to lowest terms before factoring. For example,

$$\frac{4 + 8\sqrt{5}}{4} \neq 1 + 8\sqrt{5}.$$

The correct simplification is $1 + 2\sqrt{5}$. Do you see why?

8.5 EXERCISES

Simplify each expression. Use the five rules given in the text. Assume all variables represent nonnegative real numbers. See Examples 1–4.

1. $3\sqrt{5} + 8\sqrt{45}$
2. $6\sqrt{2} + 4\sqrt{18}$
3. $9\sqrt{50} - 4\sqrt{72}$
4. $3\sqrt{80} - 5\sqrt{45}$
5. $\sqrt{2}(\sqrt{8} - \sqrt{32})$
6. $\sqrt{3}(\sqrt{27} - \sqrt{3})$
7. $\sqrt{5}(\sqrt{3} + \sqrt{7})$
8. $\sqrt{7}(\sqrt{10} - \sqrt{3})$
9. $2\sqrt{5}(\sqrt{2} + \sqrt{5})$
10. $3\sqrt{7}(2\sqrt{7} - 4\sqrt{5})$
11. $-\sqrt{14} \cdot \sqrt{2} - \sqrt{28}$
12. $\sqrt{6} \cdot \sqrt{3} - 2\sqrt{50}$
13. $(2\sqrt{6} + 3)(3\sqrt{6} - 5)$
14. $(4\sqrt{5} - 2)(2\sqrt{5} + 3)$
15. $(5\sqrt{7} - 2\sqrt{3})(3\sqrt{7} + 3\sqrt{3})$
16. $(2\sqrt{10} + 5\sqrt{2})(3\sqrt{10} - 4\sqrt{2})$
17. $(3\sqrt{2} + 4)(3\sqrt{2} + 4)$
18. $(4\sqrt{5} - 1)(4\sqrt{5} - 1)$
19. $(\sqrt{2} + \sqrt{3})^2$
20. $(\sqrt{6} + \sqrt{2})^2$
21. $(3 - \sqrt{2})(3 + \sqrt{2})$
22. $(7 - \sqrt{5})(7 + \sqrt{5})$
23. $(2 + \sqrt{8})(2 - \sqrt{8})$
24. $(3 + \sqrt{11})(3 - \sqrt{11})$
25. $(\sqrt{6} - \sqrt{5})(\sqrt{6} + \sqrt{5})$
26. $(\sqrt{11} + \sqrt{10})(\sqrt{11} - \sqrt{10})$
27. $(\sqrt{8} - \sqrt{2})(\sqrt{2} + \sqrt{4})$
28. $(\sqrt{6} - \sqrt{3})(\sqrt{3} + \sqrt{18})$
29. $(\sqrt{5x} + \sqrt{30})(\sqrt{6x} + \sqrt{3})$
30. $(\sqrt{10y} - \sqrt{20})(\sqrt{2y} - \sqrt{5})$
31. $(3\sqrt{t} + \sqrt{7})(2\sqrt{t} - \sqrt{14})$
32. $(2\sqrt{z} - \sqrt{3})(\sqrt{z} - \sqrt{5})$
33. $(\sqrt{3m} + \sqrt{2n})(\sqrt{5m} - \sqrt{5n})$
34. $(\sqrt{4p} - \sqrt{3k})(\sqrt{2p} + \sqrt{9k})$

---

35. Choose the best answer to complete the sentence. The product of two conjugates is always a(n) ____________ number.
(a) rational **(b)** irrational
(c) whole **(d)** positive
(See Objective 2.)

Rationalize the denominators. See Examples 5 and 6.

36. $\dfrac{5}{2 + \sqrt{5}}$
37. $\dfrac{7}{2 - \sqrt{11}}$
38. $\dfrac{\sqrt{12}}{\sqrt{3} + 1}$
39. $\dfrac{\sqrt{18}}{\sqrt{2} - 1}$
40. $\dfrac{2\sqrt{3}}{\sqrt{3} + 5}$
41. $\dfrac{\sqrt{12}}{2 - \sqrt{10}}$
42. $\dfrac{\sqrt{2} + 3}{\sqrt{3} - 1}$
43. $\dfrac{3 + \sqrt{2}}{\sqrt{2} + 1}$

44. $\dfrac{2\sqrt{6}+1}{\sqrt{2}+5}$
45. $\dfrac{3\sqrt{2}-4}{\sqrt{3}+2}$
46. $\dfrac{\sqrt{7}+\sqrt{2}}{\sqrt{3}-\sqrt{2}}$
47. $\dfrac{\sqrt{6}+\sqrt{5}}{\sqrt{3}+\sqrt{5}}$
48. $\dfrac{3+\sqrt{3}}{\sqrt{2}}$
49. $\dfrac{2-\sqrt{5}}{\sqrt{3}}$
50. $\dfrac{\sqrt{6}+\sqrt{2}}{\sqrt{2}}$
51. $\dfrac{\sqrt{7}+\sqrt{5}}{\sqrt{5}}$

Write each quotient in lowest terms. See Example 7.

52. $\dfrac{5\sqrt{7}-10}{5}$
53. $\dfrac{6\sqrt{5}-9}{3}$
54. $\dfrac{2\sqrt{3}+10}{8}$
55. $\dfrac{4\sqrt{6}+6}{10}$
56. $\dfrac{12-2\sqrt{10}}{4}$
57. $\dfrac{9-6\sqrt{2}}{12}$
58. $\dfrac{16+8\sqrt{2}}{24}$
59. $\dfrac{25+5\sqrt{3}}{10}$

---

60. What is wrong with the following example? (See Objective 4.)

$$\frac{2\sqrt{6}+8}{4} = 2\sqrt{6}+2$$

Simplify the radical expressions in Exercises 61–68.

61. $\sqrt[3]{4}(\sqrt[3]{2}-3)$
62. $\sqrt[3]{5}(4\sqrt[3]{5}-\sqrt[3]{25})$
63. $2\sqrt[4]{2}(3\sqrt[4]{8}+5\sqrt[4]{4})$
64. $6\sqrt[4]{9}(2\sqrt[4]{9}-\sqrt[4]{27})$
65. $(\sqrt[3]{2}-1)(\sqrt[3]{4}+3)$
66. $(\sqrt[3]{9}+5)(\sqrt[3]{3}-4)$
67. $(\sqrt[3]{5}-\sqrt[3]{4})(\sqrt[3]{25}+\sqrt[3]{20}+\sqrt[3]{16})$
68. $(\sqrt[3]{4}+\sqrt[3]{2})(\sqrt[3]{16}-\sqrt[3]{8}+\sqrt[3]{4})$

Work the following problems.

69. The radius of the circular top or bottom of a tin can with a surface area A and a height h is given by

$$r = \frac{-h+\sqrt{h^2+.64A}}{2}.$$

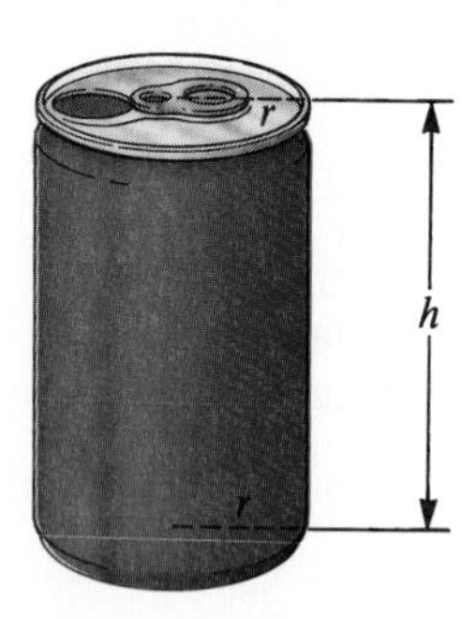

What radius should be used to make a can with a height of 12 centimeters and a surface area of 100 square centimeters? Leave the answer as a simplified radical expression.

70. If an investment of P dollars grows to A dollars in two years, the annual rate of return on the investment is given by

$$r = \frac{\sqrt{A}-\sqrt{P}}{\sqrt{P}}.$$

First rationalize this expression, then find the annual rate of return (in percent) if $50,000 increases to $58,320.

71. Why is the expression $\sqrt{3}+\sqrt{27}$ not in simplified form?

72. In your own words, explain what is meant by the conjugate of a binomial expression. (See Objective 2.)

PREVIEW EXERCISES

Solve each equation. See Section 4.5.

73. $y^2-4y+3=0$
74. $x^2-x-20=0$
75. $2m^2+m=15$
76. $3a^2=14a+5$
77. $k-1=(k-1)^2$
78. $(t+2)^2=-(t+2)$

8.6 EQUATIONS WITH RADICALS

OBJECTIVES

1 SOLVE EQUATIONS WITH RADICALS.

2 IDENTIFY EQUATIONS WITH NO SOLUTIONS.

3 SOLVE EQUATIONS THAT REQUIRE SQUARING A BINOMIAL.

1 The addition and multiplication properties of equality are not enough to solve an equation with radicals such as

$$\sqrt{x+1} = 3.$$

Solving equations that have square roots requires a new property, the **squaring property.**

SQUARING PROPERTY OF EQUALITY

If both sides of a given equation are squared, all solutions of the original equation are *among* the solutions of the squared equation.

CAUTION Be very careful with the squaring property. Using this property can give a new equation with *more* solutions than the original equation. For example, starting with the equation $y = 4$ and squaring each side gives

$$y^2 = 4^2, \qquad \text{or} \qquad y^2 = 16.$$

This last equation, $y^2 = 16$, has *two* solutions, 4 or -4, while the original equation, $y = 4$, has only *one* solution, 4. Because of this possibility, checking is more than just a guard against algebraic errors when solving an equation with radicals. It is an essential part of the solution process. *All potential solutions from the squared equation must be checked in the original equation.*

EXAMPLE 1 USING THE SQUARING PROPERTY OF EQUALITY

Solve the equation $\sqrt{p+1} = 3$.

Use the squaring property of equality to square both sides of the equation and then solve this new equation.

$$(\sqrt{p+1})^2 = 3^2$$

$$p + 1 = 9 \qquad (\sqrt{p+1})^2 = p + 1$$

$$p = 8 \qquad \text{Subtract 1.}$$

Now check this answer in the original equation.

$$\sqrt{p+1} = 3$$

$$\sqrt{8+1} = 3 \quad ? \qquad \text{Let } p = 8.$$

$$\sqrt{9} = 3 \quad ?$$

$$3 = 3 \qquad \text{True}$$

Since this statement is true, 8 is the solution of $\sqrt{p+1} = 3$. In this case the squared equation had just one solution, which also satisfied the original equation. ■

EXAMPLE 2 USING THE SQUARING PROPERTY OF EQUALITY

Solve $3\sqrt{x} = \sqrt{x+8}$.

Squaring both sides gives

$$(3\sqrt{x})^2 = (\sqrt{x+8})^2$$
$$3^2(\sqrt{x})^2 = (\sqrt{x+8})^2 \qquad (ab)^2 = a^2b^2$$
$$9x = x + 8 \qquad (\sqrt{x})^2 = x;\ (\sqrt{x+8})^2 = x + 8$$
$$8x = 8 \qquad \text{Subtract } x.$$
$$x = 1. \qquad \text{Divide by 8.}$$

Check this potential solution.

$$3\sqrt{x} = \sqrt{x+8}$$
$$3\sqrt{1} = \sqrt{1+8} \quad ? \qquad \text{Let } x = 1.$$
$$3(1) = \sqrt{9} \quad ? \qquad \sqrt{1} = 1$$
$$3 = 3 \qquad \text{True}$$

The check shows that the solution of the given equation is 1. ■

2 Not all equations with radicals have a solution, as shown by the equations in Examples 3 and 4.

EXAMPLE 3 USING THE SQUARING PROPERTY OF EQUALITY

Solve the equation $\sqrt{y} = -3$.

Square both sides.

$$(\sqrt{y})^2 = (-3)^2$$
$$y = 9$$

Check this proposed answer in the original equation.

$$\sqrt{y} = -3$$
$$\sqrt{9} = -3 \quad ? \qquad \text{Let } y = 9.$$
$$3 = -3 \qquad \text{False}$$

Since the statement $3 = -3$ is false, the number 9 is not a solution of the given equation and is said to be *extraneous*. In fact, $\sqrt{y} = -3$ has no solution. Since $\sqrt{y}$ represents the *nonnegative* square root of y, we might have seen immediately that there is no solution. ■

The steps to use when solving an equation with radicals are summarized below.

SOLVING AN EQUATION WITH RADICALS

Step 1 Arrange the terms so that one radical is alone on one side of the equation.
Step 2 Square both sides.
Step 3 Combine like terms.
Step 4 If there is still a term with a radical, repeat Steps 1–3.
Step 5 Solve the equation from Step 3 for potential solutions.
Step 6 Check all potential solutions from Step 5 in the original equation.

EXAMPLE 4
USING THE SQUARING PROPERTY OF EQUALITY

Solve $a = \sqrt{a^2 + 5a + 10}$.

Square both sides.

$$(a)^2 = (\sqrt{a^2 + 5a + 10})^2$$

$$a^2 = a^2 + 5a + 10 \qquad (\sqrt{a^2 + 5a + 10})^2 = a^2 + 5a + 10$$

$$0 = 5a + 10 \qquad \text{Subtract } a^2.$$

$$a = -2 \qquad \text{Subtract 10; divide by 5.}$$

Check this proposed solution in the original equation.

$$a = \sqrt{a^2 + 5a + 10}$$

$$-2 = \sqrt{(-2)^2 + 5(-2) + 10} \quad ? \qquad \text{Let } a = -2.$$

$$-2 = \sqrt{4 - 10 + 10} \quad ? \qquad \text{Multiply.}$$

$$-2 = 2 \qquad \text{False}$$

Since $a = -2$ leads to a false result, the equation has no solution. ■

3 The next examples use the following facts from Section 3.4.

$$(a + b)^2 = a^2 + 2ab + b^2 \qquad \text{and} \qquad (a - b)^2 = a^2 - 2ab + b^2.$$

By the second pattern, for example,

$$(y - 3)^2 = y^2 - 2(y)(3) + (3)^2$$
$$= y^2 - 6y + 9.$$

EXAMPLE 5
USING THE SQUARING PROPERTY OF EQUALITY

Solve the equation $\sqrt{2y - 3} = y - 3$.

Square each side, using the result above on the right side of the equation.

$$(\sqrt{2y - 3})^2 = (y - 3)^2$$
$$2y - 3 = y^2 - 6y + 9$$

This equation is quadratic because of the y^2 term. As shown in Section 4.5, solving this equation requires that one side be equal to 0. Subtract $2y$ and add 3, getting

$$0 = y^2 - 8y + 12.$$

Solve this equation by factoring.

$$0 = (y - 6)(y - 2)$$

Set each factor equal to 0.

$$y - 6 = 0 \quad \text{or} \quad y - 2 = 0$$
$$y = 6 \quad \text{or} \quad y = 2$$

Check both of these potential solutions in the original equation.

If $y = 6$,

$$\sqrt{2y - 3} = y - 3$$
$$\sqrt{2(6) - 3} = 6 - 3 \quad ?$$
$$\sqrt{12 - 3} = 3 \quad ?$$
$$\sqrt{9} = 3 \quad ?$$
$$3 = 3. \quad \text{True}$$

If $y = 2$,

$$\sqrt{2y - 3} = y - 3$$
$$\sqrt{2(2) - 3} = 2 - 3 \quad ?$$
$$\sqrt{4 - 3} = -1 \quad ?$$
$$\sqrt{1} = -1 \quad ?$$
$$1 = -1. \quad \text{False}$$

Only 6 is a solution of the equation. ■

Sometimes it is necessary to write an equation in a different form before squaring both sides. The next example shows why.

EXAMPLE 6
USING THE SQUARING PROPERTY OF EQUALITY

Solve the equation $3\sqrt{x} - 1 = 2x$.

Squaring both sides gives

$$(3\sqrt{x} - 1)^2 = (2x)^2$$
$$9x - 6\sqrt{x} + 1 = 4x^2,$$

an equation that is more complicated, and still contains a radical. It would be better instead to rewrite the original equation so that the radical is alone on one side of the equals sign. Get the radical alone by adding 1 to both sides to get

$$3\sqrt{x} = 2x + 1.$$

Now square both sides. $\quad (3\sqrt{x})^2 = (2x + 1)^2$

$$9x = 4x^2 + 4x + 1$$

Subtract $9x$ from both sides. $\quad 0 = 4x^2 - 5x + 1$

This quadratic equation can be solved by factoring.

$$0 = (4x - 1)(x - 1)$$
$$4x - 1 = 0 \quad \text{or} \quad x - 1 = 0$$
$$x = \frac{1}{4} \quad \text{or} \quad x = 1$$

Both of these potential solutions must be checked in the original equation.

If $x = \frac{1}{4}$,

$$3\sqrt{x} - 1 = 2x$$
$$3\sqrt{\frac{1}{4}} - 1 = 2\left(\frac{1}{4}\right) \quad ?$$
$$\frac{3}{2} - 1 = \frac{1}{2}. \quad \text{True}$$

If $x = 1$,

$$3\sqrt{x} - 1 = 2x$$
$$3\sqrt{1} - 1 = 2(1) \quad ?$$
$$3 - 1 = 2. \quad \text{True}$$

The solutions to the original equation are 1/4 and 1. ■

CAUTION Errors often occur when each side of an equation is squared. For example, in Example 6 after the equation is rewritten as

$$3\sqrt{x} = 2x + 1,$$

it would be *incorrect* to write the next step as

$$9x = 4x^2 + 1.$$

Don't forget that the binomial $2x + 1$ must be squared to get $4x^2 + 4x + 1$.

Some equations with radicals require squaring twice, as in the next example.

EXAMPLE 7
USING THE SQUARING PROPERTY TWICE

Solve $\sqrt{21 + x} = 3 + \sqrt{x}$.

Square both sides.

$$(\sqrt{21 + x})^2 = (3 + \sqrt{x})^2$$
$$21 + x = 9 + 6\sqrt{x} + x$$

Combine terms and simplify.

$$12 = 6\sqrt{x}$$
$$2 = \sqrt{x}$$

Since we still have a radical, square both sides a second time to get

$$4 = x.$$

Check the potential solution.

If $x = 4$,

$$\sqrt{21 + x} = 3 + \sqrt{x}$$
$$\sqrt{21 + 4} = 3 + \sqrt{4} \quad ?$$
$$5 = 5. \quad \text{True}$$

The solution is 4. ■

8.6 EXERCISES

Find all solutions for each equation. See Examples 1–4.

1. $\sqrt{x} = 2$
2. $\sqrt{m} = 5$
3. $\sqrt{m + 5} = 0$
4. $\sqrt{y - 4} = 0$
5. $\sqrt{z + 5} = -2$
6. $\sqrt{t - 3} = -2$
7. $\sqrt{k} - 2 = 5$
8. $\sqrt{p} - 3 = 7$
9. $\sqrt{y} + 4 = 2$
10. $\sqrt{m} + 6 = 5$
11. $\sqrt{5t - 9} = 2\sqrt{t}$
12. $\sqrt{3n + 4} = 2\sqrt{n}$
13. $3\sqrt{r} = \sqrt{8r + 16}$
14. $2\sqrt{r} = \sqrt{3r + 9}$
15. $\sqrt{5y - 5} = \sqrt{4y + 1}$
16. $\sqrt{2x + 2} = \sqrt{3x - 5}$
17. $\sqrt{x + 2} = \sqrt{2x - 5}$
18. $\sqrt{3m + 3} = \sqrt{5m - 1}$
19. $p = \sqrt{p^2 - 3p - 12}$
20. $k = \sqrt{k^2 - 2k + 10}$
21. $2r = \sqrt{4r^2 + 5r - 30}$

Find all solutions for each equation. Remember that $(a + b)^2 = a^2 + 2ab + b^2$ and $(\sqrt{a})^2 = a$. See Examples 5 and 6.

22. $\sqrt{5x + 11} = x + 3$

23. $\sqrt{2x + 1} = x - 7$

24. $\sqrt{5x + 1} = x + 1$

25. $\sqrt{3x + 10} = 2x - 5$

26. $\sqrt{4x + 13} = 2x - 1$

27. $\sqrt{x + 1} - 1 = x$

28. $\sqrt{3x + 3} + 5 = x$

29. $\sqrt{4x + 5} - 2 = 2x - 7$

30. $\sqrt{6x + 7} - 1 = x + 1$

31. $3\sqrt{x + 13} = x + 9$

32. $2\sqrt{x + 7} = x - 1$

33. $\sqrt{4x} - x + 3 = 0$

34. $\sqrt{2x} - x + 4 = 0$

35. $\sqrt{3x} - 4 = x - 10$

36. $\sqrt{x} + 9 = x + 3$

In the following two exercises, it is necessary to square both sides twice. See Example 7.

37. $\sqrt{x} = \sqrt{x - 5} + 1$

38. $\sqrt{2x} = \sqrt{x + 7} - 1$

Solve each problem.

39. The square root of the sum of a number and 4 is 5. Find the number.

40. A certain number is the same as the square root of the product of 8 and the number. Find the number.

41. Three times the square root of 2 equals the square root of the sum of some number and 10. Find the number.

42. The negative square root of a number equals that number decreased by 2. Find the number.

43. To estimate the speed at which a car was traveling at the time of an accident, police sometimes use the following procedure. A police officer drives the car involved in the accident under conditions similar to those during which the accident took place, and then skids to a stop. If the car is driven at 30 miles per hour, the speed at the time of the accident is given by

$$s = 30\sqrt{\frac{a}{p}},$$

where a is the length of the skid marks left at the time of the accident and p is the length of the skid marks in the police test. Find s if
(a) $a = 900$ feet and $p = 100$ feet;
(b) $a = 400$ feet and $p = 25$ feet;
(c) $a = 80$ feet and $p = 20$ feet;
(d) $a = 120$ feet and $p = 30$ feet.

44. Explain why $\sqrt{m} = -4$ has no solution. (See Objective 2.)

45. Why is it *necessary* to check all solutions in the original equation whenever the squaring property of equality is used? (See Objective 1.)

46. What is wrong with the following "solution"?

$-\sqrt{x - 1} = 4$	
$-(x - 1) = 16$	Square both sides.
$-x + 1 = 16$	Distributive property
$-x = 15$	Subtract 1 on each side.
$x = -15$	Multiply each side by -1.

PREVIEW EXERCISES

Use the rules for exponents to simplify each expression. Write each answer in exponential form with only positive exponents. See Sections 3.2 and 3.5.

47. $(5^2)^3$

48. $3^{-4} \cdot 3^{-1}$

49. $\dfrac{a^{-2}a^3}{a^4}$

50. $(2x^3)^{-1}$

51. $\left(\dfrac{p}{3}\right)^{-2}$

52. $\left(\dfrac{2y^3}{y^{-1}}\right)^{-2}$

53. $\dfrac{(c^3)^2c^4}{(c^{-1})^3}$

54. $\dfrac{(m^2)^4m^{-1}}{(m^3)^{-1}}$

8.7 FRACTIONAL EXPONENTS

OBJECTIVES

1 DEFINE AND USE $a^{1/n}$.

2 DEFINE AND USE $a^{m/n}$.

3 USE RULES FOR EXPONENTS WITH FRACTIONAL EXPONENTS.

4 USE FRACTIONAL EXPONENTS TO SIMPLIFY RADICALS.

In this section we introduce exponential expressions with fractional exponents such as $5^{1/2}$, $16^{3/4}$, and $8^{-2/3}$.

1 How should $5^{1/2}$ be defined? We want to define $5^{1/2}$ so that all the rules for exponents developed earlier in this book still hold. Then we should define $5^{1/2}$ so that

$$5^{1/2} \cdot 5^{1/2} = 5^{1/2+1/2} = 5^1 = 5.$$

This agrees with the product rule for exponents from Section 3.2. By definition,

$$(\sqrt{5})(\sqrt{5}) = 5.$$

Since both $5^{1/2} \cdot 5^{1/2}$ and $\sqrt{5} \cdot \sqrt{5}$ equal 5,

$$5^{1/2} \text{ should equal } \sqrt{5}.$$

Similarly,

$$5^{1/3} \cdot 5^{1/3} \cdot 5^{1/3} = 5^{1/3+1/3+1/3} = 5^{3/3} = 5,$$

and

$$\sqrt[3]{5} \cdot \sqrt[3]{5} \cdot \sqrt[3]{5} = \sqrt[3]{5^3} = 5,$$

so

$$5^{1/3} \text{ should equal } \sqrt[3]{5}.$$

These examples suggest the following definition.

$a^{1/n}$

If a is a nonnegative number and n is a positive integer,

$$a^{1/n} = \sqrt[n]{a}.$$

EXAMPLE 1 USING THE DEFINITION OF $a^{1/n}$

Simplify each expression by first writing it in radical form.

(a) $16^{1/2}$

By the definition above,

$$16^{1/2} = \sqrt{16} = 4.$$

(b) $27^{1/3} = \sqrt[3]{27} = 3$ **(c)** $64^{1/3} = \sqrt[3]{64} = 4$ **(d)** $64^{1/6} = \sqrt[6]{64} = 2$ ∎

2 Now a more general exponential expression like $16^{3/4}$ can be defined. By the power rule, $(a^m)^n = a^{mn}$, so that

$$16^{3/4} = (16^{1/4})^3 = (\sqrt[4]{16})^3 = 2^3 = 8.$$

However, $16^{3/4}$ could also be written as

$$16^{3/4} = (16^3)^{1/4} = (4096)^{1/4} = \sqrt[4]{4096} = 8.$$

The expression can be evaluated either way to get the same answer. As the example suggests, taking the root first involves smaller numbers and is often easier. This example suggests the following definition for $a^{m/n}$.

$a^{m/n}$

If a is a nonnegative number and m and n are integers, with $n > 0$,

$$a^{m/n} = \sqrt[n]{a^m} = (\sqrt[n]{a})^m.$$

EXAMPLE 2
USING THE DEFINITION OF $a^{m/n}$

Evaluate each expression.

(a) $9^{3/2}$

Use the definition to write

$$9^{3/2} = (9^{1/2})^3 = 3^3 = 27.$$

(b) $64^{2/3} = (64^{1/3})^2 = 4^2 = 16$

(c) $-32^{4/5} = -(32^{1/5})^4 = -2^4 = -16$ ■

Earlier, a^{-n} was defined as

$$a^{-n} = \frac{1}{a^n}$$

for nonzero numbers a and integers n. This same result applies for negative fractional exponents.

$a^{-m/n}$

If a is a positive number and m and n are integers, with $n > 0$,

$$a^{-m/n} = \frac{1}{a^{m/n}}.$$

EXAMPLE 3
USING THE DEFINITION OF $a^{-m/n}$

Write each expression with a positive exponent and then evaluate.

(a) $32^{-3/5} = \dfrac{1}{32^{3/5}} = \dfrac{1}{(32^{1/5})^3} = \dfrac{1}{2^3} = \dfrac{1}{8}$

(b) $27^{-4/3} = \dfrac{1}{27^{4/3}} = \dfrac{1}{(27^{1/3})^4} = \dfrac{1}{3^4} = \dfrac{1}{81}$ ■

3 All the rules for exponents given earlier still hold when the exponents are fractions. The next examples show how to use these rules to simplify expressions with fractional exponents.

EXAMPLE 4 USING THE RULES FOR EXPONENTS WITH FRACTIONAL EXPONENTS

Simplify each expression. Write each answer in exponential form with only positive exponents.

(a) $3^{2/3} \cdot 3^{5/3} = 3^{2/3+5/3} = 3^{7/3}$

(b) $\dfrac{5^{1/4}}{5^{3/4}} = 5^{1/4-3/4} = 5^{-2/4} = 5^{-1/2} = \dfrac{1}{5^{1/2}}$

(c) $(9^{1/4})^2 = 9^{2(1/4)} = 9^{2/4} = 9^{1/2} = \sqrt{9} = 3$

(d) $\dfrac{2^{1/2} \cdot 2^{-1}}{2^{-3/2}} = \dfrac{2^{1/2+(-1)}}{2^{-3/2}} = \dfrac{2^{-1/2}}{2^{-3/2}} = 2^{-1/2-(-3/2)} = 2^{2/2} = 2^1 = 2$

(e) $\left(\dfrac{9}{4}\right)^{5/2} = \dfrac{9^{5/2}}{4^{5/2}} = \dfrac{(9^{1/2})^5}{(4^{1/2})^5} = \dfrac{(\sqrt{9})^5}{(\sqrt{4})^5} = \dfrac{3^5}{2^5}$ ■

EXAMPLE 5 USING THE RULES FOR EXPONENTS WITH FRACTIONAL EXPONENTS

Simplify each expression. Write each answer in exponential form with only positive exponents. Assume that all variables represent positive numbers.

(a) $m^{1/5} \cdot m^{3/5} = m^{1/5+3/5} = m^{4/5}$

(b) $\dfrac{p^{5/3}}{p^{4/3}} = p^{5/3-4/3} = p^{1/3}$

(c) $(x^2y^{1/2})^4 = (x^2)^4(y^{1/2})^4 = x^8y^2$

(d) $\left(\dfrac{z^{1/4}}{w^{1/3}}\right)^5 = \dfrac{(z^{1/4})^5}{(w^{1/3})^5} = \dfrac{z^{5/4}}{w^{5/3}}$

(e) $\dfrac{k^{2/3} \cdot k^{-1/3}}{k^{5/3}} = k^{2/3+(-1/3)-5/3} = k^{-4/3} = \dfrac{1}{k^{4/3}}$ ■

CAUTION Errors often occur in problems like those in Examples 4 and 5 because students try to convert the expressions to radicals. Remember that the *rules of exponents* apply here.

4 Sometimes a radical can be simplified by writing it in exponential form. The next example shows how this is done.

EXAMPLE 6 SIMPLIFYING RADICALS BY USING RATIONAL EXPONENTS

Simplify each radical by writing it in exponential form.

(a) $\sqrt[6]{9^3} = (9^3)^{1/6} = 9^{3/6} = 9^{1/2} = \sqrt{9} = 3$

(b) $(\sqrt[4]{m})^2 = (m^{1/4})^2 = m^{2/4} = m^{1/2} = \sqrt{m}$

Here it is assumed that $m \geq 0$. ■

8.7 EXERCISES

Simplify each expression by first writing it in radical form. See Examples 1 and 2.

1. $36^{1/2}$ **2.** $64^{1/2}$ **3.** $25^{1/2}$ **4.** $49^{1/2}$ **5.** $8^{1/3}$ **6.** $27^{1/3}$

7. $64^{1/3}$ **8.** $125^{1/3}$ **9.** $16^{1/4}$ **10.** $81^{1/4}$ **11.** $32^{1/5}$ **12.** $243^{1/5}$

13. $4^{3/2}$ **14.** $9^{5/2}$ **15.** $27^{2/3}$ **16.** $8^{5/3}$ **17.** $16^{3/4}$ **18.** $64^{5/3}$

19. $32^{2/5}$ **20.** $144^{3/2}$ **21.** $-8^{2/3}$ **22.** $-27^{5/3}$ **23.** $-64^{1/3}$ **24.** $-125^{5/3}$

Simplify each expression. Write each answer in exponential form with only positive exponents. Assume that all variables represent positive numbers. See Examples 3–5.

25. $2^{1/2} \cdot 2^{5/2}$ **26.** $5^{2/3} \cdot 5^{4/3}$ **27.** $6^{1/4} \cdot 6^{-3/4}$ **28.** $12^{2/5} \cdot 12^{-1/5}$

29. $\dfrac{15^{3/4}}{15^{5/4}}$ **30.** $\dfrac{7^{3/5}}{7^{-1/5}}$ **31.** $\dfrac{11^{-2/7}}{11^{-3/7}}$ **32.** $\dfrac{4^{-2/3}}{4^{1/3}}$

33. $(8^{3/2})^2$ **34.** $(5^{2/5})^{10}$ **35.** $(6^{1/3})^{3/2}$ **36.** $(7^{2/5})^{5/3}$

37. $\left(\dfrac{25}{4}\right)^{3/2}$ **38.** $\left(\dfrac{8}{27}\right)^{2/3}$ **39.** $\dfrac{2^{2/5} \cdot 2^{-3/5}}{2^{7/5}}$ **40.** $\dfrac{3^{-3/4} \cdot 3^{5/4}}{3^{-1/4}}$

41. $\dfrac{6^{-2/9}}{6^{1/9} \cdot 6^{-5/9}}$ **42.** $\dfrac{8^{6/7}}{8^{2/7} \cdot 8^{-1/7}}$ **43.** $p^{2/3} \cdot p^{7/3}$ **44.** $k^{-1/4} \cdot k^{5/4}$

45. $\dfrac{z^{2/3}}{z^{-1/3}}$ **46.** $\dfrac{r^{5/4}}{r^{3/4}}$ **47.** $(m^3n^{1/4})^{2/3}$ **48.** $(p^4 \cdot q^{1/2})^{4/3}$

49. $\left(\dfrac{a^{1/2}}{b^{1/3}}\right)^{4/3}$ **50.** $\left(\dfrac{m^{2/3}}{n^{3/4}}\right)^{1/2}$ **51.** $\dfrac{c^{2/3} \cdot c^{-1/3}}{c^{5/3}}$ **52.** $\dfrac{d^{3/4} \cdot d^{5/4}}{d^{1/4}}$

Simplify each radical in Exercises 53–60 by first writing it in exponential form. Give final answers in radical form where appropriate. Assume that all variables represent positive numbers. See Example 6.

53. $\sqrt[6]{4^3}$ **54.** $\sqrt[9]{8^3}$ **55.** $\sqrt[8]{16^2}$ **56.** $\sqrt[9]{27^3}$

57. $\sqrt[4]{a^2}$ **58.** $\sqrt[9]{b^3}$ **59.** $\sqrt[6]{k^4}$ **60.** $\sqrt[8]{m^4}$

---

61. A formula for calculating the distance, d, one can see from an airplane to the horizon on a clear day is

$$d = 1.22x^{1/2},$$

where x is the altitude of the plane in feet and d is given in miles. How far can one see to the horizon in a plane flying at the following altitudes? Give answers to the nearest hundredth.

(a) 20,000 feet **(b)** 30,000 feet

62. A biologist has shown that the number of different plant species S on a Galápagos Island is related to the area of the island, A, by

$$S = 28.6A^{1/3}.$$

How many plant species would exist on such an island with the following areas?

(a) 8 square miles

(b) 27,000 square miles

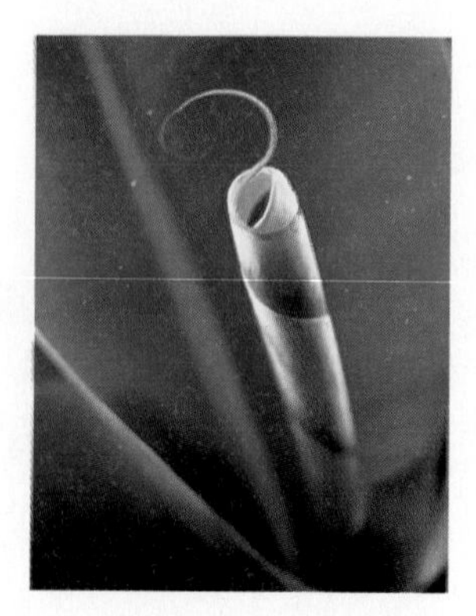

63. Why is $7^{1/2}$ defined so as to equal $\sqrt{7}$? (See Objective 1.)

64. What property of exponents is used to suggest that $6^{2/3}$ should equal $\sqrt[3]{6^2}$? (See Objective 2.)

PREVIEW EXERCISES

Find all square roots of each number. Simplify where possible. See Section 8.1.

65. 25 **66.** 49 **67.** 14 **68.** 29

69. 18 **70.** 48 **71.** 80 **72.** 75

CHAPTER 8 GLOSSARY

KEY TERMS

8.1 **square root** The square roots of a^2 are a and $-a$ (a is nonnegative).

radicand The number or expression under a radical sign is called the radicand.

radical A radical sign with a radicand is called a radical.

radical expression An algebraic expression containing a radical is called a radical expression.

perfect square A number with a rational square root is called a perfect square.

cube root The cube root of a number a is the number whose cube is a.

index (order) The index of the radical $\sqrt[n]{a}$ is n.

8.2 **perfect cube** A perfect cube is a number with a rational cube root.

8.3 **like radicals** Like radicals are multiples of the same radical.

8.4 **rationalizing the denominator** The process of changing the denominator of a fraction from a radical (irrational number) to a rational number is called rationalizing the denominator.

simplified form A square root radical is in simplified form if the radicand contains no factor (except 1) that is a perfect square, the radicand contains no fractions, and no denominator contains a radical.

8.5 **conjugate** The conjugate of $a + b$ is $a - b$.

NEW SYMBOLS

$\sqrt{}$ radical sign

$\approx$ is approximately equal to

$\sqrt[3]{a}$ cube root of a

$\sqrt[n]{a}$ nth root of a

$a^{1/m}$ mth root of a

CHAPTER 8 QUICK REVIEW

SECTION	CONCEPTS	EXAMPLES
8.1 FINDING ROOTS	If a is a positive real number,	
	$\sqrt{a}$ is the positive square root of a;	$\sqrt{49} = 7$
	$-\sqrt{a}$ is the negative square root of a; $\sqrt{0} = 0$.	$-\sqrt{81} = -9$

SECTION	CONCEPTS	EXAMPLES
	If a is a negative real number, $\sqrt{a}$ is not a real number. If a is a positive rational number, $\sqrt{a}$ is rational if a is a perfect square. $\sqrt{a}$ is irrational if a is not a perfect square. Each real number has exactly one real cube root.	$\sqrt{-25}$ is not a real number. $\sqrt{\frac{4}{9}}$, $\sqrt{16}$ are rational. $\sqrt{\frac{2}{3}}$, $\sqrt{21}$ are irrational. $\sqrt[3]{27} = 3$; $\sqrt[3]{-8} = -2$
8.2 MULTIPLICATION AND DIVISION OF RADICALS	**Product Rule for Radicals** For nonnegative real numbers x and y, $\sqrt{x} \cdot \sqrt{y} = \sqrt{xy}$ and $\sqrt{xy} = \sqrt{x} \cdot \sqrt{y}$. **Quotient Rule for Radicals** If x and y are nonnegative real numbers and y is not 0, $\frac{\sqrt{x}}{\sqrt{y}} = \sqrt{\frac{x}{y}}$ and $\sqrt{\frac{x}{y}} = \frac{\sqrt{x}}{\sqrt{y}}$. If all indicated roots are real, $\sqrt[n]{x} \cdot \sqrt[n]{y} = \sqrt[n]{xy}$ and $\frac{\sqrt[n]{x}}{\sqrt[n]{y}} = \sqrt[n]{\frac{x}{y}}$.	$\sqrt{5} \cdot \sqrt{7} = \sqrt{35}$ $\sqrt{8} \cdot \sqrt{2} = \sqrt{16} = 4$ $\sqrt{48} = \sqrt{16} \cdot \sqrt{3} = 4\sqrt{3}$ $\sqrt{\frac{25}{64}} = \frac{\sqrt{25}}{\sqrt{64}} = \frac{5}{8}$ $\frac{\sqrt{8}}{\sqrt{2}} = \sqrt{\frac{8}{2}} = \sqrt{4} = 2$ $\sqrt[3]{5} \cdot \sqrt[3]{3} = \sqrt[3]{15}$ $\frac{\sqrt[4]{12}}{\sqrt[4]{4}} = \sqrt[4]{\frac{12}{4}} = \sqrt[4]{3}$
8.3 ADDITION AND SUBTRACTION OF RADICALS	Add and subtract like radicals by using the distributive property. Only like radicals can be combined in this way.	$2\sqrt{5} + 4\sqrt{5} = (2 + 4)\sqrt{5}$ $= 6\sqrt{5}$ $\sqrt{8} + \sqrt{32} = 2\sqrt{2} + 4\sqrt{2}$ $= 6\sqrt{2}$
8.4 RATIONALIZING THE DENOMINATOR	The denominator of a radical can be rationalized by multiplying both the numerator and denominator by the same number.	$\frac{2}{\sqrt{3}} = \frac{2 \cdot \sqrt{3}}{\sqrt{3} \cdot \sqrt{3}} = \frac{2\sqrt{3}}{3}$ $\sqrt{\frac{5}{11}} = \frac{\sqrt{5} \cdot \sqrt{11}}{\sqrt{11} \cdot \sqrt{11}} = \frac{\sqrt{55}}{11}$
8.5 SIMPLIFYING RADICAL EXPRESSIONS	When appropriate, use the rules for adding and multiplying polynomials to simplify radical expressions.	$\sqrt{6}(\sqrt{5} - \sqrt{7}) = \sqrt{30} - \sqrt{42}$ $(\sqrt{5} - \sqrt{3})(\sqrt{5} + \sqrt{3}) = 5 - 3 = 2$

SECTION	CONCEPTS	EXAMPLES
	Any denominators with radicals should be rationalized.	$\frac{3}{\sqrt{6}} = \frac{3\sqrt{6}}{6} = \frac{\sqrt{6}}{2}$
	If a radical expression contains two terms in the denominator and at least one of those terms is a radical, multiply both the numerator and the denominator by the conjugate of the denominator.	$\frac{6}{\sqrt{7}-\sqrt{2}} = \frac{6}{\sqrt{7}-\sqrt{2}} \cdot \frac{\sqrt{7}+\sqrt{2}}{\sqrt{7}+\sqrt{2}}$ $= \frac{6(\sqrt{7}+\sqrt{2})}{7-2}$ Multiply fractions. $= \frac{6(\sqrt{7}+\sqrt{2})}{5}$ Simplify.
8.6 EQUATIONS WITH RADICALS	**Solving an Equation with Radicals**	Solve: $\sqrt{2x-3} + x = 3$
	Step 1 Arrange the terms so that a radical is alone on one side of the equation.	$\sqrt{2x-3} = 3 - x$ Isolate radical.
	Step 2 Square each side. (By the squaring property of equality, all solutions of the original equation are *among* the solutions of the squared equation.)	$(\sqrt{2x-3})^2 = (3-x)^2$ Square. $2x - 3 = 9 - 6x + x^2$
	Step 3 Combine like terms.	$0 = x^2 - 8x + 12$ Get one side = 0.
	Step 4 If there is still a term with a radical, repeat Steps 1–3.	$0 = (x-2)(x-6)$ Factor. $x - 2 = 0$ or $x - 6 = 0$ Set each factor = 0.
	Step 5 Solve the equation for potential solutions.	$x = 2$ or $x = 6$ Solve.
	Step 6 Check all potential solutions from Step 5 in the original equation.	Verify that 2 is the only solution (6 is extraneous).
8.7 FRACTIONAL EXPONENTS	Assume $a \geq 0$, m and n are integers, $n > 0$. $a^{1/n} = \sqrt[n]{a}$ $a^{m/n} = \sqrt[n]{a^m} = (\sqrt[n]{a})^m$	$8^{1/3} = \sqrt[3]{8} = 2$ $(81)^{3/4} = \sqrt[4]{81^3} = (\sqrt[4]{81})^3$ $= 3^3 = 27$

CHAPTER 8 REVIEW EXERCISES

[8.1] *Find all square roots of each number.*

1. 49
2. 81
3. 225
4. 729

Find each root that is a real number.

5. $\sqrt{16}$
6. $-\sqrt{4225}$
7. $\sqrt{-8100}$
8. $\sqrt[3]{-64}$
9. $\sqrt[4]{16}$
10. $\sqrt[5]{32}$

Write rational, irrational, *or* not a real number *for each radical. If a number is rational, give its exact value. If a number is irrational, use a calculator to give a decimal approximation to the nearest thousandth.*

11. $\sqrt{15}$
12. $\sqrt{64}$
13. $-\sqrt{169}$
14. $\sqrt{-170}$

[8.2] *Use the product rule to simplify each expression.*

15. $\sqrt{5} \cdot \sqrt{15}$
16. $\sqrt{12} \cdot \sqrt{3}$
17. $\sqrt{27}$
18. $\sqrt{160}$
19. $\sqrt{32} \cdot \sqrt{48}$
20. $\sqrt{98} \cdot \sqrt{50}$

Use the quotient rule and the product rule, as necessary, to simplify each expression.

21. $\sqrt{\frac{9}{4}}$
22. $\sqrt{\frac{10}{169}}$
23. $\sqrt{\frac{1}{6}} \cdot \sqrt{\frac{5}{6}}$
24. $\sqrt{\frac{2}{5}} \cdot \sqrt{\frac{2}{45}}$
25. $\frac{12\sqrt{75}}{4\sqrt{3}}$
26. $\frac{24\sqrt{12}}{16\sqrt{3}}$

Simplify each expression. Assume that all variables represent positive real numbers.

27. $\sqrt{p} \cdot \sqrt{p}$
28. $\sqrt{m^2p^4}$
29. $\sqrt{x^3}$
30. $\sqrt{r^{18}}$
31. $\sqrt{x^{10}y^{16}}$
32. $\sqrt{\frac{13}{k^4}}$
33. $\sqrt[3]{\frac{5}{8}}$
34. $\sqrt[3]{\frac{6}{125}}$
35. $\sqrt[3]{375x^4}$

[8.3] *Simplify and combine terms wherever possible. Assume that all variables represent nonnegative real numbers.*

36. $3\sqrt{2} + 5\sqrt{2}$
37. $\frac{1}{3}\sqrt{18} + \frac{1}{4}\sqrt{32}$
38. $\frac{2}{5}\sqrt{75} - \frac{3}{4}\sqrt{160}$
39. $4\sqrt{24} - 3\sqrt{54} + \sqrt{6}$
40. $3\sqrt[3]{54} - 2\sqrt[3]{16}$
41. $4\sqrt[4]{16} + 3\sqrt[4]{32} \cdot \sqrt[4]{2}$
42. $\sqrt{16p} + 3\sqrt{p} - \sqrt{49p}$
43. $\sqrt{20m^2} - m\sqrt{45}$
44. $3k\sqrt{8k^2n} + 5k^2\sqrt{2n}$

[8.4] *Perform the indicated operations, and write all answers in simplest form. Rationalize all denominators. Assume that all variables represent positive real numbers.*

45. $\frac{10}{\sqrt{3}}$
46. $\frac{5}{\sqrt{2}}$
47. $\frac{3\sqrt{2}}{\sqrt{5}}$
48. $\frac{\sqrt{2}}{\sqrt{15}}$
49. $\sqrt{\frac{3}{5}}$
50. $\sqrt{\frac{5}{14}} \cdot \sqrt{28}$

51. $\sqrt{\frac{2}{3}} \cdot \sqrt{\frac{1}{5}}$

52. $\sqrt[3]{\frac{1}{3}}$

53. $\sqrt[3]{\frac{5}{4}}$

54. $\sqrt{\frac{7}{x}}$

55. $\sqrt{\frac{r^3t}{5w^2}}$

56. $\sqrt[3]{\frac{k^4}{4n}}$

[8.5] *Simplify each expression.*

57. $3\sqrt{2}(\sqrt{3} + 2\sqrt{2})$

58. $(2\sqrt{3} - 4)(5\sqrt{3} + 2)$

59. $(\sqrt{7} + 2\sqrt{6})(\sqrt{12} - \sqrt{2})$

60. $(5\sqrt{7} + 2)^2$

61. $(\sqrt{5} - \sqrt{7})(\sqrt{5} + \sqrt{7})$

62. $(2\sqrt{3} + 5)(2\sqrt{3} - 5)$

Rationalize the denominators.

63. $\frac{\sqrt{3}}{1 + \sqrt{3}}$

64. $\frac{1}{2 + \sqrt{5}}$

65. $\frac{\sqrt{7}}{4 - \sqrt{7}}$

66. $\frac{\sqrt{8}}{\sqrt{2} + 6}$

67. $\frac{\sqrt{5} - 1}{\sqrt{2} + 3}$

68. $\frac{2 + \sqrt{6}}{\sqrt{3} - 1}$

[8.6] *Find all solutions for each equation.*

69. $\sqrt{p} = -2$

70. $\sqrt{k + 1} = 10$

71. $\sqrt{y} - 8 = 0$

72. $\sqrt{x} + 2 = 1$

73. $\sqrt{5m + 4} = 3\sqrt{m}$

74. $\sqrt{2p + 3} = \sqrt{5p - 3}$

75. $\sqrt{4y + 1} = y - 1$

76. $\sqrt{-2k - 4} = k + 2$

77. $\sqrt{2 - x} + 3 = x + 7$

[8.7] *Simplify each expression. Assume that all variables represent positive numbers.*

78. $81^{1/2}$

79. $125^{1/3}$

80. $7^{2/3} \cdot 7^{5/3}$

81. $\frac{12^{2/5}}{12^{-2/5}}$

82. $\frac{x^{1/4} \cdot x^{5/4}}{x^{3/4}}$

83. $\sqrt[8]{49^4}$

MIXED REVIEW EXERCISES

Simplify each expression. Assume all variables represent positive numbers.

84. $\sqrt[3]{2} \cdot \sqrt[3]{8}$

85. $(2^{4/3} \cdot 2^{1/3})^{1/5}$

86. $\sqrt{\frac{36}{p^2}}$

87. $-\sqrt{2}\left(\sqrt{2} - \sqrt{\frac{5}{2}}\right)$

88. $\sqrt{\frac{1}{6}} \cdot \sqrt{\frac{18}{5}}$

89. $\sqrt[3]{54x^2}$

90. $\sqrt[12]{125^4}$

91. $-\sqrt{3}(\sqrt{5} - \sqrt{27})$

92. $\sqrt{\frac{9m^3}{2p}}$

93. $27^{2/3}$

94. $3\sqrt{75} + 2\sqrt{27}$

95. $\frac{(7^{5/2})^{1/5}}{7^{3/2}}$

96. $(\sqrt{3} + 4)(\sqrt{3} - 4)$

97. $-\sqrt{36}$

98. $\sqrt{48}$

99. $\frac{12}{\sqrt{24}}$

100. $\frac{16^{3/4}}{16^{1/4}}$

101. $\sqrt{15} \cdot \sqrt{2} + 5\sqrt{30}$

102. Enrique wants to rationalize the denominator of the fraction $\frac{4}{\sqrt[3]{2}}$ by multiplying by the fraction $\frac{\sqrt[3]{2}}{\sqrt[3]{2}}$. Will this plan work? Why?

CHAPTER 8 TEST

In this test, assume that all variables represent positive numbers.

In Exercises 1–3, find the indicated root. Give a decimal approximation if necessary. Round to the nearest thousandth.

1. $\sqrt{100}$

2. $-\sqrt{190}$

3. $\sqrt[3]{-27}$

Simplify where possible.

4. $\sqrt{27}$

5. $\sqrt{\frac{128}{25}}$

6. $\sqrt[3]{-32}$

7. $\frac{20\sqrt{18}}{5\sqrt{3}}$

8. $3\sqrt{28} + \sqrt{63}$

9. $3\sqrt{27x} - 4\sqrt{48x} + 2\sqrt{3x}$

10. $\sqrt[3]{32x^2y^3}$

11. $\sqrt[4]{32m^3n^4p^6}$

12. $(6 - \sqrt{5})(6 + \sqrt{5})$

13. $(2 - \sqrt{7})(3\sqrt{2} + 1)$

14. $(\sqrt{5} + \sqrt{6})^2$

Rationalize each denominator.

15. $\frac{3\sqrt{2}}{\sqrt{6}}$

16. $\frac{4p}{\sqrt{k}}$

17. $\sqrt[3]{\frac{5}{9}}$

18. $\frac{-3}{4 - \sqrt{3}}$

Solve each equation.

19. $\sqrt{k + 2} = 5$

20. $\sqrt{2y + 8} = 2\sqrt{y}$

21. $6\sqrt{k} - 3 = k + 2$

Simplify each expression. Write answers with only positive exponents.

22. $8^{4/3}$

23. $5^{3/4} \cdot 5^{7/4}$

24. $\frac{(3^{1/4})^3}{3^{7/4}}$

25. What is wrong with the following "solution"?

$$\begin{aligned} \sqrt{2m + 1} + 5 &= 0 \\ \sqrt{2m + 1} &= -5 \\ 2m + 1 &= 25 \\ 2m &= 24 \\ m &= 12 \end{aligned}$$

The solution is 12.

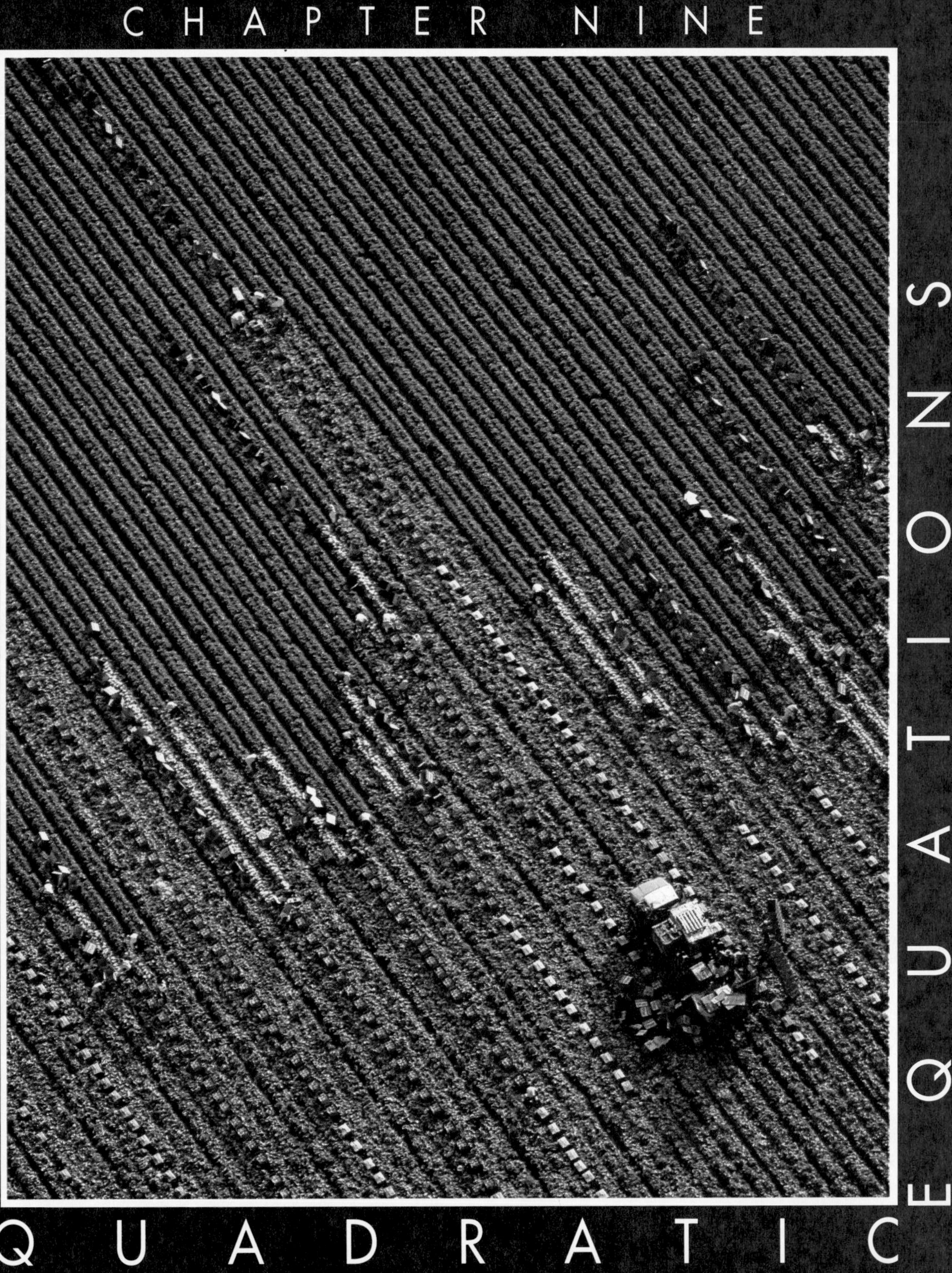
CHAPTER NINE
QUADRATIC EQUATIONS

Suppose a worker is cleaning the face of the Abraham Lincoln sculpture at Mount Rushmore. If he were to drop a tool from this height, the number of seconds that it would take for the tool to hit the ground and the distance in feet the object would travel would be related by the formula $d = 16t^2$, where t is the time and d is the distance. This is an example of a quadratic equation, where one of the variables is squared.

A method of solving quadratic equations (by factoring) was studied in Chapter 4. However, not all quadratic equations can be solved this way. In this chapter we study several other ways of solving quadratic equations.

9.1 SOLVING QUADRATIC EQUATIONS BY THE SQUARE ROOT PROPERTY

OBJECTIVES

1 SOLVE EQUATIONS OF THE FORM x^2 = A NUMBER.

2 SOLVE EQUATIONS OF THE FORM $(ax + b)^2$ = A NUMBER.

Recall that a *quadratic equation* is an equation that can be written in the form

$$ax^2 + bx + c = 0$$

for real numbers a, b, and c, with $a \neq 0$. In Chapter 4, these equations were solved by factoring. However, not all quadratic equations can be solved by factoring. Other ways to solve quadratic equations are shown in this chapter. For example, the quadratic equation

$$(x - 3)^2 = 15,$$

in which the square of a binomial is equal to some number, can be solved with square roots.

1 The **square root property of equations** justifies taking square roots of both sides of an equation.

SQUARE ROOT PROPERTY OF EQUATIONS

If b is a positive number and if $a^2 = b$, then

$$a = \sqrt{b} \quad \text{or} \quad a = -\sqrt{b}.$$

NOTE When we solve an equation, we want to find *all* values of the variable that satisfy the equation. Therefore, we want both the positive and the negative square roots of b.

EXAMPLE 1 **SOLVING A QUADRATIC EQUATION BY THE SQUARE ROOT PROPERTY**

Solve each equation. Write radicals in simplified form.

(a) $x^2 = 16$

By the square root property, since $x^2 = 16$, then

$$x = \sqrt{16} = 4 \quad \text{or} \quad x = -\sqrt{16} = -4.$$

An abbreviation for "$x = 4$ or $x = -4$" is written $x = \pm 4$, and is read "x equals positive or negative 4." Check each solution by substituting back into the original equation.

(b) $z^2 = 5$

The solutions are $z = \sqrt{5}$ or $z = -\sqrt{5}$, which may be written $z = \pm\sqrt{5}$.

(c) $m^2 = 8$

Again, use the square root property.

$$m^2 = 8$$

$$m = \sqrt{8} \quad \text{or} \quad m = -\sqrt{8} \qquad \text{Square root property}$$

$$m = 2\sqrt{2} \quad \text{or} \quad m = -2\sqrt{2} \qquad \text{Simplify } \sqrt{8}.$$

$$m = \pm 2\sqrt{2} \qquad \text{Abbreviation}$$

(d) $y^2 = -4$

Since -4 is a negative number and since the square of a real number cannot be negative, there is no real number solution for this equation. (The square root property cannot be used because of the requirement that b must be positive.)

(e) $3x^2 + 5 = 11$

First solve the equation for x^2.

$$3x^2 + 5 = 11$$

$$3x^2 = 6 \qquad \text{Subtract 5.}$$

$$x^2 = 2 \qquad \text{Divide by 3.}$$

Now use the square root property to get $x = \pm\sqrt{2}$. ■

2 In each of the equations in Example 1, the exponent 2 appeared with a single variable as its base. The square root property of equations can be extended to solve equations where the base is a binomial, as shown in the next example.

EXAMPLE 2
SOLVING A QUADRATIC EQUATION BY THE SQUARE ROOT PROPERTY

Solve the equation $(x - 3)^2 = 16$.

Apply the square root property, using $x - 3$ as the base.

$$(x - 3)^2 = 16$$

$$x - 3 = \sqrt{16} \quad \text{or} \quad x - 3 = -\sqrt{16}$$

$$x - 3 = 4 \quad \text{or} \quad x - 3 = -4 \qquad \sqrt{16} = 4$$

$$x = 7 \quad \text{or} \quad x = -1 \qquad \text{Add 3.}$$

Check both answers in the original equation.

$$(x - 3)^2 = 16 \qquad\qquad (x - 3)^2 = 16$$

$$(7 - 3)^2 = 16 \quad ? \quad \text{Let } x = 7. \qquad (-1 - 3)^2 = 16 \quad ? \quad \text{Let } x = -1.$$

$$4^2 = 16 \quad ? \qquad\qquad (-4)^2 = 16 \quad ?$$

$$16 = 16 \quad \text{True} \qquad\qquad 16 = 16 \quad \text{True}$$

Both 7 and -1 are solutions. ■

EXAMPLE 3 **SOLVING A QUADRATIC EQUATION BY THE SQUARE ROOT PROPERTY**

Solve $(x - 1)^2 = 6$.

By the square root property,

$$x - 1 = \sqrt{6} \quad \text{or} \quad x - 1 = -\sqrt{6}.$$
$$x = 1 + \sqrt{6} \quad \text{or} \quad x = 1 - \sqrt{6}.$$

Check:

$$(1 + \sqrt{6} - 1)^2 = (\sqrt{6})^2 = 6;$$
$$(1 - \sqrt{6} - 1)^2 = (-\sqrt{6})^2 = 6.$$

The solutions are $1 + \sqrt{6}$ and $1 - \sqrt{6}$. ■

NOTE The solutions in Example 3 may be written in abbreviated form as

$$1 \pm \sqrt{6}.$$

If they are written this way, keep in mind that there are *two* solutions indicated, one with the + sign and the other with the − sign.

EXAMPLE 4 **SOLVING A QUADRATIC EQUATION BY THE SQUARE ROOT PROPERTY**

Solve the equation $(3r - 2)^2 = 27$.

$3r - 2 = \sqrt{27}$	or	$3r - 2 = -\sqrt{27}$	Square root property
$3r - 2 = 3\sqrt{3}$	or	$3r - 2 = -3\sqrt{3}$	$\sqrt{27} = \sqrt{9 \cdot 3} = 3\sqrt{3}$
$3r = 2 + 3\sqrt{3}$	or	$3r = 2 - 3\sqrt{3}$	Add 2.
$r = \dfrac{2 + 3\sqrt{3}}{3}$	or	$r = \dfrac{2 - 3\sqrt{3}}{3}$	Divide by 3.

The solutions are

$$\frac{2 + 3\sqrt{3}}{3} \quad \text{and} \quad \frac{2 - 3\sqrt{3}}{3},$$

which may be abbreviated as

$$\frac{2 \pm 3\sqrt{3}}{3}. \quad ■$$

CAUTION The solutions in Example 4 are fractions that cannot be reduced, since 3 is *not* a common factor in the numerator.

EXAMPLE 5 **RECOGNIZING A QUADRATIC EQUATION WITH NO REAL SOLUTION**

Solve $(x + 3)^2 = -9$.

The square root of -9 is not a real number. There is no real number solution for this equation. ■

9.1 EXERCISES

Solve each equation by using the square root property. Express all radicals in simplest form. See Example 1.

1. $x^2 = 25$
2. $y^2 = 100$
3. $x^2 = 64$
4. $z^2 = 81$
5. $m^2 = 13$
6. $x^2 = 7$
7. $y^2 = -15$
8. $p^2 = -10$
9. $3p^2 = 6$
10. $2q^2 = 12$
11. $4k^2 = 5$
12. $5k^2 = 8$
13. $3x^2 - 8 = 64$
14. $2t^2 + 7 = 61$
15. $5a^2 + 4 = 8$
16. $4p^2 - 3 = 7$

17. Explain why the square of a real number cannot be negative. (See Objective 1.)

18. Does the equation $-x^2 = 16$ have real solutions? Explain. (See Objective 1.)

19. Which of these equations has exactly one real number solution?
(a) $x^2 = 4$
(b) $y^2 = -4$
(c) $(x - 4)^2 = 1$
(d) $t^2 = 0$

20. Which of the equations in Exercise 19 has no real solution?

Solve each equation by using the square root property. Express all radicals in simplest form. See Examples 2–5.

21. $(x - 2)^2 = 16$
22. $(r + 4)^2 = 25$
23. $(a + 4)^2 = 10$
24. $(r - 3)^2 = 15$
25. $(m - 1)^2 = -4$
26. $(t + 2)^2 = -8$
27. $(x - 1)^2 = 32$
28. $(y + 5)^2 = 28$
29. $(2m - 1)^2 = 9$
30. $(3y - 7)^2 = 4$
31. $(6m - 2)^2 = 121$
32. $(7m - 10)^2 = 144$
33. $(2a - 5)^2 = 30$
34. $(2y + 3)^2 = 45$
35. $(3p - 1)^2 = 18$
36. $(5r - 6)^2 = 75$
37. $(2k - 5)^2 = 98$
38. $(4x - 1)^2 = 48$
39. $(3m + 4)^2 = -8$
40. $(5y - 3)^2 = -50$

Use a calculator with a square root key to solve each of the following. In Exercises 45–48, round your answers to the nearest hundredth.

41. $k^2 = 2.56$
42. $z^2 = 9.61$
43. $r^2 = 77.44$
44. $y^2 = 43.56$
45. $(k + 2.14)^2 = 5.46$
46. $(r - 3.91)^2 = 9.28$
47. $(2.11p + 3.42)^2 = 9.58$
48. $(1.71m - 6.20)^2 = 5.41$

Solve the following problems.

49. One expert at marksmanship can hold a silver dollar at forehead level, drop it, draw his gun, and shoot the coin as it passes waist level. The distance traveled by a falling object is given by

$$d = 16t^2,$$

where d is the distance the object falls in t seconds. If the coin falls about 4 feet, use the formula to estimate the time that elapses between the dropping of the coin and the shot.

50. The illumination produced by a light source depends on the distance from the source. For a particular light source, this relationship can be expressed as

$$d^2 = \frac{4050}{I},$$

where d is the distance from the source (in feet) and I is the amount of illumination in footcandles. How far from the source is the illumination equal to 50 footcandles?

51. The amount A that P dollars invested at a rate of interest r will grow to in 2 years is
$$A = P(1 + r)^2.$$
At what interest rate will \$1 grow to \$1.21 in two years?

52. The area A of a circle with radius r is given by the formula
$$A = \pi r^2.$$
What radius will give a circle with area 36π square inches?

53. The surface area S of a sphere with radius r is given by the formula $S = 4\pi r^2$. What radius will give a sphere with surface area 100π square feet?

54. Becky and Brad are the owners of Cole's Baseball Cards. They have found that the price p, in dollars, of a particular Kirby Puckett baseball card depends on the demand d, in hundreds, for the card, according to the formula $p = (d - 2)^2$. What demand produces a price of \$5 for the card?

PREVIEW EXERCISES

Simplify all radicals and combine terms. See Sections 5.4 and 8.4.

55. $\frac{3}{2} + \sqrt{\frac{27}{4}}$ **56.** $-\frac{1}{4} + \sqrt{\frac{5}{16}}$ **57.** $6 + \sqrt{\frac{2}{3}}$ **58.** $5 - \sqrt{\frac{7}{2}}$

Factor each perfect square trinomial. See Section 4.4.

59. $x^2 + 8x + 16$ **60.** $m^2 - 10m + 25$ **61.** $p^2 - 5p + \frac{25}{4}$ **62.** $z^2 + 3z + \frac{9}{4}$

9.2 SOLVING QUADRATIC EQUATIONS BY COMPLETING THE SQUARE

OBJECTIVES

1 SOLVE QUADRATIC EQUATIONS BY COMPLETING THE SQUARE WHEN THE COEFFICIENT OF THE SQUARED TERM IS 1.

2 SOLVE QUADRATIC EQUATIONS BY COMPLETING THE SQUARE WHEN THE COEFFICIENT OF THE SQUARED TERM IS NOT 1.

3 SIMPLIFY AN EQUATION BEFORE SOLVING.

FOCUS ON PROBLEM SOLVING

A rule for estimating the number of board feet of lumber that can be cut from a log depends on the diameter of the log. The diameter d required to get 9 board feet is found from the equation
$$\left(\frac{d - 4}{4}\right)^2 = 9.$$
Solve this equation for d. Are both answers reasonable?

Many applications of algebra lead to quadratic equations, like the one shown in this problem. Earlier, we saw how to solve quadratic equations by factoring (Section 4.5), but not all quadratic equations can be solved this way. In this section we develop a method of solving quadratic equations called *completing the square*.

The problem above is Exercise 39 in the exercises for this section. After working through this section, you should be able to solve this problem.

1 The properties and methods studied so far are not enough to solve the equation

$$x^2 + 6x + 7 = 0.$$

For a method of solving this equation, recall the method from the preceding section for solving equations of the type

$$(x + 3)^2 = 2.$$

If the equation $x^2 + 6x + 7 = 0$ could be rewritten in the form $(x + 3)^2 = 2$, it could be solved by using the square root property. The following example shows how to rewrite the equation in this form.

EXAMPLE 1
REWRITING AN EQUATION TO USE THE SQUARE ROOT PROPERTY

Solve $x^2 + 6x + 7 = 0$.

Start by subtracting 7 from both sides of the equation to get $x^2 + 6x = -7$. If $x^2 + 6x = -7$ is to be written in the form $(x + 3)^2 = 2$, the quantity on the left-hand side of $x^2 + 6x = -7$ must be made into a perfect square trinomial. The expression $x^2 + 6x + 9$ is a perfect square, since

$$x^2 + 6x + 9 = (x + 3)^2.$$

Therefore, if 9 is added to both sides of $x^2 + 6x = -7$, the equation will have a perfect square trinomial on one side, as needed.

$$x^2 + 6x + 9 = -7 + 9 \qquad \text{Add 9 on both sides.}$$
$$(x + 3)^2 = 2 \qquad \text{Factor; combine terms.}$$

Now use the square root property to complete the solution.

$$x + 3 = \sqrt{2} \quad \text{or} \quad x + 3 = -\sqrt{2}$$
$$x = -3 + \sqrt{2} \quad \text{or} \quad x = -3 - \sqrt{2}$$

The solutions of the original equation are $-3 + \sqrt{2}$ and $-3 - \sqrt{2}$. Check this by substituting $-3 + \sqrt{2}$ and $-3 - \sqrt{2}$ for x in the original equation. ■

The process of changing the form of the equation in Example 1 from

$$x^2 + 6x + 7 = 0 \quad \text{to} \quad (x + 3)^2 = 2$$

is called **completing the square.** Completing the square changes only the form of the equation. To see this, square on the left side of $(x + 3)^2 = 2$ and combine terms; the result will be $x^2 + 6x + 7 = 0$.

EXAMPLE 2
SOLVING A QUADRATIC EQUATION BY COMPLETING THE SQUARE

Solve the quadratic equation $m^2 - 8m = 5$.

A suitable number must be added to both sides to make one side a perfect square. Find this number as follows: recall from Chapter 3 that

$$(m + a)^2 = m^2 + 2am + a^2.$$

In the equation $m^2 - 8m = 5$, the value of $2am$ is $-8m$ and a^2 must be found. Set $2am$ equal to $-8m$ to find a.

$$2am = -8m$$
$$a = -4$$

Squaring -4 gives 16, the number to be added to both sides.

$$m^2 - 8m + 16 = 5 + 16 \qquad (1)$$

The trinomial $m^2 - 8m + 16$ is a perfect square trinomial. Factor this trinomial to get

$$m^2 - 8m + 16 = (m - 4)^2.$$

Equation (1) becomes

$$(m - 4)^2 = 21.$$

Now use the square root property.

$$m - 4 = \sqrt{21} \quad \text{or} \quad m - 4 = -\sqrt{21}$$
$$m = 4 + \sqrt{21} \quad \text{or} \quad m = 4 - \sqrt{21}$$

Check that the solutions are

$$4 + \sqrt{21} \quad \text{and} \quad 4 - \sqrt{21}. \blacksquare$$

As illustrated by Example 2, the number to be added to both sides of the quadratic equation $x^2 + 2ax = b$ to complete the square is found by taking $1/2$ of $2a$, the coefficient of the x term, and squaring it. This gives $(1/2)(2a) = a$, with a^2 then added to both sides of the equation as shown in Example 2.

The steps used in solving a quadratic equation by completing the square are given below.

SOLVING A QUADRATIC EQUATION BY COMPLETING THE SQUARE

Step 1 If the coefficient of the squared term is 1, proceed to Step 2. If the coefficient of the squared term is not 1 but some other nonzero number a, divide both sides of the equation by a. This gives an equation that has 1 as coefficient of the squared term.

Step 2 Make sure that all terms with variables are on one side of the equals sign and that all numbers are on the other side.

Step 3 Take half the coefficient of x and square the result. Add the square to both sides of the equation. The side containing the variables now can be written as a perfect square.

Step 4 Apply the square root property.

2 The process of completing the square requires the coefficient of the squared variable to be 1 (Step 1). The next example shows how to solve a quadratic equation with this coefficient different from 1. The steps are numbered according to the summary above.

EXAMPLE 3
SOLVING A QUADRATIC EQUATION BY COMPLETING THE SQUARE

Solve $4y^2 + 24y - 13 = 0$.

Step 1 Divide each side by 4 to get the coefficient of y^2 to be 1.

$$4y^2 + 24y - 13 = 0$$

$$y^2 + 6y - \frac{13}{4} = 0$$

Step 2 Add 13/4 to each side to get the variable terms on the left and the number on the right.

$$y^2 + 6y = \frac{13}{4}$$

Step 3 Now take half the coefficient of y, or $(1/2)(6) = 3$, and square the result: $3^2 = 9$. Add 9 to both sides of the equation, perform the addition on the right-hand side, and factor on the left.

$$y^2 + 6y + 9 = \frac{13}{4} + 9 \qquad \text{Add 9.}$$

$$y^2 + 6y + 9 = \frac{49}{4} \qquad \text{Add on the right.}$$

$$(y + 3)^2 = \frac{49}{4} \qquad \text{Factor on the left.}$$

Step 4 Use the square root property and solve for y.

$$y + 3 = \frac{7}{2} \quad \text{or} \quad y + 3 = -\frac{7}{2} \qquad \text{Square root property}$$

$$y = -3 + \frac{7}{2} \quad \text{or} \quad y = -3 - \frac{7}{2} \qquad \text{Add } -3.$$

$$y = \frac{1}{2} \quad \text{or} \quad y = -\frac{13}{2}$$

The two solutions are 1/2 and $-13/2$. Check by substitution into the original equation. ■

EXAMPLE 4
SOLVING A QUADRATIC EQUATION BY COMPLETING THE SQUARE

Solve $4p^2 + 8p + 5 = 0$.

First divide both sides by 4 to get the coefficient 1 for the p^2 term (Step 1). The result is

$$p^2 + 2p + \frac{5}{4} = 0.$$

Subtract 5/4 from both sides (Step 2).

$$p^2 + 2p = -\frac{5}{4}$$

The coefficient of p is 2. Take half of 2, square the result, and add it to both sides. The left-hand side can then be written as a perfect square (Step 3).

$$p^2 + 2p + 1 = -\frac{5}{4} + 1 \qquad \left[\frac{1}{2}(2)\right]^2 = 1$$

$$(p + 1)^2 = -\frac{1}{4}$$

The square root of $-1/4$ is not a real number so the square root property does not apply. This equation has no real number solution. ■

3 Sometimes an equation must be simplified before applying the steps for solving by completing the square. The next example illustrates this.

EXAMPLE 5 SIMPLIFYING AN EQUATION BEFORE COMPLETING THE SQUARE

Solve $(x + 3)(x - 1) = 2$.

$$(x + 3)(x - 1) = 2$$

$$x^2 + 2x - 3 = 2 \qquad \text{Use FOIL.}$$

$$x^2 + 2x = 5 \qquad \text{Add 3.}$$

$$x^2 + 2x + 1 = 5 + 1 \qquad \text{Add 1 to get a perfect square on the left.}$$

$$(x + 1)^2 = 6 \qquad \text{Factor on the left; add on the right.}$$

$$x + 1 = \sqrt{6} \quad \text{or} \quad x + 1 = -\sqrt{6} \qquad \text{Square root property}$$

$$x = -1 + \sqrt{6} \quad \text{or} \quad x = -1 - \sqrt{6} \qquad \text{Add } -1.$$

The two solutions are $-1 + \sqrt{6}$ and $-1 - \sqrt{6}$. These can be checked by substituting into the original equation. ■

NOTE The solutions given in Example 5 are *exact*. If we were asked to find approximations using a calculator, we would need to use the square root key, finding that $\sqrt{6} \approx 2.449$. Evaluating the two solutions we find that

$$x \approx 1.449 \quad \text{and} \quad x \approx -3.449.$$

9.2 EXERCISES

Find the number that should be added to each expression to make it a perfect square.

1. $x^2 + 2x$ **2.** $y^2 - 4y$ **3.** $x^2 + 18x$ **4.** $m^2 - 3m$

5. $z^2 + 9z$ **6.** $p^2 + 22p$ **7.** $y^2 + 5y$ **8.** $r^2 + 7r$

9. What is the first step in solving the quadratic equation below? (See Objective 2.)

$$2x^2 - 4x = 9$$

(a) Add 4 to both sides of the equation.
(b) Factor the left side as $2x(x - 2)$.
(c) Factor the left side as $x(2x - 4)$.
(d) Divide both sides by 2.

10. Explain the steps used in solving a quadratic equation by completing the square. (See Objective 1.)

Solve each equation by completing the square. You may have to simplify first. Round answers to the nearest thousandth in Exercises 34–37. See Examples 2–5.

11. $x^2 + 4x = -3$ **12.** $y^2 - 4y = 0$ **13.** $a^2 + 2a = 5$

14. $m^2 + 4m = -1$ **15.** $z^2 + 6z = -8$ **16.** $q^2 - 8q = -16$

17. $x^2 - 6x + 1 = 0$ **18.** $b^2 - 2b - 2 = 0$ **19.** $c^2 + 3c = 2$

20. $(x + 2)(x - 4) = -5$ **21.** $(w + 4)(w - 2) = 16$ **22.** $(2n + 1)(n - 5) = -5$

23. $k^2 + 5k - 3 = 0$ **24.** $2m^2 + 4m = -7$ **25.** $3y^2 - 9y + 5 = 0$

26. $6q^2 - 8q + 3 = 0$ **27.** $4y^2 + 4y - 3 = 0$ **28.** $-x^2 + 6x = 4$ **29.** $-x^2 + 4 = 2x$

30. $3x^2 - 2x = 1$ **31.** $-x^2 - 4 = 2x$ **32.** $m^2 - 4m + 8 = 6m$ **33.** $2z^2 = 8z + 5 - 4z^2$

34. $3r^2 - 2 = 6r + 3$ **35.** $4p - 3 = p^2 + 2p$ **36.** $(x + 1)(x + 3) = 2$ **37.** $(x - 3)(x + 1) = 1$

Work the following problems. In Exercises 40–41, round your answer to the nearest thousandth.

38. Two cars travel at right angles to each other from an intersection until they are 17 miles apart. At that point one car has gone 7 miles farther than the other. How far did the slower car travel?

39. A rule for estimating the number of board feet of lumber that can be cut from a log depends on the diameter of the log. To find the diameter d required to get 9 board feet of lumber, we use the equation

$$\left(\frac{d - 4}{4}\right)^2 = 9.$$

Solve this equation for d. Are both answers reasonable?

40. A rancher has determined that the number of cattle in his herd has increased over a two-year period at a rate r given by the equation

$$5r^2 + 10r = 1.$$

Find r. Do both answers make sense?

41. Two painters are painting a house in a development of new homes. One of the painters takes 2 hours longer to paint a house working alone than the other painter. When they do the job together, they can complete it in 4.8 hours. How long would it take the faster painter alone to paint the house?

PREVIEW EXERCISES

Write each quotient in lowest terms. Simplify the radicals if possible. See Section 8.5.

42. $\dfrac{2 + 2\sqrt{3}}{2}$ **43.** $\dfrac{3 + 6\sqrt{5}}{3}$ **44.** $\dfrac{4 + 2\sqrt{7}}{8}$ **45.** $\dfrac{5 + 5\sqrt{2}}{10}$

46. $\dfrac{8 + 6\sqrt{3}}{4}$ **47.** $\dfrac{4 + \sqrt{28}}{6}$ **48.** $\dfrac{6 + \sqrt{45}}{12}$ **49.** $\dfrac{8 + \sqrt{32}}{8}$

9.3 SOLVING QUADRATIC EQUATIONS BY THE QUADRATIC FORMULA

OBJECTIVES

1 IDENTIFY THE VALUES OF a, b, AND c IN A QUADRATIC EQUATION.

2 USE THE QUADRATIC FORMULA TO SOLVE QUADRATIC EQUATIONS.

3 SOLVE QUADRATIC EQUATIONS WITH ONLY ONE SOLUTION.

4 SOLVE QUADRATIC EQUATIONS WITH FRACTIONS.

5 USE THE QUADRATIC FORMULA TO SOLVE AN APPLIED PROBLEM.

FOCUS ON PROBLEM SOLVING

In a bicycle race over a 12-mile route, Donnie finished 8 minutes ahead of Juan. Donnie pedaled 3 miles per hour faster than Juan. What was Donnie's speed?

A quadratic equation is needed to solve this problem. While the equation can be solved by completing the square (as shown in Section 9.2), it can be solved more quickly by the quadratic formula, which is introduced in this section.

The problem above is Exercise 67 in the exercises for this section. After working through this section, you should be able to solve this problem.

Any quadratic equation can be solved by completing the square, but the method is not very handy. This section introduces a general formula, the quadratic formula, that gives the solution for any quadratic equation.

In order to use the quadratic formula, it is necessary to put the equation in **standard form,**

$$ax^2 + bx + c = 0, \qquad a \neq 0.$$

The restriction $a \neq 0$ is important in order to make sure that the equation is quadratic. If $a = 0$, then the equation becomes $0x^2 + bx + c = 0$, or $bx + c = 0$, which is a linear, and not a quadratic, equation.

1 The first step in solving a quadratic equation by this new method is to identify the values of a, b, and c in the standard form of the quadratic equation.

EXAMPLE 1 DETERMINING VALUES OF a, b, AND c IN A QUADRATIC EQUATION

For each of the following quadratic equations, put the equation in standard form if necessary, and then identify the values of a, b, and c.

(a) $2x^2 + 3x - 5 = 0$

This equation is already in standard form. The values of a, b, and c are

$$a = 2, \qquad b = 3, \qquad \text{and} \qquad c = -5.$$

(b) $-x^2 + 2 = 6x$

First rewrite the equation with 0 on the right side to match the standard form of $ax^2 + bx + c = 0$.

$$-x^2 + 2 = 6x$$
$$-x^2 - 6x + 2 = 0$$

Now identify $a = -1$, $b = -6$, and $c = 2$. (Notice that the coefficient of x^2 is understood to be -1.)

(c) $(2x - 7)(x + 4) = -23$

Put the equation in standard form.

$$(2x - 7)(x + 4) = -23$$

$$2x^2 + x - 28 = -23 \qquad \text{Use FOIL on the left.}$$

$$2x^2 + x - 5 = 0 \qquad \text{Add 23 on each side.}$$

Now, identify the values: $a = 2$, $b = 1$, $c = -5$. ■

2 The quadratic formula is developed by solving $ax^2 + bx + c = 0$ by completing the square. To see how this is done, we can compare solving $2x^2 + x - 5 = 0$ (from Example 1(c)) by completing the square. We use the steps as given in Section 9.2.

Step 1 Make the coefficient of the squared term equal to 1.

$2x^2 + x - 5 = 0$	$ax^2 + bx + c = 0$
$x^2 + \frac{1}{2}x - \frac{5}{2} = 0$ Divide by 2.	$x^2 + \frac{b}{a}x + \frac{c}{a} = 0$ Divide by a.

Step 2 Get the variable terms alone on the left side.

$x^2 + \frac{1}{2}x = \frac{5}{2}$ Add 5/2.	$x^2 + \frac{b}{a}x = -\frac{c}{a}$ Subtract c/a.

Step 3 Add the square of half the coefficient of x, factor the left side, and combine terms on the right.

$x^2 + \frac{1}{2}x + \frac{1}{16} = \frac{5}{2} + \frac{1}{16}$ Add 1/16.	$x^2 + \frac{b}{a}x + \frac{b^2}{4a^2} = -\frac{c}{a} + \frac{b^2}{4a^2}$ Add $\frac{b^2}{4a^2}$.
$\left(x + \frac{1}{4}\right)^2 = \frac{41}{16}$ Factor; add on right.	$\left(x + \frac{b}{2a}\right)^2 = \frac{b^2 - 4ac}{4a^2}$ Factor; add on right.

Step 4 Use the square root property to complete the solution.

$x + \frac{1}{4} = \pm\sqrt{\frac{41}{16}}$	$x + \frac{b}{2a} = \pm\sqrt{\frac{b^2 - 4ac}{4a^2}}$
$x + \frac{1}{4} = \pm\frac{\sqrt{41}}{4}$	$x = -\frac{b}{2a} \pm \frac{\sqrt{b^2 - 4ac}}{2a}$
$x = -\frac{1}{4} \pm \frac{\sqrt{41}}{4}$	$x = \frac{-b \pm \sqrt{b^2 - 4ac}}{2a}$
$x = \frac{-1 \pm \sqrt{41}}{4}$	

The final result in the column on the right is called the quadratic formula, and it is a key result that should be memorized. Notice that there are two values, one for the + sign and one for the − sign.

QUADRATIC FORMULA

The solutions of the quadratic equation $ax^2 + bx + c = 0$, $a \neq 0$, are

$$x = \frac{-b + \sqrt{b^2 - 4ac}}{2a} \quad \text{and} \quad x = \frac{-b - \sqrt{b^2 - 4ac}}{2a}$$

or, in compact form,

$$x = \frac{-b \pm \sqrt{b^2 - 4ac}}{2a}.$$

CAUTION Notice that the fraction bar is under $-b$ as well as the radical. In using this formula, be sure to find the values of $-b \pm \sqrt{b^2 - 4ac}$ first, then divide those results by the value of $2a$.

EXAMPLE 2
SOLVING A QUADRATIC EQUATION BY THE QUADRATIC FORMULA

Solve $2x^2 + x - 5 = 0$ by the quadratic formula.

As found in Example 1(c), the values of a, b, and c are: $a = 2$, $b = 1$, and $c = -5$. Substitute these numbers into the quadratic formula and simplify the result.

$$x = \frac{-b \pm \sqrt{b^2 - 4ac}}{2a}$$

$$x = \frac{-1 \pm \sqrt{(1)^2 - 4(2)(-5)}}{2(2)}$$ Let $a = 2$, $b = 1$, $c = -5$.

$$x = \frac{-1 \pm \sqrt{1 + 40}}{4}$$ Perform the operations under the radical and in the denominator.

$$x = \frac{-1 \pm \sqrt{41}}{4}$$

Notice that this result agrees with the result obtained in the left column when the quadratic formula was derived. The two solutions are $\frac{-1 + \sqrt{41}}{4}$ and $\frac{-1 - \sqrt{41}}{4}$. ■

EXAMPLE 3
SOLVING A QUADRATIC EQUATION BY THE QUADRATIC FORMULA

Solve $x^2 = 2x + 1$.

One side of the equation must be 0 before a, b, and c can be found. Subtract $2x$ and 1 from both sides of the equation to get

$$x^2 - 2x - 1 = 0.$$

Then $a = 1$, $b = -2$, and $c = -1$. The solution is found by substituting these values into the quadratic formula.

$$x = \frac{-b \pm \sqrt{b^2 - 4ac}}{2a}$$

$$= \frac{-(-2) \pm \sqrt{(-2)^2 - 4(1)(-1)}}{2(1)}$$ Let $a = 1$, $b = -2$, $c = -1$.

$$= \frac{2 \pm \sqrt{4 + 4}}{2} = \frac{2 \pm \sqrt{8}}{2}$$

Since $\sqrt{8} = \sqrt{4 \cdot 2} = \sqrt{4} \cdot \sqrt{2} = 2\sqrt{2}$,

$$x = \frac{2 \pm 2\sqrt{2}}{2}.$$

Write the solutions in lowest terms by factoring $2 \pm 2\sqrt{2}$ as $2(1 \pm \sqrt{2})$ to get

$$x = \frac{2(1 \pm \sqrt{2})}{2} = 1 \pm \sqrt{2}.$$

The two solutions of the given equation are $1 + \sqrt{2}$ and $1 - \sqrt{2}$. ■

3 When the quantity under the radical, $b^2 - 4ac$, equals zero, the equation has just one rational number solution. In this case, the trinomial $ax^2 + bx + c$ is a perfect square.

EXAMPLE 4 **USING THE QUADRATIC FORMULA WHEN THERE IS ONE SOLUTION**

Solve $4x^2 + 25 = 20x$.

Write the equation as $4x^2 - 20x + 25 = 0$. Here, $a = 4$, $b = -20$, and $c = 25$. By the quadratic formula,

$$x = \frac{-(-20) \pm \sqrt{(-20)^2 - 400}}{8} = \frac{20 \pm 0}{8} = \frac{5}{2}.$$

Since there is just one solution, 5/2, the trinomial $4x^2 - 20x + 25$ is a perfect square. ■

4 The next example shows how to solve quadratic equations with fractions.

EXAMPLE 5 **SOLVING A QUADRATIC EQUATION WITH FRACTIONS**

Solve the equation $\frac{1}{10}t^2 = \frac{2}{5}t - \frac{1}{2}$.

Eliminate the denominators by multiplying both sides of the equation by the common denominator, 10.

$$10\left(\frac{1}{10}t^2\right) = 10\left(\frac{2}{5}t - \frac{1}{2}\right)$$

$$t^2 = 4t - 5 \quad \text{Distributive property}$$

$$t^2 - 4t + 5 = 0 \quad \text{Standard form}$$

From this form identify $a = 1$, $b = -4$, and $c = 5$. Use the quadratic formula to complete the solution.

$$t = \frac{-(-4) \pm \sqrt{(-4)^2 - 4(1)(5)}}{2(1)} \quad \text{Substitute into the formula.}$$

$$= \frac{4 \pm \sqrt{16 - 20}}{2} \quad \text{Perform the operations.}$$

$$= \frac{4 \pm \sqrt{-4}}{2}$$

The radical $\sqrt{-4}$ is not a real number, so the equation has no real number solution. ■

5 Quadratic equations are needed to solve some applied problems.

PROBLEM SOLVING

When a quadratic equation must be solved in an applied problem, it is important to check that the solution satisfies the physical requirements for the unknown. For example, it would be impossible to have a negative length, or a fractional number of people. ■

EXAMPLE 6 SOLVING AN APPLIED PROBLEM USING THE QUADRATIC FORMULA

If an object is thrown upward from a height of 50 feet, with an initial velocity of 32 feet per second, then its height after t seconds is given by

$$h = -16t^2 + 32t + 50, \quad \text{where } h \text{ is in feet.}$$

After how many seconds will it reach a height of 30 feet?

We must find the value of t for which $h = 30$.

$$30 = -16t^2 + 32t + 50 \qquad \text{Let } h = 30.$$
$$16t^2 - 32t - 20 = 0 \qquad \text{Put in standard form.}$$
$$4t^2 - 8t - 5 = 0 \qquad \text{Divide by 4.}$$

Use the quadratic formula, with $a = 4$, $b = -8$, and $c = -5$.

$$t = \frac{-b \pm \sqrt{b^2 - 4ac}}{2a}$$
$$t = \frac{-(-8) \pm \sqrt{(-8)^2 - 4(4)(-5)}}{2(4)}$$
$$t = \frac{8 \pm \sqrt{64 + 80}}{8}$$
$$t = \frac{8 \pm \sqrt{144}}{8} = \frac{8 \pm 12}{8}$$

The two solutions of the equation are

$$\frac{8 + 12}{8} = \frac{20}{8} = \frac{5}{2} \quad \text{and} \quad \frac{8 - 12}{8} = -\frac{4}{8} = -\frac{1}{2}.$$

Since t represents time, the solution $-1/2$ must be rejected, since time cannot be negative here. The object will reach a height of 30 feet after 5/2, or 2 1/2 seconds. ■

9.3 EXERCISES

For each equation, put it in standard form $ax^2 + bx + c = 0$ if necessary, and then identify the values of a, b, and c. (Hint: In some equations we may have $b = 0$ or $c = 0$.) Do not solve. See Example 1.

1. $3x^2 + 4x - 8 = 0$
2. $9x^2 + 2x - 3 = 0$
3. $-8x^2 - 2x - 3 = 0$
4. $-2x^2 + 3x - 8 = 0$
5. $2x^2 = 3x - 2$
6. $9x^2 - 2 = 4x$
7. $x^2 = 2$
8. $x^2 - 3 = 0$

9. $3x^2 - 8x = 0$ **10.** $5x^2 = 2x$ **11.** $(x - 3)(x + 4) = 0$ **12.** $(x + 6)^2 = 3$

13. $9(x - 1)(x + 2) = 8$ **14.** $(3x - 1)(2x + 5) = x(x - 1)$

---

15. Why is the restriction $a \neq 0$ necessary in the definition of quadratic equation?

16. Consider the equation $x^2 - 9 = 0$.
(a) Solve the equation by factoring.
(b) Solve the equation by the quadratic formula.
(c) Compare your answers. If a quadratic equation can be solved by both the factoring and the quadratic formula methods, should you always get the same results? Explain.

Use the quadratic formula to solve each equation. Write all radicals in simplified form. Reduce answers to lowest terms. Round answers to the nearest thousandth in Exercises 47–49. See Examples 2–4.

17. $z^2 + 6z + 9 = 0$ **18.** $6k^2 + 6k + 1 = 0$ **19.** $y^2 + 4y + 4 = 0$ **20.** $3r^2 - 5r + 1 = 0$

21. $z^2 = 13 - 12z$ **22.** $x^2 = 8x + 9$ **23.** $4p^2 - 12p + 9 = 0$ **24.** $k^2 = 20k - 19$

25. $5x^2 + 4x - 1 = 0$ **26.** $5n^2 + n - 1 = 0$ **27.** $2z^2 = 3z + 5$ **28.** $7r - 2r^2 + 30 = 0$

29. $p^2 + 2p - 2 = 0$ **30.** $x^2 - 2x + 1 = 0$ **31.** $2w^2 + 12w + 5 = 0$ **32.** $9r^2 + 6r + 1 = 0$

33. $5m^2 + 5m = 0$ **34.** $4y^2 - 8y = 0$ **35.** $6p^2 = 10p$ **36.** $3r^2 = 16r$

37. $m^2 - 20 = 0$ **38.** $k^2 - 5 = 0$ **39.** $9r^2 - 16 = 0$ **40.** $4y^2 - 25 = 0$

41. $2x^2 + 2x + 4 = 4 - 2x$ **42.** $3x^2 - 4x + 3 = 8x - 1$ **43.** $2x^2 + x + 7 = 0$

44. $x^2 + x + 1 = 0$ **45.** $2x^2 = 3x - 2$ **46.** $x^2 = 5x - 20$

47. $x^2 = 1 + x$ **48.** $2x^2 + 2x = 5$ **49.** $5x^2 = 3 - x$

Use the quadratic formula to solve each equation. See Example 5.

50. $\frac{3}{2}r^2 - r = \frac{4}{3}$ **51.** $\frac{1}{2}x^2 = 1 - \frac{1}{6}x$ **52.** $\frac{3}{5}x - \frac{2}{5}x^2 = -1$

53. $\frac{2}{3}m^2 - \frac{4}{9}m - \frac{1}{3} = 0$ **54.** $\frac{m^2}{4} + \frac{3m}{2} + 1 = 0$ **55.** $\frac{r^2}{2} = r + \frac{1}{2}$

56. $\frac{2y^2}{7} + \frac{10}{7}y + 1 = 0$ **57.** $k^2 = \frac{2k}{3} + \frac{2}{9}$ **58.** $9 - \frac{24}{y^2} = -\frac{17}{y}$

59. $\frac{m^2}{2} = \frac{m}{2} - 1$ **60.** $3 + \frac{1}{p^2} = \frac{6}{p}$ **61.** $2 = \frac{4}{x} - \frac{3}{x^2}$

---

62. If an applied problem leads to a quadratic equation, what must you be aware of after you have solved the equation? (See Objective 5.)

63. Suppose that a problem asks you to find the length of a rectangle, and the problem leads to a quadratic equation. Which one of the following solutions to the equation cannot be an answer to the problem, if L represents this length of the rectangle? (See Objective 5.)

(a) $L = 9$ **(b)** $L = 5\frac{1}{4}$

(c) $L = \frac{1 + \sqrt{5}}{2}$ **(d)** $L = \frac{1 - \sqrt{5}}{2}$

Work each of the following problems. See Example 6.

64. The time t in seconds under certain conditions for a ball to be 48 feet in the air is given (approximately) by

$$48 = 64t - 16t^2.$$

Solve this equation for t. Are both answers reasonable?

65. A certain projectile is located $d = 2t^2 - 5t + 2$ feet from the ground after t seconds have elapsed. How many seconds will it take the projectile to be 14 feet from the ground?

66. Karen and Jessie work at a fast-food restaurant after school. Working alone, Jessie can close up in 1 hour less time than Karen. Together they can close up in 2/3 of an hour. How long does it take Jessie to close up alone?

67. In a bicycle race over a 12-mile route, Donnie finished 8 minutes ahead of Juan. Donnie pedaled 3 miles per hour faster than Juan. What was Donnie's speed?

PREVIEW EXERCISES

Perform the indicated operations. See Sections 3.1, 3.3, and 3.4.

68. $(4 + 6z) + (-9 + 2z)$

69. $(10 - 3t) - (5 - 7t)$

70. $(4 + 3r)(6 - 5r)$

71. $(5 + 2x)(5 - 2x)$

Use the product rule for radicals to simplify each expression. See Section 8.2.

72. $\sqrt{24}$

73. $\sqrt{48}$

74. $\sqrt{72}$

75. $\sqrt{44}$

SUMMARY EXERCISES ON QUADRATIC EQUATIONS

Four methods have now been introduced for solving quadratic equations written in the form $ax^2 + bx + c = 0$. The chart below shows some advantages and some disadvantages of each method.

METHOD	ADVANTAGES	DISADVANTAGES
1. Factoring	Usually the fastest method.	Not all equations can be solved by factoring. Some factorable polynomials are hard to factor.
2. Square root property	Simplest method for solving equations of the form $(ax + b)^2 =$ a number.	Few equations are given in this form.
3. Completing the square	Can always be used (also, the procedure is useful in other areas of mathematics).	It requires more steps than other methods.
4. Quadratic formula	Can always be used.	It is more difficult than factoring because of the $\sqrt{b^2 - 4ac}$ expression.

Solve each quadratic equation by the method of your choice.

1. $y^2 + 3y + 1 = 0$

2. $p^2 + 3p + 2 = 0$

3. $2x^2 - x = 1$

4. $2a^2 + 1 = a$

5. $8m^2 = 2m + 15$

6. $8x^2 + 2x = 15$

7. $(2p - 1)^2 = 10$

8. $2q^2 + 3q = 1$

9. $5k^2 + 8 = 22k$

10. $5k^2 + 2k = 1$

11. $3z^2 = 4z + 1$

12. $2c^2 + 11c + 12 = 0$

13. $(2q + 9)^2 = 48$

14. $4x^2 + 5x = 1$

15. $15t^2 + 58t + 48 = 0$

16. $12d^2 + 19d = 21$

17. $p^2 + 5p + 5 = 0$

18. $(7m - 1)^2 = 32$

19. $3c^2 - 4c = 4$

20. $5k^2 + 17k = 12$

21. $2x^2 - 5x + 1 = 0$

22. $(5r - 7)^2 = -1$

23. $4m^2 - 11m + 10 = 0$

24. $2a^2 - 7a + 4 = 0$

25. $(3p - 1)(p + 2) = -3$

26. $(3r + 2)^2 = 5$

27. $(a + 6)^2 = 121$

28. $(3x - 1)(2x + 5) = x$

29. $(2x + 1)(x - 1) = 5$

30. $(z - 1)(z - 3) = -3z$

---

31. How would you respond to this statement? "Since I know how to solve quadratic equations by the factoring method, there is no reason for me to learn any other method of solving quadratic equations."

32. How many real solutions are there for a quadratic equation that has a negative number as its radicand in the quadratic formula?

9.4 COMPLEX NUMBERS

OBJECTIVES

1. WRITE COMPLEX NUMBERS LIKE $\sqrt{-5}$ AS MULTIPLES OF i.
2. ADD AND SUBTRACT COMPLEX NUMBERS.
3. MULTIPLY COMPLEX NUMBERS.
4. WRITE COMPLEX NUMBER QUOTIENTS IN STANDARD FORM.
5. SOLVE QUADRATIC EQUATIONS WITH COMPLEX NUMBER SOLUTIONS.

As shown earlier in this chapter, some quadratic equations have no real number solutions. For example, the solution

$$x = \frac{-3 + \sqrt{-5}}{2}$$

is not a real number because of $\sqrt{-5}$. For every quadratic equation to have a solution, a new set of numbers, which includes the real numbers, is needed.

1 This new set of numbers is defined using a new number i having the properties given below.

THE NUMBER i

$$i = \sqrt{-1} \quad \text{and} \quad i^2 = -1$$

Numbers like $\sqrt{-5}$, $\sqrt{-4}$, and $\sqrt{-8}$ can now be written as multiples of i, using a generalization of the product rule for radicals, as in the next example.

EXAMPLE 1 **SIMPLIFYING SQUARE ROOTS OF NEGATIVE NUMBERS**

Write each number as a multiple of i.

(a) $\sqrt{-5} = \sqrt{-1 \cdot 5} = \sqrt{-1} \cdot \sqrt{5} = i\sqrt{5}$

(b) $\sqrt{-4} = \sqrt{-1} \cdot \sqrt{4} = i\sqrt{4} = i \cdot 2 = 2i$

(c) $\sqrt{-8} = i\sqrt{8} = i \cdot 2 \cdot \sqrt{2} = 2i\sqrt{2}$ ■

CAUTION It is easy to mistake $\sqrt{2}\,i$ for $\sqrt{2i}$, with the i under the radical. For this reason, it is customary to write the i factor first when it is multiplied by a radical. For example, we usually write $i\sqrt{2}$ rather than $\sqrt{2}\,i$.

Numbers that are nonzero multiples of i are *imaginary numbers*. The *complex numbers* include all real numbers and all imaginary numbers.

COMPLEX NUMBER

A **complex number** is a number of the form $a + bi$, where a and b are real numbers. If $b \neq 0$, $a + bi$ also is an **imaginary number.**

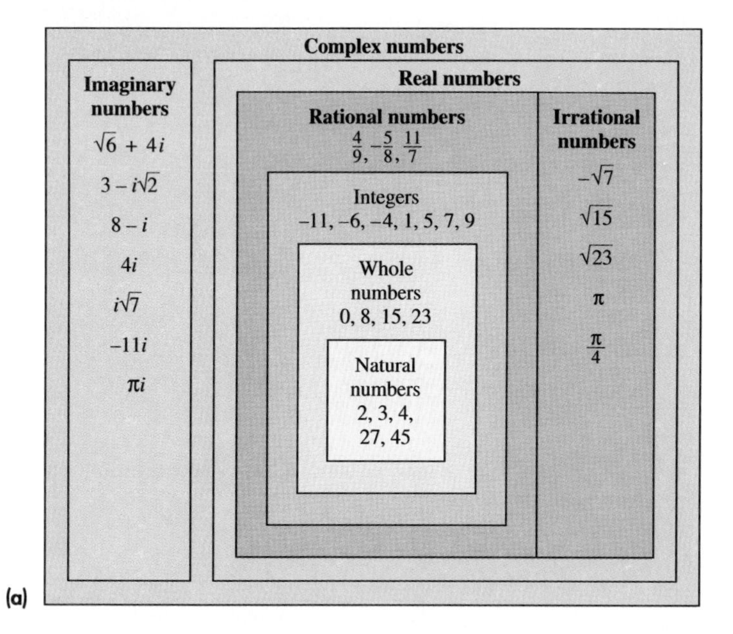

(b)
Complex numbers — Imaginary numbers; Real numbers
Real numbers — Irrational numbers; Rational numbers
Rational numbers — Integers; Non-integer rational numbers
Integers — Whole numbers; Negative integers
Whole numbers — Zero; Natural numbers

FIGURE 9.1

For example, the real number 2 is a complex number since it can be written as $2 + 0i$. Also, the imaginary number $3i = 0 + 3i$ is a complex number. Other complex numbers are

$$3 - 2i,$$
$$1 + i\sqrt{2},$$
and
$$-5 + 4i.$$

In the complex number $a + bi$, a is called the **real part** and b (*not* bi) is called the **imaginary part.**

A complex number written in the form $a + bi$ (or $a + ib$) is in **standard form.** Figure 9.1 shows the relationships among the various types of numbers discussed in this book. (Compare this figure to Figure 1.4 in Chapter 1.)

2 The operations of addition and subtraction of complex numbers are similar to these operations with binomials.

ADDITION AND SUBTRACTION OF COMPLEX NUMBERS

1. To add complex numbers, add their real parts and add their imaginary parts.
2. To subtract complex numbers, change the number following the subtraction sign to its negative, and then add.

The properties of Section 1.9 (commutative, associative, etc.) also hold for operations with complex numbers.

EXAMPLE 2 ADDING AND SUBTRACTING COMPLEX NUMBERS

Add or subtract.

(a) $(2 - 6i) + (7 + 4i) = (2 + 7) + (-6 + 4)i = 9 - 2i$

(b) $3i + (-2 - i) = -2 + (3 - 1)i = -2 + 2i$

(c) $(2 + 6i) - (-4 + i)$

Change $-4 + i$ to its negative, and then add.

$$(2 + 6i) - (-4 + i) = (2 + 6i) + (4 - i) \qquad -(-4 + i) = 4 - i$$

$$= (2 + 4) + (6 - 1)i \qquad \text{Commutative, associative, and distributive properties}$$

$$= 6 + 5i$$

(d) $(-1 + 2i) - 4 = (-1 - 4) + 2i = -5 + 2i$ ■

3 Multiplication of complex numbers is performed in the same way as multiplication of polynomials. Whenever i^2 appears, replace it with -1.

EXAMPLE 3 MULTIPLYING COMPLEX NUMBERS

Find the following products.

(a)

$$3i(2 - 5i) = 6i - 15i^2 \qquad \text{Distributive property}$$

$$= 6i - 15(-1) \qquad i^2 = -1$$

$$= 6i + 15$$

$$= 15 + 6i \qquad \text{Commutative property}$$

The last step gives the result in standard form.

(b) $(4 - 3i)(2 + 5i)$

Use FOIL.

$$(4 - 3i)(2 + 5i) = 4(2) + 4(5i) + (-3i)(2) + (-3i)(5i)$$

$$= 8 + 20i - 6i - 15i^2$$

$$= 8 + 14i - 15(-1)$$

$$= 8 + 14i + 15$$

$$= 23 + 14i$$

(c)

$$(5 - 2i)(5 + 2i) = 25 + 10i - 10i - 4i^2$$

$$= 25 - 4(-1)$$

$$= 25 + 4 = 29$$ ■

4 The quotient of two complex numbers is expressed in standard form by changing the denominator into a real number. For example, to write

$$\frac{1+3i}{5-2i}$$

in standard form, the denominator must be a real number. As seen in Example 3(c), the product $(5-2i)(5+2i)$ is 29, a real number. This suggests multiplying the numerator and denominator of the given quotient by $5+2i$, as follows.

$$\frac{1+3i}{5-2i} = \frac{1+3i}{5-2i} \cdot \frac{5+2i}{5+2i}$$

$$= \frac{5+2i+15i+6i^2}{25-4i^2}$$

$$= \frac{5+17i+6(-1)}{25-4(-1)} \qquad i^2 = -1$$

$$= \frac{5+17i-6}{25+4}$$

$$= \frac{-1+17i}{29} = -\frac{1}{29} + \frac{17}{29}i$$

The last step gives the result in standard form. Recall that this is the method used to rationalize certain expressions in Chapter 8. The complex numbers $5-2i$ and $5+2i$ are *conjugates*. That is, the **conjugate** of the complex number $a+bi$ is $a-bi$. Multiplying the complex number $a+bi$ by its conjugate $a-bi$ gives the real number a^2+b^2.

PRODUCT OF CONJUGATES

$$(a+bi)(a-bi) = a^2+b^2$$

The product of a complex number and its conjugate is the sum of the squares of the real and imaginary parts.

EXAMPLE 4 DIVIDING COMPLEX NUMBERS

Write the following quotients in standard form.

(a) $\frac{-4+i}{2-i}$

Multiply numerator and denominator by the conjugate of the denominator, $2+i$.

$$\frac{-4+i}{2-i} \cdot \frac{2+i}{2+i} = \frac{-8-4i+2i+i^2}{4-i^2}$$

$$= \frac{-8-2i-1}{4-(-1)} \qquad i^2 = -1$$

$$= \frac{-9-2i}{5} = -\frac{9}{5} - \frac{2}{5}i$$

Note that the final step is performed in order to get the complex number in standard form.

(b) $\dfrac{3 + i}{-i}$

Here, the conjugate of $0 - i$ is $0 + i$, or i.

$$\frac{3+i}{-i} \cdot \frac{i}{i} = \frac{3i + i^2}{-i^2}$$

$$= \frac{-1 + 3i}{-(-1)} \qquad i^2 = -1\text{, commutative property}$$

$$= -1 + 3i \quad \blacksquare$$

5 Quadratic equations that have no real solutions do have complex solutions, as shown in the next examples.

EXAMPLE 5 SOLVING A QUADRATIC EQUATION WITH COMPLEX SOLUTIONS (SQUARE ROOT METHOD)

Solve $(x + 3)^2 = -25$ for complex solutions.

Use the square root property.

$$(x + 3)^2 = -25$$

$$x + 3 = \sqrt{-25} \quad \text{or} \quad x + 3 = -\sqrt{-25}$$

Since $\sqrt{-25} = 5i$,

$$x + 3 = 5i \quad \text{or} \quad x + 3 = -5i$$

$$x = -3 + 5i \quad \text{or} \quad x = -3 - 5i.$$

In standard form, the two complex solutions are $-3 + 5i$ and $-3 - 5i$. ■

EXAMPLE 6 SOLVING A QUADRATIC EQUATION WITH COMPLEX SOLUTIONS (QUADRATIC FORMULA)

Solve $t^2 - 4t + 5 = 0$ for complex solutions.

The polynomial on the left cannot be factored, so use the quadratic formula, with $a = 1$, $b = -4$, and $c = 5$. The solutions are

$$t = \frac{-(-4) \pm \sqrt{(-4)^2 - 4(1)(5)}}{2(1)}$$

$$= \frac{4 \pm \sqrt{16 - 20}}{2}$$

$$= \frac{4 \pm \sqrt{-4}}{2}$$

$$= \frac{4 \pm 2i}{2}. \qquad \sqrt{-4} = i\sqrt{4} = 2i$$

Factor out the common factor of 2 in the numerator and reduce to get

$$t = \frac{2(2 \pm i)}{2} = 2 \pm i.$$

The two complex solutions are $2 + i$ and $2 - i$. ■

EXAMPLE 7

SOLVING A QUADRATIC EQUATION WITH COMPLEX SOLUTIONS (QUADRATIC FORMULA)

Solve $2p^2 = 4p - 5$ for complex solutions.

Write the equation as $2p^2 - 4p + 5 = 0$. Then $a = 2$, $b = -4$, and $c = 5$. The solutions are

$$\begin{aligned} p &= \frac{-(-4) \pm \sqrt{(-4)^2 - 4(2)(5)}}{2(2)} \\ &= \frac{4 \pm \sqrt{16 - 40}}{4} \\ &= \frac{4 \pm \sqrt{-24}}{4}. \end{aligned}$$

Since $\sqrt{-24} = i\sqrt{24} = i \cdot \sqrt{4} \cdot \sqrt{6} = i \cdot 2 \cdot \sqrt{6} = 2i\sqrt{6}$,

$$\begin{aligned} p &= \frac{4 \pm 2i\sqrt{6}}{4} \\ &= \frac{2(2 \pm i\sqrt{6})}{4} && \text{Factor out a 2.} \\ &= \frac{2 \pm i\sqrt{6}}{2}. && \text{Lowest terms} \end{aligned}$$

In standard form, the solutions are the complex numbers

$$1 + \frac{\sqrt{6}}{2}i \quad \text{and} \quad 1 - \frac{\sqrt{6}}{2}i. \quad ■$$

9.4 EXERCISES

Write the following numbers as multiples of i. See Example 1.

1. $\sqrt{-9}$ **2.** $\sqrt{-18}$ **3.** $\sqrt{-20}$ **4.** $\sqrt{-27}$

5. $\sqrt{-36}$ **6.** $\sqrt{-50}$ **7.** $\sqrt{-125}$ **8.** $\sqrt{-98}$

Add or subtract as indicated. See Example 2.

9. $(2 + 8i) + (3 - 5i)$ **10.** $(4 - 5i) + (7 - 2i)$

11. $(8 - 3i) - (2 + 6i)$ **12.** $(1 + i) - (3 - 2i)$

13. $(3 - 4i) + (6 - i) - (3 + 2i)$ **14.** $(5 + 8i) - (4 + 2i) + (3 - i)$

Find each product. See Example 3.

15. $(3 + 2i)(4 - i)$ **16.** $(9 - 2i)(3 + i)$ **17.** $(5 - 4i)(3 - 2i)$

18. $(10 + 6i)(8 - 4i)$ **19.** $(3 + 6i)(3 - 6i)$ **20.** $(11 - 2i)(11 + 2i)$

Write each quotient in standard form. See Example 4.

21. $\dfrac{-2 + i}{1 - i}$ **22.** $\dfrac{6 + i}{2 + i}$ **23.** $\dfrac{3 - 4i}{2 + 2i}$

24. $\dfrac{7-i}{3-2i}$

25. $\dfrac{4-3i}{i}$

26. $\dfrac{-i}{1+2i}$

---

27. Write an explanation of the method used to divide complex numbers. (See Objective 4.)

28. Use the fact that $i^2 = -1$ to complete each of the following. Do them in order.
(a) Since $i^3 = i^2 \cdot i$, $i^3 =$ ______.
(b) Since $i^4 = i^3 \cdot i$, $i^4 =$ ______.
(c) Since $i^{48} = (i^4)^{12}$, $i^{48} =$ ______.

Solve each quadratic equation for complex solutions by the square root property. Write solutions in standard form. See Example 5.

29. $(a+1)^2 = -4$ **30.** $(p-5)^2 = -36$ **31.** $(k-3)^2 = -5$

32. $(y+6)^2 = -12$ **33.** $(3x+2)^2 = -18$ **34.** $(4z-1)^2 = -20$

Solve each quadratic equation for complex solutions by the quadratic formula. Write solutions in standard form. See Examples 6 and 7.

35. $m^2 - 2m + 2 = 0$ **36.** $b^2 + b + 3 = 0$ **37.** $2r^2 + 3r + 5 = 0$ **38.** $3q^2 = 2q - 3$

39. $p^2 - 3p + 4 = 0$ **40.** $2a^2 = -a - 3$ **41.** $5x^2 + 3 = 2x$ **42.** $6y^2 + 2y + 1 = 0$

43. $2m^2 + 7 = -2m$ **44.** $4z^2 + 2z + 3 = 0$ **45.** $r^2 + 3 = r$ **46.** $4q^2 - 2q + 3 = 0$

Exercises 47–48 deal with quadratic equations having real number coefficients.

47. Suppose you are solving a quadratic equation by the quadratic formula. How can you tell, before completing the solution, whether the equation will have imaginary solutions? (See Objective 5.)

48. Refer to the solutions in Examples 5, 6, and 7, and complete the following statement: If a quadratic equation has imaginary solutions, they are ______ of each other. (See Objective 5.)

Answer true *or* false *to each of the following. (See Objective 1.)*

49. Every real number is a complex number.

50. Every imaginary number is a complex number.

51. Every complex number is a real number.

52. Every complex number is an imaginary number.

PREVIEW EXERCISES

Complete the ordered pairs using the given equation. Then plot the ordered pairs. See Section 6.1.

Equation	*Ordered Pairs*
53. $2x - 3y = 6$	(0,) (, 0) (, 2) (−3,)
54. $4x - y = 8$	(0,) (, 0) (4,) (, −4)
55. $3x + 5y = 15$	(0,) (, 0) (−5,) (, −3)
56. $6x + 4y = 12$	(0,) (, 0) (−2,) (, −3)
57. $y - 2x = 0$	(0,) (3,) (, 5) (, 2)
58. $x - 3y = 0$	(0,) (3,) (, 4) (4,)

9.5 GRAPHING QUADRATIC EQUATIONS IN TWO VARIABLES

OBJECTIVES

1. GRAPH QUADRATIC EQUATIONS OF THE FORM $y = ax^2 + bx + c$ $(a \neq 0)$.
2. FIND THE VERTEX OF A PARABOLA.
3. USE A GRAPH TO DETERMINE THE NUMBER OF REAL SOLUTIONS OF A QUADRATIC EQUATION.

1 In Chapter 6 we saw that the graph of a linear equation in two variables is a straight line that represents all the solutions of the equation. Quadratic equations in two variables, of the form $y = ax^2 + bx + c$, are graphed in this section. Perhaps the simplest quadratic equation is $y = x^2$ (or $y = 1x^2 + 0x + 0$). The graph of this equation cannot be a straight line since only linear equations of the form $Ax + By = C$ have graphs that are straight lines. However, $y = x^2$ can be graphed in much the same way that straight lines were graphed, by finding ordered pairs that satisfy the equation $y = x^2$.

EXAMPLE 1 GRAPHING A QUADRATIC EQUATION

Graph $y = x^2$.

Select several values for x; then find the corresponding y-values. For example, selecting $x = 2$ gives

$$y = 2^2 = 4,$$

and so the point (2, 4) is on the graph of $y = x^2$. (Recall that in an ordered pair such as (2, 4), the x-value comes first and the y-value second.) Some ordered pairs that satisfy $y = x^2$ are shown in a table to the side of Figure 9.2. If the ordered pairs from the table are plotted on a coordinate system and a smooth curve drawn through them, the graph is as shown in Figure 9.2. ■

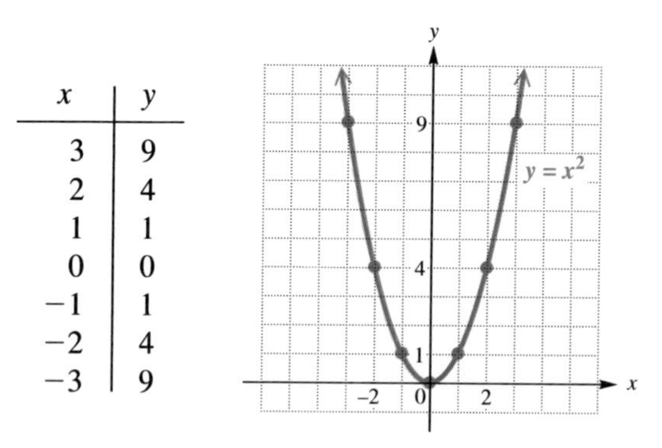

x	y
3	9
2	4
1	1
0	0
−1	1
−2	4
−3	9

FIGURE 9.2

The curve in Figure 9.2 is called a **parabola.** The point (0, 0), the lowest point on this graph, is called the **vertex** of the parabola. The vertical line through the vertex (the y-axis here) is called the **axis** of the parabola.

Every equation of the form

$$y = ax^2 + bx + c,$$

with $a \neq 0$, has a graph that is a parabola. Because of its many useful properties, the parabola occurs frequently in real-life applications. For example, if an object is thrown into the air, the path that the object follows is a parabola (ignoring wind resistance). The cross sections of radar, spotlight, and telescope reflectors also form parabolas.

EXAMPLE 2
GRAPHING A PARABOLA

Graph $y = x^2 - 3$.

Find several ordered pairs. Begin with the intercepts. If $x = 0$,

$$y = x^2 - 3 = 0^2 - 3 = -3,$$

giving the ordered pair $(0, -3)$. If $y = 0$,

$$\begin{aligned} y &= x^2 - 3 \\ 0 &= x^2 - 3 \\ x^2 &= 3 \\ x &= \sqrt{3} \quad \text{or} \quad -\sqrt{3}, \end{aligned}$$

giving the two ordered pairs $(-\sqrt{3}, 0)$ and $(\sqrt{3}, 0)$. Now choose additional x-values near the x-values of these three points. Some of these are shown in the table of values with Figure 9.3. Plot all these points and connect them with a smooth curve as shown in Figure 9.3. The vertex of this parabola is $(0, -3)$. ■

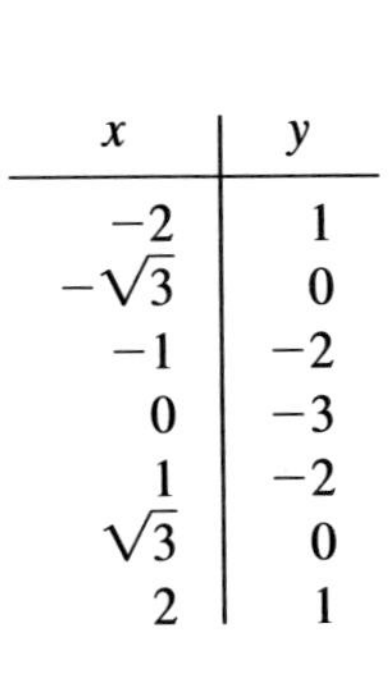

x	y
-2	1
$-\sqrt{3}$	0
-1	-2
0	-3
1	-2
$\sqrt{3}$	0
2	1

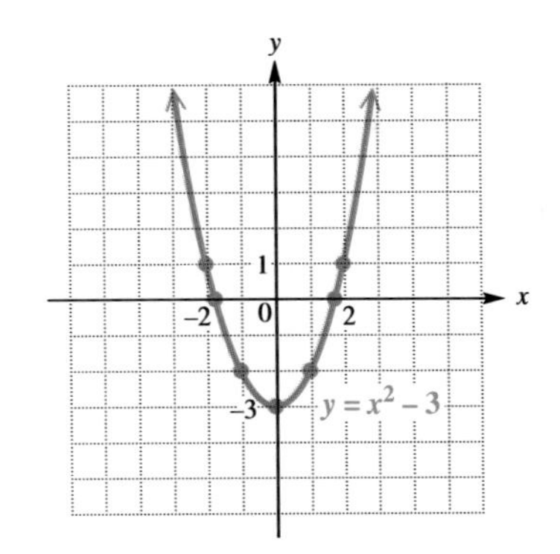

FIGURE 9.3

2 As the graphs we found above suggest, the vertex is the most important point to locate when you are graphing a quadratic equation. The next example shows how to find the vertex in a more general case.

EXAMPLE 3
FINDING THE VERTEX TO GRAPH A PARABOLA

Graph $y = x^2 - 2x - 3$.

We want to find the vertex of the graph. Note in Figure 9.3 that the vertex is exactly halfway between the x-intercepts. If a parabola has two x-intercepts this is always the case. Therefore, let us begin by finding the x-intercepts. Let $y = 0$ in the equation and solve for x.

$$0 = x^2 - 2x - 3$$

$$0 = (x + 1)(x - 3) \qquad \text{Factor.}$$

$$x + 1 = 0 \quad \text{or} \quad x - 3 = 0 \qquad \text{Set each factor equal to 0.}$$

$$x = -1 \quad \text{or} \quad x = 3$$

There are two x-intercepts, $(-1, 0)$ and $(3, 0)$. Now find any y-intercepts. Substitute $x = 0$ in the equation.

$$y = 0^2 - 2(0) - 3 = -3$$

There is one y-intercept, $(0, -3)$.

As mentioned above, the x-value of the vertex is halfway between the x-values of the two x-intercepts. Thus, it is 1/2 their sum.

$$x = \frac{1}{2}(-1 + 3) = 1$$

Find the corresponding y-value by substituting 1 for x in the equation.

$$y = 1^2 - 2(1) - 3 = -4$$

The vertex is $(1, -4)$. The axis is the line $x = 1$. Plot the three intercepts and the vertex. Find additional ordered pairs as needed. For example, if $x = 2$,

$$y = 2^2 - 2(2) - 3 = -3,$$

leading to the ordered pair $(2, -3)$. A table of values with the ordered pairs we have found is shown with the graph in Figure 9.4. ■

x	y
−2	5
−1	0
0	−3
1	−4
2	−3
3	0
4	5

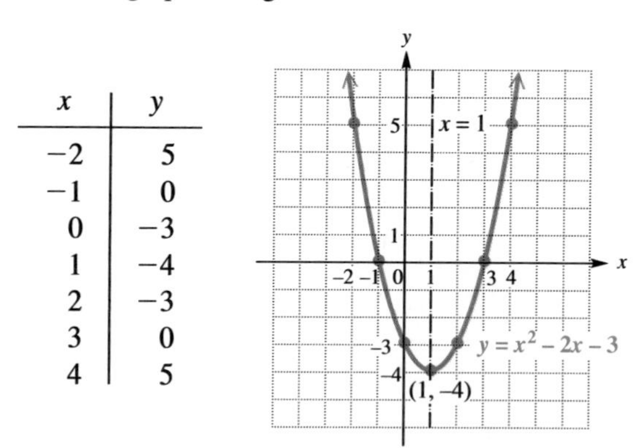

FIGURE 9.4

We can generalize from Example 3. The x-values of the x-intercepts for the equation $y = ax^2 + bx + c$, by the quadratic formula, are

$$x = \frac{-b + \sqrt{b^2 - 4ac}}{2a} \quad \text{and} \quad x = \frac{-b - \sqrt{b^2 - 4ac}}{2a}.$$

Thus, the x-value of the vertex is

$$x = \frac{1}{2}\left(\frac{-b + \sqrt{b^2 - 4ac}}{2a} + \frac{-b - \sqrt{b^2 - 4ac}}{2a}\right)$$

$$x = \frac{1}{2}\left(\frac{-b + \sqrt{b^2 - 4ac} - b - \sqrt{b^2 - 4ac}}{2a}\right)$$

$$x = \frac{1}{2}\left(\frac{-2b}{2a}\right) = -\frac{b}{2a}.$$

For the equation in Example 3, $y = x^2 - 2x - 3$, $a = 1$, and $b = -2$. Thus,

$$x = -\frac{b}{2a} = -\frac{-2}{2(1)} = 1,$$

which is the same x-value for the vertex we found in Example 3. (The x-value of the vertex is $x = -\frac{b}{2a}$ even if the graph has no x-intercepts.) A procedure for graphing quadratic equations follows.

GRAPHING THE PARABOLA $y = ax^2 + bx + c$

Step 1 Find the y-intercept.
Step 2 Find any x-intercepts.
Step 3 Find the vertex. Let $x = -\frac{b}{2a}$ and find the corresponding y-value by substituting for x in the equation.
Step 4 Plot the intercepts and the vertex.
Step 5 Find and plot additional ordered pairs near the vertex and intercepts as needed.

EXAMPLE 4 GRAPHING A PARABOLA

Graph $y = x^2 - 4x + 1$.

Find the intercepts. Let $x = 0$ in $y = x^2 - 4x + 1$ to get the y-intercept $(0, 1)$. Let $y = 0$ to get the x-intercepts. If $y = 0$, the equation is $0 = x^2 - 4x + 1$, which cannot be solved by factoring. Use the quadratic formula to solve for x.

$$x = \frac{4 \pm \sqrt{16 - 4}}{2} \qquad \text{Let } a = 1, b = -4, c = 1.$$

$$x = \frac{4 \pm \sqrt{12}}{2}$$

$$x = \frac{4 \pm 2\sqrt{3}}{2} \qquad \sqrt{12} = 2\sqrt{3}$$

$$x = \frac{2(2 \pm \sqrt{3})}{2} = 2 \pm \sqrt{3}$$

Use a calculator to find that the x-intercepts are $(3.7, 0)$ and $(.3, 0)$ to the nearest tenth. The x-value of the vertex is

$$x = -\frac{b}{2a} = -\frac{-4}{2(1)} = 2.$$

The y-value of the vertex is

$$y = 2^2 - 4(2) + 1 = -3,$$

so the vertex is $(2, -3)$. The axis is the line $x = 2$.

A table of values of the points found so far, along with some others, is shown with the graph. Join these points with a smooth curve, as shown in Figure 9.5. ■

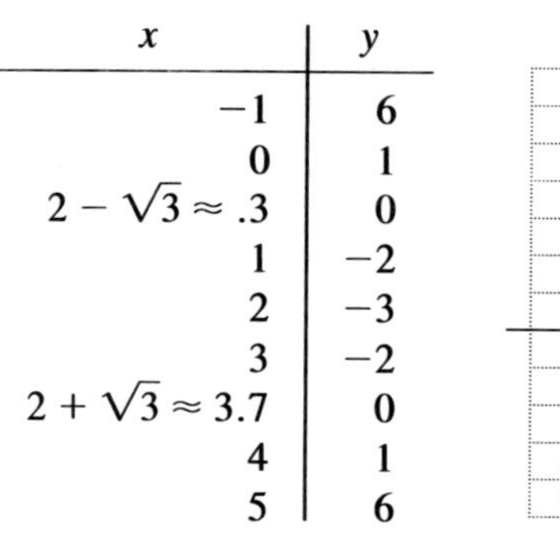

x	y
-1	6
0	1
$2 - \sqrt{3} \approx .3$	0
1	-2
2	-3
3	-2
$2 + \sqrt{3} \approx 3.7$	0
4	1
5	6

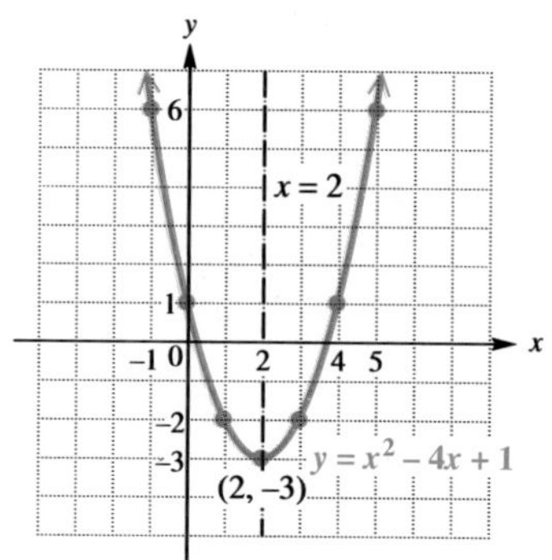

FIGURE 9.5

3 It can be verified by the vertical line test (Section 6.6) that the graph of an equation of the form $y = ax^2 + bx + c$ is the graph of a function. A function defined by an equation of the form $f(x) = ax^2 + bx + c$ $(a \neq 0)$ is called a **quadratic function.** The domain of a quadratic function is all real numbers; the range can be determined after the function is graphed.

Look again at Figure 9.5, which gives the graph of $y = x^2 - 4x + 1$. Recall that setting y equal to 0 gives the x-intercepts, where the x-values are

$$2 + \sqrt{3} \approx 3.7 \qquad \text{and} \qquad 2 - \sqrt{3} \approx .3.$$

The solutions of $0 = x^2 - 4x + 1$ are the x-values of the x-intercepts of the graph of the corresponding quadratic function.

INTERCEPTS OF THE GRAPH OF A QUADRATIC FUNCTION

The real number solutions of a quadratic equation $ax^2 + bx + c = 0$ are the x-values of the x-intercepts of the graph of the corresponding quadratic function $y = ax^2 + bx + c$.

Since the graph of a quadratic function can cross the x-axis in two, one, or no points, this result shows why some quadratic equations have two, some have one, and some have no real solutions.

EXAMPLE 5

DETERMINING THE NUMBER OF REAL SOLUTIONS FROM A GRAPH

(a) Figure 9.6 shows the graph of $y = x^2 - 3$. The equation $0 = x^2 - 3$ has two real solutions. (They are $\sqrt{3}$ and $-\sqrt{3}$.)

(b) Figure 9.7 shows the graph of $y = x^2 - 4x + 4$. The equation $0 = x^2 - 4x + 4$ has one real solution. (It is 2.)

(c) Figure 9.8 shows the graph of $y = x^2 + 2$. The equation $0 = x^2 + 2$ has no real solutions, since there are no x-intercepts. (The equation *does* have two imaginary solutions, $i\sqrt{2}$ and $-i\sqrt{2}$.) ■

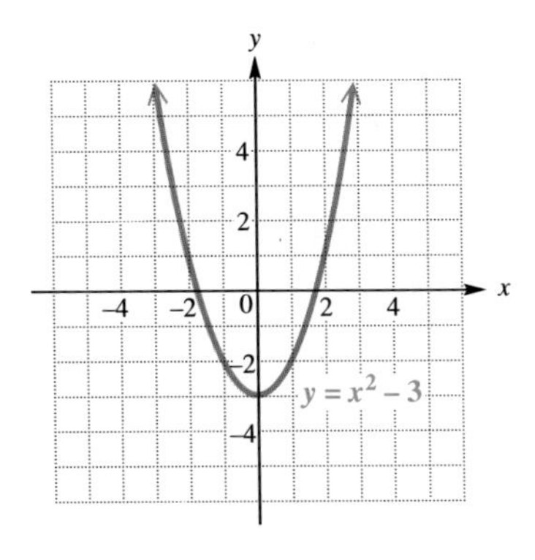

FIGURE 9.6

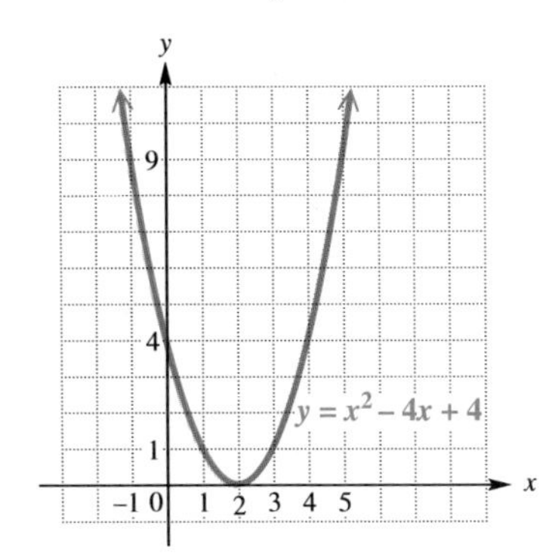

FIGURE 9.7

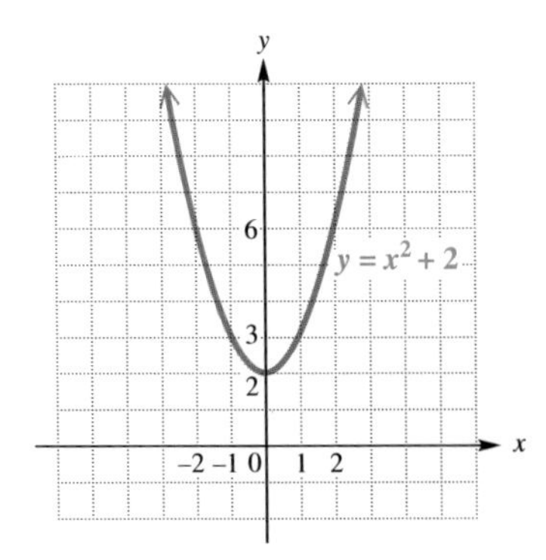

FIGURE 9.8

9.5 EXERCISES

Sketch the graph of each equation. Identify each vertex. See Examples 1–4.

1. $y = 2x^2$

2. $y = -2x^2$

3. $y = x^2 + 2x + 1$

4. $y = x^2 - 2x + 8$

5. $y = -x^2 - 2x - 1$

6. $y = -x^2 + 4x - 4$

7. $y = x^2 + 1$

8. $y = -x^2 - 2$

9. $y = 2 - x^2$

10. $y = x^2 - x - 2$

11. $y = x^2 + 6x + 8$

12. $y = x^2 - 3x - 4$

13. $y = x^2 + 2x + 3$

14. $y = x^2 - 4x + 3$

15. $y = x^2 - 8x + 15$

16. $y = x^2 + 6x + 12$

17. $y = -x^2 - 4x - 3$

18. $y = -x^2 + 6x - 5$

Decide from the graph how many real number solutions the corresponding equation has. Find any real solutions from the graph. See Example 5.

19.

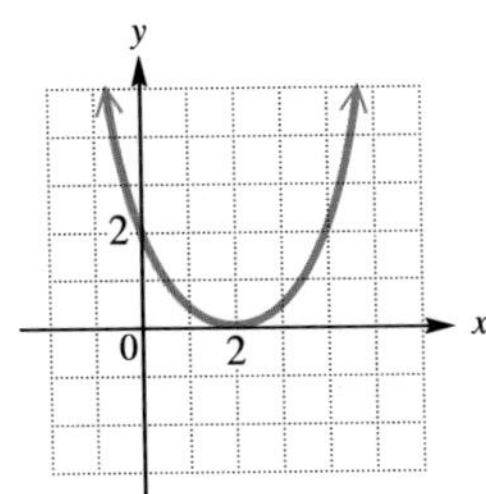

20.

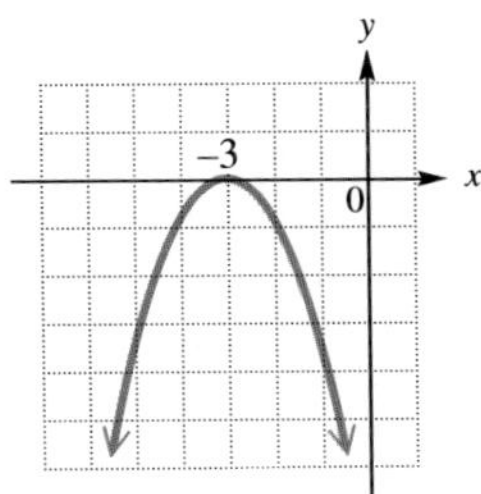

21.

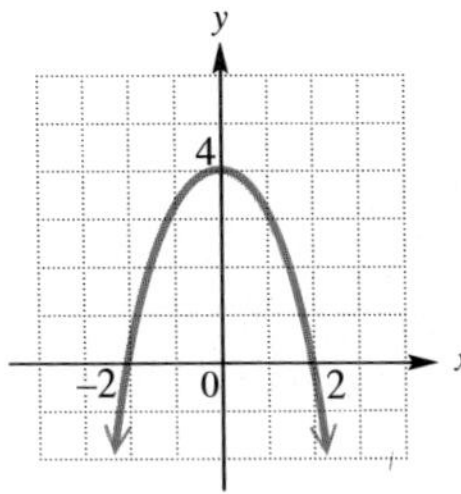

22.

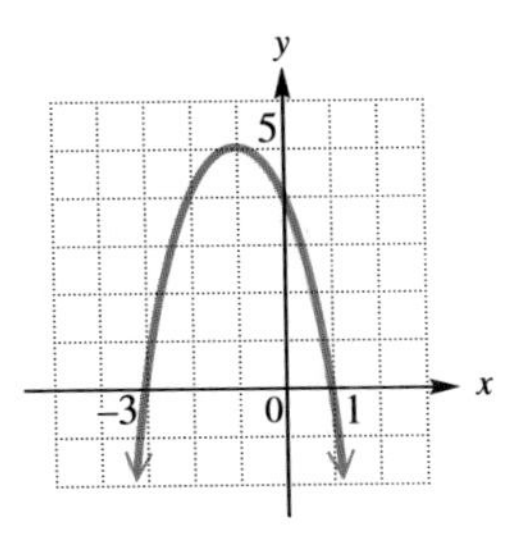

23.

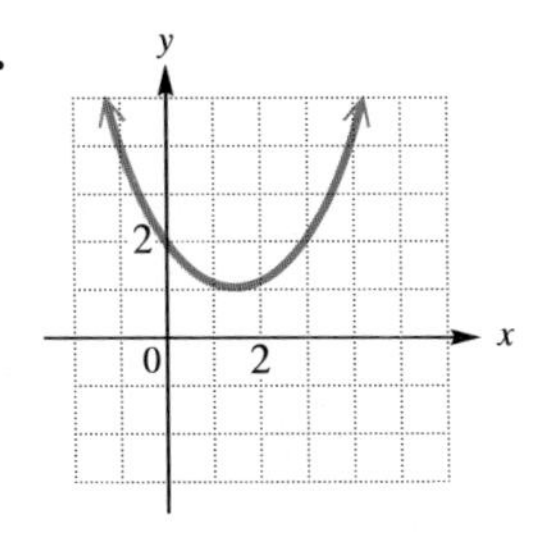

24.

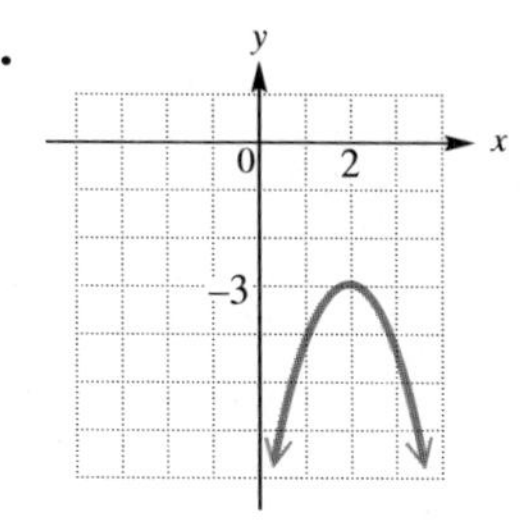

25. Explain how to find the number of real solutions of the equation $ax^2 + bx + c = 0$ by observing the graph of $y = ax^2 + bx + c$. (See Objective 3.)

26. If the graph of $y = ax^2 + bx + c$ does not intersect the x-axis, how many imaginary solutions does the equation $ax^2 + bx + c = 0$ have? (See Objective 3.)

Find the domain and range of each of the functions graphed in the indicated exercises.

27. Exercise 19 **28.** Exercise 20 **29.** Exercise 21

30. Exercise 22 **31.** Exercise 23 **32.** Exercise 24

Given $f(x) = 2x^2 - 5x + 3$, *find each of the following.*

33. $f(0)$ **34.** $f(1)$ **35.** $f(-2)$ **36.** $f(-1)$

---

37. Which of the following functions has a parabola as its graph? (See Objective 3.)
(a) $f(x) = 2x^2 - 5x + 3$ **(b)** $f(x) = -3x + 2$
(c) $f(x) = x$ **(d)** $f(x) = 4$

38. The graph of $f(x) = ax^2$ is a parabola that opens upward if $a > 0$ or downward if $a < 0$. Use this information to tell whether the graph of the given quadratic function opens upward or downward.
(a) $f(x) = \frac{1}{2}x^2$ **(b)** $f(x) = -\frac{1}{2}x^2$
(c) $f(x) = -\sqrt{2}x^2$ **(d)** $f(x) = \sqrt{2}x^2$

CHAPTER 9 GLOSSARY

KEY TERMS

9.3 standard form (of a quadratic equation) A quadratic equation written as $ax^2 + bx + c = 0$ $(a \neq 0)$ is in standard form.

9.4 complex number A complex number is a number of the form $a + bi$, where a and b are real numbers.

imaginary number The complex number $a + bi$ is imaginary if $b \neq 0$.

real part The real part of $a + bi$ is a.

imaginary part The imaginary part of $a + bi$ is b.

standard form (of a complex number) A complex number written in the form $a + bi$ (or $a + ib$) is in standard form.

conjugate The conjugate of $a + bi$ is $a - bi$.

9.5 parabola The graph of a quadratic equation of the form $y = ax^2 + bx + c$ $(a \neq 0)$ is called a parabola.

vertex The vertex of a parabola is the highest or lowest point on the graph.

axis The axis of a parabola is a vertical line through the vertex.

quadratic function A function of the form $f(x) = ax^2 + bx + c$ $(a \neq 0)$ is called a quadratic function.

NEW SYMBOLS

$\pm$ positive or negative

i $i = \sqrt{-1}$ and $i^2 = -1$

CHAPTER 9 QUICK REVIEW

SECTION	CONCEPTS	EXAMPLES
9.1 SOLVING QUADRATIC EQUATIONS BY THE SQUARE ROOT PROPERTY	**Square Root Property of Equations** If b is positive, and if $a^2 = b$, then $a = \sqrt{b}$ or $a = -\sqrt{b}$.	Solve: $(2x + 1)^2 = 5$. $2x + 1 = \pm\sqrt{5}$ $2x = -1 \pm \sqrt{5}$ $x = \dfrac{-1 \pm \sqrt{5}}{2}$
9.2 SOLVING QUADRATIC EQUATIONS BY COMPLETING THE SQUARE	**Completing the Square** **1.** If the coefficient of the squared term is 1, go to Step 2. If it is not 1, divide each side of the equation by this coefficient. **2.** Make sure that all variable terms are on one side of the equation and all constant terms are on the other. **3.** Take half the coefficient of x, square it, and add the square to each side of the equation. Factor the variable side and combine terms on the other. **4.** Use the square root property to solve the equation.	Solve: $2x^2 + 4x - 1 = 0$. **1.** $x^2 + 2x - \dfrac{1}{2} = 0$ Divide by 2. **2.** $x^2 + 2x = \dfrac{1}{2}$ Add $\dfrac{1}{2}$. **3.** $x^2 + 2x + 1 = \dfrac{1}{2} + 1$ Add 1. $(x + 1)^2 = \dfrac{3}{2}$ Factor and add. **4.** $x + 1 = \pm\sqrt{\dfrac{3}{2}} = \pm\dfrac{\sqrt{6}}{2}$ $x = -1 \pm \dfrac{\sqrt{6}}{2} = \dfrac{-2 \pm \sqrt{6}}{2}$
9.3 SOLVING QUADRATIC EQUATIONS BY THE QUADRATIC FORMULA	**Quadratic Formula** The solutions of $ax^2 + bx + c = 0$, $(a \neq 0)$, are $x = \dfrac{-b \pm \sqrt{b^2 - 4ac}}{2a}$.	Solve: $3x^2 - 4x - 2 = 0$. $x = \dfrac{-(-4) \pm \sqrt{(-4)^2 - 4(3)(-2)}}{2(3)}$ $x = \dfrac{4 \pm \sqrt{16 + 24}}{6}$ $x = \dfrac{4 \pm \sqrt{40}}{6} = \dfrac{4 \pm 2\sqrt{10}}{6}$ $x = \dfrac{2(2 \pm \sqrt{10})}{6} = \dfrac{2 \pm \sqrt{10}}{3}$

SECTION	CONCEPTS	EXAMPLES
9.4 COMPLEX NUMBERS	For the positive number b, $$\sqrt{-b} = i\sqrt{b}.$$	Simplify: $\sqrt{-19}$. $$\sqrt{-19} = \sqrt{-1 \cdot 19} = i\sqrt{19}$$
	Add complex numbers by adding the real parts and adding the imaginary parts.	Add: $(3 + 6i) + (-9 + 2i)$. $$(3 + 6i) + (-9 + 2i) = (3 - 9) + (6 + 2)i = -6 + 8i$$
	To subtract complex numbers, change the number following the subtraction sign to its negative and add.	Subtract: $(5 + 4i) - (2 - 4i)$. $$(5 + 4i) - (2 - 4i) = (5 + 4i) + (-2 + 4i) = (5 - 2) + (4 + 4)i = 3 + 8i$$
	Multiply complex numbers in the same way polynomials are multiplied. Replace i^2 with -1.	Multiply: $(7 + i)(3 - 4i)$. $$(7 + i)(3 - 4i) = 7(3) + 7(-4i) + i(3) + i(-4i)$$ FOIL $$= 21 - 28i + 3i - 4i^2$$ $$= 21 - 25i - 4(-1)$$ $i^2 = -1$ $$= 21 - 25i + 4 = 25 - 25i$$
	Divide complex numbers by multiplying the numerator and the denominator by the conjugate of the denominator.	Divide: $\frac{2}{6 + i}$. $$\frac{2}{6 + i} = \frac{2}{6 + i} \cdot \frac{6 - i}{6 - i} = \frac{2(6 - i)}{36 - i^2} = \frac{12 - 2i}{36 + 1} = \frac{12 - 2i}{37} = \frac{12}{37} - \frac{2}{37}i$$
	A quadratic equation may have complex solutions. The quadratic formula will give complex solutions in such cases.	Solve for all complex solutions of $$x^2 + x + 1 = 0.$$ Here, $a = 1$, $b = 1$, and $c = 1$.

SECTION	CONCEPTS	EXAMPLES
		$x = \frac{-1 \pm \sqrt{1^2 - 4(1)(1)}}{2(1)}$ $x = \frac{-1 \pm \sqrt{1 - 4}}{2}$ $x = \frac{-1 \pm \sqrt{-3}}{2}$ $x = \frac{-1 \pm i\sqrt{3}}{2}$ The solutions are $-\frac{1}{2} + \frac{\sqrt{3}}{2}i$ and $-\frac{1}{2} - \frac{\sqrt{3}}{2}i$.
9.5 GRAPHING QUADRATIC EQUATIONS IN TWO VARIABLES	**Graphing $y = ax^2 + bx + c$** **1.** Find the y-intercept. **2.** Find any x-intercepts. **3.** Find the vertex: $x = -\frac{b}{2a}$; find y by substituting for x in the equation.	Graph $y = 2x^2 - 5x - 3$. **1.** $y = 2(0)^2 - 5(0) - 3 = -3$ The y-intercept is $(0, -3)$. **2.** $0 = 2x^2 - 5x - 3$ $0 = (2x + 1)(x - 3)$ $2x + 1 = 0$ or $x - 3 = 0$ $2x = -1$ or $x = 3$ $x = -\frac{1}{2}$ or $x = 3$ The x-intercepts are $\left(-\frac{1}{2}, 0\right)$ and $(3, 0)$. **3.** For the vertex: $x = -\frac{b}{2a} = -\frac{-5}{2(2)} = \frac{5}{4}$ $y = 2\left(\frac{5}{4}\right)^2 - 5\left(\frac{5}{4}\right) - 3$ $y = 2\left(\frac{25}{16}\right) - \frac{25}{4} - 3$ $y = \frac{25}{8} - \frac{50}{8} - \frac{24}{8} = -\frac{49}{8}$ $y = -6\frac{1}{8}$. The vertex is $\left(\frac{5}{4}, -\frac{49}{8}\right)$.

SECTION	CONCEPTS	EXAMPLES
	4. Plot the intercepts and the vertex. **5.** Find and plot additional ordered pairs near the vertex and intercepts as needed. The number of real solutions of the equation $ax^2 + bx + c = 0$ can be determined from the number of x-intercepts of the graph of $y = ax^2 + bx + c$.	**4.** and **5.** The figure shows that the equation $2x^2 - 5x - 3 = 0$ has 2 real solutions.

CHAPTER 9 REVIEW EXERCISES

[9.1] *Solve each equation by using the square root property. Express all radicals in simplest form.*

1. $y^2 = 49$ **2.** $x^2 = 15$ **3.** $m^2 = 48$ **4.** $(k + 2)^2 = 9$

5. $(r - 3)^2 = 7$ **6.** $(2p + 1)^2 = 11$ **7.** $(3k + 2)^2 = 12$ **8.** $\left(x + \frac{1}{2}\right)^2 = \frac{3}{4}$

9. The amount A that P dollars invested at a rate of interest r will amount to in 2 years is

$$A = P(1 + r)^2.$$

At what interest rate will \$1 grow to \$1.44?

10. Which of the following equations has two real solutions?
(a) $x^2 = 0$ **(b)** $x^2 = -4$
(c) $(x + 5)^2 = -16$ **(d)** $(x + 6)^2 = 25$

[9.2] *Solve each equation for real number solutions by completing the square.*

11. $m^2 + 6m + 5 = 0$ **12.** $p^2 + 4p = 7$ **13.** $-x^2 + 5 = 2x$

14. $2y^2 + 8y = 3$ **15.** $5k^2 - 3k - 2 = 0$ **16.** $(4a + 1)(a - 1) = -3$

17. What must be added to $x^2 + kx$ so that it will become a perfect square?

18. It takes Al one hour longer than Marcia to shovel snow from the sidewalk. Together it takes them $1\frac{3}{4}$ hours. How long would it take each person to do the job working alone?

[9.3] *Use the quadratic formula to solve each equation for real number solutions.*

19. $x^2 - 2x - 4 = 0$

20. $-m^2 + 3m + 5 = 0$

21. $3k^2 + 2k + 3 = 0$

22. $5p^2 = p + 1$

23. $2p^2 - 3 = 4p$

24. $-4a^2 + 7 = 2a$

25. $\frac{c^2}{4} = 2 - \frac{3}{4}c$

26. $\frac{3}{2}r^2 = \frac{1}{2} - r$

---

27. Why is it impossible to solve the equation

$$0x^2 - 4x + 5 = 0$$

by the quadratic formula?

28. An object is moving in such a manner that at time t its position in *inches* from the origin is given by $d = t^2 + 5t + 1$, where t is in seconds. After how many seconds is the object 9 *feet* from the origin? Express the answer to the nearest hundredth of a second.

[9.4] *Perform the indicated operation.*

29. $(3 + 5i) + (2 - 6i)$

30. $(-2 - 8i) - (4 - 3i)$

31. $(-1 + i) - (2 - i)$

32. $(4 + 3i) + (-2 + 3i)$

33. $(6 - 2i)(3 + i)$

34. $(2 + 3i)(4 - 2i)$

35. $(5 + 2i)(5 - 2i)$

36. $(8 - i)(8 + i)$

37. $\frac{1 + i}{1 - i}$

38. $\frac{3 + 2i}{1 + i}$

39. $\frac{2 - 4i}{3 + i}$

40. $\frac{5 + 6i}{2 + 3i}$

---

41. What is the conjugate of the real number a?

42. Is it possible to multiply a complex number by its conjugate and get an imaginary product? Explain.

Find the complex solutions of each quadratic equation.

43. $(m + 2)^2 = -3$

44. $(x - 1)^2 = -2$

45. $(3p - 2)^2 = -8$

46. $(4p + 1)^2 = -12$

47. $3k^2 = 2k - 1$

48. $h^2 + 3h = -8$

49. $4q^2 + 2 = 3q$

50. $9z^2 + 2z + 1 = 0$

[9.5] *Sketch the graph of each equation. Identify each vertex.*

51. $y = 3x^2$

52. $y = x^2 - 2x + 1$

53. $y = -x^2 + 3$

54. $y = x^2 - 1$

55. $y = -x^2 + 2x + 3$

56. $y = x^2 + 4x + 1$

Decide from the graph how many real number solutions the corresponding equation has. Find any real solutions from the graph.

57.

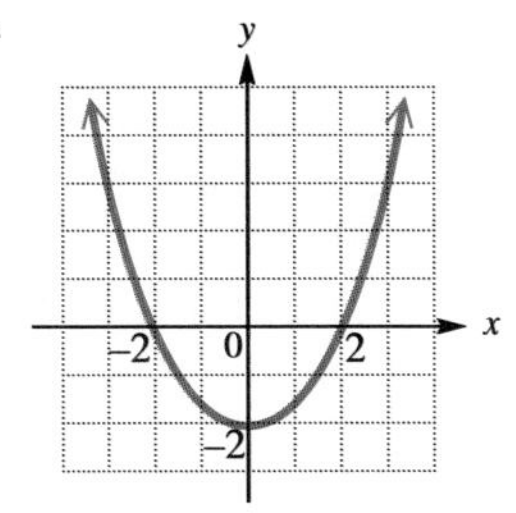

58.

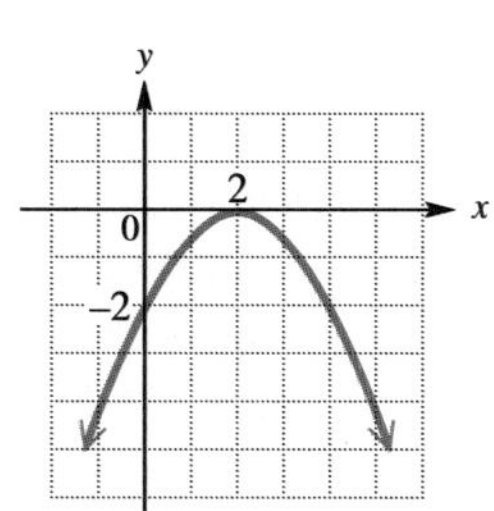

59.

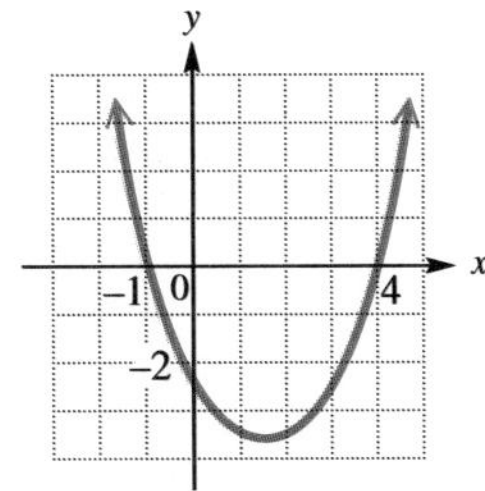

60.

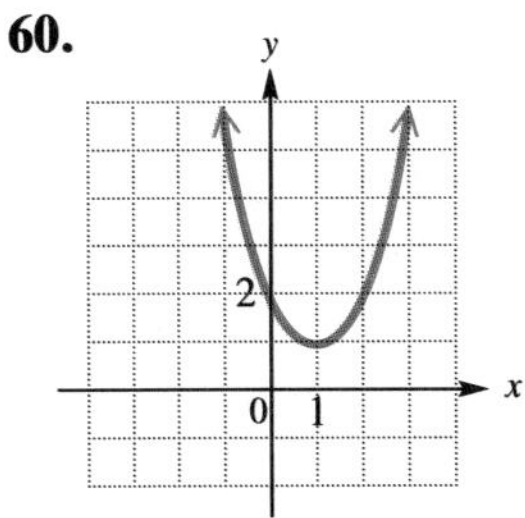

MIXED REVIEW EXERCISES

Solve each equation.

61. $(2t - 1)(t + 5) = 0$

62. $(2p + 1)^2 = 4$

63. $(k + 2)(k - 1) = 3$

64. $6t^2 + 7t - 3 = 0$

65. $2x^2 + 3x + 2 = x^2 - 2x$

66. $x^2 + 2x + 1 = 3$

67. $(3x + 5)(3x + 5) = 7$

68. $\frac{1}{2}r^2 = \frac{7}{2} - r$

69. $\frac{2}{m} - 3 = -\frac{1}{m^2}$

70. $(2y - 3)^2 = 7$

71. $x^2 + 4x - 1 = 0$

72. $7x^2 - 8 = 5x^2 + 16$

CHAPTER 9 TEST

In the equations for Exercises 1–15, find only real number solutions.

Solve by using the square root property.

1. $x^2 - 5 = 0$

2. $(k - 3)^2 = 49$

3. $(3r - 2)^2 = 35$

Solve by completing the square.

4. $x^2 + 4x - 2 = 0$

5. $5x^2 = 2 - 9x$

Solve by the quadratic formula.

6. $m^2 = 3m + 10$

7. $3z^2 + 2 = 7z$

8. $y^2 - \frac{5}{3}y + \frac{1}{3} = 0$

Solve by the method of your choice.

9. $m^2 - 2m = 1$

10. $(2x - 1)^2 = 18$

11. $(x - 5)(3x + 2) = 0$

12. $(q - 5)(3q - 2) = 4$

13. $(x - 5)(x - 5) = 8$

14. $y^2 = 6y - 2$

---

15. If an object is thrown upward from ground level with an initial velocity of 64 feet per second, then its height after t seconds is given by

$$h = -16t^2 + 64t,$$

where h is in feet. After how many seconds will the object reach a height of 64 feet?

Perform the indicated operations.

16. $(3 + i) + (5 - 2i) - (1 + i)$

17. $(4 - 3i)(6 + i)$

18. $(2 + 5i)(2 - 5i)$

19. $\frac{1 + 2i}{3 - i}$

Find the complex solutions of the following quadratic equations.

20. $p^2 = 5p - 7$

21. $3m^2 + 1 = 2m$

Sketch the graph and identify the vertex of each parabola in Exercises 22–24.

22. $y = -x^2 - 6x - 9$

23. $y = x^2 + 4$

24. $y = x^2 + 6x + 7$

---

25. From your answer to Exercise 24, determine how many real number solutions there are for the equation $x^2 + 6x + 7 = 0$.

APPENDIX A GEOMETRY REVIEW AND FORMULAS

SPECIAL ANGLES

NAME	CHARACTERISTIC	EXAMPLES
RIGHT ANGLE	Measure is 90°.	
STRAIGHT ANGLE	Measure is 180°.	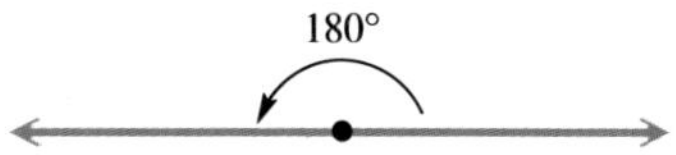 180°
COMPLEMENTARY ANGLES	The sum of the measures of two complementary angles is 90°.	Angles 1 and 2 are complementary. 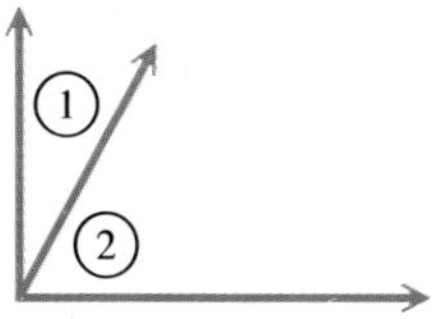
SUPPLEMENTARY ANGLES	The sum of the measures of two supplementary angles is 180°.	Angles 3 and 4 are supplementary. 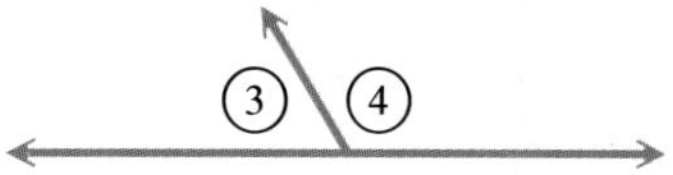
VERTICAL ANGLES	Vertical angles have equal measures.	Angle 2 = Angle 4 Angle 1 = Angle 3 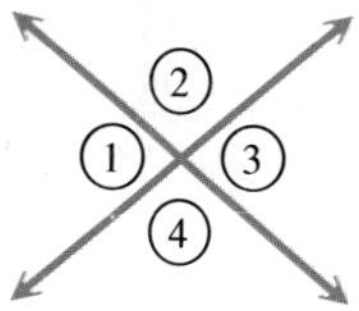
ANGLES FORMED BY PARALLEL LINES AND A TRANSVERSAL		m and n are parallel. 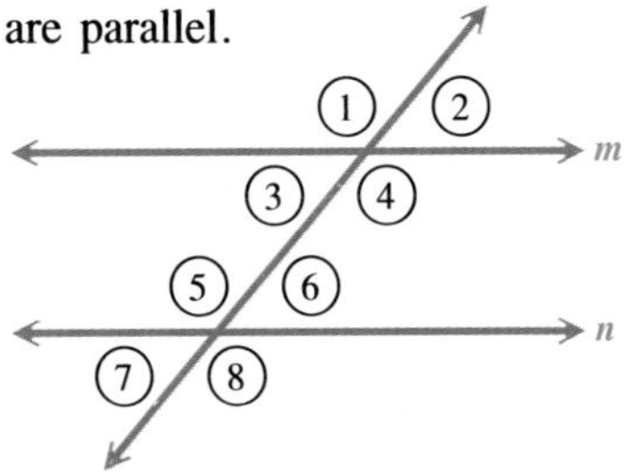
ALTERNATE INTERIOR ANGLES	Measures are equal.	Angle 3 = Angle 6
ALTERNATE EXTERIOR ANGLES	Measures are equal.	Angle 1 = Angle 8
INTERIOR ANGLES ON THE SAME SIDE	Angles are supplementary.	Angles 3 and 5 are supplementary.

SPECIAL TRIANGLES

NAME	CHARACTERISTIC	EXAMPLES
RIGHT TRIANGLE	Triangle has a right angle.	90°
ISOSCELES TRIANGLE	Triangle has two equal sides.	$AB = BC$
EQUILATERAL TRIANGLE	Triangle has three equal sides.	$AB = BC = CA$
SIMILAR TRIANGLES	Corresponding angles are equal; corresponding sides are proportional.	$A = D$, $B = E$, $C = F$ $\dfrac{AB}{DE} = \dfrac{AC}{DF} = \dfrac{BC}{EF}$

FORMULAS

FIGURE	FORMULAS	EXAMPLES
SQUARE	Perimeter: $P = 4s$ Area: $A = s^2$	

FIGURE	FORMULAS	EXAMPLES
RECTANGLE	Perimeter: $P = 2L + 2W$ Area: $A = LW$	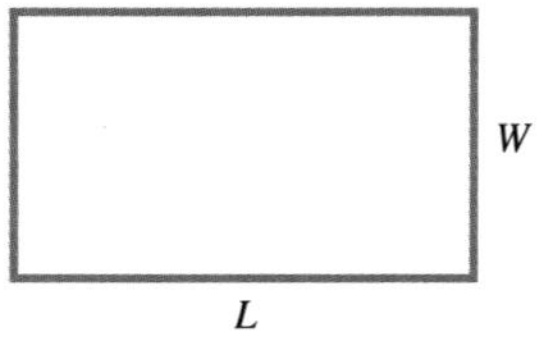 *W*, *L*
TRIANGLE	Perimeter: $P = a + b + c$ Area: $A = \frac{1}{2}bh$	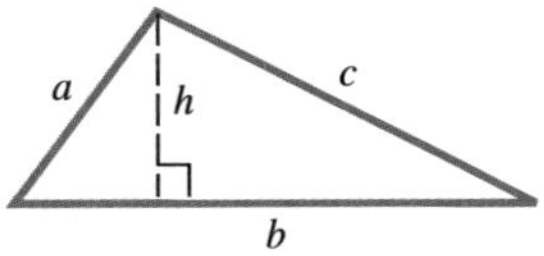 *a*, *c*, *h*, *b*
PYTHAGOREAN FORMULA (for right triangles)	In a right triangle with legs a and b and hypotenuse c, $c^2 = a^2 + b^2$.	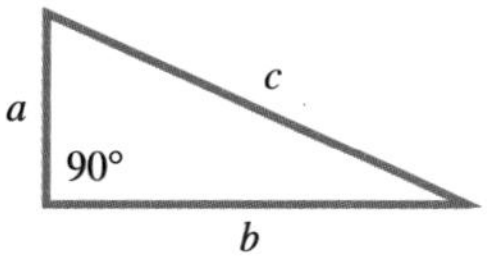 *a*, *c*, 90°, *b*
SUM OF THE ANGLES OF A TRIANGLE	$A + B + C = 180°$	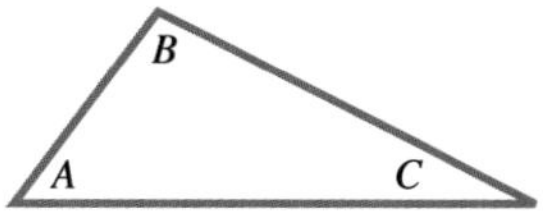 *B*, *A*, *C*
CIRCLE	Diameter: $d = 2r$ Circumference: $C = 2\pi r$ $C = \pi d$ Area: $A = \pi r^2$	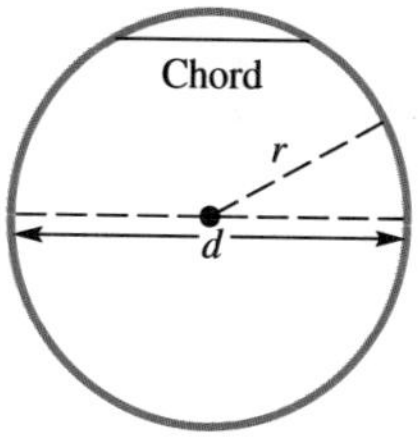 Chord, *r*, *d*
PARALLELOGRAM	Area: $A = bh$ Perimeter: $P = 2a + 2b$	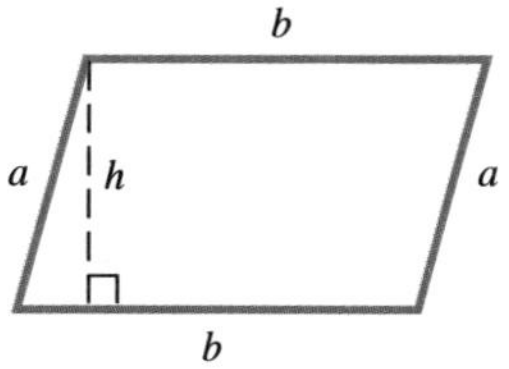 *b*, *a*, *h*, *a*, *b*
TRAPEZOID	Area: $A = \frac{1}{2}(B + b)h$ Perimeter: $P = a + b + c + B$	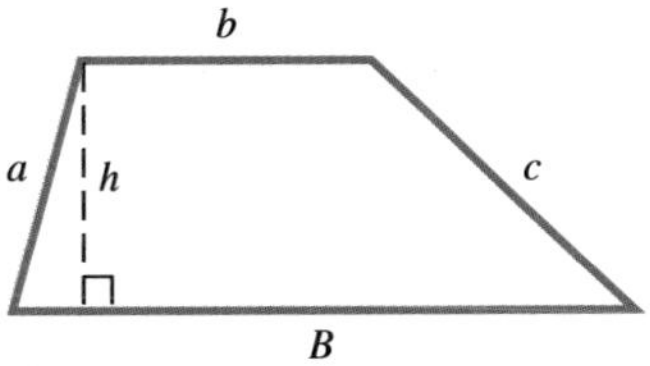 *b*, *a*, *h*, *c*, *B*

FIGURE	FORMULAS	EXAMPLES
SPHERE	Volume: $V = \frac{4}{3}\pi r^3$ Surface area: $S = 4\pi r^2$	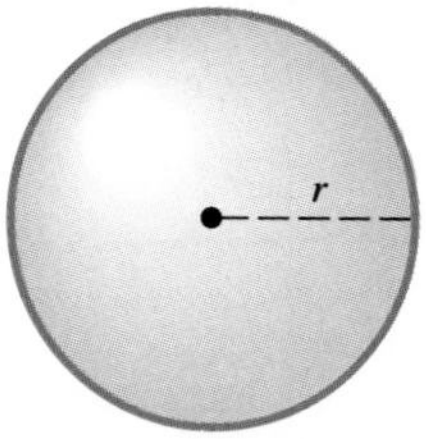 r
CONE	Volume: $V = \frac{1}{3}\pi r^2 h$ Surface area: $S = \pi r\sqrt{r^2 + h^2}$	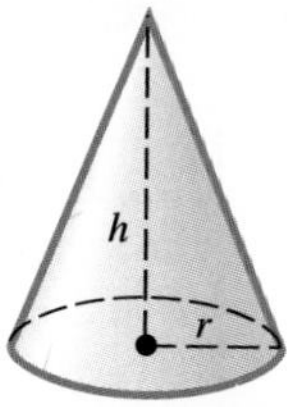 h, r
CUBE	Volume: $V = e^3$ Surface area: $S = 6e^2$	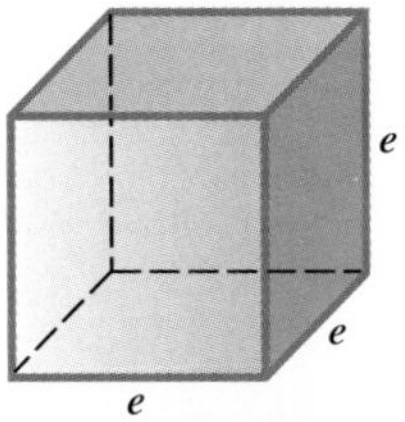 e, e, e
RECTANGULAR SOLID	Volume: $V = LWH$ Surface area: $A = 2HW + 2LW + 2LH$	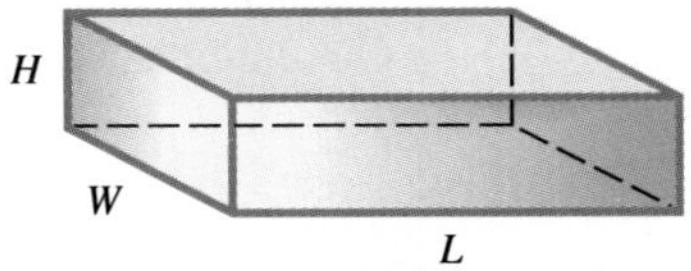 H, W, L
RIGHT CIRCULAR CYLINDER	Volume: $V = \pi r^2 h$ Surface area: $S = 2\pi rh + 2\pi r^2$	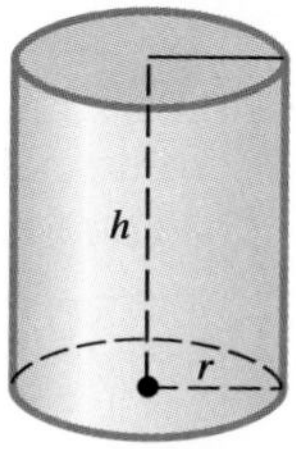 h, r
RIGHT PYRAMID	Volume: $V = \frac{1}{3}Bh$ B = area of the base	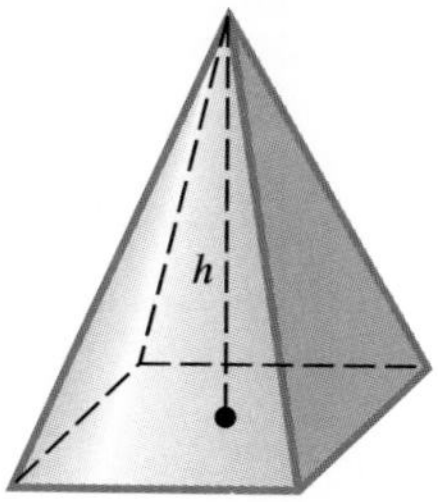 h

APPENDIX B OTHER FORMULAS

Distance: $d = rt$; r = rate or speed, t = time

Percent: $p = br$; p = percentage, b = base, r = rate

Temperature: $F = \frac{9}{5}C + 32$ $\qquad C = \frac{5}{9}(F - 32)$

Simple Interest: $I = prt$; p = principal or amount invested, r = rate or percent, t = time in years

APPENDIX C REVIEW OF DECIMALS AND PERCENTS

OBJECTIVES

1 ADD AND SUBTRACT DECIMALS.
2 MULTIPLY AND DIVIDE DECIMALS.
3 CONVERT PERCENTS TO DECIMALS AND DECIMALS TO PERCENTS.
4 FIND PERCENTAGES BY MULTIPLICATION.

1 A **decimal** is a number written with a decimal point, such as 4.2. The operations on decimals—addition, subtraction, multiplication, and division—are explained in the next examples.

EXAMPLE 1 ADDING AND SUBTRACTING DECIMALS

Add or subtract as indicated.

(a) $6.92 + 14.8 + 3.217$

Place the numbers in a column, with decimal points lined up, then add. If you like, attach 0s to make all the numbers the same length; this is a good way to avoid errors. For example,

Decimal points lined up

$$\begin{array}{r} 6.92 \\ 14.8 \\ +\ 3.217 \\ \hline 24.937 \end{array} \quad \text{or} \quad \begin{array}{r} 6.920 \\ 14.800 \\ +\ 3.217 \\ \hline 24.937. \end{array}$$

(b) $47.6 - 32.509$

Write the numbers in a column, attaching zeros to 47.6.

$$\begin{array}{r} 47.6 \\ -32.509 \\ \hline \end{array} \quad \text{becomes} \quad \begin{array}{r} 47.600 \\ -32.509 \\ \hline 15.091 \end{array}$$

(c) 3 − .253

$$\begin{array}{r} 3.000 \\ -\ \ .253 \\ \hline 2.747 \end{array}$$ ■

2 Multiplication and division of decimals are explained next.

EXAMPLE 2 MULTIPLYING DECIMALS

Multiply.

(a) 29.3 × 4.52

Multiply as if the numbers were whole numbers. The number of decimal places in the answer is found by adding the numbers of decimal places in the factors.

$$\begin{array}{r} 29.3 \\ \times 4.52 \\ \hline 5\ 86 \\ 14\ 6\ 5 \\ 117\ 2 \\ \hline 132.4\ 36 \end{array}$$

1 decimal place
2 decimal places
3 decimal places in answer

(b) 7.003 × 55.8

$$\begin{array}{r} 7.003 \\ \times\ \ 55.8 \\ \hline 5\ 602\ 4 \\ 35\ 015 \\ 350\ 15 \\ \hline 390.767\ 4 \end{array}$$

3 decimal places
1 decimal place
4 decimal places in answer ■

EXAMPLE 3 DIVIDING DECIMALS

Divide: 279.45 ÷ 24.3.

Move the decimal point in 24.3 one place to the right, to get 243. Move the decimal point the same number of places in 279.45. By doing this, 24.3 is converted into the whole number 243.

$$24\,3.\overline{)279\,4.5}$$

Bring the decimal point straight up and divide as with whole numbers.

$$\begin{array}{r} 11.5 \\ 243\overline{)2794.5} \\ 243 \\ \hline 364 \\ 243 \\ \hline 121\ 5 \\ 121\ 5 \\ \hline 0 \end{array}$$ ■

3 One of the main uses of decimals comes from percent problems. The word **percent** means "per one hundred." Percent is written with the sign %. One percent means "one per one hundred."

PERCENT

$$1\% = .01 \quad \text{or} \quad 1\% = \frac{1}{100}$$

EXAMPLE 4
CONVERTING BETWEEN DECIMALS AND PERCENTS

Convert.

(a) 75% to a decimal
Since $1\% = .01$,

$$75\% = 75 \cdot 1\% = 75 \cdot (.01) = .75.$$

The fraction form $1\% = 1/100$ can also be used to convert 75% to a decimal.

$$75\% = 75 \cdot 1\% = 75 \cdot \left(\frac{1}{100}\right) = .75$$

(b) 2.63 to a percent

$$2.63 = 263 \cdot (.01) = 263 \cdot 1\% = 263\% \quad \blacksquare$$

4 A part of a whole is called a **percentage.** For example, since 50% represents $50/100 = 1/2$ of a whole, 50% of 800 is half of 800, or 400. Percentages are found by multiplication, as in the next example.

EXAMPLE 5
FINDING PERCENTAGES

Find the percentages.

(a) 15% of 600
The word *of* indicates multiplication here. For this reason, 15% of 600 is found by multiplying.

$$15\% \cdot 600 = (.15) \cdot 600 = 90$$

(b) 125% of 80

$$125\% \cdot 80 = (1.25) \cdot 80 = 100 \quad \blacksquare$$

APPENDIX C EXERCISES

Perform the indicated operations. See Examples 1–3.

1. $14.23 + 9.81 + 74.63 + 18.715$

2. $89.416 + 21.32 + 478.91 + 298.213$

3. $19.74 - 6.53$

4. $27.96 - 8.39$

5. $219 - 68.51$

6. $283 - 12.42$

7. $\begin{array}{r} 48.96 \\ 37.421 \\ +\ 9.72 \\ \hline \end{array}$

8. $\begin{array}{r} 9.71 \\ 4.8 \\ 3.6 \\ 5.2 \\ +8.17 \\ \hline \end{array}$

9. $\begin{array}{r} 8.6 \\ -3.751 \\ \hline \end{array}$

10. $\begin{array}{r} 27.8 \\ -13.582 \\ \hline \end{array}$

11. 39.6×4.2
12. 18.7×2.3
13. 42.1×3.9
14. 19.63×4.08
15. $.042 \times 32$
16. 571×2.9
17. $24.84 \div 6$
18. $32.84 \div 4$
19. $7.6266 \div 3.42$
20. $14.9202 \div 2.43$
21. $2496 \div .52$
22. $.56984 \div .034$

Convert the following percents to decimals. See Example 4(a).

23. 53%
24. 38%
25. 129%
26. 174%
27. 96%
28. 11%
29. .9%
30. .1%

Convert the following decimals to percents. See Example 4(b).

31. .80
32. .75
33. .007
34. 1.4
35. .67
36. .003
37. .125
38. .983

Answer each of the following. Round your answers to the nearest hundredth. See Example 5.

39. What is 14% of 780?
40. Find 12% of 350.
41. Find 22% of 1086.
42. What is 20% of 1500?
43. 4 is what percent of 80?
44. 1300 is what percent of 2000?
45. What percent of 5820 is 6402?
46. What percent of 75 is 90?
47. 121 is what percent of 484?
48. What percent of 3200 is 64?
49. Find 118% of 125.8.
50. Find 3% of 128.
51. What is 91.72% of 8546.95?
52. Find 12.741% of 58.902.
53. What percent of 198.72 is 14.68?
54. 586.3 is what percent of 765.4?

Solve each problem. Formulas can be found in Appendix B.

55. A retailer has $23,000 invested in her business. She finds that she is earning 12% per year on this investment. How much money is she earning per year?

56. Harley Dabler recently bought a duplex for $144,000. He expects to earn 16% per year on the purchase price. How many dollars per year will he earn?

57. For a recent tour of the eastern United States, a travel agent figured that the trip totaled 2300 miles, with 35% of the trip by air. How many miles of the trip were by air?

58. Capitol Savings Bank pays 8.9% interest per year. What is the annual interest on an account of $3000?

59. An ad for steel-belted radial tires promises 15% better mileage when the tires are used. Alexandria's Escort now goes 420 miles on a tank of gas. If she switched to the new tires, how many extra miles could she drive on a tank of gas?

60. A home worth $77,000 is located in an area where home prices are increasing at a rate of 12% per year. By how much would the value of this home increase in one year?

61. A family of four with a monthly income of $2000 spends 90% of its earnings and saves the rest. Find the *annual* savings of this family.

APPENDIX D SETS

OBJECTIVES

1 LIST THE ELEMENTS OF A SET.

2 LEARN THE VOCABULARY AND SYMBOLS USED TO DISCUSS SETS.

3 DECIDE WHETHER A SET IS FINITE OR INFINITE.

4 DECIDE WHETHER A GIVEN SET IS A SUBSET OF ANOTHER SET.

5 FIND THE COMPLEMENT OF A SET.

6 FIND THE UNION AND THE INTERSECTION OF TWO SETS.

1 A **set** is a collection of things. The objects in a set are called the **elements** of the set. A set is represented by listing its elements between **set braces,** { }. The order in which the elements of a set are listed is unimportant.

EXAMPLE 1 **LISTING THE ELEMENTS OF A SET**

Represent the following sets by listing the elements.

(a) The set of states in the United States that border on the Pacific Ocean = {California, Oregon, Washington, Hawaii, Alaska}.

(b) The set of all counting numbers less than 6 = {1, 2, 3, 4, 5}. ■

2 Capital letters are used to name sets. To state that 5 is an element of

$$S = \{1, 2, 3, 4, 5\},$$

write $5 \in S$. The statement $6 \notin S$ means that 6 is not an element of S.

A set with no elements is called the **empty set,** or the **null set.** The symbols $\emptyset$ or { } are used for the empty set. If we let A be the set of all cats that fly, then A is the empty set.

$$A = \emptyset \qquad \text{or} \qquad A = \{\ \}$$

CAUTION Do not make the common error of writing the empty set as $\{\emptyset\}$.

In any discussion of sets, there is some set that includes all the elements under consideration. This set is called the **universal set** for that situation. For example, if the discussion is about presidents of the United States, then the set of all presidents of the United States is the universal set. The universal set is denoted U.

3 In Example 1, there are five elements in the set in part (a), and five in part (b). If the number of elements in a set is either 0 or a counting number, then the set is **finite.** On the other hand, the set of natural numbers, for example, is an **infinite set,** because there is no final number. We can list the elements of the set of natural numbers as

$$N = \{1, 2, 3, 4 \ldots\}$$

where the three dots indicate that the set continues indefinitely. Not all infinite sets can be listed in this way. For example, there is no way to list the elements in the set of all real numbers between 1 and 2.

EXAMPLE 2
DISTINGUISHING BETWEEN FINITE AND INFINITE SETS

List the elements of each set, if possible. Decide whether each set is finite or infinite.

(a) The set of all integers
One way to list the elements is $\{\ldots, -2, -1, 0, 1, 2, \ldots\}$. The set is infinite.

(b) The set of all natural numbers between 0 and 5
$\{1, 2, 3, 4\}$ The set is finite.

(c) The set of all irrational numbers
This is an infinite set whose elements cannot be listed. ■

Two sets are **equal** if they have exactly the same elements. Thus, the set of natural numbers and the set of positive integers are equal sets. Also, the sets

$$\{1, 2, 4, 7\} \quad \text{and} \quad \{4, 2, 7, 1\}$$

are equal. The order of the elements does not make a difference.

4 If all elements of a set A are also elements of a new set B, then we say A is a **subset** of B, written $A \subseteq B$. We use the symbol $A \nsubseteq B$ to mean that A is not a subset of B.

EXAMPLE 3
USING SUBSET NOTATION

Let $A = \{1, 2, 3, 4\}$, $B = \{1, 4\}$, and $C = \{1\}$. Then $B \subseteq A$, $C \subseteq A$, and $C \subseteq B$, but $A \nsubseteq B$, $A \nsubseteq C$, and $B \nsubseteq C$. ■

The set $M = \{a, b\}$ has four subsets: $\{a, b\}$, $\{a\}$, $\{b\}$, and $\emptyset$. The empty set is defined to be a subset of any set. How many subsets does $N = \{a, b, c\}$ have? There is one subset with 3 elements: $\{a, b, c\}$. There are three subsets with 2 elements:

$$\{a, b\}, \quad \{a, c\}, \quad \text{and} \quad \{b, c\}.$$

There are three subsets with 1 element:

$$\{a\}, \quad \{b\}, \quad \text{and} \quad \{c\}.$$

There is one subset with 0 elements: $\emptyset$. Thus, set N has eight subsets.

The following generalization can be made.

NUMBER OF SUBSETS OF A SET

A set with n elements has 2^n subsets.

To illustrate the relationships between sets, **Venn diagrams** are often used. A rectangle represents the universal set, U. The sets under discussion are represented by regions within the rectangle. The Venn diagram in Figure 1 shows that $B \subseteq A$.

5 For every set A, there is a set A', the **complement** of A, that contains all the elements of U that are not in A. The shaded region in the Venn diagram in Figure 2 represents A'.

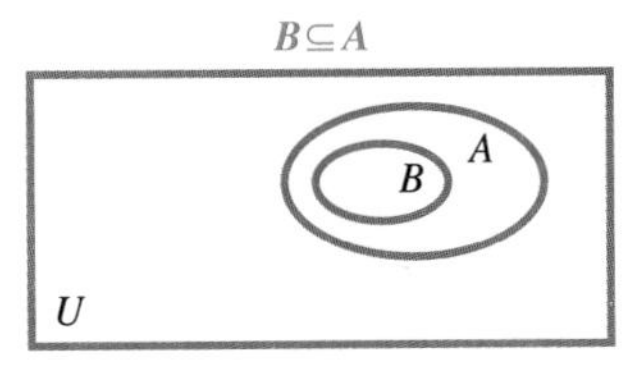

FIGURE 1

A′ is in color.

FIGURE 2

EXAMPLE 4 **DETERMINING THE COMPLEMENT OF A SET**

Given $U = \{a, b, c, d, e, f, g\}$, $A = \{a, b, c\}$, $B = \{a, d, f, g\}$, and $C = \{d, e\}$, find A', B', and C'.

(a) $A' = \{d, e, f, g\}$

(b) $B' = \{b, c, e\}$

(c) $C' = \{a, b, c, f, g\}$ ■

6 The **union** of two sets A and B, written $A \cup B$, is the set of all elements of A together with all elements of B. Thus, for the sets in Example 4,

$$A \cup B = \{a, b, c, d, f, g\}$$

and

$$A \cup C = \{a, b, c, d, e\}.$$

In Figure 3 the shaded region is the union of sets A and B.

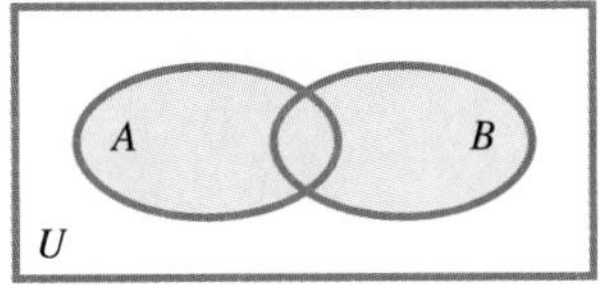

FIGURE 3

EXAMPLE 5 **FINDING THE UNION OF TWO SETS**

If $M = \{2, 5, 7\}$ and $N = \{1, 2, 3, 4, 5\}$, then

$$M \cup N = \{1, 2, 3, 4, 5, 7\}. \quad ■$$

The **intersection** of two sets A and B, written $A \cap B$, is the set of all elements that belong to both A and B. For example, if

$$A = \{\text{Jose, Ellen, Marge, Kevin}\}$$

and

$$B = \{\text{Jose, Patrick, Ellen, Sue}\},$$

then

$$A \cap B = \{\text{Jose, Ellen}\}.$$

The shaded region in Figure 4 represents the intersection of the two sets A and B.

EXAMPLE 6
FINDING THE INTERSECTION OF TWO SETS

Suppose that $P = \{3, 9, 27\}$, $Q = \{2, 3, 10, 18, 27, 28\}$, and $R = \{2, 10, 28\}$.

(a) $P \cap Q = \{3, 27\}$ **(b)** $Q \cap R = \{2, 10, 28\} = R$ **(c)** $P \cap R = \emptyset$ ■

Sets like P and R in Example 6 that have no elements in common are called **disjoint sets.** The Venn diagram in Figure 5 shows a pair of disjoint sets.

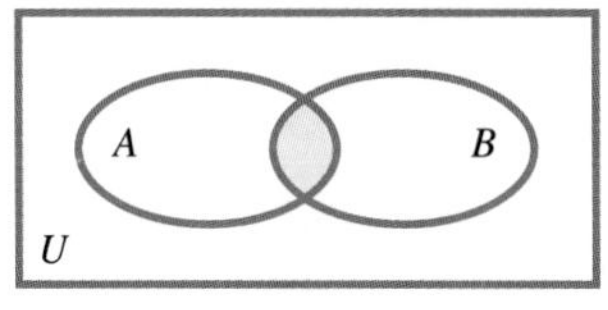

FIGURE 4

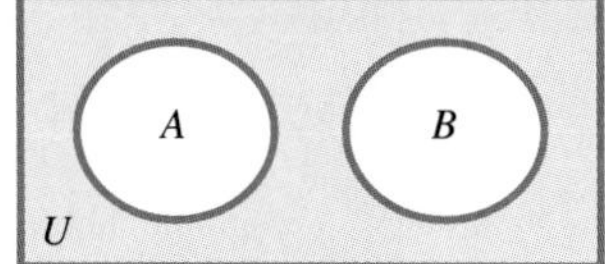

FIGURE 5

EXAMPLE 7
USING SET OPERATIONS

Let $U = \{2, 5, 7, 10, 14, 20\}$, $A = \{2, 10, 14, 20\}$, $B = \{5, 7\}$, and $C = \{2, 5, 7\}$. Find each of the following.

(a) $A \cup B = \{2, 5, 7, 10, 14, 20\} = U$ **(b)** $A \cap B = \emptyset$

(c) $B \cup C = \{2, 5, 7\} = C$ **(d)** $B \cap C = \{5, 7\} = B$

(e) $A' = \{5, 7\} = B$ ■

APPENDIX D EXERCISES

List the elements of each of the following sets. See Examples 1 and 2.

1. The set of all natural numbers less than 8

2. The set of all integers between 4 and 10

3. The set of seasons

4. The set of months of the year

5. The set of women presidents of the United States

6. The set of all living humans who are more than 200 years old

7. The set of letters of the alphabet between K and M

8. The set of letters of the alphabet between D and H

9. The set of positive even integers

10. The set of all multiples of 5

11. Which of the sets described in Exercises 1–10 are infinite sets?

12. Which of the sets described in Exercises 1–10 are finite sets?

Tell whether each of the following is true or false.

13. $5 \in \{1, 2, 5, 8\}$

14. $6 \in \{1, 2, 3, 4, 5\}$

15. $2 \in \{1, 3, 5, 7, 9\}$

16. $1 \in \{6, 2, 5, 1\}$

17. $7 \notin \{2, 4, 6, 8\}$

18. $7 \notin \{1, 3, 5, 7\}$

19. $\{2, 4, 9, 12, 13\} = \{13, 12, 9, 4, 2\}$

20. $\{7, 11, 4\} = \{7, 11, 4, 0\}$

Let $A = \{1, 3, 4, 5, 7, 8\}$
$B = \{2, 4, 6, 8\}$
$C = \{1, 3, 5, 7\}$
$D = \{1, 2, 3\}$
$E = \{3, 7\}$
$U = \{1, 2, 3, 4, 5, 6, 7, 8, 9, 10\}$.

Tell whether each of the following is true or false. See Examples 3, 5, 6, and 7.

21. $A \subseteq U$

22. $D \subseteq A$

23. $\emptyset \subseteq A$

24. $\{1, 2\} \subseteq D$

25. $C \subseteq A$

26. $A \subseteq C$

27. $D \subseteq B$

28. $E \subseteq C$

29. $D \not\subseteq E$

30. $E \not\subseteq A$

31. There are exactly 4 subsets of E.

32. There are exactly 8 subsets of D.

33. There are exactly 12 subsets of C.

34. There are exactly 16 subsets of B.

35. $\{4, 6, 8, 12\} \cap \{6, 8, 14, 17\} = \{6, 8\}$

36. $\{2, 5, 9\} \cap \{1, 2, 3, 4, 5\} = \{2, 5\}$

37. $\{3, 1, 0\} \cap \{0, 2, 4\} = \{0\}$

38. $\{4, 2, 1\} \cap \{1, 2, 3, 4\} = \{1, 2, 3\}$

39. $\{3, 9, 12\} \cap \emptyset = \{3, 9, 12\}$

40. $\{3, 9, 12\} \cup \emptyset = \emptyset$

41. $\{4, 9, 11, 7, 3\} \cup \{1, 2, 3, 4, 5\}$
$= \{1, 2, 3, 4, 5, 7, 9, 11\}$

42. $\{1, 2, 3\} \cup \{1, 2, 3\} = \{1, 2, 3\}$

43. $\{3, 5, 7, 9\} \cup \{4, 6, 8\} = \emptyset$

44. $\{5, 10, 15, 20\} \cup \{5, 15, 30\} = \{5, 15\}$

Let $U = \{a, b, c, d, e, f, g, h\}$
$A = \{a, b, c, d, e, f\}$
$B = \{a, c, e\}$
$C = \{a, f\}$
$D = \{d\}$.

List the elements in the following sets. See Examples 4–7.

45. A'

46. B'

47. C'

48. D'

49. $A \cap B$

50. $B \cap A$

51. $A \cap D$

52. $B \cap D$

53. $B \cap C$

54. $A \cup B$

55. $B \cup D$

56. $B \cup C$

57. $C \cup B$

58. $C \cup D$

59. $A \cap \emptyset$

60. $B \cup \emptyset$

61. Name every pair of disjoint sets among A–D above.

ANSWERS TO SELECTED EXERCISES

TO THE STUDENT

If you need further help with algebra, you may want to obtain a copy of the *Student's Solutions Manual* that goes with this book. It contains solutions to all the odd-numbered exercises and all the chapter test exercises. You also may want the *Student's Study Guide*. It has extra examples and exercises to complete, corresponding to each learning objective of the book. In addition, there is a practice test for each chapter. Your college bookstore either has these books or can order them for you.

In this section we provide the answers that we think most students will obtain when they work the exercises using the methods explained in the text. If your answer does not look exactly like the one given here, it is not necessarily wrong. In many cases there are equivalent forms of the answer that are correct. For example, if the answer section shows 3/4 and your answer is .75, you have obtained the right answer but written it in a different (yet equivalent) form. Unless the directions specify otherwise, .75 is just as valid an answer as 3/4.

In general, if your answer does not agree with the one given in the text, see whether it can be transformed into the other form. If it can, then it is the correct answer. If you still have doubts, talk with your instructor.

CHAPTER 1

1.1 Exercises (page 8)

1. $2 \cdot 3 \cdot 5$ **3.** $2 \cdot 5 \cdot 5$ **5.** $5 \cdot 13$ **7.** $2 \cdot 2 \cdot 5 \cdot 5$ **9.** 17 **11.** $2 \cdot 2 \cdot 31$ **13.** $\frac{1}{2}$ **15.** $\frac{5}{6}$ **17.** $\frac{8}{9}$ **19.** $\frac{2}{3}$ **21.** $\frac{2}{3}$ **23.** (c) **25.** $\frac{9}{20}$ **27.** $\frac{3}{25}$ **29.** $\frac{6}{5}$ **31.** $\frac{3}{10}$ **33.** $\frac{1}{9}$ **35.** $\frac{3}{20}$ **37.** $\frac{7}{18}$ **39.** $2\frac{1}{6}$ **41.** 8 **43.** $\frac{2}{3}$ **45.** $\frac{1}{2}$ **47.** $\frac{4}{5}$ **49.** $\frac{10}{9}$ **51.** $\frac{19}{22}$ **53.** $\frac{1}{15}$ **55.** $\frac{8}{15}$ **57.** $3\frac{3}{8}$ **59.** $8\frac{1}{6}$ **61.** $1\frac{5}{12}$ **63.** $\frac{49}{30}$ **65.** $\frac{26}{63}$ **67.** $2\left(\frac{10}{100}\right) + 3\left(\frac{10}{100}\right) = \frac{50}{100}$ **69.** $\frac{1}{3}$ cup **71.** $14\frac{7}{16}$ tons **73.** $8\frac{23}{24}$ hours **75.** $\frac{9}{16}$ inch

1.2 Exercises (page 16)

1. 36 **3.** 64 **5.** 289 **7.** 125 **9.** 1296 **11.** 32 **13.** 729 **15.** $\frac{1}{4}$ **17.** $\frac{8}{125}$ **19.** $\frac{64}{125}$ **21.** Yes. For example, $\left(\frac{1}{2}\right)^2 = \frac{1}{4}$, and $\frac{1}{4} < \frac{1}{2}$. **23.** Multiply $4 \cdot 3$ to get 12. Then add the 9. If no parentheses or other grouping symbols are present, multiplication is performed before addition. **25.** 16 **27.** 2 **29.** 0

31. 60 **33.** 65 **35.** 15 **37.** $\frac{10}{3}$ **39.** 66 **41.** $\neq, <, \leq$ **43.** $\neq, \geq, >$ **45.** $\leq, \geq$ **47.** $\neq, \geq, >$ **49.** $\neq, >, \geq$ **51.** $\neq, <, \leq$ **53.** $\neq, <, \leq$ **55.** $\neq, <, \leq$ **57.** One person *is younger than* another. **59.** true **61.** true **63.** true **65.** true **67.** false **69.** false **71.** $66 > 72$; false **73.** $2 \geq 3$; false **75.** $3 \geq 3$; true **77.** $7 = 5 + 2$ **79.** $3 < \frac{50}{5}$ **81.** $12 \neq 5$ **83.** $0 \geq 0$ **85.** $14 > 6$ **87.** $3 \leq 15$ **89.** $17 > 12$

1.3 Exercises (page 22)

1. **(a)** 12 **(b)** 24 **3.** **(a)** 15 **(b)** 75 **5.** **(a)** 14 **(b)** 38 **7.** **(a)** $\frac{4}{3}$ **(b)** $\frac{16}{3}$ **9.** **(a)** $\frac{2}{3}$ **(b)** $\frac{4}{3}$ **11.** **(a)** 30 **(b)** 690 **15.** Possible pairs: $x = 0, y = 6$; $x = 1, y = 4$; $x = 2, y = 2$; $x = 3, y = 0$. There are many others. **17.** **(a)** 24 **(b)** 33 **19.** **(a)** 6 **(b)** $\frac{9}{5}$ **21.** **(a)** $\frac{23}{6}$ **(b)** $\frac{4}{3}$ **23.** **(a)** 150 **(b)** 195 **25.** **(a)** 2 **(b)** $\frac{17}{7}$ **27.** **(a)** $\frac{5}{6}$ **(b)** $\frac{16}{27}$ **29.** **(a)** $\frac{10}{3}$ **(b)** $\frac{13}{3}$ **31.** **(a)** 18 **(b)** 21 **33.** $8x$ **35.** $\frac{5}{x}$ **37.** $8 - x$ **39.** $3x + 8$ or $8 + 3x$ **41.** $8x + 52$ or $52 + 8x$ **43.** yes **45.** no **47.** yes **49.** yes **51.** yes **53.** yes **57.** $x + 8 = 12$; 4 **59.** $2x + 2 = 10$; 4 **61.** $5 + 2x = 13$; 4 **63.** $3x = 2 + 2x$; 2 **65.** $\frac{20}{5x} = 2$; 2 **67.** expression **69.** equation **71.** expression **75.** $\frac{110}{3}$ **77.** $\frac{24}{5}$

1.4 Exercises (page 30)

1. **(a)** -8 **(b)** 8 **3.** **(a)** 9 **(b)** 9 **5.** **(a)** 2 **(b)** 2 **7.** **(a)** 0 **(b)** 0 **11.** **(a)** A **(b)** A **(c)** B **(d)** B **13.** -12 **15.** -8 **17.** 3 **19.** $|-3|$ or 3 **21.** $-|-6|$ or -6 **23.** $|5 - 3|$ or 2 **25.** true **27.** true **29.** false **31.** true **33.** true **35.** true **37.** true **39.** true **41.** false **43.** true **45.** false **47.** **(a)** 3, 7 **(b)** 0, 3, 7 **(c)** $-9, 0, 3, 7$ **(d)** $-9, -1\frac{1}{4}, -\frac{3}{5}, 0, 3, 5.9, 7$ **(e)** $-\sqrt{7}, \sqrt{5}$ **(f)** All the numbers are real numbers.

49.
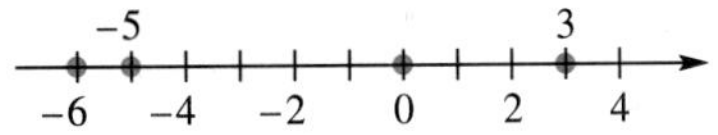

51.
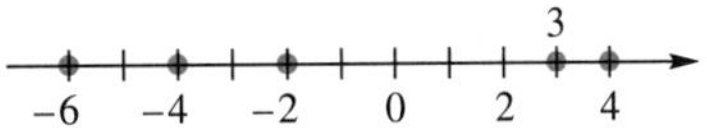

53.
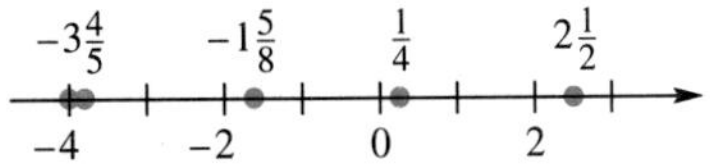

55.
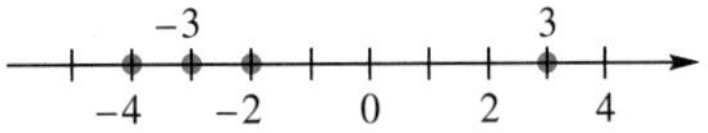

Many different answers are possible in Exercises 57–61. We give three possible answers for each exercise.
57. $\frac{1}{2}, \frac{5}{8}, 1\frac{3}{4}$ **59.** $-3\frac{1}{2}, -\frac{2}{3}, \frac{3}{7}$ **61.** $\sqrt{5}, \pi, -\sqrt{3}$ **63.** true **65.** false **67.** false (0 is a whole number, but not positive.) **69.** true **71.** false **73.** true: $a = 0$ or $b = 0$ or both $a = 0$ and $b = 0$; false: Choose any values for a and b so that not both are zero. **75.** true: a is equal to the opposite of b $(a = -b)$; false: a is not equal to the opposite of b $(a \neq -b)$.

1.5 Exercises (page 36)

1. 2 **3.** -2 **5.** -8 **7.** -12 **9.** 4 **11.** 12 **13.** 5 **15.** 2 **17.** -9 **19.** -11 **21.** $\frac{1}{2}$ **23.** $-\frac{19}{24}$ **25.** $-\frac{3}{4}$ **27.** -7.7 **29.** -8 **31.** 0 **33.** -20 **35.** no **37.** $-9 + 2 + 6$; -1 **39.** $[-17 + (-6)] + 12$; -11 **41.** $[-11 + (-4)] + (-5)$; -20 **43.** false **45.** true **47.** false **49.** true **51.** true **53.** false **55.** true **57.** true **59.** -3 **61.** -2 **63.** -2 **65.** 2 **67.** $50 + (-36) = 14$; \$14 left **69.** $34{,}000 + (-3500) = 30{,}500$; the new altitude is 30,500 feet. **71.** $-4 + 49 = 45$; the temperature was then 45° F. **73.** $4 + (-3) + (-2) = -1$; -1 yard **75.** 26.833 inches **77.** 17 **79.** -16

1.6 Exercises (page 42)

1. -3 **3.** -4 **5.** -8 **7.** -14 **9.** 9 **11.** -4 **13.** 4 **15.** $\frac{3}{4}$ **17.** $-\frac{11}{8}$ **19.** $\frac{15}{8}$ **21.** 11.6 **23.** -9.9 **25.** -1.8 **29.** 10 **31.** -5 **33.** 11 **35.** -10 **37.** -22 **39.** 2 **41.** -6 **43.** -12 **45.** positive **47.** positive **49.** $3-(-7)$; 10 **51.** $-11+[-4-2]$; -17 **53.** $[8+(-13)]-(-2)$; -3 **55.** $[-3-(-7)]-8$; -4 **57.** -2 **59.** -3 **61.** -3 **63.** $-5-10=-15$; $-15°$ F **65.** $14{,}494-(-282)=14{,}776$; 14,776 feet **67.** $-10-7=-17$; $-\$17$ **69.** $76{,}000-(-29{,}000)=105{,}000$; \$105,000 **71.** -9 **73.** -17

1.7 Exercises (page 48)

1. 12 **3.** -12 **5.** 120 **7.** 0 **9.** -165 **11.** $\frac{5}{12}$ **13.** $-\frac{1}{6}$ **15.** 6 **17.** $-.102$ **19.** -12.96 **21.** -11.9 **23.** $-36, -18, -12, -9, -6, -4, -3, -2, -1, 1, 2, 3, 4, 6, 9, 12, 18, 36$ **25.** $-25, -5, -1, 1, 5, 25$ **27.** $-17, -1, 1, 17$ **33.** -14 **35.** -3 **37.** -36 **39.** 12 **41.** 5 **43.** 12 **45.** 18 **47.** -14 **49.** -10 **51.** -28 **53.** 12 **55.** -360 **57.** 0 **59.** -24 **61.** 23 **63.** positive **65.** $(-9)(2)+6$; -12 **67.** $-9-(-1)(6)$; -3 **69.** $7(-6)-9$; -51 **71.** $\frac{3}{5}[(-8)+3]$; -3 **73.** 2 **75.** 0 **77.** -3 **79.** -2 **81.** -1 **83.** true **85.** true **87.** -34 **89.** 42 **91.** 34

1.8 Exercises (page 55)

1. $\frac{1}{9}$ **3.** $-\frac{1}{4}$ **5.** undefined **7.** $-\frac{10}{9}$ **11.** -2 **13.** -6 **15.** undefined **17.** 0 **19.** $\frac{2}{3}$ **21.** 2.1 **23.** -5 **25.** -4 **27.** -10 **29.** $-\frac{5}{2}$ **31.** -60 **33.** 2 **35.** 5 **37.** 4 **39.** -23 **41.** $\frac{52}{75}$ **43.** $\frac{7}{3}$ **45.** negative **47.** negative **49.** negative **51.** -2 **53.** 4 **55.** 0 **57.** -2 **59.** $4x=-32$; -8 **61.** $\frac{x}{-3}=-4$; 12 **63.** $\frac{x}{5}=-2$; -10 **65.** $\frac{x^2}{3}=12$; 6 or -6 **67.** $4+\frac{(-28)}{4}$; -3 **69.** $\frac{-16}{4}-3$; -7 **71.** 10 **73.** $-\frac{7}{10}$

1.9 Exercises (page 62)

1. associative **3.** commutative **5.** commutative **7.** associative **9.** inverse **11.** identity **13.** inverse **15.** identity **17.** distributive **21.** $k\cdot 9$ **23.** m **25.** $3r+3m$ **27.** 1 **29.** 0 **31.** $-5+5$; 0 **33.** $-3r-6$ **35.** 9 **37.** $5[k(-6)]$; $-30k$ **41.** $5m+10$ **43.** $-4r-8$ **45.** $-8k+16$ **47.** $-9a-27$ **49.** $4r+32$ **51.** $-16+2k$ **53.** $10r+12m$ **55.** $-12x+16y$ **57.** $5(8+9)$; 85 **59.** $7(2+8)$; 70 **61.** $9(p+q)$ **63.** $5(7z+8w)$ **65.** $-3k-5$ **67.** $-4y+8$ **69.** $4-p$ **71.** $1+15r$ **73.** No. For $a=2$, $b=3$, $c=4$, $a+(b\cdot c)=2+(3\cdot 4)=14$, whereas $(a+b)\cdot(a+c)=(2+3)(2+4)=5\cdot 6=30$. **75.** no **77.** yes **79.** Subtraction is not associative. **81.** true **83.** true

Chapter 1 Review Exercises (page 68)

1. $\frac{1}{3}$ **3.** $\frac{1}{5}$ **5.** $\frac{7}{50}$ **7.** $\frac{7}{8}$ **9.** $\frac{5}{12}$ **11.** 64 **13.** $\neq, >, \geq$ **15.** $\neq, <, \leq$ **17.** $70<-68$; false **19.** $\frac{8}{9}<1$; true **21.** $41\leq 40$; false **23.** $\frac{2}{3}$ **25.** $\frac{137}{3}$ **27.** $4-x$ **29.** $|-7|$ or 7 **31.** $-|-7|$ or -7 **33.** false **35.** true **37.** false **39.** false

41. 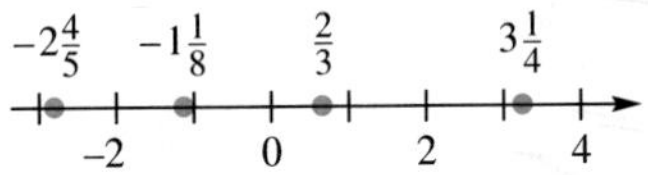

43. −13 **45.** $\frac{23}{40}$ **47.** 5.6 **49.** [12 + (−7)] + 5; 10 **51.** 18
53. $\frac{17}{12}$ **55.** 0 **57.** [9 − (−3)] − 12; 0 **59.** $\frac{8}{7}$ **61.** −28.25 **63.** −60 **65.** −2 **67.** −73
69. positive **71.** $\frac{7}{4}$ **73.** negative **75.** identity **77.** commutative **79.** inverse **81.** $18k + 30r$
83. −15 **85.** −38 **87.** −3 **89.** 2 **91.** 65 **93.** 9 **95.** −6 **97.** $\frac{5}{14}$ **101.** $5(x + 6)$; $5x + 30$

Chapter 1 Test (page 71)

[1.1] **1.** $\frac{7}{11}$ **2.** $\frac{203}{120}$ **3.** $\frac{18}{19}$ [1.2] **4.** false **5.** true **6.** true **7.** true **8.** false [1.3] **9.** 46
10. $\frac{124}{5}$ [1.4] **11.** $-|-8|$ or −8 **12.** .705 [1.8] **13.** $\frac{-6}{8(-3)}$; $\frac{1}{4}$ **14.** negative [1.3] **15.** yes
16. no [1.5–1.8] **17.** −7 **18.** $\frac{61}{20}$ **19.** 0 **20.** −1 **21.** 3 **22.** 4 **23.** undefined
[1.4] **24.** 30,500 feet [1.5] **25.** 178° F [1.6] **26.** 3 [1.9] **27.** B **28.** D **29.** E **30.** A **31.** C
32. $-3 + 4m$

CHAPTER 2

2.1 Exercises (page 77)

1. 15 **3.** −22 **5.** 35 **7.** −9 **9.** 1 **11.** −1 **13.** like **15.** unlike **17.** like **19.** like
21. $17y$ **23.** $-6a$ **25.** $13b$ **27.** $7k + 15$ **29.** $-4y$ **31.** $2x + 6$ **33.** $14 - 7m$ **35.** $-17 + x$
37. $23x$ **39.** $9y^2$ **41.** $-14p^3 + 5p^2$ **45.** $30t + 66$ **47.** $-3n - 15$ **49.** $4r + 15$ **51.** $12k - 5$
53. $-2k - 3$ **55.** $(x + 3) - 3x$; $-2x + 3$ **57.** $9x - (8x - 5x)$; $6x$ **59.** $9(6x + 3) + (3 - x)$; $53x + 30$
61. $-2x + 6y - 2$ **63.** $11p^2 + 5p + 1$ **65.** $10r - 34$ **67.** $-z^2 - 3z + 7$ **69.** −6 **71.** 4 **73.** 7
75. −6 **77.** 3 **79.** $-\frac{1}{3}$

2.2 Exercises (page 85)

1. 10 **3.** −2 **5.** 10 **7.** −8 **9.** −5 **11.** −2 **13.** 4 **15.** −5 **17.** −11 **19.** −6 **23.** 19
25. −1 **27.** −8 **29.** 7 **31.** 0 **33.** 2 **35.** 4 **37.** 2 **39.** 34 **41.** −14 **43.** 67 **45.** −65
47. 4 **49.** −12 **51.** $-\frac{18}{5}$ **53.** −4 **55.** −7 **57.** 0 **59.** 2 **61.** 6 **63.** 7 **65.** 4 **67.** 49
69. $\frac{27}{2}$ **71.** $-\frac{15}{2}$ **73.** −64 **75.** $\frac{18}{5}$ **77.** $\frac{22}{3}$ **79.** (c) **81.** $\frac{1}{2}$ **83.** 3 **85.** 4
87. $7x - 6x = -9$; −9 **89.** $4x = 6$; $\frac{3}{2}$ **91.** $\frac{2x}{5} = 4$; 10 **93.** $12m - 20$ **95.** $-6y - 30$
97. $-14 + 3y$ **99.** $-26 + 40p$

2.3 Exercises (page 93)

1. 2 **3.** 24 **5.** 9 **7.** −2 **9.** −1 **11.** −4 **13.** −3 **15.** −3 **17.** 8 **19.** −2 **21.** 6
23. −5 **25.** 0 **27.** 4 **29.** 0 **31.** $-\frac{1}{5}$ **33.** $-\frac{5}{7}$ **35.** no solution **37.** all real numbers
39. no solution **43.** 5 **45.** 0 **47.** $-\frac{7}{5}$ **49.** 120 **51.** $\frac{12}{19}$ **53.** 15,000 **55.** $11 - q$ **57.** $x + 7$
59. $a + 12$ **61.** $\frac{t}{5}$ **63.** $\frac{3}{5}x$ **65.** $x + 2x$ **67.** $\frac{x - 7}{x}$ **69.** $\frac{-9}{x + 3}$

2.4 Exercises (page 100)

1. 9 **3.** 1 **5.** Kennedy: 303; Nixon: 219 **7.** Brian: 26 sit-ups; Katherine: 54 sit-ups **9.** 9 men, 11 women **11.** Jean: 30 miles; Larry: 60 miles **13.** 4, 9, and 27 centimeters **15.** Boggs: 551 times; Evans: 541 times; Burks: 558 times **17.** 18 prescriptions **19.** peanuts: $22\frac{1}{2}$ ounces; cashews: $4\frac{1}{2}$ ounces **21.** $k - m$ **23.** no **25.** 80° **27.** 26° **29.** 55° **31.** 68 and 69 **33.** 10 and 12 **35.** \$450 **37.** 36 quart cartons **39.** 42 **41.** 640 **43.** 60

2.5 Exercises (page 109)

1. 60 **3.** 4 **5.** 8 **7.** 14 **9.** 10 **11.** 21 **13.** 1.5 **15.** 254.34 **17.** 113.04 **19.** 2 **21.** 3 **23.** 7 **25.** 1 **27.** 4 **29.** $L = \frac{A}{W}$ **31.** $H = \frac{V}{LW}$ **33.** $r = \frac{C}{2\pi}$ **35.** $W = \frac{1}{2}(P - 2L)$ or $W = \frac{P - 2L}{2}$ **37.** $b = \frac{2A}{h} - B$ or $b = \frac{2A - Bh}{h}$ **39.** $h = \frac{S - 2\pi r^2}{2\pi r}$ or $h = \frac{S}{2\pi r} - r$ **43.** area **45.** 10.2 meters **47.** 23,800.10 square feet **49.** 37.68 feet **51.** 107°, 73° **53.** 75°, 75° **55.** 139°, 139° **57.** $x = \frac{6 - y}{5}$ **59.** $x = \frac{9 - 3y}{2}$ **61.** $x = \frac{5y + 15}{3}$ **63.** $\frac{8}{3}$ **65.** $-\frac{15}{7}$ **67.** -20

2.6 Exercises (page 116)

1. $\frac{3}{2}$ **3.** $\frac{36}{55}$ **5.** $\frac{2}{5}$ **7.** $\frac{5}{8}$ **9.** $\frac{1}{10}$ **11.** $\frac{8}{5}$ **13.** $\frac{1}{6}$ **15.** $\frac{4}{15}$ **17.** true **19.** true **21.** false **23.** false **29.** 175 **31.** 4 **33.** 16 **35.** $\frac{14}{3}$ **37.** $\frac{25}{3}$ **39.** $\frac{35}{8}$ **41.** $\frac{15}{2}$ **43.** 19 **45.** $16\frac{1}{2}$ ounces **47.** \$9.90 **49.** \$338 **51.** 4.4 feet **53.** $25\frac{2}{3}$ inches **55.** $2\frac{1}{2}$ cups **57.** 8-ounce size **59.** 13-ounce size **61.** 28-ounce size **63.** -1 **65.** 10 **67.** 19.2 yards **69.** 8% per year **71.** $2\frac{1}{2}$ years **73.** 6 **75.** 4

2.7 Exercises (page 122)

1. 10 milliliters **3.** \$300 **5.** \$2.70 **7.** 0 liters **9.** 160 liters **11.** 18 barrels **13.** 7.5 gallons **15.** 25 milliliters **17.** \$10,000 **19.** \$10,000 at 10%; \$15,000 at 14% **21.** \$16,400 **23.** 10 nickels **25.** 27 twenties; 15 fifties **27.** (d) **29.** pure acid; interest; monetary value **31.** 18 **33.** 8 hamburgers; 4 bags of french fries **35.** $13\frac{1}{3}$ liters **37.** $r = \frac{d}{t}$ **39.** $L = \frac{P - 2W}{2}$ or $L = \frac{P}{2} - W$ **41.** $h = \frac{2A}{b}$ **43.** 40 miles per hour

2.8 Exercises (page 128)

1. 3.453 hours **3.** 2.260 meters per second **5.** 530 miles **9.** 5 hours **11.** 3 hours **13.** $2\frac{1}{2}$ hours **15.** 4 meters **17.** 300 feet, 400 feet, 500 feet **19.** 46 feet **21.** *A* and *B*: 52°; *C*: 76° **23.** 65 miles per hour **25.** 120 pounds **27.** 10 inches **29.** exports: 741 million dollars; imports: 426 million dollars **31.** \$17.05 **33.** 109°, 71° **35.** 25.5-ounce size **39.** $<$ **41.** $<$ **43.** $>$ **45.** $>$

2.9 Exercises (page 139)

1.
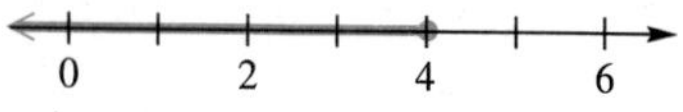

3.
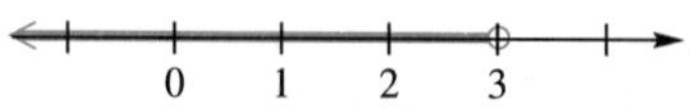

5.
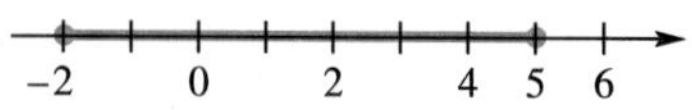

7. 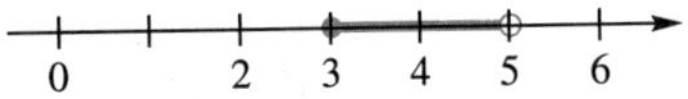 **11.** $z \geq 1$ **13.** $k \geq 5$ **15.** $x < 9$ **17.** $k \geq -6$ **19.** $y > 9$

21. $n \leq -11$ 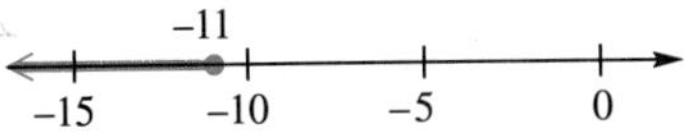 **23.** $z > 5$

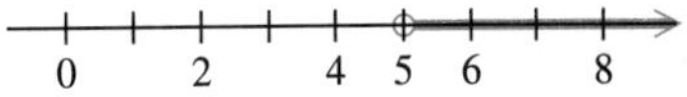

25. $k \geq -5$ 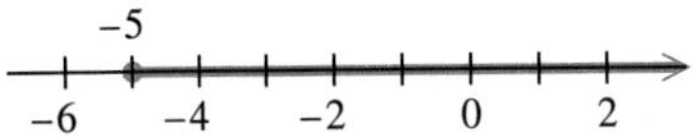 **27.** $r < 30$

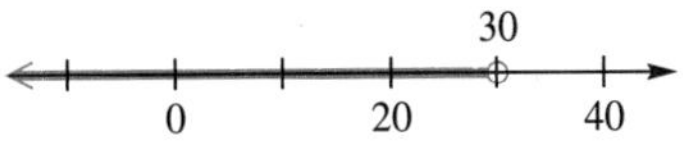

29. $q > -\frac{1}{2}$ 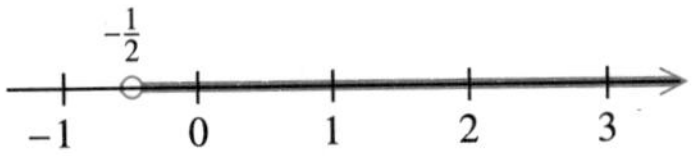 **31.** $p > -5$

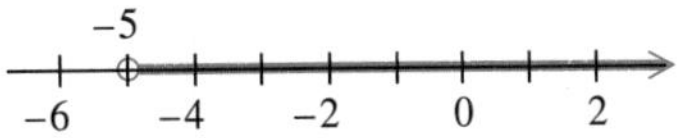

33. $k \leq 0$ 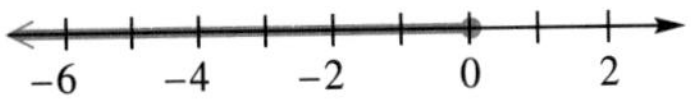 **35.** $r \geq -1$

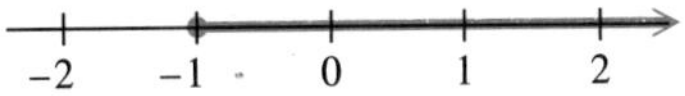

39. $-1 \leq x \leq 6$ 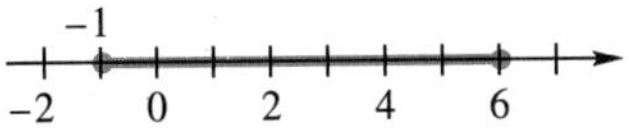 **41.** $-\frac{11}{6} < m < -\frac{2}{3}$

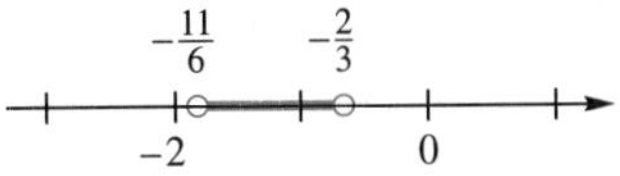

43. $1 < p < 3$ **45.** $-26 \leq z \leq 6$

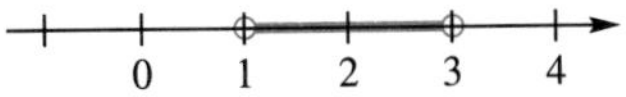

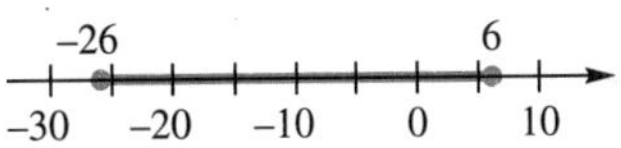

47. $-3 \leq p \leq 6$ 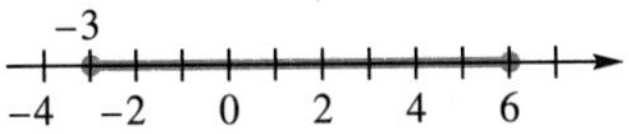 **49.** $-\frac{24}{5} \leq r \leq 0$

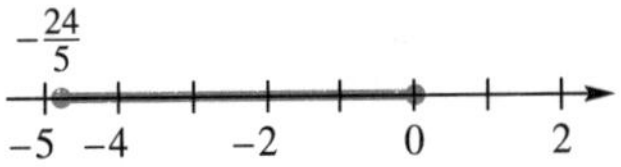

51. all numbers greater than $-\frac{7}{4}$ **53.** at least 88 **55.** at least \$275 **57.** 32° to 95° Fahrenheit **59.** $x \geq 500$ **61.** 38 meters or less **63.** $k > -21$ **65.** $p \leq \frac{8}{3}$ **67.** 11 **69.** 10 **71.** $14x - 12$ **73.** $3x^3 + 7x^2 + 2$

Chapter 2 Review Exercises (page 146)

1. $11m$ **3.** $16p^2 + 2p$ **5.** $-2m + 29$ **7.** 6 **9.** 7 **11.** 11 **13.** 5 **15.** 5 **17.** $\frac{64}{5}$ **19.** all real numbers **21.** all real numbers **23.** no solution **25.** $\frac{5}{2}$ **27.** Hawaii: 6425 square miles; Rhode Island: 1212 square miles **29.** 80° **31.** 84 **33.** 4.19 (rounded) **35.** $h = \frac{2A}{b + B}$ **37.** 100°, 100° **39.** $\frac{5}{3}$ **41.** $\frac{3}{4}$ **43.** true **45.** false **47.** $\frac{7}{2}$ **49.** 2300 miles **51.** 8 slices for \$2.19 **53.** 25 pieces **55.** 3.75 liters **57.** \$5000 at 5%; \$5000 at 9% **59.** 13 hours **61.** $\frac{5}{6}$ hour or 50 minutes **63.** 10 meters

65. 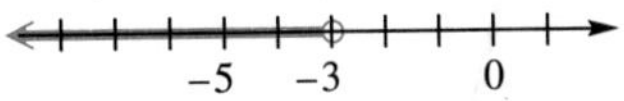 **67.** 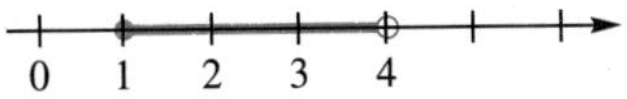 **69.** $y > 8$ **71.** $z \leq -37$

73. $7 < y < 9$ **75.** $y > -2$ **77.** $y \leq \frac{27}{10}$ **79.** $\frac{4}{3} < m \leq 5$ **81.** all numbers less than or equal to $-\frac{2}{9}$ **83.** $r = \frac{I}{pt}$ **85.** $\frac{36}{5}$ **87.** $m \leq -6$ **89.** $\frac{13}{4}$ **91.** 50 miles per hour; 80 miles per hour **93.** \$10,000 at 5%; \$16,000 at 10% **95.** 5 meters **97.** 50 meters or less

Chapter 2 Test (page 149)

[2.1] **1.** $-r$ **2.** 4 **3.** $4m + 6$ [2.2–2.3] **4.** -5 **5.** 0 **6.** $-\frac{12}{5}$ **7.** $\frac{5}{6}$ **8.** 3 **9.** no solution

[2.4] **10.** 2 **11.** Oldsmobile: \$110; Bronco: \$203 [2.5] **12.** $L = \frac{P - 2W}{2}$ or $L = \frac{P}{2} - W$ **13.** 73 yards

[2.6] **15.** 1 **16.** -26 **17.** \$17.55 [2.7] **18.** \$11,000 at 6%; \$17,000 at 7.5% **19.** 10 liters

[2.8] **20.** 4 hours [2.9] **21.** $m > 7$ **22.** $k \le 1$

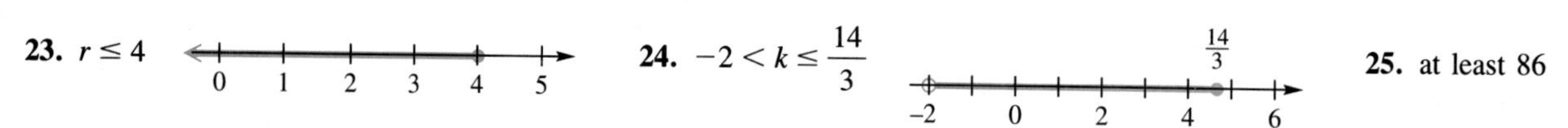

23. $r \le 4$ **24.** $-2 < k \le \frac{14}{3}$ **25.** at least 86

■ CHAPTER 3

3.1 Exercises (page 157)

1. $-r^5$ **3.** x^5 **5.** $-p^7$ **7.** $6y^2z$ **9.** 0 **11.** $9y^4 + 7y^2$ **13.** $14z^5 - 9z^3 + 8z^2$
15. $1.17q^2 + 1.40q - .25$ **17.** already simplified; degree 4; (b) binomial **19.** already simplified; degree 10; (c) trinomial **21.** already simplified; degree 8; (c) trinomial **23.** x^5z^3; degree 8; (a) monomial
25. already simplified; degree 9; (b) binomial **27.** sometimes **29.** never **31.** sometimes
33. (a) 38 (b) -1 **35.** (a) 19 (b) -2 **37.** (a) -5 (b) 1 **39.** (a) -6 (b) 0 **41.** $10a^3 + a^2$
43. $-n^5 - 12n^3 - 2$ **45.** $12m^3 + m^2 + 12m - 14$ **47.** $5a^4 - 4a^3 + 3a^2 - a + 1$ **49.** $-2r^2 + 7r - 6$
51. $-2x^2 - x + 1$ **53.** $4x^3 + 2x^2 + 5x$ **55.** $25y^4 + 8y^2 + 3y$ **57.** $10x^2 - 2x + 46$
59. $r^3 + 4r^2 + 11r + 10$ **61.** 7.652 **63.** $5m^2 + 8m - 10$ **65.** $-6x^2 + 6x - 7$ **67.** $10x^4 - 6x^2 + 10x - 1$
71. 64 **73.** 625 **75.** $\frac{8}{27}$ **77.** 128

3.2 Exercises (page 163)

1. base 5, exponent 12 **3.** base $3m$, exponent 4 **5.** base 2, exponent 4; base -2, exponent 4 **7.** base -24, exponent 2; base 24, exponent 2 **9.** base r, exponent 6; base $-r$, exponent 6 **11.** 3^5 **13.** $(-2)^5$
15. $\frac{1}{(-2)^3}$ **17.** p^5 **19.** $(-2z)^3$ **21.** 64 **23.** 16 **25.** 36 **27.** 12 **29.** 4^5 **31.** 3^{11} **33.** 4^{18}
35. $(-3)^5$ **37.** y^{14} **39.** r^{17} **41.** $-63r^9$ **43.** $10p^{13}$ **45.** $13m^3$; $36m^6$ **47.** $-p$; $-132p^2$
49. $15r$; $105r^3$ **51.** cannot be added; $-15a^5$ **53.** 6^6 **55.** 9^6 **57.** 5^8 **59.** -4^6 **61.** 5^3m^3 or $125m^3$
63. $(-2)^4p^4q^4$ or $16p^4q^4$ **65.** $\frac{(-3)^2x^{10}}{4^2}$ or $\frac{9x^{10}}{16}$ **67.** $\frac{5^3a^6b^3}{c^{12}}$ or $\frac{125a^6b^3}{c^{12}}$ **69.** $\frac{4^8}{3^5}$ **71.** 3^7m^7 **73.** 8^9z^9
75. $2^3m^7n^5$ or $8m^7n^5$ **77.** $5^7a^{11}b^{12}$ **79.** 5^{8r} **81.** 2^pm^p **83.** $\frac{4^{2r}}{3^{3r}}$ **85.** $\frac{2^np^{mn}}{q^{rn}}$ **87.** $12x^5$ **89.** $6p^7$
91. 10^3 or 1000 **95.** $15k^2$ **97.** $18r^4$ **99.** $28y^7$ **101.** $10z^{11}$

3.3 Exercises (page 169)

1. $-32x^7$ **3.** $15y^{11}$ **5.** $30a^9$ **7.** $-18r^3s^2$ **9.** $-30p + 15p^2$ **11.** $12x + 8x^2 + 20x^4$
13. $21y + 35y^3 - 14y^4$ **15.** $-6m^6 - 10m^5 - 12m^4$ **17.** $27a^{11} + 18a^9 - 72a^7$ **19.** $32z^5 + 20z^4y - 12z^3y^2$
21. $6p^4q^3 - 10p^3q + 4p^2q^3$ **23.** $72y^3 - 70y^2 + 21y - 2$ **25.** $6r^3 + 5r^2 - 12r + 4$
27. $3y^5 + 12y^4 - 2y^3 - 8y^2 + y + 4$ **29.** $2a^5 + a^4 - a^3 + a^2 - a + 3$ **31.** $2m^4 - 7m^3 + 3m^2 + 17m - 15$
33. $x^2 - 3x - 10$ **35.** $r^2 + 9r + 18$ **37.** $8y^2 - 6y - 5$ **39.** $6x^2 + 7x - 3$ **41.** $110m^2 + 21m - 110$
43. $16 - 2x - 3x^2$ **45.** $-10 + 17z - 6z^2$ **47.** $-6 + 21y + 12y^2$ **49.** $6r^2 - 7rs - 3s^2$
51. $18m^2 - 19mk - 12k^2$ **53.** $-8r^2 + 2rs + 45s^2$ **55.** $6p^2 - \frac{5}{2}pq - \frac{25}{12}q^2$ **57.** $2m^6 - 5m^3 - 12$
59. $2k^5 - 6k^3h^2 + k^2h^2 - 3h^4$ **61.** $6p^8 + 15p^7 + 12p^6 + 36p^5 + 15p^4$ **63.** $-24x^8 - 28x^7 + 32x^6 + 20x^5$
67. $9m^2$ **69.** $4r^2$ **71.** $16x^4$

3.4 Exercises (page 173)

1. $m^2 + 4m + 4$ **3.** $r^2 - 6r + 9$ **5.** $x^2 + 4xy + 4y^2$ **7.** $25p^2 + 20pq + 4q^2$ **9.** $16a^2 + 40ab + 25b^2$ **11.** $9r^2 + 12rs + 4s^2$ **13.** $a^2 - 64$ **15.** $4 - p^2$ **17.** $4m^2 - 25$ **19.** $9x^2 - 16y^2$ **21.** $25y^2 - 9x^2$ **23.** $9h^2 - 4j^2$ **25.** $m^3 - 15m^2 + 75m - 125$ **27.** $8a^3 + 12a^2 + 6a + 1$ **29.** $y^4 + 16y^3 + 96y^2 + 256y + 256$ **31.** $81r^4 - 216r^3t + 216r^2t^2 - 96rt^3 + 16t^4$ **33.** $4z^2 - 10zx + \frac{25}{4}x^2$ **35.** $4m^2 - \frac{25}{9}$ **37.** $x^4 - 1$ **39.** $49y^4 - 100z^2$ **41.** $(a + b)^2 = 49$; $a^2 + b^2 = 29$; not equal **43.** $\frac{1}{2}m^2 - 2n^2$ **45.** $9a^2 - 4$ **49.** $\frac{1}{5}$ **51.** -4 **53.** $\frac{4}{3}$ **55.** $-\frac{8}{3}$

3.5 Exercises (page 181)

1. 2 **3.** 2 **5.** $\frac{1}{27}$ **7.** $-\frac{1}{12}$ **9.** 32 **11.** 2 **13.** 1 **15.** $-\frac{3}{p}$ **17.** $\frac{5}{6}$ **19.** $\frac{9}{20}$ **21.** $\frac{3^4}{5^2}$ **23.** $\frac{b^5}{a^2}$ **25.** 4^5 **27.** $\frac{1}{8^6}$ **29.** $\frac{1}{6^6}$ **31.** 14^3 **33.** x^{15} **35.** z^2 **37.** 2^5 **39.** $2k^2$ **41.** $\frac{1}{4^9}$ **43.** 5^9 **45.** 2^6 **47.** 3^5 **49.** $\frac{5^2}{m^2}$ or $\frac{25}{m^2}$ **51.** $\frac{1}{x^3}$ **53.** $\frac{a^2}{3^3}$ or $\frac{a^2}{27}$ **55.** $4^2k^4m^8$ or $16k^4m^8$ **57.** 5^{6r} **59.** $\frac{1}{x^{11a}}$ **61.** q^{5k} **63.** $\frac{p^{3y}}{6^y}$ **65.** $\frac{8}{3}$ **67.** $-\frac{5}{8}$ **69.** $\frac{a^{11}}{2b^5}$ **71.** $\frac{108}{y^5z^3}$ **75.** 10 **77.** $x - 2 + \frac{9}{x}$ **79.** $\frac{y^3}{4} + \frac{3y}{2} + \frac{2}{y}$

3.6 Exercises (page 185)

1. $2x$ **3.** $2a^2$ **5.** $\frac{9k^3}{m}$ **7.** $\frac{-3m^2}{p^2}$ **9.** $30m^3 - 10m$ **11.** $5m^4 - 8m + 4m^2$ **13.** $4m^4 - 2m^2 + 2m$ **15.** $m^4 - 2m + 4$ **17.** $m - 1 + \frac{5}{2m}$ **19.** $-5q^2 + 2q - 1$ **21.** $8q^4 - 7q^2 + \frac{4}{q}$ **23.** $-\frac{3q^3}{5} - q + \frac{6}{5}$ **25.** $-q^2 + \frac{3q}{5} - \frac{2}{5} + \frac{7}{5q}$ **27.** $x^2 + 3x$ **29.** $4x^2 + 1 - x$ **31.** $\frac{12}{x} + 8 + x$ **33.** $\frac{x}{3} + 2 - \frac{1}{3x}$ **35.** $4k^3 - 6k^2 - k + \frac{7}{2} - \frac{3}{2k}$ **37.** $-10p^3 + 5p^2 - 3p + \frac{3}{p}$ **39.** $4y^3 - 2 + \frac{3}{y}$ **41.** $\frac{12}{x} - \frac{6}{x^2} + \frac{14}{x^3} - \frac{10}{x^4}$ **43.** 8; 13; no **45.** $12x^5 + 9x^4 - 12x^3 + 6x^2$ **47.** $-63y^4 - 21y^3 - 35y^2 + 14y$ **49.** 2 **51.** $2x^3 - 5x^2 + x$ **53.** $-20k^4 + 12k^3 - 8k^2$ **55.** $6m^{10} - 12m^8 + 3m^7$ **57.** $5x + 2$ **59.** $-2x^2 - 11$ **61.** $7x^2 - 10x + 1$

3.7 Exercises (page 189)

1. $x + 2$ **3.** $2y - 5$ **5.** $p - 4 + \frac{4}{p + 6}$ **7.** $r - 5$ **9.** $6m - 1$ **11.** $x - 2 + \frac{11}{2x + 4}$ **13.** $4r + 1 + \frac{10}{3r - 5}$ **15.** $2k + 5 + \frac{10}{7k - 8}$ **17.** $x^2 - x + 2$ **19.** $4k^3 - k + 2$ **21.** $3y^2 - 2y + 6 + \frac{-5}{y + 1}$ **23.** $3k^2 + 2k - 2 + \frac{6}{k - 2}$ **25.** $2p^3 - 6p^2 + 7p - 4 + \frac{14}{3p + 1}$ **27.** $x^2 + 1 + \frac{-6x + 2}{x^2 - 2}$ **29.** $m^3 - 2m^2 - 7 + \frac{-7m - 36}{4m^2 - m - 3}$ **31.** $y^2 - y + 1$ **33.** $a^2 - 1$ **35.** $y - \frac{9}{2} + \frac{15}{2y + 4}$ **37.** $3w + \frac{11}{3} + \frac{52/3}{3w - 2}$ **39.** $2p^2 + \frac{4}{3}p - \frac{5}{3} + \frac{(-8/3)p^2 + (19/3)p + 2}{3p^3 + 2p}$ **41.** divisor: $2x + 5$; quotient: $2x^3 - 4x^2 + 3x + 2$ **43.** 64,270 **45.** 1230 **47.** 3.4 **49.** .237

3.8 Exercises (page 193)

1. 6.835×10^9 **3.** 8.36×10^{12} **5.** 2.15×10^2 **7.** 2.5×10^4 **9.** 3.5×10^{-2} **11.** 1.01×10^{-2} **13.** 1.2×10^{-5} **15.** 8,100,000,000 **17.** 9,132,000 **19.** 324,000,000 **21.** .00032 **23.** .041 **25.** 800,000 **27.** .000004 **29.** 420 **31.** 3,000,000 **33.** .2 **35.** 1300 **37.** .00025 **39.** .000008 **41.** 7×10^{-3} **43.** 9.29×10^7 **45.** .000000001 **47.** .00023; .006 **49.** about 15,300 seconds

Chapter 3 Review Exercises (page 197)

1. $22m^2$; degree 2; monomial **3.** already in descending powers; degree 5; none of these **5.** $-8x^5 + 9x^3 - 7x$; degree 5; trinomial **7.** $10p^2 - 3p - 5$; degree 2; trinomial **9.** $a^3 + 4a^2$ **11.** $-13k^4 - 15k^2 + 18k$ **13.** $7r^4 - 4r^3$ **15.** 80 **17.** 5^{11} **19.** $27x^6$ **21.** a^4c^{12} **23.** $10x^2 - 55x$ **25.** $-22y^4 + 4y^3 + 18y^2$ **27.** $a^3 - 2a^2 - 7a + 2$ **29.** $5p^5 - 2p^4 - 3p^3 + 25p^2 + 15p$ **31.** $6k^2 - 9k - 6$ **33.** $12k^2 - 32kq - 35q^2$ **35.** $a^2 + 8a + 16$ **37.** $4r^2 + 20rt + 25t^2$ **39.** $36m^2 - 25$ **41.** $25a^2 - 36b^2$ **43.** $r^3 + 6r^2 + 12r + 8$ **45.** $\frac{64}{25}$ **47.** 5^8 **49.** $(-3)^{10}$ or 3^{10} **51.** x^2 **53.** y^7 **55.** $\frac{3^5}{p^3}$ **57.** $72r^5$ **59.** $2x^2y$ **61.** $p - 3 + \frac{5}{2p}$ **63.** $2r + 7$ **65.** $k^2 + k + \frac{7}{2} + \frac{(-9/2)k - (23/2)}{2k^2 + k + 1}$ **67.** 1.58×10^7 **69.** 9.76×10^{-5} **71.** 12,000 **73.** 90,000,000 **75.** .0003 **77.** 2 **79.** $144a^2 - 1$ **81.** $\frac{1}{8^{12}}$ **83.** $p - 3 + \frac{5}{2p}$ **85.** $6k^3 - 21k - 6$ **87.** $4r^2 + 20rs + 25s^2$ **89.** $10r^2 + 21r - 10$ **91.** $y^2 + 5y + 1$ **93.** $-11 + p$ **95.** $\frac{1}{5^{11}}$ **97.** 5^8 **99.** $\frac{1}{m^2}$

Chapter 3 Test (page 200)

[3.1] **1.** $-x^2 + 6x$; degree 2; (b) binomial **2.** $2m^4 + 11m^3 - 8m^2$; degree 4; (a) trinomial **3.** $x^5 - x^2 - 2x + 12$ **4.** $3y^2 - 2y - 2$ [3.2] **5.** $\frac{64}{27}$ [3.5] **6.** $\frac{1}{25}$ [3.2] **7.** 5^9 [3.5] **8.** $\frac{1}{8^5}$ **9.** $\frac{1}{6q}$ **10.** $8p^5$ [3.3] **12.** $6m^5 + 12m^4 - 18m^3 + 42m^2$ **13.** $15m^2 - 37m - 66$ **14.** $14t^2 + 27ts - 20s^2$ **15.** $6k^3 - 17k^2 + 10k + 16$ [3.4] **16.** $4k^2 + 20km + 25m^2$ **17.** $4r^2 + 2rt + \frac{1}{4}t^2$ **18.** $36p^4 - 25r^2$ [3.6] **19.** $-5y^3 - \frac{8y^2}{3} + 2y$ **20.** $2 + \frac{5}{r} - \frac{3}{r^2} + \frac{8}{5r^3}$ [3.7] **21.** $3x^2 + 4x + 2$ **22.** $5r^2 + 2r - 5 + \frac{5}{2r^2 - 3}$ [3.8] **23.** 3.79×10^{-4} **24.** 900 **25.** .0004

CHAPTER 4

4.1 Exercises (page 206)

1. 12 **3.** $10p^2$ **5.** 1 **7.** $6m^2n$ **9.** $4y^3x$ **11.** x **13.** $3m^2$ **15.** $-2z^4$ **17.** xy^2 **19.** $3ab$ **21.** $-4m^2n$ **23.** $7a(7a + 2)$ **25.** $11p^4(11p - 3)$ **27.** no common factor other than 1 **29.** $19y^2p^2(y + 2p)$ **31.** $13y^3(y^3 + 2y^2 - 3)$ **33.** $9qp^3(5q^3p^2 - 4p^3 + 9q)$ **35.** $(5 - x)(r + t)$ **37.** $(1 - 4p)(3p - 2q)$ **39.** $(a - 2)(a + b)$ **41.** $(7z + 3)(z + 2m)$ **43.** $2(y - 3)(y - x)$ **45.** $(a^2 + b^2)(3a + 2b)$ **47.** $(ab - 4)(a - b^3)$ **49.** $5z^3a^3(25z^2 - 12za^2 + 17a)$ **51.** $11y^3(3y^5 - 4y^9 + 7 + y)$ **53.** $4a^4b^2(9b + 8a - 12a^2b)$ **55.** no **57.** yes **59.** yes **61.** no **63.** $m^2 + 7m + 12$ **65.** $r^2 - 4r - 45$ **67.** $y^2 - 12y + 32$ **69.** $a^2 - 49$

4.2 Exercises (page 210)

1. $x + 3$ **3.** $r + 8$ **5.** $t - 12$ **7.** $x - 8$ **9.** $m + 6$ **11.** $x - 5y$ **13.** $(x + 5)(x + 1)$ **15.** $(a + 4)(a + 5)$ **17.** $(x - 7)(x - 1)$ **19.** prime **21.** $(y - 4)(y - 2)$ **23.** $(s - 5)(s + 7)$ **25.** prime **27.** $(b - 8)(b - 3)$ **29.** $(k - 5)(k - 5)$ or $(k - 5)^2$ **31.** $(x + 3a)(x + a)$ **33.** $(y + 5b)(y - 6b)$ **35.** $(x + 6y)(x - 5y)$ **37.** $(r - s)(r - s)$ or $(r - s)^2$ **39.** $(p - 5q)(p + 2q)$ **41.** $3m(m + 3)(m + 1)$

43. $6(a+2)(a-10)$ **45.** $3r(r-4)(r-6)$ **47.** $3x^2(x-6)(x+5)$ **51.** y^2-y-42 **53.** $a^3(a+4b)(a-b)$ **55.** $yz(y+3z)(y-2z)$ **57.** $z^8(z-7y)(z+3y)$ **59.** $(a+b)(x+4)(x-3)$ **61.** $(2p+q)(r-9)(r-3)$ **63.** $2y^2+y-28$ **65.** $15z^2-4z-4$ **67.** $8p^2-10p-3$

4.3 Exercises (page 216)

1. $x-1$ **3.** $b-3$ **5.** $4y-3$ **7.** $(2a+5b)(3a-4b)$ **9.** $(2x^3-5x-3)$; $(2x+1)(x-3)$ **11.** $(6m^2+7m-20)$; $(3m-4)(2m+5)$ **13.** $(2x+1)(x+3)$ **15.** $(3a+7)(a+1)$ **17.** $(4r-3)(r+1)$ **19.** $(3m-1)(5m+2)$ **21.** $(2m-3)(4m+1)$ **23.** $(5a+3)(a-2)$ **25.** $(3r-5)(r+2)$ **27.** $(y+17)(4y+1)$ **29.** $(19x+2)(2x+1)$ **31.** $(2x+3)(5x-2)$ **33.** $(2w+5)(3w+2)$ **35.** $(2q+3)(3q+7)$ **37.** $(5m-4)(2m-3)$ **39.** $(4k-5)(2k+3)$ **41.** $(5m-8)(2m+3)$ **43.** $(4x-1)(2x-3)$ **45.** $-(8m-3)(5m+2)$ **47.** $(4p-3q)(3p+4q)$ **49.** $(5a+2b)(5a+3b)$ **51.** $(3a-5b)(2a+b)$ **53.** $3n^2(5n-3)(n-2)$ **55.** $-4w^4(8z+3)(z-1)$ **57.** $2k^2(2k-3w)(k+w)$ **59.** $m^4n(3m+2n)(2m+n)$ **61.** $3zy(3z+7y)(2z-5y)$ **63.** $(5q-2)(5q+1)(m+1)^3$ **65.** $-5, -1, 1, 5$ **67.** $-10, -6, 6, 10$ **69.** $9r^2-1$ **71.** $4m^2-12m+9$ **73.** $8z^3-27$

4.4 Exercises (page 223)

1. $(x+4)(x-4)$ **3.** $(a+b)(a-b)$ **5.** $(m+1)(m-1)$ **7.** $(5m+4)(5m-4)$ **9.** $4(3t+2)(3t-2)$ **11.** $(5a+4r)(5a-4r)$ **13.** prime **15.** $(p^2+7)(p^2-7)$ **17.** $(a^2+1)(a+1)(a-1)$ **19.** $(m^2+9)(m+3)(m-3)$ **21.** $(a+2)^2$ **23.** $(x-5)^2$ **25.** $(7+a)^2$ **27.** $(k+11)^2$ **29.** prime **31.** $(4t-5)^2$ **33.** $(3x+4)^2$ **35.** $(7x+2y)^2$ **37.** $xy(x+3y)^2$ **39.** $(a+1)(a^2-a+1)$ **41.** $(x-2)(x^2+2x+4)$ **43.** $(p+q)(p^2-pq+q^2)$ **45.** $(3x-1)(9x^2+3x+1)$ **47.** $(2p+q)(4p^2-2pq+q^2)$ **49.** $(3a-4b)(9a^2+12ab+16b^2)$ **51.** $(4x+5y)(16x^2-20xy+25y^2)$ **53.** $(5m-2p)(25m^2+10mp+4p^2)$ **55.** $(4y^2+1)(16y^4-4y^2+1)$ **57.** $(2k^2-3q)(4k^4+6k^2q+9q^2)$ **59.** $4mn$ **61.** $(x-1)(x+3)$ **63.** $(m-p+2)(m+p)$ **65.** 10 **67.** 9 **71.** 2 **73.** $\frac{2}{3}$ **75.** $-\frac{9}{7}$ **77.** $-\frac{5}{8}$

Summary Exercises on Factoring (page 224)

1. $(a-6)(a+2)$ **3.** $6(y-2)(y+1)$ **5.** $6(a+2b+3c)$ **7.** $(p-11)(p-6)$ **9.** $(5z-6)(2z+1)$ **11.** $(m+n+5)(m-n)$ **13.** $8a^3(a-3)(a+2)$ **15.** $(z-5a)(z+2a)$ **17.** $(x-5)(x-4)$ **19.** $(3n-2)(2n-5)$ **21.** $4(4x+5)$ **23.** $(3y-4)(2y+1)$ **25.** $(6z+1)(z+5)$ **27.** $(2k-3)^2$ **29.** $6(3m+2z)(3m-2z)$ **31.** $(3k-2)(k+2)$ **33.** $7k(2k+5)(k-2)$ **35.** $(y^2+4)(y+2)(y-2)$ **37.** $8m(1-2m)$ **39.** $(z-2)(z^2+2z+4)$ **41.** prime **43.** $8m^3(4m^6+2m^2+3)$ **45.** $(4r+3m)^2$ **47.** $(5h+7g)(3h-2g)$ **49.** $(k-5)(k-6)$ **51.** $3k(k-5)(k+1)$ **53.** $(10p+3)(100p^2-30p+9)$ **55.** $(2+m)(3+p)$ **57.** $(4z-1)^2$ **59.** $3(6m-1)^2$ **61.** prime **63.** $8z(4z-1)(z+2)$ **65.** $(4+m)(5+3n)$ **67.** $2(3a-1)(a+2)$ **69.** $(a-b)(a^2+ab+b^2+2)$ **71.** $(8m-5n)^2$ **73.** $(4k-3h)(2k+h)$ **75.** $(m+2)(m^2+m+1)$ **77.** $(5y-6z)(2y+z)$ **79.** $(8a-b)(a+3b)$ **81.** $(3m+8)(3m-8)$ **83.** prime **85.** $(a+4)^2$

4.5 Exercises (page 231)

1. $2, -4$ **3.** $-\frac{5}{3}, \frac{1}{2}$ **5.** $-\frac{1}{5}, \frac{1}{2}$ **7.** $-\frac{9}{2}, \frac{1}{3}$ **9.** $1, -\frac{5}{3}$ **11.** $\frac{7}{3}, -4$ **13.** $-2, -3$ **15.** $-1, 6$ **17.** $-7, 4$ **19.** $-8, 3$ **21.** $-1, -2$ **23.** $\frac{1}{3}, -2$ **25.** $\frac{5}{2}, -2$ **27.** $\frac{1}{3}, -\frac{5}{2}$ **29.** $-\frac{2}{5}$ **31.** $-\frac{2}{5}, 2$ **33.** $\frac{3}{2}, -\frac{3}{2}$ **35.** $2, -2$ **37.** $0, -\frac{3}{2}$ **39.** $0, \frac{5}{3}$ **41.** $\frac{1}{2}, -4$ **43.** $\frac{2}{3}, 6$ **45.** 1 **47.** 2 **49.** $\frac{4}{3}, -\frac{2}{3}, \frac{1}{2}$ **51.** $\frac{4}{3}, -\frac{3}{2}, 1$ **53.** $0, 2, -2$ **55.** $0, \frac{3}{4}, -\frac{3}{4}$ **57.** $0, 3, -2$ **59.** $0, 4, 2$ **61.** $0, -1, 3$ **63.** $-\frac{3}{2}, \frac{4}{3}, -\frac{1}{2}$ **65.** $0, \frac{3}{4}, \frac{3}{2}$ **69.** 1 gallon **71.** 12 meters

4.6 Exercises (page 235)

1. **(a)** $A = LW$ **(b)** $64 = (x + 6)(x - 6)$ **(c)** 10, -10 (discard) **(d)** length: 16 centimeters; width: 4 centimeters
3. **(a)** $A = \frac{1}{2}bh$ **(b)** $10 = \frac{1}{2}(3x + 2)(2x - 1)$ **(c)** $\frac{11}{6}$, -2 (discard) **(d)** base: $\frac{15}{2}$ feet; height: $\frac{8}{3}$ feet
5. 6 centimeters **7.** length: 18 inches; width: 9 inches **9.** 2 inches **11.** 5 centimeters **13.** 19 feet
15. 6 feet **17.** 12 centimeters **19.** 12 miles **21.** 25 miles **23.** 4 inches **25.** 10° **27.** 7 centimeters
29. 60° **31.** 9 centimeters **33.** 256 feet **35.** 10 seconds **37.** 1 second, 3 seconds **41.** $\pi x^2 + 2xy$
43. true **45.** false **47.** true

4.7 Exercises (page 242)

1. $-2 < m < 5$
3. $t \leq -6$ or $t \geq -5$ 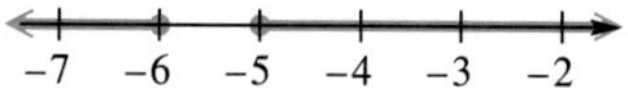
5. $-3 < a < 3$
7. $a \leq -6$ or $a \geq 7$ 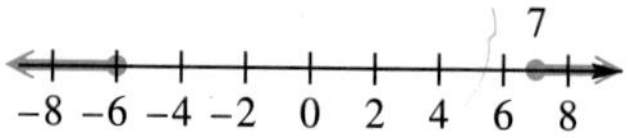
9. $m < -3$ or $m > -2$ 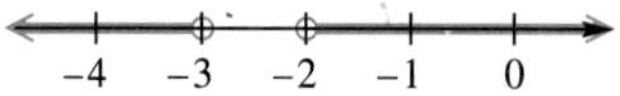
11. $-1 \leq z \leq 5$ 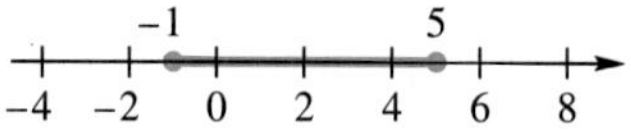
13. $-1 < m < \frac{2}{5}$
15. $-\frac{1}{2} < r < \frac{4}{3}$
17. $1 < q < 6$
19. $m < -\frac{1}{2}$ or $m > \frac{1}{3}$ 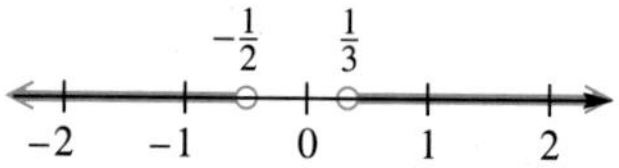
21. $-\frac{2}{3} < p < -\frac{1}{4}$
23. $m < -2$ or $m > 2$ 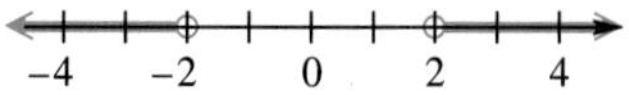
25. $r < -4$ or $r > 4$ 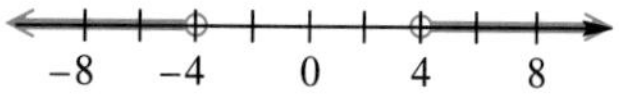
27. $-2 \leq a \leq \frac{1}{3}$ or $a \geq 4$ 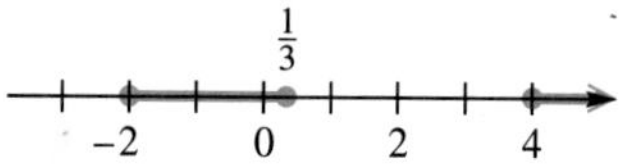
29. $r < -1$ or $2 < r < 4$
31. (d) **33.** (b) **35.** $\frac{2}{3}$ **37.** $\frac{1}{3}$ **39.** $\frac{25}{36}$
41. $\frac{1}{6}$

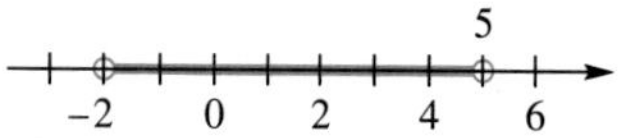

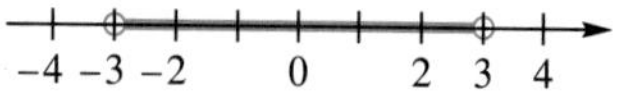

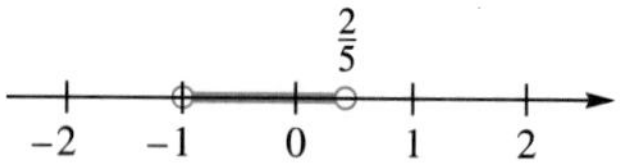

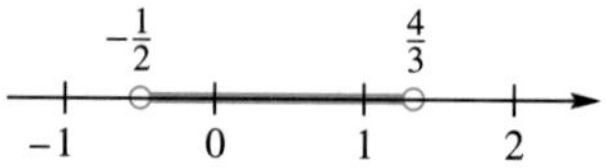

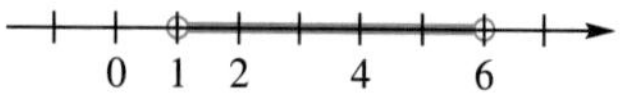

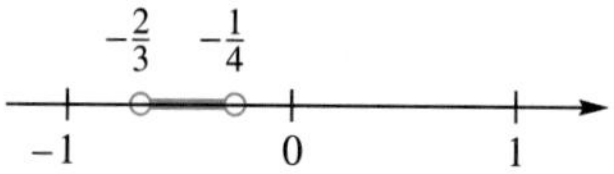

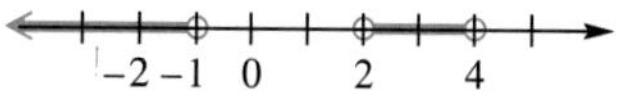

Chapter 4 Review Exercises (page 246)

1. $6(1 - 3r^5 + 2r^3)$ **3.** $(2p + 3)(3p + q)$ **5.** $(r - 9)(r + 3)$ **7.** $(z - 11)(z + 4)$ **9.** $p^5(p - 2q)(p + q)$
11. $(2k - 1)(k - 2)$ **13.** $(3r + 2)(2r - 3)$ **15.** $(7m - 2n)(m + 3n)$ **17.** $(10a + 3)(10a - 3)$
19. $36(2p + q)(2p - q)$ **21.** $(4m + 5n)^2$ **23.** $6x(3x - 2)^2$ **25.** $(7x + 4)(49x^2 - 28x + 16)$ **27.** 4, 1
29. 6 **31.** 0, 3, -3 **33.** 6 meters **35.** 8 feet **37.** $r \leq -4$ or $r \geq \frac{1}{2}$ **39.** $3 < z < 5$
41. $-3 \leq m \leq -\frac{5}{2}$ or $m \geq 1$ **43.** $4k^3(2k - 1)(3k - 1)$ **45.** $(10t - 3)(100t^2 + 30t + 9)$
47. $(4y + 3)(y + 2)$ **49.** $50y^3(2y^3 - 1 + 6y)$ **51.** $(3r - 7)^2$ **53.** $(p + 12q)(p - 10q)$ **55.** $-\frac{4}{3}$, 5

57. $\frac{7}{10}, -\frac{7}{10}$ **59.** 4 meters **61.** shorter leg: 15 meters; longer leg: 36 meters; hypotenuse: 39 meters

Chapter 4 Test (page 249)

[4.1] **1.** $6ab(1 - 6b)$ **2.** $5kt(3k + 5t)$ **3.** $8m^2(2 - 3m + 4m^2)$ **4.** $14p(2q + 1 + 4pq^3)$ **5.** prime **6.** $(2 - a)(6 + b)$ [4.2] **7.** $(x - 9)(x + 5)$ **8.** $3p^2q(p + 7)(p - 1)$ [4.3] **9.** $(3a - 2)(a + 5)$ **10.** $3z^3(5z + 1)(2z - 5)$ **11.** $p^2(4r + 5)(3r + 1)$ **12.** $(3t - 2x)(2t + x)$ [4.4] **13.** $2(5m + 7)(5m - 7)$ **14.** $(12a + 13b)(12a - 13b)$ **15.** $(a^2 + 25)(a + 5)(a - 5)$ **16.** $(2p + 3)^2$ **17.** $(5z - 1)^2$ **18.** $4y(y + 2)^2$ **19.** $(2p - 5)(4p^2 + 10p + 25)$ **20.** $(3r + 4t^2)(9r^2 - 12rt^2 + 16t^4)$ **21.** $(m - n - 4)(m + n)$ [4.5] **22.** $\frac{1}{3}, -2$ **23.** $\frac{5}{2}, -4$ **24.** $3, \frac{5}{2}, -\frac{2}{3}$ **25.** $0, 4, -4$ **26.** width: 3 feet; length: 5 feet **27.** 17 feet

29. $-3 \le p \le \frac{1}{2}$

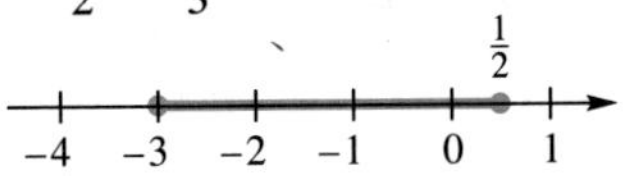

30. $m < -6$ or $m > 4$

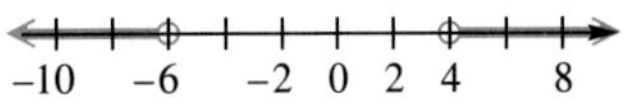

CHAPTER 5

5.1 Exercises (page 256)

1. 0 **3.** -5 **5.** 5, 3 **7.** none **9.** **(a)** 1 **(b)** $\frac{14}{9}$ **11.** **(a)** 2 **(b)** $-\frac{14}{3}$ **13.** **(a)** $\frac{256}{15}$ **(b)** undefined **15.** **(a)** -5 **(b)** $\frac{5}{23}$ **17.** **(a)** -5 **(b)** $\frac{15}{2}$ **19.** **(a)** undefined **(b)** $\frac{1}{10}$ **23.** $2k$ **25.** $-\frac{4m}{3p}$ **27.** $\frac{3}{4}$ **29.** $\frac{x - 1}{x + 1}$ **31.** $4m^2 - 3$ **33.** $\frac{8y + 5}{6}$ **35.** cannot be written in simpler form **37.** $m - n$ **39.** $\frac{m}{2}$ **41.** $2(2r + s)$ or $4r + 2s$ **43.** $\frac{m - 2}{m + 3}$ **45.** $\frac{-(x + 4)}{1 + x}$ **47.** -1 **49.** 1 **51.** $-(x + 1)$ or $-x - 1$ **53.** -1 **55.** (d) **57.** $\frac{a + 1}{b}$ **59.** $\frac{m + n}{2}$ **61.** $-\frac{b^2 + ba + a^2}{a + b}$ **63.** $\frac{z + 3}{z}$ **65.** $\frac{15}{32}$ **67.** $\frac{32}{9}$ **69.** 4 **71.** $\frac{81}{10}$

5.2 Exercises (page 263)

1. $\frac{3m}{4}$ **3.** $\frac{3}{32}$ **5.** $2a^4$ **7.** $\frac{1}{4}$ **9.** $\frac{1}{6}$ **11.** $\frac{2}{r^5}$ **15.** $\frac{6}{a + b}$ **17.** $\frac{2}{9}$ **19.** $\frac{3}{10}$ **21.** $\frac{2r}{3}$ **23.** $(y + 4)(y - 3)$ **25.** $\frac{18}{(m - 1)(m + 2)}$ **27.** $-\frac{7}{8}$ **29.** -1 **31.** $m - 4$ **33.** $\frac{k + 2}{k + 3}$ **35.** $\frac{z + 4}{z - 4}$ **37.** $\frac{m - 3}{2m - 3}$ **39.** $\frac{m + 4p}{m + p}$ **41.** $(x + 1)(x + 2)$ **43.** $\frac{10}{x + 10}$ **45.** $\frac{3 - a - b}{2a - b}$ or $\frac{a + b - 3}{b - 2a}$ **47.** $-\frac{(x + y)^2(x^2 - xy + y^2)}{3y(y - x)(x - y)}$ or $\frac{(x + y)^2(x^2 - xy + y^2)}{3y(x - y)^2}$ **49.** $2^2 \cdot 3$ **51.** $2^3 \cdot 3^2$ **53.** 3 **55.** $4p^2q^3$

5.3 Exercises (page 268)

1. 60 **3.** 120 **5.** 1800 **7.** x^5 **9.** $30p$ **11.** $180y^4$ **13.** $15a^5b^3$ **15.** $12p(p - 2)$ **17.** Multiply the denominators together. **19.** $32r^2(r - 2)$ **21.** $18(r - 2)$ **23.** $12p(p + 5)^2$ **25.** $8(y + 2)(y + 1)$ **27.** $m - 3$ or $3 - m$ **29.** $p - q$ or $q - p$ **31.** $a(a + 6)(a - 3)$ **33.** $(k + 3)(k - 5)(k + 7)(k + 8)$ **35.** $(2y - 1)(y + 4)(y - 3)$ **37.** $(3r - 5)(2r + 3)(r - 1)$ **39.** $\frac{42}{66}$ **41.** $\frac{-88}{8m}$ **43.** $\frac{24y^2}{70y^3}$ **45.** $\frac{60m^2k^3}{32k^4}$

47. $\frac{57z}{6z - 18}$ **49.** $\frac{-4a}{18a - 36}$ **51.** $\frac{6(k + 1)}{k(k - 4)(k + 1)}$ **53.** $\frac{36r(r + 1)}{(r - 3)(r + 2)(r + 1)}$ **55.** $\frac{ab(a + 2b)}{2a^3b + a^2b^2 - ab^3}$
57. $\frac{(t - r)(4r - t)}{t^3 - r^3}$ **59.** $\frac{2y(z - y)(y - z)}{y^4 - z^3y}$ or $\frac{-2y(y - z)^2}{y^4 - z^3y}$ **63.** $\frac{7}{3}$ **65.** $\frac{9}{10}$ **67.** $\frac{2}{3}$ **69.** $\frac{19}{72}$

5.4 Exercises (page 275)

1. $\frac{7}{p}$ **3.** $-\frac{3}{k}$ **5.** 1 **7.** $-\frac{2q}{3}$ **9.** m **11.** $q - 2$ **13.** 1 **17.** $\frac{2m + 3}{6}$ **19.** $\frac{4y - 3}{3y}$ **21.** $\frac{m}{6}$
23. $\frac{k + 2}{2}$ **25.** $\frac{5m + 11}{4m}$ **27.** $\frac{6 - 2y}{y^2}$ **29.** $\frac{-y + 3x}{x^2y}$ **31.** $\frac{4x + 24}{(x - 2)(x + 2)}$ **33.** $\frac{x}{2(x + y)}$
35. $\frac{m}{3(m + 3)(m - 3)}$ **37.** $\frac{3}{(m + 1)(m - 1)(m + 2)}$ **39.** $\frac{3}{q - 2}$ or $\frac{-3}{2 - q}$ **41.** $\frac{-6}{4p - 5}$ or $\frac{6}{5 - 4p}$
45. $\frac{43m + 1}{5m(m - 2)}$ **47.** $\frac{9r + 2}{r(r + 2)(r - 1)}$ **49.** $\frac{11y^2 - y - 11}{(2y - 1)(y + 3)(3y + 2)}$ **51.** $\frac{2x^2 + 6xy + 8y^2}{(x + y)(x + y)(x + 3y)}$ or
$\frac{2x^2 + 6xy + 8y^2}{(x + y)^2(x + 3y)}$ **53.** $\frac{15r^2 + 10ry - y^2}{(3r + 2y)(6r - y)(6r + y)}$ **55.** $\frac{2p^2 + 2p^2q^2 + 4q^2 - 14p + 6}{(1 + 2q^2)(1 - 2q^2)(p - 4)}$ **57.** $\frac{2k^2 - 10k + 6}{(k - 3)(k - 1)^2}$
59. $\frac{7k^2 + 31k + 92}{(k - 4)(k + 4)^2}$ **61.** (a) $\frac{7y - 2}{6}$ (b) $\frac{y^2 - y - 2}{12}$ **63.** $2\frac{5}{6}$ **65.** $-3\frac{1}{11}$

5.5 Exercises (page 281)

1. $\frac{9}{14}$ **3.** -6 **5.** $\frac{1}{pq}$ **7.** $\frac{1}{x}$ **9.** $\frac{mp}{40}$ **11.** $\frac{x(x + 1)}{4(x - 3)}$ **13.** $\frac{18q}{1 - q}$ **15.** $\frac{5(2x + 3)}{4(x - 1)}$ **17.** $\frac{31}{50}$
19. $\frac{5}{18}$ **21.** $\frac{2}{y}$ **23.** $\frac{8}{x}$ **25.** $\frac{x^2 + 1}{4 + xy}$ **27.** $\frac{5(p + 3)}{5 + p}$ **29.** $\frac{40 - 12p}{85p}$ **31.** $\frac{5y - 2x}{3 + 4xy}$ **33.** $\frac{a - 2}{2a}$
35. $\frac{z - 5}{4}$ **37.** $\frac{-m}{m + 2}$ **39.** division **43.** $\frac{3m(m - 3)}{(m - 1)(m - 8)}$ **45.** $\frac{1}{3}$ **47.** $\frac{19r}{15}$ **49.** $12x + 2$
51. $44p^2 - 27p$ **53.** -2 **55.** $-\frac{13}{3}$ **57.** -4, 1

5.6 Exercises (page 289)

1. expression; $\frac{41}{35}x$ **3.** equation; $\frac{35}{41}$ **5.** expression; $-\frac{13}{12}y$ **7.** equation; 12 **11.** $\frac{1}{2}$ **13.** 12
15. $\frac{21}{8}$ **17.** $\frac{2}{5}$ **19.** 24 **21.** 2 **23.** -7 **25.** 1 **27.** 2 **29.** -8 **31.** 12 **33.** 2 **35.** 3
37. 5 **39.** 13 **41.** 12 **43.** 5 **47.** -3 **49.** 1 **51.** no solution **53.** no solution **55.** 3, $\frac{4}{3}$
57. 3, $-\frac{6}{7}$ **59.** Factor out k on the left side. **61.** $R = \frac{kE}{I}$ **63.** $r = \frac{k}{F}$ **65.** $R = \frac{E - Ir}{I}$ **67.** $r = \frac{S - a}{S}$
or $r = \frac{a - S}{-S}$ **69.** $y = \frac{xz}{x + z}$ **71.** $z = \frac{3y}{5 - 9xy}$ or $z = \frac{-3y}{9xy - 5}$ **73.** $a = \frac{bc}{c + b}$ **75.** 3, $-\frac{1}{3}$
77. $\frac{1}{2}$, -6 **79.** 0, -4 **81.** $\frac{d}{10}$ kilometers per hour **83.** $\frac{1}{x}$

Summary on Rational Expressions (page 293)

1. $\frac{8}{m}$ **3.** $\frac{3}{4x^2(x + 3)}$ **5.** no solution **7.** $\frac{r + 4}{r + 1}$ **9.** -43 **11.** $\frac{17}{3(y - 1)}$ **13.** 2, $\frac{1}{5}$ **15.** $-\frac{19}{18z}$
17. -7 **19.** $\frac{3m + 5}{(m + 3)(m + 2)(m + 1)}$

5.7 Exercises (page 300)

1. 12 **3.** $\frac{9}{5}$ **5.** $\frac{1386}{97}$ **7.** $312 **9.** 16 pills **11.** $\frac{D}{R}=\frac{d}{r}$ **13.** 110 miles per hour **15.** 25 miles per hour **17.** 900 miles **19.** $\frac{1}{10}$ job per hour **21.** $\frac{24}{7}$ or $3\frac{3}{7}$ hours **23.** $\frac{84}{19}$ or $4\frac{8}{19}$ hours **25.** $\frac{27}{10}$ or $2\frac{7}{10}$ hours **27.** 36 hours **29.** $\frac{33}{4}$ or $8\frac{1}{4}$ hours **31.** 30 **33.** 147 **35.** $\frac{64}{25}$ **37.** 100 pounds per square inch **39.** $\frac{400}{27}$ or $14\frac{22}{27}$ footcandles **41.** $\frac{14}{3}$ **43.** 165 miles per hour **45.** 100 amps **47.** $1, -\frac{3}{2}$ **49.** $\frac{16}{3}$ or $5\frac{1}{3}$ days **51.** (a) 1 (b) -23 **53.** (a) 7 (b) -17 **55.** (a) $\frac{4}{7}$ (b) $\frac{22}{7}$ **57.** (a) $-\frac{2}{3}$ (b) $\frac{4}{3}$

Chapter 5 Review Exercises (page 308)

1. 0 **3.** 4, -2 **5.** (a) $\frac{9}{5}$ (b) 1 **7.** (a) $\frac{24}{7}$ (b) 8 **9.** $3p$ **11.** $\frac{3x-4}{2}$ **13.** $\frac{40}{3p^2}$ **15.** $\frac{7}{2}$ **17.** $\frac{2p-1}{2p+3}$ **19.** 90 **21.** $y(y+2)(y+4)$ **23.** $\frac{20}{45}$ **25.** $\frac{9m}{24m^3}$ **27.** $\frac{-10k}{15k+75}$ **29.** $\frac{1}{r}$ **31.** $\frac{35-2k}{5k}$ **33.** $\frac{-y-5}{(y+1)(y-1)}$ **35.** $\frac{8p-30}{p(p-2)(p-3)}$ **37.** 4 **39.** $\frac{16}{5}$ **41.** 20 **43.** 5 **45.** no solution **47.** 0, 3 **49.** $y=\frac{xz}{3-2x}$ or $y=\frac{-xz}{2x-3}$ **51.** $\frac{15}{11}$ **53.** 150 kilometers per hour **55.** 2 days **57.** $\frac{9}{4}$ **59.** $\frac{4p+8q}{15}$ **61.** $\frac{zx+1}{zx-1}$ **63.** $\frac{16-3r}{2r^2}$ **65.** -4 **67.** -6 **69.** $\frac{6}{5}, \frac{1}{2}$

Chapter 5 Test (page 310)

[5.1] **1.** 3, 1 **2.** (a) $\frac{1}{3}$ (b) undefined **3.** $\frac{4}{3mp^3}$ **4.** $\frac{5y(y-1)}{2}$ [5.2] **5.** a **6.** $-\frac{40}{27}$ **7.** $\frac{3m-2}{3m+2}$ **8.** $\frac{a+3}{a+4}$ [5.3] **9.** $150p^5$ **10.** $(2r+3)(r+2)(r-5)$ **11.** $\frac{77r}{49r^2}$ **12.** $\frac{15m}{24m^2-48m}$ [5.4] **13.** $-\frac{1}{x}$ **14.** $\frac{-13}{6(a+1)}$ **15.** $\frac{m^2-m-1}{m-3}$ or $\frac{-m^2+m+1}{3-m}$ **16.** $\frac{-2k^2+4k+3}{(2k-1)(k+2)(k+1)}$ [5.5] **17.** $\frac{2k}{3p}$ **18.** $\frac{-2p-7}{2p+9}$ [5.6] **19.** $-1, 4$ **20.** no solution **21.** 1 **22.** $\frac{2}{3}$ [5.7] **23.** 20 miles per hour **24.** $\frac{20}{9}$ or $2\frac{2}{9}$ hours **25.** $\frac{56}{3}$

CHAPTER 6

6.1 Exercises (page 318)

1. yes **3.** yes **5.** yes **7.** yes **9.** yes **13.** 20 **15.** 5 **17.** -7 **19.** -5

21.

x	2	0	-3
y	1	-5	-14

23.

x	4	0	-4
y	-22	-2	18

25.

m	1	0	-2
n	7	4	-2

27.

x	4	0	-4
y	-8	-8	-8

29.

x	9	-5	0
y	4	4	4

31.

x	0	12	-8
y	0	3	-2

33. $(-2, 2)$ **35.** $(-6, -2)$

39.–43.

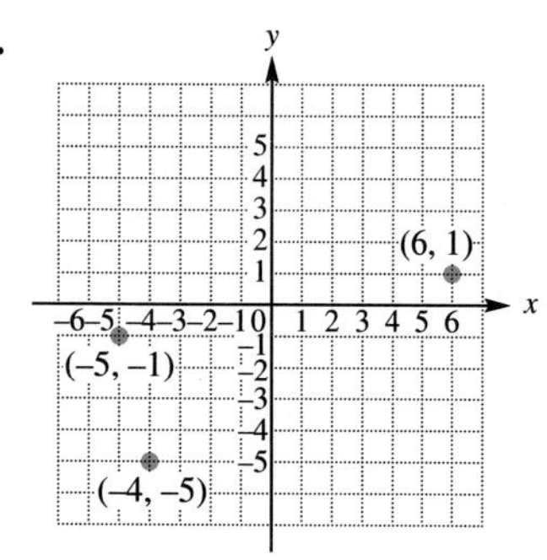

45.–49.

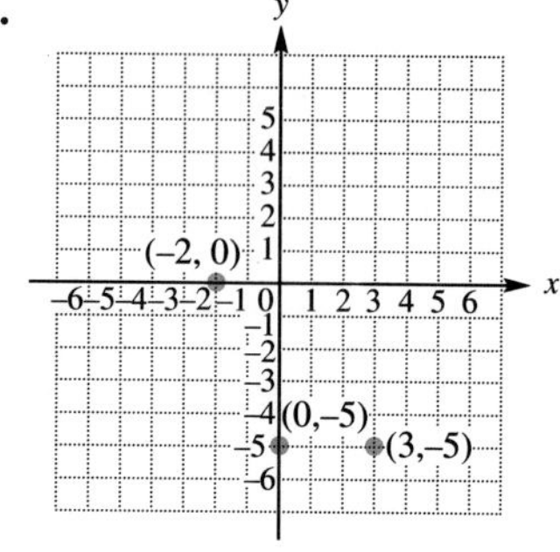

51. I **53.** II

55. III **57.** II

59. IV **61.** none

63.

x	0	2	−3	−2
y	6	10	0	2

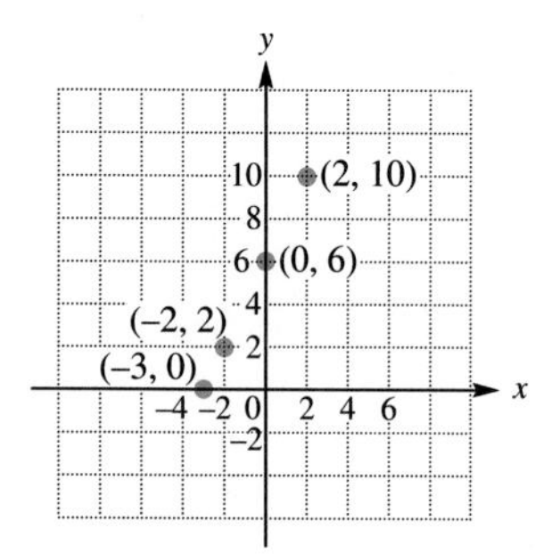

65.

x	0	10	5	−5
y	3	−3	0	6

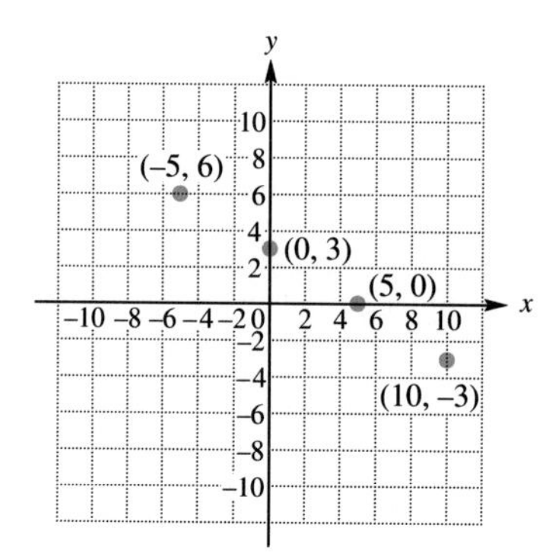

67.

x	0	−2	4	−1
y	0	−6	12	−3

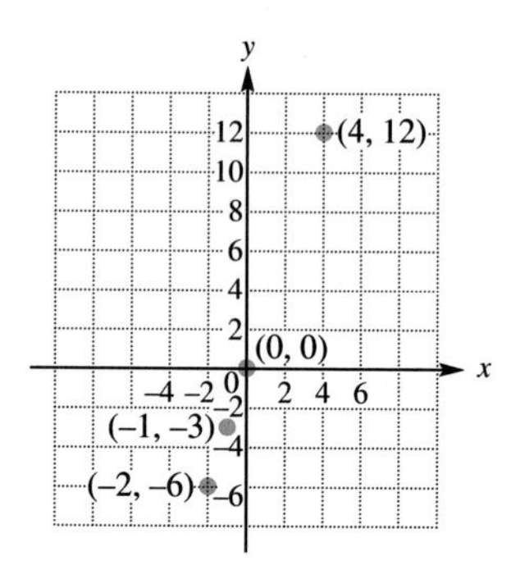

69.

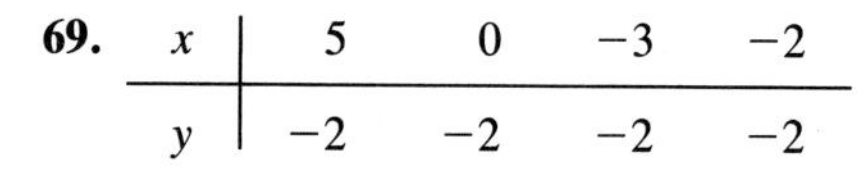

x	5	0	−3	−2
y	−2	−2	−2	−2

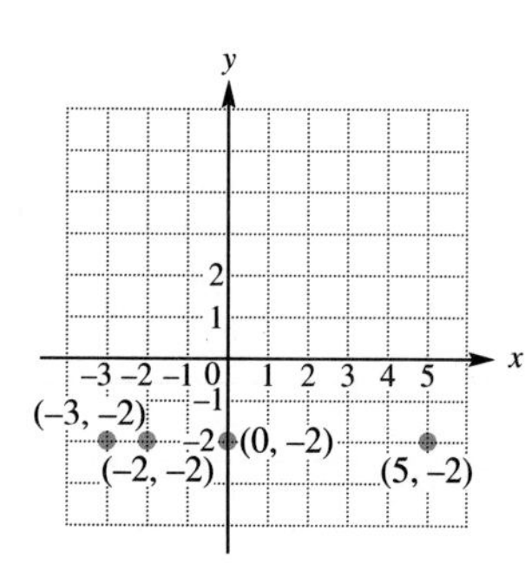

71.

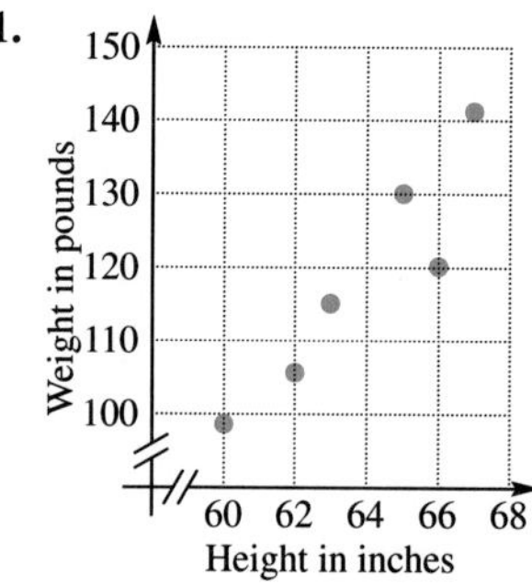

73. yes

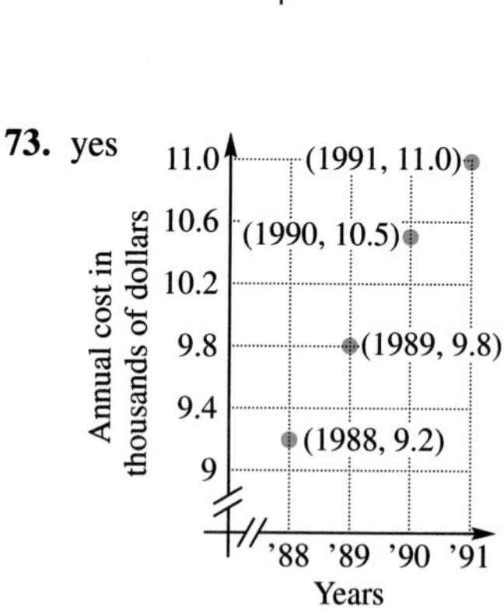

75. $\frac{8}{3}$ **77.** −9 **79.** 1 **81.** 15

6.2 Exercises (page 328)

1. (0, 5), (5, 0), (2, 3)

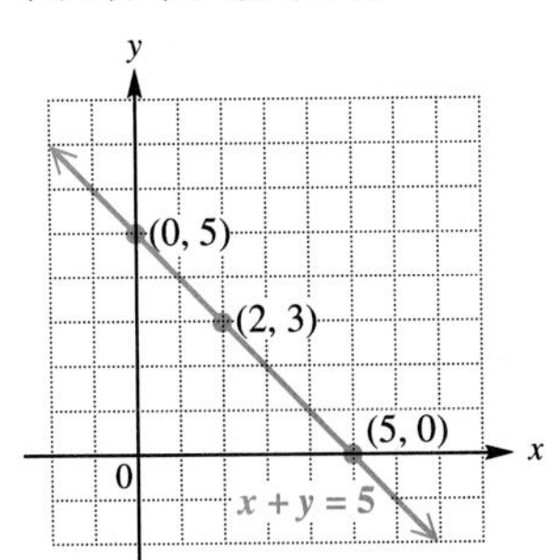

3. (0, 4), (−4, 0), (−2, 2)

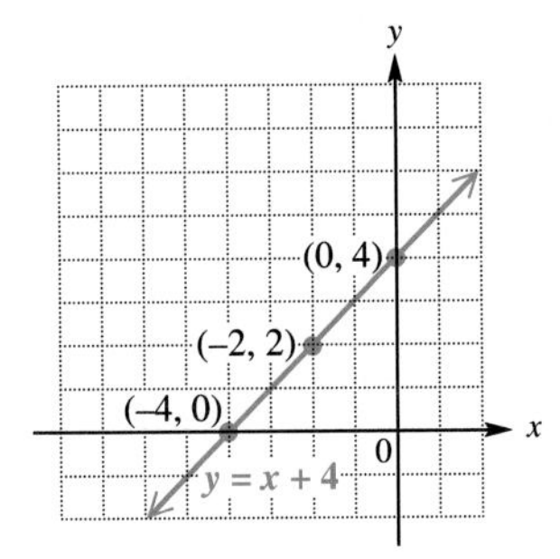

5. (0, −6), (2, 0), (3, 3)

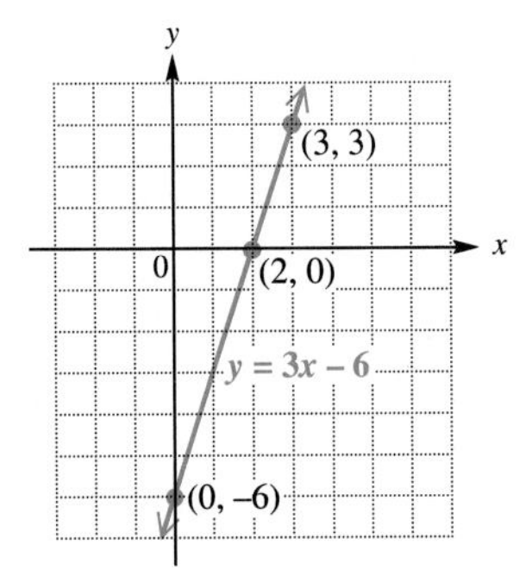

7. (0, 4), (10, 0), (5, 2)

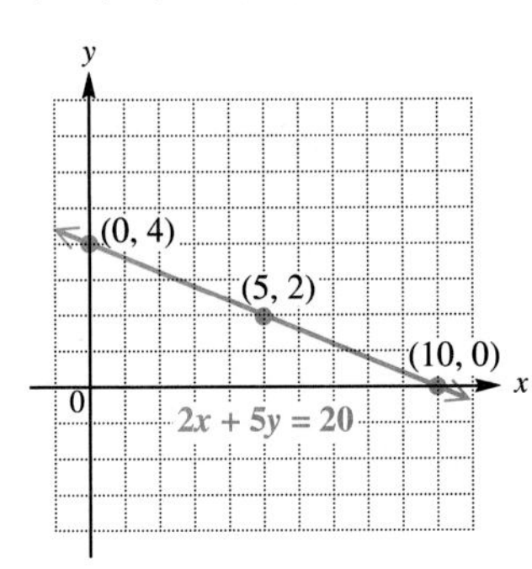

9. (−5, 2), (−5, 0), (−5, −3)

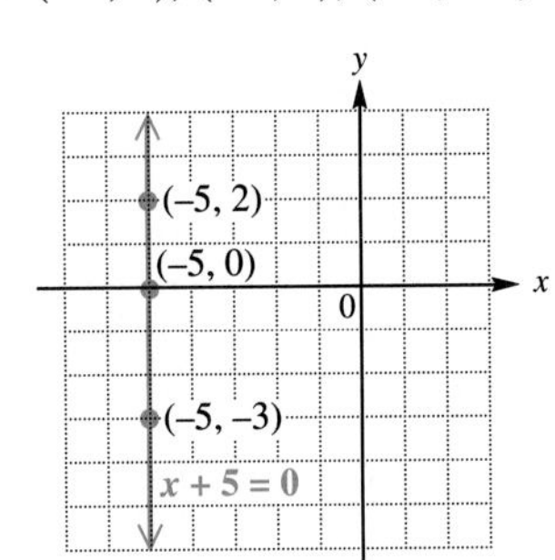

11. x-intercept: (3, 0); y-intercept: (0, 2)

13. x-intercept: (3, 0); y-intercept: $\left(0, -\frac{9}{5}\right)$

15. x-intercept: (0, 0); y-intercept: (0, 0)

17. x-intercept: (−4, 0); y-intercept: none

19.

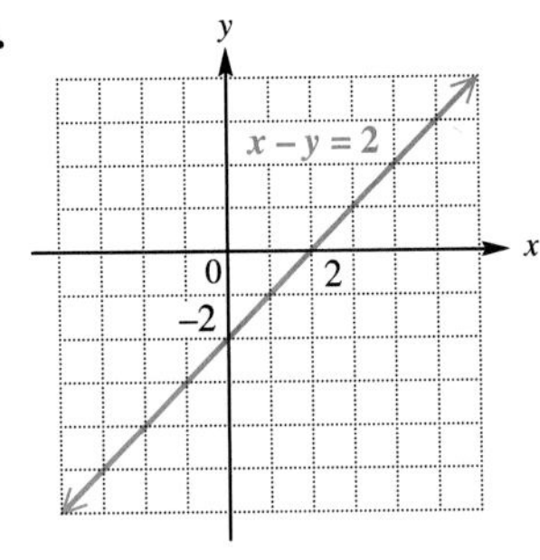

21.

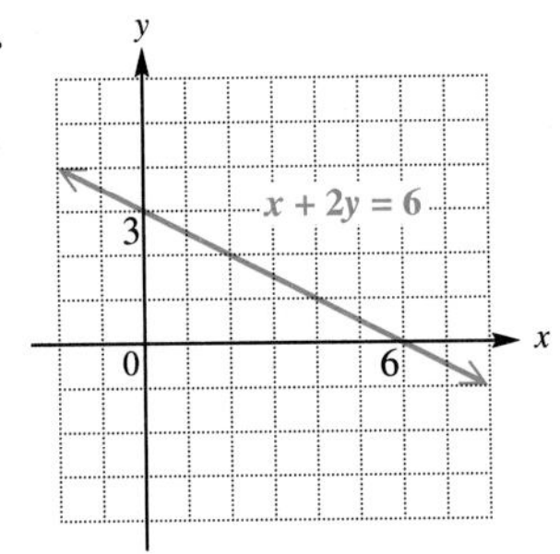

23.

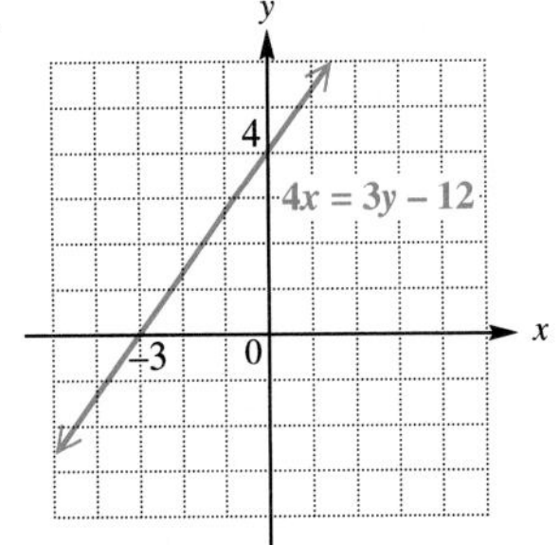

25.

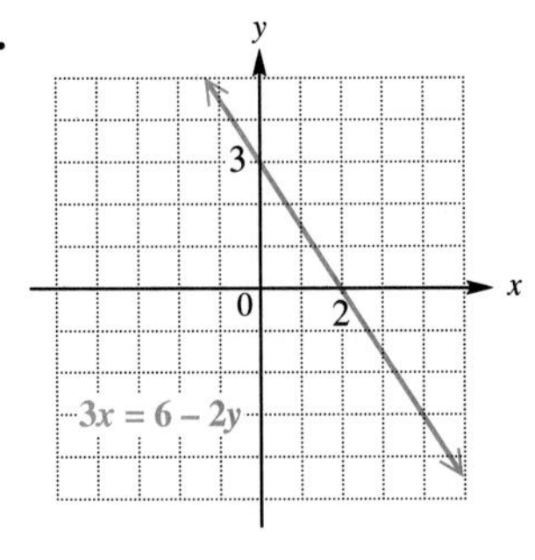

27.

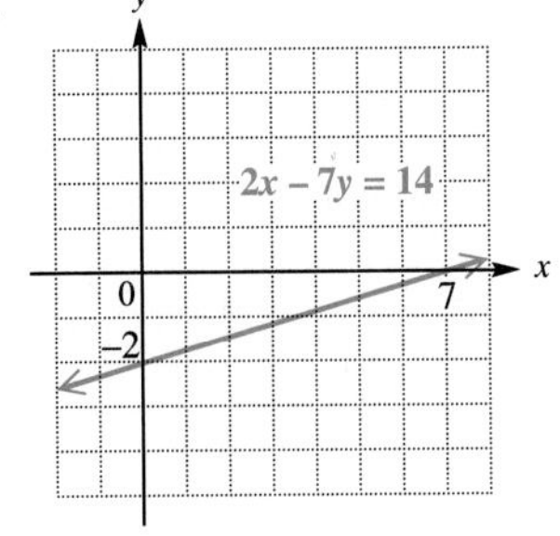

29.

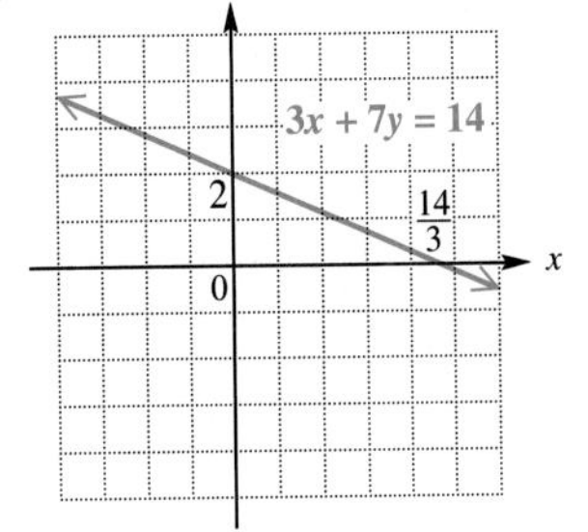

31.

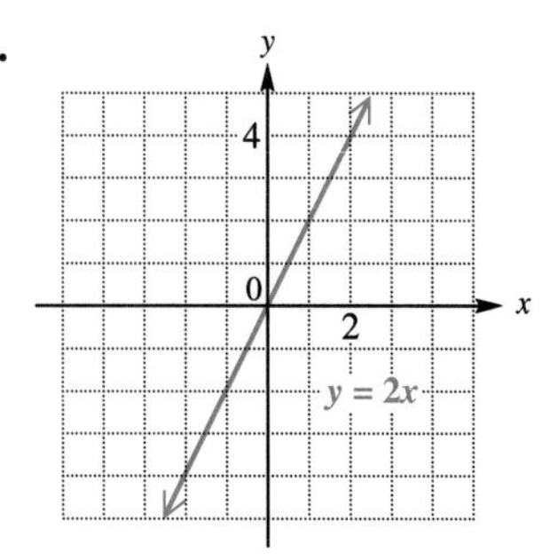

33.

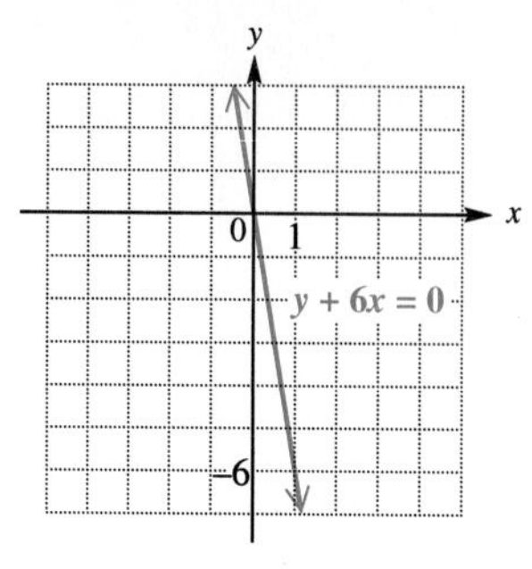

35.

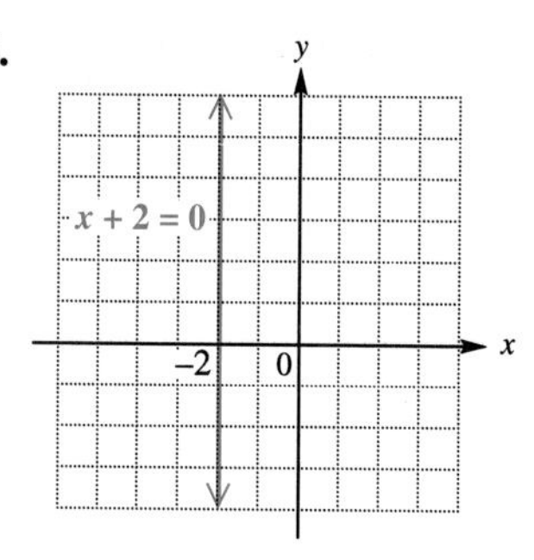

37.

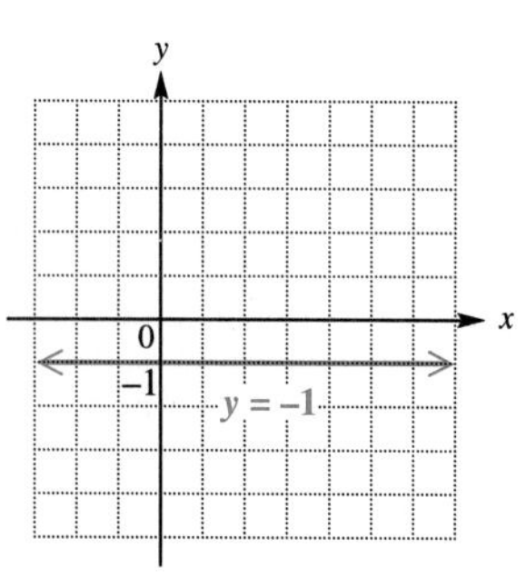

39. $x = y + 2$

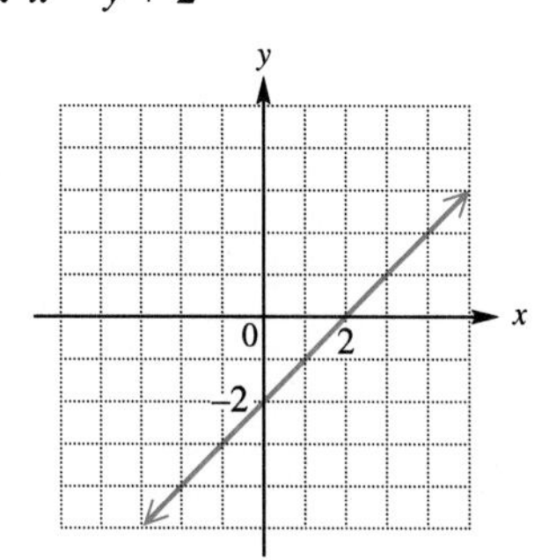

41. $y = 2x - 3$

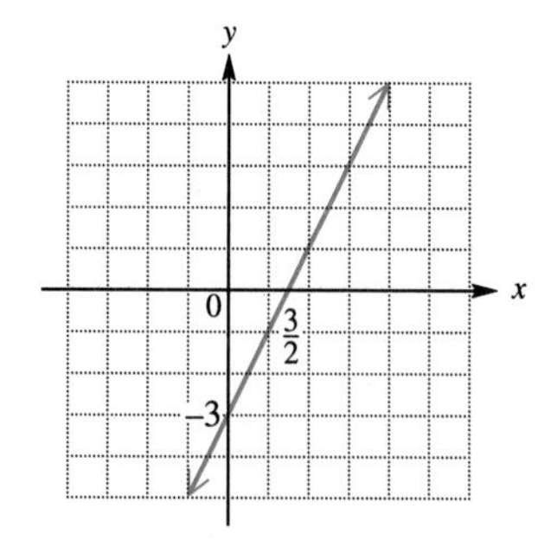

43. $3 + y = 2x - 4$ or $y = 2x - 7$

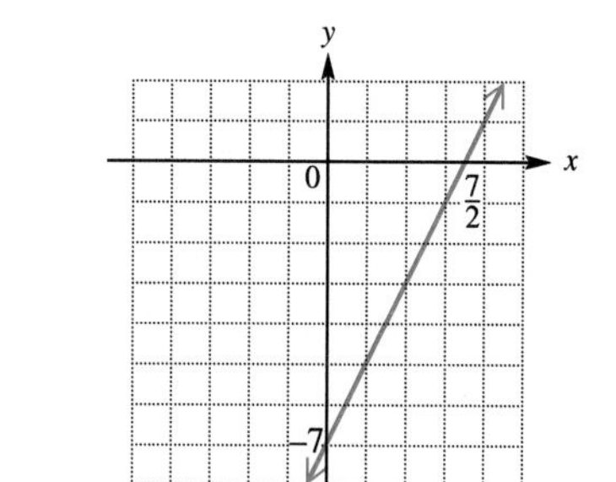

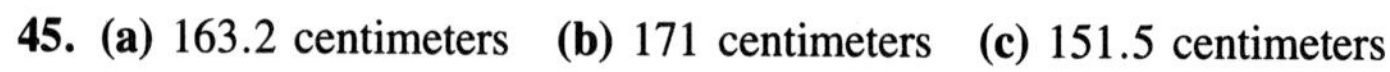
45. **(a)** 163.2 centimeters **(b)** 171 centimeters **(c)** 151.5 centimeters

(d)

h
180
170
160
150
0
20 21 22 23 24 25
r
$h = 73.5 + 3.9r$
Height in centimeters
Length of radius bone in centimeters

47. **(a)** \$1250 **(b)** \$1250 **(c)** \$1250 **(d)** \$6250 **51.** $\frac{2}{3}$ **53.** $\frac{1}{2}$ **55.** $-\frac{1}{7}$ **57.** $\frac{13}{5}$ **59.** undefined

6.3 Exercises (page 336)

1. $-\frac{1}{2}$ **3.** $\frac{2}{7}$ **5.** 0 **7.** $\frac{7}{6}$ **9.** $\frac{9}{2}$ **11.** $-\frac{5}{8}$ **13.** 0 **15.** undefined **17.** 5 **19.** 1 **21.** 9 **23.** 4 **25.** $\frac{3}{2}$ **27.** $-\frac{9}{7}$ **29.** (c) **31.** (d) **33.** perpendicular **35.** perpendicular **37.** perpendicular **39.** parallel **41.** neither **43.** perpendicular **47.** $\frac{15}{8}$ **49.** $-\frac{1}{2}$ **51.** $\frac{1}{5}$ **53.** $\frac{8}{27}$ **55.** $y = 3x - 10$ **57.** $y = -2x - 1$ **59.** $y = 4x - 1$ **61.** $y = -\frac{x}{2} - \frac{21}{10}$ or $y = -\frac{1}{2}x - \frac{21}{10}$

6.4 Exercises (page 343)

1. $y = 3x + 5$ **3.** $y = -x - 6$ **5.** $y = \frac{2}{5}x - \frac{1}{4}$

7. 

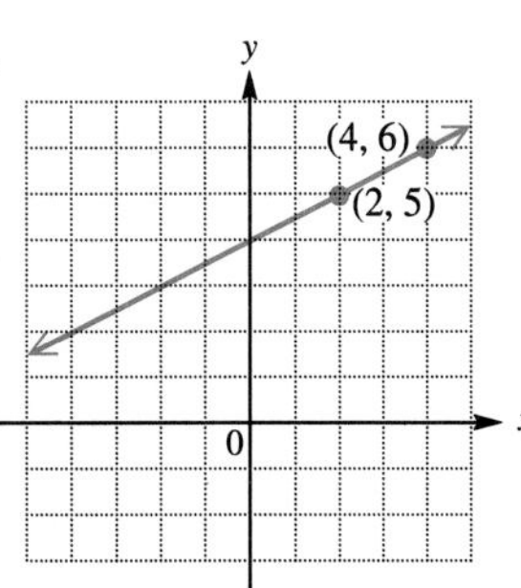

9. 

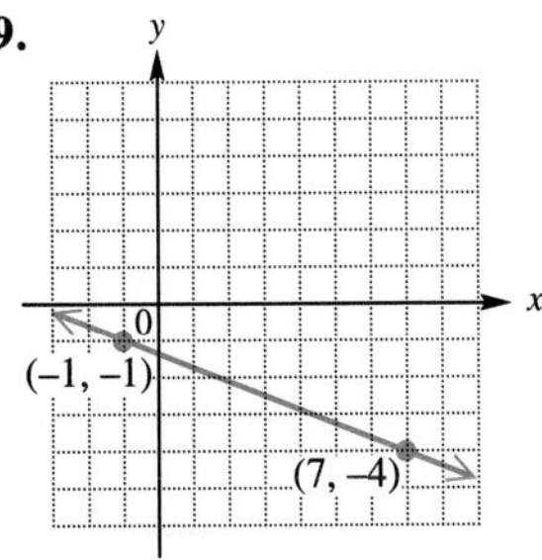

11.

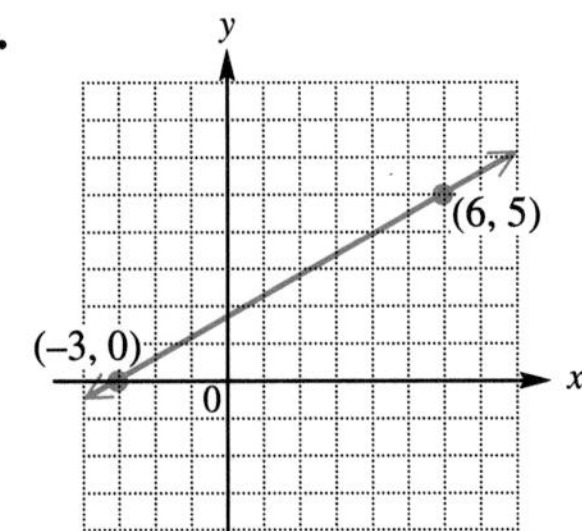

13.

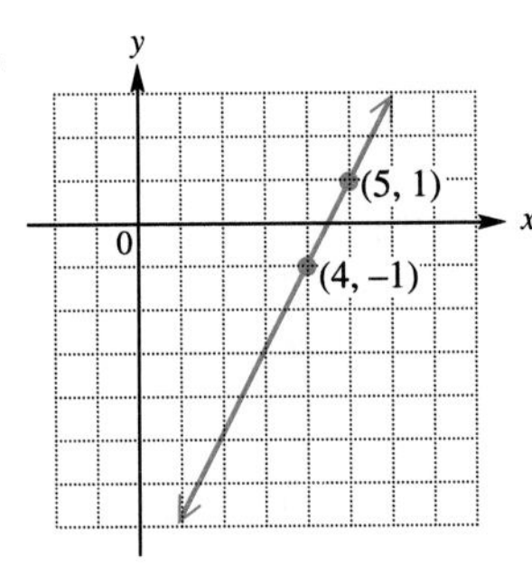

15.

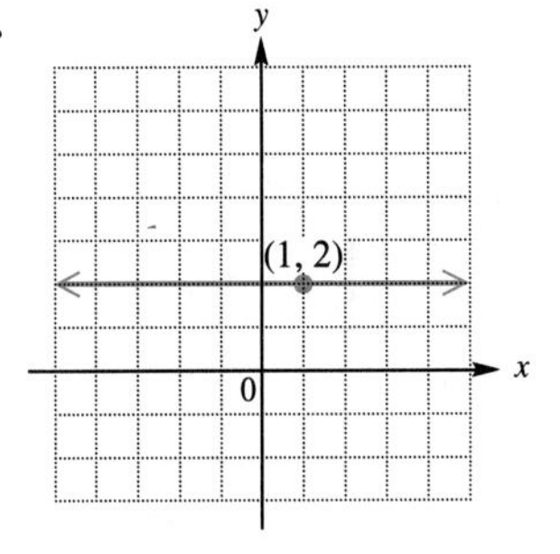

17.

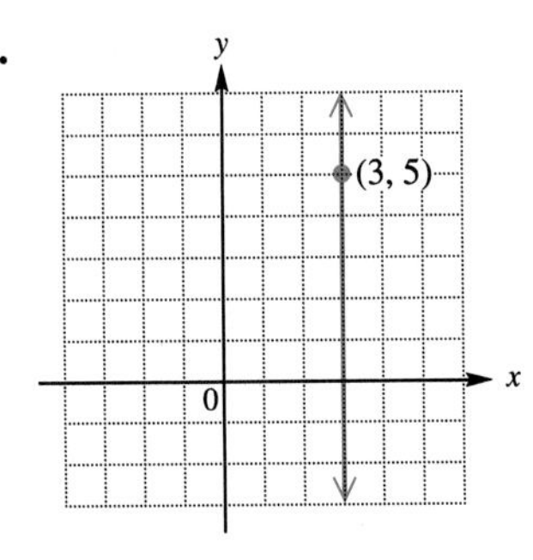

19. $y = 2x - 8$

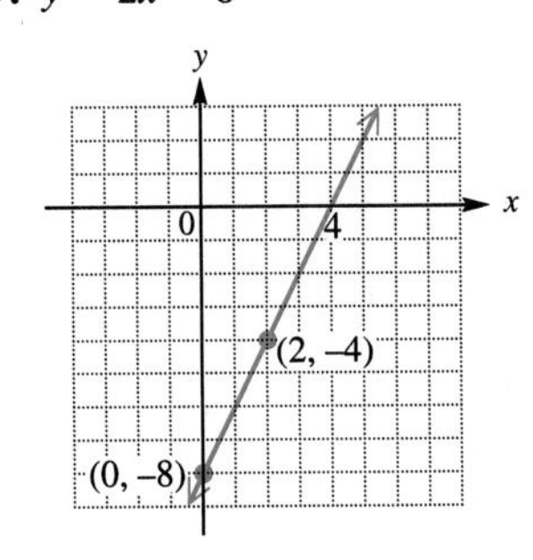

21. $y = -3x - 5$

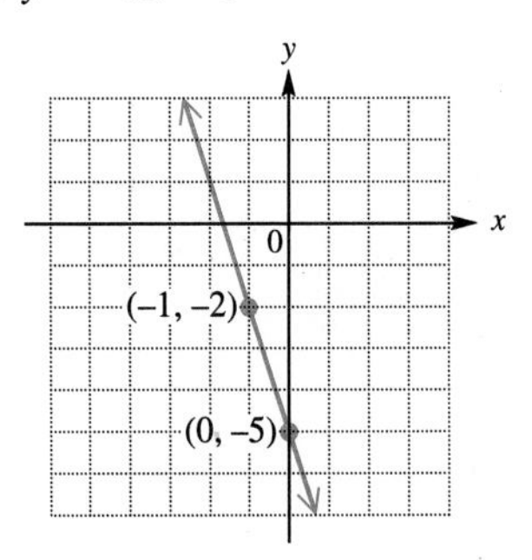

23. (d) **25.** (c) **27.** $2x - y = 7$ **29.** $2x + y = -4$ **31.** $2x - 3y = -9$ **33.** $3x + 4y = -17$
35. $8x + 11y = 48$ **37.** $3x - 5y = -11$ **39.** $3x + 5y = -11$ **41.** $2x + 3y = 6$ **43.** $2x + y = -1$
45. $x - 3y = -4$ **47.** $88x - 72y = -133$ **49.** **(a)** $3x - y = 11$ **(b)** $x + 3y = 17$ **51.** $y = 12x + 100$
53. **(a)** \$700 **(b)** \$1600 **55.** **(a)** (1, 24); (5, 48) **(b)** $6x - y = -18$

59. $m < 7$

0 1 2 3 4 5 6 7 8 9

61. $p \leq \frac{1}{2}$

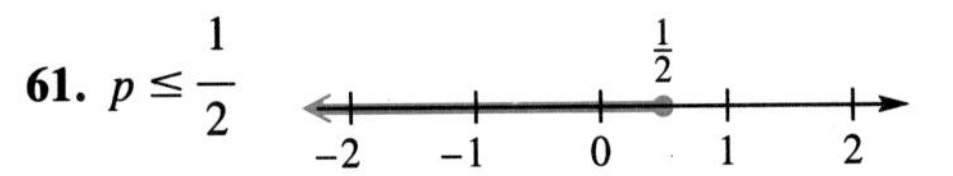

63. $p > -3$

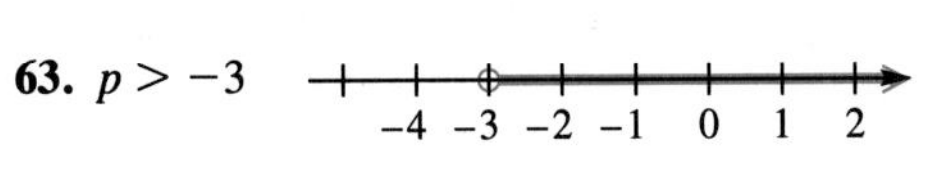

65. $p < -11$

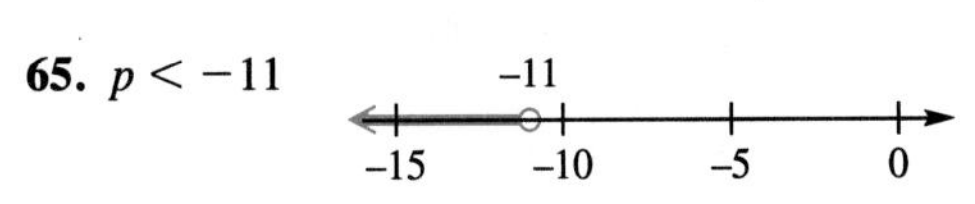

6.5 Exercises (page 349)

1.

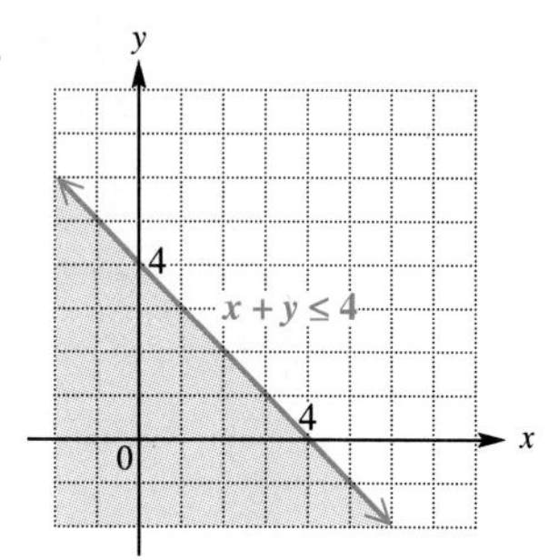

3.

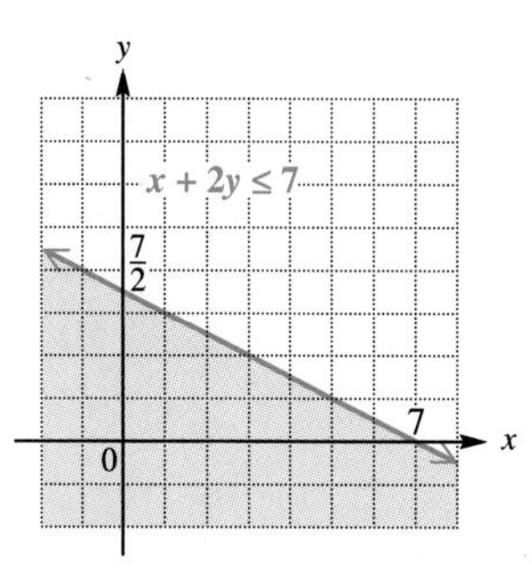

5.

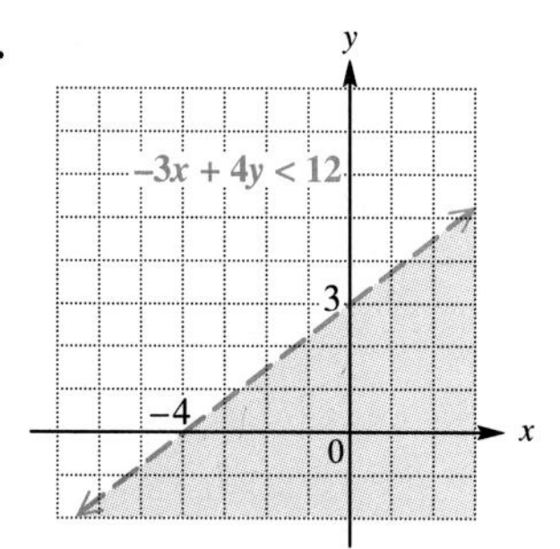

7.

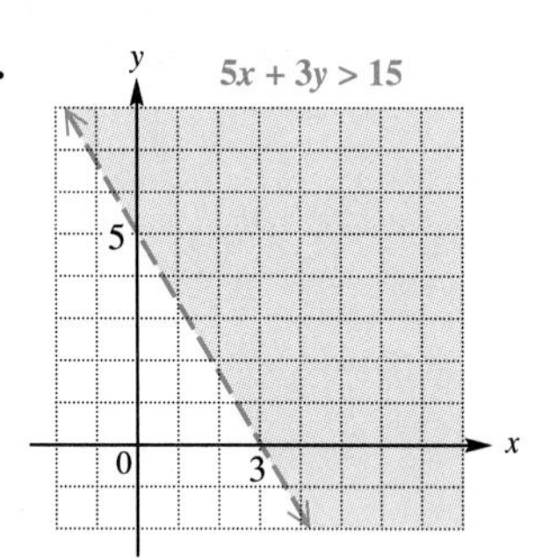

9.

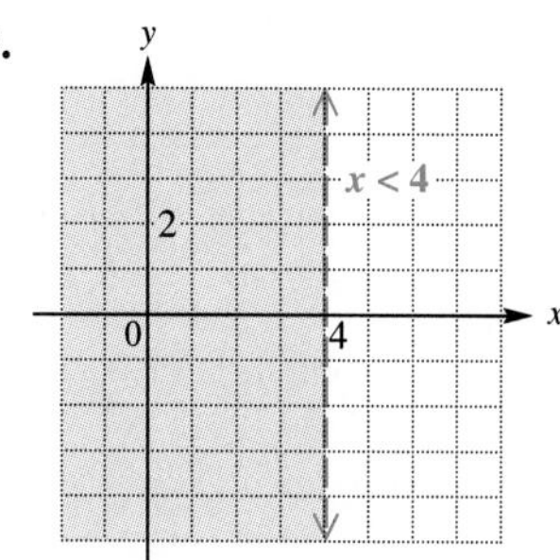

11.

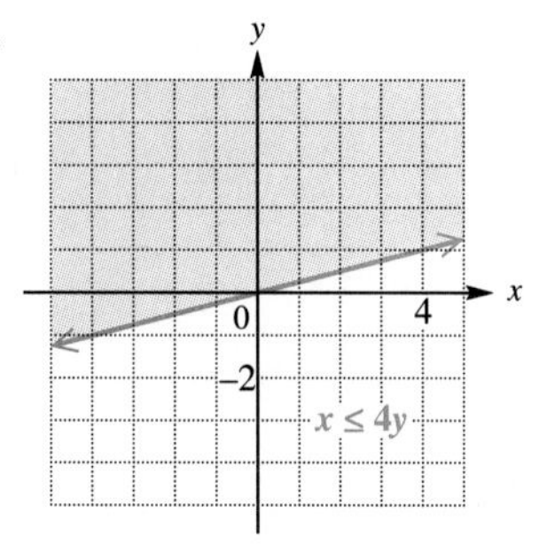

13.

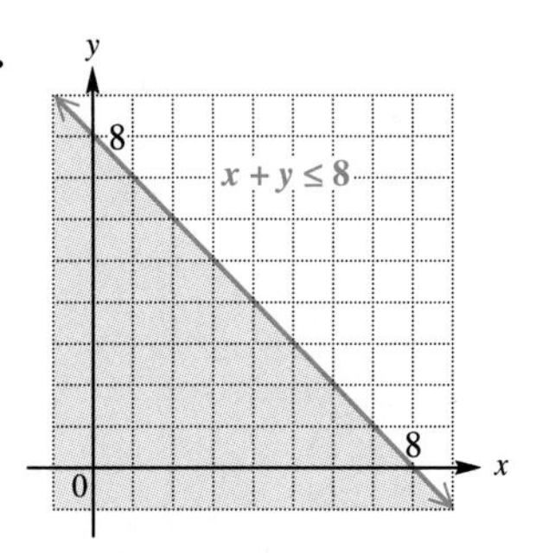

15.

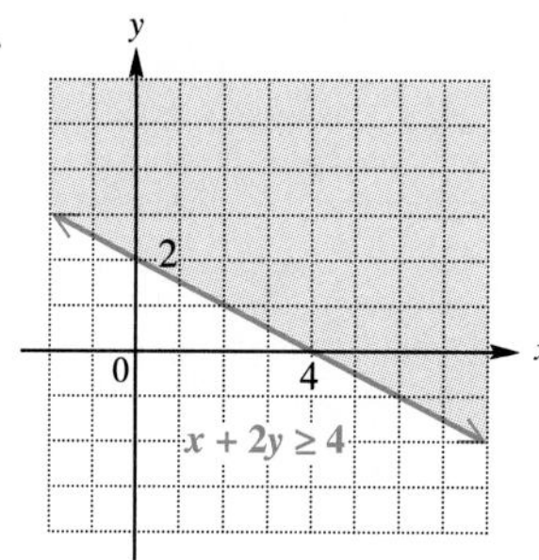

17.

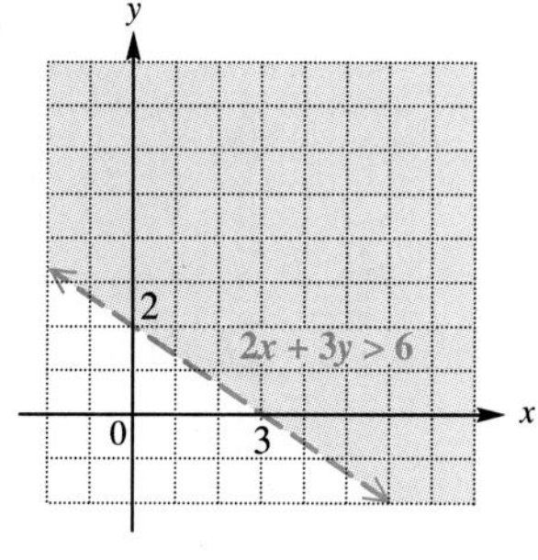

19.

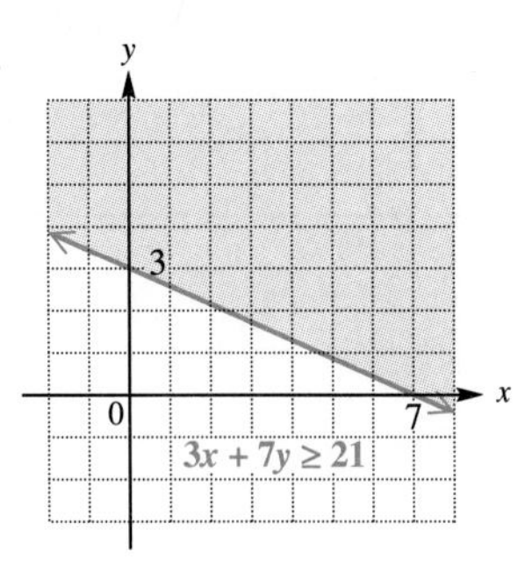

21.

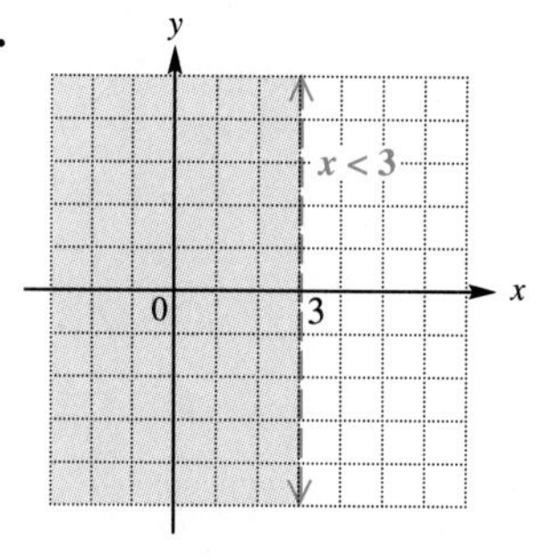

23.

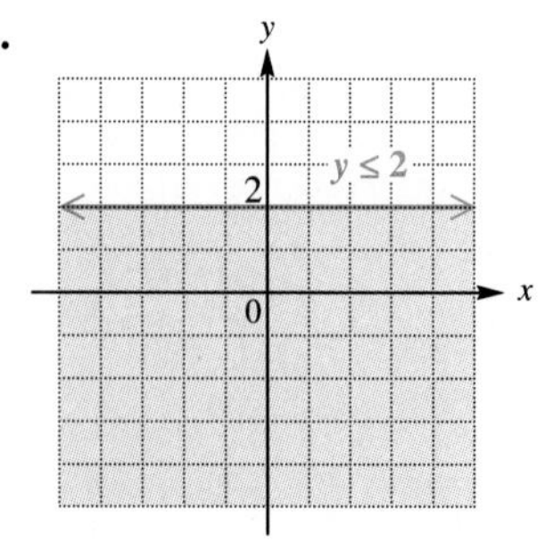

25.

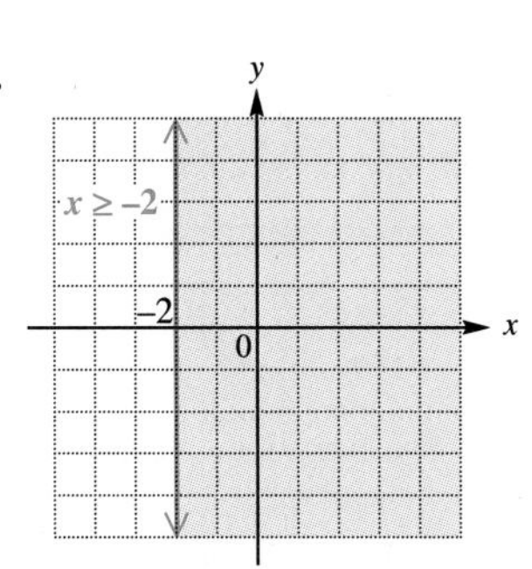

27.

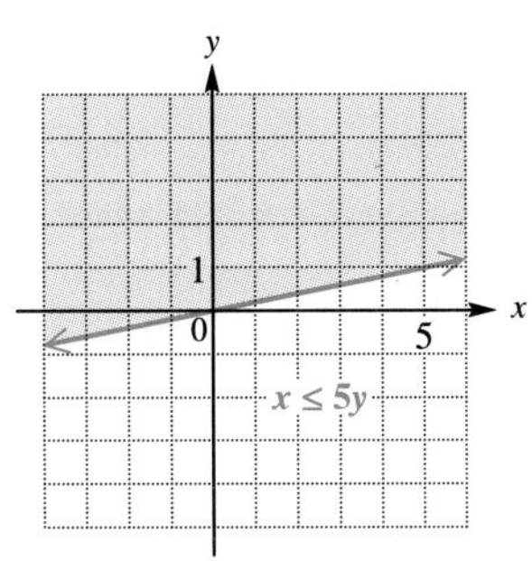

29.

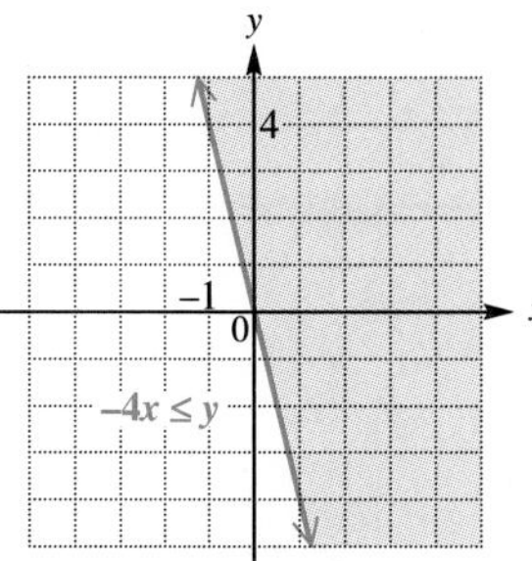

31.

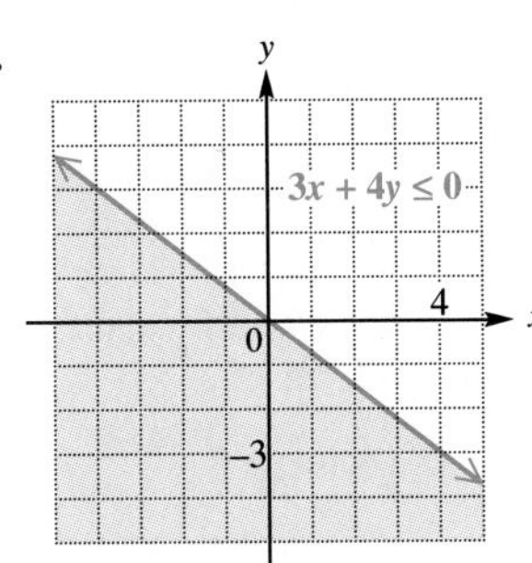

33. (c) **35.** (d) **37.** **(a)**

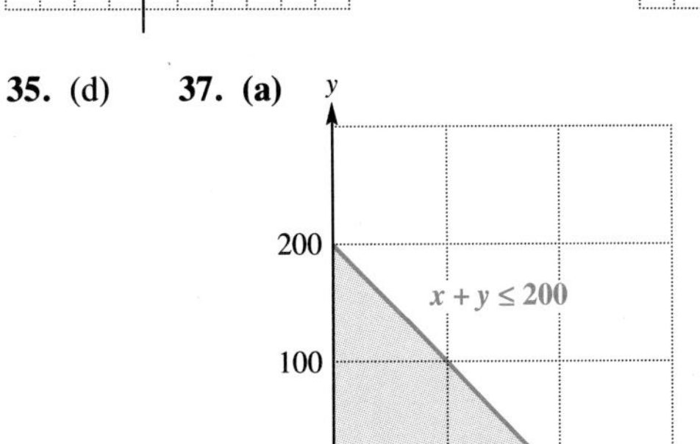

(b) (200, 0), (100, 100); other points also are possible.

39. **(a)**

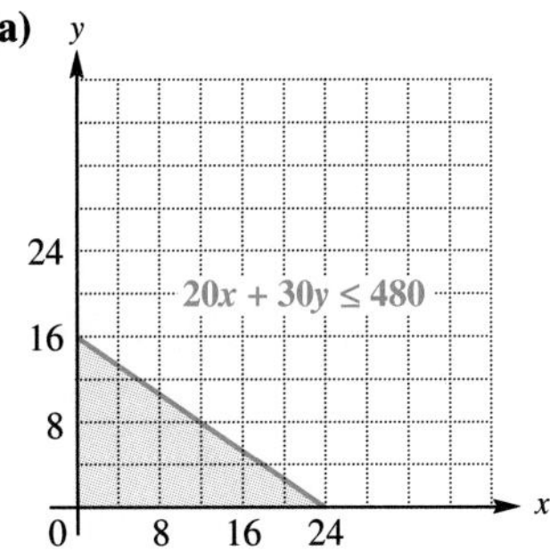

(b) (16, 1), (20, 2); other points also are possible. **43.** 13 **45.** 26 **47.** 153 **49.** 1

6.6 Exercises (page 356)

1. not a function; domain $\{-4, -2, 0\}$; range $\{3, 1, 5, -4\}$ **3.** function; domain $\{-2, -1, 0, 1, 2\}$; range $\{3, 2, 0\}$ **5.** not a function; domain $\{2, 4, 6, 8, 10\}$; range $\{-1, -2, -5, -6, -8\}$ **7.** function **9.** not a function **11.** not a function **13.** function **15.** function **17.** function **19.** not a function **21.** not a function **23.** function **25.** Both domain and range are the set of all real numbers. **27.** Both domain and range are the set of all real numbers. **29.** Domain is the set of all real numbers; range is the set of all $y \geq -3$. **31.** Domain is the set of all real numbers; range is the set of all $y \leq 6$. **33.** **(a)** 7 **(b)** -1 **(c)** -13 **35.** **(a)** 6 **(b)** 2 **(c)** 11 **37.** **(a)** 1 **(b)** 9 **(c)** 36 **39.** **(a)** -4 **(b)** -2 **(c)** -1 **41.** **(a)** 0 **(b)** 3 **(c)** 8 **43.** **(a)** 9 **(b)** 24 **(c)** -11 **45.** **(a)** 1 **(b)** -62 **(c)** -7 **47.** function **49.** not a function **51.** **(a)** 302 **(b)** 353 **(c)** 370 **55.** $17x$ **57.** $-q$

Chapter 6 Review Exercises (page 361)

1. yes **3.** no **5.** yes **7.** $(-1, -5)$, $(0, -2)$, $\left(\frac{7}{3}, 5\right)$ **9.**

x	-4	-4	-4
y	-3	0	5

11.

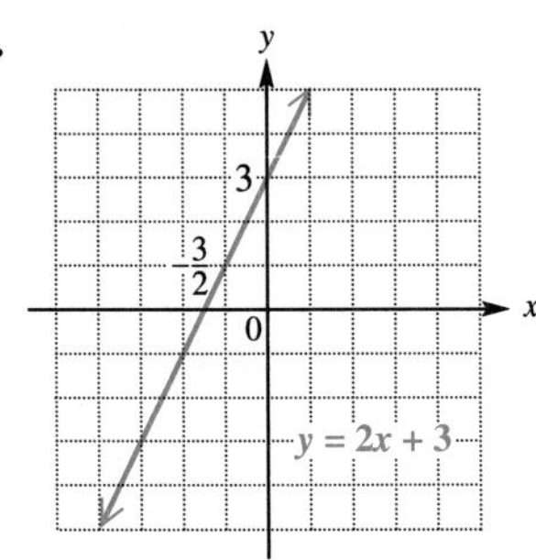

13.

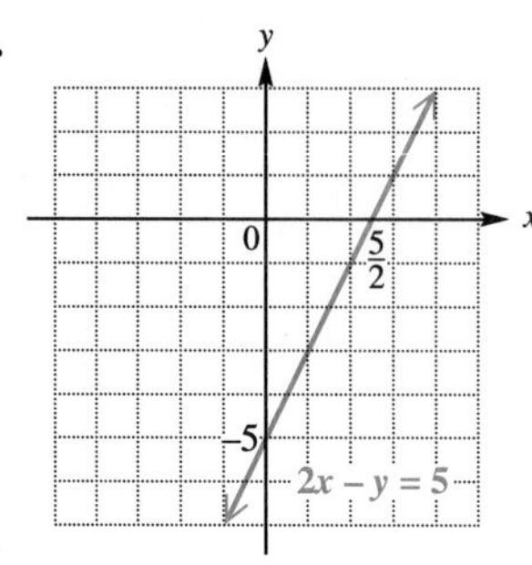

15.

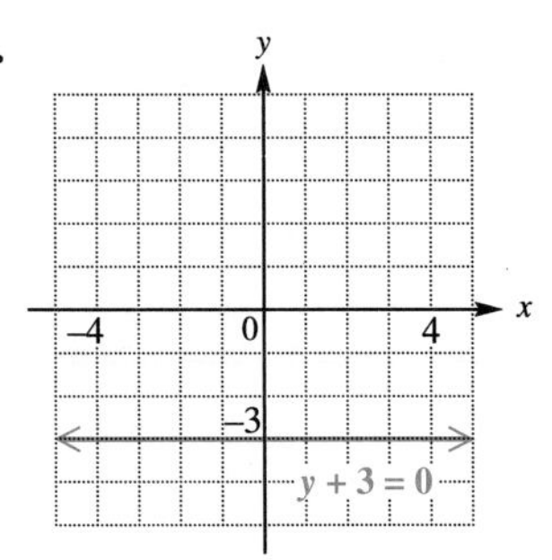

17. x-intercept: $\left(\frac{5}{2}, 0\right)$; y-intercept: $(0, -5)$ **19.** x-intercept: $(0, 0)$; y-intercept: $(0, 0)$ **21.** 2 **23.** 3

25. undefined **27.** parallel **29.** perpendicular **31.** $3x - y = 2$ **33.** $2x - 3y = -15$ **35.** $2x - 3y = -14$

37. $x + 4y = 6$ **39.**

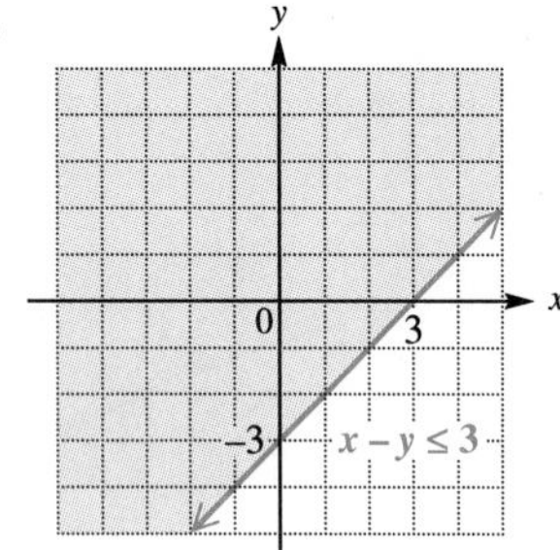

41.

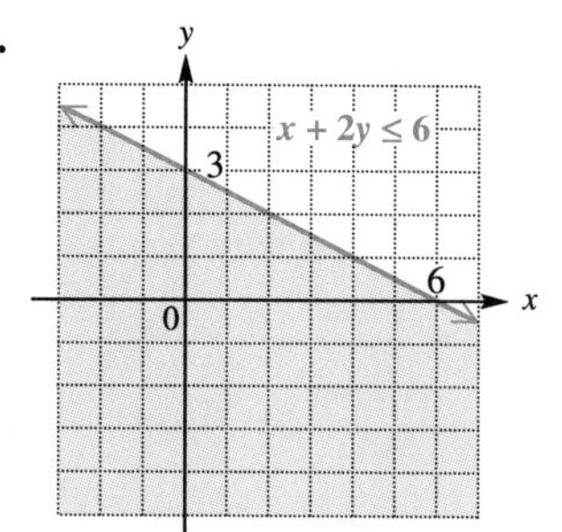

43. not a function
45. not a function
47. function
49. not a function
51. Domain is the set of all real numbers; range is the set of all $y \geq 1$.
53. **(a)** 8 **(b)** -1
55. **(a)** 5 **(b)** 2

57. $\frac{1}{3}$; x-intercept: $(7, 0)$; y-intercept: $\left(0, -\frac{7}{3}\right)$ **59.** function **61.** $(0, 3)$, $\left(\frac{9}{4}, 0\right)$, $\left(-2, \frac{17}{3}\right)$

63. $x = 5$ **65.** $3x + y = 30$ **67.** Both domain and range are the set of all real numbers.

69.

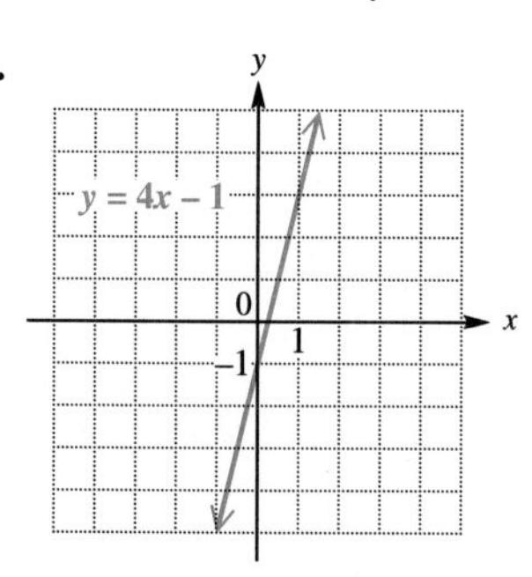

71.

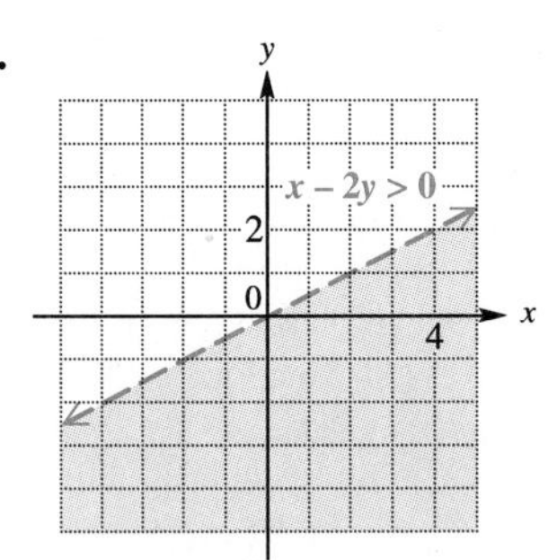

Chapter 6 Test (page 363)

[6.1] **1.** $(0, 3)$, $\left(\frac{21}{2}, 0\right)$, $\left(3, \frac{15}{7}\right)$, $\left(\frac{7}{2}, 2\right)$ **2.** $(0, 0)$, $(6, 2)$, $\left(8, \frac{8}{3}\right)$, $(-12, -4)$

[6.2] **3.** (4, 0); (0, 4) **4.** (6, 0); $\left(0, -\frac{9}{2}\right)$ **5.** (0, 0); (0, 0)

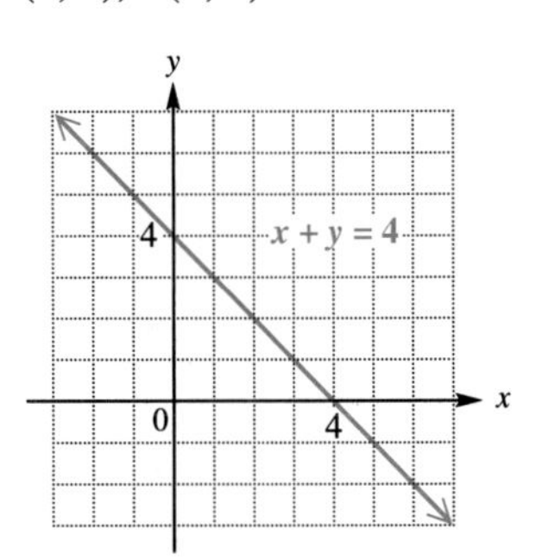

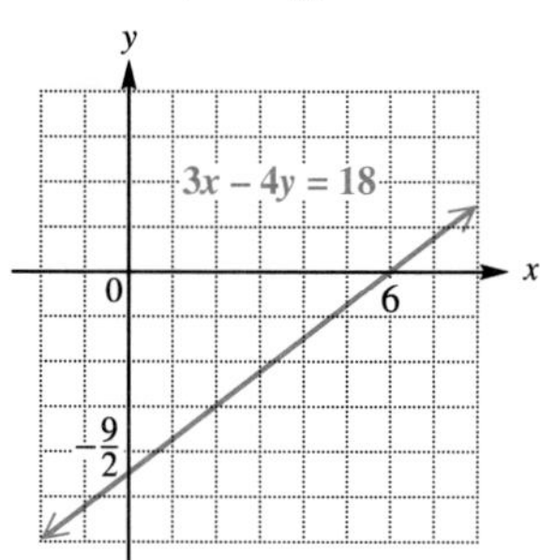

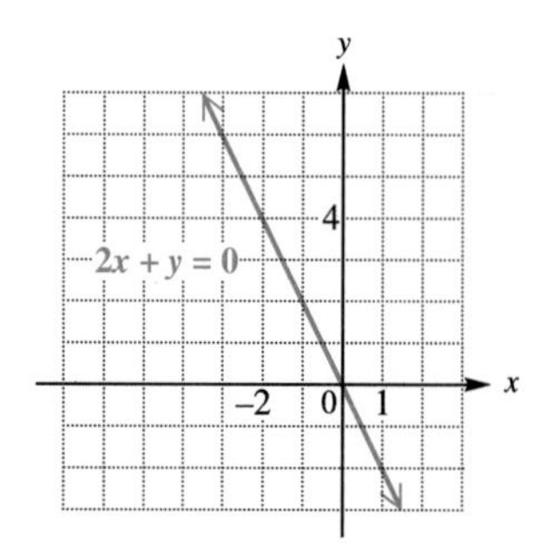

6. (−5, 0); none [6.3] **7.** $-\frac{3}{7}$ **8.** $-\frac{4}{7}$ **9.** undefined [6.4] **10.** $4x + y = 1$

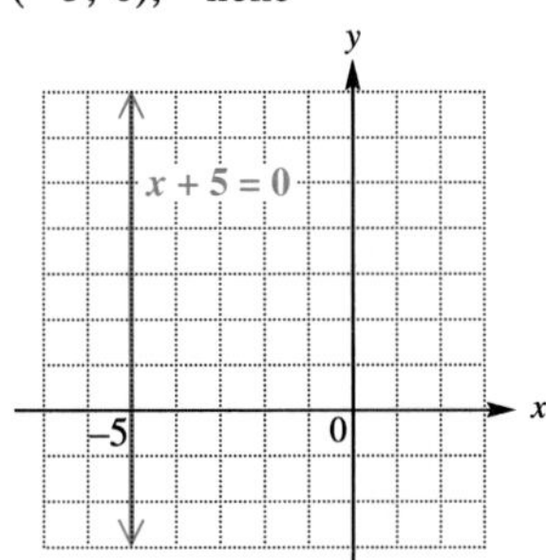

11. $3x + 4y = 8$ **12.** $9x - y = -12$ [6.5] **13.**

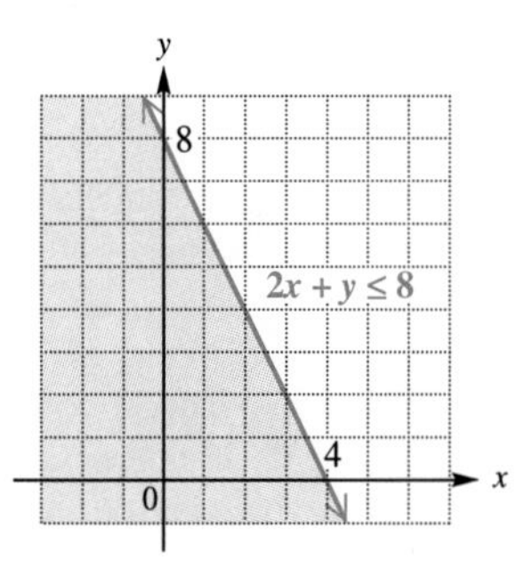

14.

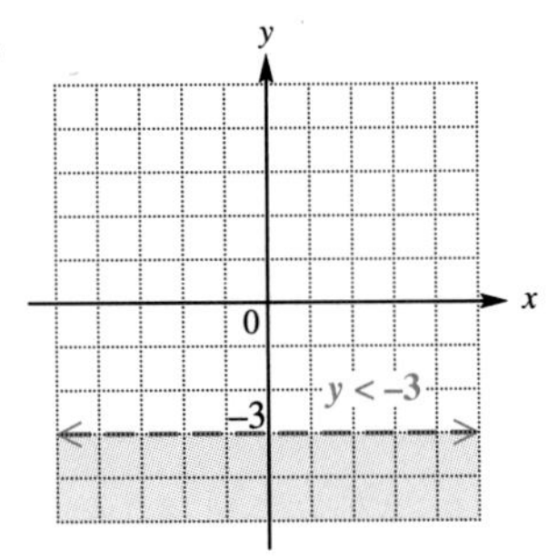

[6.6] **15.** not a function **16.** function **17.** not a function
18. Domain is the set of all real numbers; range is the set of all $y \geq 1$.
19. −20; 28

CHAPTER 7

7.1 Exercises (page 370)

1. yes **3.** no **5.** no **7.** yes **9.** no **11.** yes **15.** (4, 2) **17.** (0, 4) **19.** (4, −1) **21.** no solution (inconsistent system) **23.** infinite number of solutions (dependent equations) **25.** (0, −3) **27.** (4, −3) **29.** (2, −3) **31.** (1, 2) **33.** no solution (inconsistent system) **35.** (6, −2) **37.** (2, 7) **39.** (0, 5) **41.** **(a)** neither **(b)** intersecting lines **(c)** one solution **43.** **(a)** dependent **(b)** one line **(c)** infinite number of solutions **45.** **(a)** neither **(b)** intersecting lines **(c)** one solution **47.** **(a)** inconsistent **(b)** parallel lines **(c)** no solution **49.** 40 **51.** (40, 30) **53.** $7m + 2n$ **55.** y **57.** $3x$ **59.** 0

7.2 Exercises (page 378)

1. $(1, -2)$ **3.** $(2, 0)$ **5.** $(6, 2)$ **7.** $(-2, 6)$ **9.** $\left(\frac{1}{2}, 2\right)$ **11.** $\left(\frac{3}{2}, -2\right)$ **15.** $(2, -3)$ **17.** $(4, 4)$ **19.** $(2, -5)$ **21.** $(4, 9)$ **23.** $(-4, 0)$ **25.** $(4, -3)$ **27.** $(-9, -11)$ **29.** $(6, 3)$ **31.** $(6, -5)$ **33.** $(-6, 0)$ **35.** $\left(\frac{1}{2}, 3\right)$ **37.** no solution **39.** infinite number of solutions **41.** no solution **43.** $(2, -5)$ **45.** $\left(\frac{3}{8}, \frac{5}{6}\right)$ **47.** $\left(\frac{2}{3}, -\frac{1}{5}\right)$ **49.** $(4, 9)$ **51.** $(-3, 2)$ **53.** infinite number of solutions **55.** $y = -3x + 4$ **57.** $y = \frac{9}{2}x - 2$ **59.** 2 **61.** 3 **63.** 1

7.3 Exercises (page 385)

1. $(2, 4)$ **3.** $(6, 4)$ **5.** $(8, -1)$ **7.** no solution **9.** $(2, -4)$ **11.** $(5, 1)$ **13.** infinite number of solutions **15.** $(4, 1)$ **17.** $(-2, 5)$ **19.** $(-6, 2)$ **21.** $(-5, 3)$ **25.** $(7, 0)$ **27.** infinite number of solutions **29.** $(6, 5)$ **31.** $(4, -6)$ **33.** $(18, -12)$ **35.** $(3, 2)$ **37.** $(3, -1)$ **39.** $(12, 6)$ **41.** 21 **43.** 8 feet **45.** 13 twenties

7.4 Exercises (page 393)

1. 43 and 9 **3.** smaller: 40°; larger: 140° **5.** 72 and 24 **7.** Heathrow: 37,525,300 passengers; Haneda: 32,177,040 passengers **11.** 53 ones; 28 fives **13.** 83 student tickets; 110 non-student tickets **15.** \$5000 at 10%; \$2500 at 8% **17.** 21 large; 18 small **19.** 10,000 algebra books; 3000 calculus books **21.** 40 liters of 40% solution; 20 liters of 70% solution **23.** 30 pounds of 60% copper; 10 pounds of 80% copper **25.** 60 pounds at \$1.20 per pound; 30 pounds at \$1.80 per pound **27.** 20 bags at \$7.00 per bag; 60 bags at \$8.00 per bag **29.** faster car: 65 miles per hour; slower car: 35 miles per hour **31.** boat: 5 miles per hour; current: 3 miles per hour **33.** plane: 470 miles per hour; wind: 70 miles per hour **35.** \$7500 at 10%; \$2500 at 14% **37.** 10 inches by 20 inches **39.** 15 pounds at \$3 per pound; 30 pounds at \$1.50 per pound **41.** John: $3\frac{1}{4}$ miles per hour; Harriet: $2\frac{3}{4}$ miles per hour **43.** 4 girls; 3 boys **45.**

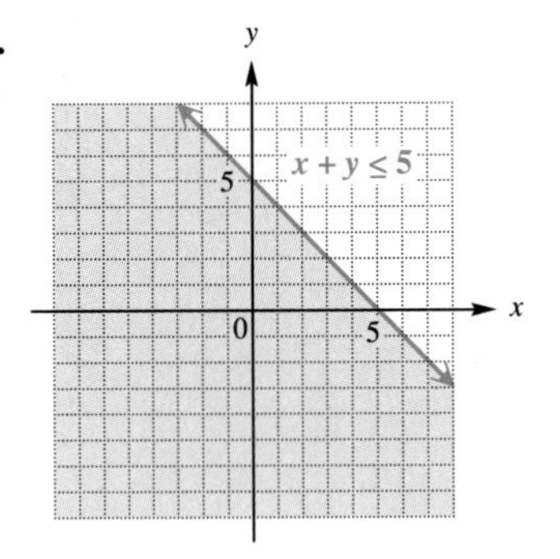

47.

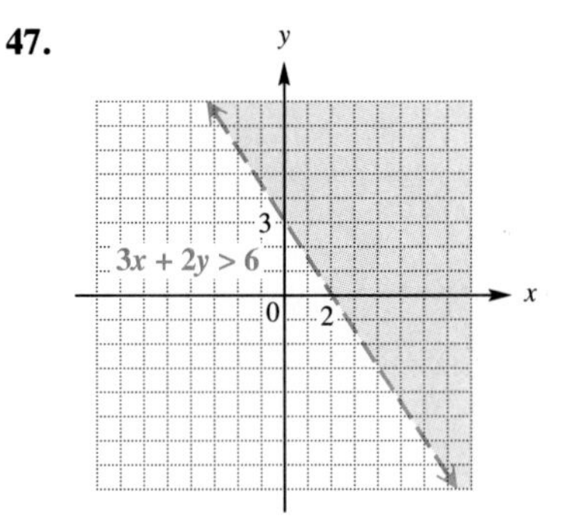

49.

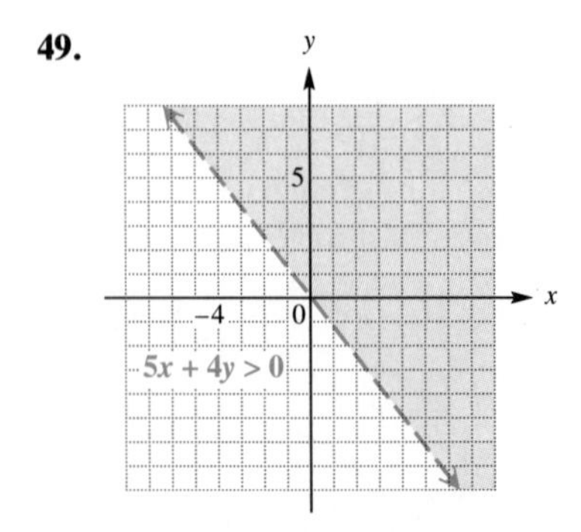

7.5 Exercises (page 399)

1.

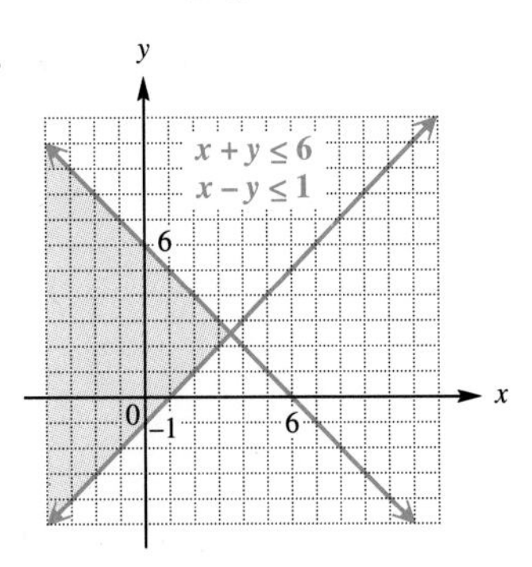

3.

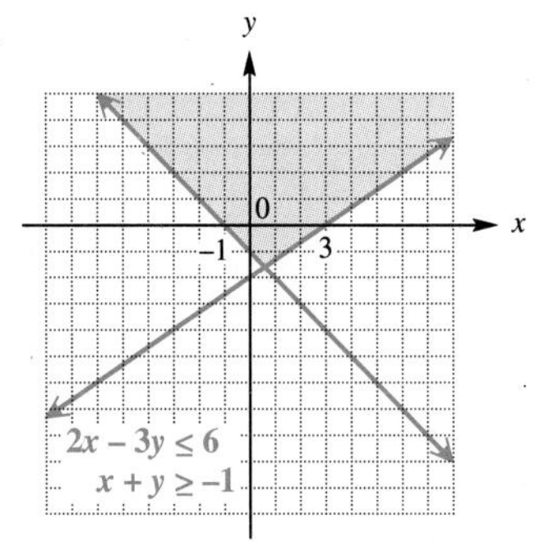

5.

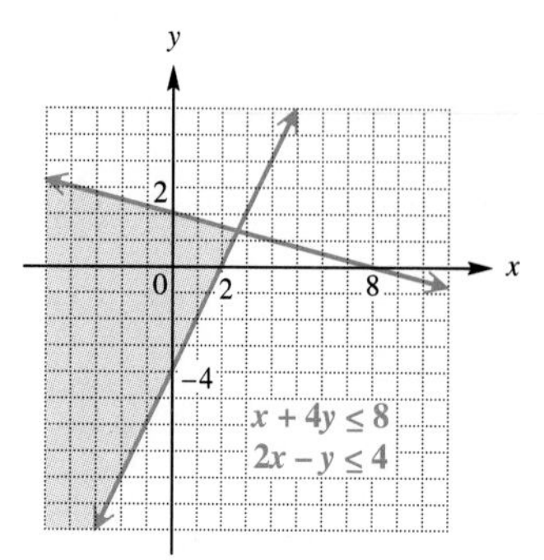

7.

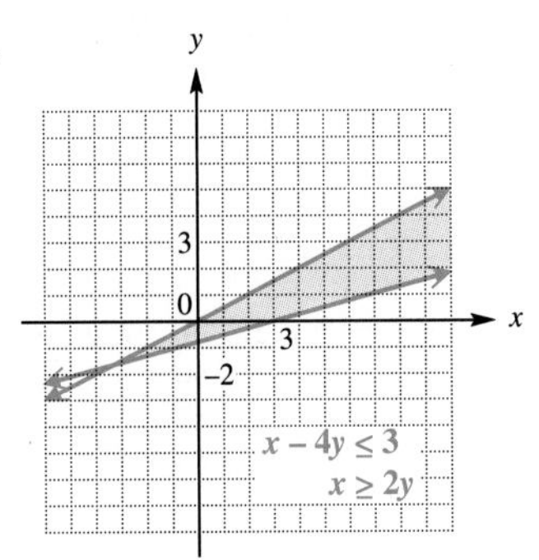

9.

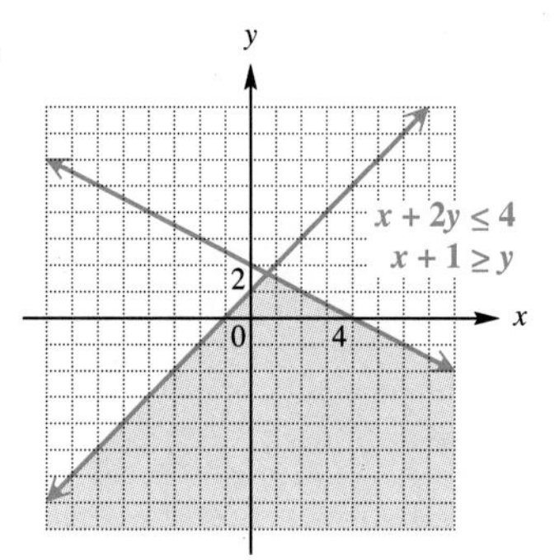

11.

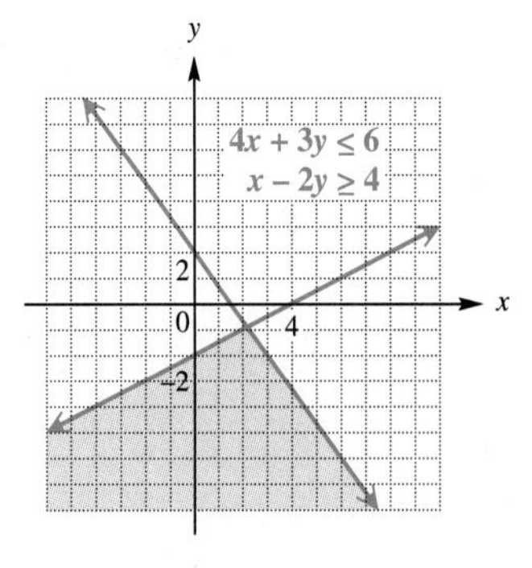

13.

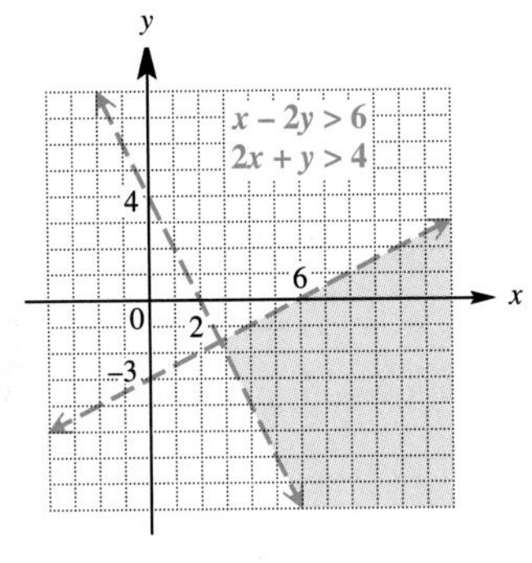

15.

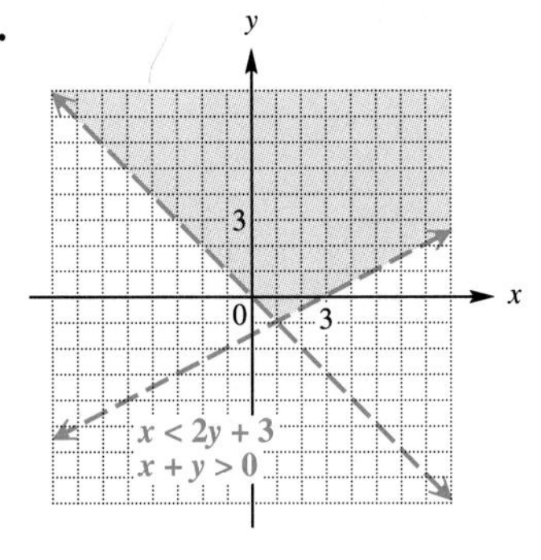

17.

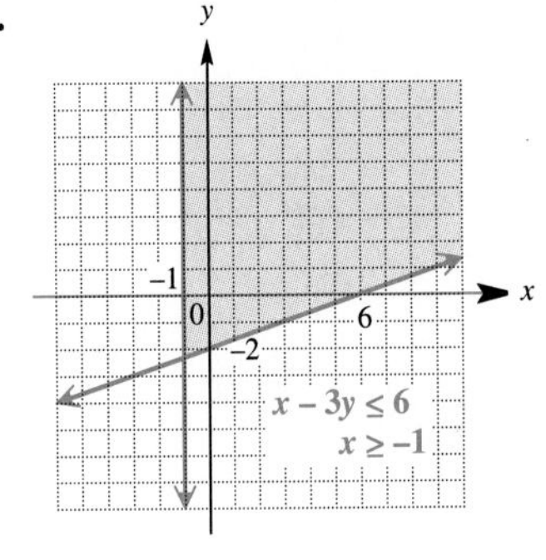

19.

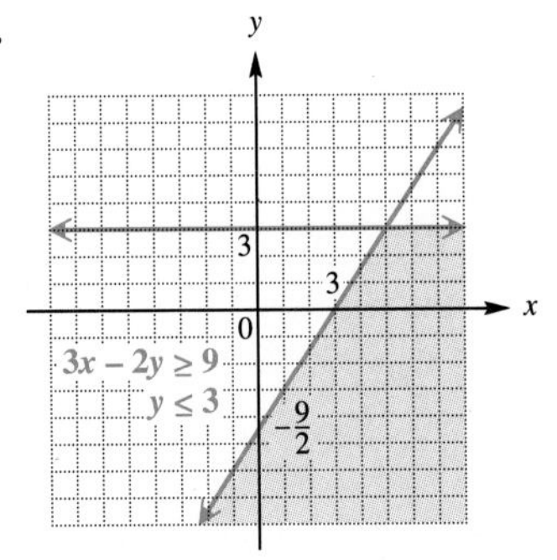

21.

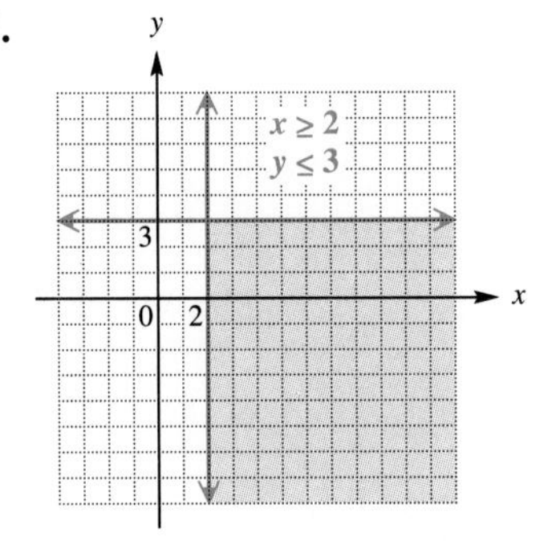

25.

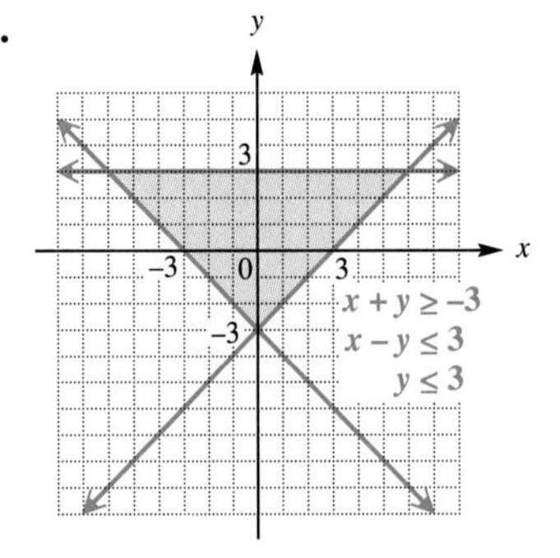

27. 25 **29.** 121 **31.** 216 **33.** 16

Chapter 7 Review Exercises (page 403)

1. no **3.** no **5.** yes **9.** (2, 1) **11.** (−1, 1) **13.** (−1, −2) **15.** (2, 4) **17.** (3, −2) **19.** infinite number of solutions (dependent equations) **21.** (4, −1) **23.** infinite number of solutions (dependent equations) **25.** (2, 2) **27.** (4, 2) **29.** (5, 3) **31.** (4, −1) **33.** (7, 1) **35.** length: 12 meters; width: 8 meters **37.** \$7000 at 12%; \$11,000 at 16% **39.** 45 miles per hour; 55 miles per hour

41.

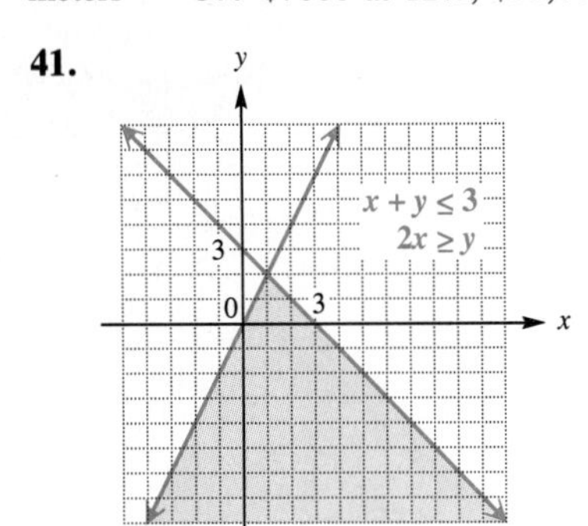

43.

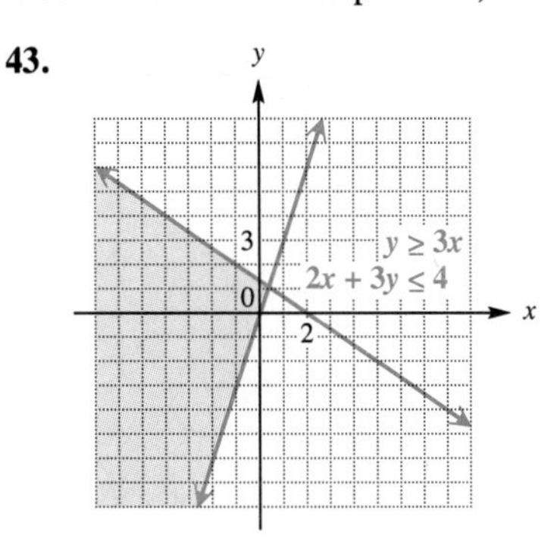

45. (9, 2) **47.** (−1, 2)

49.

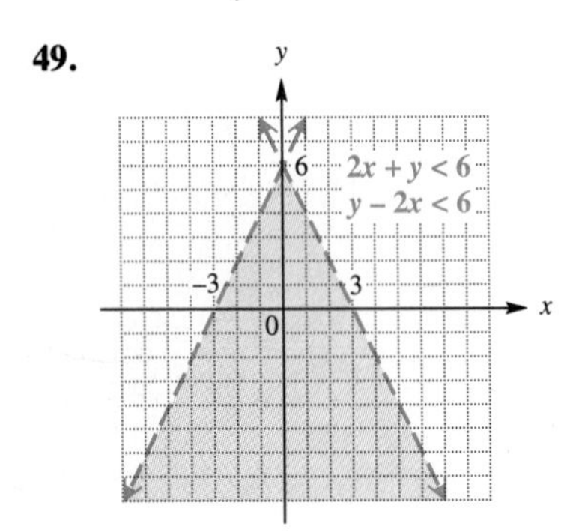

51. (3, 1) **53.** 28 and 14 **55.** (−6, 1)

Chapter 7 Test (page 406)

[7.1] **1.** (4, −3) **2.** (4, 1) [7.2] **3.** (6, 5) **4.** (4, 3) **5.** infinite number of solutions **6.** $\left(\frac{3}{2}, -2\right)$ **7.** no solution **8.** $\left(-\frac{11}{14}, -\frac{8}{7}\right)$ [7.3] **9.** (3, −5) **10.** (−6, 8) [7.1–7.3] **11.** infinite number of solutions **12.** (−6, 4) **13.** none [7.4] **14.** 15 and 29 **15.** 8 at \$9.00; 2 at \$7.00 **16.** 75 liters of 40% solution; 25 liters of 60% solution **17.** 45 miles per hour; 60 miles per hour

[7.5] **18.**

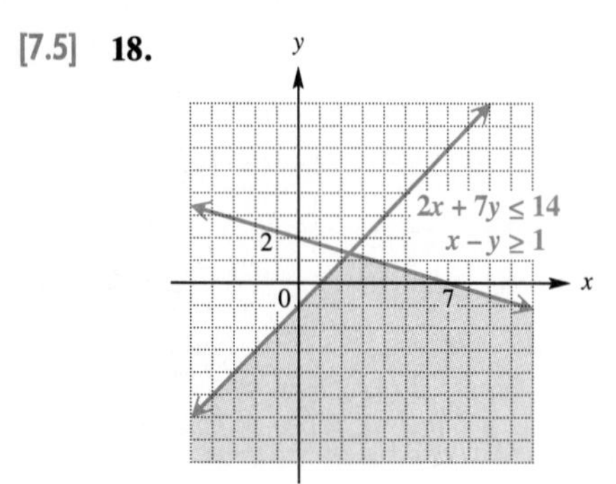

19.

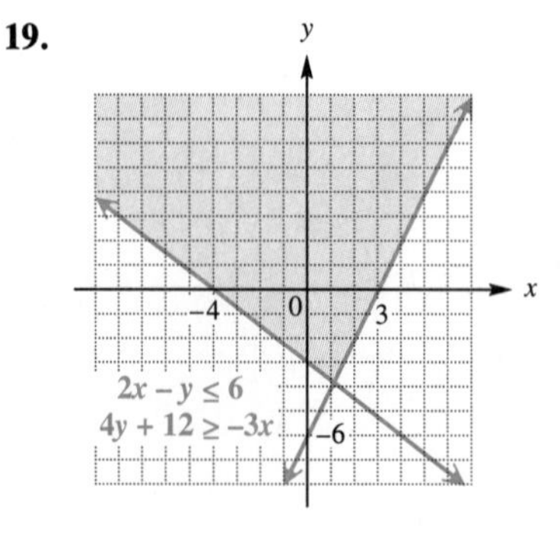

20.

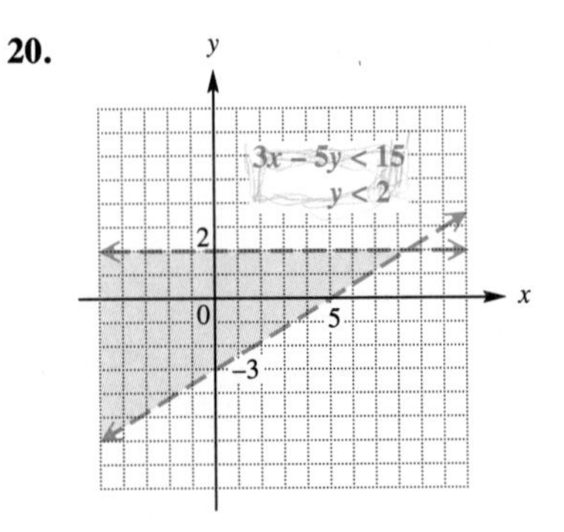

CHAPTER 8

8.1 Exercises (page 413)

1. 3, −3 **3.** 11, −11 **5.** $\frac{20}{9}, -\frac{20}{9}$ **7.** 25, −25 **9.** 16 **11.** 57 **13.** $4x^2 + 3$ **15.** 5 **17.** −8 **19.** $\frac{13}{7}$ **21.** $-\frac{7}{3}$ **23.** not a real number **25.** rational; 9 **27.** irrational; 5.568 **29.** not a real number **31.** irrational; 8.485 **33.** rational; −12 **35.** irrational; 13.038 **37.** not a real number **39.** irrational; −16.125 **41.** 26.325 **43.** 21.284 **45.** 1.015 **47.** .020 **49.** $-\sqrt{17}$ or $\sqrt{17}$; two **51.** $c = 13$ **53.** $a = 24$ **55.** $c = 11.402$ **57.** $b = 7.416$ **59.** 2 **61.** 6 **63.** −4 **65.** −3 **67.** −2 **69.** not a real number **71.** −4 **73.** −2 **75.** 17 centimeters **77.** 40 feet **79.** 130 miles **81.** **(a)** 9 **(b)** −9 or 9 **83.** 1, 2, 3, 4, 6, 8, 12, 24 **85.** 1, 2, 3, 4, 5, 6, 10, 12, 15, 20, 25, 30, 50, 60, 75, 100, 150, 300 **87.** 1, 2, 3, 4, 6, 8, 9, 12, 18, 24, 36, 72 **89.** 1, 2, 3, 5, 6, 10, 15, 25, 30, 50, 75, 150

8.2 Exercises (page 420)

1. 4 **3.** 6 **5.** 21 **7.** $\sqrt{21}$ **9.** $3\sqrt{3}$ **11.** $3\sqrt{2}$ **13.** $4\sqrt{3}$ **15.** $5\sqrt{6}$ **17.** $30\sqrt{3}$ **19.** $4\sqrt{5}$ **21.** 36 **23.** 60 **25.** $20\sqrt{3}$ **27.** $10\sqrt{10}$ **29.** $\frac{10}{3}$ **31.** $\frac{\sqrt{30}}{7}$ **33.** $\frac{2}{5}$ **35.** $\frac{4}{25}$ **37.** 5 **39.** 4 **41.** $3\sqrt{5}$ **43.** $5\sqrt{10}$ **45.** y **47.** x **49.** x^2 **51.** xy^2 **53.** $x\sqrt{x}$ **55.** $\frac{4}{x}$ **57.** $\frac{\sqrt{11}}{r^2}$ **59.** $2m\sqrt{7m}$ **61.** $2\sqrt[3]{5}$ **63.** $3\sqrt[3]{2}$ **65.** $4\sqrt[3]{2}$ **67.** $2\sqrt[4]{5}$ **69.** $\frac{2}{3}$ **71.** $\frac{10}{3}$ **73.** 2 **75.** 2 **77.** $2x\sqrt[3]{4}$ **79.** $\frac{\sqrt[3]{3m}}{2n}$ **81.** 6 centimeters **87.** $\frac{30}{66}$ **89.** $\frac{12}{56}$ **91.** $\frac{5y}{2y^2}$ **93.** $\frac{24}{3\sqrt{5}}$

8.3 Exercises (page 423)

1. $7\sqrt{3}$ **3.** $-5\sqrt{7}$ **5.** $2\sqrt{6}$ **7.** $3\sqrt{17}$ **9.** $4\sqrt{7}$ **11.** $10\sqrt{2}$ **13.** $3\sqrt{3}$ **15.** $20\sqrt{2}$ **17.** $19\sqrt{7}$ **19.** $12\sqrt{6} + 6\sqrt{5}$ **21.** $-2\sqrt{2} - 12\sqrt{3}$ **23.** $2\sqrt{2}$ **25.** $3\sqrt{3} - 2\sqrt{5}$ **27.** $6\sqrt{x}$ **29.** $3\sqrt{21}$ **31.** $5\sqrt{3}$ **33.** $-\sqrt[3]{2}$ **35.** $24\sqrt[3]{3}$ **37.** $10\sqrt[4]{2} + 4\sqrt[4]{8}$ **39.** $4x\sqrt{6}$ **41.** $-y\sqrt{5}$ **43.** $10b\sqrt{2}$ **45.** 0 **47.** $4p\sqrt{14m} - 12p\sqrt{7m}$ **49.** $21r^2\sqrt{3s}$ **51.** $21\sqrt[4]{m^3}$ **53.** $36k\sqrt[3]{2k}$ **55.** $27\sqrt{3}$ meters ≈ 46.8 meters **61.** $m^2 + 8m + 15$ **63.** $6k^2 - 17k + 12$ **65.** $9x^2 - 6x + 1$ **67.** $a^2 - 9$ **69.** $16k^2 - 25$

8.4 Exercises (page 427)

1. $\frac{6\sqrt{5}}{5}$ **3.** $\frac{3\sqrt{7}}{7}$ **5.** $\frac{8\sqrt{15}}{5}$ **7.** $\frac{\sqrt{30}}{2}$ **9.** $\frac{8\sqrt{3}}{9}$ **11.** $\sqrt{2}$ **13.** $\sqrt{2}$ **15.** $\frac{2\sqrt{30}}{3}$ **17.** $\frac{\sqrt{2}}{2}$ **19.** $\frac{3\sqrt{5}}{5}$ **21.** $\frac{\sqrt{15}}{10}$ **23.** $\frac{3\sqrt{14}}{4}$ **25.** $\frac{\sqrt{3}}{5}$ **27.** $\frac{4\sqrt{3}}{27}$ **29.** $\frac{\sqrt{6p}}{p}$ **31.** $\frac{\sqrt{3y}}{y}$ **33.** $\frac{z\sqrt{2x}}{x}$ **35.** $\frac{a\sqrt{5ab}}{b}$ **37.** $\frac{p\sqrt{2pm}}{m}$ **39.** $\frac{3a\sqrt{5r}}{5}$ **41.** $\frac{\sqrt[3]{4}}{2}$ **43.** $\frac{\sqrt[3]{2}}{4}$ **45.** $\frac{\sqrt[3]{121}}{11}$ **47.** $\frac{\sqrt[3]{50}}{5}$ **49.** $\frac{\sqrt[3]{6y}}{2y}$ **51.** $\frac{\sqrt[3]{42mn^2}}{6n}$ **53.** $\frac{\sqrt[4]{2^3}}{\sqrt[4]{2^3}}$ **55.** $\frac{9\sqrt{2}}{4}$ seconds **59.** $4x$ **61.** $-2m - 3$ **63.** $3y + 8z$

8.5 Exercises (page 433)

1. $27\sqrt{5}$ **3.** $21\sqrt{2}$ **5.** −4 **7.** $\sqrt{15} + \sqrt{35}$ **9.** $2\sqrt{10} + 10$ **11.** $-4\sqrt{7}$ **13.** $21 - \sqrt{6}$ **15.** $87 + 9\sqrt{21}$ **17.** $34 + 24\sqrt{2}$ **19.** $5 + 2\sqrt{6}$ **21.** 7 **23.** −4 **25.** 1 **27.** $2 + 2\sqrt{2}$ **29.** $x\sqrt{30} + \sqrt{15x} + 6\sqrt{5x} + 3\sqrt{10}$ **31.** $6t - 3\sqrt{14t} + 2\sqrt{7t} - 7\sqrt{2}$ **33.** $m\sqrt{15} + \sqrt{10mn} - \sqrt{15mn} - n\sqrt{10}$ **35.** (a) **37.** $-2 - \sqrt{11}$ **39.** $6 + 3\sqrt{2}$ **41.** $\frac{-2\sqrt{3} - \sqrt{30}}{3}$ **43.** $2\sqrt{2} - 1$ **45.** $-3\sqrt{6} + 6\sqrt{2} + 4\sqrt{3} - 8$ **47.** $\frac{-3\sqrt{2} - \sqrt{15} + \sqrt{30} + 5}{2}$ **49.** $\frac{2\sqrt{3} - \sqrt{15}}{3}$ **51.** $\frac{\sqrt{35} + 5}{5}$ **53.** $2\sqrt{5} - 3$ **55.** $\frac{2\sqrt{6} + 3}{5}$ **57.** $\frac{3 - 2\sqrt{2}}{4}$ **59.** $\frac{5 + \sqrt{3}}{2}$ **61.** $2 - 3\sqrt[3]{4}$ **63.** $12 + 10\sqrt[4]{8}$

65. $-1 + 3\sqrt[3]{2} - \sqrt[3]{4}$ **67.** 1 **69.** $-6 + 2\sqrt{13}$ centimeters **73.** 1, 3 **75.** $\frac{5}{2}, -3$ **77.** 1, 2

8.6 Exercises (page 439)

1. 4 **3.** -5 **5.** no solution **7.** 49 **9.** no solution **11.** 9 **13.** 16 **15.** 6 **17.** 7 **19.** no solution **21.** 6 **23.** 12 **25.** 5 **27.** 0, -1 **29.** 5 **31.** 3 **33.** 9 **35.** 12 **37.** 9 **39.** 21 **41.** 8 **43.** **(a)** 90 miles per hour **(b)** 120 miles per hour **(c)** 60 miles per hour **(d)** 60 miles per hour **47.** 5^6 **49.** $\frac{1}{a^3}$ **51.** $\frac{3^2}{p^2}$ **53.** c^{13}

8.7 Exercises (page 444)

1. 6 **3.** 5 **5.** 2 **7.** 4 **9.** 2 **11.** 2 **13.** 8 **15.** 9 **17.** 8 **19.** 4 **21.** -4 **23.** -4 **25.** 2^3 **27.** $\frac{1}{6^{1/2}}$ **29.** $\frac{1}{15^{1/2}}$ **31.** $11^{1/7}$ **33.** 8^3 **35.** $6^{1/2}$ **37.** $\frac{5^3}{2^3}$ **39.** $\frac{1}{2^{8/5}}$ **41.** $6^{2/9}$ **43.** p^3 **45.** z **47.** $m^2n^{1/6}$ **49.** $\frac{a^{2/3}}{b^{4/9}}$ **51.** $\frac{1}{c^{4/3}}$ **53.** 2 **55.** 2 **57.** $\sqrt{a}$ **59.** $\sqrt[3]{k^2}$ **61.** **(a)** 172.53 miles **(b)** 211.31 miles **65.** 5, -5 **67.** $\sqrt{14}, -\sqrt{14}$ **69.** $3\sqrt{2}, -3\sqrt{2}$ **71.** $4\sqrt{5}, -4\sqrt{5}$

Chapter 8 Review Exercises (page 448)

1. 7, -7 **3.** 15, -15 **5.** 4 **7.** not a real number **9.** 2 **11.** irrational; 3.873 **13.** rational; -13 **15.** $5\sqrt{3}$ **17.** $3\sqrt{3}$ **19.** $16\sqrt{6}$ **21.** $\frac{3}{2}$ **23.** $\frac{\sqrt{5}}{6}$ **25.** 15 **27.** p **29.** $x\sqrt{x}$ **31.** x^5y^8 **33.** $\frac{\sqrt[3]{5}}{2}$ **35.** $5x\sqrt[3]{3x}$ **37.** $2\sqrt{2}$ **39.** 0 **41.** $8 + 6\sqrt[4]{4}$ **43.** $-m\sqrt{5}$ **45.** $\frac{10\sqrt{3}}{3}$ **47.** $\frac{3\sqrt{10}}{5}$ **49.** $\frac{\sqrt{15}}{5}$ **51.** $\frac{\sqrt{30}}{15}$ **53.** $\frac{\sqrt[3]{10}}{2}$ **55.** $\frac{r\sqrt{5rt}}{5w}$ **57.** $3\sqrt{6} + 12$ **59.** $2\sqrt{21} - \sqrt{14} + 12\sqrt{2} - 4\sqrt{3}$ **61.** -2 **63.** $\frac{-\sqrt{3} + 3}{2}$ or $\frac{\sqrt{3} - 3}{-2}$ **65.** $\frac{4\sqrt{7} + 7}{9}$ **67.** $\frac{-\sqrt{10} + 3\sqrt{5} + \sqrt{2} - 3}{7}$ **69.** no solution **71.** 8 **73.** 1 **75.** 6 **77.** -2 **79.** 5 **81.** $12^{4/5}$ **83.** 7 **85.** $2^{1/3}$ **87.** $-2 + \sqrt{5}$ **89.** $3\sqrt[3]{2x^2}$ **91.** $-\sqrt{15} + 9$ **93.** 9 **95.** $\frac{1}{7}$ **97.** -6 **99.** $\sqrt{6}$ **101.** $6\sqrt{30}$

Chapter 8 Test (page 450)

[8.1] **1.** 10 **2.** -13.784 **3.** -3 [8.2] **4.** $3\sqrt{3}$ **5.** $\frac{8\sqrt{2}}{5}$ **6.** $-2\sqrt[3]{4}$ **7.** $4\sqrt{6}$ [8.3] **8.** $9\sqrt{7}$ **9.** $-5\sqrt{3x}$ [8.5] **10.** $2y\sqrt[3]{4x^2}$ **11.** $2np\sqrt[4]{2m^3p^2}$ **12.** 31 **13.** $6\sqrt{2} + 2 - 3\sqrt{14} - \sqrt{7}$ **14.** $11 + 2\sqrt{30}$ [8.4] **15.** $\sqrt{3}$ **16.** $\frac{4p\sqrt{k}}{k}$ **17.** $\frac{\sqrt[3]{15}}{3}$ **18.** $\frac{-12 - 3\sqrt{3}}{13}$ [8.6] **19.** 23 **20.** 4 **21.** 1, 25 [8.7] **22.** 16 **23.** $5^{5/2}$ **24.** $\frac{1}{3}$

CHAPTER 9

9.1 Exercises (page 455)

1. 5, -5 **3.** 8, -8 **5.** $\sqrt{13}, -\sqrt{13}$ **7.** no real number solution **9.** $\sqrt{2}, -\sqrt{2}$ **11.** $\frac{\sqrt{5}}{2}, -\frac{\sqrt{5}}{2}$ **13.** $2\sqrt{6}, -2\sqrt{6}$ **15.** $\frac{2\sqrt{5}}{5}, -\frac{2\sqrt{5}}{5}$ **19.** (d) **21.** 6, -2 **23.** $-4 + \sqrt{10}, -4 - \sqrt{10}$ **25.** no real number solution **27.** $1 + 4\sqrt{2}, 1 - 4\sqrt{2}$ **29.** 2, -1 **31.** $\frac{13}{6}, -\frac{3}{2}$ **33.** $\frac{5 + \sqrt{30}}{2}, \frac{5 - \sqrt{30}}{2}$ **35.** $\frac{1 + 3\sqrt{2}}{3}, \frac{1 - 3\sqrt{2}}{3}$ **37.** $\frac{5 + 7\sqrt{2}}{2}, \frac{5 - 7\sqrt{2}}{2}$ **39.** no real number solution **41.** 1.6, -1.6

43. 8.8, −8.8 **45.** .20, −4.48 **47.** −.15, −3.09 **49.** about $\frac{1}{2}$ second **51.** 10% **53.** 5 feet
55. $\frac{3+3\sqrt{3}}{2}$ **57.** $\frac{18+\sqrt{6}}{3}$ **59.** $(x+4)^2$ **61.** $\left(p-\frac{5}{2}\right)^2$

9.2 Exercises (page 460)

1. 1 **3.** 81 **5.** $\frac{81}{4}$ **7.** $\frac{25}{4}$ **9.** (d) **11.** −1, −3 **13.** $-1+\sqrt{6}, -1-\sqrt{6}$ **15.** −2, −4
17. $3+2\sqrt{2}, 3-2\sqrt{2}$ **19.** $\frac{-3+\sqrt{17}}{2}, \frac{-3-\sqrt{17}}{2}$ **21.** 4, −6 **23.** $\frac{-5+\sqrt{37}}{2}, \frac{-5-\sqrt{37}}{2}$
25. $\frac{9+\sqrt{21}}{6}, \frac{9-\sqrt{21}}{6}$ **27.** $\frac{1}{2}, -\frac{3}{2}$ **29.** $-1+\sqrt{5}, -1-\sqrt{5}$ **31.** no real number solution
33. $\frac{4+\sqrt{46}}{6}, \frac{4-\sqrt{46}}{6}$ **35.** no real number solution **37.** 3.236, −1.236 **39.** 16, −8; only 16 feet is a reasonable answer. **41.** 8.703 hours **43.** $1+2\sqrt{5}$ **45.** $\frac{1+\sqrt{2}}{2}$ **47.** $\frac{2+\sqrt{7}}{3}$ **49.** $\frac{2+\sqrt{2}}{2}$

9.3 Exercises (page 466)

Answers are given in the order *a*, *b*, *c* in Exercises 1–13.
1. 3, 4, −8 **3.** −8, −2, −3 **5.** 2, −3, 2 **7.** 1, 0, −2 **9.** 3, −8, 0 **11.** 1, 1, −12 **13.** 9, 9, −26
17. −3 **19.** −2 **21.** 1, −13 **23.** $\frac{3}{2}$ **25.** $\frac{1}{5}$, −1 **27.** $\frac{5}{2}$, −1 **29.** $-1+\sqrt{3}, -1-\sqrt{3}$
31. $\frac{-6+\sqrt{26}}{2}, \frac{-6-\sqrt{26}}{2}$ **33.** 0, −1 **35.** 0, $\frac{5}{3}$ **37.** $2\sqrt{5}, -2\sqrt{5}$ **39.** $\frac{4}{3}, -\frac{4}{3}$ **41.** 0, −2
43. no real number solution **45.** no real number solution **47.** 1.618, −.618 **49.** .681, −.881
51. $\frac{-1+\sqrt{73}}{6}, \frac{-1-\sqrt{73}}{6}$ **53.** $\frac{2+\sqrt{22}}{6}, \frac{2-\sqrt{22}}{6}$ **55.** $1+\sqrt{2}, 1-\sqrt{2}$ **57.** $\frac{1+\sqrt{3}}{3}, \frac{1-\sqrt{3}}{3}$
59. no real number solution **61.** no real number solution **63.** (d) **65.** 4 seconds **67.** 18 miles per hour
69. $5+4t$ **71.** $25-4x^2$ **73.** $4\sqrt{3}$ **75.** $2\sqrt{11}$

Summary Exercises on Quadratic Equations (page 468)

1. $\frac{-3+\sqrt{5}}{2}, \frac{-3-\sqrt{5}}{2}$ **3.** $-\frac{1}{2}$, 1 **5.** $-\frac{5}{4}, \frac{3}{2}$ **7.** $\frac{1+\sqrt{10}}{2}, \frac{1-\sqrt{10}}{2}$ **9.** $\frac{2}{5}$, 4
11. $\frac{2+\sqrt{7}}{3}, \frac{2-\sqrt{7}}{3}$ **13.** $\frac{-9+4\sqrt{3}}{2}, \frac{-9-4\sqrt{3}}{2}$ **15.** $-\frac{6}{5}, -\frac{8}{3}$ **17.** $\frac{-5+\sqrt{5}}{2}, \frac{-5-\sqrt{5}}{2}$
19. $-\frac{2}{3}$, 2 **21.** $\frac{5+\sqrt{17}}{4}, \frac{5-\sqrt{17}}{4}$ **23.** no real number solution **25.** $\frac{-5+\sqrt{13}}{6}, \frac{-5-\sqrt{13}}{6}$
27. 5, −17 **29.** 2, $-\frac{3}{2}$

9.4 Exercises (page 474)

1. $3i$ **3.** $2i\sqrt{5}$ **5.** $6i$ **7.** $5i\sqrt{5}$ **9.** $5+3i$ **11.** $6-9i$ **13.** $6-7i$ **15.** $14+5i$ **17.** $7-22i$
19. 45 **21.** $-\frac{3}{2}-\frac{1}{2}i$ **23.** $-\frac{1}{4}-\frac{7}{4}i$ **25.** $-3-4i$ **29.** $-1+2i, \ -1-2i$
31. $3+i\sqrt{5}, \ 3-i\sqrt{5}$ **33.** $-\frac{2}{3}+i\sqrt{2}, \ -\frac{2}{3}-i\sqrt{2}$ **35.** $1+i, \ 1-i$
37. $-\frac{3}{4}+\frac{\sqrt{31}}{4}i, \ -\frac{3}{4}-\frac{\sqrt{31}}{4}i$ **39.** $\frac{3}{2}+\frac{\sqrt{7}}{2}i, \ \frac{3}{2}-\frac{\sqrt{7}}{2}i$ **41.** $\frac{1}{5}+\frac{\sqrt{14}}{5}i, \ \frac{1}{5}-\frac{\sqrt{14}}{5}i$
43. $-\frac{1}{2}+\frac{\sqrt{13}}{2}i, \ -\frac{1}{2}-\frac{\sqrt{13}}{2}i$ **45.** $\frac{1}{2}+\frac{\sqrt{11}}{2}i, \ \frac{1}{2}-\frac{\sqrt{11}}{2}i$ **49.** true **51.** false

53. $-2, 3, 6, -4$

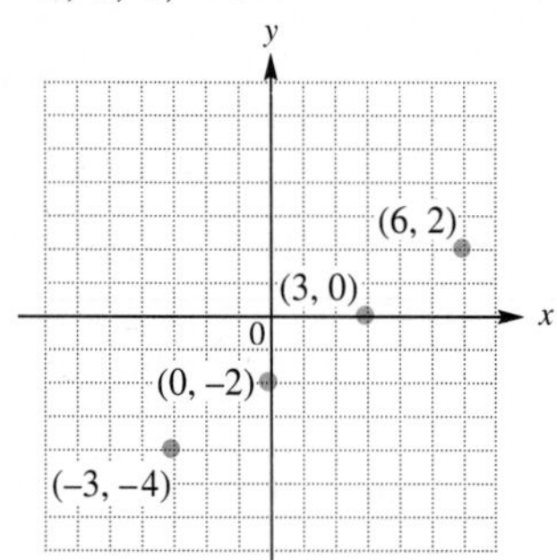

55. $3, 5, 6, 10$

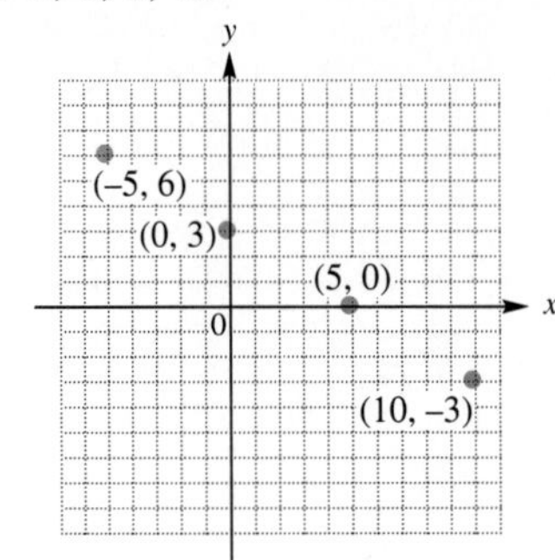

57. $0, 6, \frac{5}{2}, 1$

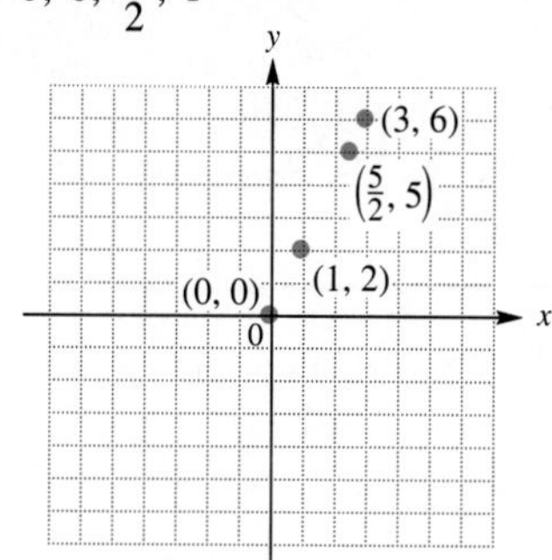

9.5 Exercises (page 481)

1. $(0, 0)$

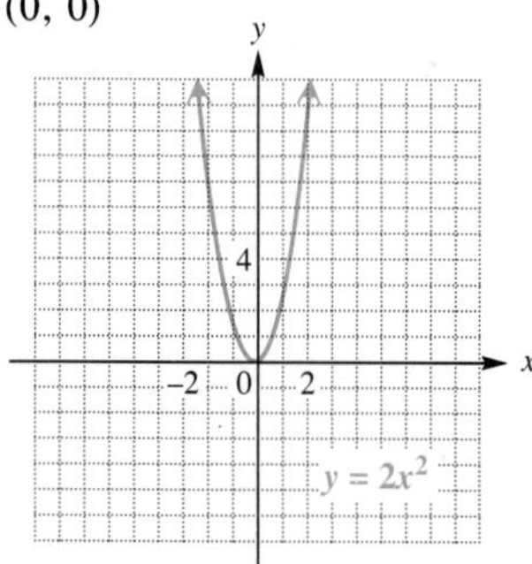

3. $(-1, 0)$

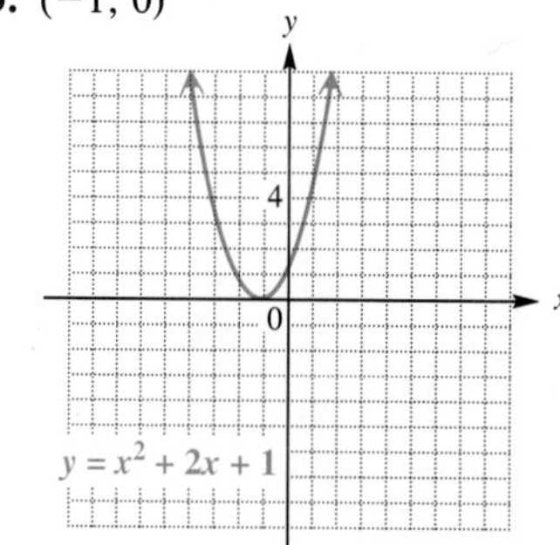

5. $(-1, 0)$

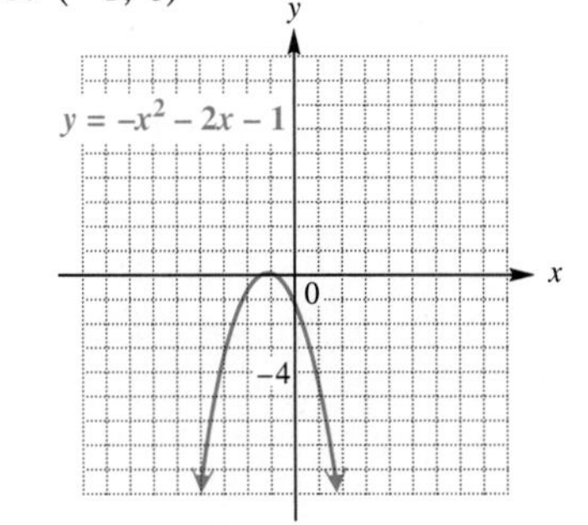

7. $(0, 1)$

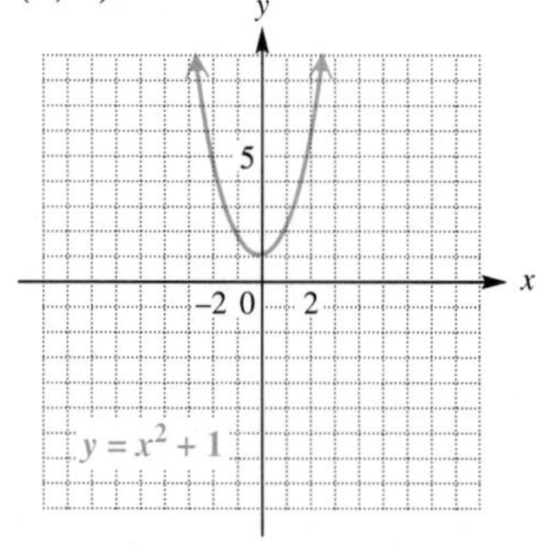

9. $(0, 2)$

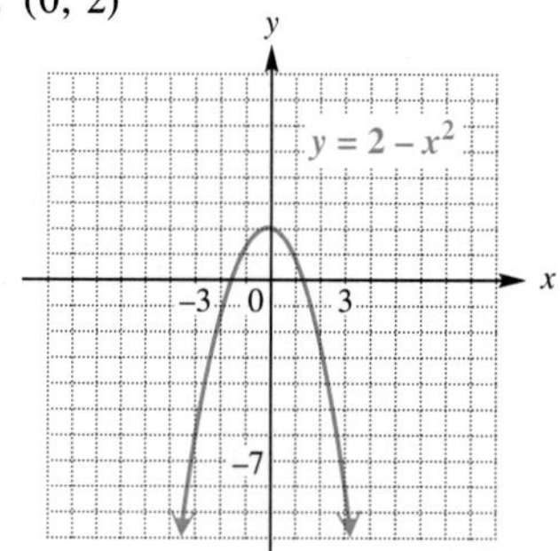

11. $(-3, -1)$

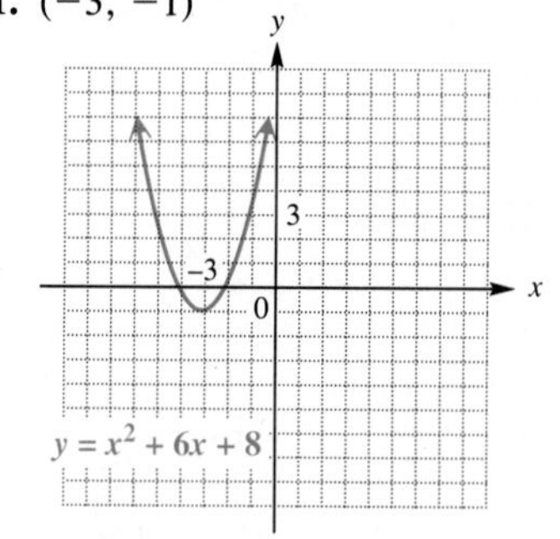

13. $(-1, 2)$

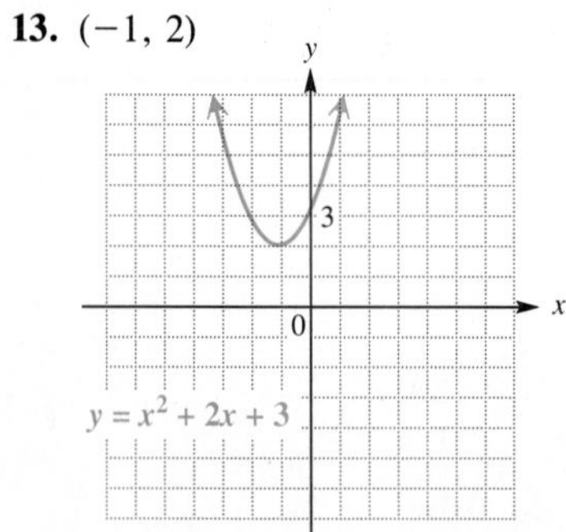

15. $(4, -1)$

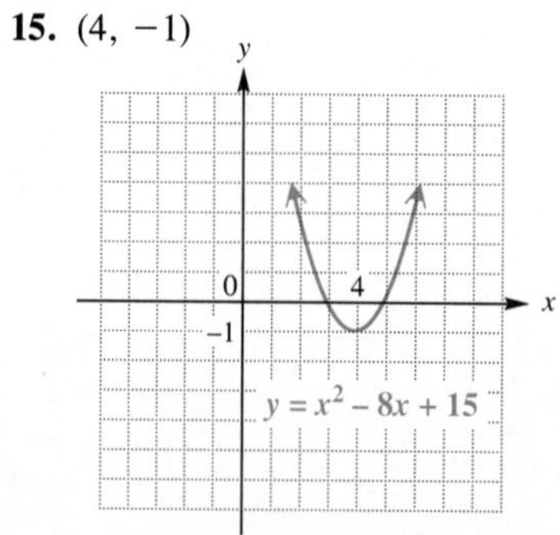

17. $(-2, 1)$

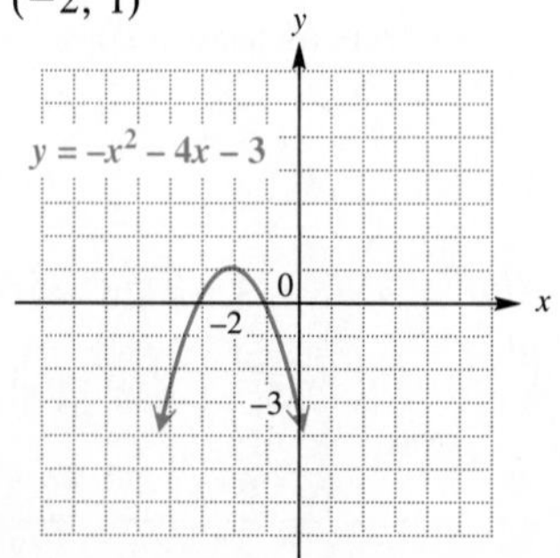

19. one solution: 2 **21.** two solutions: -2 and 2 **23.** no real solution **27.** domain: the set of all real numbers; range: the set of all $y \geq 0$ **29.** domain: the set of all real numbers; range: the set of all $y \leq 4$ **31.** domain: the set of all real numbers; range: the set of all $y \geq 1$ **33.** 3 **35.** 21 **37.** (a)

Chapter 9 Review Exercises (page 486)

1. $7, -7$ **3.** $4\sqrt{3}, -4\sqrt{3}$ **5.** $3 + \sqrt{7},\ 3 - \sqrt{7}$ **7.** $\frac{-2 + 2\sqrt{3}}{3}, \frac{-2 - 2\sqrt{3}}{3}$ **9.** 20% **11.** $-5, -1$ **13.** $-1 + \sqrt{6},\ -1 - \sqrt{6}$ **15.** $1, -\frac{2}{5}$ **17.** $\left(\frac{k}{2}\right)^2$ or $\frac{k^2}{4}$ **19.** $1 + \sqrt{5},\ 1 - \sqrt{5}$ **21.** no real number solution **23.** $\frac{2 + \sqrt{10}}{2}, \frac{2 - \sqrt{10}}{2}$ **25.** $\frac{-3 + \sqrt{41}}{2}, \frac{-3 - \sqrt{41}}{2}$ **27.** Because the coefficient of x^2 is 0, it is not a quadratic equation. **29.** $5 - i$ **31.** $-3 + 2i$ **33.** 20 **35.** 29 **37.** i **39.** $\frac{1}{5} - \frac{7}{5}i$ **41.** a **43.** $-2 + i\sqrt{3},\ -2 - i\sqrt{3}$ **45.** $\frac{2}{3} + \frac{2\sqrt{2}}{3}i, \frac{2}{3} - \frac{2\sqrt{2}}{3}i$ **47.** $\frac{1}{3} + \frac{\sqrt{2}}{3}i, \frac{1}{3} - \frac{\sqrt{2}}{3}i$ **49.** $\frac{3}{8} + \frac{\sqrt{23}}{8}i, \frac{3}{8} - \frac{\sqrt{23}}{8}i$ **51.** (0, 0) **53.** (0, 3)

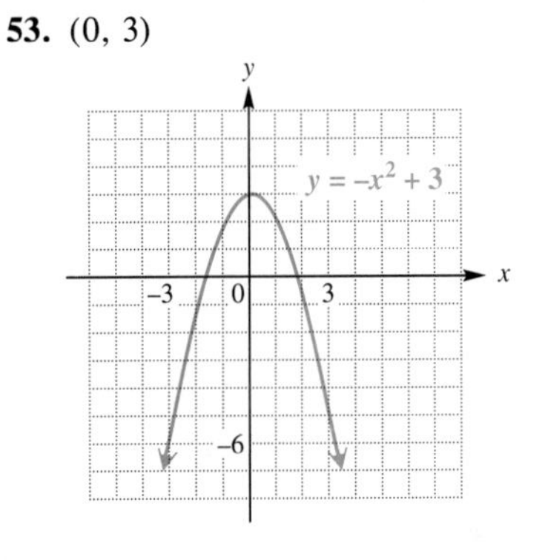

55. (1, 4)

57. two real number solutions: -2 and 2 **59.** two real number solutions: -1 and 4 **61.** $\frac{1}{2}, -5$ **63.** $\frac{-1 + \sqrt{21}}{2}, \frac{-1 - \sqrt{21}}{2}$ **65.** $\frac{-5 + \sqrt{17}}{2}, \frac{-5 - \sqrt{17}}{2}$ **67.** $\frac{-5 + \sqrt{7}}{3}, \frac{-5 - \sqrt{7}}{3}$ **69.** $1, -\frac{1}{3}$ **71.** $-2 + \sqrt{5},\ -2 - \sqrt{5}$

Chapter 9 Test (page 488)

[9.1] **1.** $\sqrt{5}, -\sqrt{5}$ **2.** $10, -4$ **3.** $\frac{2 + \sqrt{35}}{3}, \frac{2 - \sqrt{35}}{3}$ [9.2] **4.** $-2 + \sqrt{6},\ -2 - \sqrt{6}$ **5.** $\frac{1}{5}, -2$ [9.3] **6.** $5, -2$ **7.** $2, \frac{1}{3}$ **8.** $\frac{5 + \sqrt{13}}{6}, \frac{5 - \sqrt{13}}{6}$ [9.1–9.3] **9.** $1 + \sqrt{2},\ 1 - \sqrt{2}$

10. $\dfrac{1 + 3\sqrt{2}}{2}, \ \dfrac{1 - 3\sqrt{2}}{2}$ **11.** $5, -\dfrac{2}{3}$ **12.** $\dfrac{17 + \sqrt{217}}{6}, \ \dfrac{17 - \sqrt{217}}{6}$ **13.** $5 + 2\sqrt{2}, \ 5 - 2\sqrt{2}$
14. $3 + \sqrt{7}, \ 3 - \sqrt{7}$ **15.** 2 seconds [9.4] **16.** $7 - 2i$ **17.** $27 - 14i$ **18.** 29 **19.** $\dfrac{1}{10} + \dfrac{7}{10}i$
20. $\dfrac{5}{2} + \dfrac{\sqrt{3}}{2}i, \ \dfrac{5}{2} - \dfrac{\sqrt{3}}{2}i$ **21.** $\dfrac{1}{3} + \dfrac{\sqrt{2}}{3}i, \ \dfrac{1}{3} - \dfrac{\sqrt{2}}{3}i$
[9.5] **22.** $(-3, 0)$ **23.** $(0, 4)$ **24.** $(-3, -2)$ **25.** 2

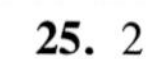

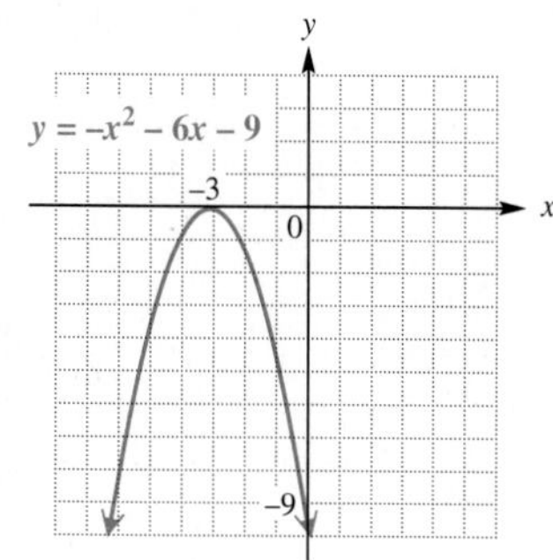

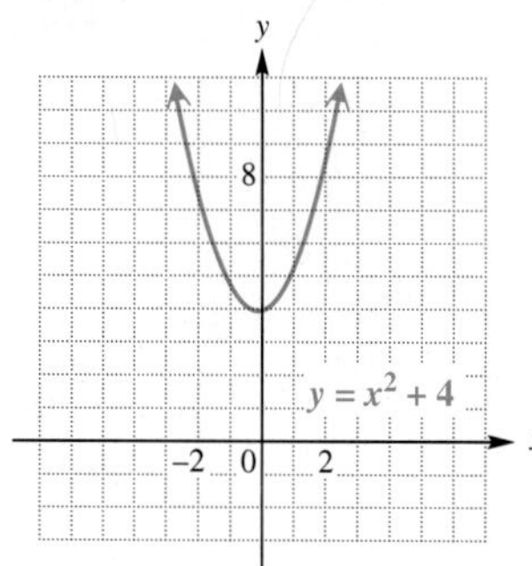

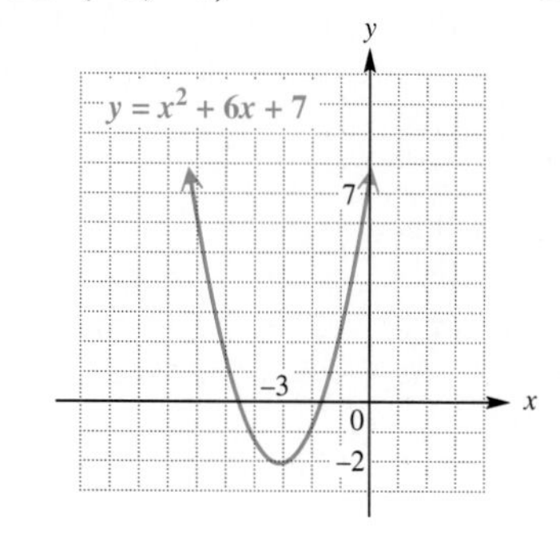

APPENDICES

Appendix C (page 495)

1. 117.385 **3.** 13.21 **5.** 150.49 **7.** 96.101 **9.** 4.849 **11.** 166.32 **13.** 164.19 **15.** 1.344 **17.** 4.14 **19.** 2.23 **21.** 4800 **23.** .53 **25.** 1.29 **27.** .96 **29.** .009 **31.** 80% **33.** .7% **35.** 67% **37.** 12.5% **39.** 109.2 **41.** 238.92 **43.** 5% **45.** 110% **47.** 25% **49.** 148.44 **51.** 7839.26 **53.** 7.39% **55.** $2760 **57.** 805 miles **59.** 63 miles **61.** $2400

Appendix D (page 500)

1. {1, 2, 3, 4, 5, 6, 7} **3.** {winter, spring, summer, fall} **5.** ∅ **7.** {L} **9.** {2, 4, 6, 8, . . .} **11.** The sets of Exercises 9 and 10 are infinite, since each contains an unlimited number of elements.

GLOSSARY

Absolute value: The absolute value of a real number can be defined as the distance between 0 and the number on the number line. [1.4]

Additive inverse (opposite): The additive inverse of a number x is the number that is the same distance from 0 on the number line as x, but on the opposite side of 0. [1.4]

Algebraic expression: An algebraic expression is a collection of numbers, variables, symbols for operations, and symbols for grouping (such as parentheses). [1.3]

Axis: The axis of a parabola is a vertical line through the vertex. [9.5]

Base: The base is the number that is a repeated factor when written with an exponent. [1.2]

Binomial: A binomial is a polynomial with exactly two terms. [3.1]

Common factor: An expression that is a factor of two or more expressions is a common factor of those expressions. [4.1]

Complementary angles: Two angles whose sum is 90° are called complementary angles. [2.4]

Complex fraction: A rational expression with fractions in the numerator, denominator, or both, is called a complex fraction. [5.5]

Complex number: A complex number is a number of the form $a + bi$, where a and b are real numbers. [9.4]

Components: In an ordered pair (x, y), x and y are called the components of the ordered pair. [6.6]

Conjugate: The conjugate of $a + b$ is $a - b$. [8.5]

Conjugate: The conjugate of $a + bi$ is $a - bi$. [9.4]

Consistent system: A system of equations with a solution is a consistent system. [7.1]

Coordinates: The numbers in an ordered pair are called the coordinates of the corresponding point. [6.1]

Coordinate system: An x-axis and y-axis at right angles form a coordinate system. [6.1]

Cube root: The cube root of a number a is the number whose cube is a. [8.1]

Degree: A degree is a unit of measure of an angle. [2.4]

Degree of a polynomial: The degree of a polynomial in one variable is the highest exponent found in any term of the polynomial. [3.1]

Degree of a term: The degree of a term is the sum of the exponents on the variables. [3.1]

Denominator: The denominator of a fraction is the number below the fraction bar. [1.1]

Dependent equations: Equations of a system that have the same graph (because they are different forms of the same equation) are called dependent equations. [7.1]

Descending powers: A polynomial in x is written in descending powers if the exponents on x decrease left to right. [3.1]

Difference: The answer to a subtraction problem is called the difference. [1.1]

Direct variation: y varies directly as x if there is a constant k such that $y = kx$. [5.7]

Domain: The domain of a relation is the set of all first components of the ordered pairs. [6.6]

Element: An object that belongs to a set is an element of the set. [1.3]

Equation: An equation is a statement that two algebraic expressions are equal. [1.3]

Equivalent equations: If two equations have exactly the same solutions, they are equivalent equations. [2.2]

Exponent: A natural number exponent is a number that indicates how many times a base is used as a factor. [1.2]

Exponential expression: An exponential expression is a base with an exponent. [3.2]

Factor: If $a \cdot b = c$, then a and b are called factors of c. [1.1]

Factor: An expression A is a factor of an expression B if B can be divided by A with zero remainder. [4.1]

Factored: A number is factored when it is written as the product of two or more numbers. [1.1]

Factored form: The indicated product of two polynomials is called the factored form of the product polynomial. [4.2]

Factoring: The process of finding the factored form of a polynomial is called factoring. [4.2]

Function: A function is a set of ordered pairs in which each first component corresponds to exactly one second component. [6.6]

Graph: The graph of a number is a point on the number line that indicates the number. [1.4]

Graph: The graph of an equation is the set of points that correspond to all ordered pairs that satisfy the equation. [6.2]

Greatest common factor: The greatest common factor is the largest quantity that is a factor of each of a group of quantities. [4.1]

Grouping symbols: Grouping symbols, such as parentheses and brackets, are symbols that are used to indicate the order in which operations should be performed. [1.2]

Hypotenuse: The longest side of a right triangle, opposite the right angle, is called the hypotenuse. [4.6]

Identity element for addition: 0 is called the identity element for addition. [1.8]

Identity element for multiplication: 1 is called the identity element for multiplication. [1.8]

Imaginary number: A complex number $a + bi$ is imaginary if $b \neq 0$. [9.4]

Imaginary part: The imaginary part of $a + bi$ is b. [9.4]

Inconsistent system: An inconsistent system of equations is a system with no solution. [7.1]

Independent equations: Equations of a system that have different graphs are called independent equations. [7.1]

Index (order): The index of the radical $\sqrt[n]{a}$ is n. [8.1]

Integers: The set of integers is $\{\ldots, -3, -2, -1, 0, 1, 2, 3, \ldots\}$. [1.4]

Inverse variation: y varies inversely as x if there is a constant k such that $y = \frac{k}{x}$. [5.7]

Irrational numbers: The set of irrational numbers is $\{x \mid x$ is a real number that is not rational$\}$. [1.4]

Least common denominator (LCD): The smallest expression that all denominators divide into evenly without remainder is called the least common denominator. [5.3]

Legs: The two shorter sides of a right triangle are called the legs of the triangle. [4.6]

Like radicals: Like radicals are multiples of the same radical. [8.3]

Like terms: Terms with exactly the same variables that have the same exponents are like terms. [2.1]

Linear equation: A linear equation can be written in the form $ax + b = 0$, for real numbers a and b, with $a \neq 0$. [2.2]

Linear equation: An equation that can be written in the form $Ax + By = C$ is a linear equation in two variables. [6.1]

Linear inequality: A linear inequality is an inequality that can be written in the form $ax + b < 0$, for real numbers a and b, with $a \neq 0$. ($<$ may be replaced with $>$, $\leq$, or $\geq$.) [2.9]

Linear inequality: An inequality that can be written in the form $Ax + By < C$ or $Ax + By > C$ is a linear inequality in two variables. [6.5]

Lowest terms: A fraction is in lowest terms when the numerator and denominator have no factors in common (other than 1). [1.1]

Lowest terms: A rational expression is written in lowest terms if the greatest common factor of its numerator and denominator is 1. [5.1]

Mixed number: A mixed number is understood to be the sum of a whole number and a fraction. For example, $3\frac{1}{4} = 3 + \frac{1}{4}$. [1.1]

Monomial: A monomial is a polynomial with exactly one term. [3.1]

Multiplicative inverse (reciprocal): Pairs of numbers whose product is 1 are multiplicative inverses, or reciprocals, of each other. [1.8]

Natural numbers: The set of natural numbers is $\{1, 2, 3, 4, \ldots\}$. [1.1]

Negative numbers: The numbers to the left of 0 on a number line are negative numbers. [1.4]

Number line: A line used to picture a set of numbers is called a number line. [1.4]

Numerator: The numerator of a fraction is the number above the fraction bar. [1.1]

Numerical coefficient: The numerical factor in a term is its numerical coefficient. [2.1]

Ordered pair: A pair of numbers written between parentheses in which the order is important is called an ordered pair. [6.1]

Origin: The point at which the x-axis and y-axis intersect is called the origin. [6.1]

Parabola: The graph of a quadratic equation of the form $y = ax^2 + bx + c$ $(a \neq 0)$ is called a parabola. [9.5]

Parallel lines: Two lines in a plane that never intersect are parallel. [6.3]

Perfect cube: A perfect cube is a number with a rational cube root. [8.2]

Perfect square: A number with a rational square root is called a perfect square. [8.1]
Perfect square trinomial: A perfect square trinomial is a trinomial that can be factored as the square of a binomial. [4.4]
Perimeter: The distance around a geometric figure is called its perimeter. [2.5]
Perpendicular lines: Perpendicular lines intersect at a 90° angle. [6.3]
Plot: To plot an ordered pair is to find the corresponding point on a coordinate system. [6.1]
Polynomial: A polynomial is a term or the sum of a finite number of terms. [3.1]
Positive numbers: The numbers to the right of 0 on a number line are positive numbers. [1.4]
Prime: A natural number (except 1) is prime if it has only itself and 1 as natural number factors. [1.1]
Prime polynomial: A prime polynomial is a polynomial that cannot be factored. [4.2]
Product: The answer to a multiplication problem is called the product. [1.1]
Proportion: A proportion is a statement that two ratios are equal. [2.6]
Quadrants: A coordinate system divides the plane into four regions called quadrants. [6.1]
Quadratic equation: A quadratic equation is an equation that can be written in the form $ax^2 + bx + c = 0$, with $a \neq 0$. [4.5]
Quadratic function: A function of the form $f(x) = ax^2 + bx + c$ $(a \neq 0)$ is called a quadratic function. [9.5]
Quadratic inequality: A quadratic inequality is an inequality that involves a second-degree polynomial. [4.7]
Quotient: The answer to a division problem is called the quotient. [1.1]
Radical: A radical sign with a radicand is called a radical. [8.1]
Radical expression: An algebraic expression containing a radical is called a radical expression. [8.1]
Radicand: The number or expression under a radical sign is called the radicand. [8.1]
Range: The range of a relation is the set of all second components of the ordered pairs. [6.6]
Ratio: A ratio is a quotient of two quantities. [2.6]
Rational expression: The quotient of two polynomials with denominator not equal to zero is called a rational expression. [5.1]
Rationalizing the denominator: The process of changing the denominator of a fraction from a radical (irrational number) to a rational number is called rationalizing the denominator. [8.4]
Rational numbers: The set of rational numbers is $\{x \mid x$ is a quotient of integers, with denominator not $0\}$. [1.4]
Real numbers: The set of real numbers is $\{x \mid x$ is a number that can be represented by a point on the number line$\}$. [1.4]
Real part: The real part of $a + bi$ is a. [9.4]
Relation: A relation is a set of ordered pairs. [6.6]
Scientific notation: A number written as $a \times 10^n$, where $1 \leq |a| < 10$ and n is an integer, is in scientific notation. [3.8]
Set: A set is a collection of objects, such as numbers. [1.3]
Signed numbers: Positive numbers and negative numbers are called signed numbers. [1.4]
Simplified form: A square root radical is in simplified form if the radicand contains no factor (except 1) that is a perfect square, the radicand contains no fractions, and no denominator contains a radical. [8.4]
Slope: The slope of a line is the ratio of the change in y compared to the change in x when moving along the line. [6.3]
Solution: A solution of an equation is any replacement for the variable that makes the equation true. [1.3]
Solution of a system: The solution of a system of linear equations includes all the ordered pairs that make all the equations of the system true at the same time. [7.1]
Solution of a system of linear inequalities: The solution of a system of linear inequalities consists of all points that make all inequalities of the system true at the same time. [7.5]
Square root: The square roots of a^2 are a and $-a$ (a is nonnegative). [8.1]
Standard form (of a complex number): A complex number written in the form $a + bi$ (or $a + ib$) is in standard form. [9.4]
Standard form (of a quadratic equation): A quadratic equation written as $ax^2 + bx + c = 0$ $(a \neq 0)$ is in standard form. [4.5, 9.3]
Straight angle: An angle that measures 180° is called a straight angle. [2.5]
Sum: The answer to an addition problem is called the sum. [1.1]
Supplementary angles: Two angles whose sum is 180° are called supplementary angles. [2.4]
System of linear equations: A system of linear equations consists of two or more linear equations with the same variables. [7.1]

System of linear inequalities: A system of linear inequalities contains two or more linear inequalities (and no other kinds of inequalities). [7.5]

Table of values: A table showing ordered pairs of numbers is called a table of values. [6.1]

Term: A term is a number, a variable, or a product or quotient of numbers and variables raised to powers. [2.1]

Trinomial: A trinomial is a polynomial with exactly three terms. [3.1]

Variable: A variable is a symbol, usually a letter, such as x, y, or z, used to represent any unknown number. [1.3]

Vertex: The vertex of a parabola is the highest or lowest point on the graph. [9.5]

Vertical angles: Vertical angles are formed by intersecting lines, and they have the same measures (see Figure 2.6). [2.5]

Whole numbers: The set of whole numbers is $\{0, 1, 2, 3, 4, \ldots\}$. [1.1]

***x*-intercept:** If a graph crosses the x-axis at k, then the x-intercept is $(k, 0)$. [6.2]

***y*-intercept:** If a graph crosses the y-axis at k, then the y-intercept is $(0, k)$. [6.2]

ACKNOWLEDGMENTS

Diagram on page 8 reprinted with permission from book #3051 ''Woodworker's 39 Sure-Fire Projects'' by the Editors of Woodworker's Magazine. Copyright ©1989 by Davis Publications. Published by TAB BOOKS, a division of McGraw-Hill, Blue Ridge Summit, PA 17294.

ILLUSTRATION CREDITS

All technical line art prepared by Precision Graphics.

All creative illustrations prepared by Rolin Graphics.

PHOTO CREDITS

Unless otherwise acknowledged, all photographs are the property of Scott, Foresman and Company (a subsidiary of HarperCollins Publishers). Page numbers are boldface in the following list. Position abbreviations are as follows: (L) left, (R) right. The photos on the title page, in the table of contents, and on chapter opener pages are courtesy of George Gerster/Comstock. Text accompanying chapter opener photos supplied by Weldon Owen, Inc.

43: Dave Brown/The Stock Market

101: P. Salovios/The Stock Market

123: Obremski/The Image Bank

128: Jacques Cochin/The Image Bank

148: Four By Five/Superstock

194: NASA

237: Gabe Palmer/The Stock Market

300: Courtesy of the Trustees of the British Museum

350: Superstock

350(R): Superstock

386: The Stock Market

387: Superstock

415(L): John Maher/The Stock Market

415(R): Four By Five/Superstock

444: The Image Bank

461: David Sailors/The Stock Market

468: The Stock Market

INDEX

INDEX OF APPLICATIONS

3.3, 3.4 PRODUCTS OF BINOMIALS

To multiply two binomials, use the FOIL method: multiply the **F**irst terms, multiply the **O**utside terms, multiply the **I**nside terms, and multiply the **L**ast terms.

Special products $(a + b)^2 = a^2 + 2ab + b^2$
$(a - b)^2 = a^2 - 2ab + b^2$
$(a + b)(a - b) = a^2 - b^2$

3.6, 3.7 QUOTIENTS OF POLYNOMIALS

To divide a polynomial by a monomial, divide each term of the polynomial by the monomial:

$$\frac{a + b}{c} = \frac{a}{c} + \frac{b}{c} \quad (c \neq 0).$$

To divide a polynomial by a polynomial, use the method of "long division."

4.1–4.4 SPECIAL FACTORIZATIONS

$x^2 - y^2 = (x + y)(x - y)$
$x^2 + 2xy + y^2 = (x + y)^2$
$x^2 - 2xy + y^2 = (x - y)^2$
$x^3 - y^3 = (x - y)(x^2 + xy + y^2)$
$x^3 + y^3 = (x + y)(x^2 - xy + y^2)$

4.5 QUADRATIC EQUATIONS

Quadratic equations can be written in the form $ax^2 + bx + c = 0$, where a, b, and c are real numbers, with $a \neq 0$.

If the polynomial can be factored, use the following property:

Zero-factor property If a and b represent real numbers and if $ab = 0$, then $a = 0$ or $b = 0$.

5.1 RATIONAL EXPRESSIONS

A rational expression is an expression of the form $\frac{P}{Q}$, where P and Q are polynomials, with $Q \neq 0$.

Fundamental property of rational expressions If $\frac{P}{Q}$ is a rational expression and if K represents any factor, where $K \neq 0$, then

$$\frac{PK}{QK} = \frac{P}{Q}.$$

5.2 MULTIPLICATION AND DIVISION OF RATIONAL EXPRESSIONS

For the rational expressions $\frac{P}{Q}$ and $\frac{R}{S}$,

$$\frac{P}{Q} \cdot \frac{R}{S} = \frac{PR}{QS} \quad \text{and} \quad \frac{P}{Q} \div \frac{R}{S} = \frac{P}{Q} \cdot \frac{S}{R} = \frac{PS}{QR}$$

where no divisors are 0.

KEY DEFINITIONS AND PROPERTIES (continued)

5.4 ADDITION AND SUBTRACTION OF RATIONAL EXPRESSIONS

If $\frac{P}{Q}$ and $\frac{R}{Q}$ are rational expressions, then

$$\frac{P}{Q} + \frac{R}{Q} = \frac{P+R}{Q} \quad \text{and} \quad \frac{P}{Q} - \frac{R}{Q} = \frac{P-R}{Q}.$$

To add or subtract rational expressions with different denominators, first find the least common denominator.

6.1 LINEAR EQUATIONS IN TWO VARIABLES

A linear equation in two variables is an equation that can be put in the form $Ax + By = C$, where A, B, and C are real numbers and A and B are not both 0.

6.2 GRAPHING LINEAR EQUATIONS

Equation	*Graphing Methods*
$y = k$	Draw a horizontal line through $(0, k)$.
$x = k$	Draw a vertical line through $(k, 0)$.
$Ax + By = 0$	Graph goes through $(0, 0)$. Find and plot another point on the graph by choosing any nonzero value of x or y. Draw a line through these points.
$Ax + By = C$ $(A, B, C \neq 0)$	Find at least two ordered pairs that satisfy the equation. Plot the corresponding points. Draw a line through these points.

6.3 SLOPE

The slope of the line through the points (x_1, y_1) and (x_2, y_2) is

$$m = \frac{y_2 - y_1}{x_2 - x_1} \quad \text{if } x_2 \neq x_1.$$

A horizontal line has slope equal to 0; a vertical line has undefined slope. A line with positive slope rises from left to right. A line with negative slope falls from left to right.

6.4 EQUATIONS OF A LINE

The slope-intercept form of the equation of a line with slope m and y-intercept $(0, b)$ is $y = mx + b$.

The point-slope form of the equation of a line with slope m going through (x_1, y_1) is $y - y_1 = m(x - x_1)$.

6.6 FUNCTIONS

A function is a set of ordered pairs in which each first component corresponds to exactly one second component.

If a vertical line intersects a graph in more than one point, then it is not the graph of a function.

7.1 SOLVING SYSTEMS OF LINEAR EQUATIONS BY GRAPHING

Graph both equations of the system on the same axes. Any points of intersection are solutions of the system.

7.2 SOLVING SYSTEMS OF LINEAR EQUATIONS BY ADDITION

Multiply one or both equations by appropriate numbers so that the coefficients of x (or y) are negatives of each other. Add the equations to get an equation with only one variable. Solve the equation and substitute the solution into one of the original equations to find the value of the remaining variable.